典型冶金固废有价金属清洁提取与资源化利用

Clean Extraction and Resource Utilization of
Valuable Metals from Typical Metallurgical Solid Waste

郭学益　王亲猛　田庆华　谢铿　管建红　徐志峰　著
Guo Xueyi　Wang Qinmeng　Tian Qinghua　Xie Keng　Guan Jianhong　Xu Zhifeng

上　册

中南大学出版社
www.csupress.com.cn
·长沙·

前 言

冶金固废产生量大、资源化利用率低是制约有色金属行业可持续发展的瓶颈问题，冶金固废中有价金属清洁提取和资源化利用是实现固废"减量化、资源化、无害化"的必然要求和有效途径，是有色金属冶金重要发展方向。

镍、钴、钨、锑是支撑国防建设和经济发展的重要战略金属，但其矿产资源复杂，有价金属与有害元素伴生。由于缺乏固废源头减量及全过程控制技术，镍、钴、钨、锑金属提取过程易产生大量多源固废，如红土镍矿冶金渣、硫基镍钴渣、钨碱煮渣、锑砷碱渣等，导致冶炼过程有价/有害组元分散、存在潜在环境风险；同时，常规镍、钴、钨、锑冶金固废处理工艺侧重于有价金属提取，对有害组分迁移及演变规律认知不深，无法解决有害物质深度脱除与安全处置难题，导致处理过程存在流程长、金属回收率低、二次污染大等问题。因此，亟须开发典型冶金固废有价金属清洁提取与资源化利用关键技术，以实现固废源头减量、无害化处置和有价金属清洁高效提取，促进我国镍、钴、钨、锑战略金属产业可持续发展。

本书作者及其团队一直从事固废资源化领域人才培养、科学研究和工程实践工作，承担了国家重点研发计划项目"镍钴/钨/锑战略金属冶金固废清洁提取与无害化技术"，提出了固废源头减量-有价金属深度提取-有害组分安全处置-尾渣资源化利用系统性解决新思路，从科学理论-关键技术-集成示范全创新链开展研究。通过技术集成及工艺优化，建立工程示范，形成可推广应用技术体系。以该研究项目为主体，作者将最新研究成果归纳总结形成此书。

本书涵盖了典型冶金固废有价金属清洁提取与资源化利用，内容丰富，创新性强，反映了作者及其科研团队近年来在有色固废处置领域的学术成就和研究成果，是团队集体智慧的结晶。

全书由郭学益、王亲猛、田庆华、谢铿、管建红和徐志峰牵头，共同完成了本书研究内容。本书是作者研究团队集体智慧的结晶，研究过程中得到了中南大学、锡矿山闪星锑业有限责任公司、北京科技大学、江西理工大学、北京矿冶科技集团有限公司、中国恩菲工程技术有限公司、中冶瑞木新能源科技有限公司、江西钨业控股集团有限公司、新疆新鑫矿业股份有限公司和长沙矿冶研究院有限责任公司等单位的支持。对本书出版做出贡献的还有金贵忠、廖春发、黄凯、王雪亮、严康、李忠岐、马育新、李渭鹏、卢东昱、张金祥、朱红斌、黎敏、霍广生、雷湘、李玉虎、金承永、姚芾、田磊、苏华、李国、董波、张加美、王青鹜、王松松、陈远林、李中臣、田苗、黄柱、邓卫华、黄少波、杨必文、邓彤、谢岁、朱佳俊、刘召波、吕东等人，对他们的支持表示感谢。

由于作者水平有限，书中难免有疏漏和不妥之处，敬请批评指正。

目 录

锑冶金固废处置现状 …………………………………………………………………（1）

红土镍矿资源现状及冶炼技术研究进展 ………………………………………（15）

红土镍矿高温硫化熔炼镍锍 ……………………………………………………（41）

改性腐殖土型红土镍矿中和高压浸出液-中和渣高压酸浸工艺 ……………（59）

氢冶金理论与方法研究进展 ……………………………………………………（72）

镍熔炼渣贫化工艺现状与展望 …………………………………………………（90）

氢氧化物沉淀法制备层状结构氧化钪的研究 …………………………………（97）

草酸盐沉淀法制备亚微米级 Sc_2O_3 粉体的研究 ………………………………（105）

新能源战略金属镍钴锂资源清洁提取研究进展 ………………………………（114）

采用亚硫酸钠浸提水淬冰铜渣中的银 …………………………………………（143）

负载纳米 ZnS 阳离子树脂选择性去除浓锌溶液中痕量级 Cu^{2+} 和 Cd^{2+} ……（151）

高冰镍浸出渣冶金过程多金属走向行为探究 …………………………………（167）

水淬焙烧渣中金铂钯的氯化浸出 ………………………………………………（178）

用石榴皮自溴化盐溶液中回收金研究 …………………………………………（189）

碱煮黑钨渣还原熔炼回收有价金属的研究 ……………………………………（197）

我国钨渣处理现状与研究进展 …………………………………………………（205）

铁热还原法处理钙砷渣及金属砷的制备工艺 …………………………………（216）

氯化焙烧铜熔炼渣回收铅工艺及动力学研究 …………………………………（228）

铜熔炼渣制备铁精矿研究 ………………………………………………………（236）

铜阳极泥典型处理工艺研究进展 ………………………………………………（249）

镓的分离提取及高纯化制备方法 ………………………………………（275）

从砷碱渣中回收锑、碱并固化砷的研究 …………………………………（287）

碳热焙烧还原砷酸钙制备金属砷 …………………………………………（296）

含砷固废碱性浸出脱砷工艺研究 …………………………………………（308）

高镍锍浸出渣高效清洁利用工艺 …………………………………………（317）

锑冶金固废处置现状

摘　要：目前锑火法冶金主要采用传统鼓风炉和反射炉技术，冶炼过程产生大量冶金固废，如熔炼渣、砷碱渣、除铅渣，其危害性大，难以资源化回收和无害化处理，已成制约锑冶炼企业发展的瓶颈问题。本文溯源了锑冶金固废产生途径，分析了固废处理工艺现状：熔炼渣堆存量大，烟化处理能耗高，分离效果不理想；砷碱渣以湿法处理为主，浸出实现锑、砷分离，含砷浸出液固化处理后堆存或填埋；除铅渣火法处理工艺铅、锑元素分离不彻底，未实现对磷酸盐的回收，湿法处理工艺流程长，工业化应用较少。因此迫切需要开发锑绿色冶金与固废资源化新方法，实现锑清洁高效提取及冶金固废源头减量与资源化回收。

关键词：锑冶金；熔炼渣；砷碱渣；除铅渣

Status of solid waste disposal in antimony metallurgy

Abstract: At present, antimony pyrometallurgy mainly adopts the traditional blast furnace and reflection furnace technology. A large number of metallurgical solid wastes are produced in the smelting process, such as smelting slag, arsenic-alkali residue and lead removal slag, which are hazardous and difficult to realize resource recovery and harmless treatment, and have become a bottleneck restricting the development of antimony smelting enterprises. This paper traces the origin of solid wastes in antimony metallurgy, and analyzes the current status of solid waste disposal. The stock of smelting slag pile is large, the energy consumption of fuming treatment is high, and the separation effect is not ideal. Arsenic alkali residue is mainly treated by wet method, antimony and arsenic are separated by leaching, and arsenic containing leaching solution is solidified and stacked or landfilled. The pyrometallurgical treatment process of lead removal slag does not completely separate lead and antimony, and does not realize the recovery of phosphate; The wet treatment has long process flow and less industrial application. Therefore, it is urgent to develop new methods for antimony clean metallurgy and solid waste recycling to achieve clean and efficient extraction of antimony, as well as source reduction and resource recovery of metallurgical solid waste.

Keywords: antimony metallurgy; smelting slag; arsenic-alkali residue; lead removal slag

锑是重要的战略金属,锑及其化合物主要用于生产半导体、远红外材料、锑铅合金、阻燃剂、颜料、催化剂、玻璃澄清剂等产品,广泛应用于军工、电子、航天等领域,是真正的"工业味精"[1-2]。锑是我国优势资源,美国地质调查局2021年调查数据显示,2020年全球锑储量190万t,我国储量48万t,占全球总储量的25%,居世界首位[3];同时,我国也是锑生产大国,2019年我国承担了世界上63%的锑供应量[4]。

锑冶炼行业中95%以上的企业采用火法冶炼生产金属锑,目前主要采用传统鼓风炉和反射炉技术,冶炼过程产生大量熔炼渣、砷碱渣、除铅渣[5],有高含量的砷、锑、铅、碱等,危害大,难以资源化回收和无害化处理。为实现对锑冶金固废中有价资源的回收利用,同时减轻对环境造成的危害,相关科研院所及冶炼企业开展了深入研究。

1 固废的产生源头

自然界中含锑矿物多达120多种,我国锑矿产资源中具有较高经济价值的有辉锑矿(Sb_2S_3)、脆硫铅锑矿($Pb_4FeSb_6S_{14}$)、锑金矿、黝铜矿($Cu_{12}Sb_4S_{13}$)等[6-7],其中以辉锑矿、脆硫铅锑矿为主。

1.1 辉锑矿冶炼

辉锑矿是我国最重要的锑矿物,其分子式是Sb_2S_3,精矿含锑为30%~50%(质量分数)。目前辉锑矿主要冶炼技术是鼓风炉挥发熔炼——反射炉还原熔炼工艺。在挥发熔炼阶段,Sb_2S_3经挥发、氧化、收尘后以Sb_2O_3的形式被收集,杂质、脉石成分与配入的石灰石、铁矿石等造渣,经前床分离后得到的熔炼渣锑质量分数为1%~2%,水淬处理后堆存或作制造水泥的原料;粗氧化锑在反射炉中通过还原熔炼后得到粗锑,在此过程中会产生含锑、砷等成分的炉渣,因其在固态时呈现蜂窝状,故称之为泡渣,其主要成分如表1所示,泡渣含锑较高,冶炼厂会将其按比例与炉料混合返回鼓风炉回收锑[8]。

表1 泡渣化学成分[5]

Table 1 Chemical composition of frothy slag %

组分	Sb	As	Fe	S	Pb	SiO_2	CaO	Al_2O_3	MgO
质量分数	36~40	0.15~0.35	2.5~3.8	0.1~0.7	0.02	23~28	2.1~3.6	3~8.1	1.4

还原熔炼得到的粗锑再经精炼脱除砷、铅等杂质元素后得到精锑产品。图1为辉锑矿冶炼工艺流程图,在精炼除砷步骤:加入纯碱,在碱性熔剂的存在下,砷首先被氧所氧化,再与纯碱发生反应生成砷酸钠和亚砷酸钠(主要为砷酸钠),进入渣相,发生如下反应:

$$4As + 5O_2 + 6Na_2CO_3 = 4Na_3AsO_4 + 6CO_2 \quad (1)$$
$$4As + 3O_2 + 6Na_2CO_3 = 4Na_3AsO_3 + 6CO_2 \quad (2)$$

同时在此精炼过程中,部分锑也被氧化,并与纯碱反应生成锑酸钠和亚锑酸钠,为砷碱渣中锑的主要来源,发生如下反应[9]:

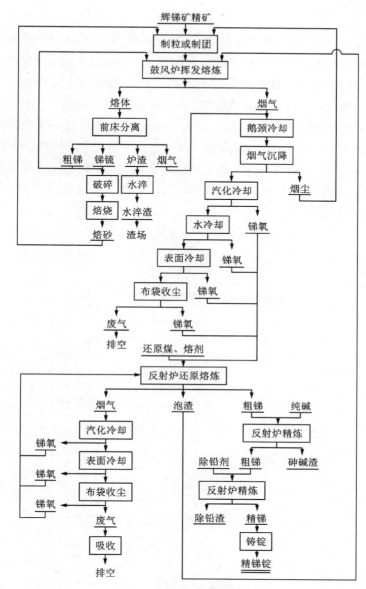

图 1 辉锑矿冶炼工艺流程

Fig. 1　Flow sheet of stibnite smelting process

$$4Sb + 5O_2 + 6Na_2CO_3 = 4Na_3SbO_4 + 6CO_2 \quad (3)$$
$$4Sb + 3O_2 + 6Na_2CO_3 = 4Na_3SbO_3 + 6CO_2 \quad (4)$$

除去砷后得到的粗锑，在同一反射炉内加入除铅剂（H_3PO_4），同时鼓入压缩空气搅动熔池，Pb 被氧化，再与除铅剂发生反应生成磷酸铅等物质进入浮渣中，同时部分锑也会与除铅剂发生反应，进入渣中，一定时间后扒去铅、锑含量较高浮渣得到精锑[10]。

1.2 脆硫铅锑矿冶炼

脆硫铅锑矿是铅和锑的复杂矿物,分子式为 $Pb_4FeSb_6S_{14}$,脆硫铅锑矿主要成分包括:Pb 为 28%~30%,Sb 为 20%~25%,Zn 为 5%~9%,S 为 21%~26%,同时还含有铟、银等有价金属[11]。目前,应用于脆硫铅锑矿的主要是沸腾焙烧—烧结—鼓风炉熔炼—吹炼的火法冶炼工艺[12](图 2)。脆硫铅锑精矿经焙烧脱硫、配料烧结后得到烧结块,再经鼓风炉挥发熔炼得到铅锑合金、含锑烟尘和鼓风炉熔炼渣,含锑烟尘经反射炉熔炼产出铅锑合金;鼓风炉熔炼渣中除含有部分铅锑元素外,还有较多锌、铟等有价金属元素。铅锑合金在反射炉中吹炼分离得到底铅和锑氧粉,与处理辉锑矿的后续工艺流程相同,对中间产品锑氧粉的处理仍采用反射炉还原熔炼、精炼生产精锑,在此精炼过程中会加入纯碱和除铅剂,脱除杂质砷、铅元素,产出砷碱渣和除铅渣,得到精锑产品[13]。

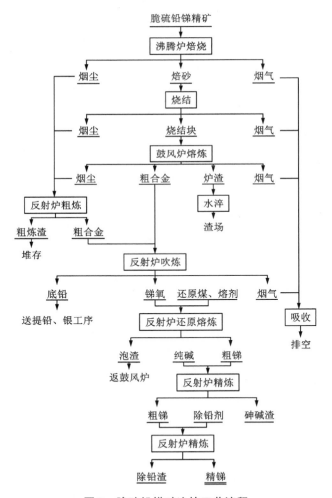

图 2 脆硫铅锑矿冶炼工艺流程

Fig. 2 Flow sheet of jamesonite smelting process

以广西某冶炼厂为例，其年产 10 kt 铅锑合金，产生 20 kt 的鼓风炉熔炼渣，由于过去资源化利用意识薄弱，仅对其进行了简单的水淬处理，得到的水淬渣除少部分用作生产水泥的原料，大部分仍处于堆存状态，广西地区鼓风炉水淬渣堆存量就有百万吨，既造成铅、锑、锌等金属资源的浪费，同时还占用大量土地，污染环境[14]。砷碱渣中砷的平均质量分数为 1%~15%，并且大多数以可溶性砷酸钠形式存在，有剧毒，加之其中还含大量的残碱，对环境造成严重的污染。目前，我国砷碱渣的堆存总量已达到 20 多万 t，且每年以 5000 t 左右的量增加[15]。同时国内每年会产生 15000 t 左右的除铅渣，其中铅锑含量约为 4000 t[16]。因此如何通过工业化处理的方法，使鼓风炉熔炼渣、砷碱渣和除铅渣中的锌、锑及铅等有价金属资源得到综合回收利用，同时减少二次污染物排放，减轻对环境造成的污染，实现对渣的资源化、无害化处理，已成为当前锑冶炼企业面临的重大难题。

2 熔炼渣回收处理方法

熔炼渣常用的处理方法是水淬后作为制造水泥的原料，但受运输成本等条件的限制，大部分水淬渣仍处于渣场堆存的状态，不仅占用土地资源，还容易造成环境污染。不同类型锑矿物在冶炼过程中产生的鼓风炉水淬渣，其化学成分质量分数不同，其典型成分如表 2 所示：

表 2 典型鼓风炉水淬渣化学成分
Table 2 Chemical composition of blast furnace slag quenched by water %

序号	Pb	Zn	Sb	In*	FeO	SiO$_2$	CaO
1[17]	3.01	5.04	0.76	103	13.54	25.52	16.73
2[17]	0.91	6.56	3.21	110	16.97	20.44	14.82
3[14]	0.10	—	1.22	—	26.64	42.03	16.11

（1，2 为脆硫铅锑矿鼓风炉熔炼渣，*单位为 g/t；3 为辉锑矿鼓风炉熔炼渣，—未检测出该元素）。

近年来，锑矿产资源不断减少，对大量堆存的鼓风炉水淬渣进行资源化处理，回收利用其中的铅、锑、锌、铟、铁等金属元素，减轻对环境的污染，具有十分重要的意义。

2.1 火法处理

火法处理主要针对脆硫铅锑矿冶炼过程中产生的热熔炼渣，通过烟化处理使得渣中铅、锑、锌和铟易挥发有价金属元素得到回收利用。但因历史原因，大量鼓风炉渣经水淬后直接堆存[18]。与烟化处理熔融渣不同，在对水淬渣进行烟化处理时需要加入还原剂煤和燃料。铅、锑、锌等金属氧化物首先被还原为金属单质，进入气相后又被氧化为金属氧化物，最终以烟尘的形态被收集。烟化产物以 ZnO、PbO、Sb$_2$O$_3$ 等金属氧化物为主，还包含部分未反应的料渣和粉煤，铅、锑、锌金属的最终回收率均为 85% 以上，烟化处理后的渣中铅锑质量分数小于 0.05%，锌质量分数小于 2%，但渣中含铁量较高[19]。除烟化处理外，还有回转窑处理，其作用原理与烟化处理相同，使用回转窑处理水淬渣时，铅锌的挥发率较高，铟也得到了进一步富集，但锑的挥发率较低，因此工厂处理一般都选用烟化炉以提高锑的回收率。

得到的含锌烟尘可进一步进行分离处理，使用锌电积废液为溶剂对烟尘进行中性浸出，分别得到含锌浸出液和含铅、锑、铟等有价金属的浸出渣，浸出液经净化除杂，送去电解生产电锌或经蒸发、浓缩、结晶处理生产一水硫酸锌产品；浸出渣再经硫酸浸出，得到含铟浸出液和含铅锑浸出渣，渣送往冶炼厂作生产铅锑的原料；对含铟浸出液进行萃取、置换得到海绵铟，再经熔铸、电解处理可得到99.99%的阴极铟[20]。

火法处理可实现对鼓风炉水淬渣中铅、锌、锑及铟金属的有效回收，使水淬渣得到资源化利用，减轻了对环境的污染，回收处理过程中产生的二次废渣可做建筑材料，是目前大规模处理鼓风炉水淬渣的有效途径。但存在的主要问题是金属分离不彻底，仅能得到混合的金属氧化物，还需要进行分离处理；处理锌含量小于10%的炉渣时，能源消耗大，挥发富集经济成本高。

2.2 湿法处理

湿法处理主要利用酸溶解或碱浸出的手段实现对渣中有价金属元素的回收，辉锑矿冶炼过程中产生的鼓风炉渣主要含锑1%~2%，铁的品位可达20%~30%。陈珍娥等对酸浸提取鼓风炉水淬渣中的铁进行了试验研究。使用盐酸硫酸双酸浸出，废渣量为3 g，1∶1 硫酸为10 mL、浓盐酸为6 mL，在$T=80$ ℃下回流反应2 h的实验条件下，铁浸出率可达87.89%；在超声波辅助条件下，达到相同浸出率只需0.5 h。得到的酸浸出液控制相应pH，可分别水解沉淀锑和铁；金则进一步在浸出渣中富集，可通过渣氰化浸出处理回收金。该方法实现对鼓风炉渣中铁、锑、金有价金属元素的富集回收，但鼓风炉渣含有大量的CaO、Al_2O_3、MgO，整体呈碱性，规模化工业处理时会消耗大量的酸，处理成本高[21]。

烟化挥发法处理含Zn小于10%鼓风炉水淬渣时，存在挥发富集成本高的问题，采用酸性浸出时酸耗大，且会有大量杂质金属元素铁、钙、镁、铝进入浸出液，为后续分离操作带来困难。韦岩松等对鼓风炉渣进行碱性浸出提锌的研究：以NaOH作浸出剂，利用锌与OH^-的配位反应，而铁、钙、锰、砷等杂质元素基本不溶，从而实现对锌的选择性沉淀[22]。

$$ZnO + 2NaOH \Longrightarrow Na_2ZnO_2 + H_2O \tag{5}$$

但该方法在实验条件下对锌的浸出率偏低，只有50%左右，考虑原因是锌在渣中的赋存状态除氧化锌外，还有硫化锌、铁酸锌、硅酸锌等，在NaOH碱性体系下仍可稳定存在，从而导致锌的浸出率较低。

现尚无规模化湿法处理鼓风炉水淬渣的应用，实验室条件下湿法处理鼓风炉水淬渣主要存在试剂消耗量大、浸出率偏低、后续元素分离困难等问题。

3 砷碱渣回收处理方法

砷碱渣中的砷主要以砷酸钠形式存在、锑以亚锑酸钠和部分金属锑形式存在，此外还有过量的碳酸钠和少量的硅酸盐成分。

3.1 火法处理

初次冶炼产生的砷碱渣锑金属含量较高，锑质量分数20%~30%，砷质量分数5%~10%，企业一般会将此种一次砷碱渣返回冶炼系统进行回收锑处理，在此过程中经反射炉还原收锑后产出的含砷渣称为二次砷碱渣。二次砷碱渣锑质量分数降至1%左右，但使砷进一步富集，其含量在10%～15%，碳酸钠为40%～60%，是目前大量堆存的砷碱渣类型。因其锑含量较低而砷含量较高，成为砷碱渣处理中产生的更难以回收处理的危险固废，现有技术对该类型砷碱渣处理的规模化应用尚属空白；且利用现有冶炼系统回收处理砷碱渣会造成砷元素在系统中的恶性循环，使操作环境进一步恶化，损害健康[23]。

刘维等提出一种砷碱渣还原熔炼处理方法，该方法是将砷碱渣与碳质还原剂混合，置于惰性或还原气氛中，在 $P \leqslant 101325$ Pa 及 $T \geqslant 800$ ℃ 的条件下进行还原熔炼，得到粗锑和还原渣及含砷烟气；含砷烟气通过冷却回收单质砷；还原渣因包含碳酸钠成分可作为除砷剂用于粗锑精炼除砷工序，或者进行水浸出及浓缩结晶处理，得到碳酸钠产品[24]。但此法得到的还原渣或碳酸钠产品中还含有少量的砷元素，用作脱砷剂使用会造成砷在炉内的富集循环，且该方法的处理对象是一次砷碱渣（含锑26.2%、砷9.68%），未涉及二次砷碱渣的处理。

金承永等提出一种从二次砷碱渣中回收锑、砷、碱的方法，将二次砷碱渣和还原剂混合后置于反应炉中，通入惰性气体，搅拌，加热；在反应炉上部出口接收含锑的砷蒸气，经冷却、结晶、收尘，得含锑的粗砷产品，反应炉下部出口排出反应碱渣，冷却，得氧化钠产品[25]。该法在实验室条件下完成了对二次砷碱渣中锑、砷、碱的回收试验，但未实现砷锑分离，没有工业化规模应用。

现有的砷碱渣火法处理主要是利用已有生产线对渣中的锑资源进行回收利用，但会产生含较低锑的二次砷碱渣，还造成了砷在冶炼系统的恶性循环；部分使用碳质还原剂对砷碱渣进行火法处理仍处于实验研究阶段，尚无更好的二次砷碱渣规模化火法处理手段。

3.2 湿法处理

砷碱渣中的砷酸钠能溶于水，而亚锑酸钠、锑酸钠难溶于水，湿法处理的主要原理就是利用两种盐类在水中的溶解性差异，实现锑与砷的分离；再对含有砷碱溶液和含锑的浸出渣进行处理[26-28]。

3.2.1 钙盐沉淀法

钙盐沉淀法是利用砷酸钙的溶解度低的特点，向含砷浸出液中加入消石灰（$Ca(OH)_2$），生成砷酸钙沉淀而除去砷，分离得到的含碱液经蒸发浓缩得到液碱或片碱，可用于粗锑精炼脱砷，其沉淀反应式如下[29]：

$$3Ca^{2+} + 2AsO_4^{3-} =\!=\!= Ca_3(AsO_4)_2 \downarrow \quad (6)$$

钙盐沉淀法处理砷碱渣，其最大优点是成本较低、处理工艺简单，具体流程如图3所示。

钙渣法虽然可以解决砷碱渣中砷锑基本分离的问题，回收其中的锑，但是

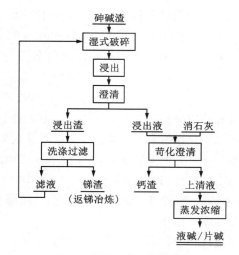

图3 钙盐沉淀工艺流程

Fig. 3　Flow sheet of causticization precipitate process

所产钙渣中的砷在水中溶解度为 6~126 mg/L,在酸性环境下溶解度更大,属于有毒的危险固废,只能堆存处理[30];同时由于空气中 CO_2 和 H_2O 的存在,砷酸钙和亚砷酸钙会分解为碳酸钙和砷酸、亚砷酸,其稳定性较差[31-32];且烧碱溶液经蒸发、浓缩后得到的液碱或片碱,砷含量较高,作为粗锑精炼脱砷剂会造成砷在冶炼系统的循环,有较大的安全隐患。

3.2.2 铁盐沉淀法

铁盐沉淀法的作用原理是砷铁共沉形成砷酸铁化合物,从而达到沉砷、固砷的目的,其反应方程式如下所示[33]:

$$Fe^{3+} + AsO_4^{3-} = FeAsO_4 \downarrow \tag{7}$$

对砷铁化合物稳定性方面的研究表明:在 pH=3~7 时,Fe/As 摩尔比>4 的碱式砷酸铁,砷的溶解度是很低的[34];且铁盐的价格低廉,使得该方法的运行成本较低。

在使用铁盐处理含砷废液时,不同的反应条件下会有一系列砷酸铁沉淀的产生,其中晶型臭葱石($FeAsO_4 \cdot 2H_2O$)相较于其他无定型砷酸铁盐沉淀,具有稳定性好、毒性低、易储存等优点。因此在对含砷废水进行处理时,控制合适的反应条件,使产生的沉淀为晶型臭葱石为更优固砷方法[35-36]。张楠等采用水热法将砷碱渣中的砷转化合成臭葱石晶体,在 pH 值为 1.5、Fe/As 摩尔比为 1.0 及温度为 150 ℃的条件下,沉砷率达到 83.12%,析出的臭葱石颗粒粒径可达 20 μm,As 浸出质量浓度为 0.08 mg/L,低于浸出毒性限值(5 mg/L),可安全储存,实现了对砷碱渣中砷的安全固定,有效地降低了砷碱渣对环境的危害[37]。但上述技术未涉及砷碱分离,造成了渣中碱资源的浪费,SU 对砷碱渣进行水浸处理,控制 T=25 ℃、L/S=2 的条件下浸出 60 min,再向浸出液中通 CO_2 处理,碱以 $NaHCO_3$ 的形式析出,得到的 $NaHCO_3$ 产品含砷低于 0.04 g/kg;再向含砷浸出液中加入 $Fe_2(SO_4)_3$ 溶液,控制 Fe/As 摩尔比=3,最终 As 去除率可达 99.9%,在实现安全有效固砷的同时,对渣中的碱资源进行了回收利用[38]。

但铁盐沉砷受限因素较多,如受各因素影响的吸附共沉淀反应、沉砷时反应条件的精准控制等[39-41]。

3.2.3 硫化沉淀法

硫化沉淀法的处理流程为:砷碱渣经浸出分离后,砷酸钠、亚砷酸钠、碳酸钠、硫酸钠等可溶性物质进入浸出液,再向浸出液中加入酸中和残碱,调节溶液 pH 至酸性,再加硫化物将溶液中的砷沉淀出来。常用的硫化剂有 H_2S、Na_2S,发生如下反应[42]:

$$2As^{3+} + 3S^{2-} = As_2S_3 \tag{8}$$

$$2As^{5+} + 5S^{2-} = As_2S_5 \tag{9}$$

仇勇海等以砷碱渣为原料,在 T=80 ℃条件下,使用热水搅拌约 2 h 浸出脱锑;在脱锑后液中通入二氧化碳气体,脱除碳酸盐;调整脱碱后液的 pH 值,在酸性条件下加入适量的硫化钠脱除砷。该"无污染砷碱渣处理技术"工业试验结果表明:锑回收率达到 99.0%;砷、碱和硫酸钠的浸出率分别达到 90%,99% 和 100%;碳酸盐中碱质量分数达到 95%,砷质量分数在 1% 左右[43]。整个工艺流程实现了水的闭路循环,无新的废渣产生,具有良好的环境效益。具体流程如图 4 所示。

但此过程产出的碳酸盐仍含有较高的砷,会造成砷在冶炼系统的循环;脱砷步骤得到的 As_2S_3、As_2S_5 沉淀的稳定性较差,仍需进一步妥善处置[44-45]。

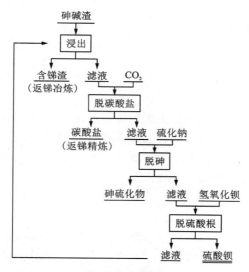

图 4　硫化沉淀法工艺流程
Fig. 4　Flow sheet of sulfide precipitation process

3.2.4　水浸—酸浸法

王建强等使用水浸—酸浸对砷碱渣进行处理，在液固比为 6∶1，温度 40 ℃，浸出时间 40 min 的条件下，可使水浸过程中锑的浸出率低于 3%，砷的浸出率达到 99%；再对渣进行盐酸浸出处理，控制酸浓度为 1∶1，液固比为 10∶1，温度 60 ℃，浸出时间 30 min，能使锑的浸出率为 88% 以上[46]。经过水浸和盐酸浸出，锑的直接回收率为 85.36%；但该方法仅对渣中锑进行回收处理，未实现对砷的无害化处理。

3.2.5　其他方法

Long 首先通过热水浸出脱锑实现砷碱渣中砷、碱与锑的分离，锑、砷、硒和碱的浸出率分别达到 1.82%、98.84%、97.9% 和 100%，再对含砷、碱浸出液通 CO_2 处理，析出的 $NaHCO_3$ 产物在 150 ℃ 下加热 2 h 得到 Na_2CO_3 产品；向结晶母液中加入 H_2SO_4 酸化处理，得到粗硒和除硒后液；再向除硒后液中通入 SO_2 气体，As^{5+} 被还原为 As^{3+}，经浓缩、冷却、过滤处理得到 As_2O_3 和脱砷后液[47-48]。该工艺实现了砷碱渣中锑、砷、硒和碱的高效回收利用，无废水和砷碱混合盐的产出。

现阶段砷碱渣湿法处理工艺技术已能较彻底地实现锑和砷、碱资源的分离，主要研究方向为针对含砷、碱浸出液的处理，其中以钙渣法、硫化沉淀法、铁盐沉淀法等为主的无害化处理技术旨在实现砷的安全高效固定，对得到的含砷固化物采取堆存或填埋处理；但其中通过钙渣法和硫化沉淀法得到的含砷固化物稳定性较差，不能达到毒性浸出要求，仍需二次处理，在设法提高含砷固化物稳定性的同时，还应兼顾规模化处理的可操作性和企业的运营成本等因素。关于砷碱渣资源化处理的研究，在回收利用锑金属资源的同时，对渣中的砷、碱资源进行了回收利用：锡矿山闪星锑业公司曾运行的砷碱渣处理系统产出砷酸钠盐和碳酸钠产品，但砷酸钠盐产品缺少利用渠道，未能真正实现砷的资源化利用[23]，已有关于砷酸钠盐制备单质砷的研究，通过还原焙烧实现砷脱除率 99.84%，得到的砷单质可进一步提纯，但尚处于实验室小试阶段[49]；另有研究以 As_2O_3 的形式回收渣中的砷资源，得到的 As_2O_3 产品可用于农药或高纯 As 的制备[50]。

4 除铅渣回收处理方法

除铅渣中的主要成分为磷酸盐与铅生成磷酸铅或偏磷酸铅，以及部分锑与磷酸盐反应生成磷酸锑或偏磷酸锑。由于操作条件的不同或者原料中的铅含量不一样，所得除铅渣中的锑、铅含量有较大波动，一般质量分数为锑15%~45%、铅5%~15%、磷10%~15%。目前对除铅渣的处置方法有火法处理与湿法处理。

4.1 火法处理

除铅渣火法处理有鼓风炉还原熔炼法，该处理方法是将除铅渣返回鼓风炉中，回收其中的铅、锑金属。但是鼓风炉中的碳质还原剂无法还原渣中的铅锑磷酸盐，导致铅锑的回收率较低，只有65%~70%，同时又产生了含铅、锑较高的渣，未实现对除铅渣的资源化高效利用[16]。

此外还有反射炉铁屑置换法，该方法以单质铁作还原剂还原置换除铅渣中的铅和锑，得到铅锑合金，其主要反应方程如下所示：

$$3Fe + 2SbPO_4 = 2Sb + Fe_3(PO_4)_2 \quad (10)$$

$$3Fe + Sb_2(HPO_4)_3 = 2Sb + 3FeHPO_4 \quad (11)$$

$$3Fe + Pb_3(PO_4)_2 = 3Pb + Fe_3(PO_4)_2 \quad (12)$$

$$Fe + PbHPO_4 = Sb + FeHPO_4 \quad (13)$$

但受炉型结构等因素影响，该法产出的炉渣中铅锑含量较高，导致铅锑金属回收率较低。张玉良等使用改进的回炼炉回收处理除铅渣，通过对反射炉的结构进行了调整，控制熔池温度为1200~1300 ℃，分次加入铁屑和纯碱，经扒渣、出炉、收尘处理工序后除铅渣中的锑的回收率达到95%，铅的回收率达到90%。此方法的缺点是同样产生了含较多铁的二次废渣，且未实现对铅、锑金属的分离回收及渣中磷酸盐的回收利用[51]。

闰方兴等发明了一种联合处理锑冶炼鼓风炉渣和除铅渣的方法，该方法主要作用原理是利用鼓风炉渣中的铁还原置换除铅渣中铅锑金属。先将锑冶炼鼓风炉渣进行破碎、造球、烘干处理，得到干燥物料球团后再还原焙烧，分别得到含有氧化锑、氧化锌和氧化铅的粉尘和金属化球团；再将金属化球团与除铅渣进行熔炼处理，利用金属化球团中的铁单质对除铅渣中的铅、锑进行还原置换，最终得到铅锑合金[52]。该方法对除铅渣中铅、锑金属的回收率均为95%以上，在解决锑冶炼鼓风炉渣堆存占用土地和污染环境问题的同时使锑冶炼鼓风炉渣和除铅渣得到综合利用。

火法回收处理除铅渣时，使用碳质还原剂的还原性效率较低，铁屑的还原效果较好，但最终是以铅锑合金的形式回收了铅、锑资源，未能实现分离回收，且存在有二次废渣产出、能源消耗较高等问题。

4.2 湿法处理

锡矿山闪星锑业公司对除铅渣进行了湿法综合回收处理[53]，工艺流程如图5所示：首先用水和铵盐对除铅渣进行浸出处理；对一次滤渣进行二次浸出，降低渣中的磷含量，滤液返回一次浸出；接着对含磷滤液通入氨气处理，得到含磷酸铵盐滤液；最后向磷酸铵盐滤液加入可溶性硫化物（如硫化钠、硫化氢等），对其中的铅、砷进行沉淀处理，沉铅、砷后的滤液通过再次过滤、澄清、浓缩、结晶处理，得到磷酸盐产品，滤液返回用于一次浸出；过滤得到的铅渣含砷小于1%，含铅5%左右，返回冶炼系统回收铅[54]。该方法回收了除铅渣中的除铅剂、铅锑金属，降低了除铅渣对环境的危害，但是还存在以下问题：未能实现对渣中铅锑、金属的分离回收；处理过程中使用了氨气，使车间内操作环境较恶劣，损害操作工人的健康；流程长、工序多、成本高。

刘鹊鸣等提出一种分离除铅渣中铅锑金属的方法，其流程如图6所示：将除铅渣破碎后，用碳酸氢铵溶液进行碱性浸出，过滤后得到磷酸盐混合溶液和高铅锑氧；对高铅锑氧用盐酸溶液浸出，过滤得到氯化铅渣和三氯化锑溶液。氯化铅渣用作炼铅原料，三氯化锑用碳酸钙粉进行中和，过滤得到氯化钙溶液和三氧化二锑[55]。

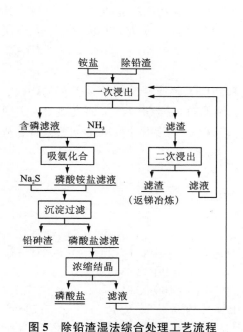

图 5　除铅渣湿法综合处理工艺流程

Fig. 5　Flow sheet of wet comprehensive treatment of lead removal slag

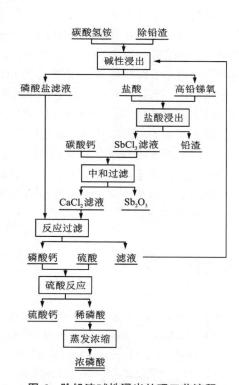

图 6　除铅渣碱性浸出处理工艺流程

Fig. 6　Flow sheet of lead removal slag alkaline leaching process

$$3CaCO_3 + 2SbCl_3 == Sb_2O_3\downarrow + 3CaCl_2 + 3CO_2\uparrow \quad (14)$$

氯化钙滤液与碱性浸出中的磷酸盐混合溶液反应生成磷酸钙和铵盐溶液（该溶液返回碱性浸出）。

$$3Ca^{2+} + 2PO_4^{3-} == Ca_3(PO_4)_2\downarrow \quad (15)$$

磷酸钙用硫酸反应生成稀磷酸和硫酸钙；稀磷酸蒸发浓缩得到浓磷酸，用于粗锑精炼除铅，硫酸钙用作建筑材料使用。

$$3CaCO_3 + 2SbCl_3 = Sb_2O_3\downarrow + 3CaCl_2 + 3CO_2\uparrow \quad (16)$$

该工艺流程能够较彻底分离回收渣中的锑、铅、磷物质元素，并分别得到三氧化二锑、氯化铅和磷酸产品，实现了对除铅渣中铅、锑金属及磷酸盐的资源化回收利用。但该方法处理工序较多，试剂使用量大，尚未规模化工业应用。

5 结论

熔炼渣、砷碱渣、除铅渣是锑冶炼过程中产生的典型固废，因其包含锑、铅、锌、砷等有价有害元素，为达到资源化回收和无害化处理的目的，需要对其进行妥善的回收处置。熔炼渣经水淬处理后的水淬渣堆存量很大，烟化处理水淬渣时能源消耗大，且得到的氧化锌烟尘杂质元素太多，分离处理效果一般，还需结合湿法浸出处理才能实现对锌、铅、锑、铟金属的分离回收；砷碱渣处理方法以湿法为主，通过浸出处理实现锑、砷元素分离，锑和砷分别进入渣和浸出液中，再对锑渣进行资源化回收，由于砷及其化合物的高毒性特点，限制了其相关产品的应用，已有针对产出砷酸钠盐和氧化砷产品的深度加工研究，但系统性、大规模处理仍以砷无害固化为主；除铅渣火法处理时，铅、锑元素分离不彻底，且未实现对磷酸盐的回收，湿法处理工艺流程长，工业化应用较少。

针对以上固废处理工艺存在的问题，在改进现有处理工艺或开发新处理工艺的同时，更应从冶炼生产源头上调控铅、锑、砷等元素的流向，例如对含锑烟尘进行梯级冷凝，实现不同氧化物的分类富集，可为后续的冶炼过程中固废（砷碱渣、除铅渣）的减排创造条件，从而实现固废源头减量。

参考文献

[1] 李增达,张福良,胡永达,等.锑矿开发利用现状及发展趋势[J].中国矿业,2014,23(4):11-15.
[2] ANDERSON C G. The metallurgy of antimony[J]. Geochemistry, 2012, 72: 3-8.
[3] U. S. Geological Survey, Mineral commodity summaries[R]. U. S. Geological Survey, 2021.
[4] 李中平.中国锑行业发展现状及高质量发展建议[J].中国国土资源经济,2021,34(3):17-20+68.
[5] 邓卫华,戴永俊.我国锑火法冶金技术现状及发展方向[J].湖南有色金属,2017,33(4):20-23.
[6] 赵天从.锑[M].北京:冶金工业出版社,1987.
[7] MULTANI R S, FELDMANN T, DEMOPOULOS G P. Antimony in the metallurgical industry: A review of its chemistry and environmental stabilization options[J]. Hydrometallurgy, 2016, 164: 141-153.
[8] 刘勇,陈芳斌,刘共元.中国锑冶炼技术的现状与发展[J].黄金,2018,39(5):55-60.
[9] 陈伟.砷碱渣中砷锑分离并选择性回收锑的工艺研究[D].昆明:昆明理工大学,2016.
[10] 吴文伟,吴学航,赖水彬,等.磷酸二氢铵与铅锑氧化物的高温反应行为[J].有色金属,2008,60(4):84-87.
[11] 戴伟明,雷禄,范庆丰,等.脆硫铅锑矿冶炼工艺研发进展综述[J].黄金科学技术,2015,23(2):98-

102.
[12] 李良斌,徐兴亮,陈晓晨.锑冶炼技术现状及研究进展与建议[J].湖南有色金属,2015,31(3):45-50+60.
[13] 马登,李东波,陈学刚,等.锑精矿冶炼技术研究进展[J].中国有色冶金,2020,49(2):49-54.
[14] 罗燊,马登.锑精矿鼓风炉挥发熔炼炉渣成分对渣含锑影响的研究[J].湖南有色金属,2021,37(1):28-30+44.
[15] 邓卫华,柴立元,戴永俊.锑冶炼砷碱渣有价资源综合回收工业试验研究[J].湖南有色金属,2014,30(3):24-27.
[16] 单桃云.火法炼锑除铅渣中锑回收工艺研究[J].湖南有色金属,2014,30(2):36-38.
[17] 何启贤,覃毅力.烟化处理铅锑鼓风炉渣回收锌铟的生产实践[J].江西有色金属,2008,22(2):29-32.
[18] 王振东,雷霆,施哲,等.烟化法处理鼓风炉炼铅炉渣试验研究[J].云南冶金,2007,36(1):45-47+55.
[19] 莫蔚.烟化法综合处理铅锑鼓风炉渣的工艺实践[J].广州化工,2012,40(18):134-136.
[20] 冉俊铭,史文革,郑燕琼,等.铅锑冶炼水淬渣综合回收有价金属工艺实践[J].有色金属(冶炼部分),2008(5):10-12.
[21] 陈珍娥,帅显泽.酸浸提取锑鼓风炉渣中铁的研究[J].矿冶工程,2018,38(1):92-94.
[22] 韦岩松,潘恒开.从铅锑冶炼鼓风炉水淬渣中碱浸锌的试验研究[J].湿法冶金,2013,32(3):158-160.
[23] 邓卫华.锑冶炼砷碱渣有价资源综合回收研究[D].长沙:中南大学,2014.
[24] 刘维,梁超,覃文庆,等.一种砷碱渣还原熔炼处理方法[P].中国:CN108220626B,2020-01-17.
[25] 金承永,金贵忠,廖光荣,等.一种从二次砷碱渣中回收锑、砷、碱的方法[P].中国:CN110541078A,2019-12-06.
[26] 石靖,易宇,郭学益.湿法冶金处理含砷固废的研究进展[J].有色金属科学与工程,2015,6(2):14-20.
[27] DB43/T 578—2016.锑冶炼砷碱渣无害化处理技术规范[S].
[28] NAZARI A M, RADZINSKI R, GHAHREMAN A. Review of arsenic metallurgy: Treatment of arsenical minerals and the immobilization of arsenic[J]. Hydrometallurgy, 2017, 174: 258-281.
[29] 金哲男,蒋开喜,魏绪钧,等.处理炼锑砷碱渣的新工艺[J].有色金属(冶炼部分),1999(5):11-14.
[30] FEI J C, MA J J, YANG J Q, et al. Effect of simulated acid rain on stability of arsenic calcium residue in residue field. [J]. Environmental Geochemistry and Health, 2020, 42(3): 769-780.
[31] ROBINS R G, JAYAWEERA L D. Arsenic in gold processing[J]. Mineral Processing and Extractive Metallurgy Review, 1992, 9(1-4): 255-271.
[32] ZHANG D N, WANG S F, WANG Y, et al. The long-term stability of calcium arsenates: Implications for phase transformation and arsenic mobilization[J]. Journal of Environmental Sciences, 2019, 84(10): 29-41.
[33] 徐蕾,郑雅杰,彭映林,等.砷酸盐的溶解理论在含砷废水处理中的应用[J].中国有色金属学报,2021,31(3):724-735.
[34] LANGMUIR D, MAHONEY J, ROWSON J. Solubility products of amorphous ferric arsenate and crystalline scorodite ($FeAsO_4 \cdot 2H_2O$) and their application to arsenic behavior in buried mine tailings[J]. Geochimica et Cosmochimica Acta, 2006, 70(12): 2942-2956.
[35] 张楠,方紫薇,郑雅杰,等.臭葱石固砷法处理含砷废水研究进展[J].广州化工,2019,47(2):20-

[36] LI X Z, CAI G Y, LI Y K, et al. Limonite as a source of solid iron in the crystallization of scorodite aiming at arsenic removal from smelting wastewater[J]. Journal of Cleaner Production, 2021, 278: 123552.

[37] 张楠, 方紫薇, 龙华, 等. 砷碱渣稳定化处理合成臭葱石晶体固砷[J]. 中国有色金属学报, 2020, 30(1): 203-213.

[38] SU R, MA X, LIN J R, et al. An alternative method for the treatment of metallurgical arsenic-alkali residue and recovery of high-purity sodium bicarbonate[J]. Hydrometallurgy, 2021, 202: 105590.

[39] 游洋, 闵小波, 彭兵, 等. 碱性高砷渣晶化稳定处理技术研究[J]. 有色金属科学与工程, 2015, 6(6): 24-28.

[40] ESCOBAR K V, VILLALOBOS M, PUIG T P, et al. Approaching the geochemical complexity of As(Ⅴ)-contaminated systems through thermodynamic modeling[J]. Chemical Geology, 2015, 410: 162-173.

[41] FERNANDEZ R L, HERY M, LE P P, et al. Biological attenuation of arsenic and iron in a continuous flow bioreactor treating acid mine drainage (AMD)[J]. Water Research, 2017, 123: 594-606.

[42] 韦岩松, 邓晓雯. 锑冶炼砷碱渣水热硫化沉淀脱砷过程的动力学[J]. 有色金属(冶炼部分), 2014(1): 8-11.

[43] 仇勇海, 卢炳强, 陈白珍, 等. 无污染砷碱渣处理技术工业试验[J]. 中南大学学报(自然科学版), 2005, 36(2): 234-237.

[44] 李倩, 成伟芳. 硫化砷渣的综合利用研究[J]. 广州化工, 2013, 41(13): 17-19.

[45] LU H B, LIU X M, FENG L, et al. Visible-light photocatalysis accelerates As(Ⅲ) release and oxidation from arsenic-containing sludge[J]. Applied Catalysis B: Environmental, 2019, 250: 1-9.

[46] 王建强, 王云燕, 王欣, 等. 湿法回收砷碱渣中锑的工艺研究[J]. 环境污染治理技术与设备, 2006, 7(1): 64-67.

[47] LONG H, ZHENG Y J, PENG Y L, et al. Recovery of alkali, selenium and arsenic from antimony smelting arsenic-alkali residue[J]. Journal of Cleaner Production, 2020, 251: 119673.

[48] LONG H, ZHENG Y J, PENG Y L, et al. Separation and recovery of arsenic and alkali products during the treatment of antimony smelting residues[J]. Minerals Engineering, 2020, 153: 106379.

[49] YANG K, QIN W Q, LIU W. Extraction of metal arsenic from waste sodium arsenate by roasting with charcoal powder[J]. Metals, 2018, 8(7): 542-555.

[50] LONG H, HUANG X Z, ZHENG Y J, et. Purification of crude As_2O_3 recovered from antimony smelting arsenic-alkali residue[J]. Process Safety and Environmental Protection, 2020, 139: 201-209.

[51] 张玉良, 谭大国, 姚光合. 一种对火法炼锑产生的除铅渣进行工业化处理的方法[P]. 中国: CN104962758A, 2015-10-07.

[52] 闫方兴, 任中山, 曹志成, 等. 联合处理锑冶炼鼓风炉渣和除铅渣的方法和系统[P]. 中国: CN107674987A, 2018-02-09.

[53] 邓卫华. 我国锑冶金技术现状及发展方向[J]. 矿冶, 2017, 26(5): 50-54.

[54] 谈应顺, 周高阳. 锑火法精炼除铅渣的处理方法[P]. 中国: CN101265520, 2008-09-17.

[55] 刘鹊鸣, 单桃云, 廖光荣, 等. 一种炼锑产生的含磷酸盐除铅渣分离锑铅磷工艺[P]. 中国: CN106065437B, 2019-02-26.

红土镍矿资源现状及冶炼技术研究进展

摘　要：镍是重要的战略金属，广泛应用于不锈钢、新能源等领域。随着新能源产业快速发展，动力电池需求急剧增长，对镍需求更加迫切，镍资源高效提取对推动我国新能源产业可持续性发展具有重要意义。由于优质硫化镍矿日趋匮乏，红土镍矿产镍已超过硫化镍矿，成重要提镍资源，并呈逐渐上升趋势。本文分析了红土镍矿资源禀赋特征，系统综述了现阶段红土镍矿常规火法和湿法提取工艺研究进展和存在的问题，展望了红土镍矿中镍提取技术发展趋势，指出红土镍矿高温硫化转型高效提镍是未来重要发展方向。

关键词：红土镍矿；火法冶金；湿法冶金；硫化熔炼

Present situation of laterite nickel ore resources and research progress of smelting technology

Abstract：Nickel is an important strategic metal, widely used in stainless steel, new energy and other fields. With the rapid development of new energy industry, the demand for power battery increases sharply, and the demand for nickel is more urgent. The efficient extraction of nickel resources is of great significance to promote the sustainable development of China's new energy industry. Due to the shortage of high quality nickel sulphide ore, the proportion of nickel in laterite ore has exceeded that of nickel sulphide ore and is gradually increasing. In this paper, the resource endowment characteristics of laterite nickel ore are analyzed, the research progress and existing problems of conventional fire and wet extraction technology of laterite nickel ore are systematically reviewed, the development trend of nickel extraction technology in laterite nickel ore is forecasted, and it is pointed out that high temperature sulfide transformation and efficient nickel extraction of laterite nickel ore is an important development direction in the future.

Keyword：laterite nickel ore；pyrometallurgy；hydrometallurgy；sulphidizing smelting

镍具有机械强度高、延展性好、难熔耐高温、化学稳定性高等特征，因而被广泛应用于不锈钢、新能源等领域，已成为建设现代化体系中不可或缺的金属。镍资源主要分布于红土镍矿和硫化镍矿中，2021年美国地质调查局公布数据显示，全球已探明镍品位高于0.5%的

本文发表在《中国有色金属学报》，2023，33(9)：2975−2997。作者：田庆华，李中臣，王亲猛，王松松，郭学益。

陆基镍资源总量(金属量)约3亿吨,其中硫化镍矿占40%,红土镍矿占60%。全球原生镍消费量大且持续增长[1],我国是世界上原生镍消费量最大的国家。国家"十四五"规划聚焦新能源汽车、绿色环保等战略性新兴产业发展,镍作为新能源汽车动力电池的关键金属,加之"高镍"正极材料的发展趋势,其需求量日益攀升[2]。

我国镍资源对外依存度持续维持在85%以上,需大量从印度尼西亚和菲律宾等国进口,供需矛盾日益突出,镍资源持续稳定供应存在挑战[3]。受印尼颁布镍矿出口禁令影响,限制了镍矿进口,我国镍冶炼企业正逐渐向印尼转移。由于硫化镍矿勘探未有突破,传统矿山资源禀赋变差,优质硫化镍矿供给维持收缩态势,红土镍矿资源逐年成为重要提镍原料[4,5]。红土镍矿资源高效提取,可缓解镍供需矛盾,对推动我国新能源产业健康可持续性发展具有重要意义。

本文分析了镍矿资源分布及红土镍矿资源禀赋特征,系统梳理总结了红土镍矿中镍等有价金属提取技术的研究进展,重点分析了红土镍矿典型处理工艺的特点,最后展望了红土镍矿中镍提取技术的发展趋势,以对开发清洁高效的镍提取新技术提供借鉴和指导。

1 镍矿资源概况

从原生镍资源上看,镍在地球上储量较为丰富,仅次于硅、氧、铁、镁。据2021年美国地质调查局(USGS)公布数据,全球陆基镍金属储量约为9400万吨,各国具体占比如图1所示,主要分布于印度尼西亚、澳大利亚、巴西、俄罗斯、古巴等国;2021年我国矿产资源报告数据,我国镍金属储量主要分布于甘肃、青海、新疆,合计储量占全国总储量的90%以上,但以硫化型镍矿为主,占全国总储量的93%;红土镍矿主要分布在云南和四川等地,占全国总储量的5%[6,7]。

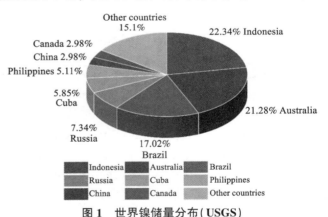

图1 世界镍储量分布(USGS)

Fig.1 World nickel reserves distribution (USUG)

全球储量40%的硫化镍矿已开发上百年,长期开采导致硫化镍矿资源储量、矿石品位等逐渐下降,加上近年来未有大型硫化镍矿发现,硫化镍矿产量呈不断下降趋势[8]。由图2(a)可知,2010年开始至今,红土镍矿产量占比已超过硫化镍矿,并呈逐渐上升趋势,红土镍矿逐年成为重要提镍原料。由图2(b)可知,红土镍矿主要分布于印度尼西亚、澳大利亚、菲律宾、古巴、新喀里多尼亚等。

据2021年国际镍研究组织(INSG)公布数据可知,2016—2020年各大洲镍产量如图3(a)所示,表明由于印度尼西亚和菲律宾镍矿开采量增加,亚洲镍产量持续增长;欧洲、美洲、大洋洲保持稳定,但占比持续下降;非洲镍产量和占比持续下降。图3(b)显示2021年印度尼西亚、菲律宾、俄罗斯和新喀里多尼亚四国镍产量占比为62.4%。图3(c)和图3(d)分别显示了全球和中国镍金属产量,2021年全球镍金属产量约为270万吨,其中俄罗斯诺里尔斯克

红土镍矿资源现状及冶炼技术研究进展 / 17

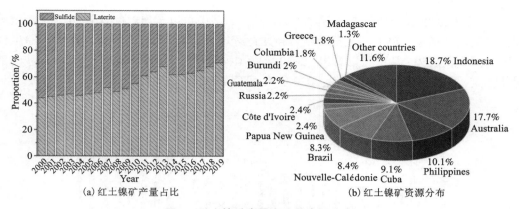

(a) 红土镍矿产量占比

(b) 红土镍矿资源分布

图 2 红土镍矿产量占比及资源分布

Fig. 2 Production proportion and resource distribution of laterite nickel ore
(a: production proportion of laterite nickel ore; b: Laterite nickel ore resource distribution)

镍业的 Kola Division 镍矿以 17.2 万吨镍产量领先全球[9]；2021 年中国镍产量约为 12 万吨，仅占比 4.8%。

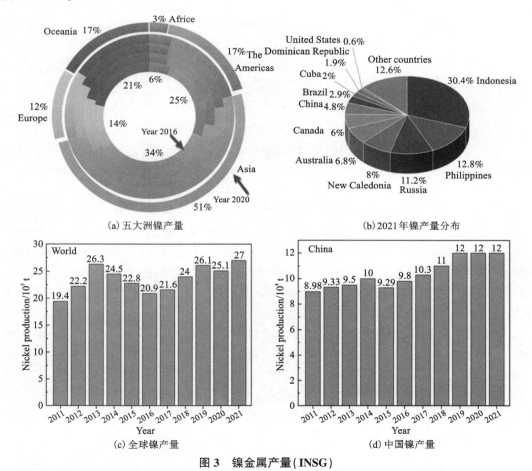

(a) 五大洲镍产量

(b) 2021年镍产量分布

(c) 全球镍产量

(d) 中国镍产量

图 3 镍金属产量（INSG）

Fig. 3 Nickel metal production: (a) Nickel production in five continents; (b) Nickel production distribution in 2021 year; (c) Global nickel production; (d) Nickel production in China (INSG)

图4(a)为全球原生镍消费量随时间变化趋势，从中可以看出，自2010年以来，全球原生镍消费量增长加速，2016年超过200万吨并持续增长。图4(b)为各大洲及中国原生镍消费量随时间变化趋势，从中可以看出，2020年我国镍消费量保持持续增长；美洲和欧洲消费量保持稳定，但由于新冠肺炎影响，2020年有所下降；非洲镍消费量减少至1.2万吨。

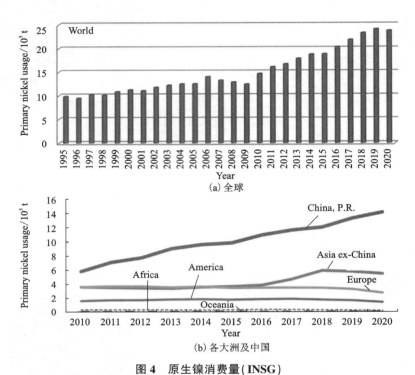

图4　原生镍消费量(INSG)

Fig. 4　Primary nickel consumption：(a) global；(b) Continents and China (INSG)

镍因其具有优异物理化学特性，广泛用于新能源汽车动力电池领域，被誉为"21世纪白色石油"，是新能源产业发展重要物质基础，决定着新能源产业链的发展命脉。从消费结构看，不锈钢消费基本稳定，电池领域用镍占比不断提高[10, 11]。根据INSG(国际镍研究组织)和中国有色金属工业协会的数据(图5)，2021年全球原生镍消费量达276万吨，其中全球不锈钢用镍占比约为71%，电池用镍占比上升至12%；我国原生镍消费量达154.2万吨，不锈钢用镍占比约为74%，电池用镍占比上升到15%。

欧盟委员会计划到2030和2035年，欧盟汽车碳排放量相较于2021年分别下降55%和100%，即到2035年仅销售零排放车辆。美国总统签署行政令要求2030年零排放车辆销量占50%。未来三元前驱体对镍需求量将持续增长，保障镍资源供应稳定将成为三元前驱体企业的重要命题[12]。由于含镍正极的能量密度高，镍可用于各种类型电池，包括镍镉电池、镍氢电池、镍铁电池、镍锌电池和镍氢电池等。

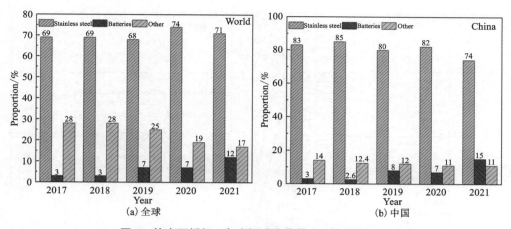

图 5 镍在不锈钢、电池领域消费量占比随时间变化

Fig. 5 Change of the proportion of nickel in stainless steel and battery consumption over time：(a) Global；(b) China

2 红土镍矿资源禀赋特征

红土镍矿是在热带或亚热带气候条件影响下，由超基性岩在地下浅层经风化-淋滤-沉积形成的地表风化壳性矿床[13]。红土镍矿成矿过程复杂，各矿层元素含量、矿相组成均有较大差异，如图 6 和表 1 所示。按照元素含量通常分为褐铁矿型、黏土型和腐殖土型三类；褐铁矿型具有 Fe、Co 含量高，Ni 含量低等特点，其主要物相为针铁矿和赤铁矿，脉石成分主要为尖晶石、滑石和闪石，镍主要以晶格取代或吸附形式赋存于针铁矿中；腐殖土型具有 Si、Mg、Ni 含量高，Co 含量低等特点，其主要物相为利蛇纹石、镍绿泥石和蒙脱石，镍主要以吸附状态或类质同相形态存在于镁或铁硅酸盐中；黏土型中各元素含量介于褐铁矿型和腐殖土型之间，其主要物相为绿脱石和斜绿泥石[3]。

表 1 红土镍矿类型与成分

Table 1 Types and compositions of laterite nickel ore

Ore type	Mass fraction/%						Characteristic	Extraction process
	Ni	TFe	MgO	SiO$_2$	Co	Cr$_2$O$_3$		
Limonite	0.8~1.4	36~50	0.5~5.0	10~30	0.1~0.2	2~5	Low Ni and Mg, high Fe	Hydrometallurgy
Nontronite	1.2~1.8	25~40	5~15	10~30	0.02~0.1	1~2	—	Pyrometallurgy/Hydrometallurgy
Saprolite	1.4~3.0	10~25	15~35	30~50	0.02~0.1	1~2	High Ni, Mg and Si; Low Fe	Pyrometallurgy

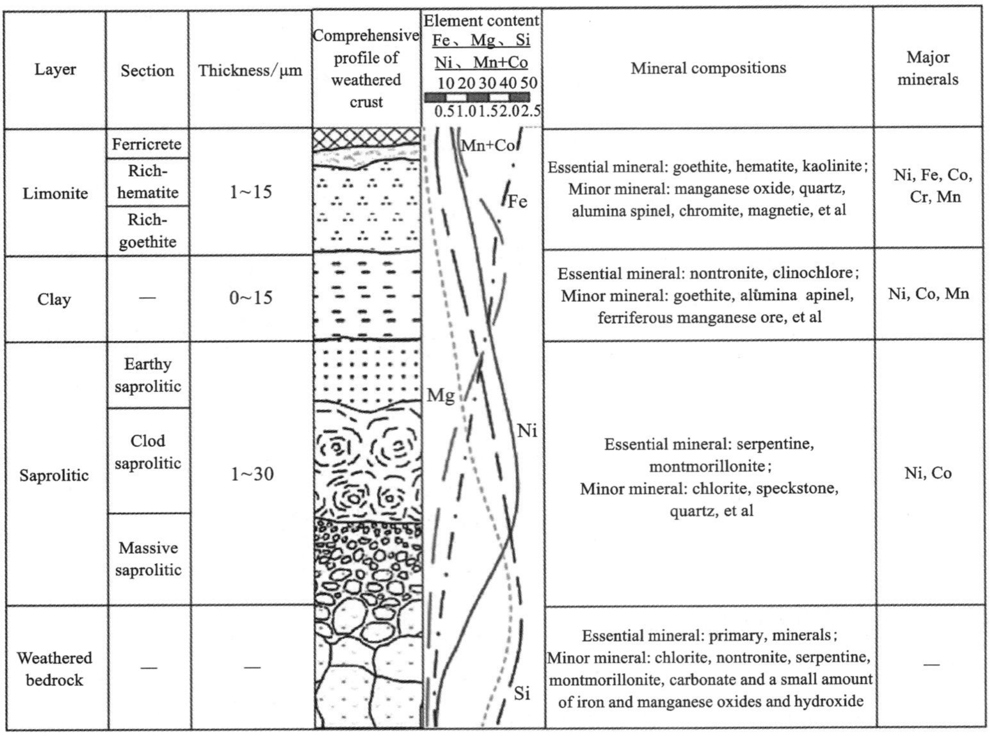

Layer	Section	Thickness/μm	Comprehensive profile of weathered crust	Element content Fe、Mg、Si Ni、Mn+Co	Mineral compositions	Major minerals
Limonite	Ferricrete	1~15		Mn+Co / Fe	Essential mineral: goethite, hematite, kaolinite; Minor mineral: manganese oxide, quartz, alumina spinel, chromite, magnetie, et al	Ni, Fe, Co, Cr, Mn
	Rich-hematite					
	Rich-goethite					
Clay	—	0~15			Essential mineral: nontronite, clinochlore; Minor mineral: goethite, alumina apinel, ferriferous manganese ore, et al	Ni, Co, Mn
Saprolitic	Earthy saprolitic	1~30		Mg / Ni	Essential mineral: serpentine, montmorillonite; Minor mineral: chlorite, speckstone, quartz, et al	Ni, Co
	Clod saprolitic					
	Massive saprolitic					
Weathered bedrock	—	—		Si	Essential mineral: primary, minerals; Minor mineral: chlorite, nontronite, serpentine, montmorillonite, carbonate and a small amount of iron and manganese oxides and hydroxide	—

图 6 红土镍矿风化壳的刨面特征[3]

Fig. 6 Planing surface characteristics of laterite weathering crust

红土镍矿含水约 30%,主要包括吸附水、结晶水和结构水,红土镍矿进入冶炼系统之前,必须对其进行干燥处理。

(1)对于褐铁矿型红土镍矿,其主要物相为针铁矿,针铁矿加热脱羟基过程是一个复杂过程,如下所示[14]:

$$\alpha - FeOOH(geothite) \rightarrow Fe_{5/3}(OH)O_2(protohematite)$$
$$\rightarrow Fe_{11/6}(OH)_{1/2}O_{5/2}(hydrohematite) \rightarrow \alpha - Fe_2O_3 \quad (1)$$

在 25~140 ℃,脱除矿中吸附水;在 200~480 ℃,脱除矿中结晶水;在 500~800 ℃下,脱除矿中羟基。褐铁矿型红土镍矿中针铁矿脱羟基过程是脱羟基程度最高的过程。在 795 ℃以上,褐铁矿型红土镍矿主要物相为 Fe_2O_3、$MgSiO_3$ 和 Mg_2SiO_4。

(2)对于腐殖土型红土镍矿,其主要物相为蛇纹石,蛇纹石物相脱羟基作用发生在 650~810 ℃,如下所示[15]:

$$(Mg, Fe, Ni)_3Si_2O_5(OH) \longrightarrow (Mg, Fe, Ni)SiO_3 + (Mg, Fe, Ni)_2SiO_4 + 2H_2O \quad (2)$$

腐殖土型红土镍矿脱羟基过程比褐铁矿型红土矿更复杂,最终形成橄榄石相和辉石相。除了上述反应,Rhamdhani 等[16]报道称,蛇纹石物相加热过程中可发生如下分解反应:

$$(Mg, Fe, Ni)_3Si_2O_5(OH) \longrightarrow 3(Mg, Fe, Ni)_2SiO_4 + SiO_2 + 4H_2O \quad (3)$$

在这种情况下,脱羟基过程不生成辉石相,而是形成橄榄石相和游离二氧化硅。

Sungging Pintowantoro 等[1]认为 700 ℃时，硅镁镍矿(garnierite)会发生分解反应，分解反应为：

$$Ni_3Mg_3Si_4O_{10}(OH)_8(s) \longrightarrow 3NiO(s) + 3MgO(s) + 4SiO_2(s) + 4H_2O(g) \qquad (4)$$

红土镍矿提镍过程主要包括火法工艺和湿法工艺。腐殖土型由于镍含量高、铁含量低，更适用于火法工艺，其冶炼工艺包括回转窑粒铁法、高炉熔炼镍铁法和还原熔炼镍铁法等；褐铁矿型由于镁、硅含量低、钴含量高，更适用于湿法工艺，其冶炼工艺包括高压酸浸法、常压酸浸法、还原焙烧-氨浸法等。

3 红土镍矿典型处理工艺

3.1 红土镍矿湿法提取工艺

3.1.1 还原焙烧-氨浸法

还原焙烧-氨浸(Caron)法能够处理 MgO 质量分数大于 10% 红土镍矿，具有试剂可循环利用、成本低等优点，但存在镍钴回收率低(镍回收率 75%~80%、钴回收率低于 50%)等缺点，工艺流程如图 7 所示。主要流程为①破碎筛分；②选择性还原焙烧：还原焙烧后原料中镍、钴主要以金属态形式存在，铁大部分以 Fe_3O_4 形式存在，少量以 FeO 或 Fe 形式存在；③氨浸：镍、钴以络氨离子形式进入浸出液，铁以 $Fe(OH)_3$ 形式进入浸出渣；④氨蒸煅烧提取镍钴。全球只有几家工厂采用该法处理红土镍矿，如澳大利亚的 Yabulu、菲律宾的 Surigao 和 Berong、印度的 Sukhinda[17]。

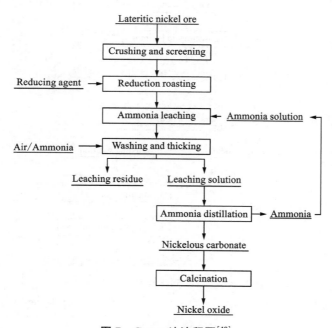

图 7 Caron 法流程图[18]

Fig. 7 Flow chart of Caron method

氨浸过程原理可表示为[17]：

$$2Ni + O_2 + 2(n-2)NH_3 + 2(NH_4)_2CO_3 \Longrightarrow 2[Ni(NH_3)_n]CO_3 + 2H_2O \quad (5)$$

$$2Co + O_2 + 2(n-2)NH_3 + 2(NH_4)_2CO_3 \Longrightarrow 2[Co(NH_3)_n]CO_3 + 2H_2O \quad (6)$$

$$2Fe + O_2 + 2(n-2)NH_3 + (NH_4)_2CO_3 \Longrightarrow 2[Fe(NH_3)_n]CO_3 + 2H_2O \quad (7)$$

$$FeO + (n-2)NH_3 + (NH_4)_2CO_3 \Longrightarrow [Fe(NH_3)_n]CO_3 + H_2O \quad (8)$$

$$4[Fe(NH_3)_n]CO_3 + 10H_2O + O_2 \Longrightarrow 4Fe(OH)_3 \downarrow + 4(n-2)NH_3 + 4(NH_4)_2CO_3 \quad (9)$$

Caron 工艺能够从褐铁型和腐殖土型红土镍矿中提取镍。巴西镍企业（Niquelândia）采用 Caron 法从原矿中提取镍，镍回收率为 70%～75%[19]。随着技术发展，相继开发了加硫还原焙烧-氨浸[20,21]、还原焙烧-酸浸、氧化焙烧-酸浸、氯化焙烧-水浸、硫酸化焙烧-水浸[22-24]、碱性焙烧-加压酸浸等工艺，但尚未实现产业化应用[25]。

3.1.2 高压酸浸法

高压酸浸法（HPAL）主要用于处理含铁高、含硅镁较低的褐铁型红土镍矿，具有能耗低、碳排放量少、镍钴回收率高（>90%）等优点[26]，是国内外处理褐铁型红土镍矿的主要技术，被广泛应用于中冶瑞木新能源、格林美股份有限公司、宁波力勤矿业公司、澳大利亚必和必拓公司（BHPB）、巴西国有矿业公司（CVRD）、加拿大鹰桥公司（Falconbridge）等。然而，该工艺对原矿中镁（<5%）、铝含量要求较高，镁、铝易沉淀结垢导致设备腐蚀严重[27]。工艺流程如图 8 所示。

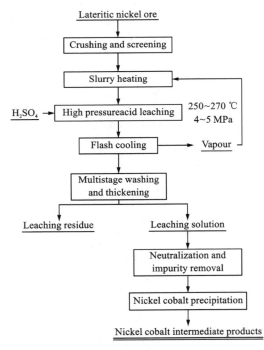

图 8　红土镍矿高压酸浸法流程图[29]

Fig. 8　Flow chart of high pressure acid leaching of laterite nickel ore

高压酸浸过程原理可表示为[28]：

$$NiO + H_2SO_4 \Longleftrightarrow NiSO_4 + H_2O \tag{10}$$

$$CoO + H_2SO_4 \Longleftrightarrow CoSO_4 + H_2O \tag{11}$$

$$2FeO \cdot OH + 3H_2SO_4 \Longleftrightarrow Fe_2(SO_4)_3 + 4H_2O \tag{12}$$

硫酸铁在高温下水解生成赤铁矿沉淀：

$$Fe_2(SO_4)_3 + H_2O \Longleftrightarrow Fe_2O_3 \downarrow + 3H_2SO_4 \tag{13}$$

针铁矿总反应：

$$2FeO \cdot OH \Longleftrightarrow Fe_2O_3 \downarrow + H_2O \tag{14}$$

因此，褐铁型红土镍矿中铁在硫酸介质中发生分解和转化反应生成赤铁矿，反应过程不消耗硫酸。

Guo 等[30]采用高压酸浸法从褐铁矿型红土镍矿中提取镍和钴。在硫酸加入量 250 g/t、浸出温度 250 ℃、浸出时间 1 h 和液固比 3∶1 条件下，Ni、Co、Mn、Mg 浸出率分别为 97%、96%、93% 和 95% 以上，而铁浸出率小于 1%。高压浸出渣检测结果表明，渣中铁、硫主要分别以赤铁矿和明矾石形式存在。

针对云南元江贫镍氧化矿的开发利用，王成彦和马保中等[31-34]提出低品位红土镍矿硝酸加压浸出(NAPL)创新工艺，2008 年该技术获得专利授权，其工艺流程如图 9 所示。该工艺处理褐铁型红土镍矿时，Ni、Co 浸出率均大于 85%、Mg 浸出率为 80%、Al 浸出率大于 60%、Fe 浸出率低于 1%，产出含铁 55% 的富铁渣。在处理高镁型红土镍矿时，Ni、Co 浸出率分别为 98% 以上、约 99%，Fe 浸出率不到 1.5%[35]。

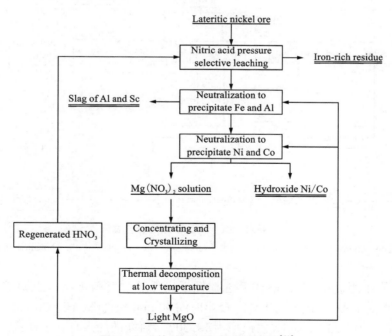

图 9 红土镍矿 NAPL 技术流程图[32]

Fig. 9 The general flow sheet of the NAPL technology for laterite ores

Luo 等[36]提出了一种用磷酸处理褐铁型红土镍矿的方法。基于磷酸盐溶解度差异，在浸出过程中，Fe^{3+} 与 PO_4^{3-} 反应生成磷酸铁，而 Ni^{2+} 和 Co^{2+} 在溶液中，从而实现了 Ni/Co 优先于 Fe 的选择性浸出。在 H_3PO_4 溶液 3 mol/L、130 ℃、恒温 2 h 条件下，镍、钴浸出率分别为 99.2%、81.8%，铁浸出率为 0.54%。与高压浸出相比，所需温度降低了 115~125 ℃，压力仅为 0.2~0.27 MPa。此外，由于 Fe^{3+} 和 Al^{3+} 的同步沉淀，在 (Fe+Al)/P 摩尔比为 1.0 时获得了层状铝掺杂磷酸铁，过程几乎没有残渣排放。

3.1.3 常压酸浸法

由于高压酸浸法对设备、规模等要求严格，因此研究者们开发了常压酸浸工艺以解决高压酸浸法的一些弊端。常压酸浸中使用的酸主要包括硫酸[37]、盐酸、硝酸[38]、有机酸和微生物衍生酸等[39]。

郭学益等[40]采用盐酸在常压下浸出高铁型红土镍矿，在矿石粒度为 0.125~0.15 mm、酸料比为 3∶1、反应温度 80 ℃、反应时间 1 h、固液比为 1∶4、不外加氯化盐下，Ni、Co、Fe、Mn、Mg 和 Cr 金属的浸出率分别为 86.9%、67.8%、86.5%、80.1%、58.5% 和 72.6%。Ni 和 Fe 的浸出过程符合 Avrami 方程。

Guo 等[41]提出了一种盐酸常压浸出和喷雾水解处理菲律宾腐殖土型红土镍矿的创新技术，其工艺流程如图 10 所示。最佳酸浸条件下，Ni、Fe 和 Mg 浸出率分别为 98.9%、97.8% 和 80.9%。用盐酸常压浸出法处理红土镍矿，破坏了矿物晶格结构，浸出残渣主要产物为二氧化硅。水解液中镍品位为 4.55%，可用于镍铁生产。

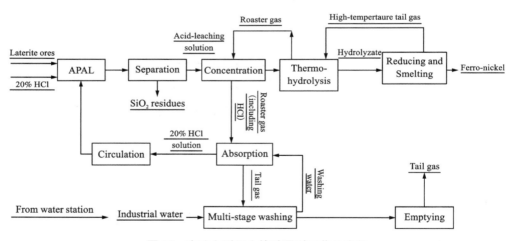

图 10 腐殖土型红土镍矿原则工艺总流程

Fig. 10 The general process flow sheet of the technology for processing the saprolitic laterite ores

李金辉等[42]采用氯化铵-盐酸体系选择性浸出低品位腐殖土型红土镍矿，在浸出温度 90 ℃、盐酸浓度 2 mol/L、固液比 1∶6、浸出时间 90 min 条件下，Ni、Co、Mn 浸出率分别为 89.45%、88.56%、90.23%，而 Fe 浸出率为 19.30%，氯盐的加入有利于针铁矿相溶解，但对其他铁矿相影响不显著。

在浸出过程中使用合适的表面活性剂和微波强化，可提高红土镍矿中有价金属的浸出率[43,44]。常压酸浸法具有能耗低、工艺简单、设备小、操作易于控制等优点，但存在浸出率低、浸出液分离困难、浸出渣中镍含量高等缺点。

3.2 红土镍矿火法工艺生产镍铁

3.2.1 回转窑粒铁法

回转窑粒铁法基于德国"Krupp-Renn"直接还原炼铁工艺发展而来,工艺流程如图 11 所示。红土镍矿在约 1400 ℃下以半熔融状态熔炼生成镍铁颗粒,随后从回转窑中排出熟料进行水淬,接着破碎和磁选,磁选后镍铁含铁 75%~80%、含镍 20%~25%、粒度≥0.1 mm[45]。回转窑粒铁法具有流程短、能耗低(85%能源由煤提供,吨矿耗煤 160~180 kg)、镍回收率高、原料适应强等优点。日本冶金公司大江山冶炼厂最早将回转窑粒铁法实现工业化。典型镍铁、尾渣组分如表 2 所示。由于熟料中微小镍铁颗粒被赋存于渣中而不能磁选回收,因此需要保证镍铁颗粒在回转窑中充分聚集长大。在工业应用中发现回转窑粒铁法存在回转窑结圈现象严重、设备作业率低等问题[46]。

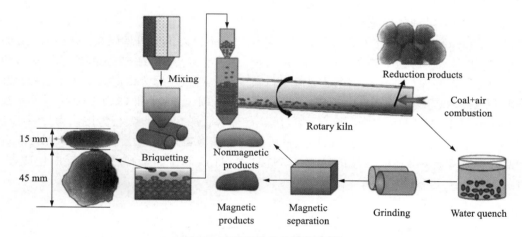

图 11 回转窑粒铁法流程图[47]

Fig. 11 Schematic diagram of rotary kiln reduction process

表 2 典型镍铁/渣成分[46]

Table 2 Typical Ferro-nickel/slag composition %

		Component	Ni+Co	C	P	S	Cr	Si	—	—
Oyama Smelter	Ferro-nickel	Mass fraction/%	21.9	0.03	0.019	0.44	0.19	0.01	—	—
	Slag	Component	Ni+Co	SiO₂	Fe	Al₂O₃	MgO	CaO	C	S
		Mass fraction/%	0.2	53.4	6.0	2.5	28.4	5.7	0.2	0.07
Jiangsu Province	Ferro-nickel	Component	Ni	Fe	C	S	P	Si	—	—
		Mass fraction/%	10.36	82.35	2.15	0.44	0.057	1.67	—	—
	Slag	Component	Ni	Fe	C	S	—	—	—	—
		Mass fraction/%	0.1	5.72	11.13	0.11	—	—	—	—

杨超等[48]研究了以 Fe_T 21.70%、Ni 1.92% 的低品位红土镍矿为原料,采用回转窑选择性还原-磁选工艺制备镍铁合金,在 1150 ℃、细磨时间 3 min、磁场强度 150 mT 条件下,所得镍铁合金中镍品位 7.26%、镍回收率 96.06%、铁品位 85.15%、铁回收率 89.23%。

针对此工艺,一些研究人员将重点放在红土镍矿还原焙烧工艺,通过使用不同还原剂如碳、CO、CH_4[5]、H_2[49]、生物质炭[50] 等,然后进行磁选工序回收镍。丁志广等[51]以含镍 0.82%、含铁 9.67% 的硅镁型红土镍矿为原料开展氢气低温还原研究,在还原温度为 600 ℃、还原时间 90 min 及氢气浓度为 60%(体积分数)的条件下,红土镍矿中镍、铁金属化率分别达到 95% 和 42%,通过氢气低温还原,矿物中的氧化镍几乎完全还原,大部分铁以低价氧化物形式存在。刘兴阳等[52]研究了 CH_4 与黏土型红土镍矿(Ni 1.65%、Fe 25.37%)中铁镍氧化物的还原行为。在 900 ℃,还原时间 60 min,CH_4 浓度为 50% 条件下,得到镍品位 3.06%,回收率为 52.9%。镍品位和回收率较低的原因主要为 CH_4 还原过程存在大量的镍仍赋存于硅酸盐中和部分还原的镍铁合金粒度过细。

3.2.2 添加剂强化还原-磁选法

自从 Okamoto[53,54]研究了硅镁镍矿在氯化钠作用下分离镍的机理以来,添加剂强化红土镍矿还原-磁选工艺得到了越来越多的关注。添加剂强化还原镍目的是大幅度地还原镍氧化物转变为金属镍并保持铁氧化物未还原或尽量减少氧化亚铁还原为金属铁。用于选择性还原的添加剂包括 Na_2SO_4、Na_2CO_3、$CaSO_4$、$Na_2S_2O_3$、$MgCl_2$、NaCl 和 $CaCl \cdot H_2O$ 等[55],反应机理如图 12 和图 13 所示。红土镍矿还原过程中镍铁颗粒的聚集长大行为被认为与液相烧结机制有关,包裹在镍铁颗粒周围的低熔点物质驱动镍铁颗粒迁移[56]。当 Na_2CO_3 或 Na_2O 作添加剂时,钠离子可破坏硅酸盐的结构,焙烧过程 Na_2O 与硅酸盐矿物发生反应生成霞石,霞石的生成加速了金属颗粒的迁移[57]。

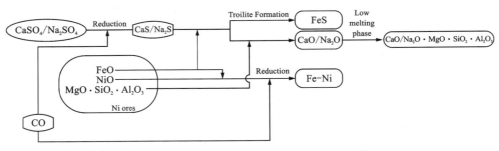

图 12 含硫添加剂促进镍选择性还原机理[14]

Fig. 12 Mechanism of sulfo compound additives on promoting selective reduction of nickel

由于添加碱金属盐可促进红土镍矿还原,因此众多研究者研究了红土镍矿与赤泥共还原焙烧的可行性。赤泥是铝企业生产氧化铝过程中产生的一种强碱性固体废弃物,包含大量的 Fe_2O_3、CaO、Na_2O、Al_2O_3、SiO_2 等成分。赤泥资源化利用已开发出有效工艺,然而这些应用仍无法解决大量赤泥问题(全球每年生成 120 Mt 的赤泥)[58]。Wang 等[59]研究发现赤泥的加入有助于镍铁聚集长大,在无烟煤 10%、赤泥 35%、1300 ℃ 下焙烧 1 h 条件下,镍铁中镍和铁的品位分别为 4.90%(红土镍矿中镍 1.62%)和 71.00%,镍和铁的回收率分别为 95%、93.77%。Guo 等[60]以生物质碳为还原剂,对含镍 1.16% 红土镍矿和赤泥进行了共还原焙烧

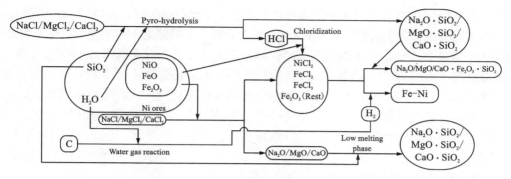

图 13 氯化物促进镍选择性还原机理[14]

Fig. 13 Mechanism of chloride compound additives on promoting selective reduction of nickel

研究,在生物质炭15%、赤泥25%、1250 ℃下焙烧80 min 条件下,镍铁中镍和铁的品位分别为1.81%和81.40%,镍和铁的回收率分别为97.21%和98.87%,制备镍铁粒径大,主要是生物质碳中硫(0.14%)和灰分(3.80%)较低,不会阻碍镍铁颗粒的团聚;且熔点低,可促进镍铁颗粒聚集长大,给贫煤地区的红土镍矿和赤泥的共还原提供一种新解决方案。

3.2.3 回转窑-矿热电炉法

20世纪50年代,回转窑-矿热电炉法(简称"RKEF")被新喀里多尼亚安博厂开发成功,是目前红土镍矿火法冶炼企业普遍采用的技术。2011年4月,国家发展和改革委员会发布了《产业结构调整指导目录(2011版)》,其中含镍红土矿高效利用RKEF技术被列为鼓励项目。

中国恩菲公司通过对RKEF工艺进行优化升级,在世界范围内进行推广应用;目前全球采用RKEF工艺的公司有几十家,遍布于东南亚、欧洲、美洲等。RKEF工艺原则流程如图14所示,红土镍矿先在干燥窑内脱除游离水;然后在回转窑内于850~1000 ℃的温度下焙烧预还原,并脱除结晶水;高温炉料直接送入电炉中在1500~1600 ℃的温度下进行还原熔炼生产镍铁[61]。

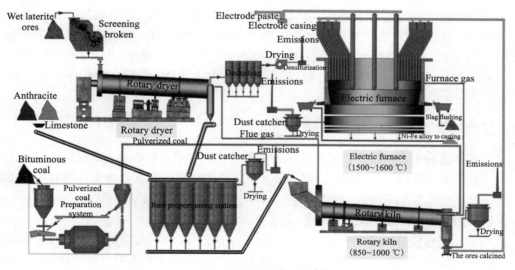

图 14 RKEF工艺流程图[66]

Fig. 14 Schematic diagram of RKEF

2011年，缅甸达贡山镍矿项目建成投产，是迄今为止中缅两国矿业领域最大的合作项目之一。缅甸达贡山镍矿拥有镍资源量（金属量）70万吨，基础储量46.74万吨，镍平均品位在1.9%左右；镍铁中镍稳定在35%左右，年产量稳定在2.2万吨镍金属[62]。

侯俊京等[63]研究了混合煤配比对回转窑预还原以及配碳量和温度对电炉熔炼的影响。优化条件下，产出的镍铁含镍23.13%、镍回收率为95.21%、铁回收率为91.97%。

该工艺具有镍铁品位高（镍15%~35%）[64]、有害元素少、原料适应性强、生产效率高、工艺成熟等优点，但也存在能耗高[65]、熔炼渣量大、粉尘污染严重、无法回收红土镍矿中钴资源等问题。

3.2.4 高炉熔炼镍铁法

1909年，高炉公司在多尼安博利用高炉冶炼工艺生产镍铁，是世界上第一家生产镍铁的厂家；我国21世纪初，采用高炉熔炼方法生产镍铁得到了快速发展[67]。烧结高炉法工艺流程为褐铁矿型红土镍矿经过干燥破碎后，通过烧结成块-高炉冶炼生产镍品位3%~5%的镍生铁[68]。此工艺具有产品质量不稳定、镍铁中S、P含量高、运营成本高、焦炭消耗量大、污染严重等缺点。另外也存在一些其他技术难题，如红土镍矿软熔区间宽、渣量大，造成高炉透气性差。

由于褐铁型红土镍矿中含大量游离水和结晶水，该工艺在红土镍矿烧结前需采用生石灰搅拌脱水，其原理是利用生石灰的吸湿性和放热达到脱水的目的。生石灰脱水一般要求5~10天，脱水时间不宜太短，若过短则会影响烧结矿质量[69]。在炉料烧结过程中存在烧损大、烧结矿强度低、燃料用量大等问题。针对以上问题，朱德庆等[65]提出低品位块矿铺底料-热风烧结-加压烧结多技术协同强化红土镍矿烧结工艺，改善了烧结矿微观结构和矿物组成，使固体能耗降低24.56%，形成了优质烧结炉料。

3.2.5 其他工艺

在分析红土镍矿火法冶炼工业化的基础上，神雾公司创新性地开发出了蓄热式转底炉-蓄热式高温熔分工艺处理红土镍矿生产镍铁合金[70]。生产表明：当石灰石粒度为-74μm，石灰石配入量约6%，还原煤粒度为1mm，还原煤配入量约7%，转底炉高温还原区温度控制在约1350℃时，转底炉金属化率约61%，熔分后可获得含镍20%以上的镍铁合金，镍回收率大于93%[71]。昆明理工大学真空冶金团队基于真空热还原原理，创新性地提出红土镍矿真空碳热还原工艺，即在真空、高温条件下，还原剂碳与红土镍矿中镍、铁和镁发生还原反应，使镁以蒸汽形式挥发进入气相中，通过冷凝系统回收金属镁，然后再采用磁选工艺生产镍铁，此工艺为高镁型红土镍矿中镁的高值化利用提供理论依据[39]。

3.3 红土镍矿火法工艺生产镍锍

红土镍矿火法工艺生产镍锍是指将矿石中的镍、钴和部分铁硫化，形成金属硫化物共熔体。常用硫化剂为硫磺、黄铁矿、石膏、含硫镍原料等物质，熔炼产物为低镍锍、炉渣和烟气，低镍锍经吹炼生产高镍锍。

3.3.1 鼓风炉硫化熔炼生产镍锍

鼓风炉还原硫化熔炼是红土镍矿最早处理工艺之一。将红土镍矿进行破碎、干燥和筛分，然后加入焦粉、石膏、黄铁矿等配料并制团，最后将团料在鼓风炉中进行熔炼，得到低镍锍，此流程镍回收率约85%[68]；低镍锍经转炉吹炼生产高镍锍，高镍锍产品一般含镍70%、硫 19.5%；全流程镍回收率约 70%[72]。该方法具有工艺成熟、操作简单、镍产品形式多样等优点，但也存在能耗高、污染大、矿石适应性差等缺点[73]。

新喀里多尼亚的红土镍矿生产镍锍过程首先选择石膏作硫化剂，加入炉体内石膏先发生脱水反应，然后与还原剂碳反应生成硫化钙，这就说明添加石膏需要加入额外的还原剂碳。当黄铁矿作为硫化剂时，可减少焦炭消耗，这是因为黄铁矿中硫含量高，其消耗量比用石膏作硫化剂时少一半；黄铁矿可调节镍锍组成，黄铁矿分解产生的 FeS 直接进入镍锍中。实践指出，用黄铁矿代替部分石膏可降低焦炭消耗 15%~20%[74]。

鼓风炉内可分为烘干区、硫化还原区、物料融化和焦炭燃烧区，每个区发生反应如下所示：

(1) 烘干区

在烘干区域，原料脱掉吸附水、结晶水和结构水。

$$(Ni, Mg)O \cdot SiO_2 \cdot H_2O = (Ni, Mg)O \cdot SiO_2 + H_2O(g) \quad (15)$$

$$CaSO_4 \cdot 2H_2O = CaSO_4 + 2H_2O(g) \quad (16)$$

(2) 还原硫化区

在此区域，炉体中物质发生还原反应：

$$2C + O_2(g) = 2CO(g) \quad (17)$$

$$3Fe_2O_3 + CO(g) = 2Fe_3O_4 + CO_2(g) \quad (18)$$

$$Fe_3O_4 + CO(g) = 3FeO + CO_2(g) \quad (19)$$

$$FeO + CO(g) = Fe + CO_2(g) \quad (20)$$

$$NiO + CO(g) = Ni + CO_2(g) \quad (21)$$

在此区域内也会进行着还原石膏和石灰石热分解：

$$CaSO_4 + 4C = CaS + 4CO(g) \quad (22)$$

$$CaSO_4 + 4CO(g) = CaS + 4CO_2(g) \quad (23)$$

$$CaCO_3 = CaO + CO_2(g) \quad (24)$$

固体的 CaS 可与 FeO 和 NiO 反应生成 FeS 和 Ni_3S_2：

$$CaS + FeO = FeS + CaO \quad (25)$$

$$2CaS + 3NiO + C = Ni_3S_2 + 2CaO + CO(g) \quad (26)$$

体系内产生的硫蒸气可与金属态镍、铁反应：

$$2Fe + S_2(g) = 2FeS \quad (27)$$

$$3Ni + S_2(g) = Ni_3S_2 \quad (28)$$

(3) 物料融化和焦炭燃烧区域

炉内存在的 SiO_2 促使石灰石与石膏分解以及造渣反应：

$$CaCO_3 + SiO_2 = CaO \cdot SiO_2 + CO_2(g) \quad (29)$$

$$CaSO_4 + SiO_2 + CO(g) = CaO \cdot SiO_2 + CO_2(g) + SiO_2(g) \quad (30)$$

$$2FeO + SiO_2 = 2FeO \cdot SiO_2 \quad (31)$$

二氧化硫还原成硫蒸气：

$$2SO_2(g) + 4CO(g) = S_2(g) + 4CO_2(g) \quad (32)$$

硫化亚铁可与氧化镍发生硫化反应：

$$FeS + NiO = NiS + FeO \quad (33)$$

图 15 为顿尼安波冶炼厂生产高冰镍流程图，即将矿石(含镍 2.6%)经过烧结机-回转窑

预焙烧，焙烧后粒矿含镍 2.9%；配料后在鼓风炉内进行硫化熔炼，低镍锍及炉渣进入炉缸，低镍锍间断出炉，温度约 1370 ℃。其低冰镍平均含镍+钴 27%、铁 63%、硫 10%，镍回收率约 90%。

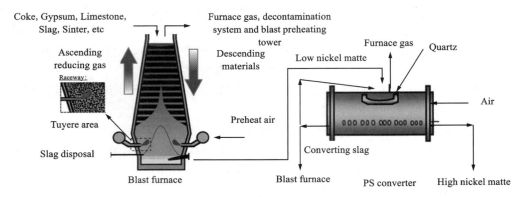

图 15　顿尼安波厂生产冰镍流程图[74]

Fig. 15　Flow chart of nickel matte production at Toniambo plant

表 3 为全球采用鼓风炉生产镍锍的部分企业及其基本情况[75]。

表 3　全球采用鼓风炉生产镍锍的部分企业及其基本情况[75]

Table 3　Some enterprises using blast furnace to produce nickel matte in the world and their basic situation

Country	Smelter	Parameters	Low nickel matte	Smelting slag	Nickel recovery/%	Cobalt recovery/%
New caledonia	Toniambo plant	Ni 2.9%, Coke 22.7%, Limestone 20.4%, Gypsum 8%	Ni+Co 27% Fe 63% S 10%	Ni+Co 0.3%; SiO_2 ~50%; CaO 10%~15%	~90%	—
Russia	Leishi plant	Ni 1.02%, Coke 30%, Limestone 31.2%, Pyrite(Gypsum) 4.6%	Ni 10% Co 0.4% Fe 65% S 20%	Ni 0.15%; SiO_2 42.5%; FeO 16.1%; CaO 17.2%; MgO 17.2%; Al_2O_3 9.5%	81.35%	60.0%
	Ufale plant	Ni 1.03%, Coke 25%, Limestone 22.9%, Pyrite(Gypsum) 10.8%	Ni 17% Co 0.7% Fe 56% S 18%	Ni 0.20%; SiO_2 43.4%; FeO 23.5%; CaO 13.8%; MgO 9.5%; Al_2O_3 9.1%	75.31%	49.5%
	Orsk plant	Ni 1.17%, Coke 20%, Limestone 35.3%, Pyrite(Gypsum) 7.6%	Ni 14%~18% Co 0.7% Fe 56% S 18%	Ni 0.17%; SiO_2 40%~46%; FeO 17%~22%; CaO 14%~18%; MgO 8%~9%; Al_2O_3 8%	67.04%	48.4%

续表3

Country	Smelter	Parameters	Low nickel matte	Smelting slag	Nickel recovery/%	Cobalt recovery/%
Japan	Sumitomo	Ni 2.6%~3.0%, Nickel sulfide concentrate 30%, Coke 13%	Ni 27%	Ni 0.17% MgO 6.8%	—	—

Su 等[76]研究了多金属结核和红土镍矿还原硫化生产镍锍工艺。利用 FactSage 热力学软件预测渣系液相线温度和镍锍形成条件。研究结果表明锍中镍浓度随着硫化剂 FeS 添加量增加而降低；还原剂加入显著影响镍品位。通过向红土镍矿和多金属结核中分别添加 0.5% 和 3% 碳，锍相中镍≥30% 和铁≤40%。在多金属结核冶炼过程中用 $CaSO_4$ 代替硫化剂 FeS，锍相中铁≤15%、镍≥50%。

Wang 等[77]提出了一种在 1500 ℃下用 CaS 作为硫化剂从腐殖土型红土镍矿中生产镍锍方法。在 Ar 气氛下，用石墨还原 $CaSO_4$ 制备 CaS。发现随着 CaS 加入量增加，镍回收率增加，镍品位降低。当 CaS 加入量为 NiO 硫化所需化学计量的 6 倍时，Ni 回收率为 93.63%、Ni 品位为 8.84%、CaS 利用率 92.10%。与已报道的以 S_2 为硫化剂的方法相比，使用 CaS 可以减少 SO_2 排放，提高 Ni 回收率。

3.3.2 回转窑硫化法生产镍锍

20 世纪 70 年代淡水河谷印尼公司 PTVI 成功开发回转窑硫化法生产镍锍工艺，1973 年在苏拉威西地区建设一条火法冶金生产线，1978 年开始生产，产品为高冰镍[78]，其工艺流程图如图 16 所示。其处理流程为红土镍矿经干燥等处理后，入回转窑进行选择性还原，硫磺先加热熔化，将熔融硫磺喷入窑尾使金属镍和铁硫化，然后加入至矿热电炉还原硫化造锍熔炼生产镍锍。该工艺在 RKEF 生产镍铁工艺上进行了一定改进，主要是在干燥窑和回转窑处加入高硫材料，并在回转窑出料口喷入熔融硫进行硫化，在电炉中冶炼得到了低冰镍。回转窑出料口处的硫化是放热反应，减少电炉和转炉的能耗，相比于青山系的镍铁硫化有一定成本优势，但是回转窑硫化时硫利用率较低，需要增加回转窑烟气脱硫系统。

危地马拉红土镍矿成分：Ni 2.1%，Fe 18.6%，SiO_2 32.5%，MgO 20.8%。用干燥窑干燥至含水 18%~20%；在回转窑中喷入熔融硫磺进行还原硫化，还原硫化温度 900 ℃；再将热焙砂送至矿热电炉中进行还原硫化熔炼，产出低镍锍、熔炼渣和烟气；将低镍锍送至转炉吹炼，得到高镍锍和转炉渣。低镍锍成分为：32.4% Ni、57.2% Fe、9.5% S；熔炼渣成分为 0.2% Ni、18% Fe、0.3% S、42% SiO_2、27.7% MgO；高镍锍成分为 76.7% Ni、0.6% Fe、21.7% S[75]。

表 4 为国际镍公司操作参数与指标[80]。

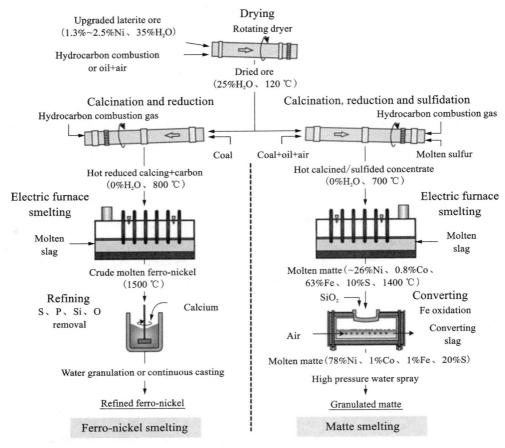

图 16 回转窑硫化法生产镍锍流程图[79]

Fig. 16 Flow chart of sulphide production of nickel matte by rotary kiln

表 4 国际镍公司操作参数与指标[80]

Table 4 Inco operating parameters and indicators

Projects	Unit	Parameters
Feed composition	%	Ni 1.8~1.9; Co 0.06 Fe 20; Fe/Ni 10.5
Ore moisture in	%	29~34
Ore moisture out	%	20
Calcine discharge temperature	℃	700
Average reductant consumption	kg/t of Dry ore	35~40
Dusting rate	%	15~17
Low nickel matte temperature	℃	1350~1400

续表4

Projects	Unit	Parameters
Smelting slag temperature	℃	1500~1550
Low nickel matte composition	%	Ni 26、Co 0.8、S 10、Fe 63
Smelting slag composition	%	SiO_2 47.6、MgO 22.7、Fe 18.4 SiO_2/MgO 2.1
Average blowing	m^3/h	18000
High nickel matte composition	%	Ni 78、Co 1、Fe 1、S 18~22
Converting slag composition	%	SiO_2 25、Fe 53
Nickel recovery	%	90

Selivanov 等[81]研究了铜硫化矿氧化焙烧产物作为硫化剂在腐殖土型红土镍矿冶炼镍锍中的应用。焙烧产物、腐殖土型红土镍矿、氧化钙和碳的混合物以 60∶100∶10∶2.5 的质量比加热焙烧可获得锍相(4.1%Ni、2.5%Cu、0.38%Co、2.1 g/t Au 和 3.6 g/t Ag)。硫化铜矿氧化焙烧产物可作为腐殖土型矿冶炼过程中硫化剂和有价金属的捕收剂。

3.3.3 镍铁合金硫化法生产镍锍

镍铁合金硫化法被新喀里多尼亚 Eramet SLN 公司开发成功,其工艺流程图如图 17 所示。镍铁合金硫化是在红土镍矿完成镍铁生产的基础上,在后续添加一个转炉,将镍铁送入转炉并加入少量硅石、液态硫,硫化得到低冰镍,之后二次转炉吹炼硫化后得高冰镍。由于传统回转窑硫化具有硫利用率低、需增加额外脱硫系统,因此目前新建设的 RKEF 生产高冰镍项目一般以镍铁硫化工艺为主。青山集团、友山镍业的高冰镍项目是非常典型的 RKEF 镍铁硫

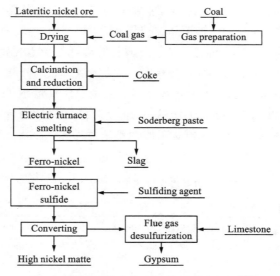

图 17 镍铁硫化法生产镍锍原则流程图[83]

Fig. 17 Process flow chart of nickel matte production by ferric sulfide method

化工艺。2021年，青山集团一年内向中伟股份供应4万吨高冰镍，向华友钴业供应6万吨高冰镍。镍铁硫化法生产镍锍可将不锈钢产业与新能源产业连接，实现红土镍矿到镍铁和镍锍产品的自由切换[82]，但镍铁硫化法仍需要电炉，导致成本高、能耗高、过程复杂等缺点。

通过电弧炉熔炼生产Fe-Ni-Cu-Co合金难以粉碎和浸出，为了克服上述问题，研究者们提出将Fe-Ni-Cu-Co合金转化为Fe-Ni-Cu-Co-S锍相（w_s>20%）。传统硫化方法是将元素硫加入熔融合金中，但具有硫利用效率低、硫成本高等问题。

Jeong等[84]提出硫酸钙（$CaSO_4$）作为硫源将Fe-Ni-Cu-Co合金转变为锍相，在1400℃、CO-CO_2-SO_2-Ar气氛，以固体碳为还原剂的条件下，研究了氧化物熔剂对硫化效率的影响规律。在未添加还原剂时，$CaSO_4$被液态合金Fe-Ni-Cu-Co中Fe还原，形成FeS，但硫化效率约为56%；添加固体碳可将反应平衡时间从36小时缩短到3.5小时，并将硫化效率从56%提高到91%。加入5%Al_2O_3-5%SiO_2和10%Al_2O_3熔剂对反应平衡时间和硫化效率均无影响，但加入5%Al_2O_3-5%Fe_2O_3熔剂后，反应平衡时间降低至2.5小时，因为前者产生具有相对高熔点的硅酸钙和铝酸钙，而后者产生具有较低熔点的铁酸钙。硫酸钙（废石膏）可以代替纯硫作为Fe-Ni-Cu-Co合金硫化的原料。由于前期探索得知加入5%Al_2O_3-5%Fe_2O_3熔剂对硫化熔炼效果有较高硫化率，因此Heo等[85]在此基础上探索氧化铁皮、赤泥、铝浮渣等工业废料对合金硫化的影响。结果发现单独添加赤泥和铝浮渣时，硫化效率为76(±2)%；在添加5%氧化铁皮和5%铝浮渣时，硫化效率约为89%，相当于化学试剂"5%Al_2O_3+5%Fe_2O_3"熔剂的硫化效果。

3.3.4 直接硫化熔炼生产镍锍

直接硫化熔炼采用富氧侧吹炉，富氧侧吹熔炼技术是在俄罗斯瓦纽科夫熔炼法的基础上由我国自主研发的[86]。富氧侧吹熔炼技术在铅、铜等有色金属领域的成功应用，为其在镍冶炼方面应用积累了丰富的工业经验。富氧侧吹还原熔炼技术以多通道侧吹喷枪以亚音速向熔池内喷入富氧空气、硫化剂和燃料（天然气、煤气、粉煤），在强烈鼓风作用下，使氧化镍矿中镍钴等金属发生还原硫化反应。具体步骤主要分为红土镍矿干燥脱水，破碎筛分，富氧侧吹炉内还原硫化等工序，具有成本低、镍锍品位控制灵活、热利用率高、对燃料要求低等优势。

2006年俄罗斯南乌拉尔镍厂利用瓦纽科夫工艺处理低品位氧化镍矿，镍回收率为88%，低镍锍含镍12.4%，炉渣含镍0.17%。董晓伟[87]等介绍了氧气侧吹熔炼技术用于处理红土镍矿的工程设计实例，主要工艺技术指标见表5。某企业设计规模为处理红土镍矿21 kt/a，产出含镍12.5%的低镍锍3100 t/a。中伟新材料股份有限公司（简称"中伟新材"）和印尼恒生新能源材料有限公司计划采用富氧侧吹熔炼技术生产高镍锍，其中中伟新材预年产高镍锍含镍金属6万吨。

表5 氧气侧吹熔炼工艺技术指标[87]

Table 5 Technical index of oxygen side blowing smelting process

Enterprise	Ore handling capacity/(t·a^{-1})	Smelting oxygen concentration/%	Coal consumption rate/%	Dust yield/%	Direct nickel yield/%	Low nickel matte/%	Slag/%
Factors	21000	90	~25	~0.75	~90	Ni 12.5 S 25.95	Ni 0.18 Fe 12.78 SiO_2 40.64

针对传统硫化过程中硫利用率低、操作环境恶劣等问题，王亲猛等[88]提出"一种红土镍矿循环硫化提取镍钴的方法"创新方法，流程如图18所示。该工艺首先将红土镍矿破碎后进行焙烧得焙烧产物；焙烧产物与硫化剂（循环硫化介质）、还原剂、渣型调质剂配料进行硫化熔炼，得低品位镍锍，实现镍钴富集；对低镍锍进行吹炼，得富钴高品位镍锍，然后再通过湿法分离提取镍钴；热态吹炼渣排出后加入硫化剂和还原剂，进行再熔炼得到富钴镍锍，返回吹炼工序；将冶炼系统产生的高温烟气收集，对红土镍矿焙烧，焙烧后烟气经乳化脱硫制备循环硫化介质，返回冶炼系统。本方法通过红土镍矿循环硫化提取镍钴，具有镍钴回收率高、硫循环利用、碳减排及实现高温烟气余热利用的优点，过程低碳环保。

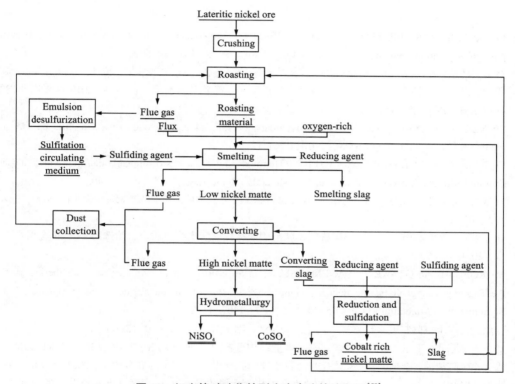

图18 红土镍矿硫化转型生产高冰镍流程图[88]

Fig. 18 Flow chart of sulphide transformation of laterite nickel ore to produce high matte

4 结论与展望

随着新能源产业快速发展，动力电池需求急剧增长，镍需求日渐迫切，因此镍资源的高效提取，对推动我国新能源产业可持续性发展具有重要意义。但现有RKEF法采用碳还原生产镍铁合金，过程能耗高、碳排放量大，原料中钴未有效利用，无法直接与新能源产业衔接；高压浸出法处理红土镍矿投资大，产生大量湿法酸性渣，属于危废，处置困难；传统鼓风炉和电炉硫化技术又存在硫利用率低、镍钴回收率低、操作环境恶劣等问题。基于上述分析，提出如下展望：

（1）全球新能源产业快速发展刺激了镍消费模式的转变，镍在电池领域中消费比例将逐年增加。

（2）镍铁合金硫化法生产镍锍可将RKEF工艺与新能源产业打通，但如何降低成本、减少碳排放，降低能耗、缩短流程是制约其广泛应用的瓶颈问题。

（3）红土镍矿直接硫化冶炼提取镍钴技术具有显著优势，是未来重要发展方向。

参考文献

[1] FARROKHPAY S, FILIPPOV L, FORNASIERO D. Pre-concentration of nickel in laterite ores using physical separation methods[J]. Minerals Engineering, 2019, 141: 105892.

[2] 饶富, 马恩, 郑晓洪, 等. 硫化镍矿中镍提取技术研究进展[J]. 化工学报, 2021, 72(1): 495-507.

[3] 潘建, 田宏宇, 朱德庆, 等. 镍矿资源供需分析及红土镍矿开发利用现状[C]. 2019年镍产业发展高峰论坛暨APOL年会会刊, 南京, 2019: 22-31.

[4] PICKLES C A, ANTHONY W. Thermodynamic modelling of the reduction of a saprolitic laterite ore by methane[J]. Minerals Engineering, 2018, 120: 47-59.

[5] PICKLES C A, FORSTER J, ELLIOTT R. Thermodynamic analysis of the carbothermic reduction roasting of a nickeliferous limonitic laterite ore[J]. Minerals Engineering, 2014, 65: 33-40.

[6] 刘贵清, 张邦胜, 张帆, 等. 中国镍矿资源与市场分析[J]. 中国资源综合利用, 2020, 38(7): 102-105.

[7] 孔令湖, 邓文兵, 尚磊. 中国镍矿资源现状与国家级镍矿床实物地质资料筛选[J]. 有色金属：矿山部分, 2021, 73(2): 79-86.

[8] MUROFUSHI A, OTAKE T, SANEMATSU K, et al. Mineralogical evolution of a weathering profile in the Tagaung Taung Ni laterite deposit: significance of smectite in the formation of highgrade Ni ore in Myanmar[J]. Mineralium Deposita, 2022, 57(7): 1107-1122.

[9] 于培云. 需求空间打开全球镍资源站上风口[N]. 期货日报, 2021-10-13(003).

[10] 唐萍芝, 陈欣, 王京. 全球镍资源供需和产业结构分析[J]. 矿产勘查, 2022, 13(1): 152-156.

[11] 徐爱东, 陈瑞瑞, 李烁, 等. 镍钴行业发展形势分析及建议[J]. 中国有色冶金, 2021, 50(6): 9-15.

[12] 程续. 层状结构高镍三元正极材料的改性及电化学性能的研究[D]. 北京：北京科技大学, 2022.

[13] 石文堂. 低品位镍红土矿硫酸浸出及浸出渣综合利用理论及工艺研究[D]. 长沙：中南大学, 2011.

[14] PINTOWANTORO S, ABDUL F. Selective reduction of laterite nickel ore[J]. Materials Transactions, 2019, 60(11): 2245-2254.

[15] ELLIOTT R, PICKLES C A, FORSTER J. Thermodynamics of the reduction roasting of nickeliferous laterite ores[J]. Journal of Minerals & Materials Characterization & Engineering. 2016, 4(6): 320-346.

[16] RHAMDHANI M A, HAYES P C, JAK E. Nickel laterite Part 2-thermodynamic analysis of phase transformations occurring during reduction roasting[J]. Mineral Processing & Extractive Metallurgy, 2009, 118(3): 146-155.

[17] 李栋, 郭学益. 低品位镍红土矿湿法冶金提取基础理论及工艺研究[M]. 北京：冶金工业出版社, 2015.

[18] 朱宇平. 红土镍矿湿法冶金工艺综述及进展[J]. 世界有色金属, 2020(18): 5-7.

[19] MANO E S, CANER L, PETIT S, et al. Ni-smectitic ore behaviour during the Caron process[J]. Hydrometallurgy, 2019, 186: 200-209.

[20] CHEN J, JAK E, HAYES P C. Investigation of the reduction roasting of saprolite ores in the Caron process: Microstructure and thermodynamic analysis[J]. Mineral Processing and Extractive Metallurgy, 2021, 130(2): 160-169.

[21] CHEN J, JAK E, HAYES P C. Investigation of the reduction roasting of saprolite ores in the Caron process: Effect of sulphur addition[J]. Mineral Processing and Extractive Metallurgy, 2021, 130(2): 170-179.

[22] 牟文宁, 崔富晖, 黄志鹏, 等. 红土镍矿硫酸焙烧-浸出溶液中铁、镍回收的研究[J]. 矿产综合利用, 2018(1): 22-25.

[23] 张云芳, 李金辉, 高岩, 等. 红土镍矿的硫酸铵焙烧过程[J]. 中国有色金属学报, 2017, 27(1): 155-161.

[24] RIBEIRO P P M, SANTOS I D, NEUMANN R, et al. Roasting and leaching behavior of nickel laterite ore[J]. Metallurgical and Materials Transactions B, 2021, 52(3): 1739-1754.

[25] GUO Qiang, Qu Jing-kui, QI Tao, et al. Activation pretreatment of limonitic laterite ores by alkali-roasting method using sodium carbonate[J]. Minerals Engineering, 2011, 24(8): 825-832.

[26] 傅建国, 刘诚. 红土镍矿高压酸浸工艺现状及关键技术[J]. 中国有色冶金, 2013, 42(2): 6-13.

[27] JOHNSON J A, CASHMORE B C, HOCKRIDGE R J. Optimisation of nickel extraction from laterite ores by high pressure acid leaching with addition of sodium sulphate[J]. Minerals Engineering, 2005, 18(13/14): 1297-1303.

[28] 皮关华, 孔凡祥, 贾露萍, 等. 瑞木红土镍矿高压酸浸的生产实践[J]. 中国有色冶金, 2015, 44(6): 11-14.

[29] 邢姜, 冷红光, 韩百岁, 等. 红土镍矿湿法冶金工艺现状及研究进展[J]. 有色矿冶, 2021, 37(5): 26-32.

[30] GUO Xue-yi, SHI Wen-tang, LI Dong, et al. Leaching behavior of metals from limonitic laterite ore by high pressure acid leaching[J]. Transactions of Nonferrous Metals Society of China. 2011, 21(1): 191-195.

[31] 马保中, 王成彦, 杨卜, 等. 硝酸加压浸出红土镍矿的中试研究[J]. 过程工程学报, 2011, 11(4): 561-566.

[32] 王成彦, 曹志河, 马保中, 等. 红土镍矿硝酸加压浸出工艺[J]. 过程工程学报. 2019(S1): 51-57.

[33] SHAO Shuang, MA Bao-zhong, WANG Xin, et al. Nitric acid pressure leaching of limonitic laterite ores: Regeneration of HNO_3 and simultaneous synthesis of fibrous $CaSO_4 \cdot 2H_2O$ byproducts[J]. Journal of Central South University, 2020, 27(11): 3249-3258.

[34] HE Fei, MA Bao-zhong, WANG Cheng-yan, et al. Microwave pretreatment for enhanced selective nitric acid pressure leaching of limonitic laterite[J]. Journal of Central South University, 2021, 28(10): 3050-3060.

[35] MA Bao-zhong, YANG Wei-jiao, YANG Bo, et al. Pilot-scale plant study on the innovative nitric acid pressure leaching technology for laterite ores[J]. Hydrometallurgy, 2015, 155: 88-94.

[36] LUO Jun, RAO Ming-jun, LI Guang-hui, et al. Self-driven and efficient leaching of limonitic laterite with phosphoric acid[J]. Minerals Engineering, 2021, 169: 106979.

[37] 郭欢, 付海阔, 靖青秀, 等. 用硫酸从红土镍矿中常压浸出镍钴铁试验研究[J]. 湿法冶金, 2020, 39(3): 190-193+202.

[38] 黄诗汉, 吴浩, 郑江峰, 等. 硝酸常压浸出红土镍矿特性及镍浸出动力学[J]. 矿冶, 2021, 30(5): 70-76.

[39] 曲涛, 谷旭鹏, 施磊, 等. 高镁硅红土镍矿开发利用研究现状[J]. 材料导报, 2020, 34(S1): 261-267.

[40] 郭学益, 吴展, 李栋, 等. 红土镍矿常压盐酸浸出工艺及其动力学研究[J]. 矿冶工程, 2011(4): 69-72+76.

[41] GUO Qiang, QU Jing-kui, HAN Bing-bing, et al. Innovative technology for processing saprolitic laterite ores by hydrochloric acid atmospheric pressure leaching[J]. Minerals Engineering, 2015, 71: 1-6.

[42] 李金辉, 徐志峰, 高岩, 等. 氯化铵选择性浸出红土镍矿有价金属[J]. 中国有色金属学报, 2019, 29(5): 1049-1057.

[43] CHE Xiao-kui, SU Xiu-zhu, CHI Ruan, et al. Microwave assisted atmospheric acid leaching of nickel from laterite ore[J]. Rare Metals, 2010, 29(3): 327-332.

[44] ZHANG Pei-yun, SUN Lin-quan, WANG Hai-rui, et al. Surfactant-assistant atmospheric acid leaching of laterite ore for the improvement of leaching efficiency of nickel and cobalt[J]. Journal of Cleaner Production, 2019, 228: 1-7.

[45] TSUJI H. Behavior of reduction and growth of metal in smelting of saprolite ni-ore in a rotary kiln for production of ferro-nickel alloy[J]. Transactions of the Iron & Steel Institute of Japan, 2012, 52(6): 1000-1009.

[46] 陶高驰, 肖峰, 蒋伟. 国内采用回转窑生产镍铁的实践[J]. 有色金属(冶炼部分), 2014(8): 51-54.

[47] GAO Li-hua, LIU Zheng-gen, PAN Yu-zhu, et al. Separation and recovery of iron and nickel from low-grade laterite nickel ore using reduction roasting at rotary kiln followed by magnetic separation technique[J]. Mining Metallurgy & Exploration, 2019, 36(2): 375-384.

[48] 杨超. 低品质红土镍矿选择性还原-磁选制备镍铁合金[J]. 矿冶工程, 2021, 41(2): 99-101.

[49] LIU Shou-jun, YANG Chao, YANG Song, et al. A robust recovery of Ni from laterite ore promoted by sodium thiosulfate through hydrogen-thermal reduction[J]. Frontiers in Chemistry, 2021, 9: 704012.

[50] SUPRIYATNA Y I, SIHOTANG I H, SUDIBYO. Preliminary study of smelting of indonesian nickel laterite ore using an electric arc furnace[J]. Materials Today: Proceedings, 2019, 13: 127-131.

[51] 丁志广, 李博, 魏永刚. 氢气作用下硅镁型红土镍矿的低温还原特性[J]. 中国有色金属学报, 2018, 28(8): 1669-1675.

[52] 刘兴阳, 刘守军, 杨颂, 等. 甲烷作用下红土镍矿中铁镍氧化物的反应行为[J]. 中国有色金属学报, 2022, 32(6): 1759-1771.

[53] OKAMOTO K, UEDA Y, NOGUCHI F. Extraction of nickel from garnierite ore by the segregation-magnetic separation process[J]. Memoirs of the Kyushu Institute of Technology Engineering, 1971.

[54] OKAMOTO K, UEDA Y, NOGUCHI F. Mechanism of nickel segregation from garnierite ore[J]. Journal of the Mining and Metallurgical Institute of Japan, 1971, 87(995): 103-108.

[55] SUHARNO B, NURJAMAN F, RAMADINI C, et al. Additives in selective reduction of lateritic nickel ores: Sodium sulfate, sodium carbonate, and sodium chloride[J]. Mining, Metallurgy & Exploration, 2021, 38(5): 2145-2159.

[56] 杭桂华. 低品位腐泥土型红土镍矿制备镍铁相关基础研究[D]. 武汉: 武汉科技大学, 2020.

[57] 赵剑波, 马东来, 吕学明, 等. Na_2CO_3作用下红土镍矿非等温碳热还原动力学研究[J]. 中国有色金属学报, 2022, 32(4): 1088-1097.

[58] HE Ao-ping, ZENG Jian-min, LIU Shi-hong. Influence of laterite nickel ore on extracting iron from Bayer

red mud by carbothermal smelting reduction[J]. Journal of Iron and Steel Research International, 2021, 28(6): 661-668.

[59] WANG Xiao-ping, SUN Ti-chang, KOU Jue, et al. Feasibility of co-reduction roasting of a saprolitic laterite ore and waste red mud[J]. International Journal of Minerals Metallurgy and Materials, 2018, 25(6): 591-597.

[60] GUO Xiao-shuang, LI Zheng-yao, HAN Ji-cai, et al. Study of straw charcoal as reductant in Co-reduction roasting of laterite ore and red mud to prepare powdered ferronickel[J]. Mining Metallurgy & Exploration, 2021, 38(5): 2217-2228.

[61] KING M G. Nickel laterite technology-Finally a new dawn[J]. JOM, 2005, 57(7): 35-39.

[62] 金永新. 缅甸达贡山镍矿项目生产实践[C]. 2018 红土镍矿行业大会暨 APOL 年会, 成都, 2018: 13-16.

[63] 侯俊京, 贾彦忠, 梁德兰, 等. 红土镍矿回转窑-电炉熔炼生产镍铁的工艺研究[J]. 中国有色冶金, 2014, 43(3): 70-73.

[64] RAO Ming-jun, LI Guang-hui, JIANG Tao, et al. Carbothermic reduction of nickeliferous laterite ores for nickel pig iron production in china: A review[J]. JOM, 2013, 65(11): 1573-1583.

[65] 朱德庆, 田宏宇, 潘建, 等. 低品位红土镍矿综合利用现状及进展[J]. 钢铁研究学报, 2020, 32(5): 351-362.

[66] LIU Peng, LI Bao-kuan, CHEUNG S, et al. Material and energy flows in rotary kiln-electric furnace smelting of ferronickel alloy with energy saving[J]. Applied Thermal Engineering, 2016, 109: 542-559.

[67] 刘继军, 胡国荣, 彭忠东. 红土镍矿处理工艺的现状及发展方向[J]. 稀有金属与硬质合金, 2011, 39(3): 62-66.

[68] 刘安治, 李韩璞. 红土镍矿冶炼工艺分析[J]. 现代冶金, 2013, 41(1): 1-4.

[69] 董训祥, 秦涔. 红土镍矿高炉工艺技术及发展趋势[J]. 炼铁, 2017, 36(3): 60-62.

[70] 王静静, 曹志成, 高建勇, 等. 转底炉处理红土镍矿的试验研究[J]. 工业加热, 2018, 47(6): 9-11.

[71] 季爱兵, 尹鑫平, 李欣. 红土镍矿蓄热式转底炉-蓄热式高温熔分冶炼新工艺[J]. 现代冶金, 2016, 44(1): 14-16.

[72] 李博. 硅镁型红土镍矿干燥特性及预还原基础研究[D]. 昆明: 昆明理工大学, 2012.

[73] 王帅. 红土镍矿火法冶炼技术现状与研究进展[J]. 中国冶金, 2021, 31(10): 1-7.

[74] 别列果夫斯基. 镍冶金学[M]. 北京: 中国工业出版社, 1962.

[75] 周建男, 周天时. 利用红土镍矿冶炼镍铁合金及不锈钢[M]. 北京: 化学工业出版社, 2016.

[76] SU Kun, WANG Feng, PARIANOS J, et al. Alternative resources for producing nickel matte-laterite ores and polymetallic nodules[J]. Mineral Processing and Extractive Metallurgy Review, 2021, 43(5): 584-597.

[77] WANG Hong-yang, HOU Yong, CHANG He-qiang, et al. Preparation of Ni-Fe-S matte from nickeliferous laterite ore using CaS as the sulfurization agent[J]. Metallurgical and Materials Transactions B, 2022, 53(2): 1136-1147.

[78] 高承君. 湿法高压酸浸和火法高冰镍未来前景最新进展[C]. 2021 年 APOL 镍与不锈钢产业链年会, 济南, 2021: 71-77.

[79] WARNER A, DÍAZ C M, DALVI A D, et al. World nonferrous smelter survey, Part IV: Nickel: Sulfide[J]. JOM, 2007, 59(4): 58-72.

[80] WARNER A, DÍAZ C M, DALVI A D, et al. World nonferrous smelter survey, part III: Nickel: Laterite[J]. JOM, 2006, 58(4): 11-20.

[81] SELIVANOV E N, KLYUSHNIKOV A M, GULYAEVA R I. Application of sulfide copper ores oxidizing roasting products as sulfidizing agent during melting nickel raw materials to matte[J]. Metallurgist, 2019, 63(7-8): 867-877.

[82] 张振芳, 陈秀法, 李仰春, 等. 双碳目标下镍资源的综合利用发展趋势[J]. 矿产综合利用, 2022(2): 31-39.

[83] 吴琦, 马文军. 碳中和背景下电池镍行业发展趋势及应对措施[J]. 中国有色冶金, 2021, 50(5): 7-11.

[84] JEONG E H, NAM C W, PARK K H, et al. Sulfurization of Fe-Ni-Cu-Co alloy to matte phase by carbothermic reduction of calcium sulfate[J]. Metallurgical & Materials Transactions B, 2016, 47(2): 1103-1112.

[85] HEO J H, JEONG E H, NAM C W, et al. Use of industrial waste (al-dross, red mud, mill scale) as fluxing agents in the sulfurization of Fe-Ni-Cu-Co alloy by carbothermic reduction of calcium sulfate[J]. Metallurgical & Materials Transactions B, 2018, 49(3): 939-943.

[86] 袁精华. 侧吹炉的现状与展望[J]. 有色金属(冶炼部分), 2022(1): 31-35.

[87] 董晓伟, 李有刚. 我国氧气侧吹炼镍技术的工程应用[J]. 湖南有色金属, 2011, 27(3): 26-27.

[88] 王亲猛, 李中臣, 郭学益, 等. 一种红土镍矿循环硫化提取镍钴的方法: 中国, CN202111522791.0[P]. 2021.

红土镍矿高温硫化熔炼镍锍

摘　要：随着新能源产业快速发展，能源金属镍需求量急剧增长，红土镍矿已成为镍冶炼重要原料。红土镍矿高温硫化冶炼镍锍工艺是未来重要发展方向，红土镍矿经硫化熔炼产低镍锍，再经吹炼产高镍锍，后经湿法浸出净化产硫酸镍，可直接为新能源行业提供原料。但该工艺硫化冶炼理论基础薄弱，亟须开展相关基础研究。本文通过热力学分析，揭示了红土镍矿硫化熔炼低镍锍过程物相演变规律，阐明了硫化熔炼过程机理。结果表明：在一定硫化熔炼条件下，红土镍矿中镍氧化物转变历程为 $NiO \rightarrow Ni \rightarrow Ni_3S_2$，钴氧化物转变历程为 $CoO \rightarrow Co \rightarrow Co_9S_8$，铁氧化物转变途径为 $Fe_2O_3 \rightarrow Fe_2O_4 \rightarrow FeO \rightarrow FeS$ 或 $Fe_2O_3 \rightarrow Fe_3O_4 \rightarrow FeO \rightarrow Fe \rightarrow FeS$；金属与 S 亲和力强弱顺序：$Ni \approx Fe > Co$；金属与 O 亲和力强弱顺序：$Fe > Co > Ni$。经理论计算：硫化熔炼过程，当硫磺添加量为矿料质量的2%，碳添加量为矿料质量的4%时，产出镍锍品位为21.45%，镍、钴回收率分别为99.43%、87.58%，硫直接利用率为62.68%。目前，红土镍矿高温硫化熔炼镍锍，已初步实现工业应用，与常规RKEF技术相比，过程绿色低碳，是具有里程碑意义的技术变革。

关键词：红土镍矿；硫化熔炼；热力学机理；生产实践

Production of nickel matte via high-temperature sulphidizing smelting from laterite nickel ore

Abstract: With the rapid development of the new energy industry, the demand for energy metal nickel has increased rapidly, and laterite nickel ore has become an important raw material for nickel smelting. The technology of producing nickel matte via high-temperature sulphidizing smelting from laterite nickel ore is an important development direction in the future. Laterite nickel ore produces low nickel matte by sulphidizing smelting, low nickel matte produces high nickel matte by converting, and high nickel matte produces nickel sulfate by leaching of hydrometallurgy and purification, which can directly provide raw materials for the new energy industry. However, the theoretical basis of this process was weak, so it was urgent to carry out relevant basic research. In this paper, the phase evolution of producing nickel matte via sulphidizing smelting from laterite nickel ore was revealed by thermodynamic analysis, and the mechanism of sulphidizing smelting

---

本文发表在《中国有色金属学报》，网络首发。作者：李中臣，王亲猛，王松松，田庆华，郭学益。

was elucidated. The results showed that under certain sulphidizing smelting conditions, the transformation course of nickel oxide in laterite ore was NiO→Ni→Ni$_3$S$_2$, the transformation course of cobalt oxide was CoO→Co→Co$_9$S$_8$, and the transformation path of iron oxide was Fe$_2$O$_3$→Fe$_3$O$_4$→FeO→FeS or Fe$_2$O$_3$→Fe$_3$O$_4$→FeO→Fe→FeS. The strength order of affinity between metal and S was Ni≈Fe>Co. The strength order of affinity between metal and O was Fe>Co>Ni. According to the theoretical calculation, when the sulfur content was 2% of the ore mass and the carbon content was 4% of the ore mass, the nickel grade was 21.45%, the recovery of nickel and cobalt were 99.43% and 87.58%, respectively, and the direct utilization rate of sulfur was 62.68%. At present, the technology of producing nickel matte via high-temperature sulphidizing smelting from laterite nickel ore has been preliminarily applied in industry. Compared with conventional RKEF technology, the process is a green and low carbon, which is a milestone technical change.

Keywords: laterite nickel ore; sulphidizing smelting; mechanism of thermodynamics; practice of production

镍是一种重要的战略金属,广泛应用于不锈钢、合金及新能源等领域[1,2]。自1950年至2021年,全球原生镍金属消费量从不足20万吨增加到270余万吨,且持续增长。原生镍资源主要为红土镍矿(氧化镍矿)和硫化镍矿[3]。据2021年美国地质调查局公布数据显示,全球已探明镍品位高于0.5%的陆基镍资源总量(金属量)约3亿吨,其中硫化镍矿占40%,红土镍矿占60%。随着优质硫化镍矿资源日趋枯竭和镍消费需求持续增长,红土镍矿已成为重要镍生产原料,其需求量逐渐增加[4-6]。红土镍矿是由含镍橄榄石经过长期风化和淋滤过程形成,主要分布在赤道附近的热带和亚热带地区[7,8]。一般来说,红土镍矿主要分为褐铁矿型、黏土型和腐殖土型[9]。褐铁矿型分布于红土镍矿上层,主要由针铁矿(FeO·OH)组成,镍主要取代氢氧化亚铁晶格中的铁形成(Fe,Ni)O·OH[10],一般含0.8%~1.5% Ni、36.0%~50.0% Fe、10.0%~30.0% SiO$_2$和0.5%~5.0% MgO。腐殖土型分布于红土镍矿下层,主要由蛇纹石[(Mg,Fe)$_3$Si$_2$O$_5$(OH)$_4$]组成,镍主要以吸附状态或类质同相形态存在于镁或铁的硅酸盐矿物中形成(Mg,Fe,Ni)$_3$Si$_2$O$_5$(OH)$_4$;一般含1.8%~3.0% Ni、10.0%~25.0% Fe、30.0%~50.0% SiO$_2$和15.0%~35.0% MgO[11]。黏土型位于褐铁矿型和腐殖土型中间,一般含1.2%~1.8% Ni、25.0%~40.0% Fe、10.0%~30.0% SiO$_2$和5%~15% MgO。

由于红土镍矿矿物学特性复杂、结晶性差等,红土镍矿中有价金属难以通过物理分选富集[12,13],主要通过火法工艺和湿法工艺[14,15]提取。常规火法工艺[16,17]主要为还原熔炼镍铁法、还原-磁选镍铁法、高炉熔炼镍铁法等,常规湿法工艺[18,19]主要为高压酸浸法、常压酸浸法[20]、还原焙烧-氨浸法等,目前世界范围内应用较为广泛的技术为还原熔炼镍铁法(RKEF)和高压酸浸法(HPAL)。还原熔炼镍铁法具有工艺成熟、流程短、投资少等优点,但存在能耗高、碳排放量大、熔炼渣量大、钴回收率低等缺点,产品主要用于不锈钢,难以满足新能源产业发展需求;高压酸浸法具有可综合回收镍钴钪、镍钴回收率高等优点,但存在酸浸渣环境危害大、投资高、建设周期长等缺点,因此亟须开发红土镍矿处理新工艺。

红土镍矿硫化冶炼法是将矿物原料与硫化剂、还原剂、造渣剂配料后在1300~1500 ℃下熔炼生产低镍锍,然后向低镍锍中吹氧、造渣,在 1200~1300 ℃下吹炼生产高镍锍。该方法具有原料适应性好、镍钴回收率高、产品可直接与新能源产业衔接等优点。但针对该工艺基础理论研究薄弱,亟须开展相关基础研究。本文结合热力学软件 Factsage 7.1 和 HSC Chemistry 6.0,研究了红土镍矿硫化熔炼镍锍过程物相演变规律,揭示了硫化熔炼机理,为红土镍矿硫化冶炼技术发展与生产实践提供理论指导。

1 红土镍矿原料分析

1.1 原料成分

本文红土镍矿取自于印度尼西亚苏拉威西岛。原料在 80 ℃下干燥后先用颚式破碎机(武汉探矿机械厂,MPE-100×250)破碎,然后用球磨机(鹤壁精中科技有限公司,JZGJ100-10)细磨后筛分至粒度 0.074 mm 以下,红土镍矿化学成分见表 1,其含镍 Ni 1.73%、Fe 22.38%;主要脉石成分为 MgO 和 SiO_2,另有少量的 Al_2O_3、MnO、CaO、Cr_2O_3,这是一种典型的腐殖土型红土镍矿。根据镍不同物相在溶剂中溶解度和溶解速度不同,镍化学物相分析见表 2。从表 2 可以看出,87.51%的镍分布于硅酸盐中,9.83%的镍分布于铁矿石中。

表 1 红土镍矿化学组成
Table 1 Chemical composition of the laterite nickel ore %

Compositions	Ni	Fe_T	SiO_2	MgO	Co	Al_2O_3	Cr_2O_3	CaO	MnO
Content	1.73	22.38	29.1	24.22	0.09	1.96	0.28	0.28	0.46

表 2 镍在各物相中分布
Table 2 Distribution of Nickel in associated minerals %

Mineral	Nickel sulfide	Bearing silicate	Bearing iron ore	Nitotal
Content	0.046	1.514	0.170	1.730
Fraction	2.66	87.51	9.83	100

图 1 为矿物解离度分析仪(MLA)检测结果图,MLA 用于分析矿物样品的物质组成、成分定量、矿物的嵌布特征等重要参数。图 1 表明原料主要矿物成分为蛇纹石、针铁矿化蛇纹石、针铁矿、磁铁矿、铬铁矿、石英等;从表 3 可以看出,红土镍矿中蛇纹石相、针铁矿化蛇纹石相和针铁矿相之和占比 90.02%;其中蛇纹石相含量最高,分布比例为 57.10%;其次为针铁矿化蛇纹石相,分布比例为 18.96%。94.14% Ni、83.42% Co、87.92% Fe 分布于蛇纹石相、针铁矿化蛇纹石相、针铁矿相中;4.62% Ni 和 6.43% Co 分布于软锰矿(含铁钴镍镁)中;Ni 基本未分布在磁铁矿相中,而 7.30% Co 和 8.07% Fe 分布在磁铁矿中。

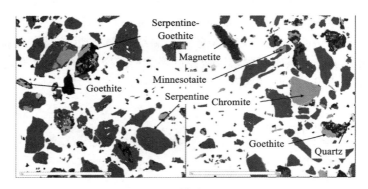

图 1 红土镍矿 MLA 图

Fig. 1　Images of MLA of the laterite nickel ore

表 3　镍、钴、铁在各矿物中的平衡分配

Table 3　The equilibrium distribution of nickel, cobalt and iron in each mineral　　%

Mineral	Content	Distribution		
		Ni	Co	Fe
Serpentine $(Mg, Fe)_3Si_2O_5(OH)_4$	57.10	64.94	19.06	18.32
Goethite $FeO \cdot OH$	13.96	8.65	32.71	36.71
Serpentine-Goethite	18.96	20.55	31.65	32.89
Magnetite Fe_3O_4	2.60	0.32	7.30	8.07
Pyrolusite MnO_2	0.71	4.62	6.43	0.30
Minnesotaite $(Fe, Mg)_3Si_4O_{10}(OH)_2$	0.85	0.59	0.48	0.75
Quartz SiO_2	1.27	0.07	0.21	0.07
Chromite $(Fe, Mg)Cr_2O_4$	2.17	0.02	1.66	2.19
Others	2.38	0.24	0.5	0.7
Total	100	100	100	100

1.2　热重-示差扫描量热分析(TG-DSC)

采用美国 TA 公司的 SDTQ-600 型热分析仪对红土镍矿进行热重-示差扫描量热(TGDSC)分析,在氩气气氛下由室温升温至 1200 ℃,升温速率为 10 ℃/min。红土镍矿中水主要以吸附水、结晶水和结构水三种形态存在[21]。结合图 2 红土镍矿热重-示差扫描量热分析可知,样品失重率约为 15.84%。DSC 曲线上有三个吸热峰,分别位于 97.4 ℃、262 ℃ 和 591.5 ℃ 处。97.4 ℃ 的吸热峰主要是因为吸附水的脱除,并未发生物相转变;升高至 262 ℃ 过程中,参与组成矿物晶体结构的结晶水被脱除;升温至 591.2 ℃ 过程中,原料脱除大部分结构水,主要发生蛇纹石的脱羟基作用,此反应为吸热反应;815.6 ℃ 放热峰主要是因为蛇纹石物相转变,引起蛇纹石的晶格破坏,由非晶态物质转变为结晶度良好的橄榄石相和顽火辉石相,此反应为放热反应[22]。橄榄石相由于晶体结构排列紧密,反应活性较差,一旦镍被包裹,将增加还原难度。

2 研究方法

在红土镍矿硫化熔炼过程中,通过配料向熔炼体系中加入硫化剂、还原剂、造渣剂后,在1300~1500 ℃下熔炼

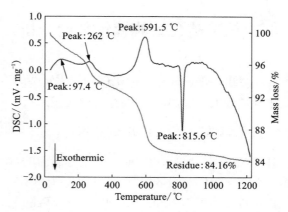

图2 红土镍矿 TG-DTA 曲线

Fig. 2 TG-DSC curves of the laterite nickel ore

生产低镍锍;低镍锍氧化吹炼过程中,向低镍锍中吹氧和加入石英造渣,在1200~1300 ℃下吹炼生产高镍锍,其工艺流程图如图3所示。基于表1中红土镍矿元素及化学物相分析检测结果,利用 HSC Chemistry 6.0 和 Factsage 7.1 热力学软件,模拟红土镍矿硫化熔炼过程物相演变。HSC 的平衡组成模块及其纯组分数据库可以清晰揭示物相的转变路径;使用 FactSage 7.1 中 Predom 模块及 FactPS(纯化合物)、FToxide(固体和液体氧化物溶液)和 FTmisc(液体合金)数据库模拟计算物质稳定存在区域,揭示体系平衡时,冶炼产物与温度和气氛(氧势和硫势)对应关系。

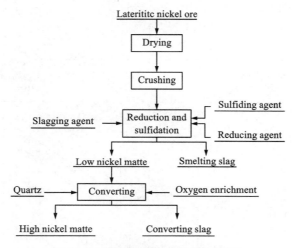

图3 红土镍矿硫化冶炼工艺流程图

Fig. 3 Flow chart of sulphide smelting of laterite nickel ore

3 红土镍矿硫化熔炼镍锍理论分析

为了尽可能准确表达反应机理和简化复杂的冶金反应过程，用 NiO、CoO 和 Fe_2O_3 来表示红土镍矿中的 Ni、Co、Fe 的存在形式。

3.1 硫化熔炼过程物相转变路径

利用 Factsage 7.1 中 Predom 模块绘制了 1500 ℃下 Ni/Co/Fe-S-O 优势区域图，如图 4 所示。从 Ni-S-O 优势区域图[图 4(a)]中可看出，在 p_{SO_2}<1 atm①、p_{O_2}<1.10×10^{-5} atm 时，镍氧化物转变历程为 NiO→Ni→Ni_3S_2（即路径 1）；从 Co-S-O 优势区域图[图 4(b)]中可以看出，在 p_{SO_2}<0.76 atm、p_{O_2}<3.70×10^{-7} atm 时，钴氧化物转变历程为 CoO→Co→Co_9S_8（即路径 1）；从 Fe-S-O 优势区域图[图 4(c)]中可看出，在 p_{SO_2} 为 1.30×10^{-4}～5.03 atm、p_{O_2} 为 9.80×10^{-10}～1.10×10^{-6} atm 时，铁氧化物转变途径为 Fe_2O_3→Fe_3O_4→FeO→FeS（即路径 1）；

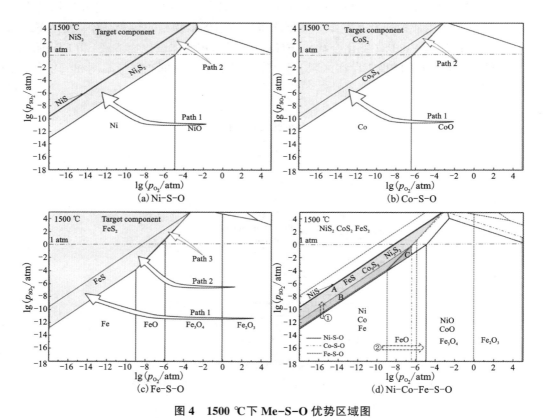

图 4　1500 ℃下 Me-S-O 优势区域图

Fig. 4　Predominant region diagram of Me-S-O at 1500 ℃

---

① 1 atm=1.01×10^5 Pa。

在 $p_{SO_2}<1.30\times10^{-4}$ atm、$p_{O_2}<9.80\times10^{-10}$ atm 时,铁氧化物转变历程为 $Fe_2O_3\rightarrow Fe_3O_4\rightarrow FeO\rightarrow Fe\rightarrow FeS$(即路径2)。

图4(d)为1500 ℃下 Ni/Co/Fe-S-O 优势区域叠加图,图中 A、B、C 区域分别表示为 $Ni_3S_2+Co_9S_8+FeS$、$Ni_3S_2+Co+FeS$、$Ni_3S_2+Co+FeO$,若想实现红土镍矿中 NiO、CoO、Fe_2O_3 的选择性硫化,需要将熔炼氧分压和硫分压控制在 A、B、C 区域内。从图4(d①)中也可得到 Ni、Co、Fe 与 S 亲和力的差异,表明了金属 Ni 和 Fe 与 S 亲和力接近,均大于 Co;从图4(d②)中也可得到 Ni、Co、Fe 与 O 亲和力的差异,表明金属 Fe 与 O 亲和力最强、Ni 与 O 亲和力最弱。

3.2 硫化熔炼过程还原硫化反应

根据上述优势区域图可知,红土镍矿中主要化合物在还原硫化熔炼过程中可能发生的化学反应如下所示:

$$4NiO + S_2(g) = 4Ni + 2SO_2(g) \quad (1)$$

$$3Ni + S_2(g) = Ni_3S_2 \quad (2)$$

$$4CoO + S_2(g) = 4Co + 2SO_2(g) \quad (3)$$

$$\frac{9}{4}Co + S_2(g) = \frac{1}{4}Co_9S_8 \quad (4)$$

$$12Fe_2O_3 + S_2(g) = 8Fe_3O_4 + 2SO_2(g) \quad (5)$$

$$4Fe_3O_4 + S_2(g) = 12FeO + 2SO_2(g) \quad (6)$$

$$4FeO + S_2(g) = 4Fe + 2SO_2(g) \quad (7)$$

$$\frac{4}{3}FeO + S_2(g) = \frac{4}{3}FeS + \frac{2}{3}SO_2(g) \quad (8)$$

$$2Fe + S_2(g) = 2FeS \quad (9)$$

图5为上述反应的吉布斯自由能(ΔG)与温度(t)关系。从图5可以看出,在500~800 ℃ 内,NiO 不能被 S_2 还原为 Ni 单质;在800~1600 ℃内,NiO 可被 S_2 还原为 Ni 单质[式(1)]。在500~1300 ℃内,CoO 不能被 S_2 还原为 Co 单质;在1300~1600 ℃内,CoO 可被 S_2 还原为 Co 单质[式(3)]。Fe_2O_3 可被 S_2 还原至 Fe_3O_4[式(5)]或 FeO[式(6)]状态,不能被 S_2 还原成 Fe 单质[式(7)];从式(2)、(4)、(9)可以得出,金属 Ni、Fe 与 S 亲和力接近,均大于 Co。

图6为镍、钴、铁氧化物碳还原平衡图。从图6可以看出,NiO 易被还原,在低温和低 CO 浓度下即可还原为 Ni 单质[23];随着温度升高,NiO 还原为 Ni 单质所需 CO 浓度升高;Fe_2O_3 比 NiO 易被还原,NiO 比 Fe_3O_4 易被还原。CoO 还原为 Co 单质所需 CO 浓度大于 NiO 还原为 Ni 单质,即 NiO 比 CoO 易被还原;反应温度小于1200 ℃时,CoO 比 Fe_3O_4 易被还原;反应温度大于1200 ℃时,Fe_3O_4 比 CoO 易被还原。反应温度小于570 ℃时,铁氧化物转变历程为 $Fe_2O_3\rightarrow Fe_3O_4\rightarrow Fe$;反应温度大于570 ℃时,铁氧化物转变历程为 $Fe_2O_3\rightarrow Fe_3O_4\rightarrow FeO\rightarrow Fe$[24]。

$$C + CO_2(g) = 2CO(g) \quad (10)$$

$$NiO + CO(g) = Ni + CO_2(g) \quad (11)$$

$$CoO + CO(g) = Co + CO_2(g) \quad (12)$$

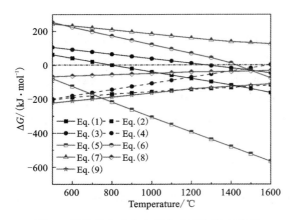

图 5 相关反应式吉布斯自由能与温度关系

Fig. 5 The Gibbs free energy change as a function of different temperatures for the chemical reactions

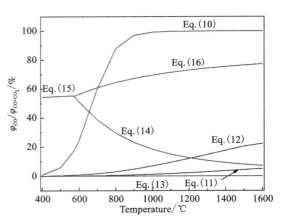

图 6 C 还原镍/钴/铁氧化物平衡图

Fig. 6 Equilibrium diagram of nickel/cobalt/iron oxides reduced by C

$$3Fe_2O_3 + CO(g) = 2Fe_3O_4 + CO_2(g) \quad (13)$$

$$Fe_3O_4 + CO(g) = 3FeO + CO_2(g) \quad (14)$$

$$\frac{1}{4}Fe_3O_4 + CO(g) = \frac{3}{4}Fe + CO_2(g) \quad (15)$$

$$FeO + CO(g) = Fe + CO_2(g) \quad (16)$$

3.3 硫化熔炼过程物相演变

通过热力学优势区图得 NiO、CoO 和 Fe_2O_3 转变为硫化物可能发生的途径，为了更加清晰地了解氧化物与硫化剂、还原剂的反应机理，利用 HSC Chemistry 6.0 的平衡组成模块计算反应产物的平衡组成。平衡组成由等温、等压和固定质量条件下的吉布斯自由能最小化方法确定[25]。

3.3.1 NiO 与 S_2、C 高温还原硫化平衡组成

1500 ℃下，初始 NiO 摩尔量为 3 kmol，S_2 以 0.1 kmol 的速率加入体系内，结果如图 7(a)所示。由图 7(a)可以看出，随着 S_2 加入，NiO 先还原成 Ni，然后再硫化成 Ni_3S_2；由图 7(c)可以看出，在 S_2 添加量为 NiO 质量的 31.3%时，体系中镍主要以镍单质形式存在；在 S_2 添加量为氧化镍质量的 70.3%时，体系中镍主要以硫化镍形式存在。在体系中额外加入 1 kmol C，结果如图 7(b)所示。由图 7(b)可以看出，碳存在条件下，SO_2 气体量减少，说明硫作为还原剂的量减少，即加入还原剂碳可减少熔炼体系中 SO_2 气体排放；由图 7(d)可以看出，在 S_2 添加量为 NiO 质量 11.7%时，体系中镍主要以镍单质形式存在；在 S_2 添加量为氧化镍质量的 50.8%时，体系中镍主要以硫化镍形式存在。在碳存在条件下，可减少熔炼体系中硫添加量。

3.3.2 CoO 与 S_2、C 高温还原硫化平衡组成

图 8(a)为 1500 ℃下，初始 CoO 摩尔量为 3 kmol，S_2 以 0.1 kmol 的速率加入体系内。

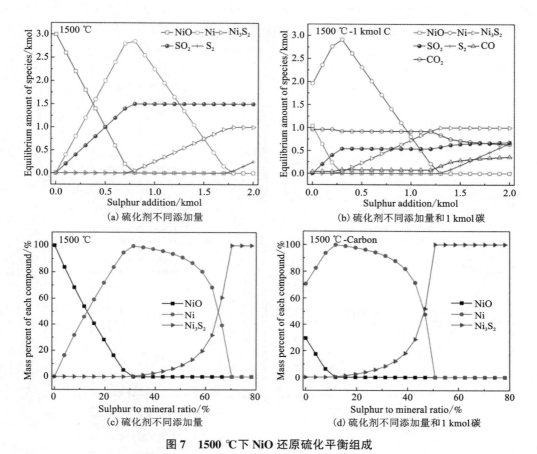

图 7 1500 ℃下 NiO 还原硫化平衡组成

Fig. 7 Equilibrium phase diagrams of NiO reduction-sulfidation at 1500 ℃ with (a、c) sulfur different addition, (b、d) sulfur different addition and 1 kmol carbon

由图 8(a)可以看出,在 S_2 添加量 0~0.8 kmol,CoO 还原成 Co;在 S_2 添加量 0.8~1.5 kmol,体系内未发生钴物相转变;在 S_2 添加量大于 1.5 kmol,Co 逐渐硫化成 Co_9S_8。由图 8(c)可以看出,在 S_2 添加量为 NiO 质量 31.3%时,体系中钴主要以钴单质形式存在;在 S_2 添加量为氧化钴质量的 62.5%时,体系中钴单质开始硫化为硫化钴。图 8(b)为体系中额外加入 1 kmol C,由图 8(b)可以看出,CoO 转变为 Co_9S_8 的规律与不添加还原剂时相一致,在 S_2 添加量大于 1.2 kmol,Co 逐渐硫化成 Co_9S_8;由图 8(d)可以看出,在 S_2 添加量为 NiO 质量 11.7%时,体系中镍主要以镍单质形式存在;在 S_2 添加量为氧化钴质量的 50.8%时,体系中开始生成硫化钴。与 NiO 硫化相比,CoO 完全硫化所需硫化剂添加量较大。

3.3.3 Fe_2O_3 与 S_2、C 高温还原硫化平衡组成

1500 ℃下,初始 Fe_2O_3 为 3 kmol,S_2 以 0.1 kmol 的速率加入体系内,结果如图 9(a)所示。由图 9(a)可以看出,随着 S_2 加入,Fe_2O_3 先还原成 Fe_3O_4/FeO,然后 Fe_3O_4 还原为 FeO,最后 FeO 硫化为 FeS;结合优势区域图可知,Fe_3O_4 被硫化为 FeS 需要 $p_{SO_2}>1$ atm,在熔炼过程中难发生,说明 Fe_3O_4 量减少的原因是 Fe_3O_4 被 S_2 还原为 FeO,然后 FeO 被硫化为 FeS。

图9(b)为在体系中额外加入 1 kmol C。由图9(b)可以看出，Fe_2O_3 转变为 FeS 规律与不加还原剂时相一致；在碳存在条件下，减少了 SO_2 生成，即提高了硫直接利用率。

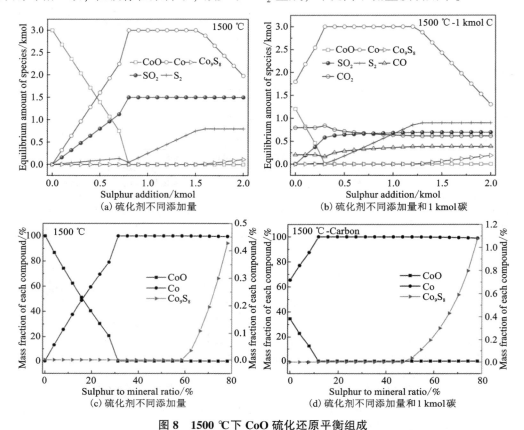

图8 1500 ℃下 CoO 硫化还原平衡组成

Fig. 8 Equilibrium phase diagrams of CoO sulfidation-reduction at 1500 ℃ with (a、c) sulfur different addition, (b、d) sulfur different addition and 1 kmol carbon

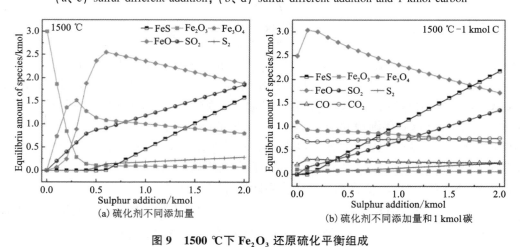

图9 1500 ℃下 Fe_2O_3 还原硫化平衡组成

Fig. 9 Equilibrium phase diagrams of Fe_2O_3 reduction-sulfidation at 1500 ℃ with (a) sulfur different addition, (b) sulfur different addition and 1 kmol carbon

从图9(b)可以看出,在加入 1 kmol C 时,熔炼体系中未出现铁单质;当加入 3 kmol C 时(图10),体系中还原气氛强,将铁氧化物还原为铁单质,并且随着硫化剂加入,铁单质硫化生成 FeS,说明铁单质可发生硫化反应(Eq.9)。

3.3.4 硫化熔炼机理模型

基于上述分析,建立了红土镍矿硫化熔炼机理模型,如图11所示。红土镍矿硫化熔炼过程中,在硫化熔炼温度条件下,固体硫会迅速气化,形成富硫环境;结合图7可知,镍/钴氧化物优先与还原剂碳发生还原反应生成镍/钴单质,若还原剂加入量不足时,镍/钴氧化物将被硫还原成镍/钴单质;最后,镍/钴单质与硫化剂反应生成 Ni_3S_2/Co_9S_8。结合图9和

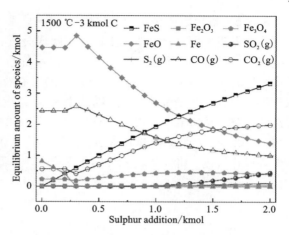

图10 1500 ℃下 Fe_2O_3 还原硫化平衡组成
（硫化剂不同添加量和 3 kmol 碳）

Fig. 10 Equilibrium phase diagrams of Fe_2O_3 reduction-sulfidation at 1500 ℃ with (a) sulfur different sulfur different addition and 3 kmol carbon

图10可知,铁氧化物优先发生还原反应生成 Fe_3O_4 和 FeO;在还原剂加入量不足以将 FeO 还原为 Fe 单质时,硫化反应为 FeO 与硫化剂反应生成 FeS;当还原剂加入量可将部分 FeO 还原为 Fe 单质时,硫化反应为 FeO/Fe 与硫化剂反应生成 FeS。

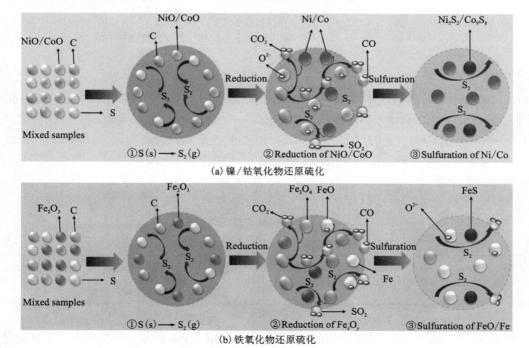

图11 红土镍矿硫化熔炼转型生产低镍锍机理

Fig. 11 Mechanism of low nickel matte production from laterite nickel
(a: reduction-sulfidation of Nickel/cobalt oxide; b: reduction-sulfidation of Iron oxide)

3.3.5 红土镍矿全组分硫化熔炼

为更贴近实际生产，按红土镍矿实际成分进行计算，其成分见表1。为了降低熔炼渣液相线温度，对熔炼渣型进行了调整。SiO_2 添加量对熔炼渣液相线温度影响如图12(a)所示。从中可以看出，不额外加入 SiO_2 助熔剂的熔渣液相线温度约为 1625 ℃，这需要高于 1650 ℃ 操作温度以保持炉渣完全熔融；随渣中 SiO_2 含量增加，熔渣液相线温度呈现先降低再稳定后升高的趋势，且初生相区由单一氧化物转变为橄榄石相区。通过控制渣中 SiO_2 质量分数为 31.5%~42.5%，使渣型 SiO_2/MgO 值为 2.5~3.0，熔渣液相线温度约 1420 ℃，低于 1500 ℃ 的熔炼操作温度，将使炉料充分熔化反应，并获得良好的渣流动性。

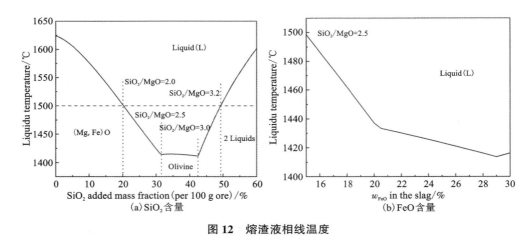

图 12 熔渣液相线温度

Fig. 12 Liquidus temperatures (a: SiO_2 content; b: FeO content)

基于表1中 NiO、CoO 完全还原，利用 FactSage 7.1 计算了 MgO-SiO_2-"FeO"-Al_2O_3-Cr_2O_3-MnO-CaO 系熔渣液相线温度，炉渣成分如表4所示。图12(b)为 SiO_2/MgO 值为 2.5 时，FeO 含量变化对熔渣液相线温度影响规律。从图12(b)可以看出，在 FeO 质量分数为 15%~30%，随 FeO 含量的增加，熔渣液相线温度逐渐降低，且均低于 1500 ℃。由此可见，在硫化熔炼过程中，随着部分铁元素硫化后进入镍锍相中，将导致渣中铁含量降低，熔渣液相线温度升高。

表 4 用于 Factsage 计算的炉渣成分

Table 4 The composition of the slag used for FactSage calculation %

"FeO"	Si_2O/MgO	Al_2O_3	CaO	MnO	Cr_2O_3
15~30	2.5	1.96	0.28	0.46	0.28

在 1500 ℃、SiO_2/MgO 为 2.5、硫磺添加量变化范围为 0%~4%，还原剂碳添加量变化范围为 0%~4.5% 条件下，研究了硫磺添加量和还原剂碳添加量对镍钴回收率、镍品位和硫直接利用率的影响，如图13所示。

图13(a)和(b)分别显示了镍、钴回收率随碳和硫磺添加量变化规律。在硫磺添加量为

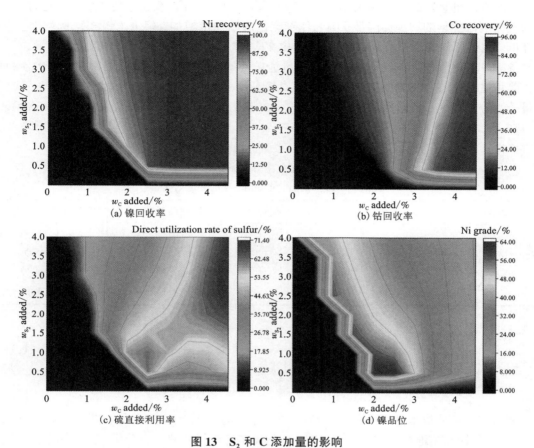

图 13 S₂ 和 C 添加量的影响

Fig. 13 The effects of S₂ and C additions

(a: Ni recovery; b: Co recovery; c: Direct utilization rate of sulfur; d: Ni grade)

矿料的 1%~4% 时,镍、钴回收率随还原剂添加量增加而升高;在硫化剂添加量为矿料的 2% 时,镍、钴回收率分别由 85.54%、17.82% 升高至 99.64%、93.51%。在还原剂添加量为矿料的 2%~2.5% 时,镍、钴回收率随硫化剂添加量增大而升高;在还原剂添加量为 3%~4.5% 时,镍回收率随硫化剂添加量增大呈现微小变化,钴回收率随硫化剂添加量增大而降低,这是因为钴分配进入锍相和渣相的比例与锍相中含铁量成正比例关系[26,27],锍相中硫含量增加,导致锍相中铁含量降低,即钴在镍锍和炉渣之间的分配系数减小,钴回收率降低。在还原剂添加量为 4% 时,钴回收率由 91.55% 降低至 70.89%。相同条件下,镍回收率均大于钴,这是因为镍与硫亲和力大于钴。

图 13(c) 显示了硫直接利用率随碳和硫磺添加量变化规律。在硫磺添加量为矿料的 2% 时,硫直接利用率随还原剂添加量增加而增加,因为还原剂碳存在条件下,减少了硫磺作还原剂比例,提高了硫直接利用率,与分析氧化镍和氧化钴纯物质硫化过程相一致。图 13(d) 显示了镍品位随碳和硫磺添加量变化规律。在硫磺添加量为矿料的 0.5%~4% 时,镍品位随还原剂添加量的增加而降低,因为随着还原剂添加量增加,铁进入锍相比例增加,镍品位降

低；硫化剂添加量为2%时，镍品位由47.59%降低至16.48%。在还原剂添加量一定时，镍品位随硫化剂添加量的增加而降低，因为随着硫化剂添加量增加，硫进入锍相比例增加，镍品位降低；在还原剂添加量为4%时，镍品位由22.45%降低至20.29%。

从表5可以看出，当硫化剂硫磺添加量为矿料的2%，还原剂碳添加量为矿料的4%时，镍、钴回收率分别为99.43%、87.58%；硫直接利用率为62.68%；产物镍锍中镍、钴、铁、硫质量分数分别为21.45%、1.03%、61.89%、15.63%；熔炼渣含55.53%SiO_2、22.20%MgO、20.47%FeO、Al_2O_3 1.80%。

表5 硫化熔炼体系参数

Table 5　Parameters of sulfide smelting system

熔炼条件	镍回收率/%	钴回收率/%	硫利用率/%	镍锍质量分数/%	熔炼渣质量分数/%
1500 ℃ $SiO_2/MgO=2.5$ 2%硫磺 4%碳	99.43	87.58	62.68	Ni 21.45 Co 1.03 Fe 61.89 S 15.63	SiO_2 55.53 MgO 22.20 FeO 20.47 Al_2O_3 1.80

4 红土镍矿侧吹硫化冶炼生产实践

侧吹熔池熔炼技术是在俄罗斯瓦纽科夫熔炼法的基础上发展的[28]，其侧吹炉示意图如图14所示。侧吹熔池熔炼技术在铅、铜、硫化镍矿炼镍等领域已成功应用，为其在红土镍矿侧吹硫化冶炼镍锍提供了借鉴经验。

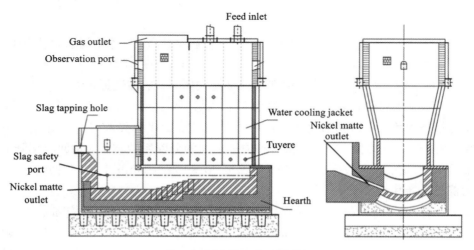

图14 侧吹炉示意图[29, 30]

Fig. 14　Schematic diagram of the side-blown furnace

红土镍矿侧吹熔炼镍锍模型如图 15 所示，结合图 4 和图 11 可知，红土镍矿在 1300～1500 ℃下经硫化熔炼生产低镍锍，向熔炼体系中加入还原剂和硫化剂，使体系的氧分压和硫分压位于低镍锍区域。我国某公司印尼产业园采用侧吹炉进行红土镍矿硫化熔炼生产镍锍，已投产运行[31]，其生产过程如图 16 所示。富氧侧吹技术是多通道侧吹喷枪以亚音速向熔池内喷入富氧空气、硫化剂和还原剂，使红土镍矿中镍、钴等有价金属发生还原硫化反应，具有镍钴回收率高、成本低、镍锍品位控制灵活、热利用率高、对原料适应性强等优点，打通了从红土镍矿到新能源电池原料的新技术路径，实现红土镍矿冶炼技术的里程碑式突破。

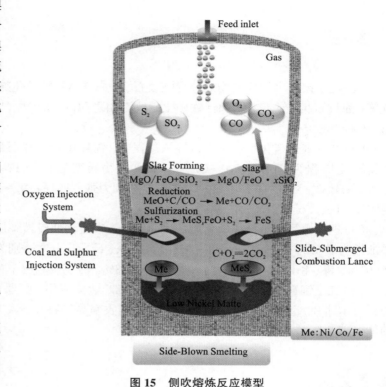

图 15　侧吹熔炼反应模型

Fig. 15　Reaction model of side-blown smelting

图 16　红土镍矿硫化熔炼生产实践[31]

Fig. 16　Smelting process of laterite nickel ore: (a) and (b) Inside of the side-blown furnace; (c) low nickel matte ingot; (d) low nickel matte product

5 结论

(1) 红土镍矿中主要矿物成分为蛇纹石、针铁矿、针铁矿化蛇纹石、磁铁矿、铬铁矿、石英等;蛇纹石相、针铁矿相和针铁矿化蛇纹石相之和占 90.02%,其分布比例分别为 57.10%、13.96%、18.96%。

(2) 红土镍矿硫化熔炼生产低镍锍过程中,镍氧化物转变历程为 $NiO \rightarrow Ni \rightarrow Ni_3S_2$;钴氧化物转变历程为 $CoO \rightarrow Co \rightarrow Co_9S_8$;铁氧化物转变途径为 $Fe_2O_3 \rightarrow Fe_3O_4 \rightarrow FeO \rightarrow FeS$ 或 $Fe_2O_3 \rightarrow Fe_3O_4 \rightarrow FeO \rightarrow Fe \rightarrow FeS$;金属与 S 亲和力强弱顺序:$Ni \approx Fe$、Co;金属与 O 亲和力强弱顺序:Fe、Co、Ni。

(3) 理论计算红土镍矿硫化熔炼优化条件:当硫化剂硫磺添加量为矿料的 2%,还原剂碳添加量为矿料的 4% 时,产品镍锍中镍、钴、铁、硫质量分数分别为 21.45%、1.03%、61.89%、15.63%,镍、钴回收率分别为 99.43%、87.58%,硫直接利用率为 62.68%。

(4) 红土镍矿侧吹熔炼生产镍锍,已初步产业化应用,打通了从红土镍矿到新能源产业的新技术路径,该技术具有低碳环保、镍钴回收率高等特点,实现世界性红土镍矿冶炼技术的新突破。

参考文献

[1] MA Bao-zhong, LI Xiang, YANG Wei-jiao, et al. Nonmolten state metalized reduction of saprolitic laterite ores: Effective extraction and process optimization of nickel and iron[J]. Journal of Cleaner Production, 2020, 256: 120415.

[2] TIAN Qing-hua, DONG Bo, GUO Xue-yi, et al. Comparative atmospheric leaching characteristics of scandium in two different types of laterite nickel ore from Indonesia[J]. Minerals Engineering, 2021, 173: 107212.

[3] ELLIOTT R, PICKLES C A, FORSTER J. Thermodynamics of the reduction roasting of nickeliferous laterite ores[J]. Journal of Minerals and Materials Characterization and Engineering, 2016, 4(6): 320-346.

[4] PICKLES C A, ANTHONY W. Thermodynamic modelling of the reduction of a saprolitic laterite ore by methane[J]. Minerals Engineering, 2018, 120: 47-59.

[5] JOWITT S. Mudd and Jowitt 2014-Detailed assessment of global nickel resource trends and endowments database[J]. 2014.

[6] SU K, WANG F, PARIANOS J, et al. Alternative resources for producing nickel matte-laterite ores and polymetallic nodules[J]. Mineral Processing and Extractive Metallurgy Review, 2021, 43(5): 584-597.

[7] SANTORO L, PUTZOLU F, Mondillo N, et al. Trace element geochemistry of iron-(oxy)-hydroxides in Ni(Co)-laterites: Review, new data and implications for ore forming processes[J]. Ore Geology Reviews, 2022, 140: 104501.

[8] GAO Jian-min, LI Wen-jie, MA Shu-jia, et al. Spinel ferrite transformation for enhanced upgrading nickel

grade in laterite ore of various types[J]. Minerals Engineering, 2021, 163(4): 106795.
[9] MUDD G M. Global trends and environmental issues in nickel mining: Sulfides versus laterites[J]. Ore Geology Reviews, 2010, 38(1/2): 9-26.
[10] PICKLES C A, ANTHONY W. A thermodynamic study of the reduction of a limonitic laterite ore by methane[J]. High Temperature Materials and Processes, 2018(9-10): 909-919.
[11] ZHOU Shi-wei, WEI Yong-gang, LI Bo, et al. Mineralogical characterization and design of a treatment process for Yunnan nickel laterite ore, China[J]. International Journal of Mineral Processing, 2017, 159: 51-59.
[12] RAO Ming-jun, LI Guang-hui, JIANG Tao, et al. Carbothermic reduction of nickeliferous laterite ores for nickel pig iron production in china: A review[J]. JOM, 2013, 65(11): 1573-1583.
[13] ZHU De-qing, CUI Yu, HAPUGODA S, et al. Mineralogy and crystal chemistry of a low grade nickel laterite ore[J]. Transactions of Nonferrous Metals Society of China, 2012, 22(4): 907-916.
[14] KESKINKILIC E. Nickel laterite smelting processes and some examples of recent possible modifications to the conventional route[J]. Metals, 2019, 9(9): 974.
[15] LYU Xue-ming, WANG Lun-wei, YOU Zhi-xiong, et al. A novel method of smelting a mixture of two types of laterite ore to prepare ferronickel[J]. JOM, 2019, 71(11): 4191-4197.
[16] MESHRAM P, ABHILASH, PANDEY B D. Advanced review on extraction of nickel from primary and secondary sources[J]. Mineral Processing and Extractive Metallurgy Review, 2019, 40(3): 157-193.
[17] JIANG Xin, HE Liang, WANG Lin, et al. Effects of reducing parameters on the size of ferronickel particles in the reduced laterite nickel ores[J]. Metallurgical and Materials Transactions B, 2020, 51(6): 1-10.
[18] STANKOVI S, STOPI S, SOKI M, et al. Review of the past, present, and future of the hydrometallurgical production of nickel and cobalt from lateritic ores[J]. Metallurgical and Materials Engineering, 2020, 26(2): 199-208.
[19] ZHANG Pei-yu, WANG Hai-rui, HAO Jing-cheng, et al. Reinforcement of the two-stage leaching of laterite ores using surfactants[J]. Frontiers of Chemical Science and Engineering, 2021, 15(3): 562-570.
[20] MCDONALD R G, WHITTINGTON B I. Atmospheric acid leaching of nickel laterites review: Part I. sulphuric acid technologies[J]. Hydrometallurgy, 2008, 91(1-4): 35-55.
[21] XU Peng-yun, WANG Qiang, LI Chuang, et al. Relationship between process mineralogical characterization and beneficiability of low-grade laterite nickel ore[J]. Journal of Central South University, 2021, 28(10): 3061-3073.
[22] 王宇鲲, 魏永刚, 彭博, 等. 镁质贫镍红土矿热分解理论计算与实验研究[J]. 材料导报, 2019, 33(8): 6.
[23] 李光辉, 饶明军, 姜涛, 等. 红土镍矿钠盐还原焙烧-磁选的机理[J]. 中国有色金属学报, 2012, 22(1): 7.
[24] 郭学益, 陈远林, 田庆华, 等. 氢冶金理论与方法研究进展[J]. 中国有色金属学报, 2021, 31(7): 1891-1906.
[25] HAN Jun-wei, LIU Wei, ZHANG Tian-fu, et al. Mechanism study on the sulfidation of ZnO with sulfur and iron oxide at high temperature[J]. Scientific Reports, 2017, 7: 42536.
[26] 何焕华, 蔡乔方. 中国镍钴冶金[M]. 北京: 冶金工业出版社, 2000.
[27] 魏寿昆, 洪彦若. 镍锍选择性氧化的热力学及动力学[J]. 有色金属, 1981(3): 50-60.
[28] 田庆华, 李中臣, 王亲猛, 等. 红土镍矿资源现状及冶炼技术研究进展[J/OL]. 中国有色金属学报.

https：//kns.cnki.net/kcms/detail/43.1238.tg.20220901.0851.003.html.
[29] XU Lei, CHEN Min, WANG Nan, et al. Degradation mechanisms of magnesia-chromite refractory bricks used in oxygen side-blown reducing furnace[J]. Ceramics International, 2020, 46：17315-17324.
[30] CHEN Lin, YANG Tian-zu, BIN Shu, et al. An efficient reactor for high-lead slag reduction process：Oxygenrich side blow furnace[J]. JOM, 2014, 66(9)：1664-1669.
[31] 中伟新材料股份有限公司. 印尼莫罗瓦利产业基地 OESBF 产线正式投产[R]. 长沙：中伟新材料股份有限公司, 2022.

改性腐殖土型红土镍矿中和高压浸出液−中和渣高压酸浸工艺

摘　要：以改性腐殖土型红土镍矿作中和剂，在改性腐殖土型红土镍矿添加量 25 g/100 mL、中和时间 60 min、中和温度 40 ℃ 的条件下对褐铁型红土镍矿高压浸出液进行中和处理，高压浸出液中杂质铁、铝脱除率分别达到 98.8%、96.8%，实现了铁、铝等杂质的高效脱除。采用高压酸浸工艺处理中和反应产物中和渣，在浸出温度 250 ℃、浸出时间 1 h、硫酸浓度 1.2 mol/L、液固比 8∶1 的条件下进行高压浸出，中和渣中镍、钴浸出率均达到 98%。改性腐殖土型红土镍矿中和高压浸出液−中和渣高压酸浸工艺全流程的镍、钴回收率分别达 99.5%、96.8%，实现了物料中镍、钴的高效回收以及中和过程"零排放"。所开发工艺物料利用率高、环境效益显著、普适性强，符合当前镍、钴金属湿法生产的工艺需求，应用前景广阔。

关键词：改性腐殖土型红土镍矿；中和；高压酸浸；镍钴提取

Neutralized high pressure leaching solution-neutralized residue high pressure acid leaching of modified saprolite

Abstract：The high pressure leaching solution of limonite laterite nickel ore was neutralized with modified saprolite as neutralizer under the conditions of 25 g/100 mL modified saprolite, 60 min neutralization time and 40 ℃ neutralization temperature. The removal rates of iron and aluminum impurities in the high pressure leaching solution reached 98.8% and 96.8%, which realized the efficient removal of iron, aluminum and other impurities. The neutralization reaction product neutralization slag was treated by high pressure acid leaching process. Under the conditions of leaching temperature 250 ℃, leaching time 1 h, sulfuric acid concentration 1.2 mol/L and liquid-solid ratio 8∶1, the leaching rate of nickel and cobalt in neutralization slag reached 98%. The recovery rates of nickel and cobalt in the whole process of neutralization high pressure leaching solution-neutralization slag high pressure acid leaching process of modified saprolite reached 99.5% and 96.8%, which realized the efficient recovery of nickel and cobalt in the

---

本文发表在《中国有色金属学报》，网络首发。作者：田庆华，王青鹙，董波，郭学益，曾奎彰，许志鹏。

material and the 'zero emission' of neutralization process. The developed process has high material utilization rate, significant environmental benefits and strong universality, which meets the current process requirements of nickel and cobalt metal wet production and has broad application prospects.

Keywords: modified saprolite; neutralization; high pressure acid leaching; nickel and cobalt extraction

镍因其具备出色的理化性能,是国家生产发展不可或缺的战略金属,被广泛应用于冶金、军工、航空航天、新能源等诸多领域[1]。镍作为电池生产的重要原料,随着新能源汽车行业的迅猛扩张以及电池器件产业的蓬勃发展,其需求量逐年激增[2-3]。镍资源主要蕴藏于硫化镍矿与红土镍矿中,由于硫化镍矿经长期开采消耗,其矿产资源趋于枯竭,红土镍矿逐渐成为主要的提镍原料[4-5]。根据红土镍矿中铁、镁含量差异,其可被分为褐铁型、腐殖土型、黏土型三大类[6]。腐殖土型红土镍矿因其"高镁低铁"的成分特性[7],通常以火法工艺处理,用于生产镍铁合金[8-11]或镍硫[12-14],但该类工艺能耗较高且污染严重,与新能源行业衔接困难[15-16]。褐铁型红土镍矿具有"高铁、低硅镁"的成分特点[7],适用于以高压酸浸为代表的湿法处理工艺[17-20],但传统的高压浸出工艺采用石灰石中和高压浸出液,产生大量硫酸盐渣,严重危害环境[21-22];近年来,有研究表明腐殖土型红土镍矿可被用于中和褐铁型红土镍矿高压浸出液,但存在中和效果较差、反应周期较长等问题[23-25]。腐殖土型红土镍矿中蛇纹石等含镁矿相溶出困难,是影响其中和效果的主要原因[26];焙烧处理可破坏腐殖土型红土镍矿中蛇纹石矿相,增强镁的溶出,从而改善其中和效果[27]。

本课题组已针对腐殖土型红土镍矿焙烧改性开展了系统研究,发现腐殖土型红土镍经600 ℃焙烧处理后,即可破坏其中的蛇纹石类含镁矿相,但未进一步考察其对高压浸出液的中和效果[28-29]。本文针对改性腐殖土型红土镍矿(以下简称"改性腐殖土"),开发了如图1所示的工艺。利用改性腐殖土作中和剂,对褐铁型红土镍矿高压浸出液进行中和处理,实现

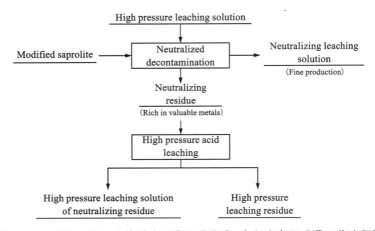

图 1 改性腐殖土型红土镍矿中和高压浸出液-中和渣高压酸浸工艺流程图
Fig. 1 Process flow chart of neutralized high-pressure leaching solution-neutralized residue high pressure acid leaching of modified saprolite

高压浸出液中铁、铝等杂质的高效脱除；改性腐殖土中未反应的有价金属进入中和产物中和渣中，采用高压酸浸工艺对中和渣进行浸出，进一步回收其中的有价金属。较之传统工艺，新工艺以矿物替代石灰石作中和剂，避免了钙盐废渣的产生，减少了药剂的使用；传统中和工艺产生的钙盐废渣难以被利用，而新工艺产生的中和渣的成分与高品位褐铁型红土镍矿的成分相近，在生产中可作原料直接投入高压酸浸工序，实现了中和过程的"零排放"。同时新工艺拓展了腐殖土型红土镍矿的湿法处理手段，解决了其难以与新能源产业衔接的问题。研究结果为红土镍矿湿法处理工艺提供了新思路，对于红土镍矿资源高效利用及清洁生产具有借鉴意义。

1 实验

1.1 实验原料

本文涉及的实验原料改性腐殖土与高压浸出液，均源自本课题组前期实验研究过程。改性腐殖土为印度尼西亚产腐殖土型红土镍矿的焙烧处理产物，其铁、镁含量较高，镍品位较高，主要成分见表1。改性腐殖土的XRD图谱见图2，主要物相为Fe_2O_3；由于低于850 ℃焙烧制备的改性腐殖土中镁钙类矿相的特征峰不明显，此图谱中未呈现该类别矿相。高压浸出液为褐铁型红土镍矿的高压酸浸产物，其酸度高、成分复杂，富含铁、镍，其他主要成分见表2。

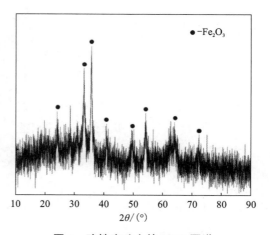

图 2　改性腐殖土的 XRD 图谱

Fig. 2　XRD pattern of modified saprolite

表 1　改性腐殖土的主要成分

Table 1　Main composition of modified saprolite

Composition	Fe	Mg	Ni	Al	Mn	Co	Ca	Cr
Mass fraction/%	26.68	10.06	1.67	1.28	0.41	0.05	0.38	0.47

表 2　高压浸出液的主要成分

Table 2　Main composition of high pressure leaching solution

Composition	Fe	Mg	Ni	Al	Mn	Co	Ca	Cr	pH
Mass concentration /(mg·L^{-1})	2956	881.9	2212	215.2	1200	71.16	30.08	0.385	0.45

1.2 实验步骤

1.2.1 高压浸出液改性腐殖土中和实验

取 100 mL 高压浸出液置于三孔烧瓶中，利用恒温水浴锅进行预热；向其中添加一定量的改性腐殖土，通过装配搅拌桨的电动搅拌器将物料混合，进行浸出反应，以 pH 计实时监测反应体系的 pH 值。对浸出后所得悬浊液进行抽滤，测量滤液体积，再次检测滤液 pH 值，并进行成分分析；所得滤渣即中和渣，将其收集并用于后续实验。

1.2.2 中和渣高压浸出实验

将中和渣置于石英胆中，加入硫酸溶液混合，将石英胆置于高压反应釜中进行高压浸出。对浸出后所得悬浊液进行抽滤，测量滤液体积并进行成分分析；将所得滤渣烘干后，进行物相分析。

1.3 数据处理

改性腐殖土中和高压浸出液过程中，高压浸出液中杂质脱除率根据式(1)进行计算，公式中 m_1 为中和后液中杂质质量分数，m_2 为高压浸出液中杂质质量分数。

$$杂质脱除率(\%) = \frac{m_2 - m_1}{m_2} \times 100\% \tag{1}$$

中和渣高压浸出过程中，中和渣中金属浸出率根据式(2)进行计算，公式中 m_1 为中和渣中金属质量分数，m_2 为中和渣的高压浸出液中金属质量分数。

$$金属浸出率(\%) = \frac{m_2}{m_1} \times 100\% \tag{2}$$

全流程中金属回收率根据式(3)进行计算，公式中 m_1 为中和渣的高压浸出液中金属质量分数，m_2 为中和后液中金属质量分数，m_3 为改性腐殖土中金属质量分数，m_4 为高压浸出液中金属质量分数。

$$金属回收率(\%) = \frac{m_1 + m_2}{m_3 + m_4} \times 100\% \tag{3}$$

2 试验结果与讨论

2.1 高压浸出液改性腐殖土中和研究

Fe^{3+} 在溶液中的水解反应如式(4)所示，该式表明溶液中 H^+ 浓度是限制水解反应的关键因素，Fe^{3+} 仅能在强酸性溶液中稳定存在[30]。根据 $Fe(OH)_3$ 的溶度积，可知当溶液 pH 值高于 1.9 时，Fe^{3+} 开始形成沉淀；当 pH=3.7 左右时，溶液中 Fe^{3+} 基本沉淀完全，该条件下可达到良好的除铁效果。结合上述原理，选取 pH=3.7 作中和除铁反应体系的终点 pH。改性腐殖土中存在大量的含镁耗酸物质，在溶出时通过大量消耗高压浸出液中 H^+ 达到中和效果。

$$Fe^{3+} + 3H_2O \rightleftharpoons Fe(OH)_3\downarrow + 3H^+ \tag{4}$$

2.1.1 改性腐殖土添加量/中和反应时间对除杂效果的影响

中和反应时间及改性腐殖土添加量对高压浸出液除杂效果的影响见图3。由图3可知，当反应时间一定时，随着改性腐殖土添加量的增加，进入反应体系的含镁耗酸物质增加，与游离酸发生反应后，使得中和后液pH值升高。反应时间低于60 min时，含镁耗酸物质与游离酸迅速反应，中和后液pH值的增幅明显；反应时间超过60 min后，改性腐殖土中易溶出的含镁耗酸物质被消耗殆尽，残余含镁物质溶出缓慢，因此中和后液pH值的增幅减缓并逐渐趋于稳定。若过度延长反应时间，将致使生产周期增长，降低生产效率。

改性腐殖土添加量低于25 g/100 mL时，中和后液终点pH值均低于3.00，未达到最佳中和除铁、铝的条件；改性腐殖土添加量高于25 g/100 mL时，中和后液终点pH值为4.94，pH值过高造成中和剂浪费，增加生产成本；改性腐殖土添加量为25 g/100 mL时，中和后液终点pH为3.74，且中和60 min时的中和后液pH值达到3.56，接近最佳的中和除铁铝条件。综上，选取最佳改性腐殖土添加量为25 g/100 mL、最佳中和反应时间为60 min。

2.1.2 中和反应温度对除杂效果的影响

中和反应温度对高压浸出液除杂效果的影响见图4。由图4可知，随着中和反应温度升高，中和后液的pH值逐渐升高，且当中和反应温度高于40 ℃后，中和后液的pH值增幅明显，可知升高中和反应温度有利于改善中和效果。中和反应温度为40 ℃时，中和后液pH值达到3.70；继续升高中和反应温度，则中和后液的pH值过高，对改善中和效果无意义且增加能耗；中和反应温度过低则无法保证中和后液达到最佳除铁pH值。因此选取最佳中和反应温度为40 ℃。

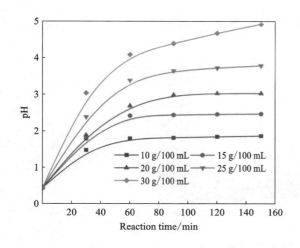

图3 改性腐殖土添加量及反应时间对除杂效果的影响（反应温度30 ℃）

Fig. 3 Effect of modified saprolite addition and reaction time on edulcoration (Reaction temperature: 30 ℃)

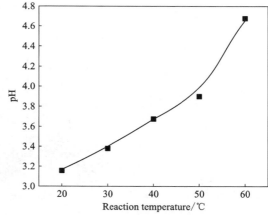

（改性腐殖土添加量25 g/100 mL、中和反应时间60 min）

图4 中和反应温度对除杂效果的影响

Fig. 4 Effect of neutralization reaction temperature on edulcoration (Modified saprolite addition: 25 g/100 mL, Reaction time: 60 min)

2.1.3 中和产物分析

最佳反应条件下的中和渣XRD图谱见图5，可知中和渣中存在Fe_2O_3、斜绿泥石

[(Mg，Fe^{2+})$_5$Al(Si$_3$Al)O$_{10}$(OH)$_8$]与Fe(OH)$_3$。由于斜绿泥石与酸反应困难，在中和过程中基本未被溶出[4]；中和时溶液中的Fe^{3+}发生如式(4)所示的水解反应，并以Fe(OH)$_3$形式沉淀进入渣中。中和后液成分见表3，对比表2中高压浸出液成分可知，中和后液中铁、铝等杂质含量明显降低，铁、铝脱除率分别达99.5%、96.8%。中和渣成分见表4，对比改性腐殖土成分(表1)可知，中和渣中镁减量明显，证明改性腐殖土中含镁耗酸物质已被大量消耗；铁、铝等杂质发生水解反应后形成沉淀，由于其沉淀物胶体具有吸附性，部分镍、钴等金属被吸附并进入中和渣中，使得中和渣中铁、铝及镍、钴等金属含量升高。上述实验证明，利用改性腐殖土中和高压浸出液，可获得良好的中和效果。

表3 中和后液的主要成分
Table 3 Main composition of leaching solution after neutralization

Composition	Fe	Mg	Ni	Al	Mn	Co	Ca	Cr
Mass concentration/(mg·L^{-1})	14.44	9073	1284	7.66	557.9	33.63	28.9	10.9

表4 中和渣的主要成分
Table 4 Main composition of neutralizing residue

Composition	Fe	Mg	Ni	Al	Mn	Co	Ca	Cr
Mass fraction/%	28.04	6.69	2.02	1.28	0.65	0.062	0.32	0.39

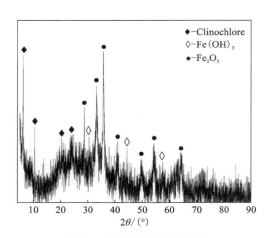

图5 中和渣的XRD图谱
Fig.5 XRD pattern of neutralizing residue

2.2 中和渣高压浸出研究

中和除铁的过程中，高压浸出液中部分镍、钴等有价金属伴随铁、铝等杂质一并沉淀进入中和渣中，渣中镍含量较高，具有回收价值；且中和渣中镁含量明显低于腐殖土型红土镍矿，适用高压浸出工艺对其进行处理。

高压酸浸工艺因其能耗较低、金属回收率高、产品品质良好等显著优势,已成为当前处理高铁、低硅镁的褐铁型红土镍矿的主流工艺技术。高压酸浸过程中的基本原理如式(5)~式(7)所示;反应中生成的硫酸铁在高温高压条件下以赤铁矿形式被沉淀,如式(8)所示,该式进一步阐释了高压酸浸对铁具有良好选择性[31-32]。因此以中和渣为原料进行高压酸浸,研究各反应条件对中和渣中金属浸出效果的影响。

$$NiO + H_2SO_4 \rightleftharpoons NiSO_4 + H_2O \quad (5)$$
$$CoO + H_2SO_4 \rightleftharpoons CoSO_4 + H_2O \quad (6)$$
$$2FeO \cdot OH + 3H_2SO_4 \rightleftharpoons Fe_2(SO_4)_3 + 4H_2O \quad (7)$$
$$Fe_2(SO_4)_3 + 3H_2O \rightleftharpoons Fe_2O_3 \downarrow + 3H_2SO_4 \quad (8)$$

2.2.1 浸出温度对高压浸出效果的影响

浸出温度对中和渣高压浸出效果的影响见图6。由图6可知,随着浸出温度的升高,镍、钴等金属的浸出率逐渐升高,铁的浸出率逐渐降低;当浸出温度为250 ℃时,镍、钴浸出率均高达98%、铁浸出率仅为1.22%,镍、铁分离效果良好。不同浸出温度下各高压浸出渣的物相分析见图7,可知浸出渣的主要物相均为Fe_2O_3;且随着浸出温度的升高,图谱中Fe_2O_3的峰强增加,晶型逐渐变好。

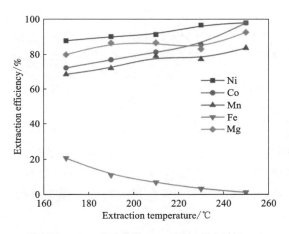 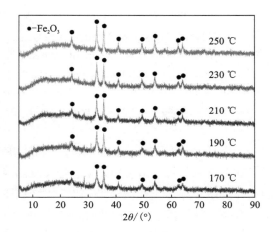

(液固比10∶1、硫酸浓度1.2 mol/L、浸出时间1 h)

图6 浸出温度对中和渣高压浸出效果的影响

Fig. 6 Effect of extraction temperature on high pressure leaching of neutralized residue
(Liquid-solid ratio: 10∶1, Sulfuric acid concentration: 1.2 mol/L, Extraction time: 1 h)

图7 不同浸出温度所得高压浸出渣的 XRD 图谱

Fig. 7 XRD pattern of high pressure leaching residue obtained at different extraction temperatures

由于提高浸出温度有利于增强金属的浸出反应,同时由式(8)可知 Fe^{3+} 仅在高温条件下才能直接转化为 Fe_2O_3 沉淀,达到除铁效果;结合实验结果可知提高浸出温度利于增强铁的选择性,强化镍、铁分离效果,最佳浸出温度为250 ℃。

2.2.2 浸出时间对高压浸出效果的影响

浸出时间对中和渣高压浸出效果的影响见图8。由图8可知,随着浸出时间延长,镍浸出率有小幅波动，但整体浸出效果仍较好，浸出率均高于 95%；钴浸出率小幅下降，铁浸出率仍保持在较低水平；锰、镁的浸出率略有提高。不同浸出时间下各高压浸出渣的物相分析见图 9，可知浸出渣的主要物相均为 Fe_2O_3，虽在浸出时间为 2 h 的渣样中检出 $[Fe_{0.6}Cr_{0.4}]O_3$，但其对镍、钴浸出率无明显影响，推测其来源于原料中的少量含铬杂质。

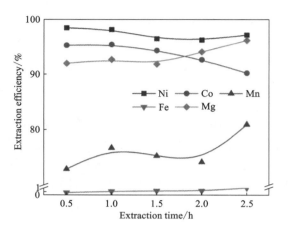 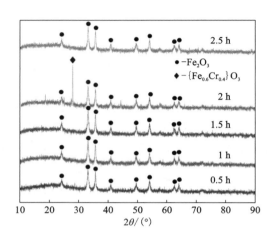

（液固比 10∶1，硫酸 1.2 mol/L，浸出温度为 250 ℃）

图 8　浸出时间对中和渣高压浸出效果的影响

Fig. 8　Effect of extraction time on high pressure leaching of neutralized residue
(Liquid-solid ratio：10∶1, Sulfuric acid concentration：1.2 mol/L, Extraction temperature：250 ℃)

图 9　不同浸出时间所得高压浸出渣的 XRD 图谱

Fig. 9　XRD pattern of high pressure leaching residue obtained at different extraction time

由上述结果可知，浸出时间对中和渣高压浸出效果的影响较小，浸出时间为 0.5 h 时，即可获得良好的镍、钴浸出效果；延长浸出时间对增强有价金属浸出效果的意义不大，且增加能耗。由于高压釜在温度调控时耗时较长，浸出时间过短不能保证反应过程的稳定进行，综合考虑上述原因后，选取最佳浸出时间为 1 h。

2.2.3　硫酸浓度对高压浸出效果的影响

硫酸浓度对中和渣高压浸出效果的影响见图 10。由图 10 可知，随着硫酸浓度升高，镍、钴、锰、镁等金属的浸出率均有小幅波动；体系中过量的硫酸会与已生成的 Fe_2O_3 进行反应，使铁的浸出率有所提高。当硫酸浓度为 1.2 mol/L 时，镍、钴浸出率为 98% 以上；继续增加硫酸浓度，会导致铁的浸出率上升，增加浸出液中杂质铁的含量。不同浓度的硫酸浸出下高压浸出渣的物相分析见图 11，可知当硫酸浓度高于 1.2 mol/L 时，高压浸出渣中的 Fe_2O_3 转变为 $Fe(OH)SO_4$，而 $Fe(OH)SO_4$ 可与体系中的游离酸反应，继续增加硫酸浓度反而会影响镍、钴的有效浸出。因此高压浸出所用最佳硫酸物质的量浓度为 1.2 mol/L。

2.2.4　液固比对高压浸出效果的影响

液固比对中和渣高压浸出效果的影响见图 12。由图 12 可知，随着液固比的增加，镍、钴、锰、镁等金属的浸出率均有小幅上升，铁浸出率无明显变化。当液固比为 8∶1 时，镍、钴浸出率最高，均达到 98%；继续增加液固比对提高镍、钴浸出率无明显影响。不同液固比浸出下高压浸出渣的物相分析见图 13，可知各浸出渣的主要成分均为 Fe_2O_3，浸出液固比变化对浸出效果的整体影响较小。结合上述实验结果，可知高压浸出时最佳液固比为 8∶1。

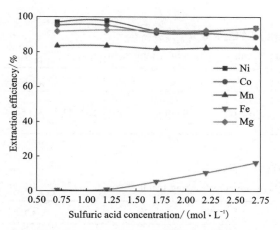

（液固比 10∶1、浸出时间为 1 h，浸出温度为 250 ℃）

图 10　硫酸浓度对中和渣高压浸出效果的影响

Fig. 10　Effect of sulfuric acid concentration on high pressure leaching of neutralized residue
(Liquid-solid ratio: 10∶1, Extraction time: 1 h, Extraction temperature: 250 ℃)

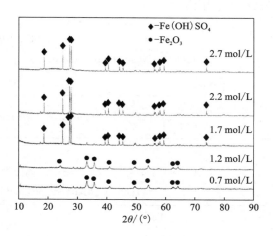

图 11　不同浓度硫酸所得高压浸出渣的 XRD 图谱

Fig. 11　XRD pattern of high pressure leaching residue obtained at different sulfuric acid concentration

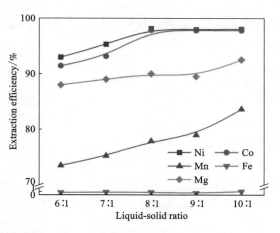

（浸出温度为 250 ℃，浸出时间为 1 h，硫酸浓度为 1.2 mol/L）

图 12　液固比对中和渣高压浸出效果的影响

Fig. 12　Effect of liquid-solid ratio on high pressure leaching of neutralized residue
(Extraction temperature: 250 ℃, Extraction time: 1 h, Sulfuric acid concentration: 1.2 mol/L)

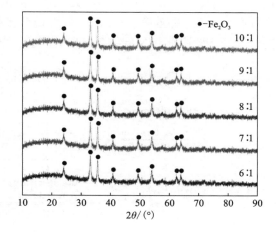

图 13　不同液固比所得高压浸出渣的 XRD 图谱

Fig. 13　XRD pattern of high pressure leaching residue obtained at different liquid-solid ratio

中和渣在最佳浸出条件下的高压浸出液与高压浸出渣成分分别见表5、表6,可知中和渣经高压浸出后,该高压浸出液中杂质铁的含量更低,浸出效果良好;浸出渣富含Fe_2O_3,可作提铁物料用于后续生产。

表5 中和渣高压浸出液的主要成分

Table 5　Main composition of high pressure leaching solution of neutralized residue

Composition	Fe	Mg	Ni	Al	Mn	Co	Ca	Cr
Mass concentration /($mg \cdot L^{-1}$)	1424.40	6093.89	1962.70	133.62	519.19	61.66	164.53	31.54

表6 高压浸出渣的主要成分

Table 6　Main composition of high pressure leaching residue

Composition	Fe	Mg	Ni	Al	Mn	Co	Ca	Cr
Mass fraction/%	29.32	4.24	0.57	1.25	0.344	0.015	0.151	0.392

2.3　工艺效果评估

针对上述实验的整体流程进行物质流分析,其结果见图14,由图可知流程中各金属来源于改性腐殖土与高压浸出液,经中和除铁后部分镍、钴留存于中和后液中,剩余的镍、钴均进入中和渣;中和渣经高压酸浸处理后,大部分镍、钴进入中和渣的高压浸出液中,浸出渣中仅有少量镍、钴残存,根据式(3)可得全流程中镍、钴回收率分别达到91.84%、90.88%。同时发现实验中存在少量的金属损失问题,均属于正常损耗。

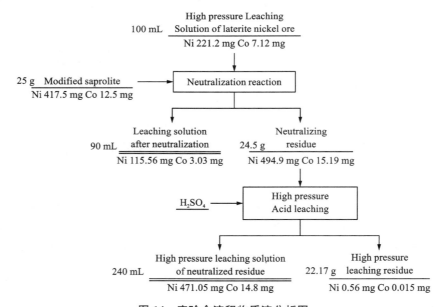

图14　实验全流程物质流分析图

Fig. 14　Material flow analysis diagram of whole experimental process

本文所开发的工艺中，采用改性腐殖土作中和剂，替代了传统工艺所用的石灰中和剂，避免了硫酸盐废渣等危废的产生；所产出的中和渣以原料的形式进入高压酸浸工序，进一步实现了物料中镍、钴金属的有效回收，解决了沉淀法除杂时有价金属损失严重的问题；高压酸浸产物高压浸出渣是良好的铁资源，具有一定的回收处理价值。本工艺所涉及的物料利用率较高、金属回收率高，实现了矿产资源的最大化运用，具有良好的经济效益；有效降低了固废排放水平，符合国家治污减排的环保政策，环境效益显著；整体工艺流程简洁，无需大规模的设备改造与技术改进即可嫁接于多数现有的湿法生产线，具有普适性强、引进成本低、衔接速度快等突出优势，应用前景广阔。

3 结论

本文针对改性腐殖土型红土镍矿富含耗酸物质的特点，开发了改性腐殖土中和高压浸出液——中和渣高压浸出工艺，通过实验研究不同工艺条件对物料中金属的浸出及回收效果的影响：

（1）以改性腐殖土作中和剂，在改性腐殖土添加量 25 g/100 mL、反应时间 60 min、反应温度 50 ℃的条件下，对高压浸出液进行中和除杂，所得中和后液 pH 达到 3.7，铁、铝脱除率分别达到 99.5%、96.8%，除杂效果良好；所得中和渣以原料形式进入高压酸浸工序处理，中和除杂过程无固废产生。

（2）以中和渣为原料，在浸出温度 250 ℃、浸出时间 1 h、硫酸物质的量浓度 1.2 mol/L、液固比 8∶1 的条件下，对其进行高压酸浸，中和渣中镍、钴浸出率均达到 98%；高压浸出渣中富含 Fe_2O_3，可作提铁物料用于后续生产。

（3）所开发的工艺以改性腐殖土替代传统的石灰中和剂，避免了危险固废的产生，环境效益显著；实现了物料的高效利用，有效控制金属损失，工艺全流程的镍、钴回收率分别达到 91.84%、90.88%，经济效益良好；工艺流程简洁，普适性强，可快速嫁接于现有的镍钴湿法生产线，符合行业的生产需求，应用前景广阔。

参考文献

[1] 田庆华，李中臣，王亲猛，等. 红土镍矿资源现状及冶炼技术研究进展[J]. 中国有色金属学报，2022，23(9)：2975-2997.
[2] 席儒恒，李园园，张建茹，等. 高镍无钴层状正极材料的研究进展[J]. 电源技术，2021，45(12)：1641-1645.
[3] 饶富，马恩，郑晓洪，等. 硫化镍矿中镍提取技术研究进展[J]. 化工学报，2021，72(1)：495-507.
[4] TIAN Qing-hua, DONG Bo, GUO Xue-yi, et al. Comparative atmospheric leaching characteristics of scandium in two different types of laterite nickel ore from Indonesia. Minerals Engineering，2021，173：107212.
[5] DONG Bo, TIAN Qing-hua, GUO Xue-yi, et al. Leaching behavior of scandium from limonitic laterite

ores under sulfation roasting-water leaching[J]. Journal of Sustainable Metallurgy, 2022, 8(3): 1078-1089.

[6] 潘建, 田宏宇, 朱德庆, 等. 镍矿资源供需分析及红土镍矿开发利用现状[C]. 2019年镍产业发展高峰论坛暨APOL年会会刊, 南京, 2019: 22-31.

[7] 李金辉, 徐志峰, 高岩, 等. 氯化铵选择性浸出红土镍矿有价金属[J]. 中国有色金属学报, 2019, 29(5): 1049-1057.

[8] 白亚东, 杨颂, 刘守军, 等. Na_2CO_3对红土镍矿的H_2还原影响规律与作用机理研究[J]. 中国有色金属学报, 2022, 32(9): 2774-2786.

[9] 丁志广, 李博, 魏永刚. 氢气作用下硅镁型红土镍矿的低温还原特性[J]. 中国有色金属学报, 2018, 28(8): 1669-1675.

[10] 刘兴阳, 刘守军, 杨颂, 等. 甲烷作用下红土镍矿中铁镍氧化物的反应行为[J]. 中国有色金属学报, 2022, 32(6): 1759-1771.

[11] 赵剑波, 马东来, 吕学明, 等. Na_2CO_3作用下红土镍矿非等温碳热还原动力学研究[J]. 中国有色金属学报, 2022, 32(4): 1088-1097.

[12] 王帅. 红土镍矿火法冶炼技术现状与研究进展[J]. 中国冶金, 2021, 31(10): 1-7.

[13] SU Kun, WANG Feng, PARIANOS J, et al. Alternative resources for producing nickel matte-laterite ores and polymetallic nodules[J]. Mineral Processing and Extractive Metallurgy Review, 2022, 43(5): 584-597.

[14] WANG Hong-yang, HOU Yong, CHANG He-qiang, et al. Preparation of Ni-Fe-S matte from nickeliferous laterite ore using CaS as the sulfurization agent[J]. Metallurgical and Materials Transactions B, 2022, 53(2): 1136-1147.

[15] 朱德庆, 田宏宇, 潘建, 等. 低品位红土镍矿综合利用现状及进展[J]. 钢铁研究学报, 2020, 32(5): 351-362.

[16] 张云芳, 李金辉, 高岩, 等. 红土镍矿的硫酸铵焙烧过程[J]. 中国有色金属学报, 2017, 27(1): 155-161.

[17] 马保中. 褐铁型红土镍矿硝酸加压浸出新技术进展[C]. 2020年APOL镍与不锈钢产业链年会会刊, 上海, 2020: 3-7.

[18] GUO Xue-Yi, SHI Wen-tang, LI Dong, et al. Leaching behavior of metals from limonitic laterite ore by high pressure acid leaching[J]. Transactions of Nonferrous Metals Society of China. 2011, 21(1): 191-195.

[19] SHAO Shuang, MA Bao-zhong, WANG Xin, et al. Nitric acid pressure leaching of limonitic laterite ores: Regeneration of HNO_3 and simultaneous synthesis of fibrous $CaSO_4 \cdot 2H_2O$ byproducts[J]. Journal of Central South University, 2020, 27(11): 3249-3258.

[20] HE Fei, MA Bao-zhong, WANG Cheng-yan, et al. Microwave pretreatment for enhanced selective nitric acid pressure leaching of limonitic laterite[J]. Journal of Central South University, 2021, 28(10): 3050-3060.

[21] 孙宁磊, 刘苏宁, 李勇, 等. 氢氧化镍钴生产工艺中碱基活化控制技术的研发及应用[J]. 中国有色冶金, 2020, 49(4): 7-10.

[22] 刘苏宁, 丁剑, 李诺, 等. 红土镍矿湿法冶炼石灰乳沉淀镍钴工艺[J]. 中国有色冶金, 2021, 50(4): 49-52.

[23] ADAMS M, VAN D M D, CZERNY C, et al. Piloting of the beneficiation and EPAL circuits for Ravensthorpe Nickel Operations[C]// IMRIE W P, et al. International Laterite Nickel Symposium-2004. Warrendale, PA: The Minerals, Metals and Materials Society, 2004: 347-367.

[24] LOVEDAY B K. The use of oxygen in high pressure acid leaching of nickel laterites[J]. Miner. Eng., 2008, 21(7): 533-538.

[25] NEUDORF D, HUGGINS D. A method for nickel and cobalt recovery from laterite ores by combination of atmospheric and moderate pressure leaching: US 2006 0024224[P]. 2006-02-02.

[26] ELLIOTT R, PICKLES C A, FORSTER J. Thermodynamics of the reduction roasting of nickeliferous laterite ores[J]. Journal of Minerals & Materials Characterization & Engineering. 2016, 4(6): 320-346.

[27] RHAMDHANI M A, HAYES P C, JAK E. Nickel laterite Part 2-thermodynamic analysis of phase transformations occurring during reduction roasting[J]. Mineral Processing & Extractive Metallurgy, 2009, 118(3): 146-155.

[28] 田庆华, 董波, 郭学益, 等. 一种低成本、低酸耗的红土镍矿的浸出方法: CN114507780A[P]. 2022-05-17.

[29] 田庆华, 董波, 郭学益, 等. 一种红土镍矿的资源综合回收方法: 中国, CN114480877A[P]. 2022.

[30] 李国卿, 周海燕, 戴向荣, 等. 不同铁浓度的含酸矿井水的中和沉淀试验[J]. 化工管理, 2021(14): 75-76+177.

[31] 皮关华, 孔凡祥, 贾露萍, 等. 瑞木红土镍矿高压酸浸的生产实践[J]. 中国有色冶金, 2015, 44(6): 11-14.

[32] 邢姜, 冷红光, 韩百岁, 等. 红土镍矿湿法冶金工艺现状及研究进展[J]. 有色矿冶, 2021, 37(5): 26-32.

氢冶金理论与方法研究进展

摘　要：冶金行业是我国重要的经济支撑行业，也是主要的碳排放行业。氢冶金是当前冶金领域低碳发展的重要方向，已受到国内外广泛关注。本文对氢冶金技术在钢铁冶金、有色金属冶金以及二次资源利用领域的基础理论研究、工艺应用研究进展进行了系统综述。作为气体还原剂，在温度大于810 ℃的条件下，氢气还原能力强于一氧化碳，且氢气的还原反应速率比碳还原剂高1到2个数量级。基于氢冶金的直接还原炼铁技术已处于技术成熟、稳步发展阶段，其应用包括典型的 Midrex、HYL-Ⅲ 工艺、欧洲的 ULCOS、瑞典的 HYBRT、日本的 COURSE50 项目以及我国的中晋矿业等。氢冶金在有色金属冶金以及二次资源利用领域的发展处于基础研究阶段，有待进一步技术突破。氢气大规模、低成本制备以及氢冶金过程的热量平衡是发展氢冶金技术亟待解决的关键问题。

关键词：氢冶金；钢铁冶金；有色金属冶金；二次资源利用；热力学；动力学

Research progress on hydrogen metallurgy theory and method

Abstract: The metallurgy industry is a strong support for economic and a major carbon emission industry in China. Hydrogen metallurgy is regarded as a vibrant research branch in developing efficient metallurgical technologies with low-carbon emission at present. Therefore, hydrogen metallurgical technology has been concerned worldwide. In the present paper, the recent developments of theory researches and applications of hydrogen metallurgical technology in the fields of extractive metallurgy of ferrous metals and nonferrous metals, and secondary resources utilization are systematically reviewed. As a gaseous reducing agent, H_2 possesses a stronger reduction capacity than CO at the temperature higher than 810 ℃. And the reduction reaction rate of H_2 is higher than that of carbon reducing agent with 1 to 2 orders of magnitude. The direct reduction iron making technology based on hydrogen metallurgy is in the stage of steady development, its application includes the typical processes Midrex and HYL-Ⅲ, and some other new projects, such as Europe's ULCOS, Sweden's HYBRT, Japan's COURSE50, and China's Zhongjin Mining. While the researches of hydrogen metallurgy in the fields of extractive metallurgy of nonferrous metals and secondary resource utilization are in theoretical study stage,

---

本文发表在《中国有色金属学报》，2021, 31(7)：1891-1906。作者：郭学益、陈远林、田庆华、王亲猛。

and further technical breakthrough is in needed. The production of hydrogen with large scale and low cost and the thermal balance in hydrogen metallurgy process are the key problems to be solved in the future.

Keywords: hydrogen metallurgy; extractive metallurgy of ferrous metals; extractive metallurgy of nonferrous metals; secondary resource utilization; thermodynamics; kinetics

随着世界工业的快速发展,由碳排放引起的温室效应等环境问题日益突出。冶金行业是我国重要的经济发展支撑行业,同时,也是主要的碳排放行业,我国钢铁冶金年碳排放量超过 $2.2×10^9$ t CO_2 [1]、有色金属冶金年碳排放也高达 $3.0×10^9$ t CO_2 [2]。在当前发展低碳经济的迫切形势下,亟须开发清洁冶金技术,保障冶金行业的可持续发展。氢冶金作为绿色冶金新技术,是当前冶金领域低碳发展的重要方向,已受到国内外科研工作者广泛关注。

1995 年,EDELSON[3] 提出了采用 H_2 熔融还原炼铁的工艺,并申请了美国专利。我国徐匡迪院士在 1999 年北京第 125 次香山科学会议等学术会议上多次提出发展 H_2 冶金的倡议,并认为可通过向反应体系施加物理场的方法来实现 H_2 冶金过程强化[4]。郑少波[5] 和 GERMESHUIZEN 等[6] 对应用 H_2 的钢铁冶金流程进行了技术、经济评价,认为采用基于核能的 H_2 替代碳作为炼铁还原剂,生产成本提高约 12.8%,但是,CO_2 排放量可降低 63%。VOGL 等[7]认为,H_2 直接还原炼铁的能量消耗与传统高炉炼铁相当,在电力成本低和碳排放要求严格的情况下,H_2 直接还原炼铁工艺则具有广阔前景。

21 世纪是氢时代,氢冶金以氢代替碳还原,不但碳排放低,而且反应速度极快[8-9]。目前,国内外冶金领域均提出了氢冶金的战略规划,包括欧洲的 ULCOS、瑞典的 HYBRT、日本的 COURSE50 项目以及我国的中晋矿业等[9]。TANG 等[9]综述了氢冶金在国内外钢铁冶金领域的研发进展,并提出了适合我国的氢冶金路线。随着制氢技术的发展,氢冶金将成为未来冶金行业新的竞争领域[10-11]。本文将对氢冶金技术在钢铁冶金、有色金属冶金及二次资源利用领域的基础理论研究、工艺应用研究进展进行系统综述,并对氢冶金技术未来发展前景及亟待解决的关键问题进行分析,为氢冶金领域的研究提供参考。

1 钢铁冶金领域氢冶金研究进展

传统高炉炼铁工艺强烈依赖冶金焦,能耗高、污染重,为了摆脱高炉工艺的固有缺点,开发清洁的钢铁冶金工艺,基于氢冶金的炼铁技术应运而生。发展氢冶金是炼铁技术的一场革命,将有效推动钢铁工业的可持续发展[12]。目前,钢铁冶金领域的氢冶金研究主要包括气基直接还原和熔融还原(见表 1)[13-14]。

表 1 氢冶金在炼铁工艺中的应用现状[13-14]

Table 1 Application status of hydrogen metallurgy in iron-making

Iron-making process	Representative process		Application progress
Direct reduction	Shaft furnace process	Midrex process	Industrialization application, annual output of about 45 million tons
		HYL-Ⅲ process	Industrialization application, annual output of about 12 million tons
	Fluidized bed process	Finmen process	Industrialization application, annual output of about 2 million tons
		Circored process	Industrialization application, annual output of about 500000 tons
Smelting reduction	—		Laboratory research

1.1 H_2 还原铁氧化物的基础研究

采用纯 H_2 代替碳作为炼铁还原剂时,产物为 H_2O,避免了碳还原产生的 CO_2,理论上可实现温室气体的零排放,为钢铁冶金的绿色发展提供了可能。因此,国内外对 H_2 还原铁氧化物的行为及机理进行了大量研究。

1.1.1 热力学研究

目前,气基还原炼铁主要采用 H_2、CO 混合气体作为还原气,H_2、CO 还原铁氧化物时,热力学平衡图如图 1 所示[15-16],还原过程分为 Fe_3O_4 稳定区、FeO 稳定区和金属铁稳定区。反应温度小于 570 ℃时,铁氧化物的还原历程为 $Fe_2O_3 \rightarrow Fe_3O_4 \rightarrow Fe$;反应温度大于 570 ℃时,铁氧化物的还原历程则为 $Fe_2O_3 \rightarrow Fe_3O_4 \rightarrow FeO \rightarrow Fe$。

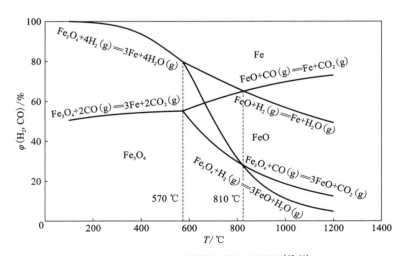

图 1 H_2、CO 还原铁氧化物平衡图[15-16]

Fig. 1 Equilibrium diagram of iron oxides reduced by H_2, CO[15-16]

在反应温度小于 810 ℃时，CO 还原平衡曲线位于 H_2 还原平衡曲线下方，相同温度条件下，还原铁氧化物生成金属铁所需 CO 平衡分压小于 H_2，表明此温度范围内 CO 还原能力强于 H_2；而反应温度大于 810 ℃时，则 H_2 还原能力强于 CO。由于实际反应温度一般大于 810 ℃，因此，采用富氢或纯氢还原气体进行铁矿还原在热力学上具有一定的优势。

1.1.2 动力学研究

H_2 对固态铁氧化物的还原过程，主要包括以下环节[17]：

① H_2 从气流层向固-气界面扩散并被界面吸附；
② 发生界面还原反应，生成气体 H_2O 和相应的固体产物；
③ 气体 H_2O 从反应界面脱附；
④ 随着反应的进行，固体产物成核、生长，并形成产物层，H_2O 需要穿过固体产物层从反应界面向气流层扩散，还原反应速率取决于速率最低的环节。

H_2 还原熔融铁氧化物为气-液反应过程，整体反应速率则主要由三个环节控制[18]：

① 气相中的质量传递速率；
② 气-液界面反应速率；
③ 液相中的质量传递速率。

近年来，一些关于 H_2 还原铁氧化物的动力学研究结果如表 2 所示。

表 2 关于 H_2 还原铁氧化物的动力学研究结果[19-27]

Table 2　Results of kinetic study on reducing of iron oxides by hydrogen[19-27]

Material	Reaction temperature/℃	Main result	Ref.
Magnetite single crystal	900~1100	The reduction rate of magnetite increased with the increasing of reaction temperature; the reduction rate was controlled by gas diffusion	[19]
Iron ore pellets	760~1000	The reduction degree and reduction rate increased with the increasing of temperature; the reduction rate is controlled by chemical reaction and the apparent activation energy is 40.95 kJ/mol; the reaction rate with 75% H_2 ~ 25% N_2 was lower than that under pure hydrogen	[20]
Fine iron ore	500~900	The reduction rate increased with the increasing of temperature and gas flow, while was not affected by the absolute pressure; particles with size of 0.5~4.0 mm showed lower reduction rate than that with size of 0.125~0.5 mm	[21]
Hematite pellets	—	Using pure H_2 as the reducing agent could reduce the size of current industrial furnace using ($CO+H_2$) owed to a faster reaction rate; the reduction rate increased with the decrease of pellet diameter	[22]

续表2

Material	Reaction temperature/℃	Main result	Ref.
Magnetite concentrates	300~570	The reduction rate of micron magnetite increased with decreasing particle size, while it is opposite for nanoscale magnetite	[23]
Precipitated iron oxide	Room temperature~900	Fe_2O_3 was reduced in two stage: $Fe_2O_3 \rightarrow Fe_3O_4 \rightarrow Fe$ with the activation energy of 89.13 and 70.412 (kJ/mol) respectively; the reaction $Fe_2O_3 \rightarrow Fe_3O_4$ followed unimolecular model and the reaction $Fe_3O_4 \rightarrow Fe$ followed two-dimensional nucleation model	[24]
Hematite ore fines	700~900	Both the reduction degree and reduction rate of hematite ore increased with the increasing of H_2 content in mixed gas and temperature; the reaction showed fitting of first-order model below 850 ℃, then shifted to follow diffusion controlled at 900 ℃ due to the formation of dense iron layer	[25]
Iron ore	400~750	In the temperature range of 400~550 ℃, the two-stage reactions $Fe_2O_3 \rightarrow Fe_3O_4$, $Fe_3O_4 \rightarrow Fe$ followed chemical reaction and three dimensional diffusion model respectively; in 600~750 ℃, the reduction reaction was in three-stage: $Fe_2O_3 \rightarrow Fe_3O_4 \rightarrow FeO \rightarrow Fe$, the reactions followed chemical reaction model, nucleation and subsequent generation model, and three dimensional diffusion model	[26]
Iron-silica magnetically stabilized porous structure	800~1000	There was a transition point during the reduction reaction, both reactions, $Fe_3O_4 \rightarrow FeO$ and $FeO \rightarrow Fe$ followed contracting sphere model until the transition point, the reaction models were $(1-x)^{2/3}$, $(1-x)^{2/3}$ respectively; after the transition point, the reaction $Fe_3O_4 \rightarrow FeO$ shifted to followed unimolecular decay model, the reaction model was $(1-x)$, and the reaction $FeO \rightarrow Fe$ shifted to followed contracting cylinder model, the reaction model was $(1-x)^{1/3}$	[27]

(1) H_2 还原铁氧化物的影响因素

H_2 对固态铁矿物的还原反应速率受多方面条件影响,在动力学上,提高温度有利于加快 H_2 扩散及界面反应速率;热力学上,铁氧化物还原的总体反应($Fe_2O_3 + 3H_2(g) = 2Fe + 3H_2O(g)$,$\Delta H_{298}^{\ominus} = 95.8$ kJ/mol)是吸热效应,提高温度,可增强反应趋势。因此,一定范围内提高温度,有利于 H_2 还原铁氧化物,这与文献研究结果一致[19-21]。

不同矿物粒度导致反应过程速率控制环节的差异,也影响 H_2 对铁矿物的还原反应速率。HABERMANN 等[21]提出,随着赤铁矿粒度增大,H_2 还原反应速率降低,最终还原率则基本不变。COSTA 等[22]对竖炉内 H_2 直接还原铁矿球团进行了模拟分析,也认为在最优反应温度 800 ℃条件下，球团粒径越小还原反应速率越快。TEPLOV[23]则认为，对于微米级磁铁矿，还原速率随粒度减小显著提高，而纳米级磁铁矿还原速率则随粒度减小显著降低。此外，气体压力、铁氧化物孔隙率等因素对 H_2 还原反应也有影响[17]。

（2）H_2 还原铁氧化物的动力学机制

对于 H_2 还原铁氧化物的反应动力学机制，国内外研究者提出了不同的数学模型。SPREITZER 等[17]认为，H_2 对固态铁氧化物的还原过程符合未反应核模型，反应速率主要由外扩散、内扩散、化学反应控制。由于铁氧化物的还原分步进行，且不同温度范围内反应历程不同，因此，不同温度条件下 H_2 还原铁氧化物的动力学机制存在差异[24-27]。如陈赓[26]提出，H_2 还原铁氧化物，在 400 至 550 ℃范围内，第一阶段发生还原反应 $Fe_2O_3 \rightarrow Fe_3O_4$，符合化学反应模型，第二阶段发生还原反应 $Fe_3O_4 \rightarrow Fe$，符合三维扩散模型；在 600 至 750 ℃范围内，第一阶段发生还原反应 $Fe_2O_3 \rightarrow Fe_3O_4$，符合化学反应模型，第二阶段则发生还原反应 $Fe_3O_4 \rightarrow FeO$，符合随机成核和随后生长模型，第三阶段发生还原反应 $FeO \rightarrow Fe$，符合三维扩散模型。

在大多数条件下，H_2 对铁氧化物的熔融还原反应速率则非常快，反应速率主要受气体层中的传质速率控制[28]。NAGASAKA 等[18]提出，在 1673 K 条件下，H_2 对纯铁氧化物熔体的还原反应是关于 H_2 分压的一级反应，其速率函数可表示为：

$$r = k_a[H_2] \cdot [p(H_2) - p(H_2O)/K_H]$$

式中：速率常数 $k_a[H_2] = 1.6 \times 10^{-6}$；$K_H$ 为气体中 H_2O 与 H_2 的分压比 $p(H_2O)/p(H_2)$。

不同渣系中，H_2 对熔融铁氧化物的还原反应速率也存在差异，在一定范围内随着渣中 CaO 含量增加，反应速率提高，而 SiO_2 含量增加，反应速率则降低[18]。

不同研究者对于 H_2 还原铁氧化物动力学机制的研究结果存在一定差异，主要原因是采用的反应条件（如温度）以及铁矿原料成分、形态存在差异。XU 等[29]和 WANG 等[30]通过实验证明了矿物种类（褐铁矿、赤铁矿、鲕状赤铁矿）、铁矿成分对铁矿中铁氧化物的还原速率有显著影响。

1.1.3 H_2 作为气基还原剂的优势

热力学研究表明，在反应温度大于 810 ℃时，H_2 对铁氧化物的还原能力强于 CO。从动力学角度看，在相同温度条件下，H_2-H_2O 的互扩散系数（1000 K 时 7.330 cm^2/s）大于 CO-CO_2 的互扩散系数（1000 K 时 1.342 cm^2/s），因此，H_2 比 CO 更容易向铁矿颗粒或球团内部扩散，对固态铁氧化物的还原速率也更快[31]。

TATEO 等[32]和 HIDEKI 等[33]采用 CO-H_2 混合还原气体还原铁氧化物表明，在温度为 1093 K 条件下，随着煤气中 H_2 含量的增加，还原反应表观速率常数均显著增大，采用 CO、H_2-CO 混合气（体积比 1∶1）及 H_2 为还原剂时的表观速率常数分别为 8.6、15.0 和 33.0 mol/(s·m^3·atm)。ZUO 等[34]也证明了，采用 CO-H_2 混合气体还原铁氧化矿球团时，提高还原气体中 H_2 含量，可显著提高气体的有效扩散系数和反应速率常数。

在熔融状态，H_2 还原铁氧化物的反应速率常数也比固态 C、熔解态[C]和 CO 均高 1 到 2 个数量级，因此，H_2 是高效的炼铁还原剂[18]。

1.2　富氢气基直接还原炼铁技术研究

气基直接还原是在低于铁矿石熔点的温度下，采用还原气体将铁氧化物还原成高品位金属铁的方法,由于直接还原铁脱氧过程中形成许多微孔,在显微镜下观看状似海绵,又称为海绵铁。目前,气基直接还原炼铁已形成工业化应用,规模最大的 Midrex 工艺年产海绵铁达到 4500 万 t,采用的还原气体为含 H_2 和 CO 的富氢混合气体,因此,气基直接还原是一种基于氢冶金的炼铁技术。富氢气基直接还原流程生产的海绵铁约占世界海绵铁总产量 75%,主要包括竖炉工艺、流化床工艺等,其中,基于竖炉法的 Midrex 工艺、HYL-Ⅲ 工艺是最成功的两种富氢气基直接还原工艺[35]。

1.2.1 气基竖炉法

(1)Midrex 工艺,由 Midrex 公司开发成功,流程如图 2 所示[36]。将铁矿氧化球团或块矿原料从炉顶加入,从竖炉中部通入富氢热还原气,炉料与热风的逆向运动中被热还原气加热还原成海绵铁。富氢还原气由天然气经催化裂化制取,裂化剂为炉顶煤气。炉顶煤气经洗涤后部分与一定比例天然气混合经催化裂化反应转化成还原气,剩余部分则与天然气混合用作热能供应,催化裂化反应主要包括[35, 37-39]:

$$CH_4 + CO_2 \longrightarrow 2CO + 2H_2 \qquad (1)$$

$$CH_4 + H_2O \longrightarrow CO + 3H_2 \qquad (2)$$

产生的富氢还原气中,$V(H_2)/V(CO) \approx 1.5$,温度为 850~900 ℃[35, 37-39]。Midrex 工艺还原气中 H_2 含量较低,竖炉中还原气和铁矿石的反应表现为放热效应[40]。

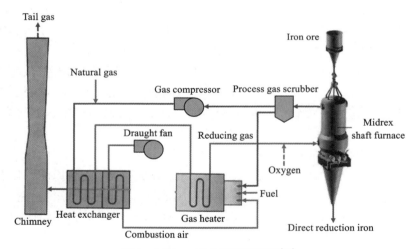

图 2 Midrex 工艺流程示意图[36]

Fig. 2 Process diagram of Midrex[36]

Midrex 工艺可获得最优铁金属化率达 100%,但产量大幅降低,铁金属化率约为 96% 时是最优生产条件,增加还原气体的 CO 比例可以提高产量[41]。对 Midrex 工艺竖炉的模拟研究表明,在竖炉 7.0 m 深度以下,随着还原气上行,H_2 体积分数迅速减少,而 CO 体积分数变化很小,直至 2.0 m 深度以上 CO 的体积分数才显著降低,铁矿原料在炉内运行约 2.0 m 即可完全变成浮氏体氧化亚铁[42]。GHADI 等[43] 则提出:赤铁矿在 Midrex 竖炉还原区上部完全转变为磁铁矿,运行到中部时被还原为方铁矿,在炉底部时方铁矿才被还原为海绵铁;采用双气体喷嘴,可提高铁矿还原率,每生产 1 t 海绵铁可减少 H_2 用量 100 m³。

（2）HYL-Ⅲ工艺，由墨西哥 Hylsa 公司开发，工艺流程如图 3 所示[36]。HYL-Ⅲ工艺使用球团矿或天然块矿为原料，原料在预热段内与上升的富氢还原气作用，迅速升温完成预热，随着温度的升高，矿石的还原反应逐渐加速，形成海绵铁。富氢还原气采用天然气为原料，水蒸气为裂化剂，经催化裂化反应制取[35,37-39]：

$$CH_4 + H_2O \longrightarrow CO + 3H_2 \tag{3}$$

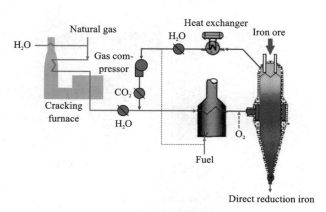

图 3　HYL-Ⅲ工艺流程示意图[36]

Fig. 3　The process diagram of HYL-Ⅲ

富氢还原气中 $V(H_2)/V(CO) = 5 \sim 6$，温度高达 930 ℃[35,37-39]。该工艺还原气中 H_2 含量较 Midrex 工艺高，竖炉中还原反应则表现为吸热效应，因此对入炉还原气温度要求较高[40]。

而后，Hylsa 公司又基于 HYL-Ⅲ工艺开发了 HYL-ZR 工艺，该工艺可以直接使用焦炉煤气、煤制气等富氢气体，为富煤缺气的地区发展气基直接还原工艺开辟了新路径。

与 Midrex 工艺相比，采用 $V(H_2)/V(CO) = 1$ 的煤制气生产 1 t 海绵铁产品，还原气消耗量增加约 100 m³，而能量利用率提高约 3.3%[35]。周渝生等[44]认为，中国的能源结构适合发展以煤气为气源的气基直接还原炼铁工艺，并提出了现有煤气化设备与 HYL 竖炉结合的工艺方案。王兆才[37]对煤制气-HYL 直接还原工艺进行了系统研究，结果表明：采用 H_2 含量 30%~75% 的煤制合成气，还原反应 2 h 内，铁矿球团金属化率均可达 95% 左右；还原气中 H_2 含量增加有利于还原反应的进行，但 H_2 含量达到 50% 后，H_2 对还原反应的增强作用逐渐减弱。

1.2.2　气基流化床法

基于流化床法的直接还原炼铁工艺主要有 Finmet 和 Circored 工艺，Finmet 工艺由委内瑞拉 Orinoco Iron 公司和奥地利 Siemens VAI Metals Technologies 公司联合开发并运营[14]。Circored 工艺由 Outotec 公司开发，采用天然气重整产生的 H_2 作为还原气体[14]。相比于竖炉法，基于流化床法的直接还原铁生产规模较小。流化床法可直接采用铁矿粉原料，在高温还原气流中进行还原，反应速度快，理论上是气基法中最合理的工艺方法。但生产实践中，使物料处于流化态所需的气体流量远大于理论还原所需的气量，造成还原气利用率极低，气体循环能耗高；其次，"失流"等生产问题难以解决，阻碍了流化床法的进一步发展[35]。

1.2.3 富氢气基直接还原炼铁技术新发展

2004 年,来自欧洲 15 个国家的 48 个企业、组织联合启动了 ULCOS(Ultra-low CO$_2$ Steelmaking)项目,并提出了基于氢冶金的钢铁冶金路线,通过电解水产氢,供给直接还原竖炉,工艺路线如图 4 所示[45]。该项目实现了氢冶金技术的突破,可使钢铁冶炼碳排放从 1850 kg CO$_2$/t 粗钢降低 84% 至 300 kg CO$_2$/t 粗钢[45]。

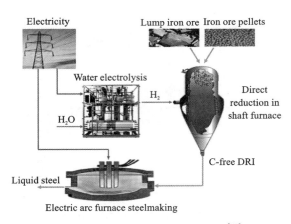

图 4 基于氢冶金的钢铁冶金路线[45]

Fig. 4 Iron and steel making route based on hydrogen metallurgy

2008 年,日本启动了创新性炼铁工艺技术开发项目(COURSE50),研究内容包括 H$_2$ 还原炼铁技术开发,提高 H$_2$ 还原效应[46-47]。目前,研究人员在 12 m^3 的试验高炉上进行了多次试验,对吹入 H$_2$ 带来的影响及 CO$_2$ 减排效果等进行了验证,确立了 H$_2$ 还原效果最大化的工艺条件。COURSE50 项目计划在 2030 年投入运行,此后将开展钢铁厂外部供氢技术的开发,最终实现"零碳钢"目标[47]。

2017 年,瑞典钢铁公司、LKAB 铁矿石公司和 Vattenfall 电力公司联合成立了合资企业,旨在推动 HYBRIT(Hydrogen breakthrough ironmaking technology)项目,开发基于 H$_2$ 直接还原的炼铁技术,替代传统的焦炭和天然气,减少瑞典钢铁行业碳排放[48]。经估算,采用纯 H$_2$ 还原,考虑间接碳排放量,可降低至 53 kg CO$_2$/t 粗钢[7]。HYBRIT 项目计划于 2018—2024 年进行中试试验,2025—2035 年建立 H$_2$ 直接还原炼铁示范厂,并依托新建的 H$_2$ 储存设施,到 2045 年实现无化石能源炼铁的目标[48]。

2019 年,德国蒂森克虏伯集团与液化空气公司联合,正式启动了高炉 H$_2$ 炼铁试验,并计划从 2022 年开始,杜伊斯堡地区其他高炉均使用 H$_2$ 进行钢铁冶炼,可使生产过程中 CO$_2$ 排放量降低 20%[47]。

2019 年 10 月,我国山西中晋矿业年产 30 万 t 氢气直接还原炼铁项目调试投产,该项目针对国内"富煤缺气"的资源特点,自主研发了"焦炉煤气干重整还原气"工艺,突破了气基竖炉直接还原技术在我国产业化的瓶颈,CO$_2$ 排放量比传统高炉炼铁降低 31.7%[49]。

由此可见,富氢气基直接还原炼铁已经进入技术成熟、稳步发展的阶段,并正在向纯氢气直接还原的方向发展。

1.3 富氢熔融还原炼铁技术研究

1995年,EDELSON[3]提出了采用H_2熔融还原铁矿制备铁水的工艺,该工艺将铁矿原料和助熔剂从炉顶加入,从还原炉中上部、中下部分别通入O_2和过量的H_2,炉料下落过程中首先通过由过量H_2与O_2燃烧产生的火焰区被完全熔化,然后熔体通过还原区被H_2还原,最后熔体在还原炉底部实现渣铁分离。从炉顶将尾气回收,并利用尾气余热对H_2和O_2进行预热。

由于纯H_2熔融还原炼铁仍存在短期内难以实现大规模、低成本制氢的问题,近年来,上海大学提出铁浴碳-氢复吹熔融还原工艺路线,其基本路线是在熔融还原反应中以H_2为主要还原剂、以碳为主要燃料,达到降低能耗和CO_2排放的目标[50]。碳-氢熔融还原工艺主要基于以下原理[5]。

还原:
$$Fe_2O_3 + 3H_2(g) = 2Fe + 3H_2O(g) \tag{4}$$

供热:
$$2C + O_2(g) = 2CO(g) \tag{5}$$

制氢:
$$CO(g) + H_2O(g) = H_2(g) + CO_2(g) \tag{6}$$

理论计算表明,采用H_2还原出1 mol Fe消耗的热量仅为碳还原的五分之一,且反应温度达到1400 ℃时,还原速率比CO高2个数量级,因此,在熔融状态下采用H_2还原铁矿具有速度快、能耗和CO_2排放均较低的优势[51]。

倪晓明等[52]进行了碳-氢熔融还原铁矿的实验研究,结果表明:几分钟之内可完成绝大部分铁氧化物的还原,且终渣Fe_T可达到1%以下;碳-氢熔融还原反应为一级反应,随着反应进行,还原速率降低,反应速率控制环节转变为铁氧化物的扩散。

曹朝真等[53]对采用H_2取代碳进行熔融还原炼铁的可行性进行了研究,认为:基于现有的熔融还原工艺,向熔炼炉喷吹H_2,通过减少喷煤量并增加H_2喷吹量,使$n(C):n(O)$小于1,可达到降低还原区所需热负荷的目的;铁矿熔融还原的吨铁能耗随$n(H_2)/n(H_2+C)$提高而降低,全碳熔融还原的理论吨铁能耗达$4×10^6$ kJ,而全H_2熔融还原理论吨铁能耗仅为约$0.8×10^6$ kJ,吨铁理论耗氢量为980 m^3。

张波[50]针对铁浴碳-氢复吹熔融还原铁矿工艺,研究建立了反应器动力学模型,认为引入H_2后还原效果优于纯碳还原,而在合适比例下的碳-氢混合还原又优于纯H_2还原;提高$n(C):n(H_2)$比例,有利于提高产能,而随着H_2量的增加,渣中铁含量降低;碳-氢混合还原以及纯碳、纯H_2还原反应速率常数分别为2.34 g Fe/($cm^2·min$)、1.7~1.85 g Fe/($cm^2·min$)、2.05 g Fe/($cm^2·min$),合适的$n(C):n(H_2)$比例为(0.5:1)~(1.2:1)。

2 有色金属冶金领域氢冶金研究进展

2.1 有色金属冶金领域发展氢冶金的可行性

我国有色金属工业发展迅速,自 2002 年以来我国有色金属产量连续 17 年居世界第一[54]。然而,我国有色金属冶金为高污染产业,每年产生大量温室效应气体。因此,发展氢冶金技术对我国有色金属冶金产业升级意义重大。

根据热力学原理,金属氧化物生成吉布斯自由能 $\Delta G^{\ominus}$(MeO) 大于水生成自由能 $\Delta G^{\ominus}$(H_2O) 时,该金属氧化物可被 H_2 还原。从 Ellingham 图[55](见图 5)可知,在一定温度条件下,铅、锡、铜、镍、钴等多种金属对应氧化物 $\Delta G^{\ominus}$(MeO) 高于 $\Delta G^{\ominus}$(H_2O),具有 H_2 还原的可行性。而且,低温条件下,$\Delta G^{\ominus}$(H_2O) 小于由碳生成 CO_2 的 $\Delta G^{\ominus}$(CO_2),即 H_2 还原能力强于碳,具备替代碳还原剂的潜能。

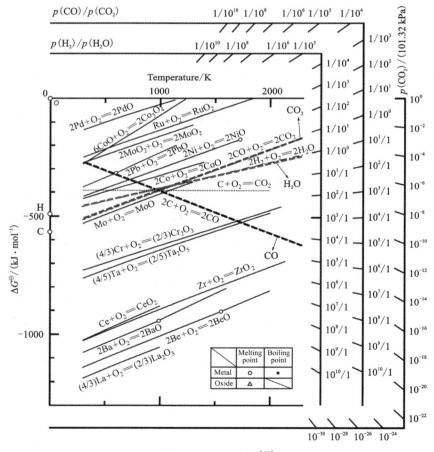

图 5 Ellingham 图[55]

Fig. 5 Ellingham diagram

近年来,在发展绿色冶金的形势驱动下,氢冶金在有色金属冶金领域也越来越受关注,特别是与钢铁冶金相关度较高的镍、钛等冶金行业。

2.2 镍冶金

基于综合利用富氢焦炉煤气降低红土镍矿冶炼能耗的设想,卢杰[56]采用焦炉煤气主要成分 H_2 作为还原剂,进行了红土镍矿还原焙烧-磁选制备镍铁的研究,结果表明:镍、铁回收率均随还原过程 H_2 分压增大而提高,在总气速为 200 L/h、还原温度 800 ℃、还原时间 220 min、硫酸钠添加量 20%、还原产物磨矿时间 10 min、磁场强度 0.156 T 的优化条件下,可获得镍品位 5.64%、镍回收率 83.59% 的镍铁产品。丁志广等[57-58]对 H_2 还原红土镍矿过程进行了详细分析,发现在小于 600 ℃ 的低温条件下,镍、铁的金属化率随温度升高呈现增加的趋势;实验所采取的氢气浓度(20%~100%)及矿物粒度(0.15~0.83 mm)条件对镍、铁金属化率的影响并不显著;对含 Ni 0.82%、Fe 9.67%、MgO 31.49% 的硅镁型红土镍矿,在温度 600 ℃、还原时间 90 min 及 H_2 浓度(体积分数)60% 的条件下,镍、铁金属化率分别为 88%、46%,其中大部分铁氧化物被还原为低价氧化物,继续提高还原温度,则红土镍矿中无定型镁硅酸盐通过重结晶生成致密的镁橄榄石相(Mg_2SiO_4),包覆镍、铁氧化物,阻碍了 Ni、Fe 的进一步还原。

对于 H_2 还原镍、铁氧化物的动力学,张海培等[59]认为:Fe_2O_3-NiO 体系在 H_2 气氛下还原过程的反应机理函数为 $G(\alpha)=[-\ln(1-\alpha)]^4$($\alpha$ 为反应分数),反应过程符合随机成核和随后生长模型,体系中 Fe_2O_3 含量越高,反应活化能越高;反应过程中,NiO 被优先还原,生成的 Ni 可作为催化剂促进 Fe_2O_3 的还原。

有研究者通过采用钠盐添加剂强化 H_2 对红土镍矿的还原过程,并取得了良好的效果。LU 等[60]研究了硫酸钠对强化 H_2 还原红土镍矿的作用,在 0%~20% 的添加范围内,增加硫酸钠添加量可提高产品镍品位与回收率,Na^+ 和 SO_4^{2-} 被吸附在红土镍矿表面,加快了界面反应速率;同时,硫酸钠被 H_2 还原形成 Fe-S 固溶体,加速了整个反应的传质过程,促进镍铁金属颗粒的传质和聚集长大。高金涛等[61]认为,焙烧过程中碳酸钠可促进红土镍矿镍、铁氧化物的释放,有利于进一步还原,还原产物中镍主要以铁为载体,并以 Fe-Ni 形式生成。

2.3 钛冶金

LI 等[62-63]采用 H_2、CO 混合气体,对钒钛磁铁矿球团进行还原,并分析反应过程的动力学,发现:随着还原温度、$V(H_2)/V(CO)$ 比例的升高以及球团粒径减小,金属总还原率呈升高趋势;还原反应表面活化能为 60.78 kJ/mol,反应过程初始阶段由表面化学反应控制,末尾阶段则由化学反应与内扩散环节联合控制;对含 FeO 26.25%、TiO_2 9.26%、V_2O_5 0.62%、Cr_2O_3 1.48% 的钒钛磁铁矿,采用焙烧-富氢气基直接还原-熔融分离工艺,进行铁、钛、钒、铬金属的提取;在焙烧温度 1200 ℃、焙烧时间 15 min、还原温度 1050 ℃、$V(H_2)/V(CO)$ 为 2.5、还原时间 30 min、熔融温度 1580 ℃、熔融分离时间 30 min 的优化条件下,铁、钛、钒、铬回收率分别达到 97.9%、89.8%、96.7%、97.8%。

SARGEANT 等[64-65]进行了 H_2 还原钛铁矿的实验研究,结果证明,钛铁矿还原过程中发生了以下反应:

$$FeTiO_3 + H_2(g) \Longleftrightarrow Fe + TiO_2 + H_2O(g) \tag{7}$$

并发现，H_2 浓度对钛铁矿的还原具有显著影响，1000 ℃ 条件下还原钛铁矿，反应初始的 1 h 时间内，低 H_2 压力下还原反应速率更高，随着反应的进行，则需要提高 H_2 压力才能保持与初始相当的还原反应速率。

ZHAGN 等[66]采用 H_2 辅助镁热还原 TiO_2 制备了钛金属粉末。由于钛对氧具有极强的亲和力，Mg 难以直接还原 TiO_2，而高温条件下 Ti-H-O 固溶体稳定性低于 Ti-O 固溶体，因此，可采用 H_2 气氛增强 Mg 与 Ti-O 反应的热力学驱动力，通过镁的进一步脱氧作用，获得钛金属粉末。

2.4 其他有色金属冶金

氢冶金技术也被应用于钨、钼金属粉末的制备。ZHU 等[67]采用 H_2 对含砷钨氧化物进行还原，制备钨金属粉末，在 800 ℃ 条件下还原 3 h，最终可获得 W-1%As 复合纳米粉末。KANG 等[68]研究了 H_2 分别还原纯 WO_3 和 WO_3-NiO 体系的动力学，结果表明：H_2 对纯 WO_3 的还原反应活化能为 94.6～117.4 kJ/mol，而对 WO_3-NiO 体系中 WO_3、NiO 的还原反应活化能分别为 87.4 kJ/mol、79.4 kJ/mol；混合物体系中，反应生成的 Ni 可促进 H_2 对 WO_3 的还原。

ZHANG 等[69]采用 H_2 对 MoO_2 粉末进行还原，并添加超细钼粉末作为成核剂，成功制备了纳米钼粉末。H_2 对 MoO_2 的还原反应速率受表面化学反应控制，活化能为 54.89～62.23 kJ/mol，在还原过程中添加 0.1%NaCl，反应活化能提高至 67.05～73.76 kJ/mol，因此，适量添加 NaCl 可限制反应生成的钼颗粒长大，从而获得微细颗粒的钼粉末[70]。

氢冶金技术在有色金属冶金领域的研究目前仍处于实验室研究阶段，而且仅针对氧化矿物的冶金过程。将氢冶金技术应用于有色金属硫化矿资源的冶金过程，如高铅渣、锑氧的还原熔炼，仍然具有很大的发展潜力。

3 二次资源利用领域氢冶金研究进展

目前，氢冶金技术在有色金属二次资源利用领域的应用也已有文献报道。

张怀伟[71]采用碳-氢复合还原对铜渣进行贫化，研究表明：在还原改性过程中，铜渣脱氧量和还原速率随还原气体中 H_2 含量的增加而提高，CO 和 H_2 最佳体积比为 $V(CO):V(H_2)=1:2$；熔融还原过程的固-液-气三相反应主要包括以下环节：

(1) 以 H_2 和 CO 为核心的气泡形成并长大；
(2) H_2 和 CO 通过气相边界层扩散到渣-气界面；
(3) 熔渣中 Fe^{3+} 向渣-气界面迁移；
(4) 在渣-气界面处，还原气体将 $Fe_3O_4(Fe^{3+})$ 还原成 $FeO(Fe^{2+})$ 和 H_2O；
(5) Fe^{2+} 向熔渣内部迁移；
(6) 气泡上升过程中，游离态的碳与 H_2O 发生反应，生产 CO 和 H_2。

计算结果表明，熔融铜渣的碳-氢复合还原反应为一级反应，表观活化能为 58.8 kJ/mol[71]。

通过分析熔融还原铜渣过程铜和铁的分布状态发现，喷吹 H_2 可为冰铜颗粒的碰撞聚集提供有利条件[72]。

EBIN 等[73]采用 H_2 还原辅助热解技术从 Zn-C 废弃电池的电极材料中回收锌，在 950 ℃的还原热解条件下，锌回收率达到 99.8%，通过分析认为：在该反应条件下，H_2 对电池黑粉的还原是主要反应，碳热还原为次要反应。

付辉龙等[74]采用铁浴碳-氢混合还原工艺对不锈钢粉尘中的铬进行回收，也获得了较好的效果，结果表明，底吹 H_2 可显著加快含铬物料的还原进程。

4 发展氢冶金亟待解决的关键问题

4.1 H_2 的大规模、低成本制备

冶金工业属于资源、能源密集型行业，要发展氢冶金技术，首先需要解决大容量、低成本制氢的问题。以钢铁冶金为例，采用纯 H_2 熔融还原炼铁技术，吨铁理论耗氢量高达 980 m^3[53]。因此，发展氢冶金需要保证 H_2 的大规模、低成本制备。此外，尽管在末端利用环节，H_2 被认为是一种清洁的能源和还原剂，但是制氢过程的清洁指数是实现整个行业清洁生产的关键。目前，比较成熟的电解水、煤焦气化、天然气裂解等制氢技术生产成本较高，且均依赖传统能源，缺乏可持续性[10]。

随着传统能源短缺问题的日益突出，基于太阳能的新型制氢技术，如光催化、光电解、热解制氢，已越来越受国内外研究者关注，并已取得大量研究成果[75]。ACAR 等[76-77]从经济性、技术可行性、环境影响等方面对不同的制氢技术进行可持续性分析认为，采用太阳能的热解制氢是基于现存电网水解制氢以外最具可持续性的制氢技术，但仍需突破大规模应用的技术障碍。

4.2 H_2 还原过程热量平衡

H_2 直接还原过程，以铁氧化物还原为例：

$$3Fe_2O_3 + H_2(g) = 2Fe_3O_4 + H_2O(g), \Delta H^{\ominus}_{298} = -12.1 \text{ kJ/mol} \tag{8}$$

$$Fe_3O_4 + H_2(g) = 3FeO + H_2O(g), \Delta H^{\ominus}_{298} = 164.0 \text{ kJ/mol} \tag{9}$$

$$FeO + H_2(g) = Fe + H_2O(g), \Delta H^{\ominus}_{298} = 135.6 \text{ kJ/mol} \tag{10}$$

总反应：

$$Fe_2O_3 + 3H_2(g) = 2Fe + 3H_2O(g), \Delta H^{\ominus}_{298} = 95.8 \text{ kJ/mol} \tag{11}$$

整个还原反应过程表现为强烈的吸热效应，导致纯 H_2 还原体系热量平衡需氢量远大于化学平衡需氢量，炉内化学能和物理能不匹配，须解决还原过程的热量平衡问题，以维持稳定的反应温度。

传统竖炉 H_2 利用率低，炉顶气的还原势较高。李路叶[78]和白明华等[79]通过模拟分析，设计了吹氧竖炉，燃烧部分未反应的 H_2，利用燃烧热对上部冷炉料进行预热，可使 H_2 用量比传统竖炉减少 22.08%、H_2 利用率提高 33.43%。因此，氢冶金过程，采用 H_2 为还原剂的

同时燃烧过剩的 H_2 补充热量，维持反应热量平衡，是提高反应效率，降低生产成本的可行方法。

5 结语

（1）作为气体还原剂，在温度大于 810 ℃的条件下，H_2 对铁氧化物的还原能力强于 CO，且 H_2 的还原反应速率比碳还原剂高 1 到 2 个数量级。因此，H_2 是高效、清洁的炼铁还原剂。

（2）富氢气基直接还原炼铁技术处于技术成熟、稳步发展的阶段。Midrex 和 HYL-Ⅲ 工艺是典型的基于氢冶金的直接还原炼铁工艺，目前，国内外冶金领域均在发展基于氢冶金的直接还原炼铁技术，如欧洲的 ULCOS、瑞典的 HYBRT、日本的 COURSE50 项目以及我国的中晋矿业等。H_2 熔融还原炼铁技术处于实验室研究阶段。

（3）氢冶金技术在镍、钛、钨、钼等有色金属冶金及二次资源利用领域的应用已越来越受关注，但目前处于实验室研究阶段，有待进一步的技术突破。

（4）由于氢冶金耗氢量大，因此，发展氢冶金须突破 H_2 的大规模、低成本、可持续制备技术。H_2 直接还原过程为吸热效应，还须解决热量平衡问题，以维持稳定的反应过程。

参考文献

[1] LIN Bo-qiang, WU Rong-xin. Designing energy policy based on dynamic change in energy and carbon dioxide emission performance of China's iron and steel industry[J]. Journal of Cleaner Production, 2020, 256: 1-14.

[2] 陈星. 中国有色金属工业全要素碳排放效率与碳排放绩效研究[D]. 厦门: 厦门大学, 2017.

[3] EDELSON J. Method for reducing particulate iron ore to molten iron with hydrogen as reductant. American, 5464464[P]. 1999-11-07.

[4] 徐匡迪, 蒋国昌. 中国钢铁工业的现状和发展[J]. 中国工程科学, 2000, 2(7): 1-8.
XU Kuang-di, JIANG Guo-chang. Present situations and development of Chinese iron and steel industry [J]. Engineering Science, 2000, 2(7): 1-8.

[5] 郑少波. 氢冶金基础研究及新工艺探索[J]. 中国冶金, 2012, 22(7): 1-6.

[6] GERMESHUIZEN L M, BLOM P W E. A techno-economic evaluation of the use of hydrogen in a steel production process, utilizing nuclear process heat[J]. International Journal of Hydrogen Energy, 2013, 38 (25): 10671-10682.

[7] VOGL V, ÅHMAN M, NILSSON L J. Assessment of hydrogen direct reduction for fossil-free steelmaking [J]. Journal of Cleaner Production, 2018, 203: 736-745.

[8] 干勇. 21 世纪是"氢"的时代[EB/OL]. [2019-09-20]. http://www.sohu.com/a/238747317_655347.

[9] TANG Jue, CHU Man-sheng, LI Feng, et al. Development and progress on hydrogen metallurgy[J]. International Journal of Minerals, Metallurgy and Materials, 2020, 27(6): 713-723.

[10] 张佩兰, 郑黎. 工业制氢技术及经济性分析[J]. 山西化工, 2014, 34(5): 54-56.

[11] DAWOOD F, ANDA M, SHAFIULLAH G M. Hydrogen production for energy: An overview[J].

International Journal of Hydrogen Energy, 2020, 45(7): 3847-3869.

[12] 王太炎, 王少立, 高成亮. 试论氢冶金工程学[J]. 鞍钢技术, 2005(1): 4-8.

[13] 张福明, 曹朝真, 徐辉. 气基竖炉直接还原技术的发展现状与展望[J]. 钢铁, 2014, 49(3): 1-10.

[14] SCHENK J L. Recent status of fluidized bed technologies for producing iron input materials for steelmaking[J]. Particuology, 2011, 9(1): 14-23.

[15] ZHOU Shi-wei, WEI Yong-gang, ZHANG Shuo-yao. Reduction of copper smelting slag using waste cooking oil[J]. Journal of Cleaner Production, 2019, 236: 1-5.

[16] 葛俊礼. 气基直接还原竖炉炉内行为与炉型关系研究[D]. 秦皇岛: 燕山大学, 2014.

[17] SPREITZER D, SCHENK J. Reduction of iron oxides with hydrogen—A review[J]. Steel Research International, 2019, 90(10): 201900108.

[18] NAGASAKA T, HINA M, BAN-YA S. Interfacial kinetics of hydrogen with liquid slag containing iron oxide[J]. Metall and Materi Trans B, 2000, 31(5): 945-955.

[19] BAHAGAT M, KHEDR M H. Reduction kinetics, magnetic behavior and morphological changes during reduction of magnetite single crystal[J]. Materials Science and Engineering B, 2007, 138(3): 251-258.

[20] BAI Ming-hua, LONG Hu, REN Su-bo, et al. Reduction behavior and kinetics of iron ore pellets under H_2-N_2 atmosphere[J]. ISIJ International, 2018, 58(6): 1034-1041.

[21] HABERMANN A, WINTER F, HOFBAUER H, et al. An experimental study on the kinetics of fluidized bed iron ore reduction[J]. ISIJ International, 2000, 40(10): 935-942.

[22] COSTA A R, WAGNER D, PATISSON F. Modelling a new, low CO_2 emissions, hydrogen steelmaking process[J]. Journal of Cleaner Production, 2013, 46: 27-35.

[23] TEPLOV O A. Kinetics of the low temperature hydrogen reduction of magnetite concentrates[J]. Russian Metallurgy (Metally), 2012(1): 8-21.

[24] LIN H Y, CHEN Y W, LI C. The mechanism of reduction of iron oxide by hydrogen[J]. Thermochimica Acta, 2003, 400(1/2): 61-67.

[25] WEI Zheng, ZHANG Jing, QIN Bao-ping, et al. Reduction kinetics of hematite ore fines with H_2 in a rotary drum reactor[J]. Powder Technology, 2018, 332: 18-26.

[26] 陈赓. 气基还原氧化铁动力学机理研究[D]. 大连: 大连理工大学, 2011.

[27] BARDE A A, KLAUSNER J F, MEI R. Solid state reaction kinetics of iron oxide reduction using hydrogen as a reducing agent[J]. International Journal of Hydrogen Energy, 2016, 41(24): 10103-10119.

[28] HAYASHI S, IGUCHI Y. Hydrogen reduction of liquid iron oxide fines in gas-conveyed systems[J]. ISIJ International, 1994, 34(7): 555-561.

[29] XU Run-sheng, DAI Bo-wen, WANG Wei, et al. Effect of iron ore type on the thermal behaviour and kinetics of coal-iron ore briquettes during coking[J]. Fuel Processing Technology, 2018, 173: 11-20.

[30] WANG H T, SOHN H Y. Effect of CaO and SiO_2 on swelling and iron whisker formation during reduction of iron oxide compact[J]. Ironmaking & Steelmaking, 2011, 38(6): 447-452.

[31] 邹宗树, 王臣. 熔融还原炼铁工艺的煤气富氢改质预还原[J]. 钢铁, 2007, 42(8): 17-20.

[32] TATEO U, TSUNEHISA N, HIDEKI O, et al. Effective use of hydrogen in gaseous reduction of iron ore agglomerates with H_2-CO[J]. Journal of Iron and Steel Research International, 2009, 16(2): 1179-1184.

[33] HIDEKI O, TOSHINARI Y, TATEO U. Effect of water-gas shift reaction on reduction of iron oxide powder packed bed with H_2-CO mixtures[J]. ISIJ International, 2003, 43(10): 1502-1511.

[34] ZUO Hai-bin, WANG Cong, DONG Jie-ji. Reduction kinetics of iron oxide pellets with H_2 and CO mixtures[J]. International Journal of Minerals, Metallurgy and Materials, 2015, 22(7): 688-696.

[35] 易凌云.铁矿球团混合气体气基直接还原基础研究[D].长沙:中南大学,2013.
[36] 刘龙.氢气直接还原竖炉还原段内温度场及流场研究[D].秦皇岛:燕山大学,2016.
[37] 王兆才.氧化球团气基竖炉直接还原的基础研究[D].沈阳:东北大学,2009.
[38] 张福明,曹朝真,徐辉.气基竖炉直接还原技术的发展现状与展望[J].钢铁,2014,49(3):1-10.
[39] 董跃,乔星星,刘改换,等.气基直接还原铁工艺还原气研究现状[J].能源与节能,2016(3):1-4.
[40] 吕建超.气基直接还原竖炉内还原过程的研究与分析[D].秦皇岛:燕山大学,2017.
[41] PARISI D R, LABORDE M A. Modeling of counter current moving bed gas-solid reactor used in direct reduction of iron ore[J]. Chemical Engineering Journal, 2004, 104(1/3):35-43.
[42] 徐辉,邹宗树,周渝生,等.竖炉生产直接还原铁过程的数值模拟[J].材料与冶金学报,2009,8(1):7-11.
[43] GHADI A Z, VALIPOUR M S, BIGLARI M. CFD simulation of two-phase gas-particle flow in the Midrex shaft furnace: The effect of twin gas injection system on the performance of the reactor[J]. International Journal of Hydrogen Energy, 2017, 42(1):103-118.
[44] 周渝生,钱晖,齐渊洪,等.煤制气生产直接还原铁的联合工艺方案[J].钢铁,2012,47(11):27-35.
[45] QUADER M A, AHMED S, DAWAL S Z. Present needs, recent progress and future trends of energy-efficient Ultra-Low Carbon Dioxide (CO_2) Steelmaking (ULCOS) program[J]. Renewable Sustainable Energy Rev., 2016, 55:537-549.
[46] 全荣.COURSE50炼铁工艺研发进展[N].世界金属导报,2016-05-17(B03).
[47] 张京萍.拥抱氢经济时代全球氢冶金技术研发亮点纷呈[N].世界金属导报,2019-11-26(F01).
[48] KARAKAYA E, NUUR C, ASSBRING L. Potential transitions in the iron and steel industry in Sweden: Towards a hydrogen-based future?[J]. Journal of Cleaner Production, 2018, 195:651-663.
[49] 中晋冶金科技有限公司.山西中晋矿业年产30万吨氢气直还铁项目即将投产[EB/OL].[2019-09-20].http://www.zjthky.com/view.asp?id=274.
[50] 张波.铁浴碳-氢复合吹终还原反应动力学研究[D].上海:上海大学,2011.
[51] 郑少波,洪新,徐建伦,等.C-H_2绿色高效铁矿熔态还原技术构想[C]//2006年非高炉炼铁年会论文集.沈阳:中国金属学会,2006:117-123.
[52] 倪晓明,骆琳,杨森龙,等.氢-碳熔融还原实验室试验研究[C]//2008年非高炉炼铁年会文集.延吉:中国金属学会,2008:310-315.
[53] 曹朝真,郭培民,赵沛,等.高温熔态氢冶金技术研究[J].钢铁钒钛,2009,30(1):1-6.
[54] 郭学益,田庆华,刘咏,等.有色金属资源循环研究应用进展[J].中国有色金属学报,2019,29(9):1859-1901.
[55] HASEGAWA M. Treatise onprocess metallurgy: Chapter 3. 3-ellingham diagram[M]. Stockholm, Sweden: Royal Institute of Technology, 2014.
[56] 卢杰.硫酸钠对红土镍矿在氢气和甲烷气氛下的还原性研究[D].太原:太原理工大学,2013.
[57] 丁志广.硅镁型红土镍矿气基固相还原的研究[D].昆明:昆明理工大学,2017.
[58] 丁志广,李博,魏永刚.氢气作用下硅镁型红土镍矿的低温还原特性[J].中国有色金属学报,2018,28(8):1669-1675.
[59] 张海培,李博,丁志广,等.非等温条件下氢气还原Fe_2O_3-NiO制备镍铁合金的反应动力学[J].中国有色金属学报,2017,27(1):171-177.
[60] LU Jie, LIU Shou-jun, SHANGGUAN Ju, et al. The effect of sodium sulphate on the hydrogen reduction process of nickel laterite ore[J]. Minerals Engineering, 2013, 49:154-164.
[61] 高金涛,张颜庭,陈培钰,等.红土镍矿富集镍和铁的焙烧、氢气还原和磁选分离[J].北京科技大学学

报，2013，35(10)：1289-1296.
[62] LI Wei, FU Gui-qin, CHU Man-sheng, et al. Reduction behavior and mechanism of Hongge vanadium titanomagnetite pellets by gas mixture of H_2 and CO[J]. Journal of Iron and Steel Research, Internationa, 2017, 24(1): 34-42.
[63] LI Wei, FU Gui-qin, CHU Man-sheng, et al. An effective and cleaner process to recovery iron, titanium, vanadium, and chromium from Hongge vanadium titanomagnetite with hydrogen-rich gases[J]. Ironmaking & Steelmaking, 2020, 1743-2812. DOI: 10.1080/03019233.2020.1721955.
[64] SARGEANT H M, ABERNETHY F A J, BARBER S J, et al. Hydrogen reduction of ilmenite: Towards an in situ resource utilization demonstration on the surface of the Moon[J]. Planetary and Space Science, 2020, 180: 104751.
[65] SARGEANT H M, ABERNETHY F A J, ANAND M, et al. Feasibility studies for hydrogen reduction of ilmenite in a static system for use as an ISRU demonstration on the lunar surface[J]. Planetary and Space Science, 2020, 180: 104759.
[66] ZHANGY, FANGZ Z, XIA Y, et al. Hydrogen assisted magnesiothermic reduction of TiO_2[J]. Chemical Engineering Journal, 2017, 308: 299-310.
[67] ZHU Hong-bo, TAN Dun-qiang, LI Ya-lei, et al. Refining mechanisms of arsenic in the hydrogen reduction processof tungsten oxide[J]. Advanced Powder Technology, 2015, 26(3): 1013-1020.
[68] KANG H, JEONG Y K, OH S T. Hydrogen reduction behavior and microstructural characteristics of WO_3 and WO_3-NiO powders[J]. International Journal of Refractory Metals & Hard Materials, 2019, 80: 69-72.
[69] ZHANG Yong, JIAO Shu-qiang, CHOU Kuo-Chih, et al. Size-controlled synthesis of Mo powders via hydrogen reduction of MoO_2 powders with the assistance of Mo nuclei[J]. International Journal of Hydrogen Energy, 2020, 45(3): 1435-1443.
[70] SUN Guo-dong, WANG Kai-fei, JI Xin-peng, et al. Preparation of ultrafine/nano Mo particles via NaCl-assisted hydrogen reduction of different-sized MoO_2 powders[J]. International Journal of Refractory Metals & Hard Materials, 2019, 80: 243-252.
[71] 张怀伟.基于直流电场和碳-氢复合还原改性的铜渣贫化过程的实验研究[D].上海：上海大学，2014.
[72] QU Guo-rui, WEI Yong-gang, LI Bo, et al. Distribution of copper and iron components with hydrogen reduction of copper slag[J]. Journal of Alloys and Compounds, 2020, 824: 1-9.
[73] EBIN B, PETRANIKOVA M, STEENARI B M, et al. Investigation of zinc recovery by hydrogen reduction assisted pyrolysisof alkaline and zinc-carbon battery waste[J]. Waste Management, 2017, 68: 508-517.
[74] 付辉龙，牛帅，陈文彬，等.含铬物料铁浴碳氢还原的实验研究[J].过程工程学报，2013，13(2)：246-249.
[75] LIU G, SHENG Y, AGER J W, et al. Research advances towards large-scale solar hydrogen production from water[J]. Energy Chem, 2019, 1(2): 1-41.
[76] ACAR C, BESKESE A, TEMUR G T. Sustainability analysis of different hydrogen production options using hesitant fuzzy AHP[J]. International Journal of Hydrogen Energy, 2018, 43(39): 18059-18076.
[77] ACAR C, DINCER I. Review and evaluation of hydrogen production options for better environment[J]. Journal of Cleaner Production, 2019, 218: 835-849.
[78] 李路叶.直接还原竖炉氢气利用率及炉内温度场研究[D].秦皇岛：燕山大学，2016.
[79] 白明华，李路叶，徐宽，等.直接还原竖炉氢气利用率的数值分析及优化[J].燕山大学学报，2016，40(6)：481-486.

镍熔炼渣贫化工艺现状与展望

摘　要：目前硫化镍矿主要采用火法冶炼工艺，降低镍熔炼渣中有价金属含量是各企业追求的目标。镍冶炼炉渣一般属于 FeO-SiO$_2$ 系和 FeO-SiO$_2$-CaO(MgO) 系，在造锍熔炼过程中应关注 Fe$_3$O$_4$ 的变化，渣含镍量与渣中 Fe$_3$O$_4$ 含量呈正相关关系。镍熔炼渣贫化工艺一般采用沉降电炉工艺，国外工艺中镍熔炼渣中镍含量一般为 0.5%~2%；国内比较先进的镍熔炼渣贫化工艺采用侧吹还原+硫化剂的方法，渣含 Ni 可降至 0.12%。硫化镍矿熔炼渣贫化工艺的关键是增强反应动力学、增加侧吹还原手段，以提高还原剂利用率，可以考虑采用多种工艺手段结合的工艺路线来降低尾渣中有价金属含量。

关键词：镍熔炼渣；贫化工艺；有价金属回收率；沉降电炉；侧吹还原

Current status and prospect of nickel smelting slag dilution process

Abstract：At present, nickel sulfide ore mainly adopts pyrometallurgical smelting process, and reducing the content of valuable metals in nickel smelting slag is the goal pursued by enterpises. Nickel smeling slag generally belongs to FeO-SiO$_2$-CaO(MgO) slag system. In the process of matte-making smelting, attention should be paid to the change of Fe$_3$O$_4$. The nickel content of the slag is positively related to the Fe$_3$O$_4$ content in the slag. The nickel smelting slag dilution process generally adopts the sedimentation electric furnace process. The nickel contet in the nickel smeling slag in foreign processes is generally 0.5% to 2%; the domestically advanced nickel smelting slag dilution process adopts the method of side blowing reduction + vulcanizing agent, which can reduce the Ni content of the slag to 0.12%. The key to the nickel sulfide smelting slag dilution process is to enhance the reaction kinetics and increase the side-blowing reduction means to increase the utilization rate of the reducing agent. A process route combining multiple process methods can be considered to reduce the valuable metal content in the slag.

Keywords：nickel smeling slag; dilution process; valuable metal recovery rate; sedimentation electric fumace; side blow reduction

随着全球经济的快速发展，作为镍主要消费领域的不锈钢需求量与产量不断增加，以镍

本文发表在《中国有色冶金》，2020，49(06)：1-4。作者：王雪亮，石润泽，李兵，顾明杰，朱荣，郭亚光。

为主要原料的燃料电池、催化剂以及电镀行业快速发展,市场对镍的需求量不断上涨。近年来,受镍铁冶炼产能提高和成本降低的影响,硫化镍矿火法冶炼企业面临激烈的竞争,如何最大限度地降低生产成本、提高有价金属回收率,成为镍熔炼系统亟待解决的问题。降低弃渣指标、提高有价金属回收率是解决问题最为直接、有效的途径,也是实施资源战略、发展循环经济的落脚点。

1 镍火法冶炼工艺

镍火法冶炼工艺分为硫化镍矿冶炼和氧化镍矿冶炼。氧化矿主要用于生产镍铁、镍锍或湿法生产氢氧化镍,典型工艺有回转窑–矿热炉联合法、回转窑直接还原法等。硫化矿选矿后主要用于生产金属镍,典型工艺主要有闪速炉法及富氧顶吹法等。硫化矿经采、选作业生产镍精矿,利用造锍熔炼和吹炼生产高镍锍。目前硫化镍矿火法冶炼工艺方法归纳见图1[1],硫化镍矿冶炼流程见图2[2]。

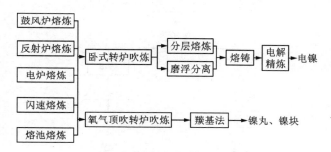

图1 硫化镍矿火法生产方法

造锍熔炼是一种常用的镍冶炼方法,其原理是利用金属镍对硫的亲和力接近铁,而对氧的亲和力远小于铁的性质,在氧化程度不同的造锍熔炼过程中,分阶段使铁的硫化物不断氧化成氧化物,随后与脉石造渣而除去。镍熔炼的主要工艺有闪速熔炼、氧气顶吹炉熔炼、电炉熔炼、反射炉熔炼和熔池熔炼等。镍闪速熔炼技术,克服了传统熔炼方法未充分利用粉状精矿的巨大表面积和矿物燃料的缺点,大大减少了能源消耗,提高了硫的利用率,改善了环境。闪速熔炼有奥托昆普闪速炉和因科纯氧闪速炉两种形式。

进行造锍熔炼时,硫化镍矿和熔剂等物料在熔炼炉中发生一系列物理化学反应,最终形成互不相溶的镍锍或铜镍锍和炉渣[1]。

镍锍吹炼的主要工艺有卧式转炉吹炼和氧气顶吹转炉吹炼。低镍锍吹炼的任务是向低镍锍熔体中鼓入空气和加入适量的石英熔剂,将低镍锍中的铁和其他杂质氧化后与石英造渣,部分硫和其他一些挥发性杂质氧化后随烟气排出,从而得到含有价金属(Ni、Cu、Co 等)较高的高镍锍和含有价金属较低的吹炼渣。低镍锍的主要成分是 FeS、Fe_3O_4、Ni_3S_2、Cu_2S、PbS、ZnS 等,在吹炼 1250 ℃左右的高温下,硫化物一般按式(1)~(2)进行氧化反应。

$$MS + 3/2O_2 = MO + SO_2 \tag{1}$$

$$MS + O_2 = M + SO_2 \tag{2}$$

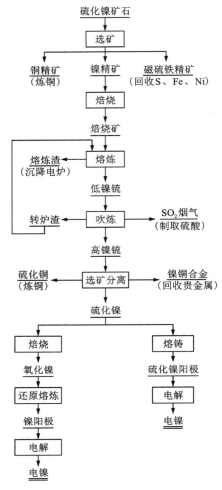

图 2 硫化镍矿冶炼流程

式中：M 表示金属；MS 表示金属硫化物；MO 表示金属氧化物。

高镍锍中 Ni、Cu 大部分仍然以金属硫化物状态存在，少部分以合金状态存在，低镍锍中的贵金属和部分钴也进入高镍流中。

2 镍熔炼渣性质

镍熔炼渣的化学成分随矿石种类和冶炼工艺的不同而存在差异。硫化镍精矿造锍熔炼产出的炉渣为铁橄榄石型炉渣，一般含 Fe 35%~40%，含 SiO_2 30%~40%。反射炉和闪速炉炉渣含 Fe_3O_4 通常高于 10%，并随镍锍品位上升而增加，而电炉炉渣含 Fe_3O_4 较低(<5%)[1]。氧化矿制备镍铁或镍锍过程中所产镍渣可用于生产铸石、矿棉纤维、人造石英石等[2]。

镍熔炼渣中的主要矿物有辉石(含镁)、橄榄石等，水碎的镍渣中还含有大量的玻璃相，玻璃相的含量与渣排出时的温度、水碎速度等有关[3]。其他镍冶炼流程产生的渣,大部分都要返回熔炼炉进行处理,因此以研究镍熔炼渣贫化工艺为主。

镍冶炼炉渣一般属于 $FeO-SiO_2$ 系和 $FeO-SiO_2-CaO(MgO)$ 系,在造锍熔炼过程中还要关注 Fe_3O_4 的变化。由于镍矿原料中往往含有较多的 MgO,所以炉渣含 MgO 也较多,渣中 MgO 低于 10% 时对渣性质无太大影响,当 MgO 质量分数超过 14% 时,炉渣的熔点迅速上升,黏度增大,单位电耗增大[1]。

炉渣的导电率与炉渣黏度有关,一般来说,黏度小的炉渣导电性好,含 FeO 高的炉渣除离子传导外,还有电子传导,所以具有很好的导电性。熔炼渣黏度受硅酸离子形状、大小的影响,一般情况下碱性氧化物能破坏硅酸离子网状结构,有降低黏度的作用,添加氟化物降低黏度非常有效。

3 国外镍熔炼渣贫化工艺现状

镍在炉渣中的损失形态为机械夹杂和化学溶解(主要以 NiO 形态)两种。化学损失随氧势的增加而递增。所以在造锍熔炼过程中镍锍品位的提高会导致更多的镍损失,熔炼渣中含 Ni 一般为 0.5%~2%,通常采用喷吹还原剂作为降低镍化学损失的方法。用炉渣贫化炉处理炉渣可提高镍回收率,焦炭是普遍使用的还原剂[1]。

奥托昆普闪速炉工艺常采用电炉进行炉渣贫化。从目前镍闪速熔炼生产企业来看,博茨瓦纳 BCL 公司和芬兰奥托昆普公司贫化过程在单独的电炉内进行。

国际镍公司汤姆森镍厂电炉熔炼原料低镍锍成分为:(Ni+Cu) 17%,Fe 47%,S 26%;炉渣成分为:Fe 36%,SiO_2 36%,(Ni+Cu) 0.3%。采用 FeS_2 洗涤、沉降、还原沉降等措施降低炉渣中镍,结果表明还原后沉降的效果最明显,镍回收率为 85%。

哈贾伐尔塔冶炼厂和奥托昆普研究中心合作开发了闪速熔炼直接产出高镍锍的新工艺,该工艺要求电炉渣含镍控制在 0.3% 以下。实际生产发现,电炉渣含镍量与渣中 Fe_3O_4 含量呈正相关关系,电炉贫化渣中 Fe_3O_4 为 2% 时,弃渣含 Ni 为 0.2%,当电炉贫化渣中 Fe_3O_4 为 7% 时,弃渣含 Ni 可达 1.0%,因此要降低渣含镍必须控制渣中 Fe_3O_4 含量[4]。

4 国内镍熔炼渣贫化工艺现状

4.1 吉恩镍渣贫化工艺试验

中国恩菲工程技术有限公司与吉林吉恩镍业股份有限公司合作进行了镍渣贫化工艺的研究,研究采用了顶吹煤粉或天然气的方法。煤粉加入采取喷粉工艺,以氮气为载气将煤粉喷入熔池底部,天然气则直接采用浸入式喷枪喷入熔池底部。

试验是以喷吹煤粉或天然气的方式选择性还原矿热炉中熔融镍转炉渣,主要验证喷粉还原工艺、天然气喷吹工艺及相关实践操作可行性,对基础性理论技术参数进行校正调整。

喷吹煤粉试验结果显示:Fe 在化渣和喷粉还原过程中含量始终保持稳定;化渣阶段温度

较高以及动力学条件良好，Ni、Co、Cu 在化渣结束后实现大部分的还原分离，还原分离率分别为 $S_{Ni}>80\%$，$S_{Co}\approx60\%$，$S_{Cu}\approx75\%$；在喷粉还原阶段 Ni、Co 还原分离率增加幅度有限，Cu 在喷粉还原阶段不发生改变，因为在化渣阶段已完成大部分分离。喷粉还原终态 $w_{Fe_T}=45\%$，$w_{Ni}=0.1\%$，$w_{Co}=0.2\%$，$w_{Cu}=0.25\%$；终态还原分离率 $w_{Ni}=95\%$，$w_{Co}=70\%$，$w_{Cu}=75\%$。

天然气喷吹试验结果显示：Fe 在化渣和天然气还原过程中含量始终保持稳定；因为化渣阶段熔池温度较高以及动力学条件良好，还原反应进行较完全，Ni、Co、Cu 在化渣结束后实现了大部分的还原分离，还原分离率分别为 $S_{Ni}>85\%$，$S_{Co}\approx60\%$，$S_{Cu}\approx75\%$；在喷吹天然气还原阶段 Ni、Co 还原分离率增加幅度有限，Cu 在喷吹天然气还原阶段不发生改变，因为在化渣阶段已完成大部分分离。喷吹天然气还原终态 $w_{Fe_T}=48.5\%$，$w_{Ni}=0.1\%$，$w_{Co}=0.15\%$，$w_{Cu}=0.25\%$，终渣含 Ni 很低；终态还原分离率 $S_{Ni}=95\%$，$S_{Co}=80\%$，$S_{Cu}=75\%$。

与煤粉喷吹相比，天然气喷吹过程稳定性更好，所得测试数据离散程度小。可以观测出渣中元素改变随着天然气量的增加可分为三个阶段：第一阶段主要是 Ni 的还原；第二阶段渣中 Ni 还原速率降低，Co 还原开始；第三阶段渣中 Ni 与 Co 都保持恒定[5-6]。

4.2 喀拉通克镍渣贫化工艺实践

新疆新鑫矿业有限责任公司喀拉通克铜镍矿采用富氧侧吹炉熔池熔炼新工艺，形成了具有年产 8000 t 镍金属的熔炼能力。新疆喀拉通克矿业有限责任公司采用富氧侧吹炉冶炼工艺的主要工艺特点是：①富氧风送至熔渣层进行造锍熔炼；②熔池深，渣与低镍锍在炉内分离完全；③炉身采用铜水套结构，生产过程中铜水套内壁形成 10~30 mm 厚的渣层，无风口砖，渣层对铜水套起到保护作用。

新疆新鑫矿业有限责任公司喀拉通克铜镍矿采用富氧侧吹炉熔池熔炼新工艺，主要工艺特点是：①富氧风送至熔渣层进行造锍熔炼；②熔池深，渣与低镍锍在炉内分离完全；③炉身采用铜水套结构，生产过程中铜水套内壁形成 10~30 mm 厚的渣层，无风口砖，渣层对铜水套起到保护作用。

侧吹炉系统处理的物料主要包括自产精矿、外购精矿、特富矿，其中自产精矿占总处理量的 60%，外购精矿及特富矿分别占 20%，其主要成分见表 1。

表 1　侧吹炉入炉物料成分　　　　　　　　　　　　　　　　%

矿类型	Ni	Cu	Fe	S	SiO_2	CaO	MgO
自产精矿	2.61	4.52	32.75	24.81	15.09	1.72	7.04
外购精矿	6.3	0.46	20.26	13.83	28.82	0.115	6.62
特富矿	2.68	4.28	45.61	28.35	6.59	1.97	1.77
混合精矿	3.99	2.28	30.06	20.61	14.79	1.53	6.66

典型的侧吹炉低镍锍及炉渣成分见表 2，侧吹炉渣含 Ni 为 0.23%。

表 2 侧吹炉低镍锍及炉渣成分 %

成分	Ni	Cu	Fe	SiO_2	CaO	MgO
低镍锍	17.34	20.46	36.37	—	—	—
侧吹炉渣	0.23	0.34	34.62	36.31	2.56	6.09

因原料混合精矿中 Ni 高 Cu 低,镍锍中 Ni 品位约为 17%,镍冶炼镍锍与炉渣中 Ni 分配比取 100 计算,炉渣含 Ni 理论上应该达到 0.17%,实际生产渣含镍为 0.15%~0.25%,因此渣贫化还有下降空间。

4.3 金川镍沉降电炉镍渣贫化试验

金川公司的沉降电炉为顶吹炉熔炼系统的配套炉窑,处理顶吹炉在高氧势气氛下形成的熔融物,其尾料含镍指标的高低,决定了顶吹炉系统运行经济效益的优劣。镍冶炼厂富氧顶吹镍熔炼系统采用的原料中 MgO 含量高,渣黏度大,再加上其他工艺条件的制约,沉降电炉尾料指标虽然逐年有所下降,但还处于较高的水平,尾渣含镍>0.35%。较高的尾料含镍,已成为制约富氧顶吹镍熔炼系统经济运行的症结所在,严重影响镍熔炼系统的经济效益和环境评价。

为拓展降低沉降电炉尾料含镍指标途径,寻求提升顶吹炉系统运行经济效益支持,研究了锍渣分离的最佳条件以及降低尾料含镍的机理,结果表明采用将还原剂强制鼓入熔池内的方式增加还原性气氛,同时通过动力学搅拌作用为渣还原提供条件,可以达到降低沉降电炉尾料含镍的目的。

根据上述思路,进行 500 kg 电炉侧吹天然气镍渣贫化扩大试验,得出以下结论:
(1)侧吹对熔池搅拌强烈,可以提供良好的化学反应动力学条件。
(2)通过扩大试验可以看出,采用侧吹连续硫化还原熔炼,通过增加动力学条件,可有效降低磁性铁含量及渣的黏度,顶吹炉渣含镍可以降到 0.12%,侧吹贫化工艺可行。
(3)试验温度与沉降电炉实际控制温度一致,但沉降电炉渣含镍远高于试验炉渣;且从连续侧吹试验数据分析可以看出,在未进行有效澄清分离的条件下,渣含有价金属依然较低。

以上说明,沉降电炉在保持现有温度及澄清时间的条件下,通过增加动力学条件,有效还原,可以实现渣中有价金属含量的降低。

5 结语

硫化镍矿熔炼渣的贫化采用沉降电炉侧吹还原贫化工艺效果较好。单独的电炉沉降和渣选矿方案都有其局限性,为达到更好的贫化效果,可以考虑多种工艺手段结合的工艺路线。在现有沉降电炉贫化工艺的基础上,要增强反应动力学,增加侧吹还原手段,以提高还原剂利用率,最终降低尾渣中有价金属含量。

国家发展和改革委员会发布的《产业结构调整指导目录(2019 年本)》中,鼓励高效、低耗、低污染、新型冶炼技术开发,高效、低耗、低污染的短流程工艺是硫化镍矿火法冶炼的发展方向,短流程镍冶炼工艺产生的高镍熔炼渣的贫化将成为新的研究热点。

参考文献

[1] 彭容秋.镍冶金[M].长沙:中南大学出版社,2005.
[2] 李小明,沈苗,王翀,等.镍渣资源化利用现状及发展趋势分析[J].材料导报,2017,31(3):100-105.
[3] 盛广宏,翟建平.镍工业冶金渣的资源化[J].材料导报,2005,19(10):68-71.
[4] 邱竹贤.冶金学·有色金属卷[M].沈阳:东北大学出版社,2001.
[5] 郭亚光,朱荣,裴忠冶,等.镍渣熔融还原提铁动力学[J].中国有色冶金,2017,46(5):75-80.
[6] 郭亚光,朱荣,王云,等.镍渣煤基熔融还原提铁工艺基础研究[J].工业加热,2015,44(6):40-43.

氢氧化物沉淀法制备层状结构氧化钪的研究

摘　要：本工作研究了氢氧化物沉淀法和焙烧制备具有层状结构的氧化钪。研究了反应温度、沉淀剂用量、搅拌速度和陈化时间等影响因素对钪沉淀回收率的影响。由 XRD 和 IR 测试分析可得出，含钪沉淀物以 ScOOH 相为主晶相。主要采用电感耦合等离子体光谱仪（ICP-OES）测试含钪沉淀物中的钪离子含量，计算钪沉淀回收率。研究结果表明，在反应温度为 60 ℃、氢氧化钠沉淀剂用量为 27.5 g/L、搅拌速度为 200 r/min、陈化时间为 2 h 时，钪沉淀回收率为 97% 以上。含钪沉淀物经过焙烧处理获得氧化钪粉体材料，采用 XRD、SEM-EDS、傅里叶红外光谱仪等对氧化钪粉体材料进行表征。结果表明，氧化钪的晶型结构好并且比较纯净，其由钪元素和氧元素组成。同时，通过对微观结构的测试可得出，氧化钪粉体材料为微米级的层状结构，粒径小于 20 μm。

关键词：沉淀法；氢氧化物；层状结构；氧化钪

Research on the preparation of layered scandium oxide by hydroxide precipitation

Abstract: The scandium oxide with layered structure prepared by hydroxide precipitation and roasting was studied in this paper. The effects of reaction temperature, precipitation, stirring rate and aging time on the recovery efficiency of scandium precipitation were studied. It was concluded that the main crystal phase of scandium sediment was ScOOH by XRD and FTIR analysis. Based on the scandium ion content value in scandiumsediment, which was measured by inductively coupled plasma spectrometer (ICP-OES). The recovery percentages of scandium in the precipitation reaction were calculated. The results indicated that the recovery percentage of scandium precipitation was up to 97%, when the sodium hydroxide precipitant dosage, reaction temperature, stirring rate and aging time were 27.5 g/L, 60 ℃, 200 r/min and 2 h, respectively. The scandium oxide powder was obtained by roasting scandium hydroxide. The scandium oxide powder material was characterized by XRD, SEM-EDS, and FTIR. The scandium oxide had good crystal structure and was relatively pure, which consisted of scandium and oxygen elements. Meanwhile, according to the test of microstructure, the scandium oxide powder material was a layer structure of micron scale and the particles less than 20 μm.

Keywords：precipitation；hydroxide；layer structure；scandium oxide

引言

钪是一种稀土元素，基于其特殊性质，被广泛应用于国防、冶金、化工、玻璃、航天、核技术、激光、电子、计算机电源、超导及医疗科学等领域[1-6]。钪产品包括金属钪、钪合金和钪化合物。其中氧化钪是很多钪系材料的原材料，例如氧化钪是优异的阴极材料添加物，其能显著提高阴极的电子发射能力[7-9]；将氧化钪加入高温防腐蚀材料中，可制成一种性能良好的防腐蚀涂料[10-12]；在普通玻璃中添加氧化钪可形成网络结构，既能降低玻璃密度，又能沟通组织结构，改善玻璃性能[13]；而氧化钪稳定的氧化锆替代传统的氧化钇稳定的氧化锆用于固体燃料电池，可大大提高固体燃料电池的功能密度[14-16]。由此可得出，氧化钪在新材料、新能源的发展中起到重要作用。材料微观结构对材料的性能具有重要的影响作用，通过调控材料微观结构和形貌往往能更有效地调节材料的性能，而制备方法和反应条件对材料的微观形貌具有决定性的影响作用，因此，对材料制备工艺条件的研究具有重要意义。

目前钪提取和制备氧化钪的方法主要包括萃取法、离子交换法、化学沉淀法和膜分离法，其中化学沉淀法是一种历史悠久的分离、提纯和净化方法[17]。它成功用于冶金过程、化工过程和材料制备等众多领域，具有操作简单、成本低和投资少等系列优势，其为制取氧化钪和纯化氧化钪的重要方法之一[18]。沉淀法能将前期钪提取工作与后续的氧化钪粉体材料制备工作相结合，实现钪提取和氧化钪粉体材料制备一体化，缩短工艺流程，降低成本，实现氧化钪粉体材料制备的短流程和绿色化。其中氢氧化物沉淀法能有效地将钪与碱金属和碱土金属分离，实现氧化钪的纯化[13]。而目前对氢氧化物沉淀法制备氧化钪及所制备的氧化钪的性质缺少深入的研究。

基于钪提取和氧化钪粉体材料制备一体化技术，本工作主要针对液碱沉淀法制备氧化钪粉体材料进行深入研究，利用液碱（氢氧化钠）作为沉淀剂，通过控制一定的制备工艺条件和焙烧条件制备出层状结构的氧化钪粉体材料，研究了沉淀剂用量、反应温度、搅拌速度和陈化时间等沉淀过程反应条件对钪沉淀回收率的影响，及在较优工艺条件下制备的氧化钪粉体材料的微观结构，为拓宽氧化钪材料的应用范围提供了数据基础。

1 实验

1.1 含钪沉淀物及氧化钪的制备

利用浓盐酸在一定温度条件下将氧化钪溶解，配制浓度为 9~12 g/L 的氯化钪溶液备用，其 pH 值为 3~4，同时，配制浓度为 30% 的氢氧化钠溶液备用。取 200 mL 氯化钪溶液于 500 mL 三口烧瓶中，并添加一定量硫酸溶液，使反应体系中硫酸根离子约为 0.2~

0.3 mol/L,升至一定反应温度(40~80 ℃)时,调节搅拌器转速为100~400 r/min,利用蠕动泵将沉淀剂缓慢加入三口烧瓶中进行沉淀反应,添加完沉淀剂后,经过一定陈化时间,进行过滤和烘干获得含钪沉淀物,取部分含钪沉淀物检测其钪含量,计算其反应条件下的钪沉淀回收率,同时将含钪沉淀物在一定焙烧工艺条件下进行焙烧处理,获得氧化钪粉体材料,之后进行氧化钪粉体材料性质的检测分析。

本工作中钪沉淀回收率主要通过式(1)和式(2)计算得出:

$$m = m_1 \times w \tag{1}$$

$$K = \frac{m}{M} \times 100\% \tag{2}$$

式中:K 代表沉淀回收率;M 为实验溶液中钪离子的质量;m 为含钪沉淀物中钪的质量;m_1 为获得的沉淀物质量;w 为沉淀物中钪的含量。

1.2 样品的性能及表征

用 Optima 8000 电感耦合等离子体光谱仪(ICP-OES,美国 PerkinElmer)测试含钪沉淀中钪离子含量,计算钪沉淀回收率;用 STA449 热重-差热分析仪(TG-DTA,德国 Netzsch)测试含钪沉淀物材料的热性能(测试条件为:测试温度为 10~1200 ℃,升温速率为 10 ℃/min,反应气氛为空气);采用 T27 傅里叶显微红外仪(FTIR,德国 Bruker)针对含钪沉淀物和氧化钪粉体材料进行定性分析;利用 Smartlab-201307 型 X 射线衍射仪(XRD,日本 Rigaku Corporation)测试粉体材料的物相组成;用 JSM-6700F 扫描电镜能谱仪(SEM-EDS,日本 JEOL)检测 Sc_2O_3 粉体材料的微观形貌及结构。

2 结果与讨论

2.1 钪沉淀回收率

图 1 为不同实验影响因素条件对钪沉淀回收率作用的结果,其中图 1(a)为沉淀剂用量对钪沉淀回收率的作用结果。通过分析图 1(a)可得出,使用氢氧化钠作为沉淀剂时,沉淀剂用量在 23~28.5 g/L 范围内,随着沉淀剂用量的增加,钪沉淀回收率逐渐增加,当沉淀剂用量为 23 g/L 时,钪沉淀回收率约为 95%,而当沉淀剂用量为 27.5 g/L 时,钪沉淀回收率约为 97.8%,当沉淀剂用量为 28.5 g/L 时,钪沉淀回收率约为 98%。由此可得出,当沉淀剂用量增加为 27.5 g/L 后,随着沉淀剂用量的持续增加,钪沉淀回收率逐渐趋于稳定。因此,在实验研究中,适宜的氢氧化钠溶液沉淀剂用量可选为 27.5 g/L。

图 1(b)为反应温度对钪沉淀回收率作用的结果。通过分析图中结果可得出,在反应温度为 40~80 ℃ 范围内,刚开始随着反应温度的提高,钪沉淀回收率增加,当反应温度为 60 ℃ 时,钪沉淀回收率最大,约为 98%,而后随着反应温度的持续升高,钪沉淀回收率反而略有下降。这可能是因为开始反应温度比较低时,反应速度较慢,形成含钪沉淀物颗粒粒径小,含钪沉淀物溶解度提高,钪沉淀回收率比较低,而当沉淀反应温度过于高时,含钪沉淀盐的溶解度提高,钪沉淀率降低。通过本实验研究可得出在 40~80 ℃ 温度范围内,适宜的反

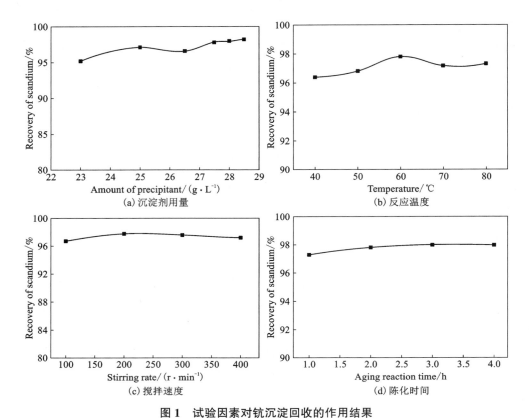

图 1 试验因素对钪沉淀回收的作用结果

Fig. 1 The effect of experimental factors on scandium precipitation recovery:
(a) amount of precipitant; (b) reaction temperature; (c) stirring rate; (d) aging time

应温度为 60 ℃。

沉淀反应过程中搅拌激烈程度直接决定着络合及沉淀反应效果。图 1(c) 为沉淀反应时搅拌速度对钪沉淀回收率的作用结果。通过分析图中结果可得出,在 100~400 r/min 搅拌速度范围内,当搅拌速度为 200 r/min 时,钪沉淀回收率最高,为 97% 左右,而随着搅拌速度增大,钪沉淀回收率反而降低。这可能是由于搅拌速度比较慢时,钪离子不能与沉淀剂充分接触,致使沉淀反应不完全,钪沉淀回收率比较低;而当搅拌速度过大时,在沉淀颗粒成核生长过程中,由于受到强烈的搅拌作用,晶粒难以成长,粒度减小,含钪沉淀物溶解度增加,钪沉淀回收率降低。因此,在本研究的搅拌速度范围内,适宜的搅拌速度可选为 200 r/min。

图 1(d) 为不同陈化时间条件下钪沉淀回收率的结果。通过分析可得出,在 1~4 h 陈化时间范围内,随着陈化时间的延长,钪沉淀回收率逐渐提高,这主要是由于随着陈化时间延长,沉淀反应体系中细小悬浮颗粒发生沉降反应,上清液中的钪离子发生沉淀发生,吸附在钪沉淀物表面,致使钪沉淀回收率提高。同时,由图 1(d) 中结果也可得出,陈化时间为 2~4 h 时,随着陈化时间的延长,钪沉淀回收率略有提升,但提高比较小,陈化时间越长,所需能耗越多。故在本研究陈化时间范围内,适宜的陈化时间可选为 2 h。

2.2 钪沉淀物及氧化钪的性质

图 2 为氢氧化钠溶液作为沉淀剂制备的含钪沉淀物粉体材料的 TG-DTA 结果。分析图 2 中 TG 曲线可以得出，粉体材料的失重大约分为三个阶段：第一个阶段是 20~200 ℃，失重率为 12.80%；第二个阶段是 200~450 ℃，失重率为 6.36%；第三个阶段为 450~800 ℃，失重率为 10.54%。第一个阶段和第二个阶段的失重主要是发生脱水反应和脱羟基反应；而第三个阶段的失重主要是含钪沉淀物的燃烧反应，生成氧化钪。同时，由图 2 可得出，焙烧温度超过 800 ℃后，其 TG 曲线逐渐趋于稳定，由此可得出，800 ℃后含钪沉淀物趋于分解完全，为获取较纯净的氧化钪粉体材料，其焙烧温度应大于 800 ℃。

将含钪沉淀物在 1100 ℃条件下焙烧获得氧化钪粉体材料。图 3 和图 4 分别为含钪沉淀物和焙烧含钪沉淀物获得的粉体材料的 XRD 结果。通过分析图 3 中结果可得出，氢氧化钠作为沉淀剂时，含钪沉淀物主要由 ScOOH 相组成，由于沉淀溶液中含有一定量的硫酸根离子，因此其沉淀物中含有少量的 $NaHSO_4$ 相。分析图 4 含钪沉淀物焙烧获得的粉体材料的 XRD 结果可得出，含钪沉淀物经过焙烧处理后获得物质的衍射峰是氧化钪物相的衍射峰，由此含钪沉淀物经过焙烧处理后得到氧化钪。同时，由图 4 中结果可得出，氧化钪的衍射峰比较尖锐，说明其晶形结构比较好。因此，利用氢氧化钠作为沉淀剂时，焙烧含钪沉淀物可获得较纯净、晶形结构较好的氧化钪粉体材料。

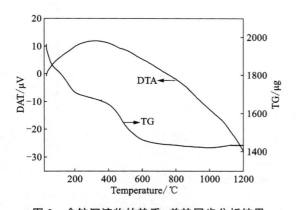

图 2　含钪沉淀物的热重-差热同步分析结果
Fig. 2　The TG-DTA result of scandium sediment

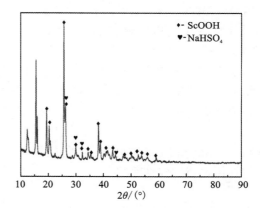

图 3　含钪沉淀物粉体材料的 XRD 结果
Fig. 3　XRD result of scandium-containing sediment powder material

物质的红外谱图可以对物质进行定性分析，图 5 和图 6 分别为含钪沉淀物和含钪沉淀物焙烧后获得的氧化钪粉体材料的红外分析结果。通过分析图 5 中结果可得出，含钪沉淀物在 3200 cm^{-1} 附近具有一个明显的吸收宽峰，其是由前驱体试样中自由—OH 的伸缩振动引起的，而试样中的水分子也会对其产生一定影响作用；前驱体试样在 1641 cm^{-1} 和 1437 cm^{-1} 处具有明显的吸收峰，此处的吸收峰主要是纯水分子的—OH 吸收峰；920~1210 cm^{-1} 处的多频带为硫酸根的振动峰；而 600 cm^{-1} 处附近为 Sc—O 键的吸收峰，由此可得出其在 665 cm^{-1} 附近的吸收峰主要是 Sc—O 键的振动峰[19]。由图 6 可得出，含钪沉淀物经过焙烧处理后样品中只剩下 634 cm^{-1} 处的吸收峰，该处的吸收峰是 Sc—O 键的吸收峰，说明含钪沉淀物经过焙烧处理后，发生分解反应，生成比较纯净的氧化钪粉体材料。

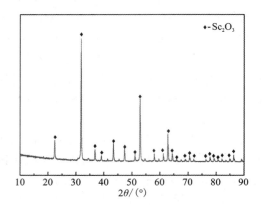

图 4　氧化钪粉体材料的 XRD 结果

Fig. 4　XRD result of scandium oxide powder material

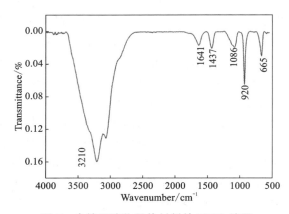

图 5　含钪沉淀物粉体材料的 FTIR 结果

Fig. 5　FTIR result of scandium-containing sediment powder material

图 7 和图 8 分别为氧化钪粉体材料的 SEM-EDS 点扫描和面扫描结果。通过分析氧化钪粉体材料的扫描结果可得出，利用液碱作为沉淀剂时制备的氧化钪粉体材料的显微结构主要是层状结构，其粒径小于 20 μm。同时，由点扫描和面扫描结果可得出，粉体材料由 Sc 和 O 元素组成，结合氧化钪粉体材料的 XRD 和 IR 结果可知，制备的氧化钪粉体材料比较纯净。

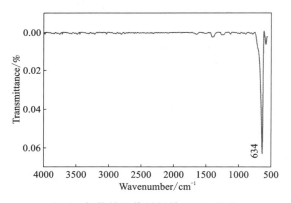

图 6　氧化钪粉体材料的 FTIR 结果

Fig. 6　FTIR result of scandium oxide powder material

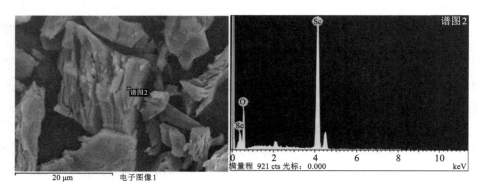

图 7　氧化钪粉体材料的点扫描结果

Fig. 7　EDS spot scan result of scandium oxide powder material

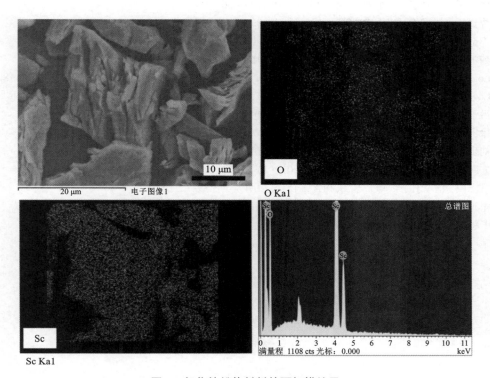

图 8 氧化钪粉体材料的面扫描结果

Fig. 8 EDS mapping micrograph result of scandium oxide powder material

3 结论

通过对氢氧化物沉淀法制备层状氧化钪粉体材料的研究可得出：利用氢氧化钠作为沉淀剂时，在研究范围内最佳反应条件为：温度为 60 ℃，沉淀剂用量为 27.5 g/L，搅拌速度为 200 r/min，陈化时间为 2 h 时，钪沉淀回收率可达 97%；制备的氧化钪粉体材料比较纯净，晶型结构较好，其具有微米级的层状结构，粒径小于 20 μm。

参考文献

[1] Ghiem V N, Atsushi I, Etsuro S, et al. Hydrometallurgy, 2016, 165：51.
[2] Kristiansen R. Norsk Bergverksmuseum, Skrifter, 2009, 41：75.
[3] 黄鑫, 张晓燕, 高红选, 等. 功能材料, 2016, 47(7)：7149.
[4] 张吉, 桑晓光, 王毅, 等. 稀有金属, 2012, 36(2)：282.
[5] 崔云涛, 王金淑, 刘伟, 等. 无机材料学报, 2012, 27(5)：480.
[6] 马平, 陈松林, 潘峰, 等. 光学学报, 2009, 29(6)：1729.

[7] 曹贵川, 祁康成, 王小菊. 光源与照明, 2018(2): 36.
[8] 曹贵川, 林祖伦, 祁康成, 等. 电子元件与材料, 2010, 29(2): 54.
[9] 王金淑, 陶斯武, 王亦曼, 等. 稀有金属材料与工程, 2004, 33(2): 324.
[10] 魏星, 方延勇. 中国专利, CN105693242A, 2016.
[11] Loghman-Estarki M R, Shoja Razavi R, Jamali H, et al. Ceramics Inter-national, 2016, 42, 11118.
[12] Loghman-Estarki M R, Razavi R S, Edris H. Current Nanoscience, 2012, 8: 767.
[13] 吕子剑, 翟秀静. 钪冶金. 化学工业出版社, 2015.
[14] 徐宏, 薛倩楠, 张建星, 等. 中国稀土学报, 2016, 34(6): 739.
[15] 刘琪, 张爱华, 顾幸勇. 中国陶瓷工业, 2013, 20(3): 14.
[16] 方海燕, 程继贵, 杨俊芳. 科技广场, 2011(1): 217.
[17] 尹志民, 潘清林, 姜锋, 等. 钪和含钪合金. 中南大学出版社, 2007.
[18] 翟秀静, 肖碧君, 李乃军. 还原与沉淀. 冶金工业出版社, 2008.
[19] 王毅, 孙旭东, 徐鸿. 功能材料, 2007, 38(6): 1007.

草酸盐沉淀法制备亚微米级 Sc_2O_3 粉体的研究

摘　要：采用草酸沉淀法制备了 Sc_2O_3 亚微米粉体材料。研究了反应温度、沉淀剂用量、反应 pH 值等实验因素对钪沉淀回收率的影响，并利用 XRD、SEM-EDS、激光粒度分析仪、傅里叶红外光谱仪(FTIR)、比表面积仪等对氧化钪粉体材料性质进行表征。研究结果表明，反应温度为 80 ℃、草酸与氯化钪的摩尔比为 1.5 及反应体系的 pH 值为 1.3 时，钪沉淀回收率为 97% 以上。结合 TG-DTA 结果及氧化钪粉体材料的 FTIR 和 XRD 结果可知，草酸钪前驱体在 1100 ℃ 条件下焙烧获得氧化钪较纯净。氧化钪粒度细小均匀，体积中心径 D_{50} 为 7.586 μm，90% 的颗粒粒径小于 16.050 μm，基本上呈正态分布；比表面积为 17.819 m^2/g，具有良好的晶型结构；微观结构为形状完整的亚微米结构层状物，结合 BET 测试结果可得出，其表面含有一定的孔洞二级结构。

关键词：沉淀法；氧化钪；草酸；粉体材料

Preparation of submicron Sc_2O_3 powder through oxalate precipitation

Abstract: Submicron Sc_2O_3 powder was prepared by using oxalic acid precipitation. The effects of reaction temperature, precipitation dosage and pH value on the recovery rate of scandium precipitation were studied. Meanwhile, the Sc_2O_3 powder was characterized by using XRD, SEM-EDS, laser particle size analyzer, Fourier infrared spectrometer (FTIR) and BET surface area measurement. It is showed that the recovery rate of scandium precipitation could be up to 97%, when the reaction temperature was 80 ℃, the molar ratio of oxalic acid to scandium chloride was 1.5 and the pH value was 1.3. According to the TG-DTA, FTIR and XRD results, it is demonstrated that the Sc_2O_3 was relative pure after the oxalate precursor was calcined at 1100 ℃, with pretty high crystallinity. The Sc_2O_3 powder has fine and uniform particles. Volume center diameter (D_{50}) of the powder was 7.586 μm, while 90% of the particles have a size of less than 16.050 μm, with a normal size distribution profile. The powder exhibits a specific surface area of 17.819 m^2/g, while it has a submicron layered structure, with surface containing certain secondary pores.

Keywords: precipitation; scandium; oxalate; powder

本文发表在《陶瓷学报》，2020, 41(4): 508-514。作者：付国燕，王玮玮，吕东，刘诚。

引言

钪属于稀土元素,在自然界中极其分散,但并不稀少,地壳中钪丰度为 $5×10^{-4} \sim 6×10^{-4}$,我国钪资源储量居世界第一,已知含钪矿物种类多达 800 多种[1]。钪及其化合物以其优异的性能在冶金、国防、电子、医学、航天、超导体等尖端技术领域具有不可替代的重要用途,是我国重要的战略储备资源[1-3]。氧化钪是含钪材料的基础材料,主要应用于固体燃料电池、催化剂、铝合金、激光材料等领域[4-8]。例如,随着经济和社会的发展、能源和环境问题的日益严重,固体燃料电池的研究成为材料领域的研究热点之一,电解质材料是整个固体燃料电池的核心部件,对固体燃料电池的商业化发展起着关键的作用。而氧化钪稳定氧化锆是目前锆基固体电解质中离子电导率最高的电解质材料,其能有效地弥补氧化钇稳定氧化锆电解质材料在中温情况下导电率比较低的问题,因此,氧化钪对固体燃料电池的发展至关重要[9-13]。而粉体材料的微观形貌结构和粒径可以直接影响材料的性能,通过控制制备工艺条件实现粒径分布均匀和形貌可控,对后续制备含钪材料,发挥其特殊性能,拓宽其应用范围具有重要的作用[4,15]。近年来,国内外针对粉体材料的制备进行了大量的研究,如溶胶-凝胶法、直接沉淀法、均相沉淀法、微乳液法等。其中,沉淀法具有一系列的技术和经济上的优势,被广泛应用于研制各种粉体材料。而草酸沉淀法是工业化制备 Sc_2O_3 常用的方法,特点是工艺简单、成本低廉、沉淀杂质少、沉淀物过滤性能好。采用沉淀法时,粉体材料的成分、粒度分布和形貌会受到溶液的成分、沉淀剂种类与用量、技术参数和操作方法等的影响[16]。

目前,针对草酸沉淀法制备氧化钪并没有系统的研究,草酸钪沉淀的分析数据亦不足,对采用此方法制备氧化钪粉体材料的工业化应用产生了不利影响,工业生产中出现的沉淀效率、过滤性能不稳定等现象和问题没有确切的数据支撑。因此,本文针对沉淀法制备氧化钪粉体材料进行了研究,通过相关实验研究发现,采用草酸沉淀法时,控制一定的制备工艺条件,可制备出粒度分布均匀的 Sc_2O_3 粉体材料,本文主要研究了沉淀过程反应条件和相关试验参数对沉淀率的影响,并研究了在较优工艺条件下氧化钪粉体材料的粒度、形貌,为采用草酸法制备氧化钪粉体材料的工业化及拓宽氧化钪材料的应用范围提供了数据基础。

1 实验

1.1 试剂

氧化钪纯度为 99.9%,东方钪业股份有限公司生产;草酸纯度为 99%,上海麦克林生化科技有限公司生产;盐酸浓度为 36% ~ 38%,北京化工厂有限责任公司生产。

1.2 实验设备与分析仪器

用 Optima 8000 电感耦合等离子体光谱仪(ICP-OES,美国 PerkinElmer)测试含钪沉淀中钪离子含量；用 STA449 热重-差热分析仪（TG-DTA，德国 Netzsch）测试 Sc_2O_3 粉体材料的热性能（测试温度为 10~1200 ℃，升温速率为 10 ℃/min，反应气氛为空气）；用 LS13320 激光粒度仪（LPSA，美国 Beckman kurman）测试氧化钪粉体材料的粒径分布及大小；采用 T27 傅里叶显微红外仪（FTIR，德国 Bruker）针对前驱体和氧化钪粉体材料的进行定性分析；利用 Smartlab-201307X 射线衍仪（XRD，日本 Rigaku Corporation）测试粉体材料的物相组成；利用 ASAP2020HD88 比表面积仪（BET，美国 Micromeritics）测定粉体的比表面积数值；用 JSM-6700F 扫描电镜能谱仪（SEM-EDS，日本 JEOL）观察 Sc_2O_3 粉体材料的微观形貌及结构。

1.3 实验方法

将氧化钪用盐酸加热溶解，将其配制成 3~5 mol/L 的氯化钪溶液备用，同时，配制浓度为 10% 的草酸溶液备用。取配制好的氯化钪溶液稀释至所需浓度，取 200 mL 的氯化钪溶液放置于 500 mL 的三孔烧瓶中，将反应体系升至一定反应温度，在 200~300 r/min 的搅拌条件下，利用蠕动泵缓慢加入草酸沉淀剂溶液进行沉淀反应，经过陈化反应后，获得含钪沉淀物溶液体系，再经过过滤、洗涤、干燥处理后获得含钪沉淀物，其主要组成物质为 $Sc_2(C_2O_4)_3 \cdot 6H_2O$，取样分析并计算钪沉淀回收率，取部分含钪沉淀物样品经过一定焙烧工艺处理后，得到较纯净的亚微米级 Sc_2O_3 粉体材料。本实验研究中制备氧化钪时主要的反应为式（1）、（2）和（3）。

$$Sc_2O_3 + 6HCl \Longrightarrow 2ScCl_3 + 3H_2O \tag{1}$$

$$2ScCl_3 + 3H_2C_2O_4 \Longrightarrow Sc_2(C_2O_4)_3 + 6HCl \tag{2}$$

$$Sc_2(C_2O_4)_3 \cdot 6H_2O + 1.5O_2 \xrightarrow{Heating} Sc_2O_3 + 6CO_2 + 6H_2O \tag{3}$$

钪沉淀回收率主要通过式（4）和式（5）计算得出。

$$m = m_1 \times w \tag{4}$$

$$K = \frac{m}{M} \times 100\% \tag{5}$$

式中：K 代表沉淀回收率；M 为实验溶液中钪离子的质量；m 为草酸盐前驱体中钪质量；m_1 为获得的沉淀物质量；w 为沉淀物中钪的含量。

2 结果与讨论

2.1 反应温度对钪沉淀回收率的影响

图 1 为不同反应温度条件下钪沉淀的回收率。由图可知，在 70~95 ℃ 温度范围内，随反应温度升高，钪沉淀回收率先升高后降低。当反应温度为 80 ℃ 时，钪沉淀回收率最高，为 95% 左右，随着反应温度的持续升高，钪沉淀回收率反而下降。这可能是由于反应温度比较低时，沉淀反应速度小，形成含钪沉淀物颗粒粒径比较小，致使含钪沉淀物溶解度提高，钪沉淀回收率下降；随着反应温度的持续升高，含钪沉淀物的溶解度增大，致使钪沉淀回收率下降[17]。因此，选择沉淀反应温度为 80 ℃。

2.2 沉淀剂用量对钪沉淀回收率的影响

图 2 为不同沉淀剂用量条件下钪沉淀的回收率。由图可知,初始阶段,随着沉淀剂用量的增加,钪沉淀回收率升高,当草酸与氯化钪摩尔比为 1.5 时,钪回收率最高,为 95% 左右;而后,随着沉淀剂用量的增加,钪沉淀回收率反而下降。这主要是因为沉淀剂用量较少时,沉淀剂不足以使钪离子完全沉淀,随着沉淀剂用量的增加,钪离子逐渐趋于完全沉淀,钪沉淀回收率提高;其次,钪与铈、钇等稀土元素化学性质相似,其草酸盐的溶解度随草酸根活度的增加而降低,通过最低点后又由小变大,而草酸根在溶液中的活度主要取决于溶液中草酸根离子的含量[17],因此,钪沉淀率开始随着沉淀剂用量的增加时逐渐提高,而后,沉淀率反而下降;同时,草酸过量浓度过高,草酸钪沉淀转化为可溶性络合物,如 $[Sc(C_2O_4)]^+$、$[Sc(C_2O_4)_2]^-$、$[Sc(C_2O_4)_3]^{3-}$ 等离子,致使沉淀剂用量过大时钪沉淀回收率降低[14,17]。

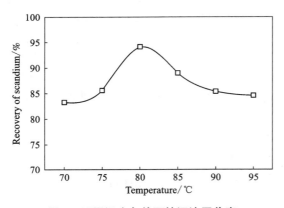

图 1 不同温度条件下钪沉淀回收率

Fig. 1 Scandium recovery versus reaction temperature

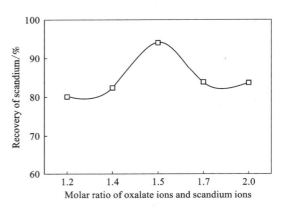

图 2 不同沉淀剂用量条件下钪回收率

Fig. 2 Scandium recovery rate versus the molar ratio of xalate to scandium

2.3 反应 pH 值对钪沉淀回收率的影响

图 3 为不同 pH 值条件下钪沉淀的回收率。由图可知,在 pH 值为 1.2~1.9 范围内,随着沉淀反应体系 pH 值的增加,钪沉淀回收率先升高后降低,当 pH 值为 1.3 时,钪回收率最高,可达到 97.68%。这可能是当沉淀反应体系 pH 值低于 1.3 时,含钪草酸盐的溶解度比较大,致使钪沉淀回收率较低;而当沉淀反应体系 pH 值比较高时,反应体系中部分杂质离子与钪离子发生共沉淀反应,致使草酸沉淀剂需求量增加,同时,反应体系中的 Fe^{3+}、Al^{3+}、Mn^{2+} 等杂质离子会与草酸反应形成草酸络合物,这也会影响草酸钪的完全沉淀,致使钪沉淀回收率下降。因此,沉淀反应选择 pH 值为 1.3。

2.4 草酸盐前驱体的热分解过程和红外分析

草酸盐前驱体在加热过程中会发生分解反应生成氧化钪。本实验研究中草酸盐前驱体主要是 $Sc_2(C_2O_4)_3 \cdot 6H_2O$,加热分解反应如式(3)所示,反应生成 Sc_2O_3。图 4 为草酸作为沉淀剂制备的草酸盐前驱体粉体材料的 TG-DTA 结果。由 TG 曲线可知,其草酸盐前驱体的失重大约分为三个阶段：第一个阶段为 200~350 ℃，失重率为 27.51%；第二个阶段为 350~550 ℃，失重率为 31.53%；第三个阶段为 550~850 ℃，失重率为 77.01%。第一个阶段和第二阶段的失重主要是发生脱水反应和脱羟基反应，生成草酸钪；而第三阶段的失重反应主要是草酸根离子的燃烧反应，进而生成氧化钪。同时，通过分析 DTA 曲线可知，其在 250 ℃、350 ℃和 850 ℃左右时有 3 个明显的吸热峰。在 250 ℃时主要是前驱体中水分的蒸发，350 ℃左右的吸收峰主要是前驱体脱羟基形成草酸钪沉淀，850 ℃左右主要是由于前驱体脱草酸根形成氧化钪而形成吸收峰，超过 850 ℃后不再有热分解发生，这与 TG 分析结果相吻合。

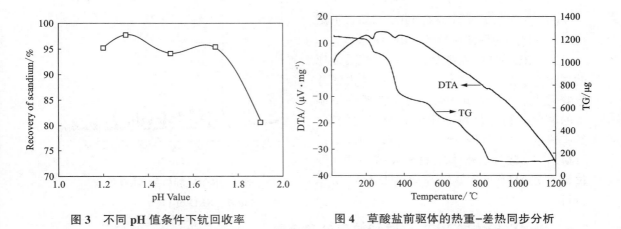

图 3　不同 pH 值条件下钪回收率
Fig. 3　Scandium recovery versus pH value

图 4　草酸盐前驱体的热重-差热同步分析
Fig. 4　TG-DTA curves of the oxalate precursor

通过查阅相关文献可得出，物质的红外谱图结果可以对物质进行定性分析。经过前期实验研究选择 1100 ℃温度条件下进行焙烧。图 5 和图 6 分别为草酸盐前驱体和草酸盐前驱体 1100 ℃焙烧后获得的氧化钪粉体材料的傅里叶显微红外分析结果。分析图 5 可知，草酸盐前

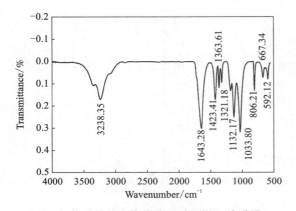

图 5　草酸盐前驱体的傅里叶显微红外谱图
Fig. 5　FTIR spectrum of the oxalate precursor

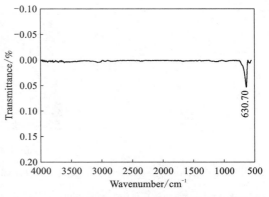

图 6　氧化钪粉体材料的傅里叶显微红外谱图
Fig. 6　FTIR spectrum of the Sc_2O_3 power

驱体在 3238 cm^{-1} 附近具有一个明显的吸收宽峰，这主要是由前驱体试样中自由 OH—的伸缩振动引起的，而试样中的水分子也会对其产生一定影响作用；同时，前驱体试样在 1643 cm^{-1} 和 1423 cm^{-1} 处具有明显的吸收峰，其吸收峰主要是纯水分子的 OH—吸收峰；1300~1000 cm^{-1} 区域是 C—O 的伸缩振动，而 600 cm^{-1} 处附近为 Sc—O 键的吸收峰[18]。对比图 5 和图 6 可知，草酸盐前驱体样品焙烧产品的吸收峰消失，只剩下 630 cm^{-1} 处的吸收峰，该处的吸收峰是 Sc—O 键的吸收峰，说明草酸盐前驱体经过焙烧处理后，前驱体发生分解反应，生成比较纯净的氧化钪粉体材料。

2.5 Sc$_2$O$_3$ 粉体的 XRD 测试

图 7 为草酸盐前驱体在 1100 ℃ 条件下焙烧 3 h 后得到的粉体材料的 XRD 图谱。

由 XRD 可知，草酸盐前驱体焙烧后形成的物质主要是氧化钪，其衍射峰比较尖锐并且峰强比较大，说明在该焙烧工艺条件下制备的氧化钪粉体材料晶化度和纯度比较高。结合热重-差热分析、红外分析和 XRD 结果可得出，草酸盐前驱体在 1100 ℃ 条件下，经过焙烧处理后可获得较纯净的氧化钪粉体材料。

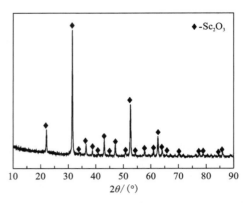

图 7　Sc$_2$O$_3$ 粉体的 XRD 图谱

Fig. 7　XRD of the Sc$_2$O$_3$ power

2.6 Sc$_2$O$_3$ 粉体的 BET 测试和粒度分布结果

图 8 为氧化钪粉体的激光粒度分布，表 1 为氧化钪粉体粒径累计分布。综合图 8 和表 1 中结果可知，制备的氧化钪粉体材料 D_{50} 为 7.586 μm，其粒径主要分布 10 μm 左右，90% 的粒度小于 16.050 μm。因此，制备出的氧化钪粉体材料应属于亚微米级材料。

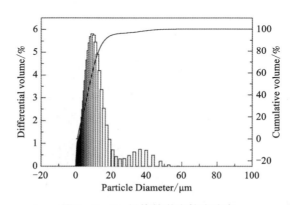

图 8　Sc$_2$O$_3$ 粉体的激光粒度分布

Fig. 8　Laser particle size distribution profile of the Sc$_2$O$_3$ power

表 1 氧化钪粉体累积粒径分布

Tab. 1　Laser particle sizes of the Sc_2O_3 powder

Cumulative distribution/%	<10	<25	<50	<75	<90
Partical size/μm	1.841	4.253	7.586	11.470	16.050

图 9 和 10 为氧化钪粉体材料的 BET 分析测试结果,其中,图 9 为吸附和脱附曲线,图 10 为孔径分布。由图 9 和图 10 可知,氧化钪粉体材料的比表面积为 17.819 m^2/g,具有相对较大的比表面积,与其他材料复合时反应活性比较大;同时,由氧化钪粉体材料的孔径分布可知,氧化钪粉体材料中具有一定的微孔结构,其单点平均孔半径为 15.15 nm,进而为材料提供更多的比表面积,增强其反应活性,使之能更加有效地与其他材料发生反应,形成复合材料,进而拓宽氧化钪粉体材料的应用范围。

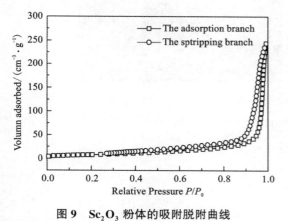 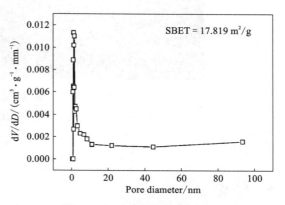

图 9　Sc_2O_3 粉体的吸附脱附曲线　　　　图 10　Sc_2O_3 粉体的孔径分布

Fig. 9　Adsorption-desorption curves of the Sc_2O_3 power　　Fig. 10　Pore size distribution of the Sc_2O_3 power

2.7　Sc_2O_3 粉体的微观扫描电镜

图 11 为氧化钪粉体的 SEM-EDS 图。由图可知,制备出的氧化钪粉体材料主要是由层片组成的块状体。由面扫描结果可知,其由 Sc 和 O 元素组成,结合氧化钪粉体材料的 XRD 结果可得出,制备的氧化钪粉体材料物质比较纯净。同时,结合氧化钪粉体材料的粒径分布结果可得出,氧化钪粉体材料的粒径为 10 μm 左右。而材料性能极大地依赖于其形貌和结构,层状材料由于其特殊的结构特性,在交换反应过程中可以形成结晶度高但是不完全交换的中间产物,进而通过控制制备工艺参数可制备出不同层间距的层状材料,与其他材料进行复合反应制备复合材料时,可以更加方便地控制反应条件,实现材料的特殊性能[19]。

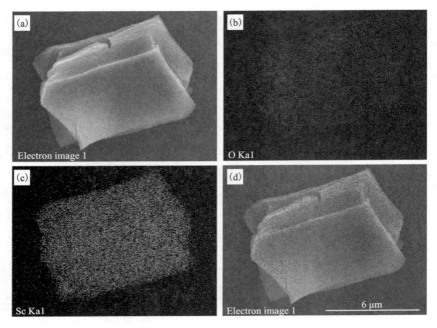

图 11　Sc_2O_3 粉体的 SEM-EDS

Fig. 11　SEM images and EDS profiles of the Sc_2O_3 power

3　结论

采用草酸沉淀法制备了亚微米级氧化钪粉体材料。

(1) 草酸沉淀法制备氧化钪粉体材料的最佳反应温度为 80 ℃，草酸根与氯化钪摩尔比为 1.5，反应体系的 pH 值为 1.3 时，钪沉淀回收率为 97% 以上。

(2) 制备出的氧化钪粉体材料晶化程度高，其粉体粒径 D_{50} 为 7.586 μm，90% 的粒度小于 16.050 μm，比表面积为 17.819 m^2/g，且颗粒表面具有一定的孔洞结构，反应活性大。

(3) 获得的氧化钪粉体材料为由片层组成的块状体，层状结构比表面积大，与其他材料复合可实现材料的特殊性能。

参考文献

[1]　日子剑, 翟秀静. 钪冶金[M]. 北京: 化学工业出版社, 2015: 5-30.

[2]　NGUYEN N V, IIZUKA A, ETSUROS, et al. Study of adsorption behavior of a new synthesized resin containing glycol amic acid group for separation of scandium from aqueous solutions[J]. Hydrometallurgy, 2016, 165: 51-56.

[3] 王普蓉,戴惠新,高利坤,等.钪的回收及提取现状[J].稀有金属,2012,36(3):501-506.
[4] 谭令,陈海清,刘俊,等.中温固体氧化物燃料电池电解质的研制[J].湖南有色金属,2015,31(6):55-58.
[5] 崔仁源,刘元钟,李相勋,等.固体氧化物燃料电池离子-电子混合传导性 Ni/ScSz 燃料极的制备以及性能分析[D].集成技术,2018,7(6):81-87.
[6] 张吉,桑晓光,王毅,等.掺钕氧化钪激光透明陶瓷纳米粉体的合成及研究[J].稀有金属,2012,36(2):282-285.
[7] 李亮星,王涛胜,黄茜琳,等.熔盐电解法制备铝钪中间合金研究进展[J].材料导报,2018,32(11):3768-3773.
[8] 王鹏,李凝,林连义,等.稀土氧化物掺杂对 $SO_4^{2-}/ZrO/Al_2O_3$ 催化剂催化性能的影响[J].中国稀土学报,2009,27(2):213-217.
[9] SHI H G, RAN R, SHAO Z P. Wet powder spraying fabrication and performance optimization of IT-SOFCs with thin-film ScSz electrolyte[J]. International Journal of Hydrogen Energy, 2012, 37(1): 1125-1132.
[10] 徐宏,薛倩楠,张建星,等. Se_2O_3 稳定 ZrO_2 电解质材料及其研究进展[J].中国稀土学报,2016,34(6):739-747.
[11] XUE Q N, HUANG X W, WANGL G, et al. Effects of Sc doping on phase stability of $Zr_{1-x}Sc_xO_2$ and phase transition mechanism: First-principles calculations and rietveld refinement[J]. Materials and Design, 2017, 114(15): 297-302.
[12] 徐宏,赵娜,张赫,等.固体氧化物燃料电池用锆基电解质材料研究概述[J].稀有金属,2017,41(4):437-444.
[13] 刘琪,张爱华,顾幸勇.固体氧化物燃料电池用 ScSz 粉体的研究进展[J].陶瓷学报,2013,34(4):538-542.
[14] 尹志民,潘清林,姜峰,等.钪和含钪合金[M].长沙:中南大学出版社,2007:1-60.
[15] 李凤生.超细粉体技术[M].北京:国防工业出版社,2000:5-100.
[16] WANG Y, SUN X D, QIU G N. Synthesis of scandium oxide nano power and fabrication of transparent scandium oxide ceramics[J]. Journal of Rare Earths, 2007, 25: 68-71.
[17] 翟秀静,肖碧君,李乃军.还原与沉淀[M].北京:冶金工业出版社,2008:290-360.
[18] 王毅,孙旭东,徐鸿均匀沉淀法制备 Sc_2O_3 纳米粉[J].功能材料,2007,38(6):1007-1008.
[19] 雷立旭,张卫锋,胡猛,等.层状复合金属氢氧化物:结构、性质及其应用[J].无机化学学报,2005,21(4):451-463.

新能源战略金属镍钴锂资源清洁提取研究进展

摘　要：镍钴锂资源提取作为新能源汽车产业链开端环节，决定着其发展命脉。然而我国镍、钴原生矿产资源严重匮乏，且随着硫化镍矿日趋枯竭，国外红土镍矿已成为镍钴主要提取原料。火法高冰镍及湿法高压酸浸作为目前红土镍矿生产电池级硫酸镍的两大主流工艺，各有利弊，但两大工艺的并行发展，能够为硫酸镍的供应增加更多确定性，进而满足新能源领域对镍钴的庞大需求。我国锂资源丰富但禀赋不佳，80%锂赋存于盐湖卤水中，长期依赖进口锂矿石。现行盐湖提锂技术中，纳滤法、反渗透、电渗析及吸附法等技术集成耦合所形成的新型提锂工艺，已通过多个盐湖的产业化验证，具有广阔的推广应用前景。此外，新型协萃体系 TBP/P507-FeCl$_3$ 的研发，基本解决了原萃取体系存在的问题，有望产业化推广应用。新能源汽车行业的蓬勃发展导致退役电池量呈指数上升，再生金属有望成为原矿金属资源的重要补充，但由于锂电正极材料种类繁多，目前回收系统并不完善。在正极材料的深度回收和处理过程中，浸出体系多采用酸碱体系，易造成二次污染。因此，应进一步研发新型浸出体系，以找到一种温和实用的浸出方法。综上，能源金属提取应以产品为导向，构建从资源到材料到电池应用及回收全生命周期技术体系，以先进技术控制紧缺资源，保障我国新能源行业可持续发展。

关键词：新能源镍；钴；锂；清洁提取

Advances in clean extraction of nickel, cobalt and lithium to produce strategic metals for new energy industry

Abstract: As the beginning of the new energy vehicle industry chain, nickel, cobalt, and lithium resource extraction determines its development lifeline. However, the primary mineral resources of nickel and cobalt in China are seriously lacking, and with the exhaustion of nickel sulfide ore, foreign laterite nickel ore has become the main raw material for nickel and cobalt extraction. At present, reduction sulfidation smelting and high-pressure acid leaching are the two main processes for producing battery grade nickel sulfate from laterite nickel ore. They have their own advantages and disadvantages. However, the parallel development of the two processes can increase the certainty of nickel sulfate supply and meet the huge demand for nickel and cobalt in

---

本文发表在《材料导报》，网络首发。作者：董波，田庆华，许志鹏，李栋，王青鹭，郭学益。

the new energy field. Lithium resources are rich, but the endowment is not good, 80% lithium occurs in salt lake brine, and there is a long-term dependence on imported lithium ore. Among the existing lithium extraction technologies from salt lakes, the new lithium extraction process formed by the integration and coupling of nanofiltration, reverse osmosis, electrodialysis and adsorption has been verified by industrialization in several salt lakes and has broad prospects for popularization and application. In addition, the research and development of the new synergistic extraction system (TBP/P507-$FeCl_3$) has basically solved the problems existing in the original extraction system, which is expected to be industrialized and applied. The rapid development of the new energy vehicle industry has led to an exponential increase in the output of retired batteries, and recycled metals are expected to become an important supplement to raw ore metal resources. However, due to a wide variety of lithium cathode materials, the current recovery system is not perfect. In the process of deep recovery and treatment of cathode materials, an acid-base system is used in the leaching system, which is easily causes secondary pollution. Therefore, a new leaching system should be further developed to find a mild and practical leaching method. In summary, energy metal extraction should be oriented by products, and the whole life cycle technology system from resources to materials to battery application and recycling should be constructed. The resources in short supply should be controlled by advanced technology, and the sustainable development of our new energy industry should be guaranteed.

Keywords: new energy; nickel; cobalt; lithium; clean extraction

引言

新能源汽车是指以非常规车用燃料作为唯一或主要驱动力的汽车,作为战略性新兴产业,是实现我国"碳达峰、碳中和"的重要途径,已成为我国国民经济发展重要支柱产业[1]。目前,新能源汽车车用动力电池基本是锂电池。锂电池作为新能源汽车的"心脏",它的能量密度、充放电倍率和稳定性等决定新能源车的续航里程、充放电效率以及安全性等多方面特性。主流动力电池的电芯由四大关键材料组成:正极、负极、电解液和隔膜,其中正极材料性质决定电势差和比容量,间接决定动力电池的能量密度,其重要性不言而喻[2-3]。因此,通常以正极材料来命名动力电池,比如三元电池(镍钴锰酸锂)、磷酸铁锂电池等[4-5]。2020年,我国新能源汽车总销量为136.7万辆,装机容量为63.6 GW·h,其中三元正极材料锂离子电池为39.7 GW·h,磷酸铁锂正极材料锂离子电池为23.2 GW·h,二者之和占比在98%以上[6]。三元锂电新能源汽车的产业链结构如图1所示,大致可划分为以下三部分:上游资源提取——中游电池材料制备——下游新能源关键部件及整车制造[7]。资源提取作为产业链开端环节,决定着新能源汽车发展命脉。镍、钴、锂等有色金属作为制备动力电池的基础材料,被称为"新能源战略金属",在低碳经济转型过程中发挥着关键作用。

我国镍、钴原生矿产资源严重匮乏,对外依存度高达80%,资源"卡脖子"风险突出。目前,随着硫化镍矿日趋枯竭,国外红土镍矿已成为镍钴主要提取原料,亟须突破红土镍矿清

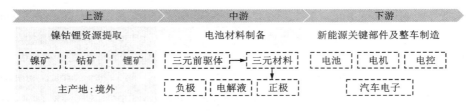

图 1 三元锂电新能源汽车的产业链结构[8]

Fig. 1 Industrial chain structure of ternary lithium new energy vehicles

洁材料化冶金新技术,以先进技术控制紧缺资源。我国锂资源虽然丰富但禀赋不佳,锂辉石和锂云母等矿石资源极度匮乏,80%锂赋存于盐湖卤水中,但镁锂比高、锂浓度低,传统技术难以实现经济性开发利用,亟须突破盐湖提锂新技术,以提高我国锂资源保障能力。截至2019 年,我国新能源汽车动力电池大规模报废的浪潮已到来,退役动力锂电池资源循环利用市场空间大,亟待开发废旧动力电池清洁材料化循环利用新技术,以有效缓解资源环境安全风险。因此,本文针对红土镍矿、盐湖卤水及退役锂电池等主要含镍、钴、锂资源,对其采储量及现行工艺进行系统评述,以求解决资源瓶颈、助力我国新能源行业低碳可持续发展。

1 镍钴锂消费及资源储备现状

1.1 镍钴锂在新能源领域的消费现状及趋势

根据我国汽车工业协会 2019 年统计数据可知[9],新能源汽车产销量分别达 125 万辆和 121 万辆,且预计到 2030 年新能源汽车年销量将超过 1520 万辆[10]。随着我国新能源汽车产销量逐年递增,动力电池消费领域对含镍、钴、锂资源及其化合物的需求量会进一步增加。近 5 年,镍、钴、锂在我国新能源领域消费占比变化趋势如图 2 所示[11-13]。由图 2 可知,在 2021 年,镍的供需格局发生了质变,但由于电池领域需求占比小、基数低,整体需求增长呈缓慢上升趋势。2021 年,中国原生镍总消费量达到 154.2 万吨,同比增长 14%,电池用镍占比上升到 15%[14]。根据预测,到 2030 年,全球动力电池对镍需求量将大幅提升至约 140 万吨镍金属量,占比提升至约 30%[14]。2020 年,我国钴在新能源领域消费量为 1.07 万吨,2021 年消费量已达到 2.09 万吨,预计 2022 年将达到 3.28 万吨,相关专家预测 2025 年国内新能源行业对钴的需求量将超过 8.6 万吨[12]。新能源行业长期以来都是碳酸锂消费最大的领域,据有关研究机构统计[10],2021 年全球动力电池对碳酸锂的需求量已达到 43 万吨,预测 2025 年全球新能源汽车产业对碳酸锂的需求将达到 154 万吨。未来几十年内,我国动力电池产销量将呈现井喷式增长,对战略金属镍、钴、锂及其化合物的需求必将持续上升。因此,亟须以产品为导向,优化升级产业结构,保障新能源汽车产业可持续发展。但我国镍钴锂资源禀赋不佳,主要依赖国外进口,亟须聚焦产业原材料供给薄弱环节,以"一带一路"为纽带,建立能源金属矿产资源优势。

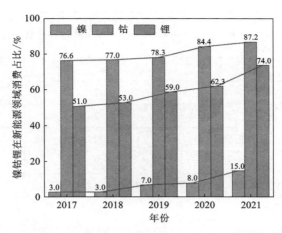

图 2 镍钴锂在新能源领域消费占比变化趋势图[11-13]

Fig. 2　The consumption proportion of nickel, cobalt and lithium in the new energy field

1.2　镍钴锂资源储备现状

1.2.1　红土镍矿资源

全球镍矿资源丰富，陆基储量为 9063 万吨，主要分为硫化物型、红土型和海底多金属结核/结壳三种类型[15]，目前开发的为硫化物型和红土型。红土型镍矿储量 5740 万吨，占陆基矿床储量的 63%；而硫化镍型储量 3318 万吨，占比 37%[15]。镍矿资源分布相对集中，主要分布在印度尼西亚、澳大利亚、俄罗斯、古巴、巴西、菲律宾、新喀里多尼亚、加拿大和中国等国家，具体分布国家如表 1 所示[15]。红土镍矿资源集中分布于印度尼西亚、澳大利亚、菲律宾、古巴、巴西、新喀里多尼亚、巴布亚新几内亚等国家；而硫化镍矿资源主要分布于南非、加拿大、俄罗斯、澳大利亚、中国等国家。近年来，随着硫化镍矿资源日趋枯竭，红土镍矿产镍占比逐年剧增，目前已成为提镍主要原材料，且矿中除含主金属镍外，还赋存大量能源金属钴[16-17]。我国镍矿资源储量有限，2020 年国内镍矿资源储量为 398 万吨，占全球 4.39%，对外依存度超 80%，是世界上镍矿、镍中间品及电解镍主要进口国[10,18]。我国钴资源同样匮乏，无独立钴矿床，多伴生于铜、镍、铁等原生矿产资源中，但钴品位极低（平均 0.02%左右），无法满足自需。全球钴矿储量主要分布国家如表 2 所示。由表 2 可知，2020 年我国钴矿资源探明储量仅 13 万吨，占全球 1.95%，原材料对外依存度超过 90%[10,19-21]。

表 1　全球镍储量主要分布国家（2020 年）[22]

Table 1　Major countries with nickel reserves in the world (2020)

排名	国家	储量/万吨	全球占比/%
1	印度尼西亚	2875	31.72
2	澳大利亚	1265	13.96
3	俄罗斯	770	8.50

续表1

排名	国家	储量/万吨	全球占比/%
4	古巴	647	7.14
5	巴西	567	6.26
6	加拿大	542	5.98
7	菲律宾	472	5.21
8	新喀里多尼亚	409	4.51
9	中国	398	4.39
10	南非	154	1.70
11	危地马拉	98	1.08
12	多米尼加	93	1.03
13	马达加斯加	79	0.87
14	其他	694	7.66
	合计	9063	100.00

表2 全球钴矿储量主要分布国家(2020年)[22]

Table 2　Major countries with cobalt reserves in the world (2020)

排名	国家	储量/万吨	全球占比/%
1	刚果(金)	297	44.46
2	印度尼西亚	107	16.02
3	澳大利亚	65	9.73
4	古巴	25	3.74
5	加拿大	19	2.84
6	喀麦隆	18	2.69
7	中国	13	1.95
8	新喀里多尼亚	12	1.80
9	菲律宾	10	1.50
10	马达加斯加	10	1.50
11	其他	92	13.77
	合计	668	100.00

红土镍矿是由铁氧化物和硅酸盐的多种水合物在热带及亚热带等雨水丰富的地区经风化所形成的[23]。成矿过程中，含镍的铁氧化物和硅酸盐矿物被酸性地表水或植物酸溶解后，镍、铁、镁、硅等金属离子随水往地下渗透，并逐渐氧化沉积[24-27]。由于Ni^{2+}与Fe^{2+}、Mg^{2+}等离子半径相似，镍通常以类质同相形式赋存于橄榄石相[$(Mg,Fe)_2SiO_4$]及针铁矿相

[FeO(OH)]晶格中[28]。红土镍矿通常被划分为三个区域：①褐铁型红土矿层；②过渡层；③腐殖土型红土镍矿层，具体如图3所示[29]。褐铁型红土镍矿具有高铁低镁的特性，其主要化学物相为针铁矿和赤铁矿等含铁矿相，脉石成分主要由滑石、尖晶石和角闪石等组成；腐殖土型红土镍矿具有高镁低铁的特性，主要物相为蛇纹石和利蛇纹石等硅酸盐矿物；过渡层中各金属元素含量介于褐铁型与腐殖土型之间，主要物相为绿脱石和二氧化硅[30]。表3为世界红土镍矿型矿床具体的分布情况，如表3可知，我国红土镍矿资源极度匮乏，长期受制于其他国家。因此，亟须突破红土镍矿清洁材料化冶金新技术，以先进技术控制海外紧缺资源。

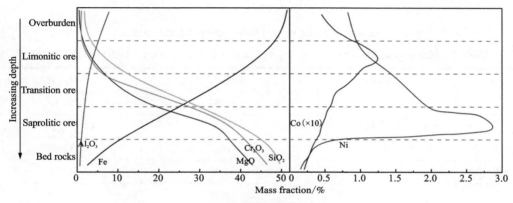

图3 红土镍矿床典型分布[31]

Fig. 3 Typical concentration profiles in a laterite nickel ore deposit

表3 全球红土镍矿资源分布概况[31]

Table 3 The distribution of the global laterite nickel ore resources

国家/地区	资源储量/10^6 t	镍品位/%	镍金属量/10^6 t	占比/%
新喀里多尼亚	2559	1.44	37	22.9
菲律宾	2189	1.28	28	17.4
印度尼西亚	1576	1.61	25	15.7
澳大利亚	2452	0.86	21	13.1
中南美洲	1131	1.51	17	10.6
非洲	996	1.31	13	8.1
加勒比海地区	944	1.17	11	6.9
亚洲和欧洲	506	1.04	5	3.3
其他	269	1.18	3	2
总计	12622	—	160	100

1.2.2 盐湖卤水资源

全球锂矿类型丰富,大致有盐湖卤水型、伟晶岩型(包括相关的花岗岩及云英岩型)、黏土型、锂沸石型、油气田卤水型和地热卤水型等。目前工业开采主要以盐湖卤水型和伟晶岩型锂矿为主。全球锂矿储量为1.28亿吨(碳酸锂当量),但资源分布相对集中,主要分布于南美锂三角地区(阿根廷、玻利维亚和智利三国毗邻区域),澳大利亚、加拿大、中国、美国及少数其他国家,具体分布情况如表4所示。澳大利亚和加拿大以开发硬岩型锂矿为主,而智利、阿根廷及我国以盐湖卤水型锂矿开发为主[32-33]。我国锂矿资源储量有限且品位较低,无法满足自需。因此,近年来我国锂矿冶炼企业所需的矿石主要来源于国外进口。2020年,我国锂矿储量为810万吨,占全球总储量6.31%,对外依存度超70%[10, 32]。

表4 全球锂矿(以 Li_2CO_3 计)储量主要分布国家(2020年)[22]

Table 4 Major countries with lithium reserves (calculated by Li_2CO_3) in the world (2020)

排名	国家	储量/万 t	全球占比/%
1	智利	5267	41.06
2	澳大利亚	1839	14.34
3	阿根廷	1693	13.20
4	中国	810	6.31
5	美国	570	4.44
6	加拿大	369	2.88
7	刚果(金)	363	2.83
8	津巴布韦	243	1.89
9	墨西哥	173	1.35
10	西班牙	79	0.62
11	其他	1422	11.09
	合计	12828	100.00

盐湖卤水型锂矿,即溶解大量锂的含盐地下水,锂主要赋存于晶间卤水、孔隙卤水及地表卤水中[34]。盐湖卤水型锂矿成矿过程主要受气候地理、地质构造和成矿物质来源等多种因素影响,多形成于干旱地区的封闭盆地。全球盐湖卤水资源储量颇丰,分布相对集中[34]。根据 SNL Metals Economics Group 统计[36],盐湖卤水型锂矿约占全球锂矿总储量的78%,从经济可开采储量角度分析,其占比则高达91%,是主要提锂原料。表5列出了全球主要盐湖卤水型锂矿床,地域不同导致各地区盐湖卤水成分差异较大,根据化学物相组成大概可将其划分为硫酸盐型、氯化物型和碳酸盐型盐湖[37]。我国盐湖卤水型锂矿资源丰富,约占全国锂资源总储量的71%以上,分布也相对集中,主要分布于青藏高原地区[38]。我国西藏的盐湖卤水型锂矿床主要为碳酸盐型,较为典型的矿床为扎布耶盐湖,其特点为 Mg/Li 比低,通常采用太阳池蒸发法直接沉淀碳酸锂,工艺相对简单[39]。而青海柴达木盆地一带,大都为高 Mg/Li 比(Mg/Li>20)的硫酸亚盐型盐湖,目前已探明该地区至少拥有11个工业级品位的硫

酸亚盐型锂盐湖[40]。由于镁锂物化性质相似，导致从高 Mg/Li 比盐湖中提锂相对困难，现行蒸发-转化工艺生产成本较高、锂回收率偏低，不利于大规模开采利用[41]。因此，亟须突破盐湖提锂新技术以提高我国锂资源保障能力。

表5　全球主要盐湖卤水型锂矿床(资源量以 Li_2CO_3 计)[42]

Table 5　Major salt lake brine resources in the world (calculated by Li_2CO_3)

盐湖	位置	海拔/m	面积/km²	锂浓度/10⁻⁶	镁锂比	类型	储量/万 t	所有者
乌尤尼	玻利维亚	3650	10582	321	9.28	硫酸镁亚型	1800	Comibol
阿塔卡玛	智利	2300	3000	1500	6.225	硫酸镁亚型	530	SQM
扎布耶	西藏	4422	247	632	0.01	强度碳酸盐型	139.7	西藏矿业
察尔汗	青海	2670	5856	310	1837	氯化物型	119.2	蓝科锂业
翁布雷穆埃尔托	阿根廷	4300	565	521	1.37	硫酸钠亚型	85	FMC
西台吉乃尔	青海	2680	570	220	65.57	硫酸镁亚型	44.1	中信国安
一里坪	青海	2600	250	210	92.3	硫酸镁亚型	29.2	五矿盐湖
当雄措	西藏	4475	318	211	0.22	中度碳酸盐型	14	中川矿业
银峰	美国	1800	32	163	1.43	硫酸钠亚型	10	Rockwood
东台吉乃尔	青海	2683	210	300	40.32	硫酸镁亚型	9.1	青海锂业

1.2.3　退役锂电池二次资源

2014 年以来，我国新能源汽车行业蓬勃发展，其产销量快速增长[43]。一般情况下，动力电池报废周期为 5~8 年，因此我国自 2019 年起便迎来了动力电池退役潮。退役锂电池中含大量战略金属镍、钴、锂，其资源回收利用价值较高，对缓解我国镍、钴、锂高消费量与低储量、高进口依存度的矛盾有重要意义。据中国汽车技术研究中心数据可知，截至 2020 年，我国退役锂电池累计超 20 万吨，市场规模高达 100 亿元[44]。2022 年起退役三元锂电池占据主要回收市场，而部分符合梯次利用标准的磷酸铁锂电池则首先进入梯次利用场景。因此，三元电池报废量占比高于退役磷酸铁锂电池。但随时间推移磷酸铁锂电池报废量将逐年递增，占比贡献不断提升[45]。据动力锂电池应用分会预测[46]，2021—2025 年，我国三元退役锂电池量合计约为 200 GW·h，镍、钴、锂回收量分别达 13.9、2.88、2.36 万 t，预计金属回收市场价值超 400 亿元。未来退役电池量将呈指数上升，预计到 2030 年，市场规模将超千亿元[14]，再生金属有望成为原矿金属资源的重要补充，支撑新能源产业发展需求。因此，亟待开发废旧动力电池清洁材料化循环利用新技术，以有效缓解资源环境安全风险。

2 镍钴锂资源提取工艺现状

2.1 红土镍矿资源提取工艺现状

红土镍矿成矿过程复杂,导致各矿层元素种类及含量、矿相组成均有较大差异。因此各矿层的提取工艺也有所不同,红土镍矿各矿层化学成分及提取工艺如表6所示[47]。由表6可知,国内外红土镍矿现行处理工艺大致可分为火法和湿法两种,其中,火法工艺适用于处理腐殖土型红土镍矿及过渡层矿,工艺成熟、生产规模较大,是目前处理高镁低铁型红土镍矿的主流工艺,但能耗及碳排放量较高、产品应用领域单一,此外矿中钴无法回收利用;湿法工艺适于处理褐铁矿型红土镍矿及过渡层矿,能耗相对较低,但工艺流程冗长复杂、设备投资成本高、废渣废水产生量大、生产规模较小[48-49]。目前,红土镍矿生产电池级硫酸镍的方法主要分为两大类:火法高冰镍及湿法高压酸浸,火法高冰镍适于处理镍品位较高且具有高镁低铁特性的腐殖土型红土镍矿;湿法高压酸浸适于处理红土镍矿中镁、镍含量均较低的褐铁型红土镍矿,上述两种工艺具体流程如图4所示。

表6 红土镍矿床不同层位的化学成分及提取工艺[47]

Table 6 Chemical composition and extraction process of different layers of laterite nickel ore

矿层	化学成分/(质量分数)%					提取工艺
	Ni	Co	Fe	Cr	MgO	
表层	<0.8	<0.1	>50	<0.8	<0.5	废弃
褐铁矿层	0.8~1.5	0.1~0.2	40~50	0.8~3.0	0.5~5.0	湿法冶金
过渡层	1.5~1.8	0.02~0.1	25~40	0.8~1.5	5~15	火法或湿法
腐殖土层	1.8~3.0	0.02~0.1	10~25	0.8~1.6	15~35	火法冶金
基层	0.25	0.01~0.02	<5	0.1~0.8	35~45	不开采

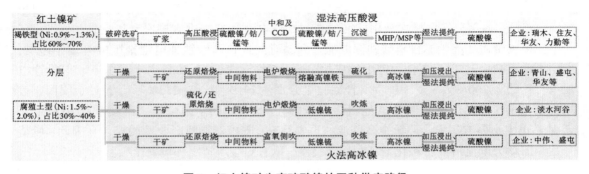

图4 红土镍矿生产硫酸镍的四种供应路径

Fig. 4 Four supply paths of producing nickel sulfate from laterite nickel ore

2.1.1 火法冶金工艺

火法冶金是指利用高温(燃料燃烧或电能产生的热或某种化学反应所放出的热)使矿石中金属与脉石或其他杂质分离的过程。据统计[50],火法冶炼是世界上处理红土镍矿的主要工艺。目前工业化的火法冶炼工艺主要有：回转窑-矿热炉还原熔炼镍铁工艺、回转窑直接还原-磁选工艺、还原硫化熔炼镍锍工艺。根据冶炼产品的不同可分为还原熔炼镍铁和还原熔炼镍锍,镍铁合金主要应用于不锈钢生产,而镍锍则需经转炉吹炼生产高冰镍产品用于新能源领域。

(1) 回转窑-矿热炉还原熔炼镍铁工艺(RKEF)

RKEF 工艺,20 世纪 50 年代在新喀里多尼亚多尼安博厂被开发,由我国引进、吸收、改进该工艺,并将其应用于镍铁生产[51],是当前应用最广泛的红土镍矿火法冶炼工艺。在中国资本的助推下,相继建成了近百座 RKEF 冶炼厂,如：中国东盟联合青山集团的印尼苏拉威西岛镍铁项目,中色集团缅甸达贡山镍铁项目以及金川集团的广西防城港镍铁项目等[52],全世界采用 RKEF 工艺冶炼镍铁的部分企业及其基本生产情况如表 7 所示。

表 7 全世界采用回转窑-矿热炉还原熔炼镍铁工艺的部分企业及其基本情况[53]

Table 7 RKEF process used in several enterprises in the world

企业名称	地点	镍平均品位/%	年产量/万 t	每吨焙砂电炉能耗/(kW·h)
塞罗马托莎厂	南美哥伦比亚	2.2	—	520
多尼安博冶炼厂	新喀里多尼亚	2.7	6	475
日向冶炼厂	日本	2.1~2.5	2.2	430~440
拉里姆纳冶炼厂	希腊	1.1~1.2	2.4	480
帕布什镍厂	乌克兰	2.3~2.4	1.75	400~600
鹰桥多米尼加镍厂	多米尼加共和国	1.56	2.9	379
中宝滨海镍业有限公司	中国河北省沧州市	20	8	500~700
福建某镍业有限公司	中国福建省福安市	9~12	30	—
中色镍业有限公司	缅甸达贡山	1.65	8.5	—

RKEF 工艺主要适合于处理镍品位较高的腐殖土型红土镍矿,其产品为镍质量分数为 8%~12% 的镍铁,镍回收率超 90%,但钴金属无法回收,造成资源浪费[54-55]。该工艺具体流程如图 5 所示,大致流程为：腐殖土经破碎筛分后混料输送入回转窑中,经焙烧预还原后直接热输送入矿热炉中直接还原冶炼得到镍铁产品[56]。

回转窑预还原过程反应机理如下[57]：

$$2C + O_2 =\!=\!= 2CO \tag{1}$$

$$NiO + C =\!=\!= Ni + CO \tag{2}$$

$$NiO + CO =\!=\!= Ni + CO_2 \tag{3}$$

$$3Fe_2O_3 + CO =\!=\!= 2Fe_3O_4 + CO_2 \tag{4}$$

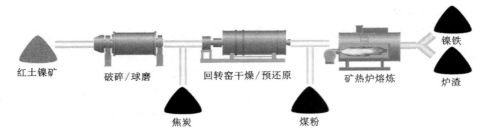

图 5　回转窑-矿热炉还原熔炼镍铁工艺流程图[56]

Fig. 5　Flow chart of RKEF process

$$Fe_3O_4 + CO = 3FeO + CO_2 \quad (5)$$

电炉还原熔炼过程反应机理如下[57]：

$$C + O_2 = CO \quad (6)$$

$$FeO + CO = Fe + CO_2 \quad (7)$$

$$SiO_2 + C = Si + CO_2 \quad (8)$$

RKEF 工艺具体优势如下[58]：①产出的镍铁品位高，镍回收率高、有害杂质元素少；②回转窑产生的烟气以及矿热炉产生的含 CO 废气均可在体系内实现循环再利用，例如：烟气可干燥炉料，而 CO 废气净化后可作为还原剂；③相较于其他火法冶炼工艺，RKEF 工艺采用热焙砂直接输送入矿热炉，一定程度上降低了能耗。但该工艺能耗及碳排量仍然相对较高、环境污染严重，镍铁产品无法应用于新能源领域[59]。而且为保证渣铁分离，在熔炼过程中，冶炼温度需维持在 1550 ℃ 以上，甚至需要更高的熔炼温度[53]。究其原因在于腐殖土型红土镍矿中 Al_2O_3、MgO、SiO_2 等易形成高熔点的物相含量较高，进而对熔体性质产生了不良的影响[60]。中南大学李光辉教授团队[61]通过渣系的合理选择和调控，在预还原及熔炼过程中强化还原气氛，降低了红土镍矿在电炉内的初渣液相生成温度，进而降低熔炼温度 50~90 ℃，有效降低了熔炼镍铁所需电耗。中南大学郭学益教授团队[62-63]采用氢气直接进行液相还原熔炼得镍铁合金，随后使用锌、镁一元或二元金属熔体高温下选择萃取镍铁合金中镍，最后利用不同金属饱和蒸气压差异，对共熔体进行真空蒸馏分离萃取介质与金属镍，得到镍粉产品，具体工艺流程如图 6 所示。该方法有效解决了现行 RKEF 工艺碳排放量大、能耗高的难题，实现了红土镍矿低碳清洁冶炼，并扩展了火法产品的应用领域。

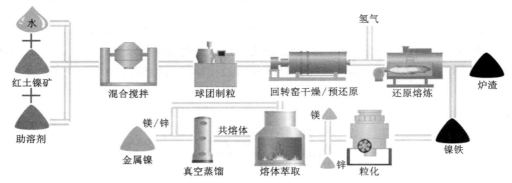

图 6　红土镍矿氢气直接还原冶炼镍铁-熔体萃取短流程提镍工艺流程图[62-63]

Fig. 6　Flow chart of nickel extraction from laterite nickel ore by hydrogen reduction-melt extraction process

（2）回转窑直接还原-磁选工艺

直接还原-磁选工艺，是1930年由德国克虏伯公司基于回转窑粒铁法（Krupp-Renn）发展而来的，首家采用该工艺大规模生产的企业为日本大江山冶炼厂[64]。目前全世界采用直接还原-磁选工艺冶炼镍铁的部分企业及其基本情况如表8所示。

表8 全世界采用直接还原-磁选工艺冶炼镍铁的部分企业及其基本情况[53]
Table 8 Direct reduction and magnetic process used in several enterprises in the world

企业名称	地点	粗镍铁品位/%	年产量/万 t
日本大江冶炼厂	日本	21.9	1.5
北海诚德镍业有限公司	中国广西北海	7.8	—
辽宁凯圣锻冶有限公司	中国辽宁铁岭	>10	1.3
南通某公司	中国江苏南通	9~12	—

直接还原-磁选工艺流程如图7所示，其处理原料主要为镍品位高于1.8%的红土镍矿，大致流程如下：先将矿石破碎筛分干燥，再与无烟煤、石灰石混料后制球得球团矿，再经回转窑半熔融还原，最后经水淬、细磨、磁选得镍铁产品。该工艺能耗较低，被公认为是目前处理高品位红土镍矿较为经济可行的方法。但由于焙烧过程采用煤为主要能源，吨矿耗煤为160~180 kg/t[65]，进而导致碳排放量较高。此外，焙烧温度较高、不易操作，回转窑易结圈、不利于生产顺行，单窑生产规模小[66]。因此，该工艺一直以来尚未大规模工业推广应用。

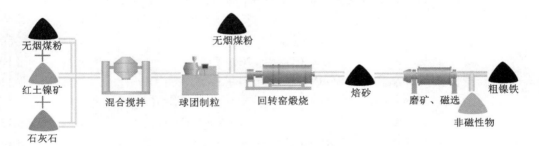

图7 直接还原-磁选工艺流程图[67]
Fig. 7 Flow chart of direct reduction and magnetic separation process

（3）还原硫化熔炼镍锍工艺

还原硫化熔炼镍锍工艺是将红土镍矿中镍、钴和部分铁还原并硫化成镍锍的方法。该工艺生产及投资成本低、产品形式多样，但存在能耗高、环境污染严重、原料适用性差、烟气难以处置等缺点[68]。全世界采用此工艺的部分企业及其基本生产情况如表9所示。近年来，采用传统硫化熔炼镍锍工艺的工厂越来越少，传统工艺流程如图8所示。首先将红土镍矿进行干燥脱除自由水和部分结晶水，添加硫化剂（石膏、黄铁矿）和还原剂（焦炭粉）进行混料后，输送入鼓风炉中熔炼，熔炼过程中通过调控硫化剂和还原剂的加入量及比例进而控制低镍锍品位，最后低镍锍经转炉吹炼后得到高镍锍产品。在此过程主要的还原反应有[50]：

表 9 全世界采用还原硫化熔炼镍锍工艺的部分企业及其基本生产情况[53]

Table 9 Reducing sulphide smelting nickel matte process used in several enterprises in the world[53]

企业名称	地点	含镍品位/%	年产量/万 t
多尼安博冶炼厂	新喀里多尼亚	2.6	6.3
俄罗斯列什厂	戈陵杜兴克斯	0.7~1.3	0.57
四阪岛别子冶炼厂	日本	2.6~3.0	—
营口鑫旺合金炉料有限公司	中国	1.66	—
友山镍业	印度尼西亚	1.85	4.5
青山实业	印度尼西亚	1.80	10.17
中青新能源	印度尼西亚	1.85	0.83

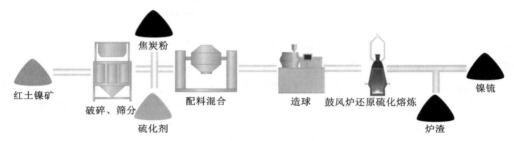

图 8 传统还原硫化熔炼镍锍工艺流程图[69]

Fig. 8 Flow chart of traditional reducing sulphide smelting nickel matte process

$$NiO + C = Ni + CO \tag{9}$$

$$NiO + CO = Ni + CO_2 \tag{10}$$

$$NiSiO_3 + 2FeO + C = Ni + Fe_2SiO_4 + CO \tag{11}$$

$$NiSiO_3 + 2FeO + CO = Ni + Fe_2SiO_4 + CO_2 \tag{12}$$

$$NiSiO_3 + 2CaO + C = Ni + Ca_2SiO_4 + CO \tag{13}$$

$$NiSiO_3 + 2CaO + CO = Ni + Ca_2SiO_4 + CO_2 \tag{14}$$

$$Fe_3O_4 + CO = 3FeO + CO_2 \tag{15}$$

$$2Fe_3O_4 + Fe_2SiO_4 + 2CO = 8FeO + 2CO_2 + SiO_2 \tag{16}$$

$$FeO + CO = Fe + CO_2 \tag{17}$$

$$Fe_2SiO_4 + 2CO = 2Fe + 2CO_2 + SiO_2 \tag{18}$$

硫化反应有[50]:

$$CaSO_4 \cdot 2H_2O = CaO + SO_3 + 2H_2O \tag{19}$$

$$3NiO + 9CO + 2SO_3 = Ni_3S_2 + 9CO_2 \tag{20}$$

$$3NiSiO_3 + 9CO + 2SO_3 = Ni_3S_2 + 9CO_2 + 3SiO_2 \tag{21}$$

$$FeO + 4CO + SO_3 = FeS + 4CO_2 \tag{22}$$

$$Fe_2SiO_4 + 8CO + 2SO_3 = 2FeS + 8CO_2 + SiO_2 \tag{23}$$

$$3NiO + 2FeS + Fe = Ni_3S_2 + 3FeO \tag{24}$$

$$6NiSiO_3 + 4FeS + 2Fe = 2Ni_3S_2 + 3Fe_2SiO_4 + 3SiO_2 \quad (25)$$
$$NiO + Fe = Ni + FeO \quad (26)$$
$$2NiSiO_3 + 2Fe = Fe_2SiO_4 + SiO_2 + 2Ni \quad (27)$$

随着新能源行业对原材料硫酸镍需求量与日俱增，新一轮红土镍矿硫化熔炼制备镍锍已成为研究的热点。现行工艺均是在传统工艺的基础上优化改进而来，主要分为两类：①回转窑干燥-回转窑预还原焙烧-电炉还原熔炼-P-S 转炉硫化-吹炼，青山、盛屯友山项目、华友华科项目均采用该工艺。镍铁硫化熔炼转产镍锍大规模量产的核心驱动力在于镍铁和硫酸镍之间的价差覆盖转产成本。据测算，当硫酸镍和镍铁的价差高于 3 万元/t 镍，才具备镍铁转产镍锍的动力[69]。虽然此工艺实现了镍铁和镍锍产品间的切换，但熔炼过程仍采用电炉，能耗较高。②以中伟为代表的富氧侧吹工艺，富氧侧吹与传统火法制高冰镍的技术路径核心区别在于用熔炼炉替代电炉，改变了硫化方式，通过富氧侧吹方式将氧气、硫化剂、煤等喷入熔炼炉中，替代了传统用电来进行加热的方式，具体工艺流程如图 9 所示。此工艺耗能低、热利用率高，但硫化过程硫化剂利用率低、操作环境恶劣[70]。针对上述问题，中南大学郭学益教授团队提出"一种红土镍矿硫化熔炼注入式强化硫化的方法"及"一种红土镍矿循环硫化提取镍和钴的方法"等创新技术[71]，实现了硫及高温烟气余热的循环利用。

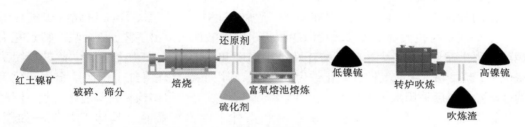

图 9 富氧侧吹硫化熔炼镍锍工艺流程图[69]

Fig. 9 Flow chart of oxygen-enriched side-blowing sulphide smelting nickel matte process

2.1.2 湿法冶金工艺

近年来，高品位红土镍矿被大规模开采利用而逐渐匮乏。相较于高品位矿石，含镍量低于 1% 的低品位红土镍矿更适于采用湿法冶金工艺处理。湿法冶金即采用特定的浸出剂将矿石中有价金属溶解至浸出液中，后续再通过沉淀法或溶剂萃取法对浸出液中有价金属进行分离提纯的过程。该工艺环境友好、过程能耗及碳排放量低、镍钴综合回收率较高。湿法冶金按浸出剂性质不同，可分为酸法与碱法；按浸出条件又可将其划分为常压、高压和生物浸出。虽然湿法工艺繁杂，但大多研究仍处于实验阶段，目前实现工业化应用的工艺仅有高压酸浸工艺（HPAL）。

高压酸浸（HPAL）是在高温高压的特殊条件下采用酸性溶液浸出红土镍矿中有价金属，相比常压酸浸（AL），此方法有价金属镍钴等浸出率高、镍钴浸出率均可达到 90%~95%，酸耗较低、一般情况下酸耗可控制在 250~520 kg 硫酸/t 矿石[72]。高压酸浸过程中主要反应如下[73]：

$$FeO(OH)(s) + 3H^+(aq) = Fe^{3+}(aq) + 2H_2O(l) \quad (28)$$
$$Mg_3Si_2O_5(OH)_4(s) + 6H^+(aq) = 3Mg^{2+}(aq) + 2SiO_2(s) + 5H_2O(l) \quad (29)$$

当体系酸度降低时，会发生式(30)的反应，且温度在 160 ℃以上尤为明显：
$$2Fe^{3+}(aq) + 3H_2O(l) \Longleftrightarrow Fe_2O_3(s) + 6H^+(aq) \quad (30)$$

古巴莫亚湾(MoaBay)冶炼厂是最早应用高压酸浸工艺的企业，其工艺流程如图 10 所示。具体包括：①破碎、研磨、筛分并制浆；②按配比将矿浆和酸溶液混合均匀并加热，在高压下浸出；③洗涤并固液分离；④沉淀法处理浸出液。该工艺投资、生产及维护成本高，高压釜易结垢而导致生产效率低，工艺操作条件苛刻，处理镁质量分数超过 4%的红土镍矿时、酸耗严重、经济性差[74-77]。为充分利用高压浸出液中的游离酸，有学者开发了 HPAL-AL 联合工艺，即将高压酸浸(HPAL)和常压酸浸(AL)结合，利用 HPAL 段浸出液中游离酸常压浸出(AL)腐殖土型红土镍矿，进而降低酸耗[78]。在 HPAL-AL 工艺的基础上，澳大利亚 BHP Billiton 公司开发了 EPAL 工艺(enhance pressure acid leach)，此方法创新点为在 AL 段混入 Na^+、K^+、NH_4^+，使浸出液中铁转化为黄铁矾沉淀，实现浸出液中铁质量浓度低于 3 g/L[79]。北京矿冶研究总院在上述工艺的基础上提出了 AL-HPAL 法，即在 AL 段先用高酸浸出褐铁型红土镍矿，在 HPAL 段用腐殖土型红土镍矿中和 AL 段余酸，使浸出液中铁以赤铁矿形式沉淀[78]。然而，研究表明，利用腐殖土型红土镍矿调节褐铁型红土镍矿高压浸出液 pH 值的过程中，中和效率低，因此，中南大学郭学益教授团队采用热处理改性后的腐殖土型红土镍矿作为中和剂中和褐铁型红土镍矿高压浸出液中的游离酸，大幅提升了中和效率，降低了单位镍产品的酸耗[80]，具体工艺流程如图 11 所示。针对浸出体系，不少研究者也进行了研究[81-83]，如北京科技大学王成彦教授课题组提出的硝酸加压浸出工艺，该方法的特点是大部分硝酸可再生循环利用，大幅降低酸耗及浸出温度和压力[84]。中南大学郭学益教授团队采用浓硫酸、硫酸铵、硫酸氢铵或氯化铵为改性剂，利用低温焙烧将红土镍矿中金属转化为相应的硫酸盐；随后采用水为浸出剂，利用水浸/水解法选择性浸出镍钴钪等有价金属，同时去除铁铝铬等贱金属，实现了红土镍矿在温和条件下的高效提取，具体工艺流程如图 12 所示[85]。

图 10 红土镍矿高压酸浸工艺流程图[86]

Fig. 10 Flow chart of high pressure acid leaching process of laterite nickel ore

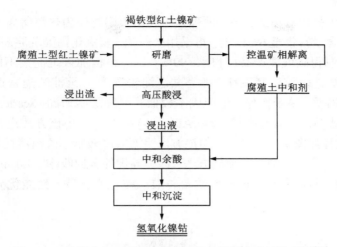

图 11 一种低成本、低酸耗的红土镍矿的浸出方法流程图[80]

Fig. 11 Flow chart of a leaching method of laterite nickel ore with low cost and low acid consumption

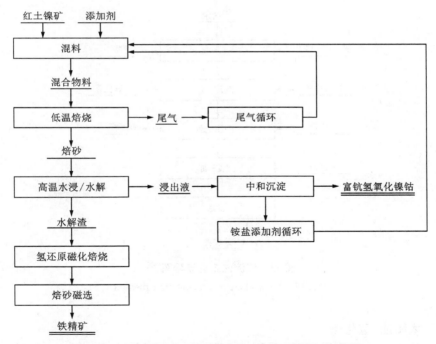

图 12 一种红土镍矿资源综合回收的方法流程图[85]

Fig. 12 Flow chart of a comprehensive recovery method for laterite nickel ore

2.2 盐湖卤水资源提取工艺现状

不同地区盐湖卤水化学成分及物相组成差别较大，导致提取工艺繁杂，包括沉淀法、太阳池法、煅烧浸取法、膜分离法、吸附法、溶剂萃取法和电化学脱嵌法等[87-88]。

2.2.1 沉淀法

沉淀法，也称为盐田富集法，基本原理是先利用太阳能作为热源将卤水在多级盐田中进行蒸发浓缩，使锂含量富集至6%以上，再采用碳酸钠、氢氧化钙等沉淀剂进行逐级除杂，最后采用碳酸钠沉淀制得碳酸锂产品（纯度≥99%）[89]。根据加入沉淀剂不同，沉淀法可分为碳酸盐沉淀法、铝酸盐沉淀法和硼镁、硼锂共沉淀法等[90]。沉淀法通常适于处理低 Mg/Li 比盐湖卤水，流程简便、易于操作，生产成本较低。美国瑟尔斯湖(Searles)盐湖、内华达州银峰(Silver peak)盐湖以及智利阿塔卡玛(Atacama)盐湖都使用该方法进行工业化量产碳酸锂[91]，具体工艺流程如图13所示[92]。但该方法生产效率较低，试剂消耗量大，生产过程锂损失严重[87-88]，亟待开发精细化沉淀新方法，以提高锂综合回收率。Zhang 等[93-94]开发了两种新型的镁沉淀剂，分别为九水偏硅酸钠及磷酸三铵三水合物，性能优异，锂回收率可达86.73%。

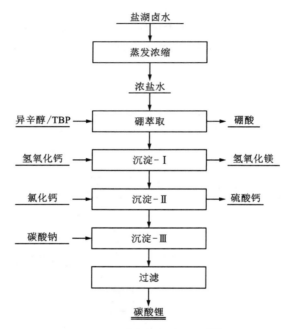

图 13　沉淀法工艺流程图[92]

Fig. 13 Flow chart of precipitation method

2.2.2 太阳池-碳化法

太阳池-碳化法通常适于处理镁含量较低的碳酸盐型盐湖卤水，是目前最适合扎布耶湖区现场条件(无电力、无矿物能源)的工艺路线[87-88]。其根本原理是利用碳酸锂溶解度的负温度效应，先将盐湖卤水分级滩晒浓缩，再在盐梯度太阳池中加热使卤水中锂以碳酸锂形式结晶析出，最后经碳化法制得工业级或电池级碳酸锂产品[95]，具体工艺流程如图14所示[96]。此方法操作简便、生产成本低，但原料适用性差、锂回收率低、生产周期长[87-88]。近年来，有研究者提出了太阳池立体结晶优化工艺，即引入辅助碳酸锂结晶的成核基体，降低成核势能，缩短生产周期，该工艺已成功应用于扎布耶矿区，增产效果显著[97]。

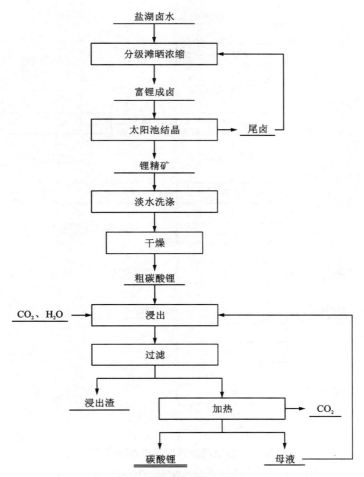

图 14 太阳池-碳化法工艺流程图[96]

Fig. 14 Flow chart of solar pond-carbonization method

2.2.3 煅烧浸取法

煅烧浸取法适用于处理我国青海的硫酸盐型盐湖,于 2005 年由青海中信国安锂业根据西台吉乃尔盐湖卤水的特性所开发,具体流程如图 15 所示[98-99]。根本原理是将溶于水的含锂氯化镁中镁焙烧转化为不溶于水的氧化镁,进而实现镁锂分离[87-88]。该法氧化镁、硼酸等副产品纯度较高、盐酸可循环利用,但能耗较高、环境污染严重、投资及生产成本高,进而制约了该工艺的推广应用[98-99]。目前,中信国安锂业二期新产线采用"纳滤膜反渗透+MVR 蒸发浓缩沉锂工艺",放弃了煅烧浸取法,并在 2021 年底新工艺中试生产线已成功运行[87-88]。

2.2.4 膜分离法

膜分离法是指利用特殊薄膜并在外力驱动下使液体中的某些成分选择性透过的方法,近年来快速发展并成功应用于盐湖卤水提锂,主要分为纳滤膜和电渗析[100]。其中纳滤膜分离法是在压力驱动下,采用一种功能性半透膜选择性分离镁[87-88]。而电渗析膜法则是采用离子选择性交换膜,利用其带电膜表面阻止二价离子(如镁离子)通过,而单价离子会顺利通过

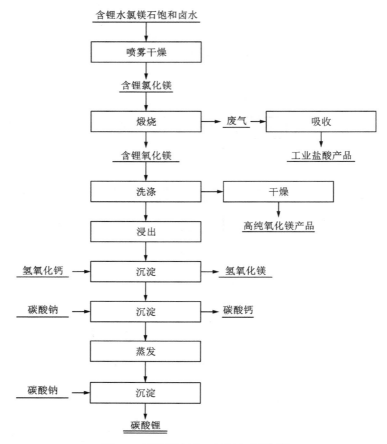

图 15 煅烧浸取法工艺流程图[98-99]

Fig. 15 Flow chart of calcination impregnation method

膜孔,进而实现镁锂分离[101]。实际生产过程中,独立使用纳滤膜或电渗析法时均存在较多缺点,因此,一般将纳滤法、反渗透、电渗析及吸附法等技术集成耦合,形成耦合膜分离法,以此提升膜法提锂效率。目前,耦合膜法提锂工艺正在国内多个盐湖推广应用[101]。

2.2.5 吸附法

吸附法,即利用对锂有选择性的吸附剂提取卤水中的锂,洗脱后再经除杂、蒸发浓缩、碱法沉淀等工序制备碳酸锂产品[34]。吸附剂大致分为无机吸附剂和有机吸附剂,其中无机吸附剂主要有铝系吸附剂、锰系和钛系尖晶石型氧化物吸附剂等,而有机吸附剂大多为离子交换树脂型[87-88]。众多吸附剂中,铝系吸附剂因其选择性强、洗脱效果好,是目前唯一得到工业化应用的吸附剂[102]。但在实际生产过程中,铝系吸附剂存在吸附容量小、杂质含量高等缺点,蓝科锂业与启迪清源科技有限公司合作开发的吸附加膜分离法耦合的提锂工艺很好地解决了上述问题,具体工艺路线如图16所示,此方法已通过多个盐湖的产业化验证,具有广阔的产业化应用前景[87]。

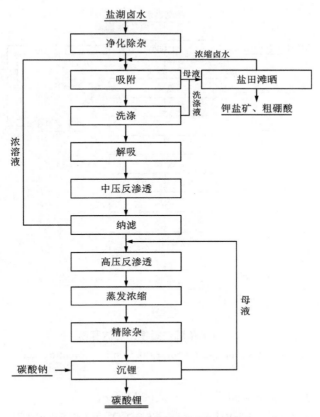

图 16 吸附加膜法提锂工艺流程图[87]

Fig. 16 Flow chart of lithium extraction by adsorption and membrane separation method

2.2.6 溶剂萃取法

溶剂萃取法与吸附法类似，即采用对锂具有高选择性的萃取体系使锂与卤水中杂质分离，富锂有机相经洗涤、反萃后得到富锂溶液[103]。目前研究最多且已工业化应用的萃取体系为磷酸三丁酯（TBP）协同萃取体系（TBP 为萃取剂，Fe^{3+} 为共萃离子）[104-105]。青海柴达木兴华锂盐采用此萃取体系在大柴旦盐湖已建成了年产 1 万吨碳酸锂产线，但产品含油量大，反萃困难[42]。针对上述问题，中科院过程所齐涛研究员团队优化了 TBP 协同萃取体系，研发出 TBP/P507-$FeCl_3$ 新型协萃体系（工艺流程见图17），实现了高镁锂比盐湖卤水中高效提锂，有望产业化推广应用[42]。

2.2.7 电化学脱嵌法

电化学脱嵌法是指在不断充电放电过程中锂从电极中脱出的过程，此概念始于锂离子电池。电化学脱嵌法提锂是目前盐湖提锂领域的一个新型开创性方法，利用水溶液锂电池脱嵌的反向工作原理，采用对锂离子具有"记忆效应"的脱锂电池正极材料为电极材料，盐湖卤水为阴极电解液，不含镁的支持电解质为阳极电解液，组成一个电化学脱嵌体系[106]。在电解过程中，阴极得电子发生还原反应，使得盐湖卤水中锂离子嵌入阴极材料中，再通过后续分离工序，实现锂的高效分离提取[107]。中南大学赵中伟教授团队[108]开发了 $LiFePO_4/FePO_4$

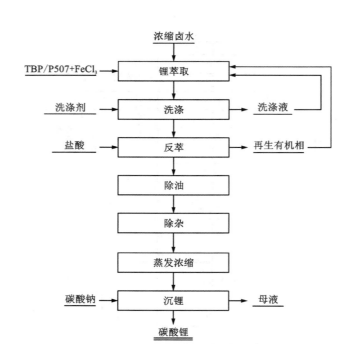

图 17　溶剂萃取法提锂工艺流程图[42]

Fig. 17　Flow chart of lithium extraction by solvent extraction

电极体系（图18），其基本原理为如式（31）和式（32）所示。

$$阳极室：LiFePO_4 - e^- \longrightarrow FePO_4 + Li^+（富集液） \tag{31}$$

$$阴极室：FePO_4 + Li^+（盐湖卤水） + e^- \longrightarrow LiFePO_4 \tag{32}$$

据报道，该技术正在西藏捌千错盐湖进行产业化试验，现已经建成年产 2000 t 碳酸锂生产线[87]。电化学脱嵌法适用于处理低品位、复杂盐湖卤水，原料适应性较强、锂回收率高。但能耗较高、电极循环性能差，需进一步优化[109]。

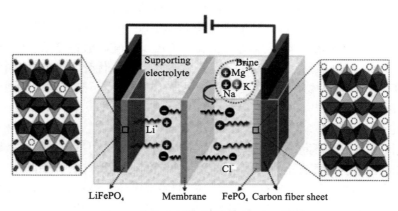

图 18　电化学脱嵌法提锂工艺原理[109]

Fig. 18　Process principle of electrochemical extraction of lithium

2.3 退役锂电池正极材料深度回收和处理现状

退役锂电池正极材料经一系列预处理后，通常需要通过火法或湿法工艺回收其中镍钴锂等有价金属[110]，主要方法如图19所示[111]。

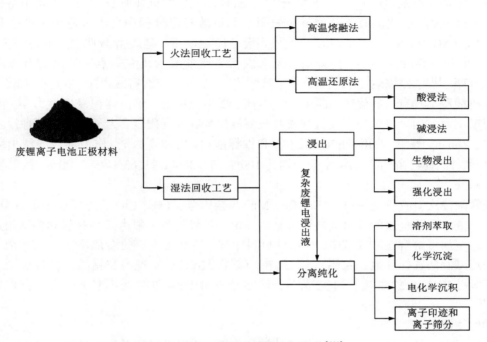

图19 废旧锂电池处理过程与方法[111]

Fig. 19 Recycling processes and methods of spent LIBs

2.3.1 火法回收工艺

火法回收工艺主要有两种方法：①高温熔融法：即利用高温（通常高于1000 ℃）将电池中有机物燃烧并去除，而电极材料则被熔化，镍、钴、锰等有价金属以合金形式被回收[112-113]。②高温还原法：即在高温还原气氛中，将电池中活性物质直接从氧化物还原为金属，以供后续回收利用[114-118]。目前，火法回收工艺已成功商业化应用，此方法效率高、工艺简便。然而，高温处理过程能源消耗严重，易产生有毒气体污染环境、需配套烟气净化设备，投资成本较高。

2.3.2 湿法回收工艺

湿法回收是目前应用最为广泛且研究较多的回收方法，回收过程大致分为：①浸出（酸浸、碱浸和生物浸出）；②有价金属的分离纯化（溶剂萃取、化学沉淀、电化学沉积、离子印迹和离子筛等）。

（1）浸出

浸出是指利用酸或碱等浸出剂将预处理后的正极活性物质转化为溶液中金属离子的过程，此工序对正极材料的回收至关重要。浸出主要有酸浸、碱浸和生物浸出，此外，还会采用超声、机械活化等手段提高浸出效果[119-122]。

①酸浸

酸浸是指采用无机或有机酸将正极材料中有价金属氧化物溶解形成离子态的过程。在研究初期，通常采用强酸（H_2SO_4[123]、HCl[124]、HNO_3[125]）浸出废锂电正极材料，且由于正极活性材料具有相对稳定的结构，与酸反应不完全，导致有价金属浸出率较低。因此，常在浸出过程中添加还原剂（H_2O_2[126]、葡萄糖[127]、$NaHSO_3$[128]和抗坏血酸[129]），以将不溶性离子还原为易溶性离子，提高有价金属浸出效果。无机酸浸出过程中会不可避免地产生有毒气体，如Cl_2、SO_2和NO_x，污染环境。此外，设备腐蚀严重，且废液酸性强、不易安全处置。近年来，有机酸因酸度适中、易降解、无二次污染逐渐成为研究热点。常用的有机酸有乙酸[130]、草酸[131]、柠檬酸[132]、甲酸[133]、乳酸[134]、苹果酸和酒石酸[110]。尽管有机酸浸出相比于无机酸浸出具有一些优势，但生产成本高、浸出效率低、单位体积浸出能力小，限制了其大规模工业应用。单独使用无机酸或有机酸浸出均存在局限性，为了结合这两种浸出体系的优点，Zhuang等[135]使用磷酸（浸出剂）和柠檬酸（浸出和还原剂）作为混合酸浸出废三元正极材料，锂、镍、钴和锰的浸出率分别达100%、93.38%、91.63%和92.00%，效果显著。

②碱浸

通常酸浸法对有价金属（Li、Ni、Co、Mn）与杂质金属（Fe、Cu）的选择性浸出效果较差，导致浸出液中有价金属的分离提纯较为复杂。由于正极材料一般由活性材料和铝箔组成，铝作为两性金属可溶解在酸或碱中，而活性材料中镍、钴和锰等有价金属不溶于碱溶液，因此通常采用碱浸法选择性地去除铝[136]。此外，碱性浸出有望实现镍和钴的选择性提取，从而降低后续分离提纯的难度，研究表明，氨溶液是铜、镍和钴选择性浸出较为理想的浸出剂[137]。

③生物浸出

生物浸出是借助某些微生物或细菌的催化作用，使正极材料中金属氧化物转化为水溶性金属硫酸盐的过程，其性能主要取决于微生物将不溶性固体化合物转化为可溶或可提取形式的能力[138]。与传统方法相比，生物浸出法具有过程温和、能耗低、环境友好等优点，但浸出效率低、有价金属浸出效果不佳是制约其工业化应用的难题。

④强化浸出

无论是酸浸还是碱浸，通常都需要较高的浸出温度和较长的反应时间，研究表明，超声、预处理等手段一定程度上可提高浸出效率。超声是一种常用的强化浸出的辅助方法，实验表明：超声波可促进固体和液体间的连续接触和相互分散，且在固液界面释放大量能量，从而有助于提高浸出效率[139]。此外，适当的研磨预处理对提高浸出效率也有显著的影响，这是由于粒度的减小、比表面积的增加及机械活化引起的颗粒晶体结构变化所致[124]。

综上，使用合适的浸出体系可有效地浸出报废锂离子电池正极材料中的金属元素，浸出产物为锂、镍、钴和锰等多种元素的混合溶液。此溶液经后续分离提纯工序可得到相应金属盐化合物，以实现循环利用。

(2) 有价金属的分离提纯

废锂电正极材料被浸出后，大部分金属（如Li、Ni、Co、Mn、Cu、Al和Fe）进入浸出液中。为从复杂浸出液中分离提纯有价金属，研究者提出了许多方法，如溶剂萃取、化学沉淀、电化学沉积以及离子印迹和离子筛等。

①溶剂萃取

溶剂萃取是遴选合适的萃取剂从浸出液中通过液-液分离提取有价金属的过程。目前常用的萃取剂有二(2-乙基己基)磷酸(D_2EHPA/P204)[140]、三正辛胺(TOA)[110]、Acorga M5640[141]、2-乙基己基膦酸单-2-乙基己基酯(PC-88A/P507)[142]、双(2,4,4-三甲基戊基)膦酸(Cyanex272)[143]等。溶剂萃取法具有能耗低、分离效果好、操作简便等优点。然而,萃取剂价格昂贵,导致生产成本偏高。

②化学沉淀

化学沉淀是使用沉淀剂将浸出液中目标金属沉淀分离的方法[110],生产成本低、能耗低,但分离不彻底,可作为初步除杂使用。

③电化学沉积

电化学沉积是从浸出液中回收纯金属或金属氢氧化物的方法[144]。电化学沉积具有操作简单、净化效率高的特点,但其能耗较高。

④离子印迹和离子筛分

吸附分离法具有工艺简便、经济环保等优点,具有良好的应用前景。由于常规吸附剂无法实现对各种金属离子的特定识别和分离,因此很难从复杂废锂电浸出液中分离和回收单一目标金属[145]。为提高吸附剂识别能力,研究者利用表面离子印迹法合成了具有选择性吸附能力的表面离子印迹材料,可选择性识别和分离目标金属离子[146]。此方法早已被研究者用于从盐湖卤水中分离锂离子,近年来,Li 等[147]也成功应用此法从废锂电浸出液中选择性提取锂,效果较佳。同时,离子筛也被证明是从废锂电浸出液中选择性分离和提取各种有价金属的可行方法[148]。因此,使用离子印迹和离子筛分有望实现废旧锂离子电池浸出液中各种有价金属的选择性分离和提取。

3 结论与展望

当前,我国正在大力发展绿色低碳循环经济,促进经济社会发展全面绿色转型,力争2030年前二氧化碳排放达峰和2060年前实现碳中和愿景。镍、钴、锂战略金属作为新能源产业发展重要物质基础,决定着其产业链的发展命脉。因此,开展新能源战略金属资源清洁提取研究,是建立健全绿色低碳循环经济的重要途径和紧迫任务,是我国有色金属产业结构优化升级和绿色低碳转型发展的需要,对我国实现"碳达峰"及"碳中和"目标具有重要意义。

(1)我国镍、钴原生矿产资源严重匮乏,目前国外红土镍矿已成为主要镍、钴提取原料。国内外处理红土镍矿的主流工艺主要为 RKEF、火法高冰镍及湿法高压酸浸工艺,其中 RKEF 工艺是当前应用最广泛的红土镍矿火法冶炼工艺,但其生产产品无法应用于新能源领域,如何高效短流程低成本提取镍铁合金中镍将是未来研究的重点。火法高冰镍及湿法高压酸浸作为目前红土镍矿生产电池级硫酸镍的两大主流工艺,各有利弊。火法高冰镍工艺的成熟,打通了红土镍矿火法工艺到硫酸镍的产业路线,缓解了目前因电池材料项目快速扩建而造成的镍原料结构性紧缺局面,对目前新能源电动车行业发展有着重要的实际意义。然而,湿法高压酸浸工艺具有金属浸出率高、过程能耗及碳排放量低、环境友好等优势,是目前发展较快的技术路线,国内多家企业已在印尼布局湿法冶炼项目。湿法高压酸浸和火法高冰镍技术的

（2）我国锂资源虽然丰富但禀赋不佳，80%锂赋存于盐湖卤水中，但因 Mg/Li 比高，分离提取难度较大、现行技术成本高，导致我国锂盐生产长期依赖于进口锂矿石。现行盐湖提锂技术中，沉淀法适于处理低 Mg/Li 比盐湖卤水，流程简便、易于操作、生产成本较低，但该法生产效率较低、试剂消耗量大、生产过程锂损失严重，亟待开发精细化沉淀新方法，以提高锂综合回收率。太阳池-碳化法适于处理镁含量较低的碳酸盐型盐湖卤水，操作简便、生产成本低，但原料适用性差、锂回收率低、生产周期长。近年来，有研究者引入辅助碳酸锂结晶的成核基体，降低了成核势能，缩短了生产周期，效果显著。煅烧浸取法由于能耗较高、环境污染严重、投资及生产成本高等缺点已逐步丧失了竞争力。通过将纳滤法、反渗透、电渗析及吸附法等技术集成耦合，形成的耦合提锂工艺，已通过多个盐湖的产业化验证，具有广阔的推广应用前景。中科院过程所研发的 TBP/P507-FeCl$_3$ 新型协萃体系，基本解决了原萃取体系存在的问题，有望产业化推广应用。电化学脱嵌法适于处理低品位、复杂盐湖卤水，原料适应性较强、锂回收率高，但能耗较高、电极循环性能差，需进一步优化升级，并亟须产业化验证。

（3）目前退役锂电池资源回收市场规模较大，但相对分散，且由于锂电正极材料种类繁多，目前的回收系统不完善，并非所有回收工艺都在经济上可行。目前，火法回收工艺主要适用于处理镍钴含量高的废电池，而湿法冶金可被应用于不同类型的废锂电正极材料的回收。由于磷酸铁锂电池正极材料中有价金属含量较低，上述两种回收工艺均不适于从此类正极材料中回收有价金属。针对已失去梯次利用价值的磷酸铁锂正极材料，一般建议利用正极材料中仍有使用价值的组分，将其直接合成为新材料，实现回收价值最大化。

参考文献

[1] 李红，刘旭升，张宜生，等. 材料导报，2019，33(23)：3853-3861.
[2] 邢宝林，鲍俐傲，李旭升，等. 材料导报，2020，34(15)：15063-15068.
[3] 赵立敏，王惠亚，解启飞，等. 材料导报，2020，34(7)：7026-7035.
[4] 袁梅梅，徐汝辉，姚耀春. 材料导报，2020，34(19)：19061-19066.
[5] 吴子彬，宋森森，董安，等. 材料导报，2019，33(1)：135-142.
[6] 高桂兰. 有机酸还原性体系浸出回收废弃锂离子电池正极材料的研究[D]. 上海：上海大学，2020.
[7] 李文辉. 新能源汽车产业链构建研究. 硕士学位论文，郑州大学，2012.
[8] 汪淑芳. 中国新能源汽车产业链优化研究. 硕士学位论文，东华大学，2015.
[9] 于占波. 商用汽车. 2015(6)：23-26.
[10] 潘伟，黎宇科，李震彪. 汽车与配件，2019(15)：34-36.
[11] 马琼，郑海军. 上海汽车，2022(4)：1-3.
[12] 王晓明. 新材料产业，2021(3)：50-54.
[13] 钟财富，刘坚，吕斌，等. 中国能源，2018，40(10)：12-15.
[14] 杨俊峰，潘寻. 有色金属（冶炼部分），2021(6)：37-41.
[15] Mudd G M. Ore Geology Reviews，2010，38(1-2)：9-26.

[16] 马北越, 吴桦, 高陟. 耐火与石灰, 2022, 47(1): 35-40.
[17] 李小明, 沈苗, 王翀, 等. 材料导报, 2017, 31(5): 100-105.
[18] Stankovi S, Stopi S, Soki M, et al. Metallurgical and Materials Engineering, 2020, 26(2): 199-208.
[19] Chandra M, Yu D, Tian Q, et al. Mineral Processing and Extractive Metallurgy Review, 2021, 43(6): 679-700.
[20] Dehaine Q, Tijsseling L T, Glass H J, et al. Minerals Engineering, 2021, 160: 106656.
[21] Mudd, G M, Weng Z, Jowitt S M, et al. Ore Geology Reviews, 2013, 55: 87-98.
[22] 江思宏, 刘书生, 王天刚, 等. 全球锂、钴、镍、锡、钾盐矿产资源储量评估报告(2021). 中国地质调查局全球矿产资源战略研究中心, 北京, 2021: 1-19.
[23] 曲涛, 谷旭鹏, 施磊, 等. 材料导报, 2020, 34(S1): 261-267.
[24] Zhu D Q, Cui Y, Hapugoda S, et al. Transactions of nonferrous metals society of China, 2012, 22(4): 907-916.
[25] Thorne R, Herrington R, Roberts S. Mineralium deposita, 2009, 44(5): 581-595.
[26] Thorne R L, Roberts S, Herrington R. Geology, 2012, 40(4): 331-334.
[27] Goodall G. Nickel recovery from reject laterite. Master's thesis, McGill University, Canada, 2007.
[28] Burkin B R. Extractive metallurgy of nickel, Hoboken: John Wiley & Sons, US, 1987: 30-35.
[29] Valix M, Cheung W H. Minerals Engineering, 2002, 15(8): 607-612.
[30] 冉启胜, 朱淑桢. 矿业工程, 2011, 9(2): 6-8.
[31] 吕学明. 红土镍矿半熔融态冶炼镍铁的理论基础及工艺研究. 博士学位论文, 重庆大学, 2019.
[32] Ye Z, Wei S, Rui X, et al. Journal of Cleaner Production, 2020, 285, 124905.
[33] Meshram P, Pandey B D, Mankhand T R. Hydrometallurgy, 2014, 150, 192-208.
[34] 许乃才, 史丹丹, 黎四霞, 等. 材料导报, 2017, 31(17): 116-121.
[35] 徐萍, 钱晓明, 郭昌盛, 等. 材料导报, 2019, 33(3): 410-417.
[36] 杨卉芃, 柳林, 丁国峰. 矿产保护与利用, 2019, 39(5): 26-40.
[37] An J W, Dong J K, Tran K T, et al. Hydrometallurgy, 2012, 117-118: 64-70.
[38] 李芳芳, 李忠. 江西化工, 2016(6): 11-13.
[39] 宋彭生. 盐湖研究, 2000(1): 1-16.
[40] 郑绵平, 侯献华. 科技导报, 2017, 35(12): 11-13.
[41] 周园园. 资源与产业, 2019, 21(3): 46-50.
[42] 苏慧. 多组分协同溶剂萃取体系应用于高镁盐湖卤水提锂的研究. 博士学位论文, 中国科学院大学, 2022.
[43] 程前, 张婧. 材料导报, 2018, 32(20): 3667-3672.
[44] 江友周, 王宜, 李淑珍, 等. 化学工业与工程, 2021, 38(6): 23-33.
[45] 昝文宇, 马北越, 刘国强. 材料研究与应用, 2021, 15(3): 297-305.
[46] Zeng X L, Li J H, Singh N. Critical reviews in environmental science and technology, 2014, 44(10): 1129-1165.
[47] 李栋. 低品位镍红土矿湿法冶金提取基础理论及工艺研究. 博士学位论文, 中南大学, 2011.
[48] Liu W R, Li X H, Hu Q Y, et al. Transactions of Nonferrous Metals Society of China, 2010, 20: s82-s86.
[49] Hallberg K B, Grail B M, Plessis C, et al. Minerals Engineering, 2011, 24(7): 620-624.
[50] 张培育. 红土镍矿酸浸-水解耦合新工艺选择性浸出镍钴应用基础研究. 博士学位论文, 中国科学院大学, 2016.

[51] 孙燕娟. 有色冶炼, 1977(3): 10-12.
[52] 马保中. 镁质氧化镍矿煤基非熔融金属化还原工艺及基础理论研究. 博士学位论文, 昆明理工大学, 2017.
[53] 王帅, 姜颖, 郑富强, 等. 中国冶金, 2021, 31(10): 1-7.
[54] Khoo J Z, Haque N, Bhattacharya S. International Journal of Mineral Processing, 2017, 161: 83-93.
[55] Takeda O, Lu X, Miki T, et al. Resources, Conservation and Recycling, 2018, 133: 362-368.
[56] 马明生. 中国有色冶金, 2013, 42(05): 57-60.
[57] 潘料庭, 罗会键, 肖琦, 等. 2016(首届)全国铁合金热点难点技术交流会. 中国内蒙古集宁, 2016: 112-119.
[58] Liu P, Li B, Cheng S, et al. Applied Thermal Engineering, 2016, 109: 542-559.
[59] 尉克俭, 马明生, 卢笠渔, 等. 中国专利, CN201210206679.0, 2012.
[60] Gao Y M, Wang S B, Hong C, et al. International Journal of Minerals Metallurgy and Materials, 2014, 21(4): 353-362.
[61] 李光辉, 罗骏, 姜涛, 等. 中国专利, CN201711388806.2, 2017.
[62] 王亲猛, 陈远林, 郭学益, 等. 中国专利, CN202010533112.9, 2020.
[63] 于大伟, 郭学益, 甘向栋, 等. 中国专利, CN202011210440.1, 2020.
[64] 陶高驰, 肖峰, 蒋伟, 等. 有色金属(冶炼部分), 2014(8): 51-54.
[65] Li G, Shi T, Rao M, et al. Minerals Engineering, 2012, 32: 19-26.
[66] Tsuji H, Tachino N. ISIJ international, 2012, 52(10): 1724-1729.
[67] 刘志国, 孙体昌, 王晓平. 中国有色金属学报, 2017, 27(3): 594-604.
[68] 路长远, 鲁雄刚, 邹星礼, 等. 自然杂志, 2015, 37(4): 269-277.
[69] 田庆华, 李中臣, 王亲猛, 等. 中国有色金属学报, https://kns.cnki.net/kcms/detail/43.1238.tg.20220901.0851.003.html.
[70] 刘云峰, 陈滨. 矿冶, 2014, 23(4): 70-75.
[71] 王亲猛, 李中臣, 郭学益, 等. 中国专利, CN202111522791.0, 2021.
[72] 杨泽宇, 张文, 申亚芳, 等. 中国有色冶金, 2020, 49(04): 1-6.
[73] Meshram P, Abhilash, Pandey B D. Mineral Processing and Extractive Metallurgy Review, 2019, 40(3): 157-193.
[74] Önal M A R, Topkaya Y A. Hydrometallurgy, 2014, 142: 98-107.
[75] Georgiou D, Papangelakis V G. Hydrometallurgy, 1998, 49(1): 23-46.
[76] Papangelakis G. Hydrometallurgy, 2009, 100(1-2): 35-40.
[77] Guo X Y, Shi W T, Li D, et al. Transactions of Nonferrous Metals Society of China, 2011, 21(1): 191-195.
[78] Mcdonald R G, Whittington B I. Hydrometallurgy, 2008, 91(1-4): 35-55.
[79] Adams M, Van D M D, Czerny C, et al. In: International laterite nickel symposium. TMS Warrendale, 2004: 193-202.
[80] 田庆华, 董波, 郭学益, 等. 中国专利, CN202111641571.X, 2021.
[81] Tian Q H, Dong B, Guo X Y, et al. Minerals Engineering, 2021, 17, 107212.
[82] 田庆华, 董波, 郭学益, 等. 中国专利, CN202111638469.4, 2021.
[83] 田庆华, 董波, 郭学益, 等. 中国专利, CN202111544544.0, 2021.
[84] Ma B, Wang C, Yang W, et al. Minerals Engineering, 2013, 45: 151-158.
[85] 田庆华, 董波, 郭学益, 等. 中国专利, CN202111546849.5, 2021.

[86] 张振芳, 陈秀法, 李仰春, 等. 矿产综合利用, 2022(02): 31-39.
[87] 乜贞, 伍倩, 丁涛, 等. 无机盐工业, 2022, 54(10): 1-12.
[88] 马珍. 无机盐工业, 2022, 54(10): 22-29.
[89] 刘卓, 周云峰, 柴登鹏, 等. 材料导报, 2015, 29(S2): 133-137.
[90] 苏慧, 朱兆武, 王丽娜, 等. 材料导报, 2019, 33(13): 2119-2126.
[91] 刘元会, 邓天龙. 世界科技研究与发展, 2006, 28(5): 69-75.
[92] Choubey P K, Chung K S, Kim M S, et al. Minerals Engineering, 2017, 110: 104-121.
[93] Zhang Y, Hu Y, Sun N, et al. Hydrometallurgy, 2019, 187: 125-133.
[94] Zhang Y, Xu R, Wang L, et al. Minerals Engineering, 2022, 180: 107468.
[95] 郑绵平, 郭珍旭, 张永生, 等. 中国专利, CN99105828.3, 2000.
[96] 赵春龙, 孙峙, 郑晓洪, 等. 过程工程学报, 2018, 18(1): 20-28.
[97] Wu Q, Yu J, Bu L, et al. Solar Energy, 2022, 244: 104-114.
[98] 杨建元, 夏康明. 中国专利, CN200510085832.9, 2006.
[99] 郑绵平, 张永生, 刘喜方, 等. 地质学报, 2016, 90(9): 2123-2166.
[100] Song J F, Nghiem L D, Li X M, et al. Environmental Science Water Research & Technology, 2017, 3(4): 593-597.
[101] 马培华, 邓小川, 温现民. 中国专利, CN200310122238.3, 2005.
[102] Ding T, Zheng M, Lin Y. Acs Omega, 2022, 7(13): 11430-11439.
[103] 石成龙. 离子液体体系用于盐湖卤水中提取锂的研究. 博士学位论文, 中国科学院大学, 2017.
[104] 李慧芳, 李丽娟, 时东, 等. 盐湖研究, 2015, 23(2): 51-57.
[105] 石成龙, 宋桂秀, 秦亚茹, 等. 化学工程, 2020, 48(2): 16-19.
[106] 李传波, 闫东豪, 贾峰, 等. 中国专利, CN202110062197.1, 2021.
[107] 郭志远, 纪志永, 陈华艳, 等. 化工进展, 2020, 39(6): 2294-2303.
[108] Liu G, Zhao Z, Ghahreman A. Hydrometallurgy, 2019, 187: 81-100.
[109] 张治奎, 赵忠伟, 何利华. 中国实用新型专利, CN202021954468.1, 2020.
[110] Gao S, Liu W, Fu D, et al. New Carbon Materials, 2022, 37(3): 435-460.
[111] Lai X, Huang Y, Gu H, et al. Energy Storage Materials, 2021, 40: 96-123.
[112] Georgi-Maschler T, Friedrich B, Weyhe R, et al. Journal of Power Sources, 2012, 207: 173-182.
[113] Zhang G, Yuan X, He Y, et al. Journal of Hazardous Materials, 2021, 406: 124332.
[114] Hu J, Zhang J, Li H, et al. Journal of Power Sources, 2017, 351: 192-199.
[115] Li J, Wang G, Xu Z. Journal of Hazardous Materials, 2016, 302: 97-104.
[116] 于大伟, 黄柱, 田庆华, 等. 中国专利, CN202011480154.7, 2021.
[117] Xiao J, Li J, Xu Z. Environmental Science & Technology, 2017, 51(20): 11960-11966.
[118] Xiao J, Li J, Xu Z. Journal of Hazardous Materials, 2017, 338: 124-131.
[119] Wang R, Lin Y, Wu S. Hydrometallurgy, 2009, 99(3-4): 194-201.
[120] Kang J, Sohn J, Chang H, et al. Advanced Powder Technology, 2010, 21(2): 175-179.
[121] Lee C K, Rhee K. Hydrometallurgy, 2003, 68(1-3): 5-10.
[122] Chen X, Ma H, Luo C, et al. Journal of Hazardous Materials, 2017, 326: 77-86.
[123] He L, Sun S, Song X, et al. Waste Management, 2017, 64: 171-181.
[124] Guan J, Li Y, Guo Y, et al. Acs Sustainable Chemistry & Engineering, 2017, 5(1): 1026-1032.
[125] Tanong K, Coudert L, Mercier G, et al. Journal of Environmental Management, 2016, 181: 95-107.
[126] Gratz E, Sa Q, Apelian D, et al. Journal of Power Sources, 2014, 262: 255-262.

[127] Pagnanelli F, Moscardini E, Granata G, et al. Journal of Industrial and Engineering Chemistry, 2014, 20(5): 3201-3207.

[128] Meshram P, Pandey B D, Mankhand T R. Chemical Engineering Journal, 2015, 281: 418-427.

[129] Nayaka G P, Pai K V, Santhosh G, et al. Journal of Environmental Chemical Engineering, 2016, 4(2): 2378-2383.

[130] Li L, Bian Y, Zhang X, et al. Journal of Power Sources, 2018, 377: 70-79.

[131] Chen X, Luo C, Zhang J, et al. Acs Sustainable Chemistry & Engineering, 2015, 3(12): 3104-3113.

[132] Li L, Bian Y F, Zhang X X, et al. Waste Management, 2018, 71: 362-371.

[133] Fu Y, He Y, Qu L, et al. Waste Management, 2019, 88: 191-199.

[134] Gao W, Zhang X, Zheng X, et al. Environmental Science & Technology, 2017, 51(3): 1662-1669.

[135] Zhuang L, Sun C, Zhou T, et al. Waste Management, 2019, 85: 175-185.

[136] Hu C, Guo J, Wen J, et al. Journal of Materials Science & Technology, 2013, 29(3): 215-220.

[137] Zheng X, Gao W, Zhang X, et al. Waste Management, 2017, 60: 680-688.

[138] Lv W, Wang Z, Cao H, et al. Acs Sustainable Chemistry & Engineering, 2018, 6(2): 1504-1521.

[139] Zhang K, Li B, Wu Y, et al. Waste Management, 2017, 64: 236-243.

[140] Li J, Yang X, Yin Z. Journal of Environmental Chemical Engineering, 2018, 6(5): 6407-6413.

[141] Tanong K, Tran L, Mercier G, et al. Journal of Cleaner Production, 2017, 148: 233-244.

[142] Virolainen S, Fini M F, Laitinen A, et al. Separation and Purification Technology, 2017, 179: 274-282.

[143] Chen X, Kang D, Cao L, et al. Separation and Purification Technology, 2019, 210: 690-697.

[144] Freitas M, Garcia E M. Journal of Power Sources, 2007, 171(2): 953-959.

[145] Nishide H, Deguchi J, Tsuchida E. Chemistry Letters, 1976, 5(2): 169-174.

[146] Donato L, Drioli E. Frontiers of Chemical Science and Engineering, 2021, 15(4): 775-792.

[147] Li Z, He G, Zhao G, et al. Separation and Purification Technology, 2021, 277: 119519.

[148] Wang H, Huang K, Zhang Y, et al. Acs Sustainable Chemistry & Engineering, 2017, 5(12): 11489-11495.

采用亚硫酸钠浸提水淬冰铜渣中的银

摘　要：采用亚硫酸钠溶液对新疆阜康冶炼厂造锍捕集所得贵金属合金经硫酸化焙烧—水淬渣进行浸出，系统考察了工艺参数如固液比、亚硫酸钠浓度、溶液pH值、浸出时间、粒径以及浸出温度对银浸出率的影响。结果表明，银的最优浸出条件为：固液比200∶1(g/L)、亚硫酸钠200 g/L、pH值为7.5、浸出时间120 min、浸出温度25 ℃，在此条件下，可以高效提取渣中的银，浸出率为97%。

关键词：冰铜渣；硫酸化焙烧—水淬；银；亚硫酸钠；浸出

Extraction of silver from water-quenched matte slag using sodium sulfit

Abstract: The precious metal alloy obtained from matte trapping by sulfation roasting and water quenching in Xinjiang Fukang Smelter was leached using sodium sulfite solution. The effects of process parameters such as solid-liquid ratio, sodium sulfite concentration, solution pH value, leaching time, particle size and leaching temperature on the leaching rate of silver were studied. The results show that the optimal leaching conditions for silver are: solid-liquid ratio 200 g/L, sodium sulfite concentration 200∶1(g/L), solution pH value 7.5, leaching time 120 min, and leaching temperature 25 ℃. The silver in the slag is efficiently extracted, and the obtained silver leaching rate is 97%.

Keywords: matte slag; sulfuration roasting-water quenching; silver; sodium sulfite; leaching

金属银有着优良的导电性、导热性和抗氧化性，被广泛应用于医疗、电子、航空和科研等领域[1-5]。我国银矿产地主要分布在中南区，其次是华东区，且主要以伴生矿为主，其中铅锌伴生矿占51.4%、铜伴生矿占34.9%、金伴生矿占2.7%，其他占11%[6]。

新疆卡拉喀拉通克矿业有限责任公司和新疆众鑫矿业有限责任公司针对当地的硫化镍铜矿，采用火法熔炼得到高冰镍，混合磨细后，再用硫酸选择性浸出镍钴；浸出渣经沸腾焙烧，焙砂经稀酸浸出产出含贵金属渣，再经加压氧化浸出铜镍，而得到贵金属富集渣，其中富含较高品位的金、银、铂、钯等贵金属。针对这种渣，目前采用再次火法熔炼捕集，以得到含

本文发表在《有色金属工程》，2022, 12(7)：83-88。作者：朱佳俊，黄凯，马育新，李渭鹏。

金、银、铂和钯的铜锍合金,进而对该合金进行硫酸化焙烧和水淬,即得到含有金银铂钯的水淬渣。由于金铂钯的理化性质相似,利用氯酸钠选择优先浸出,得到含有金铂钯的氯化浸出液和含银的浸出渣。

该浸出渣中银主要以 AgCl 形式存在,目前针对 AgCl 中 Ag 的回收,工业上可以采用的方法有多种,其中氰化法是利用氰根离子捕捉 AgCl 中的 Ag 离子发生络合反应,形成络合离子,从而达到浸出和提取贵金属银的目的。其技术简单、成本低、效率高、技术相对成熟。但是由于氰化物有毒,不环保,氰化法被其他很多方法替代[7-9]。硫脲法,是利用硫脲作为络合剂,通过络合反应来达到提取银的目的。该方法具有环保、高效的特点,但是药品消耗大,成本高,工艺不够成熟[10-12]。氨浸法,是氯化银溶于氨水形成络合离子,从而达到浸出和提取银的目的。氨浸法成本低、效率高,但是氨水气味较大,操作环境太恶劣,兼之国家对环境治理的政策要求日趋严格,氨挥发造成的气、水、土污染问题必须引起重视[13-15]。亚硫酸钠法是利用在亚硫酸钠溶液中氯化银溶解,从而得到银的亚硫酸络合离子,再利用还原置换或电解将溶液中的银离子制成金属银。亚硫酸钠法浸出率高,工业上工艺较为成熟。

本研究拟采用亚硫酸钠浸出法对富康冶炼厂的富银渣开展浸出提取研究,探索该方法对该渣提银的基本规律。

1 试验

原料由新疆阜康冶炼厂提供,含银 0.46%,且以 AgCl 形式存在。亚硫酸钠为分析纯,水为去离子水。

试验在烧杯中进行,水浴加热。重点考察固液比[(20~400):1(g/L)]、亚硫酸钠浓度(50~250 g/L)、pH 值(6.5~10.5)、浸出时间(30~720 min)、粒径(0.425 mm、0.25 mm、0.18 mm)以及浸出温度(25 ℃、50 ℃、75 ℃)对银浸出率的影响。浸出结束后,过滤,得浸出液。

用 ICP 测定浸出液中银离子浓度,计算银浸出率。X 射线衍射(XRD)分析样品的晶体结构,XRF 分析样品的主要成分,扫描电子显微镜及其附带的能谱仪分析浸出渣的微观形貌和元素含量。

2 结果与讨论

2.1 XRD 与 XRF 分析

原料浸出银前后的 XRD 图谱以及浸出 XRF 结果如图 1 和表 1 所示,主要成分为 SiO_2、$BaSO_4$ 和 $PbSO_4$,银的物相未显示,原因可能是原料中银含量较少,未达到 X 射线衍射分析的极限。

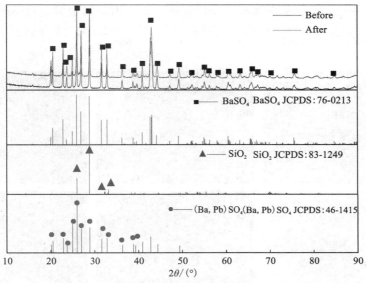

图 1　原料和浸出渣的 XRD 图谱

Fig. 1　XRD patterns of raw material and leaching residue

表 1　浸出后渣的 XRF 分析结果

Table 1　XRF analysis result of leaching residue

元素	Ba	O	Pb	S	Si	其他
质量分数/%	47.93	19.87	12.1	11.61	4.68	3.81

2.2　SEM 和 EDS 分析

浸出渣的 SEM 和 EDS 结果如图 2 所示，浸出渣的主要成分为二氧化硅，硫酸钡，硫酸铅，这一结果与 XRD 结果一致，亚硫酸钠成功将渣中的银完全浸出。

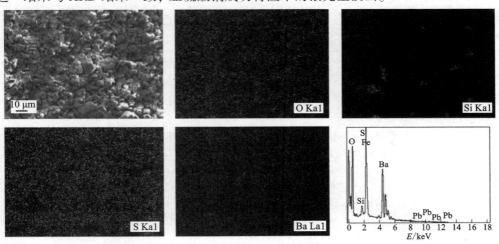

图 2　浸出渣表面的 SEM 面扫图和 EDS 谱图

Fig. 2　SEM surface scanning mappings and EDX spectrum of leaching residue

浸银后的渣颗粒,通过树脂镶嵌抛光后,对浸出渣内部的成分进行 SEM 和 EDS 分析,分析结果见图 3。从图 3 可知,浸出渣主要成分为 SiO_2 和 $BaSO_4$,渣的内部与表面成分无明显差别。

图 3　浸出渣的 SEM 面扫图像

Fig. 3　SEM surface scanning mappings of leaching residue

2.3　固液比对银浸出率的影响

在 25 ℃、亚硫酸钠 250 g/L、粒径 0.425 mm、溶液 pH=8.5、溶液体积 10 mL、浸出时间 120 min、搅拌速度 400 r/min 条件下,固液比对银浸出率的影响试验结果如图 4 所示。从图 4 可以看出,固液比在 (20~200):1(g/L) 时,随着原料用量的增加,银的浸出率变化不大,比较稳定,在 90% 左右,但随着固液比增加到 200:1(g/L) 以后,继续增加浸出渣用量,银的浸出率不断降低。随着原料量的增加,浸出药剂不足以浸出原料中的所有银,药剂不足。因此,在此浸出条件下,可保证固液比小于 200:1(g/L) 的银的浸出,且浸出率保持在 90%。

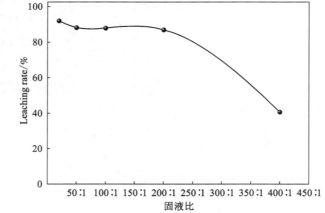

图 4　固液比对银浸出率的影响

Fig. 4　Effect of solid-liquid ratio on silver leaching rate

2.4 亚硫酸钠对银浸出率的影响

在 25 ℃、固液比 200∶1(g/L)、原料平均粒径 0.425 mm、溶液 pH=8.5、溶液体积 10 mL、浸出时间 120 min、搅拌速度 400 r/min 条件下,亚硫酸钠浓度对银浸出率的影响试验结果如图 5 所示。从图 5 可以看出,随着亚硫酸钠浓度的增加,银的浸出率不断增加,当亚硫酸钠浓度增加到 200 g/L 以后,银的浸出率几乎不变,保持在 93.66%,最佳浸出银的亚硫酸钠浓度为 200 g/L。

2.5 pH 值对银的浸出率影响

在 25 ℃、固液比 200∶1(g/L)、原料平均粒径 0.425 mm、亚硫酸钠 200 g/L、溶液体积 10 mL、浸出时间 120 min、搅拌速度 400 r/min 条件下,溶液 pH 值对银浸出率的影响试验结果如图 6 所示。从图 6 可以看出,随着溶液 pH 值的升高,银的浸出率先升高后降低,当溶液 pH 值增加到 7.5 以后,随着溶液 pH 值的升高,银的浸出率不断下降。这是因为,pH<7.5 时,亚硫酸根离子发生加氢反应可生成亚硫酸氢根,溶液中的亚硫酸根浓度降低,不利于银的浸出;当 pH>8.5 时,$Ag(SO_3)_2^{3-}$ 稳定性降低[16],$Ag(SO_3)_2^{3-}$ 氧化成沉淀,降低浸出液中银的含量,不利于银的浸出。因此,银的最佳浸出溶液 pH 值为 7.5。

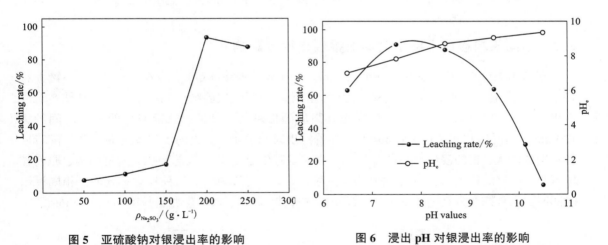

图 5　亚硫酸钠对银浸出率的影响

Fig. 5　Effect of sodium sulfite on silver leaching rate

图 6　浸出 pH 对银浸出率的影响

Fig. 6　Effect of leaching pH on silver leaching rate

2.6 不同温度下浸出时间对银浸出率的影响

在固液比 200∶1(g/L)、原料平均粒径 0.425 mm、亚硫酸钠浓度 200 g/L、溶液 pH=8.5、溶液体积 10 mL、搅拌速度 400 r/min 和不同浸出温度条件下,浸出时间对银浸出率的影响试验结果如图 7 所示。

从图 7 可以看出,不同温度下,随着浸出时间的增加,银的浸出率均呈现先升高后降低趋势,浸出率峰值对应的浸出时间约为 120 min,当浸出时间超过 120 min 后,随着浸出时间的延长,银的浸出率不断下降。这是因为,随着浸出时间的延长,亚硫酸钠被氧化为硫酸钠,降低了溶液中亚硫酸钠的浓度,同时使 $Ag(SO_3)_2^{3-}$ 氧化沉淀,降低浸出液中银的含量,所以银浸出的最佳时间为 120 min。从图 7 还可以看出，相同浸出时间时，随着浸出温度的升高，银的浸出率不断下降。这是因为，浸出温度升高，加速亚硫酸钠氧化为硫酸钠[16-17]，溶液中亚硫酸钠的浓度降低，同时使 $Ag(SO_3)_2^{3-}$ 氧化沉淀，降低了浸出液中银的含量，所以高的浸出温度不利于银的浸出。

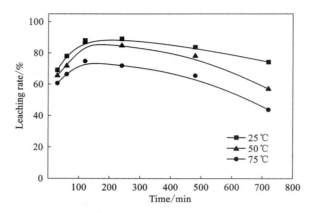

图 7 不同浸出温度下浸出时间对银浸出率的影响

Fig. 7 Effect of leaching time on silver leaching rate under different leaching temperatures

2.7 不同原料粒度下浸出时间对银浸出率的影响

在浸出温度 25 ℃、固液比 200∶1(g/L)、亚硫酸钠 200 g/L、溶液 pH = 8.5、溶液体积 10 mL、搅拌速度 400 r/min，原料粒度对银浸出率的影响试验结果如图 8 所示。从图 8 可以看出，随着浸出时间的延长，均呈现先升高后降低趋势，浸出率峰值对应的浸出时间约为 120 min，当浸出时间超过 120 min 后，随着浸出时间的延长，银的浸出率不断下降。随着原料粒度的增大，银的浸出率不断降低，说明粒度越细越有利于银的浸出，这是因为，原料粒度越小，颗粒的比表面积越大，浸出率越高。从图 8 可以看出，三种粒度相比较，相同浸出时间时，原料粒度为 0.180 mm 时的银浸出率最佳，因此，取最佳原料粒度为 0.180 mm。

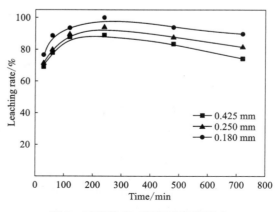

图 8 矿渣粒度对银浸出率的影响

Fig. 8 Effect of leaching time on silver leaching rate with different raw material particle size

2.8 综合条件试验

在最优浸出条件：原料粒度 0.180 mm、固液比 200∶1(g/L)、亚硫酸钠浓度 200 g/L、溶液 pH 值 7.5、浸出时间 120 min、浸出温度 25 ℃、搅拌速度 400 r/min 条件下进行浸出综合条件试验，共进行 3 次试验，试验结果如图 9 所示。从图 9 可以看出，综合条件试验所得银的浸出率保持在 97% 以上，且试验重复性好。

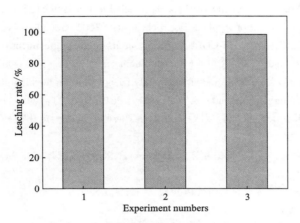

图 9　最佳条件下的综合条件试验结果
Fig. 9　Results of comprehensive condition tests

3　结论

（1）新疆阜康冶炼厂造锍捕集后得到的贵金属合金经硫酸化焙烧和水淬后的焙烧渣含有一定的银，具有较高的回收利用价值。

（2）采用亚硫酸钠为浸出剂可有效浸出焙烧渣中的银，最优浸出条件为：原料粒度 0.180 mm、固液比 200∶1(g/L)、亚硫酸钠浓度 200 g/L、溶液 pH 值 7.5、浸出时间 120 min、浸出温度 25 ℃，在此最佳浸出条件下，渣中银的浸出率可达 97%。

参考文献

[1] RAGHAVAN R, MOHANAN P K, PATNAIK S C. Innovative processing technique to produce zinc concentrate from zinc leach residue with simultaneous recovery of lead and silver[J]. Hydrometallurgy, 1998, 48(2): 225-237.

[2] APARAJITH B, MOHANTY D B, GUPTA M L. Recovery of enriched lead-silver residue from silver-rich concentrate of hydrometallurgical zinc smelter[J]. Hydrometallurgy, 2010, 105(1-2): 127-133.

[3] ZHANG Y, YU X, LI X. Kinetics of simultaneous leaching of Ag and Pb from hydrometallurgical zinc

residues by chloride[J]. Rare Metals, 2012, 31(4): 402-409.

[4] ZHANG C, WANG S, CAO Z, et al. Two-stage leaching of manganese and silver from manganese-silver ores by reduction with calcium sulfide and oxidation with copper(Ⅱ)[J]. Hydrometallurgy, 2018, 175: 240-249.

[5] FAN Y, LIU Y, NIU L, et al. Leaching of silver from silver-bearing residue by a choline chloride aqueous solution and the sustained deposition of silver on copper[J]. Hydrometallurgy, 2020, 197: 105454-105462.

[6] 卢宜源, 宾万达. 贵金属冶金学[M]. 中南工业大学出版社, 1994.

[7] CRUZ V R, OROPEZA M T, GONZÁLEZ I, et al. Electrochemical recovery of silver from cyanide leaching solutions[J]. Journal of Applied Electrochemistry, 2002, 32(5): 473-479.

[8] PORTILLA R E, HE C, JACOME-COLLAZOS M, et al. Acidic pretreatment of a copper-silver ore and its beneficial effect on cyanide leaching[J]. Minerals Engineering, 2020, 149: 106233-106240.

[9] XIAO L, QIAN P, YU Y, et al. An environmentally friendly system for high efficient silver recovery from anode slime[J]. Metallurgical Research & Technology, 2019, 116(2): 208-215.

[10] 陈洪流, 柳城, 高升, 等. 酸性硫脲浸取-石墨炉原子吸收光谱法测定化探样品中的微量银[J]. 黄金, 2018, 39(07): 80-82.

[11] 张村, 方夕辉, 夏艳圆, 等. 难浸银精矿在[Bmim]H_2SO_4—硫脲体系中的浸出[J]. 有色金属工程, 2016, 6(4): 38-40, 47.

[12] 张帅, 曾怀远, 张村, 等. 氰化法、硫代硫酸盐法、硫脲法浸出某难浸银精矿比较研究[J]. 有色金属科学与工程, 2015, 6(1): 74-78.

[13] 玉荣华, 覃祚观, 高大明, 等. Cu^{2+}催化氧化氨浸法从含银电镀污泥中浸取银[J]. 广东化工, 2016, 43(8): 134-135.

[14] 肖红新. 氨水浸取-盐酸介质火焰原子吸收光谱法测定杂铜物料中的银[J]. 黄金, 2014(9): 82-84.

[15] 王龙, 薛娜, 贾玉镯, 等. 氨浸工艺从废弃线路板中回收铜的试验研究[J]. 矿产综合利用, 2019(3): 98-101, 92.

[16] 陈小红, 赵祥麟, 楚广, 等. 用亚硫酸钠从分银渣中浸出银[J]. 中南大学学报(自然科学版), 2014, 45(2): 356-360.

[17] 程德平, 夏式均. $Ag-SO_3^{2-}-H_2O$系热力学分析[J]. 杭州大学学报(自然科学版), 1991(4): 449-454.

负载纳米 ZnS 阳离子树脂选择性去除浓锌溶液中痕量级 Cu^{2+} 和 Cd^{2+}

摘 要：通过将 ZnS 纳米粒子负载在阳离子树脂 D001 中制备负载纳米 ZnS 阳离子树脂吸附剂，即 ZnS@D001，吸附湿法冶金锌浸出液中共存杂质金属离子（如 Cu^{2+} 和 Cd^{2+}）。在 10 g/L Zn^{2+} 的合成溶液中实现了对 Cu^{2+} 的高度优先吸附，而对 Cd^{2+} 没有任何吸附效果。吸附后的 ZnS@D001 树脂可以被 1 mol/L HNO_3 有效解吸，然后再次加载 ZnS。同时利用含有 Cu^{2+} 和 Cd^{2+} 的锌离子溶液进行固定床柱吸附，经过吸附处理后，可以有效去除溶液中铜杂质，纳米 ZnS 树脂可作为锌冶炼行业深度净化锌液的潜在功能材料。

关键词：纳米粒子；离子交换树脂；D201；杂质金属离子；Cu^{2+}；Cd^{2+}；唐南膜效应

Selective removal of trace levels of Cu^{2+} and Cd^{2+} in concentrated zinc solution with loaded nano-ZnS cationic resin

Abstract: Nano-ZnS loaded cationic resin adsorbent, namely ZnS@D001, was prepared by loading ZnS nanoparticles in cationic resin D001 and used to absorb coexisting impurity metal ions (such as Cu^{2+} and Cd^{2+}) in hydrometallurgical zinc leaching solution. High priority adsorption of Cu^{2+} were achieved in 10 g/L Zn^{2+} synthetic solution, but there was no adsorption effect on Cd^{2+}. The adsorbed ZnS@D001 resin can be effectively desorbed by 1 mol/L HNO_3, and then loaded with ZnS again. At the same time, the fixed bed column adsorption of synthetic zinc ion solution containing Cu^{2+} and Cd^{2+} metal ion impurities were tested. After treatment, copper impurity content in effluent can be effectively removed, which can be used as a potential functional material for deep purification of zinc liquid in the zinc smelting industry.

Keywords: nanoparticles; ion exchange resins; impurity metal ions; Cu^{2+}; Cd^{2+}; tangnan membrane effect

从浓缩锌溶液中选择性去除痕量铜(Ⅱ)和镉(Ⅱ)离子是湿法炼锌工艺中的一项具有挑战性的任务[1-3]。目前，已经发明了很多分离方法来有效去除 1 g/L 数量级的铜(Ⅱ)和镉(Ⅱ)离子，例如置换[4-6]和溶剂萃取[7-10]。而对于铜(Ⅱ)和镉(Ⅱ)离子较少的锌(Ⅱ)溶

本文发表在《有色金属(冶炼部分)》，2022(5)：14-25。作者：朱佳俊，修祎帆，孙建刚，黄凯。

液，例如低于 1 g/L，最有效的分离方法是通过各种吸附剂吸附，如无机吸附剂[11, 12]、溶剂-浸渍树脂[13-15]和螯合树脂[16-21]。

在过去的几十年中，无机二氧化硅和有机聚合物树脂已被广泛研究用于回收铜(Ⅱ)和镉(Ⅱ)离子[22-26]。但它们对目标金属离子的选择性通常不足以满足许多相互分离的需求。因此，非常有必要制备更有效的吸附剂，以有效去除浓缩锌(Ⅱ)溶液中的铜(Ⅱ)和镉(Ⅱ)等杂质。考虑到浓锌(Ⅱ)溶液中离子强度大，从该溶液中选择性去除痕量铜(Ⅱ)和镉(Ⅱ)将是相当困难的，并且最有效的树脂和吸附剂可能会失去其强的选择性吸附能力。

因此，本研究提出了一种新型混合纳米复合树脂，即阴离子树脂负载的纳米级 ZnS 颗粒，它是将 ZnS 纳米颗粒封装到 D001 树脂。在这种混合树脂中，由于 pK_{sp} 值的差异(ZnS 23.8、CdS 26.1、CuS 35.2)，纳米尺寸的 ZnS 作为活性位点来捕获痕量铜(Ⅱ)和镉(Ⅱ)，而 D001 具有固定化带电配体($—SO_3^-$)有助于在静电引力作用下将上述痕量杂质预浓缩到树脂珠中，从而与沉积在树脂纳米孔中的纳米级 ZnS 颗粒进行化学结合。在过去的二十年里，SUZUKI 等[27, 28]、SENGUPTA 等[29-33]和 PAN 等[34-38]系统地研究了在商业树脂中负载各种无机颗粒，即有机-无机杂化纳米复合吸附剂。他们的研究充分证实了唐南膜效应对目标金属离子的预浓缩及纳米粒子巨大比表面积效应的选择性吸附能力的耦合所产生的独特增强效应。在他们的研究中使用的大多数纳米级无机颗粒是无机氢氧化物，如水合 Fe(Ⅲ)氧化物(HFO)[39-40]、磷酸锆(ZrP)、水合氧化锆(HZO)、水合 Mn(Ⅳ)氧化物(HMO)，而很少有研究提到硫化物无机材料。他们的大部分研究只集中在去除干净水中的痕量有毒元素，很少集中在浓缩电解质溶液中，这可能是由主要金属离子如 Zn(Ⅱ)在水中的强竞争吸附作用所致。关于这一挑战，我们提出了一种将纳米级 ZnS 颗粒封装在树脂中的新想法，它可能会有效分离浓锌(Ⅱ)溶液中的痕量铜(Ⅱ)和镉(Ⅱ)离子。而且，根据路易斯酸碱理论的 HSAB 定律，硫化物是典型的软碱，对 Cd^{2+}、Cu^{2+}、Ag^+ 等软酸具有很强的亲和力。因此，基于纳米金属硫化物的强吸附能力并结合唐南膜效应，为了实现对痕量铜(Ⅱ)和镉(Ⅱ)离子的优异去除，有必要利用阴离子树脂负载纳米级硫化锌，并评估其在深度净化行为中应用的可行性。

1 试验方法

1.1 试验试剂和设备

本文使用的所有化学品，如硝酸锌、硫化钠、硝酸铅、硝酸铜和硝酸镉等，均为市售分析纯试剂。

D001 树脂是一种聚苯乙烯基体的大孔强酸性阳离子交换剂，由江苏苏庆离子交换器有限公司提供。使用前，D001 用 HCl 溶液(5%质量分数)浸泡 4 h，NaOH(5%质量分数)浸泡 4 h，在上述两次操作之间用去离子水冲洗至中性。之后，将其浸泡在去离子水中进行进一步改性。

烧瓶电动摇床(Guohua102)摇动溶液样品。采用配备火焰炉的原子吸收光谱仪(4320AAS)对铜和镉元素进行定量分析。其他金属离子的测量使用 ICP/OES(iCAP7000, USA)。ZnS@D001 的孔径分布和比表面积是通过在 NOVA3000 仪器(Quantachrome)上

77 K 下进行 N_2 吸附/解吸测试来测量的。使用扫描电子显微镜（SEM）和能量色散 X 射线光谱（EDX）（ZeissSupra55）进一步观察 ZnS@D001 和 D001 树脂的结构。傅里叶变换红外（FTIR）光谱是通过使用 NicoletIS10 光谱仪（ThermoFisher）在 4 cm^{-1} 分辨率下从 4000 cm^{-1} 到 400 cm^{-1} 的波数获得的。使用配备有 CuKα 辐射（40 kV，40 mA）的 D8Advance 衍射仪以 10°/min 的扫描速度获得 X 射线衍射（XRD）图案。

1.2 ZnS 负载阳离子树脂（ZnS@D001）制备

在烧瓶中制备 500 mL 亚硝酸锌溶液（1 mol/L），在氩气气氛下向溶液中加入 100 g D001 湿树脂（水的质量分数为 68%），溶液 pH 调整为 4.5。Zn^{2+} 离子将被捕获到 D001 树脂上，3 h 后通过过滤和用去离子水洗涤直至中性 pH 来收集树脂。然后将湿树脂放入装有 400 mL Na_2S 溶液（1.0 mol/L）的烧瓶中，在 333 K 下鼓泡 3 h，负载 Zn^{2+} 的树脂将与 S^{2-} 反应形成 ZnS 相 D001 树脂，反应如下：

$$D001 - Zn^{2+} + S^{2-} (aq) \longrightarrow D001 - ZnS(s) \qquad (1)$$

由于 ZnS 的 K_{sp} 值极小（1.6×10^{-24}），纳米级的 ZnS 颗粒沉积在 D001 的纳米孔内，最后用大量去离子水冲洗，得到白色 ZnS@D001 杂化吸附剂。

1.3 批量吸附测试

吸附试验在 298 K 下通过传统的摇瓶分批法进行。将上所述制备的测试混合物溶液与一定量的 ZnS@D001 混合吸附剂或 D001 树脂一起在烧瓶中剧烈摇动 12 h，在恒温振荡器中达到平衡。使用原子吸收分光光度计分析吸附前后的金属浓度。由吸附前后的金属浓度（分别为 C_0 和 C_e）和树脂的干重（W），以及水溶液的体积（V），根据式（2）计算吸附量（Q，单位 mg/g）：

$$Q = (C_o - C_e) \times V/W \qquad (2)$$

吸附效率（A，单位%）根据式（3）计算：

$$A = (C_o - C_e)/C_o \times 100\% \qquad (3)$$

1.4 柱吸附测试

基于上述分批试验结果，在填充柱中对合成锌溶液中的铜和镉进行选择性分离试验，以检验从合成锌溶液中去除痕量铜（Ⅱ）和镉（Ⅱ）杂质的可行性。考虑到其实际条件，即在镀锌或电积工业中，少量的铜和镉作为杂质通常共存于锌溶液中，因为它们具有化学相似性。本研究中使用的填充柱是内径为 8 mm 的玻璃柱。将 3.5 mL FeS@D001 或 D001 树脂分别装在两个独立的柱子中，以进行比较。使用填充有上述树脂的柱子进行吸附测试，在开始穿透测试前，通过 pH 为 4.5（与测试溶液的 pH 相同）的无金属水溶液经过 4 h 进行预处理。使用蠕动泵（BT-100S）以 30 mL/h 的恒定流速使含有铜（Ⅱ）和镉（Ⅱ）的进料锌（Ⅱ）溶液渗透通过色谱柱。进料溶液的 pH 保持在 4.5。使用馏分收集器（HuxiModelBS-110A）每小时收集的流出物样品用于测量其残留浓度和 pH。对于洗脱测试，用去离子水洗涤柱子以去除任何残留的物理吸附金属离子，然后用 1 mol/L 硝酸作为洗脱液，同样以 30 mL/h 的恒定流速渗透通过柱子，使用 AAS4320 型原子吸收分光光度计测定初始溶液和流出物中铜（Ⅱ）和镉（Ⅱ）的浓度。

2 结果与讨论

2.1 SEM/EDS 分析

负载 ZnS 的 D001 和 D001 的树脂珠均为白色,不易从颜色或外观上区分。而通过用 SEM/EDS 观察它们的横截面,如图 1 所示,可以发现,ZnS 成功地负载到 D001 树脂的表面上,形成了一个环状区域。显然,该环状区域的厚度受锌离子通过树脂的渗透效率的影响。因此证实锌离子扩散到树脂珠中是可以通过增加压力、温度和锌浓度等控制参数来促进。

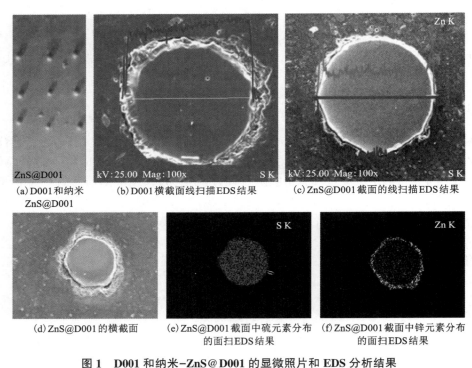

(a) D001 和纳米 ZnS@D001
(b) D001 横截面线扫描 EDS 结果
(c) ZnS@D001 截面的线扫描 EDS 结果
(d) ZnS@D001 的横截面
(e) ZnS@D001 截面中硫元素分布的面扫 EDS 结果
(f) ZnS@D001 截面中锌元素分布的面扫 EDS 结果

图 1 D001 和纳米-ZnS@D001 的显微照片和 EDS 分析结果

Fig. 1 Photomicrographs and EDS analysis results of D001 and nano-ZnS@D001

图 2 展示了一系列改性树脂或吸附前后的 SEM/EDS 分析结果。可以看出,ZnS 成功地沉积在树脂中,分散的 ZnS 颗粒大小不等,形状不规则。而在吸附后,负载在树脂上的纳米颗粒的形状改变为针状和片状。树脂形态变化表明已经发生吸附。

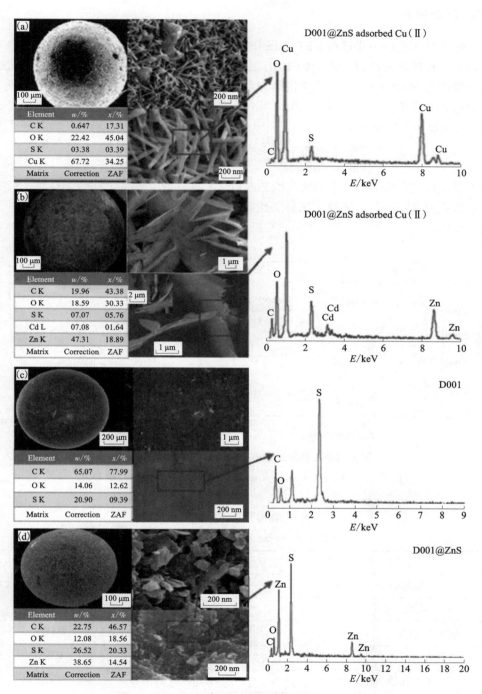

图 2 D001 和纳米 ZnS-D001 的 SEM/EDS 图
Fig. 2 SEM/EDS images of D001 and nano-ZnS-D001

2.2 N$_2$ 吸附分析

对两种树脂珠粒进一步进行 BET 测试分析,如图 3 和表 1 所示,在树脂微孔内负载 ZnS 颗粒会增加珠粒的比表面积,这可能归因于比表面积更大的纳米 ZnS 的贡献。锌在树脂上的负载量经测试为约 12.1%,这对应于混合树脂珠粒中包含约 18.0% ZnS。显然,不难理解,沉积在树脂珠微孔内壁的纳米颗粒,由于其化学亲和性强,有助于提高对目标金属离子的吸附能力,可能发生的反应如下:

$$D001@ZnS + Cu^{2+} \rightleftharpoons D001@CuS + Zn^{2+}$$

或

$$D001@ZnS + Cd^{2+} \rightleftharpoons D001@CdS + Zn^{2+}$$

由于其更好的反应表面,更细的沉积 ZnS 颗粒,将有助于提高对目标金属离子的吸附能力。树脂微孔内负载硫化锌颗粒工艺易于预成型,在金属离子溶液的相互分离中具有很大的实际应用潜力。

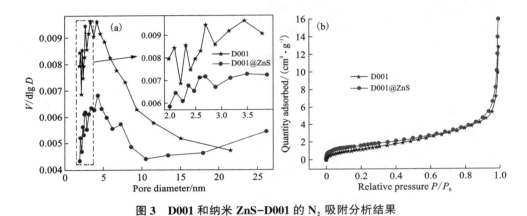

图 3 D001 和纳米 ZnS-D001 的 N$_2$ 吸附分析结果

Fig. 3 N$_2$ adsorption analysis results of D001 and nano-ZnS-D001

表 1 树脂样品的基本参数

Table 1 Basic parameters of resin samples

名称	D001@ZnS	D001
矩阵结构	聚苯乙烯二乙烯基苯	
外观、颜色	圆球(0.62~0.73 mm),白色	
BET 表面积/(m^2·g^{-1})	5.696	4.65
微孔孔容/(cm^3·g^{-1})	0.0021	0.0016
平均孔径/nm	17.272	17.004
平均密度/(g·cm^{-3})	1.648	1.467
ZnS 负载量(Zn^{2+}总量)/%	12.1	0

2.3 红外分析测试

为了进一步了解吸附过程，进行了 FTIR 测试，结果如图 4 所示，这表明 ZnS 的负载会导致 500~1500 cm^{-1} 范围内的大部分峰消失。而铜和镉离子吸附到 ZnS@D001 珠子上会带来 IR 曲线的剧烈变化，如图 4 所示，新峰出现在铜的 1418 cm^{-1} 和 1337 cm^{-1} 波数范围内，证实了 Cu^{2+} 和 Cd^{2+} 在 ZnS@D001 珠上的吸附形成了化学键，这意味着吸附过程是基于化学反应过程，而不是简单的物理过程。

2.4 XRD

图 5 是树脂珠的 XRD 测试结果，可以发现检测到 ZnS 相，这很容易理解，因为负载的 Zn^{2+} 与 S^{2-} 接触形成 ZnS。但沉积的 ZnS 的极小粒径会使 XRD 谱显示无定形或弱结晶相。这可能会使沉积的 ZnS 保持相当活跃的化学反应活性，这有利于吸附过程。

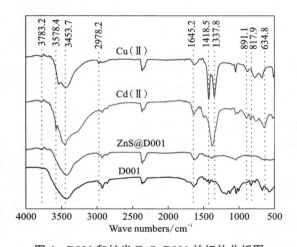

图 4　D001 和纳米 ZnS-D001 的红外分析图
Fig. 4　Infrared analysis diagrams of D001 and nano-ZnS-D001

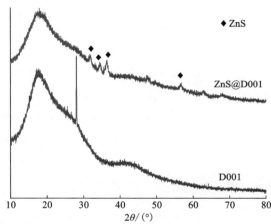

图 5　D001 和纳米 ZnS-D001 的 XRD 谱
Fig. 5　XRD patterns of D001 and nano-ZnS-D001

2.5 pH 值

图 6 显示了溶液 pH 值对 10 g/L Zn^{2+} 溶液中 Cu^{2+} 和 Cd^{2+} 在 ZnS@D001 和 D001 树脂上去除效率的影响。结果表明，分别在 ZnS@D001 和 D001 树脂上去除了 75% 和 5% 的 Cu^{2+}，ZnS@D001 和 D001 树脂上 Cd^{2+} 的去除率分别为 13% 和 3%。可以发现，在 ZnS@D001 上的去除效率远好于 D001 树脂，而且铜比镉容易去除。显然，在含有如此高浓度 Zn^{2+} 的浓溶液中，固定在 D001 市售树脂中的—SO$_3^-$ 酸性配体，在静电引力作用下会很快被 Zn^{2+} 完全占据，吸附痕量浓度的 Cu^{2+} 和 Cd^{2+} 的机会很少。相反，对于 ZnS@D001 吸附剂，也会发生同样的事情，但 ZnS 与 Cu^{2+} 和 Cd^{2+} 的强亲和力形成更稳定的 CuS 和 CdS 相，使吸附产生上述现象。另外还发现，随着 pH 的增加，两种杂质金属的去除率都会相应增加。在试验中发现，在 pH 为 1 和 2 时，对于 ZnS@D001 珠子，溶液会浑浊和变成黑色，形成 CuS，而形成 CdS 的溶液变黄，清澈并带有 H$_2$S 的难闻气味，这意味着 ZnS@D001 珠在酸性条件下存在分解过程，只

有在 pH>3.5 时,才能有效抑制 ZnS 的溶解,因此 ZnS@D001 的适宜工作 pH 值在 3.5 以上。并且该结果还表明,负载金属的 ZnS@D001 可能会被强酸性溶液洗脱(这将在下面进一步测试)。因此,杂质金属离子的去除率大可以解释为溶解在溶液中的硫化氢沉淀,在振荡操作后将硫化氢从溶液中过滤除去。因此,基于以上结果和讨论,可以推导出 ZnS@D001 吸附剂对杂质金属离子的去除机理如下(以 Cu^{2+} 为例):

$$D001@ZnS(s) + Cu^{2+}(aq) \longrightarrow D001@CuS(s) + Zn^{2+}(aq) \tag{4}$$

$$D001@ZnS(s) + 2H^+(aq) \longrightarrow D001 + Zn^{2+}(aq) + H_2S(aq) \tag{5}$$

$$Cu^{2+}(aq) + H_2S(aq) \longrightarrow 2H^+(aq) + CuS(s) \tag{6}$$

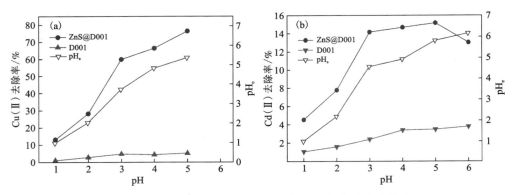

图 6 溶液 pH 对 ZnS@D001 上 Cu^{2+} 和 Cd^{2+} 去除率的影响

Fig. 6 Effects of pH value of solution on removal rate of Cu^{2+} and Cd^{2+} on ZnS@D001

2.6 杂质金属离子的选择性吸附

使用 ZnS@D001、D001 树脂和 D001+ZnS 颗粒从溶液中分离出铜和镉离子(图 7)。结果表明,对于镉离子,三种材料的吸附百分比几乎没有差异。而对于铜离子,在锌离子浓度分别为 10 g/L 和 30 g/L 的情况下,对铜的吸附率分别达到 75% 和 50%。10 g/L 是锌离子有效分离铜杂质的合适浓度。表 2 列出了铜和镉在三种吸附剂上的分离系数,可以发现,负载 ZnS 的树脂对铜和镉离子的吸附效果最好,验证了唐南膜对促进金属离子吸附的作用。

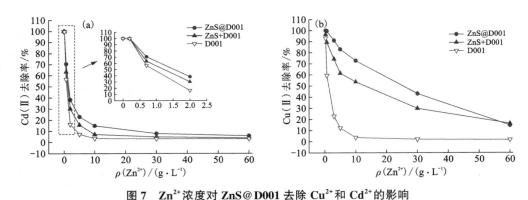

图 7 Zn^{2+} 浓度对 ZnS@D001 去除 Cu^{2+} 和 Cd^{2+} 的影响

Fig. 7 Effects of Zn^{2+} concentration on removal rate of Cu^{2+} and Cd^{2+} on ZnS@D001

表 2　Zn^{2+} 浓度对铜和镉在吸附剂上的分配系数(K_d)的影响

Table 2　Effect of Zn^{2+} concentration on partition coefficient (K_d) of copper and cadmium on adsorbent

金属(M^{2+})	吸附剂	初始锌离子质量浓度/(g·L^{-1})						
		0.3	0.7	3	5	10	30	60
Cu(Ⅱ)	D001@ZnS	∞ *	124.13	5.02	2.44	1.34	0.37	0.08
	D001+ZnS	13.66	4.19	1.44	0.79	0.58	0.21	0.96
	D001	6.69	0.73	0.15	0.07	0.02	0.01	0.008
Cd(Ⅱ)	D001@ZnS	7.99	1.19	0.25	0.15	0.09	0.04	0.03
	D001+ZnS	∞ *	5.85	0.16	0.09	0.04	0.03	0.02
	D001	5.25	0.65	0.08	0.04	0.0178	0.0159	0.016

注：* K_d 值是无穷大(∞)，因为吸附后的 M^{2+} 低于 AAS 的检测限。

2.7　选择性吸附机制

基于以上结果和讨论，改性和吸附的整个过程如图 8 所示。树脂微孔内 ZnS 颗粒的存在对于选择性吸附固定目标金属离子是非常必要的。根据 HSAB 原理，S^{2-} 是典型的软路易斯碱，因此与中间酸锌离子相比，有利于与铜、镉等软酸形成稳定的化合物。

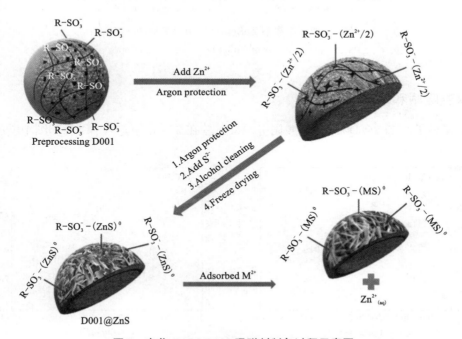

图 8　杂化 ZnS@D001 吸附剂制备过程示意图

Fig. 8　Schematic diagram of preparation process of hybrid ZnS@D001 adsorbent

前面的研究已经证实,铜和镉离子在树脂上的吸附引起了硫化锌颗粒的形态及吸附后树脂中的官能团变化。为了深入了解选择性吸附机制,对金属负载前后的 ZnS@D001 样品进行了 XPS 研究。图 9 表明,金属离子负载到 ZnS@D001 上导致更多的 Zn(2p)、Cu(2p)、Cd(3d)、S(2p) 和 O(1s) 区域的电子结合能转移,表明 ZnS 与目标金属离子铜和镉离子之间的吸附,这是分离过程的选择性吸附机制。

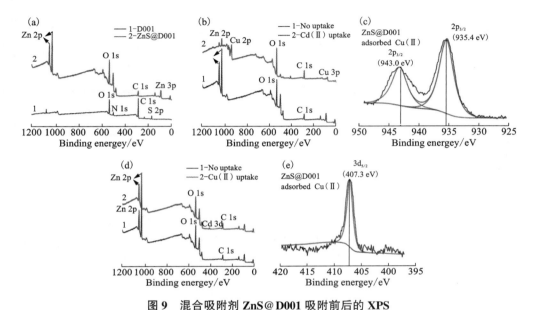

图 9 混合吸附剂 ZnS@D001 吸附前后的 XPS
Fig. 9 XPS of mixed adsorbent ZnS@D001 before and after adsorption

2.8 等温吸附和动力学

铜和镉离子在 ZnS@D001 上的吸附等温线试验在 298 K 下进行,结果如图 10(a) 所示。

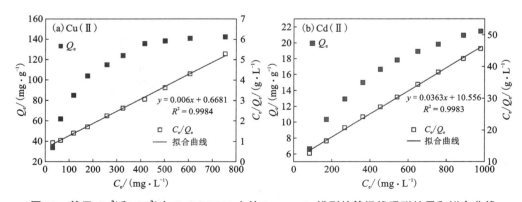

图 10 基于 Cu^{2+} 和 Cd^{2+} 在 ZnS@D001 上的 Langmuir 模型的等温线吸附结果和拟合曲线
Fig. 10 Isotherm adsorption results and corresponding fitting curves based on Langmuir model of Cu^{2+} and Cd^{2+} on ZnS@D001

根据 Langmuir 模型：

$$\frac{C_e}{Q_e} = \frac{1}{K_L Q_m} + \frac{C_e}{Q_m} \tag{7}$$

式中：C_e 为溶液中重金属离子的平衡浓度，mg/L；Q_e 为平衡浓度下的吸附量，mg/g；Q_m 为最大吸附量，mg/g；K_L 是结合常数，L/mg。

通过绘制 $1/q_e$ 和 $1/C_e$ 曲线，可以确定截距的 q_m 值和斜率的 K_L 值。图10(b)和表3结果表明，ZnS@D001 对 Cu(Ⅱ) 和 Cd(Ⅱ) 的去除可以用 Langmuir 模型表示，Cu(Ⅱ) 和 Cd(Ⅱ) 的吸附容量分别为 166.67 mg/g、27.55 mg/g。

表3 ZnS-D001 树脂上目标金属吸收的等温线吸附参数

Table 3 Adsorption isotherm parameters of target metals on ZnS-D001 resin

重金属	$Q_{e, exp}/(mg \cdot g^{-1})$	$Q_{e, cal}/(mg \cdot g^{-1})$	$K_L/(L \cdot mg^{-1})$	R^2
Cu(Ⅱ)	142.43	166.67	0.008 9	0.9984
Cd(Ⅱ)	21.45	27.55	0.0034	0.9983

铜和镉在 ZnS@D001 上的吸附动力学试验结果如图11所示。结果表明，两种杂质金属的初始吸附非常快，随后缓慢吸附接近平衡。

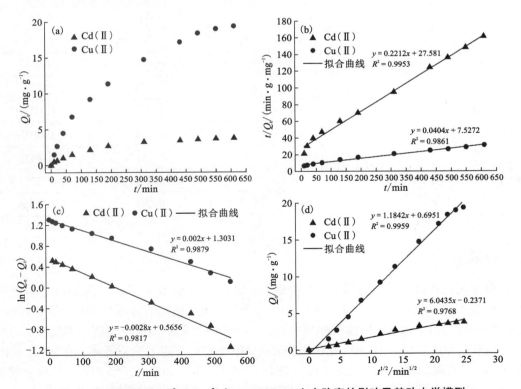

图11 吸附时间对 Cu^{2+} 和 Cd^{2+} 在 ZnS@D001 上去除率的影响及其动力学模型

Fig. 11 Effects of adsorption time on removal rate of Cu^{2+} and Cd^{2+} on ZnS@D001 and its kinetic model

两种金属离子吸附的动力学数据由下述模型表示。

伪动力学第一模型：

$$\ln(Q_e - Q_t) = \ln Q_e - \frac{k_1}{2.303}t \tag{8}$$

伪动力学第二模型：

$$\frac{t}{Q_t} = \frac{1}{k_2 Q_e^2} + \frac{t}{Q_e} \tag{9}$$

Intraparticle diffusion 模型：

$$Q_t = k_{id} t^{1/2} + C \tag{10}$$

式中：Q_t 是时间 t 时的吸附量；k_1、k_2、k_{id} 分别为对应模型的吸附动力学常数。

利用图 11(a) 的数据，按照式(8)~(10)做拟合曲线，结果如图 11(b)、图 11(c)、图 11(d) 和表 4 所示。高相关系数(表4)表明，Cu(Ⅱ)和 Cd(Ⅱ)在两种吸附剂上的吸收可以通过伪二阶模型近似表达。

表 4　目标金属吸附到 ZnS-D001 树脂上的动力学吸附参数

Table 4　Kinetic adsorption parameters of target metal adsorbed on ZnS-D001 resin

模型	模型参数	Cu(Ⅱ)	Cd(Ⅱ)
伪动力学第一模型	$Q_{e,exp}/(mg \cdot g^{-1})$	19.4	3.8
	$k_1/(10^{-2} L \cdot min^{-1})$	0.0046	0.0064
	$Q_{e,cal}/(mg \cdot g^{-1})$	20.1	3.68
	R^2	0.9879	0.9817
伪动力学第二模型	$k_2/(10^{-2} g \cdot mg^{-1} \cdot min^{-1})$	24.75	4.52
	$Q_{e,cal}/(mg \cdot g^{-1})$	0.0002	0.0017
	R^2	0.9861	0.9953
Intraparticle diffusion 模型	$k_{id}/(mg \cdot g^{-1} \cdot min^{-1/2})$	1.1842	6.0435
	C	0.6951	0.2371
	R^2	0.9959	0.9768

2.9　柱吸附

柱吸附测试在两个单独的固定床柱中进行，分别填充有 ZnS@D001 和 D001，合成水由 50 mg/L Cu^{2+} 和 50 mg/L Cd^{2+} 与 10 g/L Zn^{2+} 混合在一起，试验结果如图 12(a) 和图 12(c) 所示。由于 D001 在约 10 BV 后选择性吸附较差，两种杂质金属都迅速突破。相反，在发生突破前，ZnS@D001 可以在 50 BV 内有效去除铜[图 12(a)]。需要注意的是，在 D001 吸附的情况下，Cd^{2+} 吸附的穿透曲线上的 C/C_0 超过一个单位，即 $C/C_0>1$，可以用 Cu^{2+} 的洗脱效应来解释，即当吸附量不足时，最初吸附的一定量的 Cd^{2+} 可能被 Cu^{2+} 取代吸附位点，因为 ZnS@D001 比 Cd^{2+} 离子优先吸附 Cu^{2+}。使用 1 mol/L HNO_3 洗脱可以有效地从树脂中解吸捕获的铜和镉离子,而负载的 ZnS 会溶解成锌离子并形成 H_2S 气体。因此,ZnS@D001 杂化聚合物只能使用一次,重复吸附需要重新加载 ZnS。虽然这在实际操作中会显得相当繁琐,但在一些特殊情况下需要选择性去除浓锌溶液中的微量铜离子,本研究将提供很好的指导和帮助。

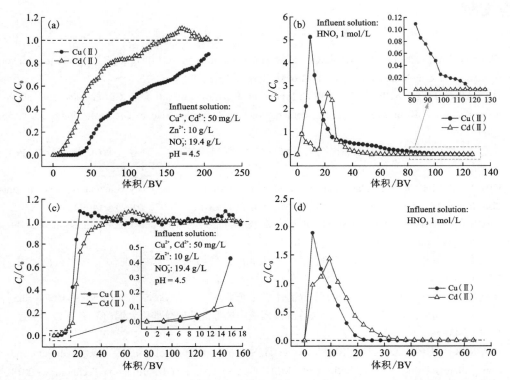

(a) Cu^{2+} 和 Cd^{2+} 吸附在 ZnS@D001 的柱吸附曲线;(b) 1 mol/L HNO_3 作为洗脱剂的 ZnS@D001 洗脱曲线;
(c) Cu^{2+} 和 Cd^{2+} 吸附在 D001 上的柱吸附曲线;(d) 1 mol/L HNO_3 作为洗脱剂的 D001 洗脱曲线。

图 12 Cu^{2+} 和 Cd^{2+} 吸附在 ZnS@D001 和 D001 上的柱吸附曲线及
以 1 mol/L HNO_3 作为洗脱剂的洗脱曲线

Fig. 12 Column adsorption curves of Cu^{2+} and Cd^{2+} adsorbed on ZnS@D001
and D001 and elution curves with 1 mol/L HNO_3 as eluant

3 结论

制备并系统评估了一种新型 ZnS-NPs@D001 聚合物负载杂化纳米复合材料,该复合材料具有良好的机械稳定性和高选择性吸附能力,可从 10 g/L Zn^{2+} 的浓锌(Ⅱ)溶液中有效去除痕量 Cu^{2+}。

ZnS@D001 对 Cu(Ⅱ)的选择性系数 $\beta_{(Cu/Zn)}$ 和平衡吸附容量均显著高于 D001。柱分离试验表明，ZnS@D001 杂化树脂在合成浓锌溶液中的除铜效率达到 75%，在 D001 树脂上没有发生镉的去除。ZnS@D001 制备简单且重现性好，有望在从浓缩锌溶液中有效去除痕量铜中进行商业应用。

参考文献

[1] ZHANG F X, YANG C H, ZHOU X J, et al. Fractional order fuzzy PID optimal control in copper removal process of zinc hydrometallurgy[J]. Hydrometallurgy, 2018, 178: 60-76.

[2] DE BARROS F C, ABREU C S, DE LEMOS L R. Separation of Cd(Ⅱ), Cu(Ⅱ) and zinc sulfate from waste produced in zinc hydrometallurgy cementation[J]. Separation Science and Technology, 2021, 56(8): 1360-1369.

[3] 秦念华, 张胜. 用低品位湿法炼锌铜渣制取海绵铜并回收锌镉[J]. 有色金属(冶炼部分), 1993(4): 24-26.

[4] DIB A, MAKHLOUFI L. Mass transfer correlation of removal of nickel by cementation onto rotating zinc disc in industrial zinc sulfate solutions[J]. Minerals Engineering, 2007, 20(2): 146-151.

[5] GROS F, BAUP S, AUROUSSEAU M. Copper cementation on zinc and iron mixtures: part 2: Fluidized bed configuration[J]. Hydrometallurgy, 2011, 106: 119-126.

[6] 牛文敏, 马高峰, 周冲冲, 等. 锰粉代替锌粉去除硫酸锌溶液中铜镉研究[J]. 湖南有色金属, 2021, 37(2): 32-34.

[7] 杨佳棋, 李立清, 冯罗, 等. 铜离子萃取剂的研究进展[J]. 电镀与涂饰, 2020, 39(9): 577-585.

[8] 李衍林, 世仙果, 李兴彬, 等. 从湿法炼锌浸出液中选择性萃取分离回收铜[J]. 有色金属工程, 2021, 11(6): 43-48.

[9] JHA M K, GUPTA D, CHOUBEY P K, et al. Solvent extraction of copper, zinc, cadmium and nickel from sulfate solution in mixer settler unit (MSU)[J]. Separation and Purification Technology, 2014, 122: 119-127.

[10] DEVI N, CHEMISTRY D O. Solvent extraction and separation of copper from base metals using bifunctional ionic liquid from sulfate medium[J]. Transactions of Nonferrous Metals Society of China, 2016, 26(3): 874-881.

[11] LAATIKAINEN K, LAHTINEN M, LAATIKAINEN M, et al. Copper removal by chelating adsorption in solution purification of hydrometallurgical zinc production[J]. Hydrometallurgy, 2010, 104: 14-19.

[12] SIROLA K, LAATIKAINEN M, LAHTINEN M, et al. Removal of copper and nickel from concentrated $ZnSO_4$, solutions with silica-supported chelating adsorbents[J]. Separation and Purification Technology, 2008, 64: 88-100.

[13] 畅如. 软化树脂吸附重金属废水的研究[J]. 化工设计通信, 2021, 47(7): 79-81.

[14] 韩义忠, 谢祥添, 贾兴州. 锑基吸附剂净化铜电解液试验研究[J]. 中国有色冶金, 2021, 50(3): 16-20.

[15] CORTINA J L, MIRALLES N, SASTRE A M, et al. Solid-liquid extraction studies of divalent metals with impregnated resins containing mixtures of organophosphorus extractants[J]. Reactive & Functional Polymers, 1997, 32: 221-229.

[16] LIU R X, TANG H X, ZHANG B W. Removal of Cu(Ⅱ), Zn(Ⅱ), Cd(Ⅱ) and Hg(Ⅱ) from waste water by poly(acrylaminophosphonic)-type chelating fiber[J]. Chemosphere, 1997, 38(13): 3169-3179.

[17] Liu D, Li Z, Zhu Y, et al. Recycled chitosan nanofibril as an effective Cu(Ⅱ), Pb(Ⅱ) and Cd(Ⅱ) ionic chelating agent: Adsorption and desorption performance[J]. Carbohydrate Polymers, 2014, 111(1): 469-476.

[18] DINU M V, DRAGAN E S, TROCHIMCZUK A W. Sorption of Pb(Ⅱ), Cd(Ⅱ) and Zn(Ⅱ) by iminodiacetate chelating resins in non-competitive and competitive conditions[J]. Desalination, 2009, 249: 374-379.

[19] IZATT R M, IZATT S R, BRUENING R L, et al. Challenges to achievement of metal sustainability in our high-tech society[J]. Chemical Society Reviews, 2014, 43: 2451-2475.

[20] TOPUZ B, MACIT M. Solid phase extraction and preconcentration of Cu(Ⅱ), Pb(Ⅱ), and Ni(Ⅱ) in environmental samples on chemically modified amberlite xad-4 with a proper schiff base[J]. Environmental Monitoring and Assessment, 2011, 173(1): 709-722.

[21] 胡耀强, 郭敏, 叶秀深, 等. 磺酸基树脂对铜、铅、镉离子的吸附行为[J]. 聊城大学学报(自然科学版), 2021, 34(3): 48-54.

[22] PLIEGER P G, TASKER P A, GALBRAITH S G. Zwitterionic macrocyclic metal sulfate extractants containing 3-dialkylaminomethylsalicylaldimine units[J]. Dalton Transactions, 2004(2): 313-318.

[23] ZHAI Y H, LIU Y W, CHANG X J, et al. Selective solid-phase extraction of trace cadmium(Ⅱ) with an ionic imprinted polymer prepared from a dual-ligand monomer[J]. Analytica Chimica Acta, 2007, 593(1): 123-128.

[24] SARKAR M, DATA P K, DAS M. Equilibrium studies on the optimization of solid-phase extraction using modified silica gel for removal, recovery, and enrichment prior to the determination of some metal ions from aqueous samples of different origin[J]. Industrial & Engineering Chemistry Research, 2005, 41(26): 307-328.

[25] 韩乐, 裴婉莹, 苏毅, 等. 生物吸附剂对污水中铜离子的吸附研究进展[J]. 化工新型材料, 2020, 48(5): 237-240.

[26] EBRAHEEM K A K, HAMDI S T. Synthesis and properties of a copper selective chelating resin containing a salicylaldoxime group[J]. Reactive and Functional Polymers, 1997, 34(1): 5-10.

[27] SUZUKI T M, BOMANI J O, MATSUNAGA H, et al. Preparation of porous resin loaded with crystalline hydrous zirconium oxide and its application to the removal of arsenic[J]. Reactive & Functional Polymers, 2000, 43: 165-172.

[28] SUZUKI T M, KOBAYASHI S, TANAKA D A P, et al. Separation and concentration of trace Pb(Ⅱ) by the porous resin loaded with α-zirconium phosphate crystals[J]. Reactive & Functional Polymers, 2004, 58: 131-138.

[29] GERMAN M, SEINGHENG H, SENGUPTA A K. Mitigating arsenic crisis in the developing world: Role of robust, reusable and selective hybrid anion exchanger (HAIX)[J]. Science of the Total Environment, 2014, 488-489: 547-553.

[30] AN B, FU Z, XIONG Z, et al. Synthesis and characterization of a new class of polymeric ligand exchangers for selective removal of arsenate from drinking water[J]. Reactive & Functional Polymers, 2010, 70(8): 497-507.

[31] PADUNGTHON S, GERMAN M, WIRIYATHAMCHAROEN S, et al. Polymeric anion exchanger supported hydrated Zr(Ⅳ) oxide nanoparticles: A reusable hybrid sorbent for selective trace arsenic removal[J]. Reactive & Functional Polymers, 2015, 93, 84-94.

[32] SENGUPTA A K, ROY T, JESSEN D. Modified anion-exchange resins for improved chromate selectivity

and increased efficiency of regeneration[J]. Reactive & Functional Polymers, 1988, 9: 293-299.

[33] SENGUPTA D, BASU B. An efficient metal-free synthesis of organic disulfides from thiocyanates using poly-ionic resin hydroxide in aqueous medium[J]. Tetrahedron Letters, 2013, 54(18): 2277-2281.

[34] CHEN L, ZHAO X, PAN B C, et al. Preferable removal of phosphate from water using hydrous zirconium oxide-based nanocomposite of high stability[J]. Journal of Hazardous Materials, 2015, 284, 35-42.

[35] SU Q, PAN B C, WAN S L, et al. Use of hydrous manganese dioxide as a potential sorbent for selective removal of lead, cadmium, and zinc ions from water[J]. Journal of Colloid and Interface Science, 2010, 349(2): 607-612.

[36] WANG J, ZHANG S J, PAN B C, et al. Hydrous ferric oxide-resin nanocomposites of tunable structure for arsenite removal: Effect of the host pore structure [J]. Journal of Hazardous Materials, 2011, 198: 241-246.

[37] XU Z W, ZHANG W M, PAN B C, et al. Application of the Polanyi potential theory to phthalates adsorption from aqueous solution with hyper-cross-linked polymer resins[J]. Journal of Colloid and Interface Science, 2008, 319(2): 392-397.

[38] JIANG Z M, LV L, ZHANG W M, et al. Nitrate reduction using nanosized zero-valent iron supported by polystyrene resins: Role of surface functional groups[J]. Water Research, 2011, 45(6): 2191-2198.

[39] PAN B J, QIU H, PAN B C, et al. Highly efficient removal of heavy metals by polymer-supported nanosized hydrated Fe(Ⅲ) oxides: Behavior and XPS study[J]. Water Research, 2010, 44(3): 815-824.

[40] 吴迪, 郭彦秀, 赵欣, 等. 铅对铁铝复合氧化物吸附铜的影响[J]. 天津农业科学, 2019, 25(3): 84-87.

高冰镍浸出渣冶金过程多金属走向行为探究

摘　要：高冰镍浸出渣是镍湿法冶金过程的重要中间产物，含有铜、镍、钴和金、银、铂、钯等有价金属，具有重要的综合回收价值。探究高冰镍浸出渣冶金过程多金属的走向行为对于提高金属综合回收率、优化系统物料平衡和推动工艺技术升级具有重要意义。本文基于新疆阜康冶炼厂高冰镍浸出渣的典型冶金工艺，结合文献调研和工业生产实践，阐释高冰镍浸出渣中主要金属铜、镍、钴、铁和贵金属金、银、铂、钯的迁移走向，揭示各金属形态变化历程，分析各金属在渣相和液相的分配行为，为工艺设计和优化提供支撑。

关键词：高冰镍浸出渣；镍；钴；铜；贵金属

Exploration on multi metal distribution behavior in the metallurgical process of leaching residue of high nickel matte

Abstract：High nickel matte leaching residue is an important intermediate product of nickel hydrometallurgy process. It contains valuable metals such as copper, nickel, cobalt, gold, silver, platinum and palladium, which has important comprehensive recovery value. Exploring distribution behavior of multimetals in metallurgical process of high nickel matte leaching residue has great significance for improving comprehensive metal recovery, optimizing material balance of the system and promoting upgrading of process technology. Based on typical metallurgical process of high nickel matteleaching residue from Fukang Refinery in Xinjiang, combined with literature investigation and industrial production practice, migration trend of main metals of copper, nickel, cobalt, iron and precious metals gold, silver, platinum and palladium in high nickel matte leaching residue were explained, morphological change process of each metal was revealed, and distribution behavior of each metal in residue and solution was analyzed. It can provide support for process design and optimization.

Keywords：leaching residue of high nickel matte；nickel；cobalt；copper；precious metals

硫化镍矿是一种重要的镍矿资源，工业上处理硫化镍矿通常采用选矿富集得到镍精矿，然后经过火法熔炼除去镍精矿中的大量脉石、铁与一部分硫，产出高冰镍，再进一步采用磨

本文发表在《有色金属（冶炼部分）》，2021（12）：20-27。作者：谢铿，王海北，马育新，李渭鹏。

浮-硫化镍阳极电解精炼工艺或湿法冶金工艺处理[1]。磨浮-硫化镍阳极电解精炼工艺是利用高冰镍熔体在铸锭缓冷时各组分相互溶解度的差异，分别生成具有不同化学成分的硫化镍和硫化铜晶粒，以及存在于这些晶粒间的镍-铜合金相，用磁选方法选出镍-铜合金，再用浮选方法分离出硫化镍精矿和硫化铜精矿，得到的二次硫化镍精矿再铸锭成阳极电解。该工艺于 20 世纪 50 年代首先在加拿大国际镍公司汤普逊镍精炼厂实现工业化，并应用于我国金川公司和成都电冶厂等[2]，生产实践显示该工艺存在镍铜分离不彻底、贵金属分散、金属回收率低、流程冗长、能耗高等缺点。与磨浮-硫化镍阳极电解精炼工艺相比，湿法冶金工艺采用选择性浸出代替高冰镍磁选-浮选、镍熔铸、一次合金硫化熔炼、酸性电解造液等生产环节，大大缩短了工艺流程，增强了对原料的适应性，提高了金属回收率，减小了溶液除杂负荷，提高了自动化程度，而且降低了对环境的污染。因此，自 20 世纪 60 年代中叶以来，硫酸浸出、氯化浸出等湿法冶金工艺得到了快速发展并广泛应用于国内外镍生产企业[3-5]，如芬兰奥托昆普公司哈贾瓦尔塔精炼厂、南非英帕拉铂公司斯普林精炼厂、加拿大鹰桥公司克里斯蒂安松厂、法国勒阿弗尔厂、日本新居滨厂、美国阿马克斯公司镍港精炼厂，以及我国的新疆阜康冶炼厂、吉林镍业公司第二精炼厂、金川公司 3 万吨镍/年加压浸出项目等，成为高冰镍处理及镍产品生产的主要工艺。

矿冶科技集团有限公司（原北京矿冶研究总院）围绕高冰镍的湿法冶金工艺进行了大量研发工作，20 世纪 80 年代，先后完成了高冰镍硫酸浸出试验研究和开发了高冰镍选择性浸出工艺，提出了硫酸常压浸出和高温加压浸出联合流程，应用于新疆喀拉通克铜镍矿冶炼-水淬高冰镍的处理。依托该工艺技术，我国第一家镍加压浸出厂-阜康冶炼厂于 1993 年在天山脚下建成，结束了新疆不产镍的历史，一举成为中国第二大电解镍生产商。阜康冶炼厂最初的工艺中，高冰镍经硫酸选择性浸出后，再通过黑镍除钴和电积等工序生产电镍[6]，选择性浸出渣富集了几乎全部的铜、铁、硫和贵金属，因渣中铜含量为 50% 以上，仅作为副产铜渣直接对外低价销售[7-9]，含有的其他有价金属和贵金属未计价，造成金属综合回收利用率低、企业资源严重浪费和经济损失。为了充分回收高冰镍浸出渣中的镍、铜和贵金属，以矿冶科技集团有限公司为代表的科研院所与阜康冶炼厂开展了精诚合作和技术攻关，通过大量的工艺试验研究、技术方案论证、可行性研究、工程设计优化等工作[10,11]，提出了浸出渣氧化焙烧-酸浸-电积的工艺流程，推动了钴回收车间和铜渣综合利用技改工程的实施，实现了企业产品从单一的电解镍生产发展为铜、钴和金、银、铂、钯等贵金属综合回收的多品种生产线，取得了显著的经济和社会效益。

近年来，随着环保政策愈趋严格以及冶炼工艺绿色低碳转型的需要，对高冰镍浸出渣综合利用技术又提出了新的要求，在该形势下，科技工作者进一步提出了氧压浸出替代回转窑焙烧、密闭加压降镍、酸碱介质循环等技术对现有工艺进行优化升级和提质增效。实现新工艺与现有镍、铜主系统的良好衔接，提高金属综合回收率，首先需要查明现有浸出渣冶金过程多金属的走向行为。为此，本文以阜康冶炼厂典型高冰镍浸出渣为对象，在文献调研和工业生产实践的基础上，探究高冰镍浸出渣中铜、镍、钴和贵金属等在冶金过程中的迁移走向，阐释有价金属形态变化历程，分析主要金属在渣相和液相的分配行为，为工艺设计和优化提供支撑。

1 高冰镍浸出渣来源及组成

新疆阜康冶炼厂的高冰镍物料主要来自喀拉通克矿和哈密众鑫矿业[12,13],不同批次高冰镍成分波动较大。比较两种高冰镍成分可知,来自喀拉通克矿的高冰镍[3]具有镍低铜高的特点,其组成大致含 Ni 29%~35%、Cu 45%~53%、Fe 0.5%~1.8%、Co 0.15%~0.28%、S 17.5%~19.5%、Au 3~4 g/t、Ag 270~310 g/t、Pt 1.5~1.8 g/t、Pd 1.5~1.8 g/t;来自哈密众鑫矿业的高冰镍[14]具有镍高铜低的特点,其组成大致含 Ni 67%~72%、Cu 6%~10%、Fe 0.7%~1.6%、Co 0.9%~1.2%、S 17%~20%、Au 2~3 g/t、Ag 200~250 g/t、Pt 1.1~1.3 g/t、Pd 1.1~1.3 g/t。高冰镍中的镍主要以 Ni_3S_2 形式存在,部分呈合金相,微量以 NiO 形式存在;铜主要以 Cu_2S 形式存在,少量呈合金相;钴和铁主要以硫化相(CoS、FeS)和合金相形式存在。将两种高冰镍按一定比例混合后,用作浸出工序的物料,混合矿组成大致含 Ni 41%~45%、Cu 32%~37%、Fe 0.8%~1.4%、Co 0.4%~0.7%、S 18.1%~18.7%、Au 2.5~4 g/t、Ag 250~300 g/t、Pt 1.2~1.6 g/t、Pd 1.2~1.6 g/t。典型高冰镍组成如表1所示。

表1 典型高冰镍主要成分/%
Table 1 Main composition of typical high nickel matte/%

	Ni	Cu	Fe	Co	S	Au*	Ag*	Pt*	Pd*
喀拉通克矿	31.82	48.97	1.17	0.22	18.43	3.62	290	1.51	1.51
哈密众鑫矿业	69.11	7.96	1.11	0.93	21.04	2.33	210	1.11	1.16
混合矿	42.77	34.06	1.13	0.48	18.13	2.85	235	1.37	1.39

*单位: g/t。

高冰镍球磨合格后,经浆化配料,进行第一段常压浸出,浸出液经过滤、净化后输送至电解车间生产电镍;第一段常压浸出渣经浆化配料,进行第二段常压浸出,浸出液返回第一段常压浸出或加压浸出配料;第二段常压浸出渣经浆化配料,进行加压浸出[15]。高冰镍硫酸选择性浸出工艺流程如图1所示。

第一段常压浸出开始后,合金相金属 Me(Ni、Co、Cu、Fe 等)与硫酸发生式(1)~(2)反应进入溶液。

$$Me + H_2SO_4 = MeSO_4 + H_2 \tag{1}$$

$$Me + H_2SO_4 + \frac{1}{2}O_2 = MeSO_4 + H_2O \tag{2}$$

$$Me + CuSO_4 = MeSO_4 + Cu \tag{3}$$

$$2Cu + \frac{1}{2}O_2 = Cu_2O \tag{4}$$

铜与合金相中的镍起反应,反应过程中有单质铜生成,但立即被氧化成氧化亚铜,如反应式(3)和(4)所示。大部分合金相被浸出后,$CuSO_4$ 与 Ni_3S_2 进行反应。同时,部分金属硫

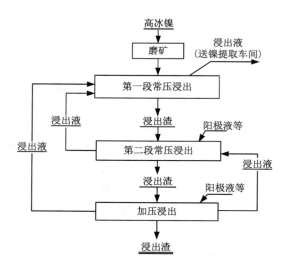

图 1　高冰镍硫酸选择性浸出工艺流程图
Figure 1　Flow sheet for sulfuric acid leaching process of high nickel matte

化物被浸出,发生式(5)~(8)反应。硫化铁在强酸的作用下产生大量的硫化氢,硫化氢继而与浸出液中的镍钴等发生反应,生成硫化物沉淀。

$$Ni_3S_2 + H_2SO_4 + \frac{1}{2}O_2 = NiSO_4 + 2NiS + H_2O \quad (5)$$

$$FeS + H_2SO_4 = FeSO_4 + H_2S \quad (6)$$

$$2FeSO_4 + H_2SO_4 + \frac{1}{2}O_2 = Fe_2(SO_4)_3 + H_2O \quad (7)$$

$$H_2S + MeSO_4 = MeS\downarrow + H_2SO_4 \quad (8)$$

随着酸不断消耗,溶液 pH 增大,当 pH 上升至 3.5~4.5,溶液中的铁离子水解,沉淀生成针铁矿,进入第一段常压浸出渣中[16]。

$$Fe_2(SO_4)_3 + 4H_2O = 3H_2SO_4 + 2FeOOH\downarrow \quad (9)$$

当 pH 继续增大至 5 以上,溶液中硫酸铜发生水解。

$$(1+m)CuSO_4 + 2mH_2O = CuSO_4 \cdot mCu(OH)_2\downarrow + mH_2SO_4 \quad (10)$$

控制第一段常压浸出温度 65~80 ℃、终点 pH 5.8~6.2,浸出渣含 Ni 29%~40%、Cu 38%~45%、Fe 0.8%~1.4%、S 15.7%~16.9%,镍以少量的合金相、Ni_3S_2、NiS、NiO 的形式存在,铜以 Cu_2O、CuO、CuS 的形式存在,铁以 FeS、FeOOH 的形式存在。浸出液中含 Ni 90~105 g/L、Co 0.4~0.75 g/L,铜铁含量很低、均小于 0.01 g/L,利于后续生产电镍。

第二段常压浸出进一步溶出剩余合金相金属以及 Ni_3S_2,同时浸出少部分铜返回至第一段常压浸出置换镍,且尽可能不浸出铁而是将其抑制于浸出渣中[17]。控制第二段常压浸出终点 pH 3~5,80% 合金相金属被浸出,镍浸出率达到 60%~70%,浸出渣含 Ni 25%~35%、Cu 40%~48%、S 15.5%~17%,主要为疏松多孔的海绵状的铜硫化物,少数包裹残余硫化镍微粒;浸出液含 Ni 75~85 g/L、ρ_{Cu}<0.5 g/L、Fe 0.2~0.7 g/L、ρ_{Co}<0.5 g/L。第二段常压浸出是在最初的一段常压浸出一段加压浸出工艺上增加的工序,它既减轻了加压浸出的负荷,

又通过提高浸出液的循环铜量和降低亚铁离子量更好地满足了生产需求。

加压浸出是用阳极液最大化地将第二段常压浸出渣中残余的合金相及硫化相中的可溶镍钴浸出，达到尽可能高的浸出率，同时浸出少部分铜返回铜至常压浸出使用，另外将大部分铜、铁、硫及贵金属等富集于终渣中[18]，从而达到选择浸出的目的。通过加压浸出，剩余镍矿物大量溶解，而铜矿物在硫酸作用下反应生成 $CuSO_4$ 并与渣中的 NiS 发生交互反应，使镍被大量浸出[19]。

$$MeS + 2O_2 = MeSO_4 \tag{11}$$

$$Cu_2O + H_2SO_4 = Cu + CuSO_4 + H_2O \tag{12}$$

$$Ni_3S_2 + H_2SO_4 + Cu_2O = NiSO_4 + 2NiS + 2Cu + H_2O \tag{13}$$

$$Cu_2S + H_2SO_4 + \frac{1}{2}O_2 = CuS + CuSO_4 + H_2O \tag{14}$$

$$MeS + CuSO_4 = CuS + MeSO_4 \tag{15}$$

由上述反应式可知，在加压浸出过程中铜的各种反应主要是由硫酸参与的反应，镍的浸出主要由铜的参加完成的，其中的铜与镍的交互反应是降低渣中镍含量较重要的一个反应。控制加压浸出温度 150~160 ℃、终点 pH 1.8~2.4，得到的加压浸出渣通常含 Ni 5%~9%、Cu 43%~53%、Fe 1.1%~1.5%、w_{Co}<0.15%、S 20%~24%、Au 4.1~5.2 g/t、Ag 350~440 g/t、Pt 2~3 g/t、Pd 2~3 g/t。浸出渣呈灰黑色，颗粒松散，含有的主要金属元素为铜和镍，其他元素为硫，同时含有少量脉石元素。渣中 70% 以上的铜和镍以硫化物存在，钴以 CoS 存在，铁以 Fe_3O_4 存在；主要矿物为 Cu_2S（包括辉铜矿、铜蓝）、Cu_2O、碱式硫酸铜、Ni_2S_3、NiS、CuS 和碱式硫酸镍。某典型高冰镍选择性浸出渣组成如表 2 所示。

表 2 典型高冰镍浸出渣主要成分

Table 2 Main composition of leaching residue of typical high nickel matte　　　　%

元素	Ni	Cu	Fe	Co	S	Au*	Ag*	Pt*	Pd*	O	V	Ti
含量	7.64	53.12	1.25	0.12	21.86	4.64	372	2.51	2.43	22.3	0.02	0.02
元素	Ba	Ca	Na	Mg	Al	Si	Se	K	Pb	As	Cr	Sr
含量	4.60	0.17	1.23	0.22	0.20	0.56	0.03	0.05	0.02	0.08	0.01	0.02

*单位：g/t。

2　高冰镍浸出渣冶金工艺

采用氧化焙烧-浸出-电积工艺[20-23]从高冰镍浸出渣中提取铜，生产电铜并得到贵金属富集渣，工艺流程如图 2 所示。高冰镍浸出渣经制粒后，加入沸腾焙烧炉，鼓入空气氧化焙烧，尽可能除去渣中的硫，使硫化物转变为氧化物。焙烧产出的二氧化硫烟气经收尘-净化-干燥-转化-吸收后制备工业硫酸，尾气碱吸后达标排放。产出的焙砂用作浸出物料，加入铜电积后液进行浸出，浸出其中的铜。大部分浸出液经板框压滤、调配槽调配后，输送、分流至电解槽生产阴极铜。浸铜后渣经干燥后送往回转窑进行硫酸化焙烧，然后进行酸化浸出，浸出其中的铜、镍、铁。酸化浸出渣富集了金、银、铂、钯，即为贵金属渣，送贵金属提取工序。酸化浸出产生的浸出液送沉银工序，沉银产生的银渣送贵金属车间；沉银后液经中和-氧化除铁工序产生铁渣，铁渣送堆场堆存并定期返回喀拉通克矿熔炼工序，除铁后液返回镍提取系统。

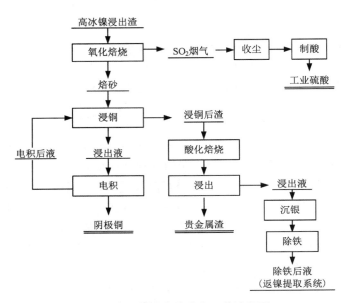

图 2　高冰镍浸出渣冶金工艺流程图

Figure 2　Flow sheet for metallurgical process of leaching residue of high nickel matte

2.1　高冰镍浸出渣氧化焙烧-浸出-电积生产阴极铜

焙烧的目的是利用空气中的氧将高冰镍中不易浸出的铜硫化物氧化成易被硫酸溶出的氧化铜[24]。高冰镍浸出渣全氧化沸腾焙烧一般在820~900 ℃的温度下进行，铜和镍的硫化物与氧气反应生成氧化铜及氧化镍[21, 25, 26]，并释放二氧化硫气体，部分金属硫化物与二氧化硫、氧气反应生成金属硫酸盐。主要发生的反应[23, 27]有：

$$Cu_2S + 2O_2 = 2CuO + SO_2 \tag{16}$$

$$CuS + \frac{3}{2}O_2 = CuO + SO_2 \tag{17}$$

$$Cu_2S + \frac{5}{2}O_2 = CuSO_4 + CuO \tag{18}$$

$$Cu_2S + 2CuSO_4 = 2Cu_2O + 3SO_2 \tag{19}$$

$$Cu_2O + 4CuSO_4 = 3CuO \cdot CuSO_4 + SO_2 \tag{20}$$

$$Ni_3S_2 + \frac{7}{2}O_2 = 3NiO + 2SO_2 \tag{21}$$

渣中氧化铁可以和主要金属氧化物反应生成具有尖晶石结构的亚铁酸盐：

$$3CuO + 2Fe_3O_4 + \frac{1}{2}O_2 = 3CuFe_2O_4 \tag{22}$$

$$3NiO + 2Fe_3O_4 + \frac{1}{2}O_2 =\!=\!= 3NiFe_2O_4 \tag{23}$$

焙烧的最终产物为铜和镍的氧化物和亚铁酸盐及 SO_2。通过控制浸铜后渣的水分、粒度、鼓风量、入炉物料量及温度等手段，可实现脱硫率大于 96%，得到焙砂组成大致含 Cu 48%~55%、Ni 7%~10%、Fe 1.2%~1.6%、S 2.9%~4.4%、Au 4.1~5.2 g/t、Ag 350~450 g/t、Pt 2~4 g/t、Pd 2~4 g/t。

浸出过程利用铜电积后液中的硫酸溶解焙砂中的氧化铜[28]：

$$CuO + H_2SO_4 =\!=\!= CuSO_4 + H_2O \tag{24}$$

控制浸出工艺参数为液固比(10~15)∶1、浸出温度 65~85 ℃、浸出时间 80 min、终酸浓度 120~150 g/L，浸出渣率低，为 7%~9%，铜浸出率为 97%~99%；大部分的镍和铁保留在浸铜后渣中，镍浸出率为 30%~40%、铁浸出率为 10%~20%。浸出液组成大致含 Ni 40~45 g/L、Cu 65~85 g/L、Fe 2.1~3.4 g/L。几乎全部贵金属富集于浸铜后渣中。得到浸铜后渣组成[29,30]大致含 Cu 7%~11%、Ni 18%~23%、w_{Co}<0.1%、Fe 1.5%~3.9%、Au 16~21 g/t、Ag 700~1550 g/t、Pt 7.9~9.5 g/t、Pd 7.9~9.5 g/t。某典型浸铜后渣组成如表 3 所示。浸铜后渣中镍 70%以上存在于铁矿物和硅酸盐中，另外还有少量以硫酸镍、氧化镍和硫化镍形式存在；铜主要存在于铁矿物中和以水溶性硫酸铜形式存在。

表 3 典型浸铜后渣主要成分

Table 3 Main composition of leaching residue after Cu extraction   %

元素	Ni	Cu	Fe	Co	S	Au*	Ag*	Pt*	Pd*	Mn	V	Ti
质量分数	20.24	7.32	3.29	0.07	7.86	17.54	1170	8.21	8.13	0.01	0.02	0.02
元素	Ba	Ca	Zn	Mg	Al	Sb	Sn	Cd	Pb	As	Cr	Sr
质量分数	7.64	0.47	0.02	0.16	0.73	0.02	0.01	0.02	0.13	0.13	0.01	0.12

*单位：g/t。

浸出液过滤后，与电积后液和硫酸铜溶解液调配成含 Cu 35~45 g/L 的电积液[31]，然后打入电解槽中，采用不溶阳极电积生产阴极铜。电积后液组成大致含 Ni 40~45 g/L、Cu 35~45 g/L、Fe 1.8~3.0 g/L、H_2SO_4 120~150 g/L。电积后液一部分返回调配，并入阴极铜生产的再循环溶液，一部分返回焙砂浸出工序。铜电积采用铅钙锶不溶阳极，种板为钛板，选用硬聚氯乙烯塑料压条封边，沉积 12~24 h 后的剥离铜皮，压纹钉耳后下入生产槽，生产周期为 5~7 d，出槽后经烫洗得到阴极铜可达到阴极铜 GB/F 467—2010 的标准。

2.2 浸铜后渣酸化焙烧-浸出提取镍铜与富集贵金属

浸铜后渣经晾晒制粒、送回转窑浆化槽配料，入回转窑在 300~420 ℃条件下进行酸化焙烧。酸化焙烧是固相与气相间的多相反应，一般采用空气或富氧空气为氧化剂。硫化镍矿的硫酸化焙烧本质是矿物中金属选择性氧化的过程，使镍、钴和铜等有价金属转化为硫酸盐，而铁则转化为氧化铁[32]，进而通过浆化浸出实现镍钴铜和铁分离。焙烧后得到酸化焙砂大致主要含 Cu 5%~8%、Ni 12%~16%、Fe 1%~2.5%、$w_{Co}<0.1\%$、S 2.5%~4.5%。

所产出的酸化焙砂进行浆化浸出，控制浸出工艺参数为液固比(5~6):1、浸出温度~80 ℃、浸出时间 2.5~3 h、终点 pH ~1.0，浸出后矿浆进入板框液固分离并洗涤产出贵金属渣。贵金属渣组成大致含 Cu 2.5%~4%、Ni 6%~9%、Fe 6.5%~11%、Au 35~45 g/t、Ag 900~3000 g/t、Pt 15~18 g/t、Pd 14~16 g/t。滤液采用沉银和除铁工序处理，沉银产生的银渣送贵金属车间；除铁产生含 Cu 2%~4%、Ni 2%~3% 和 Fe 20%~28% 的铁渣返回喀拉通克铜镍矿熔炼工序，除铁后液含 Cu 34~43 g/L、Ni 25~34 g/L、Fe 1.1~1.9 g/L，除铁后液与部分电积后液混合送镍提取系统。

2.3 贵金属渣综合回收

贵金属渣经电炉熔炼，铁、钙氧化物和石英结合形成硅酸盐熔炼渣，将大部分脉石及铁去除，同时形成镍、铜锍相，通常称为冰铜[33]。反应式如下。

$$2FeS + 3O_2 = 2FeO + 2SO_2 \quad (25)$$

$$2FeO + SiO_2 = (FeO)_2 \cdot SiO_2 \quad (26)$$

$$CaO + SiO_2 = CaO \cdot SiO_2 \quad (27)$$

$$3Fe + 3NiS + 2S = 3FeS + Ni_3S_2 \quad (28)$$

$$2CuFeS_2 = Cu_2S + 2FeS + S \quad (29)$$

得到熔炼渣含 Cu 0.1%~1%、Ni 0.3%~1.8%、Fe 2%~5%、Ag 100~200 g/t、Au 1.5~4 g/t、Pt 0.5~1.5 g/t、Pd 0.5~1.5 g/t。金、铂、钯等贵金属被捕集入冰铜，得到冰铜主要由 Cu_2S、Ni_3S_2、NiS 及 FeS 组成，含 Cu 16%~21%、Ni 35%~48%、Fe 8%~14%、Ag 5000~8000 g/t、Au 300~600 g/t、Pt 100~250 g/t、Pd 100~250 g/t。冰铜经过破碎球磨得到 180~200 目的粉矿，粉矿进行硫酸化焙烧-浸出，发生以下反应：

$$Cu + 2H_2SO_4 = CuSO_4 + 2H_2O + SO_2 \uparrow \quad (30)$$

$$Ni + 2H_2SO_4 = NiSO_4 + 2H_2O + SO_2 \uparrow \quad (31)$$

$$Cu_2S + 6H_2SO_4 = 2CuSO_4 + 6H_2O + 5SO_2 \uparrow \quad (32)$$

$$Ni_3S_2 + 10H_2SO_4 = 3NiSO_4 + 10H_2O + 9SO_2 \uparrow \quad (33)$$

$$NiS + 2H_2SO_4 = NiSO_4 + 2H_2O + SO_2 \uparrow \quad (34)$$

$$2Ag + 2H_2SO_4 = Ag_2SO_4 + 2H_2O + SO_2 \uparrow \quad (35)$$

通过多次酸化焙烧和浸出，绝大多数镍、铜、铁形成硫酸盐被浸出进入溶液，控制适当温度和料酸比，银的浸出率可达 95%，实现了银与金、铂、钯的初步分离。浸出液中银可采用铜板置换制得银品位大于 97% 的工业粗银粉，置换后液 $\rho_{Ag}<0.001$ g/L，可返回镍提取系统利用。焙烧-浸出后，浸出液含 Cu 15~60 g/L、Ni 10~50 g/L、Fe 1~15 g/L，贵金属富集渣含 Au 1500~2000 g/t、Pt 500~1500 g/t、Pd 500~1500 g/t、Ag 1.2%~2.5%，铜、镍、铁金属质量分数均为 0.8%~1.2%。贵金属富集渣再经一次氯化浸出，Au 浸出率约为 98.5%，铂、钯浸出率>92%，一次氯浸渣中 $\rho_{Au}<300$ g/t，一次氯浸液经铜板置换得到贵金属精矿，Au、Pt、Pd 三者总含量级为 40%，经洗涤后进入二次氯浸，氯浸后得到贵液。贵液过滤后用草酸选择性还原金[反应式(36)]，沉金后液用氯化铵沉铂[反应式(37)]，铂盐经煅烧得到海绵铂。沉铂后液用氨水络合沉钯[反应式(38)]，钯盐用水合肼还原为钯粉。得到金、铂和

钯产品的品位都能够达到98%以上。

$$2HAuCl_4 + 3H_2C_2O_4 = 2Au\downarrow + 8HCl + 6CO_2\uparrow \qquad (36)$$

$$H_2PtCl_6 + 2NH_4Cl = (NH_4)_2PtCl_6\downarrow + 2HCl \qquad (37)$$

$$H_2PdCl_4 + 4NH_4OH = Pd(NH_3)_4Cl_2\downarrow + 4H_2O + 2HCl \qquad (38)$$

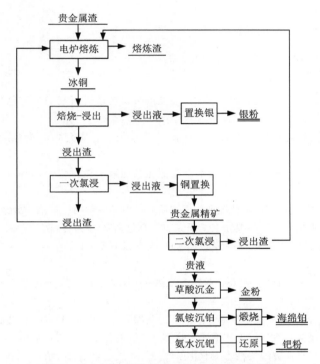

图3 贵金属渣综合回收金银铂钯工艺流程图

Figure 3 Flow sheet for metallurgical process of Au, Ag, Pt and Pd recovery from precious metals-bearing residue

3 结论

（1）高冰镍浸出渣富含铜、镍、钴和金、银、铂、钯等贵金属，经济价值巨大。新疆阜康冶炼厂采用高冰镍浸出渣氧化焙烧—浸出—电积提铜、浸铜后渣酸化焙烧—浸出提取镍铜和富集贵金属、贵金属渣电炉熔炼—焙烧—浸出—氯浸—置换—分步沉淀等组合工艺，通过控制各工序技术参数，实现了目标有价金属的定向分离、精准富集和综合回收，得到阴极铜、银粉、金粉、海绵铂、钯粉等产品，同时含镍钴的溶液可返回主系统进一步提取镍和钴。

（2）合理的物料调配和循环利用制度有利于调节体系物料平衡和酸平衡、调整目标金属的走向和实现目标金属的充分提取。例如：由于高冰镍浸出渣中含镍较高，焙烧-浸出提铜过程中，大量镍随铜一起浸出，为了回收镍与利用余酸，可将一部分铜电积后液返回镍系统配料；贵金属回收过程中，一次氯浸渣和二次氯浸渣可返回电炉熔炼，渣中残留贵金属在体系中不断被累积提取而不损失。

(3)在环保政策愈趋严格以及冶炼工艺绿色低碳转型的新形势下,迫切要求对高冰镍浸出渣冶金工艺进行技术创新,取消回转窑焙烧等污染大、能耗高的工序,综合应用加压浸出等清洁高效冶金技术最新成果,统筹优化镍、铜、钴、金、银、铂、钯提取工艺和产品结构,实现高质量发展目标。

参考文献

[1] 黄振华,林江顺,王忠,等.对我国电镍生产工艺的分析与改进研究[J].有色金属(冶炼部分),2013(6):1-5.

[2] 赵磊,刘大星.高铜硫化镍阳极电解工艺中阳极液净化工艺研究[J].有色金属(冶炼部分),2012(10):7-9.

[3] 贾玉斌.喀拉通克金属化高冰镍吹炼及水淬工业化的回顾与探讨[J].新疆有色金属,2003,26(2):23-26.

[4] 蒋开喜,王玉芳,郑朝振,等.硫化镍加压浸出研究进展与应用[J].矿冶,2018(3):45-50.

[5] 马育新.喀拉通克铜精矿加压酸浸新工艺研究[J].有色金属,2008,60(3):62-66.

[6] 游清治.阜康冶炼厂湿法流程中的镍金属总回收率及计算[J].有色金属(冶炼部分),2003(2):16-19.

[7] 陈廷扬.技术创新,完善工艺,提高企业经济效益——阜康冶炼厂铜渣处理工艺简介[J].新疆有色金属,2002,25(4):27-29.

[8] 李昌福,黄忠淼,李晔,等.关于阜康铜渣冶炼工艺的研究和选择[J].矿冶,2003,12(2):53-57.

[9] 吴涛.阜康冶炼厂尾渣技改工程工艺论证试验[J].新疆有色金属,2003,26(3):22-23.

[10] 李春雷.阜康冶炼厂高冰镍湿法精炼工业化回顾[J].中国有色冶金,2007,36(3):20-24.

[11] 张向红.十年弹指一挥间——阜康冶炼厂经济运行状况蒸蒸日上[J].新疆有色金属,2004,27(4):50-52.

[12] 木拉提.二段常压浸出在阜康冶炼厂的应用[J].新疆有色金属,2008,31(4):52-54.

[13] 姜惠云.哈密高镍矿在阜冶浸出过程的生产实践[J].新疆有色金属,2008,31(B08):120-121.

[14] 王玉珠.高冰镍中有价金属浸出过程中的研究与探讨[J].中国化工贸易,2013,5(7):263.

[15] 赵江.浸出工序中铜的行为[J].新疆有色金属,2004,27(4):31-32.

[16] 贾玉斌.阜康镍湿法冶金过程的除杂质方法机理分析[J].新疆有色金属,2003,26(3):24-26.

[17] 姜惠云.二段常压浸出在阜冶的生产实践[J].新疆有色金属,2009(A01):147-148.

[18] 王鹏伟.金属化高冰镍硫酸选择性浸出工艺中酸铜平衡及生产实践[J].新疆有色金属,1995(4):31-35.

[19] 许钢.浅析原料铜含量是影响硫酸选择性浸出终渣含镍的重要因素[J].新疆有色金属,2014,37(B07):150-151.

[20] 宋连民.阜康铜渣精炼工艺与生产实践[J].新疆有色金属,2002,25(1):33-37.

[21] 王春海.降低沸腾焙烧残硫率的生产实践[J].新疆有色金属,2014,37(3):75-78.

[22] 李杰.浅谈铜冶炼过程中影响阴极铜质量的因素[J].新疆有色金属,2014,37(B05):127-128.

[23] 马俊忠.铜渣焙烧-浸出-电积半湿法工艺中杂质控制的生产实践[J].新疆有色金属,2005,28(1):18-20.

[24] 吴涛.沸腾焙烧炉在铜冶炼生产中的应用[J].新疆有色金属,2003,26(1):18-19.

[25] 李江平.沸腾炉焙烧尾渣(铜渣)技改的应用[J].新疆有色金属,2009(A01):89-91.
[26] 李晔.阜康镍冶炼厂含镍铜渣冶炼工艺研究[J].矿冶,2000,9(3):59-62.
[27] 李江平.铜渣在沸腾焙烧炉中的生产实践[J].新疆有色金属,2004,27(4):27-28.
[28] 王春海.浅析铜冶炼过程中影响电流效率的因素[J].新疆有色金属,2008,31(4):64-66.
[29] 王英彬.降低铁渣含银浅析[J].新疆有色金属,2011(A01):94-95.
[30] 赵富平.阜康冶炼厂浸铜后渣加压浸出工艺探讨[J].新疆有色金属,2003,26(S2):43-44.
[31] 宋连民.阴极铜板面发黑、分层、长刺的工艺探讨[J].新疆有色金属,2007,30(2):29-30.
[32] 崔富晖,牟文宁,顾兴利,等.铜镍氧硫混合矿焙烧-浸出过程铜、镍、铁的转化[J].中国有色金属学报,2017(7):1471-1478.
[33] 史文峰.阜冶尾渣中贵金属的提取实践[J].新疆有色金属,2003,26(3):28-29.

水淬焙烧渣中金铂钯的氯化浸出

摘　要：为了回收水淬焙烧渣中的贵金属，采用 H_2SO_4-NaCl-$NaClO_3$ 浸液从冰铜水淬渣中浸出金、铂、钯。系统研究了固液比、硫酸浓度、氯化钠浓度、溶液氧化性、浸出温度、粒度和浸出时间等参数对浸出的影响。H_2SO_4-NaCl-$NaClO_3$ 浸出冰铜水淬渣中金、铂、钯的优化工艺参数为：硫酸浓度 0.5 mol/L、氯化钠浓度 25%、初始 ORP（氧化还原电位）1062 mV，在此条件下，金、铂、钯的浸出率分别为 90%、89%、88%。

关键词：冰铜水淬渣；浸出；金；铂；钯

Chlorinated leaching of gold, platinum and palladium from water-quenched roasting slag

Abstract: In order to recover precious metals in water-quenched roasting slag, a method of leaching gold, platinum and palladium from matte water-quenched slag was proposed using H_2SO_4-NaCl-$NaClO_3$ leaching solution system. The effects of main parameters including the solid-liquid ratio, sulfuric acid concentration, sodium chloride concentration, solution oxidation, leaching temperature, particle size and leaching time on leaching were systematically studied. The optimized process parameters of H_2SO_4-NaCl-$NaClO_3$ leaching of gold, platinum and palladium from water-quenched copper matte slag are that the concentration of sulfuric acid is 0.5 mol/L, the concentration of sodium chloride is 25%, and the initial ORP (redox potential) is 1062 mV, under the optimal conditions, the leaching rates of gold, platinum and palladium are 90%, 89% and 88%, respectively.

Keywords: copper matte water-quenching slag; leaching; gold; platinum; palladium; kinetics

贵金属以其资源稀少，用途广泛，独特的理化性质，以及价格昂贵而为人们关注[1-5]。贵金属主要包括金、银、铂族元素等，被广泛应用于军事、医疗、电子技术、催化等工业[6-10]。贵金属主要来源于贵金属矿，随着高品位矿逐年减少，人们对各类含贵金属的废弃物的关注也逐渐提升。据统计，城市矿山废弃汽车催化剂、废弃电子产品、废弃电路板中的贵金属含量远高于矿石中的贵金属含量[11-15]。目前，从城市矿山废品和冶金渣中回收贵金

本文发表在《有色金属工程》，2022，12(8)：62-69。作者：朱佳俊，黄凯，马育新，李渭鹏。

属已成为研究的重点[16-20]。

新疆卡拉通克矿业有限责任公司和新疆众鑫矿业有限责任公司针对新疆当地的铜镍矿通过造锍熔炼得到的冰铜，经硫酸选择性浸出镍钴，浸出渣经进一步沸腾焙烧、稀酸浸出和加压氧化浸出铜镍后得到贵金属渣。贵金属渣中富集了一定量的金、银、铂和钯，需要进一步冶炼回收。目前，工厂根据贵金属渣的特性，先进行火法捕集，以富集渣中的金、银、铂、钯，得到贵金属铜锍。贵金属铜锍经过硫酸化焙烧后，进行水淬，得到水淬贵金属渣。由于火法捕集工艺仅仅实现对渣中金、银、铂、钯等贵金属的富集分离，因此需要对含金、银、铂、钯的水淬渣进行专门的湿法浸出处理。可供选择的浸出方法主要有氰化[21-25]、王水浸出和氯酸钠浸出[26-30]。其中，氰化法浸出回收金、铂、钯，成本低，技术成熟，金、铂、钯的浸出率较高，但氰化法污染较大[31-33]。王水浸出回收金、铂、钯成本低，浸出率高，但是王水酸性和氧化性强，对设备腐蚀及挥发防逸都需要做专门设计。目前还提出了回收金、铂、钯的其他浸出工艺，包括臭氧、氯化钠、无机酸混合浸出。工艺污染小，浸出率高，但工艺成熟度低，不能在工业上大规模应用。氯化钠、硫酸、氯酸钠回收金、铂、钯的混合浸出工艺相对成熟，成本低，污染小，且金、铂、钯的浸出率较高，适宜回收金、铂、钯。本文针对新疆阜康冶炼厂水淬渣金、铂、钯含量高的特点，以氯化钠、硫酸、氯酸钠混合液浸出渣中金、铂、钯，考察主要工艺参数对浸出行为的影响规律，为生产提供参考。

1 试验

试验用水淬焙烧渣由新疆阜康冶炼厂提供，采用标准筛进行筛分，取筛分粒度级别分别在 0.425 mm、0.25 mm、0.18 mm 以下的物料为试验原料。试验用硫酸、氯化钠、氯酸钠为分析纯，试验用水皆为去离子水。浸出过程控制固液比（质量体积比，g/L，下同）为 100~333、硫酸浓度为 0.5~6 mol/L、氯化钠固体浓度为 10%~30%，用氯酸钠将氧化还原电位（ORP）调节至 1010~1280 mV，浸出温度为 25~80 ℃，浸出时间为 30~720 min，系统研究固液比、硫酸浓度、氯化钠浓度、氯酸钠浓度、温度、粒度和浸出时间等主要工艺参数对浸出的影响。

2 结果与讨论

2.1 XRD 和 XRF 分析

水淬焙烧矿浸出金铂钯前后的 XRD 和 XRF 结果分别如图 1 以及表 1 所示，主要成分为二氧化硅、硫酸钡、硫酸铅和少量单质金、铂、钯，金、铂、钯含量较少。

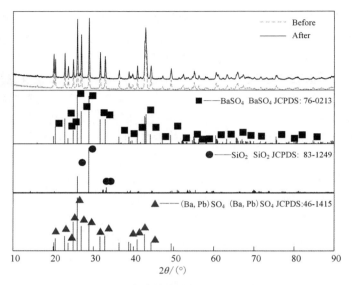

图 1 水淬焙烧渣的 XRD 图谱

Fig. 1 XRD patterns of water-quenched roasting slag

表 1 水淬焙烧渣浸出前后的主要成分

Table 1 Main contents of water-quenched roasting slag before and after leaching   %

	Ba	O	Pb	S	Si	Other
Before leaching	46.09	18.75	12.68	11.19	4.76	6.53
After leaching	47.75	21.47	12.17	11.37	3.75	3.49

2.2 SEM 与 EDS 分析

水淬焙烧渣浸金、铂、钯前后的 SEM 和 EDS 结果分别如图 2 和图 3 所示。对比图 2 和图 3 可以看出，浸出前后主要成分几乎不变化，浸出渣的主要成分为二氧化硅，硫酸钡，未检出金、铂、钯，这一结果与 XRD 的分析结果一致，进一步说明渣中的金、铂、钯已被有效浸出。

浸出渣的 SEM 和 EDS 分析结果如图 4~图 5 所示，主要成分为二氧化硅，硫酸钡，渣的内部与表面成分无差别，说明氯酸钠成功将渣的金、铂、钯浸出，这一结果与表面 SEM 和 EDS 图以及 XRD 结果一致。

2.3 固液比对金铂钯浸出率的影响

在硫酸浓度 2 mol/L、氯化钠浓度 25%、温度 25 ℃、溶液体积 10 mL、氧化还原电位（ORP）初始值 1200 mV、搅拌速度 100 r/min、浸出时间 1 h、炉渣质量 2 g、粒度 0.425 mm 条件下，研究浸出液固比对浸出的影响，结果如图 6 所示。从图 6 可以看出，浸出固液比在（100~333）∶1（g/L）时，金、铂、钯的浸出效率分别保持在 87.22%~93.67%、7.68%~18.

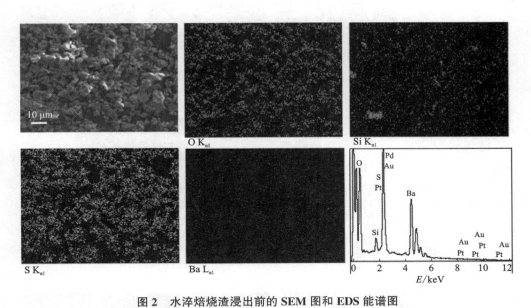

图 2 水淬焙烧渣浸出前的 SEM 图和 EDS 能谱图
Fig. 2 SEM images and EDS spectrum of water-quenched roasting slag before leaching

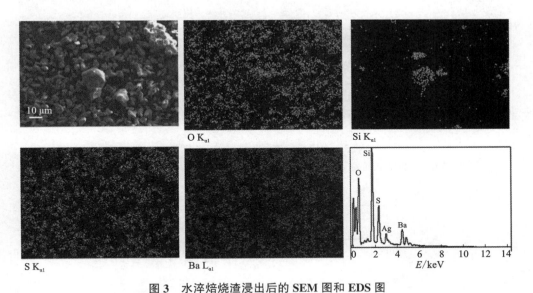

图 3 水淬焙烧渣浸出后的 SEM 图和 EDS 图
Fig. 3 SEM images and EDS spectrum of water-quenched roasting slag after leaching

68%、88.77%~93%。可见,固液比为 100∶1(g/L)时即可达到较高的金、钯浸出率,且随着固液比的增大,金、铂、钯的浸出效率相对较为稳定。从图 6 可以看出,浸出固液比对金、铂、钯的氧化还原电位(ORP)影响不大,由于浸出铂的难度较大,所以铂浸出率较低。综合考虑,选择固液比为 333∶1(g/L),对应金、铂、钯浸出率分别为 93.67%、18.68% 和 93%。

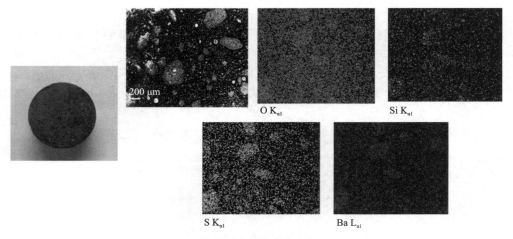

图 4　水淬焙烧渣浸出前的 SEM 图像

Fig. 4　SEM images of water-quenched roasting slag before leaching

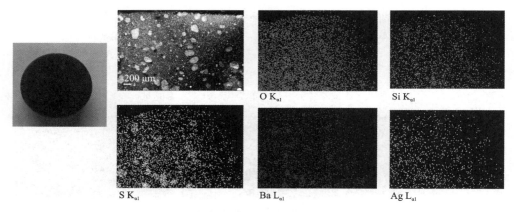

图 5　水淬焙烧渣浸出渣的 SEM 图像

Fig. 5　SEM images of leaching residue from water-quenched roasting slag

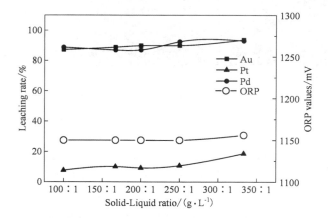

图 6　固液比对金、铂、钯浸出的影响

Fig. 6　Effects of solid-liquid ratio on leaching of gold, platinum, and palladium

2.4 硫酸浓度对金铂钯浸出率的影响

在氯化钠质量分数 25%、温度 25 ℃、溶液体积 10 mL、氧化还原电位（ORP）初始值 1200 mV、搅拌速度 100 r/min、时间 1 h、炉渣质量 3.33 g、炉渣粒度 0.425 mm 的条件下，研究硫酸浓度对金、铂、钯的影响，结果如图 7 所示。从图 7 可以看出，硫酸浓度在 0.5~6 mol/L，金、铂、钯的浸出率分别保持在 100%~90%、56.49%~8.48%、100%~81%。可见，硫酸浓度在 0.5 mol/L 即可达到高浸出金、钯的效率，由于铂浸出难度较大，所以铂浸出率较低。随着硫酸浓度的增加，金、钯的浸出效率将逐渐趋于稳定，但铂的浸出效率将逐渐降低。这是因为，随着硫酸浓度的增加，溶液中的盐酸浓度增加，加速了氯酸钠歧化反应，消耗了溶液中的氯酸钠浓度和氯化钠浓度。从图 7 不同硫酸浓度条件下金、铂、钯的氧化还原电位（ORP）数据可以看出，硫酸浓度对金、铂、钯的氧化还原电位（ORP）影响较大，说明浸出过程中氧化不是浸出过程的限制条件，而氯化钠浓度成为浸出限制环节。在氧化剂充足的条件下，起络合作用的是氯离子，由于氯离子浓度对铂浸出过程影响较大，对金、钯浸出影响较小。综合考虑选择硫酸浓度为 0.5 mol/L，此条件下金、铂、钯浸出效果最好，金和钯的浸出率接近 100%，铂浸出率为 56.49%。

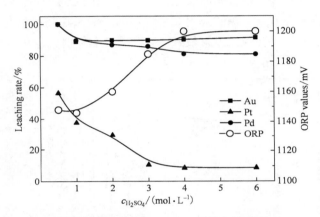

图 7　硫酸浓度对金、铂、钯浸出的影响

Fig. 7　Effects of sulfuric acid concentration on leaching of gold, platinum and palladium

2.5 氯化钠浓度对金铂钯浸出率的影响

在硫酸 0.5 mol/L、温度 25 ℃、溶液体积 10 mL、氧化还原电位（ORP）初始值 1200 mV、搅拌速度 100 r/min、时间 1 h、炉渣质量 3.33 g、炉渣粒度 0.425 mm 的条件下，氯化钠浓度对金、铂、钯浸出率的影响如图 8 所示。从图 8 可以看出，氯化钠浓度为 10%~30% 时，金、铂、钯的浸出效率分别保持在 100%~83.73%、49.93%~56.6%、94.84%~73.88%；只有在 10% 的条件下，金、钯的浸出效率才能达到最高，由于铂浸出难度较大，所以铂浸出率较低。由于随着氯化钠浓度的增加，加大氯酸钠自身歧化反应，消耗氯酸钠的浓度。从图 8 还可以看出，氯化钠浓度对氧化还原电位（ORP）影响较大，随着氯化钠浓度的增加，溶液的 ORP 值降低。由于 ORP 值越高，氯酸钠消耗量越大，浸出液氧化性不足，氯酸钠浓度成为金钯浸出的限制环节，而对于铂浸出，氯离子浓度影响更大。因此，在增加氯化钠浓度时，铂的浸出率有所增加，之后由于氧化性不足，限制铂进一步浸出。在氯化钠浓度为 10% 时，金、钯浸出效果最好，金的浸出率为 100%，钯的浸出率为 94.84%，在氯化钠浓度为 25% 时，铂浸出率最高，为 56.6%。综合考虑，后续试验氯化钠质量分数选择为 25%。

2.6 初始氧化还原电位(ORP)对金铂钯浸出率的影响

在硫酸 0.5 mol/L、氯化钠 25%、温度 25 ℃、溶液体积 10 mL、搅拌速度 100 r/min、时间 1 h、炉渣质量 3.33 g、炉渣粒度 0.425 mm 的条件下，初始氧化还原电位(ORP)对金、铂、钯的影响如图 9 所示。从图 9 可以看出，初始氧化还原电位在 1062~1370 mV 时，金、铂、钯的浸出效率分别为 97.65%~89.95%、88.56%~36.37%、76.08%~88.12%。可以推测，随着氯酸钠浓度的增加，氯酸钠的歧化反应增强，加速浸出液中氯离子的消耗，由于升高氯酸钠浓度，浸出液中的氧化还原电位(ORP)升高，氯化钠浓度成为金、铂、钯浸出的限制环节。因此只有在初始氧化还原电位(ORP)为 1062 mV 时，氯酸钠浓度较低，才能达到较高的金、钯浸出效率。随着初始氧化还原电位(ORP)的升高，氯离子浓度降低，金和钯的浸出效率将逐渐趋于稳定，铂的浸出效率将逐渐降低。在初始氧化还原电位(ORP)为 1062 mV 时，金、铂、钯的综合浸出效果最好，对应浸出率分别为金 97.65%、钯 76.08%，铂 88.56%。

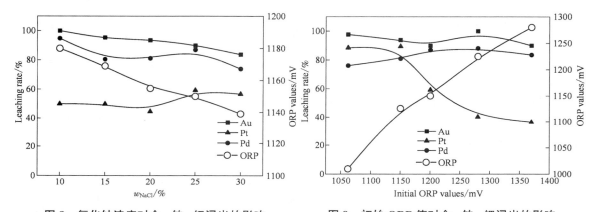

图 8 氯化钠浓度对金、铂、钯浸出的影响
Fig. 8 Effects of sodium chloride concentration on leaching of gold, platinum and palladium

图 9 初始 ORP 值对金、铂、钯浸出的影响
Fig. 9 Effects of initial ORP value on leaching of gold, platinum and palladium

2.7 不同温度下浸出时间对金铂钯浸出率的影响

在硫酸 0.5 mol/L、氯化钠 25%、溶液体积 10 mL、氧化还原电位(ORP)为 1062 mV、搅拌速度 100 r/min、炉渣质量 3.33 g、炉渣粒度 0.425 mm 条件下，研究浸出温度分别为 25、40、60、80 ℃ 条件下，浸出温度对浸出的影响，结果如图 10 所示。从图 10 可以看出，浸出时间为 120 min 时即可达到较高的金、铂、钯浸出效率。随着浸出时间的延长，金、铂、钯的浸出效率将逐渐趋于稳定。温度的升高可加速氯酸钠歧化反应，加速氯气产生，消耗浸出液中的氯酸钠和氯离子，随着温度升高，金、铂、钯的浸出效率不断降低。从图 10(d)可以看出，随着浸出温度的升高，氧化还原电位(ORP)不断降低，说明氯酸钠消耗不断增加，氯化钠浓度不断降低，限制了金铂钯的浸出。因此，常温下浸出最佳，适宜的浸出时间为 120 min，有利于金、铂、钯的浸出，此条件下对应金、铂、钯浸出率分别为 90%、82% 和 95%。

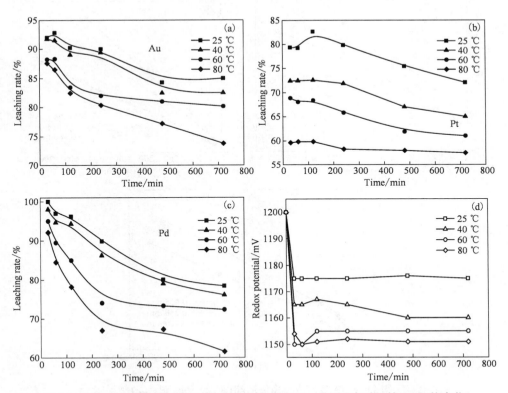

图 10 温度对浸出率的影响(a)金;(b)铂;(c)钯;(d)浸出过程的 ORP 值变化

Fig. 10 Effects of temperature on leaching rate (a) gold; (b) platinum; (c) palladium; (d) ORP value change during leaching

2.8 不同粒径下浸出时间对金铂钯浸出率的影响

在硫酸 0.5 mol/L、氯化钠 25%、温度 25 ℃、溶液体积 10 mL、氧化还原电位(ORP)值 1062 mV、搅拌速度 100 r/min、炉渣质量 3.33 g,炉渣粒度分别为 0.425、0.25、0.18 mm 条件下,原料粒度对金、铂和钯浸出率的影响如图 11 所示。从图 11 可以看出,原料粒度为 0.425~0.180 mm 时,浸出时间仅需 30 min 即可达到较高的金、铂、钯浸出效率,随着时间的延长,金、铂、钯的浸出效率变化不大,随着原料粒度的减小,金、铂、钯的浸出效率会略有提高。因此,适宜的原料粒度为 0.180 mm,浸出时间为 120 min,该条件有利于金、铂、钯的浸出,对应金、铂、钯的浸出率分别为 90%、89% 和 88.8%。

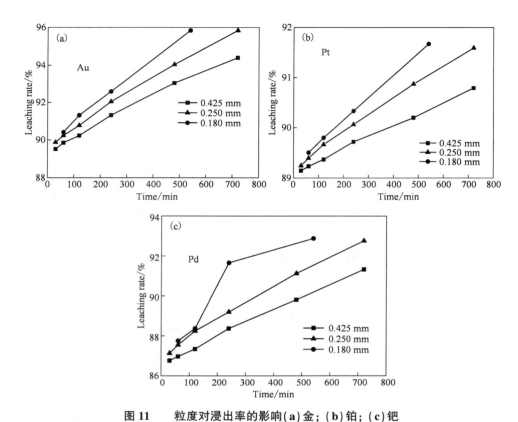

图 11　粒度对浸出率的影响 (a) 金；(b) 铂；(c) 钯

Fig. 11　Effects of particle size on leaching rate (a) gold; (b) platinum; (c) palladium

3　结论

(1) 针对主要成分为二氧化硅、硫酸钡的水淬焙烧渣，采用 H_2SO_4-NaCl-$NaClO_3$ 浸出体系，有效浸出水淬焙烧渣中的金铂钯。

(2) 该体系最佳浸出条件为：0.5 mol/L 硫酸，25% 氯化钠，初始氧化还原电位 (ORP) 1062 mV，在上述条件下金铂钯的浸出率分别为 90%、89% 和 88%。

参考文献

[1] FUJIWARA K, RAMESH A, MAKI T, et al. Adsorption of platinum (Ⅳ), palladium (Ⅱ) and gold (Ⅲ) from aqueous solutions onto l-lysine modified crosslinked chitosan resin [J]. Journal of Hazardous Materials, 2007, 146(1/2): 39-50.

[2] RAMESH A, HASEGAWA H, SUGIMOTO W, et al. Adsorption of gold (Ⅲ), platinum (Ⅳ) and

palladium（Ⅱ）onto glycine modified crosslinked chitosan resin[J]. Bioresource Technology, 2008, 99(9): 3801-3809.

[3] AKTAS S, MORCALI M H. Platinum recovery from dilute platinum solutions using activated carbon[J]. Transactions of Nonferrous Metals Society of China, 2011, 21(11): 2554-2558.

[4] 杨昆. 铜镍硫化物矿物中铂族元素的分析研究[J]. 化工管理, 2022, 33(8): 13-15.

[5] 宋飞跃, 薛泳波, 高欣, 等. 不同离子液体双水相萃取钯[J]. 应用化学, 2019, 36(3): 335-340.

[6] AGHAEI E, ALORRO R D, ENCILA A N, et al. Magnetic adsorbents for the recovery of precious metals from leach solutions and wastewater[J]. Metals, 2017, 7(12): 529-560.

[7] 谭涛. 贵金属铂、钯、铑含量检测及判定规范[J]. 化工管理, 2021, 34(30): 147-148.

[8] RZELEWSKA-PIEKUT M, REGEL-ROSOCKA M. Separation of Pt(Ⅳ), Pd(Ⅱ), Ru(Ⅲ) and Rh(Ⅲ) from model chloride solutions by liquid-liquid extraction with phosphonium ionic liquids[J]. Separation and Purification Technology, 2019, 212: 791-801.

[9] KOMENDOVA R. Recent advances in the preconcentration and determination of platinum group metals in environmental and biological samples[J]. TrAC Trends in Analytical Chemistry, 2020, 122: 115708-115719.

[10] 田茂江, 崔得锋, 张晓辉. 贵/廉金属层状复合材料的发展现状及趋势[J]. 电工材料, 2021, 49(5): 3-7.

[11] HONG H J, YU H, HONG S, et al. Modified tunicate nanocellulose liquid crystalline fiber as closed loop for recycling platinum-group metals[J]. Carbohydrate Polymers, 2020, 228: 115424-115430.

[12] 叶子豪, 李静, 蔡卫卫. 贵金属电化学固氮催化剂的研究进展[J]. 聊城大学学报(自然科学版), 2022, 35(2): 41-51.

[13] MOHAMED S K, HASSAN H M A, SHAHAT A, et al. A ligand-based conjugate solid sensor for colorimetric ultra-trace gold(Ⅲ) detection in urban mining waste[J]. Colloids and Surfaces A: Physicochemical and Engineering Aspects, 2019, 581: 123842-123848.

[14] 彭浩, 朱军, 王斌, 等. 废旧电路板中有价金属回收试验研究[J]. 矿冶工程, 2021, 41(5): 99-102.

[15] KARIM S, TING Y P. Ultrasound-assisted nitric acid pretreatment for enhanced biorecovery of platinum group metals from spent automotive catalyst[J]. Journal of Cleaner Production, 2020, 255: 120199-120212.

[16] BIATA N R, JAKAVULA S, MOUTLOALI R M, et al. Recovery of palladium and gold from PGM ore and concentrate leachates using Fe_3O_4@SiO_2@Mg-Al-LDH nanocomposite[J]. Minerals, 2021, 11(9): 917-935.

[17] SETHURAJAN M, VAN HULLEBUSCH E D, FONTANA D, et al. Recent advances on hydrometallurgical recovery of critical and precious elements from end of life electronic wastes-a review[J]. Critical Reviews in Environmental Science and Technology, 2019, 49(3): 212-275.

[18] RAO M D, SINGH K K, MORRISON C A, et al. Challenges and opportunities in the recovery of gold from electronic waste[J]. RSC Advances, 2020, 10(8): 4300-4309.

[19] WITT K, URBANIAK W, KACZOROWSKA M A, et al. Simultaneous recovery of precious and heavy metal ions from waste electrical and electronic equipment (WEEE) using polymer films containing Cyphos IL 101[J]. Polymers, 2021, 13(9): 1454-1471.

[20] 方田, 高奕吟, 王思雨, 等. Mn-Zr复合氧化物负载贵金属催化剂的氯乙烯催化燃烧性能[J]. 石油化工高等学校学报, 2021, 34(5): 1-8.

[21] BEOLCHINI F, FONTI V, FERELLA F, et al. Metal recovery from spent refinery catalysts by means of biotechnological strategies[J]. Journal of Hazardous Materials, 2010, 178(1/2/3): 529-534.

[22] 秦昌静, 杨建国. 全泥氰化提金工艺设计与生产实践[J]. 世界有色金属, 2021, 36(10): 117-118.

[23] NATARAJAN G, TING Y P. Gold biorecovery from e-waste: An improved strategy through spent medium leaching with pH modification[J]. Chemosphere, 2015, 136: 232-238.

[24] ADAMS C R, PORTER C P, ROBSHAW T J, et al. An alternative to cyanide leaching of waste activated carbon ash for gold and silver recovery via synergistic dual-lixiviant treatment[J]. Journal of Industrial and Engineering Chemistry, 2020, 92: 120-130.

[25] 王青丽, 吕超飞, 苏晨曦, 等. 国外某低品位氧化矿浮选-氰化法回收金的试验研究[J]. 有色金属(选矿部分), 2020(6): 71-76.

[26] SAEEDI M, SADEGHI N, ALAMDARI E K. Modeling of Au chlorination leaching kinetics from copper anode slime[J]. Mining, Metallurgy & Exploration, 2021, 38(6): 2559-2568.

[27] 张文岐, 熊亚东. 氯化法浸金工艺研究与实践[J]. 中国有色冶金, 2020, 49(1): 85-89.

[28] 李超, 李宏煦, 杨翀, 等. 某难浸金矿的次氯酸盐法直接浸金试验研究[J]. 黄金科学技术, 2014, 27(4): 108-112.

[29] 李晶莹, 黄璐, 徐秀丽. 废旧线路板次氯酸盐法浸金的实验研究[J]. 环境工程学报, 2011, 5(2): 453-456.

[30] TORRES R, LAPIDUS G T. Platinum, palladium and gold leaching from magnetite ore, with concentrated chloride solutions and ozone[J]. Hydrometallurgy, 2016, 166: 185-194.

[31] 王瑞祥, 刘茶香, 杨裕东, 等. 复杂金矿物氧化焙烧-硫酸浸出-氰化法回收金银模拟计算[J]. 有色金属(冶炼部分), 2020(7): 69-75.

[32] ZHANG K, LIU Z, QIU X, et al. Hydrometallurgical recovery of manganese, gold and silver from refractory Au-Ag ore by two-stage reductive acid and cyanidation leaching[J]. Hydrometallurgy, 2020, 196: 105406-105414.

[33] SCHOEMAN E, BRADSHAW S M, AKDOGAN G, et al. The extraction of platinum and palladium from a synthetic cyanide heap leach solution with strong base anion exchange resins[J]. International Journal of Mineral Processing, 2017, 162: 27-35.

用石榴皮自溴化盐溶液中回收金研究

摘 要：传统的氰化钾湿法浸金存在毒性大、污染难控制等安全隐患，开发无氰化浸金新方法具有广阔的发展前景。溴化盐溶液介质对金的浸出具有显著的优势，但是如何从其中回收金离子，仍然是一个挑战。本研究采用石榴皮来做吸附剂，考察了从配制的模拟溴化盐介质溶液中吸附金离子的规律，系统研究了pH值、时间、温度、溴离子浓度等因素对吸附金离子效率的影响规律，可为溴化盐介质溶液中回收金工艺设计提供有益的参考。

关键词：溴化钾；石榴皮；金；生物吸附

Recovery of gold from the bromide solution by pomegranate peel

Abstract：Chemical extraction of gold by the traditional cyanide leaching process was toxic and dangerous, and therefore the development of non-cyanide leaching process was quite important in industry. Bromide salt had been studied to the leaching and recovery gold for many years, while how to enrich the trace concentration of gold from the bromide solution was a big challenge. In present study, pomegranate peel was proposed as biosorbent to extrct gold from solution, and main parameters like pH value, contact time, temperature, bromide concentration on extraction of gold were investigated. The results are beneficial to improvement of gold extraction from the bromide solution system.

Keywords：potassium bromide；pomegranate peel；gold；biosorption

电子信息技术的迅速发展带来了数量众多的电子产品[1]，电子产品的更新换代周期越来越短，从而产生了大量淘汰下来的电子废弃物，如果不进行适当的处理，就会对环境产生威胁[2]。电子废弃物中含有大量的金属及其他资源[3-4]，如果直接作为垃圾进行处理，不仅会造成资源浪费，还会危害环境[5]。有必要对电子废弃物进行回收和再生利用[6,7]，其中也包括从电子废弃物中提取黄金[8]。传统氰化法提金工艺[9-11]存在毒性大、污染难控制等安全隐患[12-13]，相比之下，无毒、污染小的溴化盐溶液介质体系下对金的浸出具有显著的优势[14]，开发无氰化浸金新方法(如溴化法[15-16])具有广阔的发展前景。DADGAR等[17]认为，与氰

本文发表在《有色金属(冶炼部分)》，2021(9)：54-59。作者：王江廷，邱颢艺，索金亮，盛勇，高胜亚，陈龙翼，朱佳俊，黄凯。

化法相比，溴化法提金工艺具有无毒、浸出速度快、成本低、对 pH 变化的适应性强等优点。

近年来人们对电子垃圾中贵金属提取进行了许多研究，其中已经取得初步成效的主要有：催化氧化法提金[18, 19]以及吸附剂法提金[20]等，但其实际应用还存在问题。吸附剂法提金的关键在于吸附剂的选择，研究发现，石榴皮中含有大量鞣质[21]，而鞣质作为一种比较复杂的多元酚类化合物[22]，在溴化盐的体系中表现出了对金元素的优良吸附性。在本工作中，我们选择皂化石榴皮作为吸附剂对模拟溴化盐介质溶液进行实验，考察其对黄金等离子的吸附效果，以期促进清洁、绿色的提金新工艺发展。

1 实验原料与装置

化学试剂：分析纯溴化钾、氯金酸、硝酸铜、硝酸锌、硝酸镍、盐酸、皂化石榴皮、蒸馏水。

实验装置与仪器：空气浴振荡器、AAS 原子吸收分光光度仪、pH 仪、电子天平、加热搅拌器、温度计、移液管、锥形瓶、试管。

2 实验过程

为研究 pH、时间、温度、溴离子浓度等因素对吸附金离子效率的影响规律以及吸附剂对金、铜、镍、锌等不同元素的吸附效果，吸附试验总体采用单因素变量实验法。

皂化石榴皮吸附剂的制备过程为：取 100 g 新鲜石榴皮，与 15 g 氢氧化钠、15 mL 甲醛一起打碎，搅拌反应 24 h 后，过滤、洗涤、烘干、破碎、筛分过 40 目筛，所得粉末状吸附剂颗粒称为皂化石榴皮吸附剂。该吸附剂富含多酚化合物，可用于金属离子的吸附提取，在本研究中主要考察其对黄金离子的吸附富集效果。

2.1 初始 pH 值对吸附剂吸附金、铜、镍、锌的影响

实验条件：吸附剂 10 mg，温度 28 ℃，反应时间 2 h，0.1 mol/L 溴离子，10 mL 待吸附液，空气浴振荡。改变初始 pH 为 1.0~7.0 时，进行皂化石榴皮对溶液中金、铜、镍、锌吸附效果的对照试验。检测反应前后 pH 和吸附量，计算不同元素的吸附率。

2.2 温度对吸附剂吸附金的影响

实验条件：吸附剂 10 mg、反应时间 2 h、0.1 mol/L 溴离子浓度、10 mL 待吸附液、空气浴振荡。分别在 28 ℃、38 ℃、48 ℃进行皂化石榴皮对金吸附率影响的对照实验，研究吸附率随时间的变化规律。

2.3 时间对吸附剂吸附金的影响

实验条件：吸附剂 10 mg、温度 28 ℃、10 mL 待吸附液、0.1 mol/L 溴离子浓度、空气浴振荡。分别进行吸附时间为 10 min、20 min、30 min 和 1 h、1.5 h、2 h、4 h、8 h、16 h、24 h 的对照实验，研究不同时间下吸附剂吸附率的规律。

2.4 [Br⁻] 对吸附剂吸附金的影响

实验条件：吸附剂 10 mg、温度 28 ℃、反应时间 2 h、10 mL 待吸附液、空气浴振荡。对 [Br⁻] 分别为 0.1 mol/L、0.2 mol/L、0.5 mol/L、1.0 mol/L、2.0 mol/L 的待测含金溶液进行对照实验，研究不同 [Br⁻] 吸附率的规律。

2.5 检测方法

采用 pH 仪测待测液吸附反应前后 pH。AAS 原子吸收分光光度仪检测已吸附溶液中所测元素的浓度，计算吸附率。

3 实验结果与讨论

3.1 溶液初始 pH 值的影响

废旧电路板浸出液一般呈酸性，故在初始 pH 在 1.0~6.0 进行皂化石榴皮吸附剂对 Au^{3+}、Cu^{2+}、Zn^{2+}、Ni^{2+} 的吸附实验，结果如图 1 所示。从图 1 可见，在初始 pH<2.5 时，皂化石榴皮对 4 种离子的吸附效率都较低，均低于 30%。初始 pH 在 2.5~5 时，随着 pH 的增大，4 种离子的吸附率快速增加。当 pH 为 5.0 左右时，Au^{3+} 的吸附率达到 92.3%，pH 继续增大，吸附率反而下降。而对于 Cu^{2+}、Zn^{2+}、Ni^{2+} 在 pH>2.5 时，吸附率随 pH 增大而增大。从以上

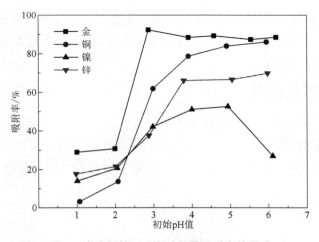

图 1 溶液初始 pH 值对金属吸附率的影响

Fig. 1 Effects of initial pH value on metal ions adsorption on saponified pomegranate peel gel

结果可见，对 Au^{3+} 吸附率最大且选择性吸附最好的 pH 为 5.0 左右。由于在低 pH 时，溶液中高浓度 H^+ 将减少吸附剂上活性吸附位点吸附的金属阳离子，降低吸附效率。随着 pH 的增大，溶液中 H^+ 浓度降低，吸附剂表面负电荷增多，有利于金属阳离子的吸附，提高吸附率。

3.2 反应温度的影响

溶液初始 pH 值为 5.0，其他条件不变，分别在不同温度（28 ℃、38 ℃、48 ℃）进行实验，考察了皂化石榴皮对 Au^{3+} 吸附率的影响，结果如图 2 所示。从图 2 可见，皂化石榴皮对 Au^{3+} 的吸附效果受温度变化的影响较小，吸附率基本保持在 95% 左右。因此实际操作中在室温下进行吸附提取即可。

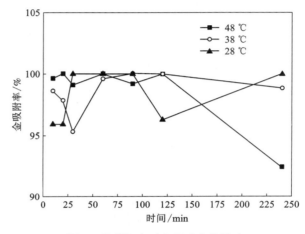

图 2　吸附温度对金吸附率的影响

Fig. 2　Effects of temperature on gold adsorption on saponified pomegranate peel gel

3.3 吸附时间对吸附效果的影响

在溶液 pH 值为 5.0，吸附温度 28 ℃，固定其他条件进行实验，反应时间对吸附的影响如图 3 所示。由图 3 可知，皂化石榴皮对金的吸附速率相当快，在吸附开始后 10 min 达到吸附平衡，吸附率达到 95% 左右。

3.4 KBr 介质浓度的影响

在 $[Au^{3+}]$ 为 0.2 mmol/L，溶液 pH 为 5.0，吸附温度 28 ℃，其他条件不变的情况下进行实验，皂化石榴皮吸附曲线如图 4 所示，可以看出，吸附率随 $[Br^-]$ 增大而缓慢下降，推测是溴离子的高浓度对吸附反应产生干扰，总体上，$[Br^-]$ 对金的吸附影响不大。

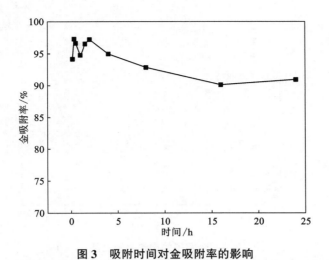

图 3　吸附时间对金吸附率的影响

Fig. 3　Effects of time on gold adsorption on saponified pomegranate peel gel

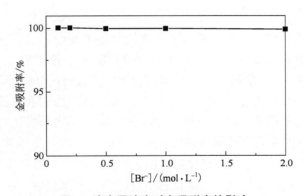

图 4　溴离子浓度对金吸附率的影响

Fig. 4　Effects of Br⁻ concentration on gold adsorption on saponified pomegranate peel gel

3.5　溶液中 Au^{3+} 初始浓度的影响

在溶液 pH 为 5.0，吸附温度 28 ℃，初始[Br⁻]分别为 0.1 mol/L、0.5 mol/L、1.0 mol/L 情况下，皂化石榴皮对金离子的饱和吸附曲线，如图 5 所示。可以看出[Br⁻]在 0.1 mol/L 情况下，吸附量随着金离子浓度的增大而增大，最后表现出平衡趋势；且从图 5 可估测出皂化石榴皮对金离子的最高吸附容量超过 550 mg/g，这是非常高的一个吸附值，具有良好的工业应用前景。在[Br⁻]增大为 0.5 mol/L 以上后，吸附率均随着金离子浓度增大而逐渐增大并趋于一个平台，但都比[Br⁻]在 0.1 mol/L 条件下的低。

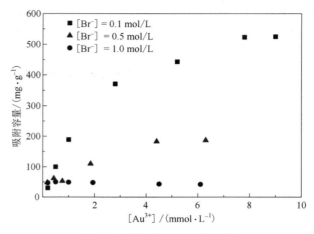

图 5　金离子的饱和吸附曲线

Fig. 5　Adsorption isotherm for gold adsorption onto saponified pomegranate peel gel

3.6　吸附机理探讨

石榴皮中富含鞣质性成分，图 6 表示交联改性鞣质的化学结构式。有文献报道[23]新鲜石榴皮含靴花单宁 10.4%，21.3%，蜡 0.8%，树脂 4.5%，甘露醇 1.8%，糖 2.7%，树胶 3.2%，没食子酸 4%等，其中的酚羟基功能团对金属离子具有良好的吸附效果[24-26]。鞣质是一类结构比较复杂的多元酚类化合物，即含有大量酚羟基。当它和甲醛、氢氧化钠混合后，酚羟基邻位和对位上的氢原子可以与甲醛进行缩聚，使分子之间发生交联，形成环状结构，在三维空间铺展开来，则生成多孔网状结构，如图 6 所示。进一步推测，酸性溶液中的金以中性分子 $H(AuBr_4)$ 的形式被这种多孔结构吸附，这些金的络合物可能是靠范德华力吸附于皂化石榴皮上。对比前面的实验结果，皂化石榴皮吸附金效果最好，对 Zn^{2+}、Ni^{2+} 吸附较少，这可能是因为 $H(AuBr_4)$ 分子较大，与环的大小相匹配，能够稳定地吸附在环状结构上。根据上述对吸附机理的探讨，认为皂化石榴皮对金以物理吸附为主，因此温度对吸附效果影响不大，这与前面的实验结果相吻合。

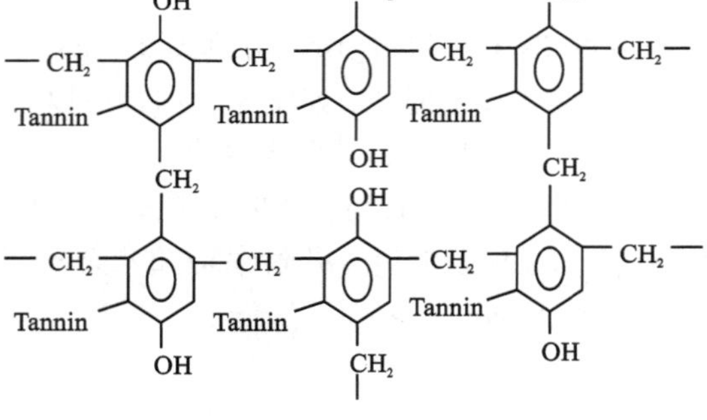

图 6　可水解单宁酸交联反应后的多孔网络结构

Fig. 6　Porous network structure after cross-linking reaction of hydrolysable tannins

4 结论

(1)在溴化盐介质溶液体系中,皂化石榴皮对金离子具有良好吸附效果。

(2)在 0.1 mol/L KBr 介质中皂化石榴皮对金、铜、镍、锌离子的吸附率随 pH 的升高均增大,在 pH 值 5.0 左右,对金的吸附率达到峰值 90% 以上,此时其他三种金属的吸附率均小于 60%,有较好的选择吸附效果。

(3)皂化石榴皮在 KBr 介质中对金的吸附速率很快,10 min 中即可达到最大值。

(4)溴离子浓度升高,会抑制皂化石榴皮对金的吸附率。

(5)温度对吸附率影响不大,仅使吸附率小范围波动,试剂吸附可在室温进行。

(6)皂化石榴皮对金离子的最高吸附容量可达 550 mg/g,即每克皂化石榴皮吸附剂可最高吸附 550 毫克黄金。

(7)石榴皮经皂化处理后形成多孔网络结构,能够高效吸附 $H(AuBr_4)$ 分子。

参考文献

[1] 查建宁.电子废弃物的环境[J].污染防治技术,2002,15(3).
[2] 罗春,蒋湛,马立实,等.电子废弃物污染现状及改善对策研究[J].安全,2008(2):14-17.
[3] 彭平安,盛国英,傅家谟.电子垃圾的污染问题[J].化学进展,2009,21(2).
[4] 周全法,刘玉海,李锋,等.金在电脑板卡中的应用[J].黄金,2003,24(8):5-7.
[5] 姜宾延,吴彩斌.电子垃圾的危害及其机械处理技术现状[J].再生资源研究,2005(3):23-26.
[6] 倪明,李冰洁,郭军华.废弃电子产品回收再利用组织模式研究[C]//第九届中国软科学学术年会论文集(上册),2013.
[7] 何阳葵.循环经济:电子垃圾治理之路[J].同济大学学报(社会科学版),2004,15(5):103-107.
[8] 夏苏湘,金成舟.浅议电子废弃物的再生利用[J].上海城市管理职业技术学院学报,2004,13(2):40-42.
[9] 石同吉.氰化提金技术发展现状评述[J].黄金科学技术,2001,9(6):22-29.
[10] Xue T, Osseo-Asare K. Heterogeneous equilibria in the Au-CN-H_2O and Ag-CN-H_2O systems[J]. Metallurgical Transactions B, 1985, 16(3):455-463.
[11] 夏光祥,涂桃枝.氨氰法从含铜金矿石中提金研究与工业实践[J].黄金,1995,16(7):26-29.
[12] 周芝兰.氰化提金废弃物污染农田的分析报告[J].考试周刊,2009(16):176-177.
[13] 李桂春,卢寿慈.非氰化提金技术的发展[J].中国矿业,2008(3):1-5.
[14] 张兴仁.溴化法提金工艺的研究及其前景[J].黄金,1993,2:009.
[15] 宋庆双,李云巍.溴化法浸出提取金和银[J].贵金属,1997,18(3):34-38.
[16] 刘建华,陈赛军.溴化法浸取硫化矿中的金[J].化工时刊,2003,17(4):38-39.
[17] DADGA A,陈炎.从难浸精矿提取金和银——氰化法与溴法的比较[J].铀矿开采,1989(4):50-56.
[18] 方兆珩,夏光祥.高砷难处理金矿的提金工艺研究[J].黄金科学技术,2004,12(2).
[19] 石伟,夏光祥,涂桃枝,等.氨性催化氧化-氰化法处理含砷难浸金矿的研究(Ⅰ)[J].过程工程学报,1996,4:2.

[20] 伍喜庆,黄志华.改性活性炭吸附金的性能[J].中国有色金属学报,2005,15(1):129-132.
[21] 王克英,左宏笛,杨理明.紫外分光光度法测定石榴皮中总鞣质含量[J].贵阳中医学院学报,2007,29(6):66-67.
[22] 刘延泽,李海霞.石榴皮中的鞣质及多元酚类成分[J].中草药,2007,38(4):502-504.
[23] Emanuele Flaeeomio. The rind of the Pomegranatefruit. Rrog. Terap[J]. Sez. Farm, 1992, 18:183-185.
[24] 孟冠华,李爱民,张全兴.活性炭的表面含氧官能团及其对吸附影响的研究进展[J].离子交换与吸附,2007,23(1):88-94.
[25] 宋立江,狄莹,石碧.植物多酚研究与利用的意义及发展趋势[J].化学进展,2000,12(2):161-170.
[26] 谢枫,樊在军,张青林,等.柿单宁在重金属吸附中的应用研究进展[J].华中农业大学学报,2012,31(3):391-396.
[27] 朱静.石榴皮中生物活性成分的提取纯化[D].北京化工大学,2006.

碱煮黑钨渣还原熔炼回收有价金属的研究

摘　要：采用还原熔炼工艺处理碱煮黑钨渣，以合金形式回收 Fe、Mn、W、Sn、Nb 有价元素，并分离 As、Pb 等有害元素。研究了还原剂（石墨粉）用量、熔炼温度和时间等对熔炼效果的影响。实验结果表明：在还原剂用量为 16%、熔炼温度为 1550 ℃、熔炼时间为 120 min 的条件下，碱煮黑钨渣中 Fe、Mn、W、Sn、Nb 元素的合金化率分别达到了 88.67%、56.28%、98.23%、67.74%、84.72%，As、Pb、Bi 基本进入烟尘。

关键词：碱煮黑钨渣；还原熔炼；有价元素；有害元素；合金化率；回收

Study on recovery of valuable metals from wolframite alkali-leaching residue by reduction smelting

Abstract: The reduction smelting process was used to process wolframite alkali-leaching residue and recover valuable elements such as Fe, Mn, W, Sn and Nb in the form of alloys, and the harmful elements such as As and Pb were separated. The effects of reducing agent (graphite powder) dosage, smelting temperature and time on smelting efficiency were studied. The experimental results show that the alloying rates of Fe, Mn, W, Sn and Nb in wolframite alkali-leaching residue are 88.67%, 56.28%, 98.23%, 67.74% and 84.72%, respectively, when reducing agent dosage is 16%, smelting temperature is 1550 ℃, and smelting time is 120 min. Meanwhile, As, Pb and Bi are in the smoke dust.

Keywords: wolframite alkali-leaching residue; reduction smelting; valuable element; harmful element; alloying rate; recovery

　　钨作为国家稀有战略金属，广泛应用于军工、航空航天和国民经济各领域。钨冶炼过程会产生大量的碱煮渣，目前碱煮渣在我国的堆存量已超百万吨，且以每年近 8 万 t 的速度递增[1]。碱煮渣中存在的 As、Pb 等重金属元素超标，严重危害生态环境，已在 2016 年被列入《国家危险废物名录》，且碱煮渣中含有 Fe、Mn、W、Sn、Nb 等有价元素，因此亟待处理并回收其中的有价元素。

　　从碱煮渣中回收有价元素的方法主要有酸法、碱法、选-冶联合法和火法。酸法主要集中于 W、Ta、Nb、Sc、Fe、Mn 的回收[2-7]。汪加军等[2]通过氟盐转型、HF-H$_2$SO$_4$ 浸出、氟盐氨转化等工序，同步提取废钨渣中的 Ta、Nb、W；苏正夫等[3]采用高温常压酸浸、离子交换工艺分离钨和其他有价金属。采用酸法处理碱煮渣，有价元素回收率高，但流程长，后续废水处理量大。碱法主要集中于 W 的回收及 Ta、Nb 的富集[8-12]。杜阳[11]采用碱熔融-水浸法回收富钨渣中 W，回收率高达 99.3%。戴艳阳等[12]采用苏打焙烧-水浸法从废钨渣中回收 W，然后进行酸浸富集 Ta、Nb、W，回收率最高达 79.45%。采用碱法处理碱煮渣，W 回收率高，但碱煮渣中 W 质量分数低于 1% 且碱消耗量大，经济价值较低。选-冶联合法先制备精矿，再进行湿法处理[13-14]，如杨俊彦等[13]采用选矿-湿法冶炼从钨渣中回收 Fe、Mn、W、Sn、Sc、Ta、Nb。选-冶联合法虽然可回收有价元素，但是回收率不高，且流程长。火法主要通过还原形成中间合金回收碱煮渣中的有价元素[15-16]，如МАСЛОВ[15]采用钨中间产物和硫酸钠在电炉中还原熔炼贫钨原料提取铁钨合金。火法的 W 回收率低，且未回收其他金属。ЗЕЛИКМАН[16]采用铝热还原钨渣制备含有 W、Ta 和 Nb 的 Fe-Mn 合金，但 Ta 和 Nb 的回收率低，仅 40%。虽然火法处理碱煮渣的 W、Ta、Nb 回收率低，但具有火法工艺流程短的特点。

本文采用还原熔炼工艺处理碱煮黑钨渣，考察还原剂用量、熔炼温度以及熔炼时间等工艺条件对还原熔炼效果的影响，以期达到有价元素 Fe、Mn、W、Sn、Nb 形成合金得以回收，As、Pb 等有害元素进入烟尘的目的。

1 实验部分

1.1 原料及设备

实验采用的原料为赣南某企业的碱煮黑钨渣，其化学组成如表 1 所示。由表 1 可知，碱煮黑钨渣的化学成分复杂，其中 O、Mn、Fe 三种元素合计 67.54%，是碱煮黑钨渣的主要成分，主要以氧化物的形式存在；其次是 Ca、Si，共占 10.79%，并含有一定量的有价元素。碱煮黑钨渣的还原熔炼设备为 GWL-1800LKQGA 型高温立式管式电炉，如图 1 所示。

表 1 碱煮黑钨渣的化学组成

Table 1 Chemical composition of wolframite alkali-leaching residue %

Fe	Mn	Ca	Si	Al	W	Sn
20.17	22.58	6.18	4.61	0.97	0.52	0.81
Nb	Ta	Pb	Cu	As	Bi	O
0.53	0.08	0.32	0.21	0.79	0.54	24.79

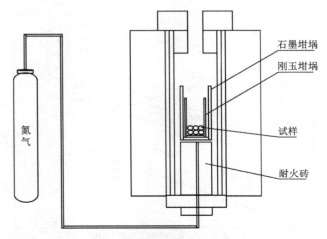

图 1 高温立式管式炉示意图
Fig. 1 Schematic diagram of high-temperature vertical tube furnace

1.2 实验方法

先将碱煮黑钨渣在 120 ℃的电热鼓风干燥箱中烘干 48 h,以去除水分,再通过磨料机进行磨碎,经筛选后得到 0.075 mm 以上粒级占 29.20% 的碱煮黑钨渣粉末。细磨、干燥后的碱煮黑钨渣粉末与还原剂(石墨粉)按实验设定的配比充分混合,压制成直径 3 cm 的圆柱置于氧化铝坩埚,放入管式炉,全程通入氩气进行保护,加热到指定温度后保温一定时间。升温速率设为 1000 ℃前 10 ℃/min,1000 ℃后 4 ℃/min。冷却至室温取出,还原产物经破碎后得到 Fe-Mn-W-Sn-Nb 合金和二次渣。合金中 Fe、Mn 分别依据 GB/T 8654.1—2007、GB/T 5686.1—2008 进行化学滴定,其他有价元素利用电感耦合等离子质谱法(ICP-MS)分析。

元素的合金化率为其在合金中的含量与其在碱煮黑钨渣中的含量之百分比。

2 结果与讨论

本文通过对碱煮黑钨渣中金属氧化物的热力学分析,查明金属氧化物的开始反应温度和金属元素的还原顺序,采用碳还原熔炼制备 Fe-Mn 合金的工艺处理碱煮黑钨渣,考察还原剂用量、熔炼温度、熔炼时间对碳热还原熔炼效果的影响。

2.1 热力学分析

吉布斯自由能是判断化学反应方向与限度的重要热力学函数。采用 FactSage 软件对碱煮黑钨渣中氧化物的吉布斯自由能进行计算,结果如图 2 所示,由此更好地分析碱煮黑钨渣中金属氧化物的吉布斯自由能与温度的关系。

从图 2 可以看出,在标准状态下,碱煮黑钨渣中各金属元素的还原顺序为 Bi>Pb>Sn>W>Fe>Nb>Mn。SnO_2、WO_3、FeO 在 800 ℃以下就已经开始与 C 发生还原反应生成金属 Sn、

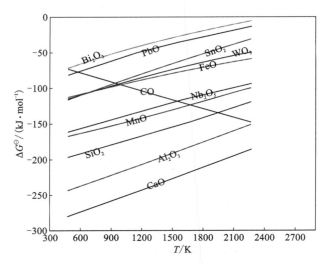

图 2 碱煮黑钨渣中主要金属氧化物的标准吉布斯自由能变化趋势

Fig. 2 Standard Gibbs free energy change trend of main metal oxides in wolframite alkali-leachingresidue

W、Fe。而 Nb_2O_5、MnO 则需在更高的温度下才能与 C 发生还原反应生成金属,其中金属 Nb 需在 1260 ℃以上的还原温度下才能生成,金属 Mn 需在 1425 ℃以上的还原温度下才能生成。SiO_2 需在 1640 ℃以上才能被还原,而 Al_2O_3、CaO 在 2000 ℃下仍不能被 C 还原成金属,留在二次渣中。

2.2 还原剂用量对金属合金化率以及元素含量的影响

在熔炼温度为 1550 ℃、熔炼时间为 120 min 的条件下,改变还原剂(石墨粉)与碱煮黑钨渣的质量比(以下简称还原剂用量),考察还原剂用量对 Fe、Mn、W、Sn、Nb 的合金化率和 As、Pb、Bi 在合金中含量的影响,结果如图 3 所示。

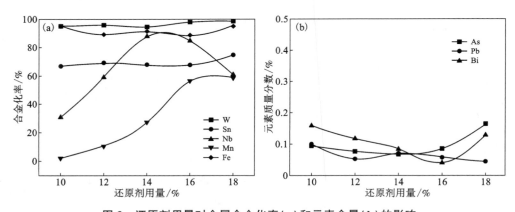

图 3 还原剂用量对金属合金化率(a)和元素含量(b)的影响

Fig. 3 Effect of reducing agent dosage on metal alloying rate (a) and element content (b)

由图3(a)可知，随着还原剂用量的增加，Fe、W、Sn 的合金化率变化较小，Nb 的合金化率呈先增大后减小的趋势，Mn 的合金化率呈先大幅增大后缓慢增大的趋势。当还原剂用量为10%时，Nb、Mn 合金化率均很低，分别仅为30.57%、2.06%，W、Fe、Sn 合金化率较高，分别为95.15%、95.10%、66.81%。当还原剂用量为16%时，Fe、Mn、W、Sn、Nb 合金化率分别为88.67%、56.28%、98.23%、67.74%、84.72%。继续加大还原剂用量，W 合金化率几乎没有变化，Fe、Sn、Mn 合金化率轻微增大，Nb 合金化率大幅下降。还原剂用量的增加促进了氧化锰的还原，使 Mn 合金化率逐渐增加。

由图3(b)可知，随着还原剂用量的增加，As、Pb、Bi 在合金中的含量先减少后增大，这是因为增加还原剂用量，还原进行得更加完全，但绝大部分 As、Pb、Bi 进入烟尘，As、Bi 在合金中的含量略有增大。综合考虑 Fe、Mn、W、Sn、Nb 的合金化率及 As、Pb、Bi 在合金中的含量，确定最适宜的还原剂用量为16%。

2.3 熔炼温度对金属合金化率及元素含量的影响

在还原剂用量为16%、保温时间为120 min 的条件下，考察熔炼温度对 Fe、Mn、W、Sn、Nb 合金化率和 As、Pb、Bi 在合金中含量的影响，结果如图4所示。

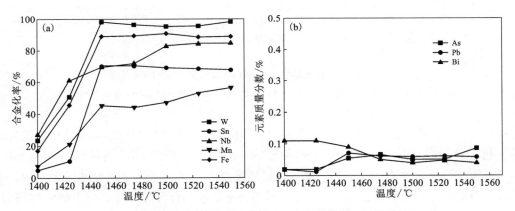

图4 熔炼温度对金属合金化率(a)和元素含量(b)的影响

Fig. 4　Effect of smelting temperature on metal alloying rate (a) and element content (b)

由图4(a)可知，随着熔炼温度的升高，Fe、W、Sn 的合金化率呈先增大后稳定的趋势，Nb、Mn 的合金化率呈逐渐增大的趋势。当温度为1400 ℃时，虽有成块的合金形成，但渣金分离效果不好，被还原的金属凝聚效果差，导致 Fe、Mn、W、Sn、Nb 的合金化率不高。当温度为1425 ℃时，获得了成块合金，但渣中金属夹杂严重，Fe、Mn、W、Sn、Nb 的合金化率仍较低。当温度为1450 ℃时，金属聚集效果好，各金属元素的合金化率均得到了较大的提高。继续升高温度，W、Fe、Sn 合金化率变化不大，Nb 和 Mn 合金化率逐渐增加，当温度为1550 ℃时，Fe、Mn、W、Sn、Nb 的合金化率分别为88.67%、56.28%、98.23%、67.74%、84.72%。升高熔炼温度有利于金属相聚集长大，增加各元素的合金化率。随着温度的升高，铌在合金中的溶解度增大，其合金化率增加[17]。同时，增加熔炼温度有助于氧化锰的还原。

由图4(b)可知，随着熔炼温度的升高，Bi 在合金中的含量呈先下降后稳定的趋势，升高温度有利于 Bi 挥发进入烟尘；Pb 含量呈先上升后稳定的趋势。而随温度升高，As 含量总体呈上升趋势，可能是因为高温时 As 与 Fe 形成了化合物进入合金，导致一部分 As 挥发不出去而留在合金中。综合考虑 Fe、Mn、W、Sn、Nb 的合金化率及 As、Pb、Bi 在合金中的含量，确定最适宜的熔炼温度为 1550 ℃。

2.4 熔炼时间对金属合金化率及元素含量的影响

在还原剂用量为 16%、熔炼温度为 1550 ℃ 的条件下，考察不同熔炼时间对 Fe、Mn、W、Sn、Nb 的合金化率和 As、Pb、Bi 在合金中含量的影响，结果如图 5 所示。

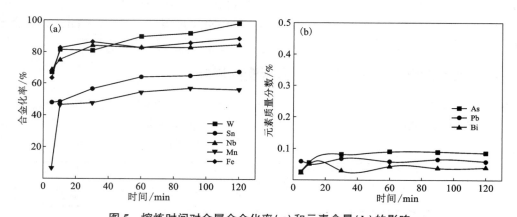

图 5 熔炼时间对金属合金化率(a)和元素含量(b)的影响

Fig. 5 Effect of smelting time on metal alloying rate (a) and element content (b)

由图 5(a)可知，随着熔炼时间的延长，Fe、Mn、W、Sn、Nb 的合金化率均呈现出先增大后趋于稳定的趋势。由于反应温度 1550 ℃ 较高，还原反应速率较快，熔炼时间为 10 min 时，各有价元素的合金化率即已达到较大值，随后缓慢增加。延长熔炼时间有助于金属相的凝聚与长大，当熔炼时间为 5 min 时，金属相凝聚不完全，导致合金化率不高，当熔炼时间为 5～10 min 时，合金化率大幅增加，表明此时金属相凝聚较好，合金化率较高。当熔炼时间在 60～120 min 时，合金化率增长缓慢，当熔炼时间为 120 min 时，Fe、Mn、W、Sn、Nb 的合金化率分别为 88.67%、56.28%、98.23%、67.74%、84.72%。

由图 5(b)可以看出，随着熔炼时间的延长，As 在合金中的含量先增加后保持稳定。熔炼 5 min 时金属相凝聚不完全，延长到 30 min 时金属相凝聚较完全，故 As 含量先增大后保持稳定。当熔炼时间为 120 min 时，As、Pb、Bi 质量分数均低于 0.1%。综合考虑 Fe、Mn、W、Sn、Nb 的合金化率及 As、Pb、Bi 在合金中的含量，确定最适宜的熔炼时间为 120 min。

2.5 合金的形貌分析

在还原剂用量 16%、熔炼温度 1550 ℃ 及熔炼时间 120 min 条件下所制备合金的 SEM-EDS 分析结果如图 6 所示。从图 6(a)可以看出，在还原剂用量为 16%、熔炼温度为 1550 ℃ 及熔炼时间为 120 min 的条件下，渣金分离效果良好。从图 6(b)可以看出，合金组织呈有规律的灰色条状，并夹杂着一些亮白色区域和黑色区域。由图 6(d_1)～图 6(d_3)可知，合金组织中的亮白区域主要为 Nb-W-C 合金相,并夹杂少量氧化硅;灰色区域主要为 Fe-Mn 合金相,伴随着少量铁锰氧化物。

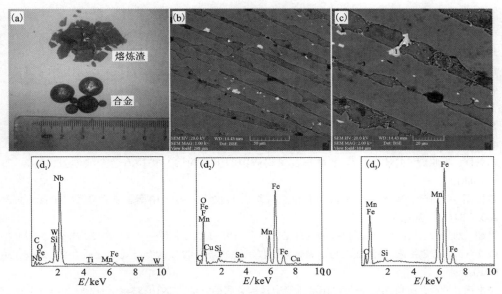

图 6　渣金宏观照片(a)、合金 SEM 图(b、c)及其区域能谱($d_1 \sim d_3$)

Fig. 6　Macroscopic photo of the slag and alloy (a), SEM images of the alloy (b, c) and its regional energy spectra ($d_1 \sim d_3$)

3　结论

(1)在碱煮黑钨渣还原熔炼过程中,各金属元素的还原顺序为 Bi>Pb>Sn>W>Fe>Nb>Mn;Al_2O_3、CaO 在 2000 ℃高温下仍不能被 C 还原,留在二次渣中。

(2)在碱煮黑钨渣还原熔炼过程中,随着还原剂用量的增加,Fe、W、Sn 的合金化率变化较小,Nb 的合金化率呈现出先增大后减小的趋势,Mn 的合金化率逐渐增加;随着熔炼温度的升高,Fe、W、Sn 的合金化率呈先增大后稳定的趋势,Nb、Mn 的合金化率呈逐渐增大的趋势;随着熔炼时间的延长,Fe、Mn、W、Sn、Nb 的合金化率均先增大后趋于稳定。

(3)碱煮黑钨渣的最佳还原熔炼条件为:还原剂用量 16%、熔炼温度 1550 ℃、熔炼时间 120 min;此条件下,Fe、Mn、W、Sn、Nb 的合金化率分别为 88.67%、56.28%、98.23%、67.74%、84.72%,As、Pb、Bi 基本进入烟尘。

参考文献

[1] LIU H, LIU H, NIE C, et al. Comprehensive treatments of tungsten slags in China: A critical review[J]. Journal of Environmental Management, 2020, 270: 110-927.

[2] 汪加军, 王晓辉, 黄波, 等. 废钨渣中钽、铌、钨高效共提新工艺研究[J]. 有色金属科学与工程, 2013, 4(5): 91-96.

[3] 苏正夫, 刘宇晖. 钨渣中钨回收利用新工艺研究[J]. 稀有金属与硬质合金, 2014, 42(4): 11-13.

[4] 范泽坤, 黄超, 徐国钻, 等. 钨冶炼渣酸分解减量资源化实验研究[J]. 稀有金属与硬质合金, 2020, 48(2): 1-4.

[5] 杨秀丽, 王晓辉, 向仕彪, 等. 盐酸法富集钨渣中的钽和铌[J]. 中国有色金属学报, 2013, 23(3): 873-881.

[6] 戴艳阳, 钟海云, 李荐, 等. 钨渣中有价金属综合回收工艺[J]. 中南工业大学学报(自然科学版), 2003, 34(1): 36-39.

[7] 肖超, 刘景槐, 吴海国. 低品位钨渣处理工艺试验研究[J]. 湖南有色金属, 2012, 28(4): 24-27.

[8] 张立, 钟晖, 戴艳阳. 钨渣酸浸与钠碱熔融回收钽铌的研究[J]. 稀有金属与硬质合金, 2008, 36(2): 6-10.

[9] 吴永谦, 吴贤, 马光, 等. 一种含钽富钨渣中钨钽分离的方法: CN103834815A[P]. 2014-06-04.

[10] 王晓辉, 向仕彪, 郑诗礼, 等. 一种从钨冶炼渣中提取钽、铌并联产氟硅酸钾的方法: CN102952951A[P]. 2013-03-06.

[11] 杜阳. 富钨渣中回收钨的研究[D]. 西安: 西安建筑科技大学, 2014.

[12] 戴艳阳, 钟晖, 钟海云. 废钨渣中同时回收铁锰的试验研究[J]. 应用化工, 2009, 38(6): 924-927.

[13] 杨俊彦, 郑其, 闫国庆, 等. 一种从钨冶炼渣回收钨的方法: CN107999274A[P]. 2018-03-27.

[14] 罗仙平, 刘北林, 唐敏康. 从钨冶炼渣中综合回收有价金属的试验研究[J]. 中国钨业, 2005, 20(3): 24-26.

[15] МАСЛОВВИ. Processing low grade material bearing tungsten by extraction of tungsten-iron alloy[J]. Nonferrous Metallurgy, 1992(10): 42-46.

[16] ЗЕЛИКМАНА H. Comprehensive treatment of tungsten concentrate[J]. Non-ferrous Metallurgy, 1993(7): 44-47.

[17] 何旭初, 范鹏, 周渝生, 等. 高炉中铌还原产生的碳化铌滞留带[J]. 北京科技大学学报, 1990(6): 504-509.

我国钨渣处理现状与研究进展

摘　要：我国是钨资源与生产大国，仲钨酸铵主要采用钨矿碱煮工艺，其生产过程产生大量的钨渣。通常钨渣中含有钨、钽铌、锡、铁、锰等多种有价金属具有极高的资源化利用价值，但同时也因含有砷等重金属有毒物质而列入危废管理。本文概述了我国钨渣基本特性、管理与处置现状，梳理了钨渣无害化与资源化工艺研究进展，提出了未来钨渣处理处置建议。

关键词：钨渣；无害化；资源化；有价金属

Current status and research progress of tungsten slag treatment in China

Abstract: With abundant tungsten resource, China is a powerhouse of tungsten industry. Ammonium paratungstate mainly adopts alkali boiling process of tungsten ore, which produces a large amount of tungsten slag. Generally, tungsten slag contains many valuable metals, such as tungsten, tantalum, niobium, tin, iron, manganese, etc., which have high resource utilization value, but at the same time, it also contains heavy metal toxic substances such as arsenic, which is included in hazardous waste management. The basic characteristics, management and disposal status of tungsten slag in China are summarized. On the basis or reviewing the research progress of harmless and resource utilization technology of tungsten slag, some suggestions for future treatment and disposal of tungsten slag are put forward.

Keywords: tungsten slag; harmless; resource recovery; valuable metals

引言

钨是我国重要的战略资源，具有高熔点、耐腐蚀、高温力学性能优异、压缩模量与弹性模量高等优良特性，可用于制造硬质合金、纯钨制品、钨合金、超合金、钢铁等多种材料，在

本文发表在《中国钨业》，2021，36(6)：62-70。作者：张金祥，罗教生，管建红，洪侃，黄叶钿，普建，李忠岐，陈远林，郭学益。

现代工业中被广泛应用于机械加工、航空航天、军事国防、电子信息等领域，其中交通运输、矿业/凿岩、工业制造是钨消费的前三大领域，占总消费量的一半以上。由于钨的稀缺性与不可替代性，目前已被各国列为重要战略金属，被誉为"高端制造的脊梁"。

钨是一种相对稀有的元素，地壳丰度为 1.0～1.5 mg/kg，地壳中含量为 0.007%[1]。我国是钨业大国，钨资源储量、生产量、贸易量和消费量均居世界第一，据美国地质调查局数据，2020 年世界钨储量 340 万 t(钨金属，下同)，钨精矿产量 8.4 万 t，其中，中国钨储量 190 万 t，占比 55.88%，钨精矿产量 6.9 万 t，占比 82.14%[2]。

仲钨酸铵(APT)是钨产业链中重要中间冶炼产品，其生产工艺按钨精矿分解方法可分为碱分解法(即碱煮工艺)、酸分解法和盐分解法，其中碱分解法是我国 APT 生产的主流工艺，分解产生的渣称之碱煮钨渣(钨渣)。钨渣中通常含有钨、钼、锡、钽、铌、钪等有价金属，具有较高的回收利用价值，同时也含有砷、氟等有害元素，浸出毒性强，存在环境污染风险，已被列入国家危废名录。几十年来我国累积的钨渣达百万吨，且每年以近 8 万 t 的速度增加，大量钨渣亟须进行无害化处理与资源化利用，由于钨渣特性差异大，常规方法难以实现无害化处置与有价金属的高效提取。因此，研究钨渣中有价元素清洁提取与有害元素安全处置技术并进行产业化推广应用，对我国钨产业绿色可持续发展具有重要意义。

1 钨渣基本特性、污染特征与管理

钨渣具有排放量大、成分及物相复杂的特点，通常含有钨、铁、锰、锡、铋、铌、钙、硅、氟、砷等多种有价、低价及有毒组分，具体化学组成随钨矿物原料成分而异，并含有碳酸钠、氢氧化钠等冶炼过程添加的药剂成分。2015 年，中国环境科学研究院杨金忠等[3]收集了 14 家仲钨酸铵(APT)生产企业生产过程中产生的钨渣，分别采用 HJT 299—2007《固体废物浸出毒性浸出方法硫酸硝酸法》和 GB 5085.3—2007《危险废物鉴别标准 浸出毒性鉴别》附录 S 推荐的方法分析了钨渣中重金属的浓度及浸出浓度，并研究了钨渣的污染特征，发现钨渣中 As、Mo 和 Hg 是特征污染物，应作为危险废物进行管理。

2016 年，钨渣被列入《国家危险废物名录》，正式作为危险废物进行管理，在此之前钨渣均按一般工业固废进行管理。2020 年我国修订发布的《国家危险废物名录(2021 版)》中钨渣仍作为危废进行管理，但满足《水泥窑协同处置固体废物污染控制标准》(GB 30485)和《水泥窑协同处置固体废物环境保护技术规范》(HJ 662)要求可进入水泥窑协同处置，且处置过程不按危险废物管理。江西是我国的钨资源与钨产业大省，其中 APT 年产量超 4 万 t，占全国一半以上，但由于企业钨渣堆场容量有限以及委外处置费用较高等问题，钨渣的处置成为钨冶炼企业面临的共性问题。为保障钨冶炼企业正常生产运营，2020 年 10 月江西省发布了两项地方标准《钨冶炼固体废物利用处置技术指南第 1 部分：水泥窑协同处置(DB36T 1295.1—2020)》和《钨冶炼固体废物利用处置技术指南第 2 部分：玻璃化处理(DB36T 1295.2—2020)》，指导相关企业采用水泥窑协同处置与玻璃化处理两种方式对钨渣进行处置，有效化解了钨渣积压的问题。

国外对钨渣的监管大都按一般固废处理办法进行管理，在美国钨精矿分解渣通常作为非有害废弃物，在被允许处置非有害工业废弃物的垃圾填埋场进行处置。欧盟则将钨精矿分解渣归类于"用物理和化学方法加工金属矿物质产生的废弃物"类别，钨精矿分解渣生产者需要确定分解渣成分，如出现某种或多种危险特性或含有毒有害成分浓度超过规定浓度限值，属于危险废物。日本将钨精矿分解渣作为产业废物进行管理[4]。

2 钨渣资源化利用

钨渣中除钨外还含有钽、铌、钪、锡、铋等多种有价元素，其品位远高于矿石中相应元素的品位，具有很高的回收利用价值[5]。针对钨渣中有价金属的回收国内外学者进行了大量研究，由于我国钨资源的绝对优势，钨渣回收的研究也主要集中在国内，国外则集中在俄罗斯，均起步于20世纪70年代。目前，钨渣的资源化利用主要分为两个方向：一是回收有价金属；二是制备新材料。

2.1 回收有价金属

2.1.1 钨的回收

钨渣中钨的回收一直是研究的热点之一，早期钨渣中钨的质量分数可达3%，近年来随着钨冶炼技术的发展与进步，钨质量分数降低到了1%左右，但仍具有很高的回收价值。从回收工艺来看，钨的回收主要有酸浸、碱焙烧-碱浸、碱焙烧-水浸等工艺。

苏正夫等[6]采用盐酸浸出-离子交换工艺回收钨渣中的钨，在优化条件下，钨浸出率为86%以上。王钦建[7]采用酸分解-萃取工艺回收钨渣中的钨，在最佳工艺条件下，最终的钨回收率为92.8%。中国专利CN102212697A[8]运用D314弱碱性阴离子树脂吸附钨渣盐酸酸浸液，解析液浓缩结晶得到钨酸钠的品位可达60%。肖超等[9]研究了一种以硫酸为浸出试剂、磷酸为添加剂的全湿法处理低品位钨渣的新工艺，在优化条件下，钨、钼的浸出率分别为69.7%与31.6%。杨利群[10]采用苏打烧结法处理低品位钨矿和废钨渣，通过添加石英粉、苏打、硝石、食盐等辅料，经磨碎混合后入回转炉烧结，再经棒磨、浸出、过滤、浓缩结晶等工序可获得粗钨酸钠溶液，滤渣中钨可降至0.5%以下。

戴艳阳[11]研究了苏打焙烧-水浸工艺回收钨渣中的钨，在优化条件下，钨的浸出率达到88.4%，水浸出渣中残余钨含量仅为0.32%，浸出液循环使用三次，浸出液中WO_3质量浓度达到16.6 g/L，钨回收率为88%以上。范泽坤等[12]采用盐酸作分解试剂处理"氢氧化钠+磷酸"冶炼黑白钨混合矿所产生的钨冶炼渣，在优化条件下，渣中钨质量分数从最初的2.56%升至9.35%，提高为之前的近4倍，并且反应过程无较大钨损，91.02%的钨仍留在渣中，酸反应渣量减至原钨渣的25%，并可与钨精矿混料进行二次冶炼。杨少华等[13]采用碳酸钠焙烧-氢氧化钠浸出的方法，从含钨1.4%的碱浸钨渣中回收钨，钨的浸出率可达90.5%。中国专利CN103103359A[14]提出一种利用低度钨渣再生APT的方法，通过改进传统苏打烧结法工艺参数，在焙烧时配入硝石并对烧结料进行湿磨浸出，得到的浸出液采用双离子交换法除杂生产APT，WO_3回收率最高达96.7%。Паланг А. А.[15]将钨渣和硫酸钠及固体还原剂一起烧结，烧结料再进行钨的水溶液浸出，在最佳条件下，钨的提取率大于90%。

表1总结对比了几种主要的提钨工艺，由表可知，钨的浸出率/回收率基本为85%以上，其中酸浸工艺中盐酸效果优于硫酸，萃取效果优于离子交换，苏打焙烧-浸出工艺中氢氧化

钠浸出率略高于水浸。

表 1　钨渣主要提钨工艺对比

Tab. 1　Comparison of main technologies for extracting tungsten from tungsten slag

提钨工艺	关键工艺与参数	原渣钨品位(WO_3)/%	回收率/%	浸出率/%	参考文献
酸浸-离子交换	酸浸：1%添加剂A、盐酸浓度20%、100 ℃、3 h、液固比3∶1	1.8~2.3		>86	苏正夫等[6]
酸浸-萃取	酸浸：盐酸浓度70%、室温、3~4 h、液固比5∶1；萃取：油水相比1∶1、10% N235、水相pH 2	2.9~3.1	92.8		王钦建[7]
酸浸	酸浸：63%浓硫酸、3.2%浓度为85%的磷酸、液固比3∶1、80 ℃、2 h	2.25		69.7	肖超等[9]
苏打焙烧-水浸	焙烧：37%苏打、850 ℃、1.5 h；浸出：以磷酸钠为添加剂、球磨水浸、浸出温度大于95 ℃、1 h、液固比5∶1	2.52	>88	88.4	戴艳阳[11]
碳酸钠焙烧-氢氧化钠浸出	焙烧：90%碳酸钠、800 ℃、1 h；浸出：80 ℃、45 min、NaOH溶液130 g/L、液固比4∶1	1.4		90.5	杨少华等[13]

2.1.2　钽铌的回收

钽、铌属于难熔稀有金属，具有熔点高、塑性好、导电导热性能好、化学稳定性高等特点，广泛应用于钢铁工业、航空航天、电子工业、超导技术等领域。我国钽铌资源相对匮乏，同时钽铌矿品位较低，开采成本高，逾85%钽铌矿依赖进口。国内部分钨渣中的(Ta_2O_5+Nb_2O_5)品位达到0.2%~0.8%，远高于钽铌原矿品位，具有极高的回收价值。杨秀丽等[16]提出稀盐酸脱硅-浓盐酸深度脱铁锰的钽铌酸法富集新工艺，在最优条件下，钽和铌回收率分别为86.57%和82.48%。向仕彪等[17]采用稀酸脱硅-浓酸脱铁锰-HF酸浸出-蒸发浓缩工艺回收钨渣中的钽铌，在最优条件下，钽和铌的回收率为80%以上。戴艳阳等[18]采用钠碱熔融-水淬浸出-盐酸浸出工艺回收钨渣中的钽铌，在最佳条件下获得Ta_2O_5和Nb_2O_5质量分数分别为3.465%和9.13%的Ta和Nb富积渣，铌和钽的回收率分别为67.6%和73.2%。张立等[19]采用酸浸-钠碱熔融法从钨渣中富集和回收钽铌，在最优条件下，Ta_2O_5由钨渣中的0.068%富集到0.48%，Nb_2O_5由0.467%富集到2.74%，钽和铌的总回收率分别为76.4%和63.3%。表2总结了钨渣中回收钽铌的主要工艺对比情况，由表2可知，酸浸工艺钽铌的回收率为80%以上，略高于碱熔工艺。

表2 钨渣中回收钽铌的主要工艺

Tab. 2 Main technologies of recovering tantalum and niobium from tungsten slag

钽铌回收工艺	关键工艺与参数	原渣钽铌品位/%		富集后钽铌品位/%		回收率/%		参考文献
		Ta_2O_5	Nb_2O_5	Ta_2O_5	Nb_2O_5	Ta_2O_5	Nb_2O_5	
稀盐酸脱硅-浓盐酸脱铁锰	脱硅：盐酸浓度6%、液固比6∶1、25 ℃、5 min；脱铁锰：盐酸浓度22%、15 min、90 ℃、理论酸用量	0.16	0.50	2.81	9.23	86.57	82.48	杨秀丽等[16]
稀酸脱硅-浓酸脱铁锰-HF酸浸出	酸浸：HF酸浓度40%、2 h、1.2倍理论酸用量、≥60 ℃	0.14	0.59	8.0*	35.4*	80.4	87.4	向仕彪等[17]
钠碱熔融-水浸-酸浸	钠碱熔融：碱渣比1.2、800 ℃、80 min；酸浸：20%盐酸加热浸出，2倍理论酸用量	0.10	0.57	3.465	9.13	73.2	67.6	戴艳阳等[18]
酸浸-钠碱熔融	酸浸：盐酸浓度5%、40 ℃、30 min、为2.5倍理论酸用量；钠碱熔融：碱渣比1.5、800 ℃、60 min	0.068	0.467	0.48	2.74	76.4	63.3	张立等[19]

*稀酸脱硅-浓酸脱铁锰-HF酸浸出钽铌回收工艺富集后钽铌品位单位为g/L。

2.1.3 铁锰的回收

采用黑钨矿或者黑白钨混合矿生产APT的过程中产生的钨渣，通常含有较多的铁锰（>15%），具有一定的回收价值。戴艳阳等[20]采用硫酸浸出-化学法除杂-共沉淀-煅烧的工艺从钨渣中回收铁、锰，制备出了锰锌铁氧体粉末，Fe、Mn浸出率分别为86.5%和88.4%。戴艳阳等[21]还研究了从钨渣中回收锰的新工艺，提出了钨渣低温硫酸化焙烧与浸出-硫化物沉淀除重金属-硫酸复盐法深度净化-中和水解除Fe-水解沉锰-H_2O_2氧化分解的工艺路径，Mn浸出率达到88.9%，制备出粒度小于0.1 μm的Mn_3O_4粉末。张建平[22]采用硫酸浸出-中和除铁-氟化剂除钙-硫化锰除重金属-碳化反应除钾钠-浓缩结晶工艺从钨渣中回收锰，制备电池级硫酸锰。刘健聪等[23]研究了钨渣硫酸浸出回收铁锰的工艺，在优化条件下，可以使铁、锰的浓度提高大约50%和38%。谭晓恒[24]研究了黑钨渣磁化焙烧回收铁锰技术，在优化工艺条件下，获得了品位为47.81%的铁精矿和品位为35.32%的锰精矿，Fe回收率为63.32%，Mn回收率为63.65%。

2.1.4 钪的回收

钪的原矿资源很少，通常伴生在钨矿、稀土矿等矿源中。我国部分钨渣中钪的氧化物品位一般在0.2%以上，可作为钪生产的重要原料之一。

周国涛等[25]研究采用P204-TPB从钨渣硫酸浸出液中萃取钪，获得的粗钪纯度为82%

以上，钪总回收率为92%以上，其主要工艺流程为：钨渣→硫酸浸出→P204、TPB萃取→硫酸洗脱→氢氧化钠反萃。刘彩云等[26]研究采用伯胺N1923萃取剂从钨渣的硫酸浸液中回收钪，通过"浸出-铁粉还原-萃取-沉淀"技术回收钨渣中的钪，钪总萃取回收率为92.33%。梁焕龙等[27]采用硫酸化焙烧-水浸工艺从钨渣中浸出钪，最佳条件下，氧化钪浸出率为93%以上。钟学明[28]采用硫酸浸出-伯胺萃取-盐酸反萃-叔胺萃取除铁-氨水和草酸两次沉淀-灼烧工艺回收钨渣中的钪，最终获得氧化钪，其纯度为90%，收率为82%。杨革[29]研究了从钨渣中提取制备高纯氧化钪的工艺，并进行了工业化试验与试生产，产品氧化钪纯度大于99.99%，实收率为45%。刘慧中等[30]研究了一种从钨渣中提取钪的方法，钨渣通过硫酸浸出、伯胺N1923萃取、盐酸反萃、草酸沉淀、加热灼烧可获得氧化钪，钪的回收率为80%~85%，纯度为99%以上。徐廷华等[31]选用酸性磷类萃取剂（P204、P507），首次提出低萃取剂浓度、大相比萃取体系，从钨渣硫酸浸出液中提取钪，钪由钨渣中的万分之几提高到72.8%。徐廷华等[32]还研究了乳状液膜法从钨渣浸出液中提取钪，一次提钪率达72.6%。聂华平等[33]对比研究了Cyanex 572、P507、Cyanex923、TBP四种萃取剂对钨渣浸出液中回收钪的影响规律，发现Cyanex 572对钪具有非常优异的萃取分离性能，可使钪浓度提高近800倍，回收率达到90.9%。丁冲等[34]研究了钨渣硫酸浸出过程中草黄铁矾法抑制铁浸出的工艺，在最佳条件下，钪浸出率为87%，铁浸出率由98%降至57%，实现了铁的抑制浸出。

总体来看，钨渣中提钪主要有盐酸浸出-萃取与硫酸浸出萃取两大类，并以硫酸浸出工艺为主，萃取的差别主要在萃取体系的选择上，主要的工艺对比情况如表3所示。

表3 钨渣中主要的提钪工艺对比

Tab. 3 Comparison of main scandium extraction processes from tungsten slag

钪回收工艺	萃取工艺参数	原渣钪品位/%	富集后钪品位/%	回收率/%	参考文献
硫酸浸出-萃取	5%P204+3%TBP+煤油、O/A=1:10	0.02	>82	>95	周国涛等[25]
	10%伯胺N1923、O/A=1:1	0.022		92.33	刘彩云等[26]
	4%伯胺N1923+0.8%ROH+煤油、O/A=1:4	0.037	90	82	钟学明等[28]
	7%P204+3%TBP+煤油、O/A=1:10	0.02~0.03	>99.99	45	杨革[29]
	15%伯胺N1923+煤油、O/A=1:10		>99	80~85	刘慧中等[30]
盐酸浸出-萃取	Cyanex 572、O/A=1:4	浸出液：0.035	洗脱液：28.623	90.9	聂华平等[33]

2.1.5 多种有价金属综合回收

戴艳阳[35]采用苏打烧结-水浸出-硫酸浸出-浸出液净化-共沉淀-烧结工艺回收钨渣中的钨、钽、铌、铁、锰，各金属的总回收率分别为88.1%、78%、56%、95.2%、68.5%。汪加军等[36]采用氟盐转型-HF-H_2SO_4浸出-氟盐氨转化循环利用过程同步提取废钨渣中的钽、铌、钨的新工艺,在最优条件下可分别获得 Ta_2O_5 和 Nb_2O_5 质量分数分别为 6.08% 和 27.29% 的钽铌富集渣及 WO_3 含量为 26.71% 的钨富集渣,钽、铌、钨的单程回收率分别达到 83.18%、88.33% 和 77.91%。罗教生等[37]采用还原熔炼法综合回收钨渣中的有价金属,提取出铁锰钨铌等多元素合金。罗仙平等[38]采用浮选-重选工艺从钨渣中回收了铋、钨和锡,获得的铋、钨、锡精矿主金属的品位分别为:8.34%、17.51%、35.39%,对应主金属回收率分别为:72.62%、53.23%、65.94%。郭超[39]研究了碳热还原法回收钨渣中有价金属工艺,铁回收率可达 93%,锰回收率约为 26%,钨回收率在 30%~70%。

中国专利 CN102212697A[8] 采用盐酸溶液低温常压浸出钨渣,并在浸出过程添加少量的钨稳定剂(碱金属氟化物和磷酸盐的混合物),钨在浸出液中富集回收,浸出渣则通过后续一系列的萃取、除杂、离子交换等工序分别回收钽铌、铁锰、钪等多种有价金属。中国专利 CN105154683B[40] 采用臭氧碱浸-氯化钙沉淀的方法回收钨渣中的钨,获得人造白钨,然后针对浸出渣运用盐酸浸出、盐酸络合浸出、萃取等工序综合回收钽铌银、镍钴铜等多种有价金属。中国专利 CN107999271A[41] 采用选-冶联合工艺综合回收钨渣中的铁、锰、钨、锡、钪、钽铌多种有价金属。中国专利 CN103614545B[42] 公开了一种低品位钨精矿、钨渣的处理方法,采用还原焙烧-中性浸出-磁选-酸浸工艺提取其中的钨、铁、锰、钽、铌等多种有价金属。湖南某循环经济技术研发中心[43]新建年处理 3 万 t 钨渣生产线,采用酸溶-碱转-萃取法,综合提取钨渣中的 W、Fe、Mn、Sc、Ta、Nb 等多种有价元素,其中 W 和 Sc 的收率分别达到 90% 和 80%。

Зеликман А. Н. 等[44]采用苏打高压浸出-盐酸浸出处理钨渣,其中 93% 的钨进浸出液,98% 的铁锰和 86%~89% 的钪进入盐酸浸出液,96% 以上的钽铌留在酸浸残渣中,残渣进一步采用硫酸盐-过氧化物处理可得到 $(Ta,Nb)_2O_5$ 40%~60% 的精矿,或者经碱液处理得 $(Ta,Nb)_2O_5$ 14%~17% 的精矿,Ta、Nb 的总回收率达到 70%~80%。Зешктан А. Н.[45-46] 不仅采用盐酸法处理钨渣,得到含 $(Ta,Nb)_2O_5$ 4%~6% 的精矿,通过萃取得到 Sc_2O_3 3%~4% 的精矿,还研究了钨渣的铝热还原,制备出含有钨、钽和铌的 Fe-Mn 合金,在小型试验中,钨入合金的回收率不超过 72%,扩大试验则大大改善了相分离,钨进入合金的回收率提高到 86.8%,但 Ta 与 Nb 的回收率仅约 40%。

综上几类钨渣资源化利用工艺,可以发现钨渣回收以多种有价元素的综合回收为主,其次是针对钪、钨、钽铌的提取,关于铁锰的回收相对较少。表 4 从回收元素种类以及回收率两个方面对比了几种主要的钨渣综合回收工艺,可以看出湿法工艺在回收元素种类以及回收率方面均优于火法工艺,但流程长且复杂。

表 4 几种主要的钨渣综合回收工艺对比

Tab. 4 Comparison of several main comprehensive recovery processes of tungsten slag

综合回收工艺	回收元素	回收率/%	参考文献
苏打烧结-水浸-硫酸浸出-净化-共沉淀-烧结	钨、钽、铌、铁、锰	88.1、78、56、95.2、68.5	戴艳阳[35]
氟盐转型-HF、H_2SO_4 浸出-氟盐氨转化循环	钽、铌、钨	83.18、88.33、77.91	汪加军等[36]

续表4

综合回收工艺	回收元素	回收率/%	参考文献
浮选-重选	铋、钨、锡	72.62、53.23、65.94	罗仙平等[38]
碳热还原	铁、锰、钨	93、26、30~70	郭超[39]
酸溶-碱转-萃取法	钨、铁、锰、钪、钙、铋、钽、铌等	钨：90，钪：90	谢建清[43]
铝热还原	钨、钽、铌	86.8、40、40	ЗеликманА. Н.[46]

2.2 制备新材料

近年来，钨渣的处理不再局限于有价金属的回收，国内部分学者开始研究利用钨渣处理氨氮废水及钨冶炼废水的"以废治废"工艺。郭欢[47]研究了以硅藻土与钨渣为主要原料烧结制备多孔陶粒的工艺及其对人工模拟的氨氮废水处理效果，通过烧结工艺优化，制取的多孔陶粒其吸水率达44.93%，孔隙率44.56%。靖青秀等[48]以硅藻土和钨渣为主要原料制备了多孔陶粒并研究其对离子型稀土矿区土壤淋滤液中氨氮的吸附去除规律，在优化条件下，陶粒对氨氮的饱和吸附量达到1.60 mg/g。邹瑜等[49]研究利用钨渣一步净化钨冶炼废水中氟、磷和砷的新方法，在优化工艺条件下，经钨渣处理后的钨冶炼废水中的残留氟浓度为9.589 mg/L，磷浓度为0.0342 mg/L，砷浓度为0.0274 mg/L。

总体来看，钨渣资源化利用工艺多种多样，既有选矿、湿法、火法或湿法-火法联用、选冶联合等传统技术，也有利用钨渣制备多孔陶粒来处理氨氮废水等新工艺；既可以针对一种或几种有价金属单独回收，也可以综合回收多种有价金属。但仍然存在回收成本高、经济效益差、二次污染等问题，导致无法进行产业化应用。因此，研发钨渣无害化与资源化处理新工艺，解决钨渣中的有毒有害物质污染问题，高效绿色回收有价金属，仍然是我国钨冶炼行业绿色发展的迫切需求。

3 钨渣无害化处理

2015年，杨金忠等[3]采集了多家APT生产企业生产过程产生的钨渣并研究其污染特性，发现钨渣浸出浓度较大的是Pb、As和Hg，其最大值分别为33.6 mg/L、26.2 mg/L和0.85 mg/L，超出GB 5085.3—2007中规定的相应限值的6.72、5.24和8.5倍，超标率分别为14.3%、21.4%和42.9%，建议钨渣应作为危险废物进行管理。2016年，仲钨酸铵生产过程中碱分解产生的碱煮渣（钨渣）被列入《国家危险废物名录》，2020年我国修订发布的《国家危险废物名录（2021版）》中钨渣仍作为危废进行管理。钨渣一旦作为危废进行管理，其资源化利用过程必须明确有毒物质的来源及走向，确保有毒物质全部转性为一般固废或者相关产品，否则钨渣资源化过程产生的二次废渣、废水、废气仍作为危废管理。目前我国关于钨渣综合回收利用的工艺虽然多种多样，但利用过程很少涉及有毒物质的转化行为及安全处置研究，而单纯的钨渣无害化处理工艺更是鲜有报道。

张钦汉[50]发明了一种钨渣无害化综合回收利用系统，通过整体的设计，回收处理效果好，防止了钨渣的随意排放对环境造成的影响，同时回收了钨；饶日荣等[51]申请了《一种APT固体废渣无害化综合回收利用系统》专利，系统包括物料回收、再循环和重复利用、收集后分离、资源价值利用和组成其他物料。

总体来看，目前我国还缺乏成熟的钨渣无害化处理技术，钨渣的处理主要以填埋、水泥窑协同处置等末端粗放技术为主，潜在污染风险，且未能回收钨渣中钨、钽、铌、锡、铋等有价金属，造成资源浪费，绝非长久之计。

4 结语与展望

2016年以前，国内外关于钨渣回收及资源化利用的研究多集中在有价金属（W、Sn、Ta、Nb、Sc等）回收，取得了许多成果，但存在经济效益较低、产生二次污染等问题，相关工艺技术在产业化及应用过程中仍面临较大的挑战。

2016年后，碱煮钨渣被列入《国家危险废物名录》，众多学者、企业及研究机构进一步研究了钨渣无害化与资源化处理工艺，但大多处于实验室研究阶段，且存在工艺复杂、成本高、二次污染、经济效益差等一种或多种问题而未能产业化推广应用，导致我国钨渣仍然以水泥窑协同处置为主，大量有价金属进入水泥后失去回收价值，造成极大的资源浪费。随着环保要求越来越严以及未来资源的紧缺，钨渣的处置应主要围绕绿色、高效、低成本、综合回收、高附加值产出等方向开展新工艺研发工作，在实现钨渣无害化的前提下，综合利用过程还必须解决工艺本身的环保问题与经济效益问题。短期来看，开发低成本高效率的钨渣无害化工艺是当务之急，解决钨渣中砷等有毒物质的安全处置问题，保障APT生产企业的稳定运营。中期来看，研发钨渣无害化-资源化协同处理新工艺势在必行，厘清钨渣中的有毒物质在资源化利用过程中的走向与转变规律，实现钨渣中有价元素的清洁提取与高值化利用，以促进钨冶炼行业的可持续发展。长远来看，基于清洁生产理念，革新钨冶炼技术，从源头上实现钨渣的减量化、无害化、资源化，是解决钨渣根本问题的最佳途径。

参考文献

[1] LIU Hu, LIU Haoling, NIE Chenxi, et al. Comprehensive treatments of tungsten slags in China: A critical review[J]. Journal of Environmental Management, 2020, 270: 1-12.

[2] 中国钨业协会.中国钨工业发展规划（2016—2020年）[J].中国钨业, 2017, 32(1): 9-15.

[3] 杨金忠, 高何凤, 王宁, 等.仲钨酸铵（APT）生产中钨渣的污染特性分析[J].环境工程技术学报, 2015, 5(6): 525-530.

[4] 吴昊, 曾欣荣, 刘宏博, 等.我国钨渣管理存在的问题及建议[J].硬质合金, 2020, 37(6): 460-465.

[5] 夏文堂.钨的二次资源及其开发前景[J].再生资源研究, 2006(1): 11-17.

[6] 苏正夫, 刘宇晖.钨渣中钨回收利用新工艺研究[J].稀有金属与硬质合金, 2014, 42(4): 11-13.

[7] 王钦建.黑钨渣的酸分解与萃取工艺优化研究[J].循环经济, 2009, 29(11): 37-39.

[8] 湖南稀土金属材料研究院.钨渣处理方法：102212697A[P].2011-10-12.
[9] 肖超,刘景槐,吴海国.低品位钨渣处理工艺试验研究[J].湖南有色金属,2012,28(4)：24-26.
[10] 杨利群.苏打烧结法处理低品位钨矿及废钨渣的研究[J].中国钼业,2008,32(4)：25-27.
[11] 戴艳阳.钨渣中有价金属综合回收新清洁工艺研究[D].长沙：中南大学,2013.
[12] 范泽坤,黄超,徐国钻,等.钨冶炼渣酸分解减量资源化实验研究[J].稀有金属与硬质合金,2020,48(2)：1-4.
[13] 杨少华,王君,谢宝如,等.低品位钨渣处理工艺[J].有色金属科学与工程,2015,6(6)：29-32.
[14] 陈泉兴,张中山.一种利用APT废低度钨渣再生APT的方法：103103359A[P].2013-05-15.
[15] Паланг А А. Recovery of tungsten from processing residue of wolframite[J]. Metallurgy, 1999(5)：19-22.
[16] 杨秀丽,王晓辉,向仕彪,等.盐酸法富集钨渣中的钽和铌[J].中国有色金属学报,2013,23(3)：873-881.
[17] 向仕彪,黄波,王晓辉,等.从废钨渣中酸法回收钽铌的研究[J].有色冶金设计与研究,2012,33(2)：5-11.
[18] 戴艳阳,钟晖,钟海云.钨渣中钽铌回收研究[J].有色金属,2009,61(3)：87-89.
[19] 张立,钟晖,戴艳阳.钨渣酸浸与钠碱熔融回收钽铌的研究[J].稀有金属与硬质合金,2008,36(2)：6-14.
[20] 戴艳阳,钟晖,钟海云.废钨渣中同时回收铁锰的实验研究[J].应用化工,2009,38(6)：924-927.
[21] 戴艳阳,钟晖,钟海云.钨渣回收制备四氧化三锰新工艺[J].中国有色金属学报,2012,22(4)：1242-1247.
[22] 张建平.钨冶炼渣制备电池级硫酸锰的工艺研究[D].赣州：江西理工大学,2018.
[23] 刘健聪,熊道陵,张建平,等.钨冶炼渣中铁、锰浸出工艺研究[J].有色金属科学与工程,2018,9(4)：14-20.
[24] 谭晓恒.黑钨渣磁化焙烧回收铁锰的技术研究[D].赣州：江西理工大学,2020.
[25] 周国涛,李青刚,刘永畅,等.从钨渣硫酸浸出液中萃取钪的研究[J].稀有金属与硬质合金,2018,46(6)：1-9.
[26] 刘彩云,符剑刚.钨渣中钪的萃取回收实验研究[J].稀有金属与硬质合金,2015,43(5)：4-11.
[27] 梁焕龙,罗东明,刘晨,等.从钨渣中浸出氧化钪的试验研究[J].湿法冶金,2015,34(2)：114-116.
[28] 钟学明.从钨渣中提取氧化钪的工艺研究[J].江西冶金,2002,22(3)：19-22.
[29] 杨革.从钨渣中提取高纯氧化钪[J].湖南有色金属,2001,17(1)：18-20.
[30] 刘慧中,汤惠民.从钨渣中提取钪的研究[J].上海环境科学,1990,9(3)：11-14.
[31] 徐廷华,邓佐国,李伟,等.从钨渣浸出液中提取钪的研究[J].江西有色金属,1997,11(4)：32-36.
[32] 徐廷华,邓佐国,傅嘉.乳状液膜法从钨渣浸出液中提取钪的研究[J].中国钨业,1998,143(2)：30-33.
[33] NIE H P, WANG Y, WANG Y L, et al. Recovery of scandium from leaching solutions of tungsten residue using solvent extraction with cyanex 572[J]. Hydrometallurgy, 2018, 175：117-123.
[34] 丁冲,王晓辉,何超然,等.黑钨渣硫酸浸钪及浸出过程中草黄铁矾法抑制铁浸出[J].过程工程学报,2014,14(6)：907-914.
[35] 戴艳阳.钨渣中有价金属综合回收新清洁工艺研究[D].长沙：中南大学,2013.
[36] 汪加军,王晓辉,黄波,等.废钨渣中钽、铌、钨高效共提新工艺研究[J].有色金属科学与工程,2013,4(5)：91-96.
[37] 罗教生,王莉莉,王冠亚.钨渣的综合利用研究[J].江西冶金,1998,18(6)：31-32.

[38] 罗仙平,刘北林,唐敏康.从钨冶炼渣中综合回收有价金属的试验研究[J].中国钨业,2005,20(3):24-26.
[39] 郭超.碱煮钨渣碳热还原过程热力学机理[D].赣州:江西理工大学,2020.
[40] 湖南世纪垠天新材料有限责任公司.钨渣中有价金属的分离回收方法:105154683B[P].2017-05-17.
[41] 北京有色金属研究总院.一种从APT钨冶炼渣综合回收有用金属的方法:107999271A[P].2017-11-17.
[42] 中南大学.一种低品位钨精矿、钨渣的处理方法:103614545B[P].2014-03-05.
[43] 谢建清.钨渣回收利用技术研究现状[J].中国钨业,2019,34(1):50-57.
[44] ЗЕЛИКМАН A H. Comprehensive treatment of tungsten concentrate[J]. Non-ferrous Metallurgy, 1993(7): 44-47.
[45] ЗЕШКТАН A H. Study on ecological technique process without pollution of residue from tungsten metallurgy and its theoretical basis[J]. Non-ferrous Metallurgy, 1995(2): 49-52.
[46] ЗЕЛИКМАН A H. Aluminothermic reduction of tungsten production waste[J]. Nonferrous Metallurgy, 1996(11): 44-46.
[47] 郭欢.硅藻土-钨渣基多孔陶粒的制备与性能研究[D].赣州:江西理工大学,2017.
[48] 靖青秀,王云燕,柴立元,等.硅藻土-钨渣基多孔陶粒对离子型稀土矿区土壤氨氮淋滤液的吸附[J].中国有色金属学报,2018,28(5):1033-1041.
[49] 邹瑜.钨冶炼渣净化钨冶炼废水新工艺及机理研究[D].赣州:江西理工大学,2018.
[50] 赣州卓越再生资源综合利用有限公司.一种钨渣无害化综合回收利用系统:205368461U[P].2016-07-06.
[51] 大余县东宏锡制品有限公司.一种APT固体废渣无害化综合回收利用系统:108160670A[P].2018-06-15.

铁热还原法处理钙砷渣及金属砷的制备工艺

摘 要：在资源与环境的双重压力下，如何解决有色冶金行业产生的冶炼钙砷渣无害化和资源化问题，成为彻底消除"砷害"的关键所在，对于整个冶金工业具有非常重要的意义。本文采用铁热还原短流程直接还原钙砷渣，砷以砷单质的形式得以回收。研究结果表明，在还原温度1050 ℃，还原时间30 min，铁配入系数为1.5 的工艺条件下，可以得到比较理想的脱砷效果，砷还原率可达96.56%。所得产物和还原后渣经 XRD、扫描电镜和能谱仪分析，生成物为单质砷，纯度达99%，反应后渣中主要成分为 $Ca_2Fe_2O_5$ 和 CaO。

关键词：钙砷渣；砷；铁热还原；热力学分析

Preparation of metal arsenic from calcium arsenic slag by ferrothermal reduction

Abstract: Under the dual pressure of resources and the environment, how to deal with the issues of harmlessness and utilization of metallurgical calcium arsenic slag produced by the nonferrous metallurgical industry has become the key to eliminating "arsenic damage" thoroughly, which has great significance to the whole metallurgical industry. In this paper, calcium arsenic slag was directly reduced by a short process of iron reduction, and arsenic was recovered in the form of arsenic. The results showed that under the conditions of a reduction temperature of 1050 ℃, holding time of 30 min, and iron blending coefficient of 1.5, the ideal arsenic removal effect could be obtained, and the reduction rate of arsenic could reach 96.56%. The product and reduced slag were analyzed by XRD and SEM-EDS. The reduction products were elemental arsenic with a purity of approximately 99%, and the main components of the slag were $Ca_2Fe_2O_5$ and CaO.

Keywords: calcium arsenic slag; arsenic; iron reduction; thermodynamic analysis

---

本文发表在《有色金属科学与工程》，2022，13（2）：22-30。作者：刘子翔，梁佳昀，孙京博，龚傲，田磊。

引言

As 元素广泛存在于自然界,共有数百种的 As 矿物已被发现[1]。As 与其化合物被广泛应用于农药、除草剂、杀虫剂和多种合金中。As 的毒性极强,且具有致癌作用[2-3]。我国有 300 万余人面临 As 中毒的威胁,属于地方性 As 中毒危害最严重的国家之一,As 对环境的危害是不可逆的[4-5]。我国矿产资源采冶活动引起的 As 害事件常见报道,在部分有色金属的开发和冶炼中,As 作为伴生金属,常常有或多或少的 As 化物排出[6-8],因此砷的无害化处置极为迫切。

在工业上处理含砷废水一般都采用操作相对简单的钙盐沉砷法,该方法是将溶液中的砷转化为钙砷渣沉淀脱除,但最终获得的钙砷渣不稳定、毒性很大,仍需进一步处理。目前,对于含砷废渣的处理也有许多研究[9]。L. G. Twidwell 等对含砷渣进行玻璃固化,并且通过实验证明可使其长期稳定保存[10]。赵宇文等研究表明,砷在玻璃化中由 As(Ⅲ)转变为 As(Ⅴ),且玻璃固化反应后固体中的 As(Ⅲ)/As(Ⅴ)摩尔比为 3∶7,砷在玻璃固化体中主要以 As(Ⅴ)的形式存在;SEM 和 XRD 结果表明高温过程的熔融反应导致非晶态玻璃固化体形成,砷被包裹在玻璃结构当中[11]。胡菁菁等研究了硼酸盐玻璃、磷酸盐玻璃和硅酸盐玻璃对含砷渣的高温玻璃固化效果。结果表明,上述三种玻璃对砷渣中的砷都有良好的固化效果,但是当下快速冶炼中未来得及完全反应的毒砂进入渣中难以避免,毒砂经历熔炼所形成的氧化砷和砷酸盐可能未来得及转变为稳定的砷硅酸盐,因此,仍然会导致严重的砷污染[12]。柴立元等研究 CaO 对钠铁硼磷玻璃体系结构及固砷效果的影响。研究表明,CaO 的加入破坏了钠铁硼磷玻璃同体系中 Q^2 桥氧结构,导致玻璃网格化程度降低,使得玻璃结构加强。对玻璃固砷体进行密度分析,CaO 进入玻璃固砷体的网络结构中,使得磷酸盐结构解聚,通过对非桥氧结构产生集聚作用且形成 Ca-O-As、Ca-O-P 等化学键,增强玻璃固砷体系的结构[13]。农泽喜等研究了含砷冶炼废渣高温烧结过程砷的迁移特性,对含砷渣,如钙砷渣、铁砷渣等进行高温煅烧。实验研究结果表明,煅烧的温度越高,煅烧后的砷渣中的砷溶解度就越低。当温度达到 1200 ℃时,砷的浸出率仅有 0.002 mg/L,砷浸出的降低率达到 99.65%;当烧结时间达到 45 min 时,砷的浸出浓度降低率可以达到 99.42%。近几年智利的几个铜冶炼厂在处理砷钙渣时就采用火法稳定化法,并取得良好的效果[14]。刘政等对处理高砷钴矿火法富集过程中产生的含砷废渣采用高温稳定化,也取得了不错的效果[15]。张洁将 SiO_2、Al_2O_3、MgO、CaO 四种添加组分掺入含砷废渣中,进行 400~1300 ℃高温烧结处理,研究烧结后烧结体中砷的稳定化程度。结果表明,SiO_2 和 Al_2O_3 对砷的稳定作用不明显,MgO 对砷有一定的稳定作用,CaO 稳定砷的作用最明显。CaO 对含砷废渣烧结体中砷的挥发和浸出有极大的抑制作用,1000 ℃高温烧结条件下,砷的稳定化程度达最大值 95.14%,浸出结果表明,砷的浸出量仅为 0.85 μg/L[16]。

虽然固化处理含砷废渣取得良好效果,但存在固化后的堆存问题,使得经济成本上升,而且长时间堆存也会有砷二次流入环境的风险,此外,砷在医药、半导体材料、合金材料等领域有重要用途,因此有学者开展砷的脱除研究,对砷进行回收利用,以求实现治理环境和资源循环利用的双重目的。曹晓恩等提出铁矾渣预氧化-煤基直接还原脱砷工艺,并按照该工艺进行脱砷热力学分析,最终脱砷率达到78.34%。预氧化铁钒渣外配15%煤粉制成含碳球团,在弱还原升温阶段(25~1075 ℃),砷总挥发率为45.90%;还原温度为1075 ℃,外配10%煤粉、还原温度30 min条件下,砷挥发率达到68.2%;调整碱度并适当提高温度后,最优脱砷率能提高到78.34%[17]。万新宇等研究了含砷铜渣在N_2-CO气氛中还原焙烧脱砷工艺的效果。实验结果表明,若还原温度在600至1200 ℃之间,调整N_2:CO到合适的比例,可以实现砷酸盐的还原且还能抑制铁橄榄石和磁铁矿的还原,达到砷以气态化合物形式挥发而铁仍以氧化物形式存在于渣中的目的。最优工艺条件为,还原温度1100 ℃、还原时间60 min、CO浓度2.5%,铜渣中砷残留为0.044%,脱砷率可以达到70.71%[18]。砷在脱除后多是以白砷等砷化合物形式进行回收,而白砷等毒性很大,若能把砷彻底制成无毒无害的单质砷,则实现了砷的彻底无害化和资源化。

本文以河南某铜业公司产生的钙砷渣为原料,探索铁热还原处理钙砷渣制备无毒无害单质砷的可行性,深入研究还原温度、铁砷比和还原时间等因素对钙砷渣还原的影响,初步实现铁热还原钙砷渣制备单质砷新工艺,对于合理、高效利用我国的砷资源及保护环境具有典型的代表性,对冶炼过程中砷的资源化回收有重大的研究意义。

1 原料与实验步骤

1.1 钙砷渣

钙砷渣主要化学成分分析结果见表1,X射线衍射(XRD)分析如图1所示。

表1 钙砷渣主要化学成分

Table 1 Main components of calcium arsenate %

成分	Ca	As	O	其他
质量分数	19.81	31.43	38.41	10.35

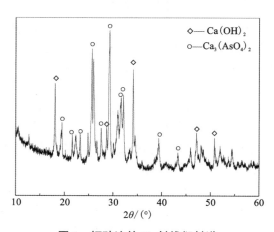

图1 钙砷渣的X-射线衍射谱

Fig. 1 X-ray diffraction pattern of calcium arsenic slag

由表 1 结果可知，钙砷渣主要元素质量分数为 Ca 19.81%，As 31.43%，O 38.41%，其他 10.35%。由表 1 结果可知，钙砷渣主要物相为 $Ca_3(AsO_4)_2$、$Ca(OH)_2$。进一步由 SEM-EDS 分析如图 2 所示。

由图 2 结果可知，钙砷渣主要以小颗粒形式存在且钙砷渣纯度较高，无其他杂质。

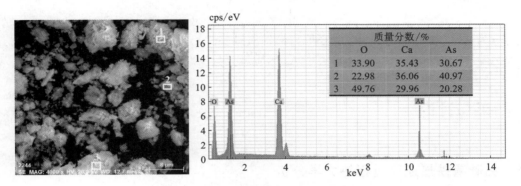

图 2 钙砷渣的 SEM 及区域 EDS 图谱

Fig. 2 SEM-EDS spectrum of calcium arsenic slag

1.2 实验流程

钙砷渣铁热还原实验在高温管式气氛炉中进行，最高温度可达 1200 ℃，加热管为耐高温的石英管，炉体结构如图 3 所示。

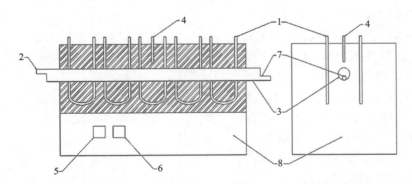

1—电阻丝；2—出气口；3—刚玉管；4—热电偶；5—电流表；6—电压表；7—进气口；8—炉体。

图 3 高温气氛管式炉

Fig. 3 Tubular furnace with high temperature atmosphere

首先，将铁粉和钙砷渣球磨并充分混匀后放入石英管中部，安装好管式气氛炉；然后设定好运行程序，通入氩气将石英管中的空气排空并逐渐升温，当升温至设定温度时开始计时，在反应过程中保持炉内氩气流量、反应温度恒定，同时保证石英管反应过程密闭以免产生泄漏，并用碱液吸收尾气以免污染空气。同时，反应结束，待炉内温度冷却后将反应物取出，送 XRD、ICP、SEM、EDS 分析。

1.3 砷还原率计算公式

铁热还原实验中,采用 ICP 电感耦合等离子体发射光谱仪分析,将钙砷渣及还原后渣溶样后,通过 ICP 测得稀释后溶液中 As 离子的浓度,公式如下:

$$\eta = \frac{M_1 \times X_1}{M_2 \times X_2} \times 100\%$$

式中: η 为 As 的还原率,%; M_1 为还原后渣的质量,g; X_1 为还原后渣中 As 的质量分数,%; M_2 为钙砷渣的质量,g; X_2 为钙砷渣中 As 的质量分数,%。

2 结果与讨论

2.1 钙砷渣铁热还原过程热力学分析

首先对钙砷渣铁热还原过程进行热力学研究,该过程分为铁热还原过程以及金属砷蒸气冷凝 2 个过程。

2.1.1 铁热还原过程

(1) $Ca_3(AsO_4)_2$ 稳定性的判断

进行铁热还原前首先判断 $Ca_3(AsO_4)_2$ 是否会发生分解,可能发生的分解反应为:

$$Ca_3(AsO_4)_2 =\!=\!= Ca(AsO_2)_2 + 2CaO + O_2(g) \tag{1}$$

$$Ca_3(AsO_4)_2 =\!=\!= As_2O_5 + 3CaO \tag{2}$$

上述反应的 $\Delta G - T$ 曲线见图 4。

由图 4 可知,在常压下 $Ca_3(AsO_4)_2$ 十分稳定,分解生成 $Ca(AsO_2)_2$ 或 As_2O_5 所需温度均超过 2500 K,因此可以判断 $Ca_3(AsO_4)_2$ 不会发生自分解。

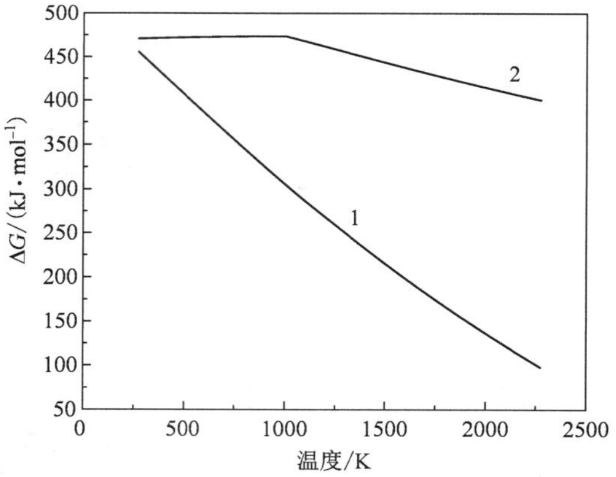

(2) $Ca_3(AsO_4)_2$ 铁热还原第一阶段

$Ca_3(AsO_4)_2$ 进行铁热还原后以 $Ca(AsO_2)_2$、As_2O_3、As_4O_6 这三种存在形式存在,可能发生的反应有:

$$Ca_3(AsO_4)_2 + 2Fe =\!=\!= Ca(AsO_2)_2 + 2FeO + 2CaO \tag{3}$$

$$3Ca_3(AsO_4)_2 + 5Fe =\!=\!= 3Ca(AsO_2)_2 + CaFe_5O_7 + 5CaO \tag{4}$$

$$2Ca_3(AsO_4)_2 + 3Fe =\!=\!= 2Ca(AsO_2)_2 + CaFe_3O_5 + 3CaO \tag{5}$$

$$1.5Ca_3(AsO_4)_2 + 2Fe =\!=\!= 1.5Ca(AsO_2)_2 + CaO \cdot Fe_2O_3 + 2CaO \tag{6}$$

$$1.5Ca_3(AsO_4)_2 + 2Fe =\!=\!= 1.5Ca(AsO_2)_2 + (CaO)_2 \cdot Fe_2O_3 + CaO \tag{7}$$

$$Ca_3(AsO_4)_2 + 2Fe =\!=\!= As_2O_3 + 2FeO + 3CaO \tag{8}$$

图 4 反应式 (1) 和反应式 (2) 的 ΔG-T 图谱

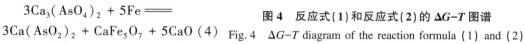

Fig. 4 ΔG-T diagram of the reaction formula (1) and (2)

$$3Ca_3(AsO_4)_2 + 5Fe \rightleftharpoons 3As_2O_3 + CaFe_5O_7 + 8CaO \quad (9)$$
$$2Ca_3(AsO_4)_2 + 3Fe \rightleftharpoons 2As_2O_3 + CaFe_3O_5 + 5CaO \quad (10)$$
$$1.5Ca_3(AsO_4)_2 + 2Fe \rightleftharpoons 1.5As_2O_3 + CaO \cdot Fe_2O_3 + 3.5CaO \quad (11)$$
$$1.5Ca_3(AsO_4)_2 + 2Fe \rightleftharpoons 1.5As_2O_3 + (CaO)_2 \cdot Fe_2O_3 + 2.5CaO \quad (12)$$
$$2Ca_3(AsO_4)_2 + 4Fe \rightleftharpoons As_4O_6 + 4FeO + 6CaO \quad (13)$$
$$3Ca_3(AsO_4)_2 + 5Fe \rightleftharpoons 1.5As_4O_6 + CaFe_5O_7 + 8CaO \quad (14)$$
$$2Ca_3(AsO_4)_2 + 3Fe \rightleftharpoons As_4O_6 + CaFe_3O_5 + 5CaO \quad (15)$$
$$2Ca_3(AsO_4)_2 + 2.667Fe \rightleftharpoons As_4O_6 + 1.333CaO \cdot Fe_2O_3 + 4.667CaO \quad (16)$$
$$3Ca_3(AsO_4)_2 + 5Fe \rightleftharpoons 1.5As_4O_6 + (CaO)_2 \cdot Fe_2O_3 + 7CaO + 3FeO \quad (17)$$

式(3)~式(17)在常压下的 $\Delta G_T - T$ 曲线见图5。

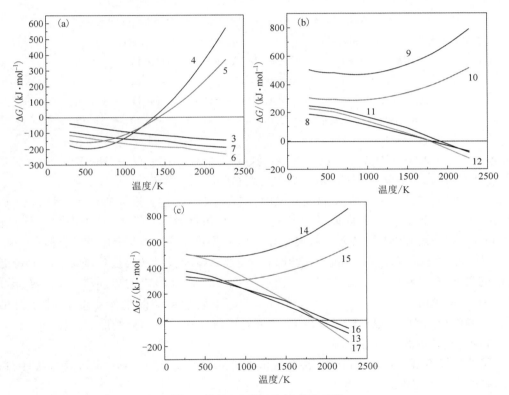

图5 式(3)~式(17)的 ΔG-T 图

Fig. 5 ΔG-T diagram of the formula (3) ~ (17)

由图5可知，$Ca_3(AsO_4)_2$ 直接反应生成 As_2O_3 或 As_4O_6 所需温度均在1750 K以上，而 $Ca(AsO_2)_2$ 的生成很容易发生。因此，可判断 $Ca_3(AsO_4)_2$ 铁热还原的第一阶段生成物为 $Ca(AsO_2)_2$。

（3）$Ca_3(AsO_4)_2$ 铁热还原第二阶段

$Ca_3(AsO_4)_2$ 铁热还原第二阶段即 $Ca(AsO_2)_2$ 的热还原过程，此过程可能发生的反应为：

$$Ca(AsO_2)_2 + 3Fe \rightleftharpoons 2As(g) + 3FeO + CaO \quad (18)$$

$$Ca(AsO_2)_2 + 3Fe \rightleftharpoons As_2(g) + 3FeO + CaO \tag{19}$$

$$1.5Ca(AsO_2)_2 + 4.5Fe \rightleftharpoons As_3(g) + 4.5FeO + 1.5CaO \tag{20}$$

$$2Ca(AsO_2)_2 + 6Fe \rightleftharpoons As_4(g) + 6FeO + 2CaO \tag{21}$$

$$2Ca(AsO_2)_2 + 5Fe \rightleftharpoons 4As(g) + CaFe_5O_7 + CaO \tag{22}$$

$$2Ca(AsO_2)_2 + 5Fe \rightleftharpoons 2As_2(g) + CaFe_5O_7 + CaO \tag{23}$$

$$2Ca(AsO_2)_2 + 5Fe \rightleftharpoons 4/3As_3(g) + CaFe_5O_7 + CaO \tag{24}$$

$$2Ca(AsO_2)_2 + 5Fe \rightleftharpoons As_4(g) + CaFe_5O_7 + CaO \tag{25}$$

$$4Ca(AsO_2)_2 + 9Fe \rightleftharpoons 8As(g) + 3CaFe_3O_5 + CaO \tag{26}$$

$$4Ca(AsO_2)_2 + 9Fe \rightleftharpoons 4As_2(g) + 3CaFe_3O_5 + CaO \tag{27}$$

$$4Ca(AsO_2)_2 + 9Fe \rightleftharpoons 8/3As_3(g) + 3CaFe_3O_5 + CaO \tag{28}$$

$$4Ca(AsO_2)_2 + 9Fe \rightleftharpoons 2As_4(g) + 3CaFe_3O_5 + CaO \tag{29}$$

$$Ca(AsO_2)_2 + 2Fe \rightleftharpoons 2As(g) + CaO \cdot Fe_2O_3 \tag{30}$$

$$Ca(AsO_2)_2 + 2Fe \rightleftharpoons As_2(g) + CaO \cdot Fe_2O_3 \tag{31}$$

$$Ca(AsO_2)_2 + 2Fe \rightleftharpoons 2/3As_3(g) + CaO \cdot Fe_2O_3 \tag{32}$$

$$Ca(AsO_2)_2 + 2Fe \rightleftharpoons 1/2As_4(g) + CaO \cdot Fe_2O_3 \tag{33}$$

$$2Ca(AsO_2)_2 + 5Fe \rightleftharpoons 4As(g) + (CaO)_2 \cdot Fe_2O_3 + 3FeO \tag{34}$$

$$2Ca(AsO_2)_2 + 5Fe \rightleftharpoons 2As_2(g) + (CaO)_2 \cdot Fe_2O_3 + 3FeO \tag{35}$$

$$2Ca(AsO_2)_2 + 5Fe \rightleftharpoons 4/3As_3(g) + (CaO)_2 \cdot Fe_2O_3 + 3FeO \tag{36}$$

$$2Ca(AsO_2)_2 + 5Fe \rightleftharpoons As_4(g) + (CaO)_2 \cdot Fe_2O_3 + 3FeO \tag{37}$$

式(18)~式(37)在常压下的 $\Delta G - T$ 图见图6。

由图6可知,在 273~2273 K 温度范围内,$Ca(AsO_2)_2$ 还原得到 $CaFe_5O_7$ 和 $CaFe_3O_5$ 的反应 ΔG 始终大于零,并且随着温度升高,其 ΔG 继续增大,反应无法进行。且常压下 $Ca(AsO_2)_2$ 还原得到单质砷以 As_4 形式存在时所需温度最低,肖若珀也指出砷蒸气很少以 As 蒸气及 As_2 蒸气形态存在,随着温度升高,As 及 As_2 的含量才会逐步增加,温度较低时,蒸气中的砷分子主要以 As_4 为主存在,由此我们可以判断形成的单质砷是以 As_4 形式存在[19]。比较式(21)、式(33)、式(37)可知在常压下发生反应的初始温度分别为 1100 K、1173 K、1005 K,因此可以推断反应后 Fe 主要是以 $(CaO)_2 \cdot Fe_2O_3$ 的形式存在。

2.1.2 金属砷蒸气冷凝过程

金属砷蒸气冷凝过程可由砷蒸气冷凝成液态砷后再冷凝成固态砷,也可直接从气态砷冷凝成固态砷,可能发生的反应有:

$$As_4(g) \rightleftharpoons 4As(l) \tag{38}$$

$$As(l) \rightleftharpoons As(s) \tag{39}$$

$$As_4(g) \rightleftharpoons 4As(s) \tag{40}$$

式(38)~式(40)在常压下的 $\Delta G - T$ 图见图7。

由图7可知砷蒸气冷凝直接生成固态砷的初始温度为 878 K,而潘崇发研究认为在常压下最适宜的冷凝温度是 573~593 K,最高不能超过 623 K,因此可以推断出冷凝过程是砷蒸气先冷凝成液态砷,而后再冷凝成固态砷[20]。

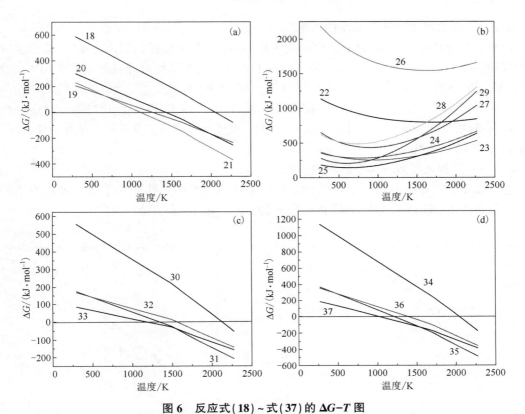

图 6 反应式(18)~式(37)的 ΔG-T 图

Fig. 6 ΔG-T diagram of the reaction formula (18) ~ (37)

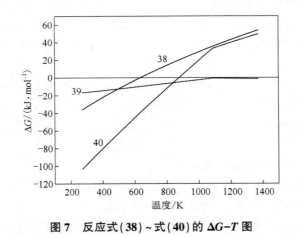

图 7 反应式(38)~式(40)的 ΔG-T 图

Fig. 7 ΔG-T diagram of the reaction formula (38) ~ (40)

2.2 工艺条件对钙砷渣铁热还原过程的影响

2.2.1 还原温度对砷还原率的影响

由热力学分析可知还原最低温度在 1023 K 左右,但经实验探索后需在 1273 K 以上才能有较好的还原效果。因此实验将探究在 1173~1373 K,还原温度对砷脱除率的影响。实验条件为:铁配入系数(铁和砷酸钙摩尔比)为 1.5,还原时间 30 min。实验结果见图 8。

由图 8 可见,在本实验条件下,随着还原温度由 1173 K 升高至 1273 K,砷还原率由 80.64% 不断增至 96.56%;随着温度进一步提升,砷还原率略有上升并趋于平稳。由此可得,还原温度取 1273 K 为宜。

2.2.2 铁配入系数对砷还原率的影响

基于 2.2.1 实验结果,在铁配入系数 0.7~1.5 内,考察了铁配入系数对砷还原率的影响。实验条件为:还原温度为 1273 K,还原时间为 60 min。实验结果见图 9。

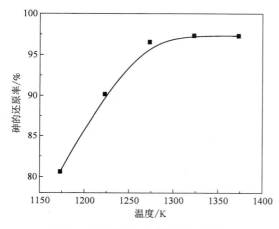

图 8 还原温度对砷还原率的影响

Fig. 8　Effect of reduction temperature on arsenic reduction ratio

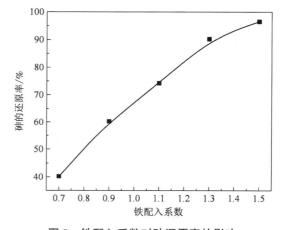

图 9 铁配入系数对砷还原率的影响

Fig. 9　Effect of iron addition coefficient on arsenic reduction ratio

由图 9 可知,在该实验条件下,随铁配入系数由 0.7 升高至 1.5,砷还原率从 40.21% 明显升高至 96.56%,由此可知,过量的铁可以使得反应 $2Ca_3(AsO_4)_2 + 7Fe \Longleftrightarrow 4As(g) + 3(CaO)_2 \cdot Fe_2O_3 + FeO$ 向正向发生,但考虑经济成本,配铁量也不能太大,故铁配入系数取 1.5 为宜。

2.2.3 还原时间对砷还原率的影响

基于 2.2.2 实验结果,在 0.5~2 h,还原温度为 1273 K,铁配入系数为 1.5 的实验条件下,考察了还原时间对砷还原率的影响。实验结果见图 10。

由图 10 可知,在该实验条件下,随着还原时间由 30 min 延长至 60 min,砷还原率有所提高,从 78.78% 升高至 96.56%。而随着时间进一步延长,砷还原率基本上保持不变。综上所述,还原时间取 60 min 为宜。

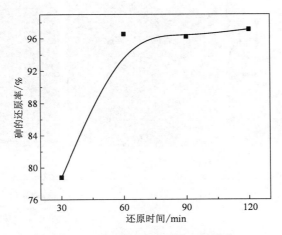

图 10　还原时间对砷还原率的影响

Fig. 10　Effect of reduction time on arsenic reduction ratio

2.3　产物与还原渣表征

实验选取了保温 60 min，铁配入系数 1.5，温度为 1273 K, 1323 K, 1373 K 等条件下的还原渣和产物送 XRD 分析，XRD 谱如图 11、图 12 所示。

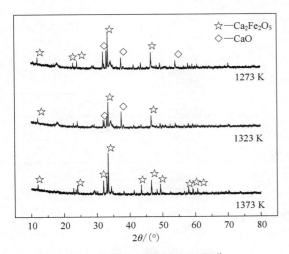

图 11　铁热还原渣 XRD 图谱

Fig. 11　XRD spectrum of reduction slag

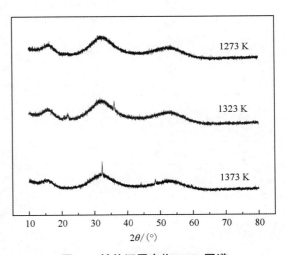

图 12　铁热还原产物 XRD 图谱

Fig. 12　XRD spectrum of reduction product

由图 11 可知，反应后还原渣主要成分为 $Ca_2Fe_2O_5$ 和 CaO，基本不含砷，可用作建筑材料。图 12 中 XRD 中无明显的特征峰，说明还原产物为无定型结构，有待进一步 SEM-EDS 分析还原产物的元素组成及含量。

图 13 和表 2 是还原产物砷的 SEM-EDS 图谱与元素含量分析，由图 13 和表 2 可知还原产物确定为片状单质砷，纯度为 99% 以上。

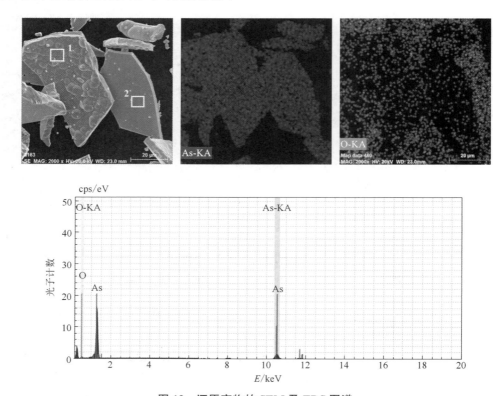

图 13 还原产物的 SEM 及 EDS 图谱

Fig. 13　SEM-EDS spectrum of reduction product

表 2　图 13 中区域 1 和 2 的 EDS 分析

Table 2　EDS analysis of areas 1 and 2 in Figure 13　　　　　　　　　%

元素	O	As	Ca
区域 1	0.75	99.23	0.02
区域 2	0.75	99.25	0

3　结论

(1)热力学分析表明,温度在 1023 K 以上时,铁热还原砷酸钙制备单质砷在理论上是可行的,反应过程为 $Ca_3(AsO_4)_2$ 先还原为 $Ca(AsO_2)_2$,再进一步还原为 $As_4(g)$。

(2)在还原温度为 1273 K,铁配入系数为 1.5,还原时间为 60 min 的条件下,可以得到比较理想的脱砷效果,砷还原率可达 96.56%。

(3)XRD 图谱表明,反应后还原渣主要成分为 $Ca_2Fe_2O_5$ 和 CaO,而生成物的 XRD 图谱中无明显的特征峰,说明还原产物为无定型结构。进一步结合 SEM-EDS 分析可知,还原产物单质砷主要为片状,纯度为 99% 以上。

参考文献

[1] 王华东,郝春曦,王建.环境中的砷[M].北京:中国环境科学出版社,1992.
[2] ABDUL K S M, JAYASINGHE S S, CHANDANA E P S, et al. Arsenic and human health effects: A review[J]. Environmental Toxicology and Pharmacology. 2015, 40(3): 828-846.
[3] 胡斌,杨天足,刘伟锋,等.基于地球化学的水热还原矿化稳定砷的技术思路[J].中国有色金属学报,2020, 30(4): 847-857.
[4] GUO L, LAN J R, DU Y G, et al. Microwave-enhanced selective leaching of arsenic from copper smelting flue dusts[J]. Journal of Hazardous Materials, 2020, 386: 121964.
[5] FRY K L, WHEELER C A, GILLINGS M M, et al. Anthropogenic contamination of residential environments from smelter As, Cu and Pb emissions: Implications for human health[J]. Environmental Pollution, 2020, 262: 114235.
[6] JAROSIKOVA A, ETTLER V, MIHALJEVIC M, et al. Characterization and pH-dependent environmental stability of arsenic trioxide-containing copper smelter flue dust[J]. Journal of Environmental Management, 2018, 209: 71-80.
[7] XU H, MIN X B, WANG Y Y, et al. Stabilization of arsenic sulfide sludge by hydrothermal treatment[J]. Hydrometallurgy, 2020, 191: 105229.
[8] 喻小强,徐家聪,易勤,等.砷锑烟灰 NaOH 常压碱浸分离砷锑的工艺[J].有色金属科学与工程,2021, 12(3): 42-49.
[9] 龚傲,陈丽杰,吴选高,等.含砷废渣处理现状及研究进展[J].有色金属科学与工程.2019, 10(4): 28-33.
[10] TWIDWELL L G, PLESSAS K O, COMBA P G, et al. Removal of arsenic from wastewaters and stabilization of arsenic bearing waste solids: Summary of experimental studies[J]. Journal of Hazardous Materials, 1994, 36(1): 69-80.
[11] 赵宇文,闵小波.As_2O_3 玻璃固化过程砷的固化过程研究[C]//中国有色金属冶金第三届学术会议论文集.沈阳:中国有色金属冶金第三届学术会议,2016: 235-239.
[12] 胡菁菁,张惠斌,曹华珍,等.有色冶炼过程砷高温熔融固化技术研究进展[J].冶金工程,2018, 5(2): 55-61.
[13] 柴立元,赵宗文,廖彦杰,等.CaO 对钠铁硼磷玻璃体系结构及固砷效果影响[J].有色金属科学与工程,2015, 6(1): 1-7.
[14] 农泽喜,王兴润,舒新前,等.含砷冶炼废渣高温烧结过程砷的迁移特性[J].环境工程学报,2013, 7(3): 1115-1120.
[15] 刘政,姚媛.高砷钴矿火法富集过程中砷的污染和治理[J].江西有色金属,2002, 16(4): 35-37.
[16] 张洁.烧结处理对含砷废渣中砷的环境释放行为的影响研究[D].西安:西北农林科技大学,2013.
[17] 曹晓恩,程相魁,张可才,等.铁矾渣预氧化-煤基直接还原过程脱砷研究[J].有色金属工程,2017, 7(5): 43-47.
[18] 万新宇,齐渊洪,高建军.含砷铜渣 N_2-CO 气氛中还原焙烧脱砷新工艺[J].矿冶工程,2017, 37(6): 80-83.
[19] 肖若珀.砷的提取、环保和应用方向[M].南宁:广西金属学会,1992.
[20] 潘崇发.提高金属砷质量的有效途径[J].有色金属(冶炼部分),1994(2): 32-34.

氯化焙烧铜熔炼渣回收铅工艺及动力学研究

摘　要：以 $CaCl_2$ 为氯化剂，进行了氯化焙烧铜熔炼渣回收铅的研究，考察了焙烧温度、保温时间、氯化剂添加量和空气流量对铅金属回收率的影响，探讨了铜熔炼渣中铅的氯化挥发动力学。结果表明，当焙烧温度为 950 ℃、焙烧时间为 12 min、$CaCl_2$ 添加量为 10%、空气流量为 100 mL/min 时，铅的金属回收率达到 92.71%。铜熔炼渣中铅的氯化挥发过程遵循界面化学反应控制的未反应核收缩模型，其反应表观活化能为 83.002 kJ/mol。

关键词：氯化焙烧；铜熔炼渣；铅；动力学；未反应核收缩模型；界面化学反应；表观活化能

Chlorination roasting for recovery of lead from copper smelting slag and its kinetics study

Abstract：With $CaCl_2$ as the chlorinating agent, the lead was recovered from copper smelting slag by chlorination roasting process. The effects of roasting temperature, holding time, the dosage of $CaCl_2$ and air flow rate on lead recovery rate were investigated, and the kinetics of the chlorination process of lead in copper smelting slag was discussed. The results show that the lead recovery rate can reach 92.71% after roasting at 950 ℃ for 12 min, with 10% of $CaCl_2$ and air flow rate at 100 mL/min. The chlorination process of lead in copper smelting slag is controlled by interface chemical reaction, following the shrinking unreacted core model, and the apparent activation energy of the reaction is calculated to be 83.002 kJ/mol.

Keywords：chlorination roasting；copper smelting slag；Pb；kinetics；shrinking unreacted core model；interface chemical reaction；apparent activation energy

铜熔炼渣是铜冶炼过程中造锍熔炼后的产物，是一种具备高回收价值的有色金属冶炼渣[1-2]。目前，对铜熔炼渣中有价元素回收时，通常只考虑回收渣中的金属铜[3]。大量铜熔炼渣经浮选、火法贫化、湿法浸出后直接堆存于渣场，渣中剩余有价金属未得到回收，严重浪费资源、污染环境[4]。铜熔炼渣中含有 0.2%~0.6% 金属铅，由于渣中铅主要以硫化物、氧化物等形式赋存，常规火法、浮选等手段难以使其分离[5]。因此，可利用铅金属氯化物沸

本文发表在《矿冶工程》，2022, 42(1)：72-76。作者：张倍恺，郭学益，王亲猛，李中臣，田庆华，李栋。

点低、挥发性高、易溶于水等特点,通过氯化焙烧过程,使铜熔炼渣中的铅元素以金属氯化物形式从伴生体系中分离并回收[6]。铜熔炼渣中其他有价元素(如 Fe、Cu、Si 等)不易被氯化,因此氯化挥发法可实现铜熔炼渣中铅的选择性回收[7]。本文以浮选提铜后的铜熔炼渣为原料,选取 $CaCl_2$ 作为氯化剂,采用氯化焙烧工艺进行铜熔炼渣中金属铅的回收,并对铅的氯化挥发动力学进行了研究。

1 实验部分

1.1 实验原料

实验原料为国内某铜冶炼厂浮选提铜后的铜熔炼渣。使用电感耦合等离子体发射光谱仪对铜熔炼渣进行化学成分分析,结果如表 1 所示。渣中 Pb 元素质量分数为 0.45%。对铜熔炼渣进行化学物相分析(见表 2)可知,渣中 Pb 主要以 PbS、PbO、$PbSO_4$ 等形式赋存。使用 X 射线衍射仪对铜熔炼渣进行物相分析,结果如图 1 所示。原料中的主要成分为磁铁矿(Fe_3O_4)和橄榄石(Fe_2SiO_4)。

表 1 铜熔炼渣化学成分(质量分数)   %

Pb	Cu	Fe	Si	Ca	Na	K	Cl	S	O
0.45	0.55	41.23	11.75	1.76	1.36	0.58	0.059	0.41	36.10

表 2 铜熔炼渣中铅物相组成

铅物相	质量分数/%	占比/%
PbS	0.178	39.56
PbO	0.141	31.33
$PbSO_4$	0.066	14.67
$PbSiO_3$	0.056	12.44
Pb	0.009	2.00
合计	0.45	100.00

1.2 实验方法

称取 10 g 经真空干燥箱烘干 12 h 后的铜熔炼渣,与一定质量比的分析纯 $CaCl_2$ 充分研磨混合后,均匀放入刚玉瓷舟中。对高温管式炉进行温控编程,以升温速度 10 ℃/min 升温,待炉温达到预设温度后,通入一定流量的空气,然后将装有混合物料的瓷舟送入炉中进行焙烧,达到焙烧时间后迅速取出瓷舟,自然冷却。将降至室温后的焙烧渣研磨成粉末,使用电感耦合等离子体发射光谱仪检测焙烧渣中 Pb 含量。实验过程中的挥发烟气通过管式炉出气口处盛有 200 mL 10%NaOH 的洗气瓶进行收集。Pb 金属回收率由式（1）确定。

$$R = \frac{w_g \times m_g - w_s \times m_s}{w_g \times m_g} \times 100\%$$

(1)

式中：R 为金属回收率，%；w_g 为铜熔炼渣中 Pb 的质量分数，%；m_g 为瓷舟中混合物料的质量，g；w_s 为焙烧渣中 Pb 的质量分数，%；m_s 为焙烧渣的质量，g。

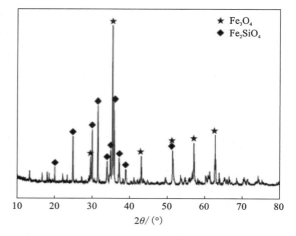

图 1　铜熔炼渣 X 射线衍射图谱

1.3　实验原理

在高温焙烧条件下，氯化钙在氧气及酸性氧化物的作用下生成氯气，含 Pb 相与氯气发生化学反应，形成低沸点、易挥发的 $PbCl_2$。最终 Pb 以气态 $PbCl_2$ 形式从焙烧渣中挥发分离，从而实现金属 Pb 的回收。体系中涉及的主要化学反应见式（2）~（5），通过 HSC chemistry 9 热力学计算软件分别计算各反应的 $\Delta G - T$ 关系（见图2）可知，当温度高于 900 ℃时，反应式（2）~（5）全部可以发生。

$$2CaCl_2 + O_2(g) + 2SiO_2 \Longrightarrow 2Cl_2(g) + 2CaSiO_3 \quad (2)$$

$$PbS + Cl_2(g) + O_2(g) \Longrightarrow PbCl_2(g) + SO_2(g) \quad (3)$$

$$2PbO + 2Cl_2(g) \Longrightarrow 2PbCl_2(g) + O_2(g) \quad (4)$$

$$PbSO_4 + Cl_2(g) \Longrightarrow PbCl_2(g) + SO_2(g) + O_2(g) \quad (5)$$

2　实验结果与讨论

2.1　焙烧温度对 Pb 回收率的影响

焙烧时间 20 min、$CaCl_2$ 添加量 10%、空气流量 100 mL/min 条件下，焙烧温度对 Pb 金属回收率的影响见图 3。从图 3 可知，焙烧温度对 Pb 金属回收率具有显著影响，在 750~1000 ℃，随着温度升高，Pb 金属回收率显著提升。这是因为温度升高会破坏铜熔炼渣结构，进

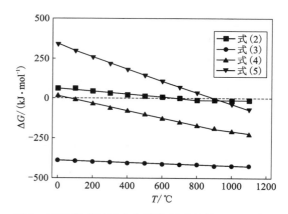

图 2　氯化焙烧过程中主要化学反应的 $\Delta G-T$ 关系

而增强氯气与含 Pb 相的接触，促进氯化反应进行。其中，在 750~950 ℃，随着温度升高，Pb 金属回收率从 58.33% 提高到 94.68%；当焙烧温度高于 950 ℃时，Pb 金属回收率增长速率下降。结合回收效率与过程能耗，选取焙烧温度为 950 ℃。

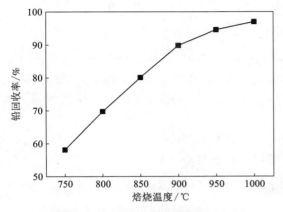

图 3 焙烧温度对 Pb 金属回收率的影响

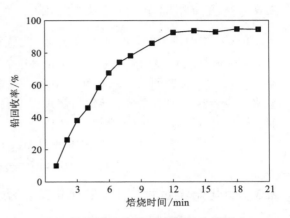

图 4 焙烧时间对 Pb 金属回收率的影响

2.2 焙烧时间对 Pb 回收率的影响

焙烧温度 950 ℃，其他条件不变，焙烧时间对 Pb 金属回收率的影响见图 4。从图 4 可知，Pb 金属回收率随焙烧时间延长而提升。由此可知，充分的焙烧时间可以提高渣中含 Pb 相氯化反应的限度，进而提升 Pb 金属回收率。在 1~12 min，随着焙烧时间延长，Pb 金属回收率增幅较大，从 10.26% 提高到 92.71%；当焙烧时间超过 12 min 后，Pb 金属回收率趋于稳定。选取焙烧时间 12 min。

2.3 $CaCl_2$ 添加量对 Pb 回收率的影响

焙烧时间 12 min，其他条件不变，$CaCl_2$ 添加量对 Pb 金属回收率的影响见图 5。由图 5 可知，$CaCl_2$ 添加量从 0 增加到 10% 时，Pb 金属回收率从 14.21% 增长到 92.71%；继续增加 $CaCl_2$ 添加量，Pb 金属回收率几乎保持不变，故选取 $CaCl_2$ 添加量为 10%。由于 Pb 化合物属于易挥发性物质，焙烧过程中无 $CaCl_2$ 时，依然会出现少量 Pb 挥发的现象。

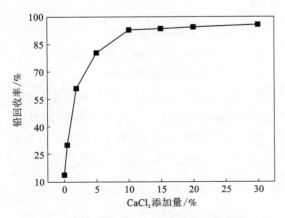

图 5 $CaCl_2$ 添加量对 Pb 金属回收率的影响

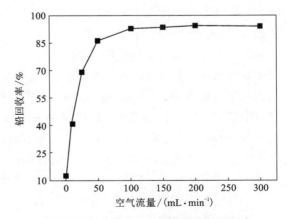

图 6 空气流量对 Pb 金属回收率的影响

2.4 空气流量对 Pb 回收率的影响

CaCl₂ 添加量 10%，其他条件不变，空气流量对 Pb 金属回收率的影响见图 6。从图 6 可知，空气流量从 0 增加到 100 mL/min，Pb 金属回收率从 12.42% 增长到 92.71%；继续增加空气流量，Pb 金属回收率几乎保持不变。故选取空气流量 100 mL/min。配料过程中 CaCl₂ 会吸收少量空气中的水分，导致焙烧过程中发生式(6)~式(7)的反应，因此在空气流量为 0 时，依然可以回收少量 Pb 金属。

$$CaCl_2 + SiO_2 + H_2O(g) = CaSiO_3 + 2HCl(g) \tag{6}$$
$$PbO + 2HCl(g) = PbCl_2(g) + H_2O(g) \tag{7}$$

2.5 氯化焙烧过程中 Pb 的氯化挥发动力学

2.5.1 氯化焙烧动力学曲线

CaCl₂ 添加量 10%、空气流量 100 mL/min 条件下，不同焙烧温度下 Pb 金属回收率与焙烧时间的关系如图 7 所示。由图 7 可知，在 800~1000 ℃，随焙烧温度升高，Pb 的氯化挥发速率不断增大。当焙烧时间低于 12 min 时，各焙烧温度下 Pb 金属回收率随焙烧时间延长而增大的规律基本一致。但当焙烧时间超过 12 min 时，900~1000 ℃，Pb 金属回收率趋于稳定。因此，选择焙烧时间 1~12 min 进行含 Pb 相氯化焙烧过程的动力学研究。

2.5.2 反应控制步骤的选择

在氯化焙烧过程中，CaCl₂ 与氧气作用的气相产物为 Cl₂，铜熔炼渣中含 Pb 相为固相，氯化反应产物主要为气态 PbCl₂，因此铜熔炼渣中含 Pb 相的氯化焙烧过程可以分为以下 4 个阶段：①氯化剂与空气中的氧气反应产生氯气，氯气向渣中含 Pb 相表面扩散，供给含 Pb 相进行氯化反应时所需氯源，为传质步骤；②在原子力场作用下，含 Pb 相吸附扩散到其表面的氯气，为扩散步骤；③高温条件下，含 Pb 相与吸附的氯气发生氯化反应，生成易挥发的 PbCl₂，为

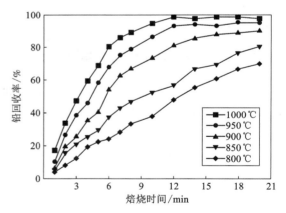

图 7 不同焙烧温度下 Pb 金属回收率
随焙烧时间变化曲线

界面化学反应步骤；④气态 PbCl₂ 由气-固接触面向空气中扩散，为传质步骤。由于氯化焙烧过程中反应物颗粒逐渐收缩，故可采用未反应核收缩模型进行处理[8]。由动力学原理可知，气-固反应的总反应速率由速度最慢的环节确定，这一环节称为反应的控制步骤[9]。假定铜熔炼渣为表面各处化学活性相同的致密固体颗粒，则渣中 Pb 的氯化反应将由其中传质、扩散、界面化学反应之一控制。

当反应由传质控制时，方程表达式为：

$$X = kt \tag{8}$$

当反应为扩散控制时，方程表达式为：

$$1 + 2(1-X) - 3(1-X)^{\frac{2}{3}} = kt \tag{9}$$

当反应为界面化学反应控制时,方程表达式为:

$$1 - (1-X)^{\frac{1}{3}} = kt \tag{10}$$

式中: k 为表观反应速率常数, min^{-1}; X 为 Pb 金属回收率,%; t 为焙烧时间, min。

将图 7 数据分别带入式(8)~(10)中,通过最小二乘法进行线性回归分析,结果如图 8 所示。由图 8(c)可知, $1-(1-X)^{1/3}$ 与 t 之间的拟合程度较好,实验数据点紧密分布在其拟合直线周围。表 3 为在不同焙烧温度下各动力学方程线性拟合相关系数 R^2 和反应速率常数 k。相关系数 R^2 可用来度量回归直线对观测值的拟合程度,其数值越接近 1,则拟合程度越好。由此可知, $1-(1-X)^{1/3}$ 与 t 之间的拟合程度最优,因此渣中 Pb 的氯化挥发过程受界面化学反应控制。且当焙烧温度升高时,反应速率常数 k 增加,Pb 氯化反应速率加快。

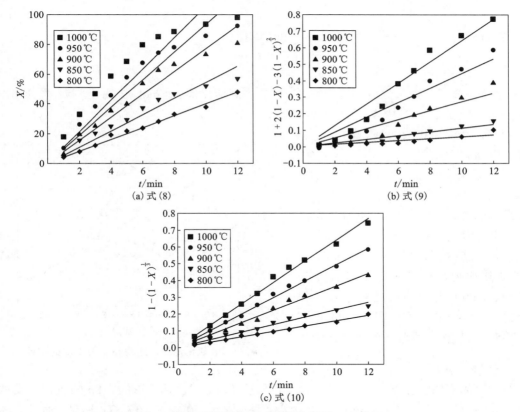

图 8 不同温度下各动力学方程线性拟合曲线

表 3 不同温度下各动力学方程线性拟合相关参数

温度/℃	式(8)		式(9)		式(10)	
	k_1	R^2	k_2	R^2	k_3	R^2
800	4.01479	0.98880	0.00579	0.79281	0.01545	0.98984
850	5.45560	0.92969	0.01109	0.90956	0.02218	0.97786

续表3

温度/℃	式(8)		式(9)		式(10)	
	k_1	R^2	k_2	R^2	k_3	R^2
900	7.75461	0.93579	0.02711	0.86725	0.03634	0.98420
950	9.33564	0.84318	0.04437	0.92057	0.04942	0.99456
1000	10.54965	0.65133	0.06381	0.94054	0.06399	0.99216

2.5.3 表观活化能计算

阿伦尼乌斯公式可以描述化学反应速率常数随温度变化的关系[10]。根据不同焙烧温度下所得 Pb 氯化反应的表观速率常数，可通过阿伦尼乌斯公式计算反应的表观活化能：

$$k = Ae^{\frac{E}{RT}} \tag{11}$$

对式(11)两边取对数可得到各温度下 $\ln k$ 的值：

$$\ln k = -\frac{E}{RT} + \ln A \tag{12}$$

式中：k 为化学反应速率常数，min^{-1}；E 为反应表观活化能，J/mol；T 为反应温度，K；A 为指前因子；R 为气体常数，8.314 J/(mol·K)。

根据表3中由式(10)所得的反应常数 k，以 $\ln k$ 对温度 $1/T$ 作直线得到图9。由式(12)可知，图9中拟合直线的斜率为 $-E/R$，由此求得焙烧过程中 Pb 氯化反应的表观活化能为 83.002 kJ/mol。表观活化能大于 40 kJ/mol 时，反应过程受化学反应控制[11]，进一步证实了 Pb 的氯化挥发过程受界面化学反应控制。通过图9中拟合直线的截距可以求得指前因子 $A = 169.37$，因此焙烧过程中 Pb 氯化反应的动力学方程可描述为：

$$1 - (1 - X)^{\frac{1}{3}} = 169.37 e^{\frac{-83002}{RT}} t \tag{13}$$

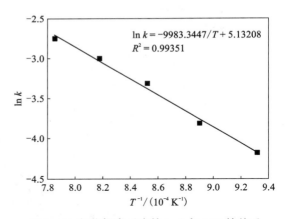

图9 Pb 氯化挥发反应的 $\ln k$ 与 $1/T$ 的关系

以上结果表明，焙烧前期 Pb 的氯化反应速率极快，然后速率减缓。Pb 的氯化挥发过程受界面化学反应控制，符合气固反应动力学模型。因此，在添加 10% $CaCl_2$ 时，通过升高焙烧温度就可以提高渣中 Pb 的氯化挥发速率，进而提升 Pb 金属回收率。

3 结论

(1)采用氯化焙烧法回收铜熔炼渣中 Pb 时，当焙烧温度 950 ℃、焙烧时间 12 min、$CaCl_2$ 添加量 10%、空气流量 100 mL/min 时，Pb 金属回收率可达 92.71%。

（2）提高焙烧温度、焙烧时间、CaCl$_2$ 添加量、空气流量，均可提高 Pb 金属回收率。由动力学方程拟合结果可知，升高焙烧温度可以提高 Pb 的氯化反应速率。

（3）在焙烧温度 800~1000 ℃、CaCl$_2$ 添加量 10%条件下，铜熔炼渣中 Pb 的氯化挥发过程符合未反应核收缩模型，受界面化学反应控制，其化学反应活化能为 83.002 kJ/mol，动力学方程为 $1-(1-X)^{\frac{1}{3}} = 169.37 e^{\frac{-83002}{RT}} t$。

参考文献

[1] 蒋亮，鄢洁，李鹏翔，等.工业铜渣固相改质后分离铁的实验研究[J].矿冶工程，2020，40(1)：96-100.

[2] 朱茂兰，肖妮，谭良春，等.铜渣还原活化制备新型胶凝材料与矿山充填的应用[J].中国有色金属学报，2020，30(11)：2736-2745.

[3] 罗仁昆，吴星琳，王俊娥，等.铜渣高温浮选药剂遴选与药剂制度优化研究[J].矿冶工程，2021，41(1)：33-36.

[4] 邱廷省，尹艳芬，崔立凤，等.磁化浮选铜冶炼废渣中铜及其他有价金属的研究[J].矿冶工程，2009，29(1)：34-36.

[5] 贺家齐，朱祖泽.现代铜冶金学[M].北京：科学出版社，2002.

[6] 伍习飞，尹周澜，李新海，等.氯化焙烧法处理宜春锂云母矿提取锂钾的研究[J].矿冶工程，2012，32(3)：95-98.

[7] QIN Hong, GUO Xue-yi, TIAN Qing-hua, et al. Pyrite enhanced chlorination roasting and its efficacy in gold and silver recovery from gold tailing[J]. Separation and Purification Technology, 2020, 250：117168.

[8] 李小斌，齐天贵，彭志宏，等.铬铁矿氧化焙烧动力学[J].中国有色金属学报，2010，20(9)：1822-1828.

[9] 袁学军，李解，李保卫，等.高硫铁精矿固硫氧化焙烧反应动力学分析[J].矿冶工程，2016，36(6)：69-74.

[10] 石美莲，华骏，颜文斌，等.含钒黏土矿直接酸浸提钒及其动力学研究[J].矿冶工程，2021，41(1)：94-97.

[11] 翟玉春.冶金动力学[M].北京：冶金工业出版社，2018.

铜熔炼渣制备铁精矿研究

摘　要：针对铜熔炼浮选尾渣中铁资源未高效利用问题，通过研究铁硅元素在低温碱性熔炼与浸出过程中的分配行为与规律，确定其优化工艺条件并制备铁精矿。研究了熔炼时间、熔炼温度、碱渣质量比对硅、铁分离效果的影响以及浸出时间、液固比、浸出温度对多元素浸出率的影响，确定优化工艺参数为熔炼温度550 ℃，熔炼时间1.5 h，碱渣质量比1.5∶1，浸出温度40 ℃，浸出时间20 min，液固比15∶1（mL/g）。在低温碱性熔炼－浸出过程中Fe、Si总回收率可分别达到99.43%与91.22%，所制铁精矿铁品位为61.82%，满足GB/T 25953—2010中三级铁精矿铁标准，且除铜外各杂质含量均低于一级标准中的限制值，可直接用于钢铁行业。

关键词：铜熔炼渣；碱性熔炼；水浸出；铁精矿

Study on preparation of iron concentrate from copper smelting slag

Abstract：Aiming at the problem of inefficient utilization of iron resources in flotation tailings of copper smelting slag, the distribution behavior and law of iron and silicon in the process of low-temperature alkaline smelting and leaching were studied, and the optimized process conditions were determined to prepare iron concentrate. The effects of melting time, melting temperature, alkali-slag ratio on the separation of silicon and iron and leaching time, liquid-solid ratio and leaching temperature on multi-element leaching rate were studied. The optimum technological parameters were the melting temperature of 550 ℃, the melting time of 1.5 h, the alkali-slag mass ratio of 1.5∶1, the leaching temperature of 40 ℃, the leaching time of 20 min and the liquid-solid ratio of 15∶1 (mL/g). In the low-temperature alkaline smelting-leaching process, the total recovery rates of Fe and Si could reach 99.43% and 91.22%, respectively. The iron grade of the iron concentrate was 61.82%, which meeted the iron standard of the third-grade iron concentrate in GB/T 25953—2010, and the content of all impurities except copper is lower than the limit value in the first-grade standard, which can be directly used in the iron and steel industry.

Keywords：Copper smelting slag；Alkaline smelting；Water leaching；iron concentrate

---

本文发表在《有色金属科学与工程》，2022，13(4)：1-9。作者：李中臣，王亲猛，田庆华，郭学益。

铜熔炼浮选尾渣为铜冶炼过程中产生的熔炼渣经过浮选贫化后的固废,简称"铜尾渣",其堆存量已高达1.4亿吨[1,2]。铜尾渣主要物相组成为铁橄榄石($2FeO·SiO_2$)、磁铁矿(Fe_3O_4)和非晶态硅石,其中铁橄榄石和磁铁矿占总渣量的90%[3-5]。由于铁橄榄石呈弱磁性、结构较稳定,很难通过常规物理或化学方法进行破坏,导致铜尾渣资源化利用难度大[6-8]。目前铜尾渣以堆弃为主,随着铜尾渣量持续增长,一方面占用大量土地;另一方面自然环境下铜尾渣中有害重金属浸出会污染水体和土壤[9]。

铜尾渣在水泥工业领域有较多应用,主要集中于作为烧制水泥熟料原料、作为矿化剂、作为水泥混凝土混合材等方面[10,11]。铜尾渣作为水泥混合材或混凝土掺合料,可改善水泥和混凝土的性能,但由于含有大量铁氧化物,掺量低杂质高,导致应用于水泥受到了限制[12-14]。Dos等利用铜尾渣作为波特兰水泥制造的细集料[15];Edwin等探讨了铜尾渣作为辅助凝胶材料在超高性能变质砂浆中的使用[16]。这些应用虽然有一定的经济价值,但铜渣中残留的有价金属资源并没有得到充分的利用,资源化利用率较低。

铜冶炼渣铁有价金属回收国内外已有大量研究,主要分为选矿法、还原焙烧-磁选法、氧化焙烧-磁选法、湿法浸出[5]。鲁兴武等采用氨水作为浸出剂从含铜品位为0.32%的铜尾渣中选择性浸出铜。结果表明,在最优化条件下:氨水浓度1.1 mol/L、浸出温度(55±2)℃、反应时间120 min、搅拌速度600 rad/min、液固比15∶1(mL/g)、尾渣粒度0.074~0.105 mm的条件下,铜浸出率为75%以上,其他杂质几乎不被浸出[17]。虽然铜回收率较高,但是由于采用浸出剂氨水浓度偏高,经济效益较低且容易产生二次污染,且渣中硅、铁元素均被废弃,未能从根本上解决铜冶炼尾渣资源化难的问题。王爽等以中国某铜渣磨矿、浮选铜尾矿为原料,以焦粉为还原剂,氧化钙为添加剂,通过高温还原反应获得了金属铁粉。但由于其过程中还原温度高达1200~1300 ℃,并需要配入大量的焦粉和氧化钙,导致能耗巨大,且渣硅组分未进行有效利用[18]。因此,找寻一种经济、环保的铜尾渣处理方法是目前铜冶炼行业所需的。

2020年我国进口铁矿石11.70亿t,对外依存度居高不下,不仅严重影响我国钢铁产业健康可持续发展,而且不利于国民经济的安全运行[19-21]。因此在我国现有资源紧缺及环境保护的严峻形势下,开发铜尾渣资源化综合利用技术,对促进循环经济和可持续发展具有重要意义[22-25]。由于铜渣储量巨大,若用来生产铁精矿,一方面能给炼铜企业带来较大经济效益,另一方面也能够达到资源循环利用、保护环境的作用[26]。本研究利用铜尾渣为原料,通过"碱性熔炼-水浸"实验,确定铜尾渣中硅、铁元素分离优化工艺条件,制备铁精矿。

1 实验

1.1 原料

铜尾渣取自于山东某公司。在85 ℃下干燥24 h,经过破碎、研磨,得到实验原料(≤0.075 mm)。对其进行元素组成进行XRF分析(XRF-1800,日本岛津公司),如表1所示。由表1可知,铜尾渣中主元素为Fe、O、Si,质量分数分别为47.9%、23.4%与17.4%,合计占总渣量的88.8%。其他元素含量较少,除两性金属Zn、Al与Ca外均低于1%。

表1 实验原料元素质量分数

Table 1 Chemical analysis of experimental raw materials %

成分	Fe	O	Si	Zn	Al	Ca	Cu	其他
质量分数	47.94	23.45	17.43	3.47	2.25	1.17	0.36	3.93

为明确实验原料物相组成,对其用 X 射线衍射仪(TTRⅢ,日本株式会学理学公司)检测,如图1所示。由 XRD 检测结果可知,尾渣中主要以铁橄榄石(Fe_2SiO_4)和四氧化三铁(Fe_3O_4)为物相组成。随后对实验原料进行扫描电镜分析(SIRION200,美国 FEI 公司),如图2所示。由图2可以看出,实验原料形貌不规则,呈黑色致密的块状,表面比较平滑,无孔状结构。

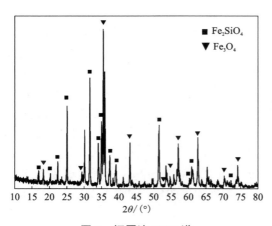

图1 铜尾渣 XRD 谱

Fig. 1 XRD patterns of the flotation tailings of copper smelting slag

图2 铜浮选尾渣 SEM 图

Fig. 2 SEM of the flotation tailings of copper smelting slag

1.2 研究方法

1.2.1 碱性熔炼实验

称取 5 g 铜冶炼尾渣置于不锈钢坩埚内,按一定量熔炼时碱加入量与尾渣量之比(碱渣质量比)称取氢氧化钠,并迅速将其置于坩埚内,随后扣置坩埚盖并将坩埚内反应物混合均匀。将坩埚放入电阻炉内,按照设定温度及设定时间进行碱性熔炼反应。反应完成后取出坩埚进行空气冷却,待坩埚表面温度降低至室温时,取出坩埚内碱性熔炼产物将其破碎并研磨后进行水浸出实验。

1.2.2 水浸出实验

将磨细后的熔炼产物放入锥形瓶中,向其中加入一定量的去离子水,超声分散 2 min,随后锥形瓶置于已经升温至指定温度的水浴振荡器中,在 4~5 次/s 的频率下震荡一定时间。待振荡结束后,取出锥形瓶,对反应产物进行离心,得到上清液与固体。液固分离后固体送入鼓风烘箱在 80 ℃下烘干 24 h。浸出液进行 ICP-MS 检测,得出含量后换算其浸出率。

1.3 数据分析

(1) 铁和硅回收率

$$\eta_{Fe} = \frac{w'_{Fe} \times m'}{w_{Fe} \times m} \tag{1}$$

$$\eta_{Si} = 1 - \frac{w'_{Si} \times m'}{w_{Si} \times m} \tag{2}$$

(2) 浸出率

通过 ICP-MS 检测液样中 Fe、Si 元素的浓度,可以计算出各金属的浸出率。

$$R_i = \frac{C_i \times V}{w_i \times m} \tag{3}$$

w_i 为元素在原料中质量分数,%;w'_i 为元素在浸出渣中质量分数,%;m 为原料质量,g;m' 为浸出渣质量,g;R_i 为某元素的浸出率,%;C_i 为液相中元素的质量浓度,g/L;V 为浸出液体积,L。

1.4 碱性熔炼原理

在熔融碱性介质中,碱能够破坏 FeO 与 SiO_2 之间的化学键,并与 SiO_2 结合生成硅酸盐,增加了 FeO 的活度从而降低了反应体系温度。

熔炼过程中可发生的反应主要化学方程式如下:

$$Fe_2SiO_4 + 2NaOH = 2FeO + Na_2SiO_3 + H_2O(g) \tag{4}$$

$$3Fe_2SiO_4 + 6NaOH + O_2(g) = 2Fe_3O_4 + 3Na_2SiO_3 + 3H_2O(g) \tag{5}$$

$$2Fe_2SiO_4 + 4NaOH + O_2(g) = 2Fe_2O_3 + 2Na_2SiO_3 + 2H_2O(g) \tag{6}$$

$$2Fe_2SiO_4 + 6NaOH(l) + O_2(g) = Na_2Si_2O_5 + 4NaFeO_2 + 3H_2O(g) \tag{7}$$

由于 NaOH 的熔点为 318.4 ℃ (591 K),故本研究选取 400 ℃ 为热力学计算起始点,进行方程式的热力学衡算。反应(4)~(7)的 ΔG 均小于零,说明在 400~1000 ℃ 的范围内,反应(4)~(7)均可自发进行。

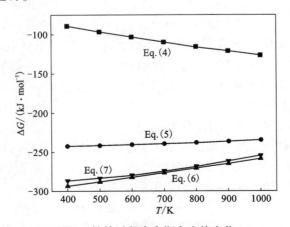

图 3 熔炼过程吉布斯自由能变化

Fig. 3 Relationship between Gibbs free energy and smelting temperature of reactions

2 结果与讨论

2.1 熔炼时间对硅铁分离的影响

熔炼时间控制着反应进行程度。若熔炼时间过短，会导致反应不完全，从而影响硅铁分离效果；若熔炼时间过长，则会造成热量浪费，增加工艺成本。

将铜尾渣与NaOH以1∶3(5 g铜尾渣，15 g NaOH)的质量比混合均匀，在550 ℃条件下进行碱性熔炼，熔炼时间对硅、铁回收率的影响如图4(a)所示。由图4(a)可以看出，随着熔炼时间增加，Si回收率随之升高，而铁元素在此条件下几乎未转化为可溶物溶于液相，故其回收率稳定在100%。在0~1.5 h，硅元素回收率升高较快，由65.89%迅速升高至86.86%，说明该阶段硅元素已大部分转为可溶性硅酸钠。反应1.5 h之后硅元素转化率虽有上升，但增量不大。

为更加直观了解硅、铁分离效果，熔炼时间对Si/Fe和渣重的影响如图4(b)所示。由图4(b)可知，随着熔炼时间增加，Si/Fe呈减小趋势。反应初期0~1.5 h时，Si/Fe下降较快，由0.37快速降低至0.04；反应1.5 h之后，曲线趋于平缓，产物的Si/Fe下降较慢，仅由0.04降低至0.03；通过产物渣重可知，随着熔炼时间增加，浸出渣质量保持4.5 g左右，在1.5 h时取得最小值4.12 g。

综合图4(a)与图4(b)可得延长熔炼时间可以增强铁硅分离效果，提高其回收率。熔炼时间超过1.5 h后虽也有一定提升但幅度并不大，且熔炼产物硬度逐渐增高，增加了后续破碎难度，综合考虑能耗与实际操作问题，选择熔炼时间1.5 h为优化条件。

2.2 熔炼温度对硅铁分离的影响

熔炼温度升高可有效提升反应速度，缩短熔炼时间，但温度升到一定程度后，反应速度的提升效果会逐渐降低，而热能消耗会急剧增加。因此，盲目提高熔炼温度并不可取。

将铜尾渣与NaOH以1∶3(5 g铜尾渣，15 g NaOH)的质量比混合，在一定温度下熔炼1.5 h，熔炼温度对硅、铁回收率的影响如图4(c)所示。由图4(c)可知，随着熔炼温度增加，Si元素回收率呈现先升高后下降的趋势，而铁元素只有少量转化为可溶物，回收率在96.50%左右。熔炼温度在350~550 ℃硅元素回收率升高较快，说明在熔炼温度低于550 ℃时，提升温度可有效提高Si回收率；在600 ℃时达到最大值82.40%，随后硅元素回收率出现小幅度降低，由82.40%下降至74.88%。

为更加直观了解硅、铁分离效果，熔炼温度对Si/Fe和渣重的影响如图4(d)所示。由图4(d)可知，Si/Fe曲线随着熔炼温度的升高而呈现先降低后升高的趋势。当熔炼温度达到550 ℃时，Si/Fe达到最小值0.06。通过分析产物渣重可知，随着温度升高，产物质量呈现先降低后升高的趋势。在550 ℃时，渣重为最小值4.39 g。

温度对熔炼反应的影响分为两个部分，一方面温度升高会导致熔炼反应生成Na_2SiO_3与铁氧化物的趋势降低，而另一方面提升熔炼温度会使反应速率增加，增强了碱与铜尾渣的作用强度。350~550 ℃后者占主导地位，从而提高了反应中硅铁元素的分离效果。温度在

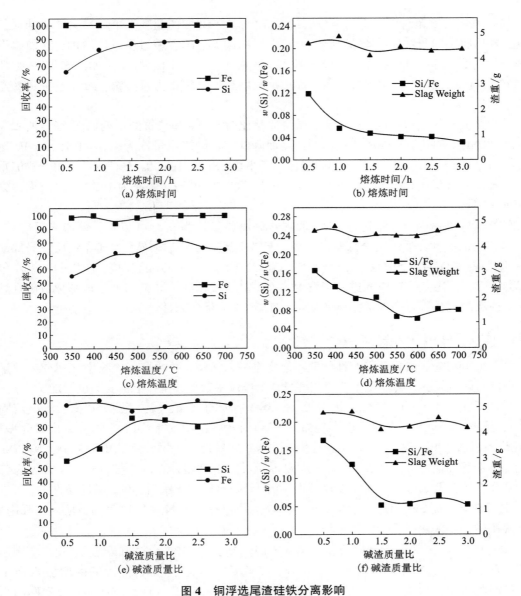

图 4 铜浮选尾渣硅铁分离影响

Fig. 4 Factors affecting separation effect of silicon and iron

550 ℃左右时,两方面作用对反应的影响达到平衡。因此,选择 550 ℃时为优化熔炼温度条件。

2.3 碱渣质量比对硅铁分离的影响

碱作为铜尾渣熔炼的反应介质,对熔炼反应影响较大。当碱渣质量比较小时,NaOH 无法充分包裹铜渣,导致反应不充分;当碱渣质量比较大时,浸出渣中 Na 残留高,需要多次水洗才能将 Na 含量降低,导致后续工艺冗杂,生产成本增加。因此选择合适碱渣质量比尤其重要。

将 NaOH 与铜尾渣以一定的碱渣质量比混合,在 550 ℃ 的温度下熔炼 1.5 h,碱渣质量比对硅、铁回收率的影响如图 4(e)所示。由图 4(e)可知,随着碱渣质量比增加,Si 回收率呈现先急剧升高后稳定的趋势,在碱渣质量比(0.5∶1)~(1.5∶1),由 55.54% 提升至 86.82%,且在 1.5∶1 时达到最大值;而铁元素只有少量转化为可溶物,回收率在 95% 左右波动。

为更加直观了解硅、铁分离效果,碱渣质量比对 Si/Fe 和渣重的影响如图 4(f)所示。由图 4(f)可知,Si/Fe 随着碱渣质量比的升高而降低。在碱渣质量比为 1.5∶1 时,Si/Fe 达到最小值 0.05;而超过 1.5∶1 后 Si/Fe 变化趋于平缓,Si/Fe 在 0.05 左右波动。由产物渣重可知,随着碱渣质量比升高,产物质量在 4.5 g 左右波动,碱渣质量比为 1.5∶1 时渣重达到最小值 4.13 g。

熔炼反应理论碱渣质量比为 0.4,但观察碱渣质量比为 0.5~1.5 时,铜渣未反应完全,可能是熔融的碱性介质无法全部浸没铜渣,上层未反应的铜渣隔绝了空气导致反应局部缺氧并发生: $3Fe_2SiO_4 + 6NaOH + O_2(g) \Longrightarrow 2Fe_3O_4 + 3Na_2SiO_3 + 3H_2O(g)$ 反应,产生的 Fe_3O_4 包裹在熔炼产物上,阻止了反应的进一步发生。而碱渣质量比高于 1.5∶1 时,能够形成碱性熔炼体系,使反应充分进行。故选择碱渣质量比 1.5∶1 为优化条件。

2.4　浸出时间对多元素浸出率的影响

浸出时间过短,会导致可溶性物质不能完全进入溶液,导致熔炼产物中元素分离效果较差;浸出时间过长,则会导致工艺流程冗长,工作效率较低,在一定程度上也增加了能耗。

碱渣质量比 1.5∶1,熔炼温度 550 ℃,熔炼时间 1.5 h,浸出温度 50 ℃,浸出液固比 20∶1,浸出时间对多元素浸出率影响如图 5(a)所示。由图 5(a)可知,随着浸出时间增加,Fe 几乎不进入液相,而 Si、Al、Zn 元素的浸出率均呈现出先增大后减小的趋势,其中 Si 元素趋势较明显而 Al、Zn 元素均较弱。推测各元素浸出过程存在两个控制过程:首先为各元素的离子扩散过程,另一个则为吸附过程。由于 $NaFeO_2$ 溶于水后,在低碱度条件下,可生成 $Fe(OH)_3$ 胶体,而 $Fe(OH)_3$ 胶体表面存在较多孔隙且具有较高的比表面,是一种吸附剂,能够吸附溶液中的离子[27]。

硅元素在 0~20 min 上升较快,在 20 min 时达到了 67.72%;20 min 之后,浸出率则逐渐减低。前 20 min 硅离子的扩散过程占主导,离子浸出率上升;而随着时间的增加 $Fe(OH)_3$ 胶体的吸附占主导地位,溶液中的离子逐渐被其吸附,故浸出率下降。Al、Zn 元素则因为其在熔炼产物中的含量较少,即浸出过程中进入液相的量较小,变化趋势不明显,但均在 20 min 时达到浸出率的最大值:37.01%、20.21%。浸出时间对渣重影响如图 5(b)所示,浸出时间对渣重无较大影响。

考虑浸出过程中离子浸出与 $Fe(OH)_3$ 胶体的吸附这两个过程的协同作用,选取 20 min 为优化浸出时间。

2.5　液固比对多元素浸出率的影响

液固比[$m(L)/m(S)$]对碱性熔炼产物的影响较复杂,理论上液固比越大,越有利于熔炼产物的浸出,但过高的液固比会抑制浸出过程。由于熔炼产物中碱过量,导致浸出液呈碱

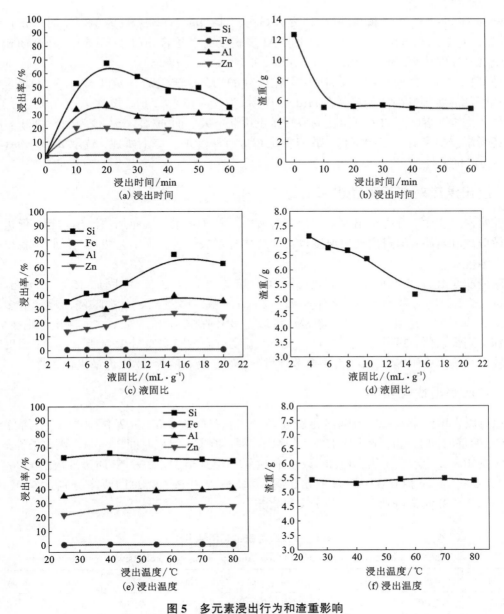

图 5　多元素浸出行为和渣重影响

Fig. 5　Factors affecting multi-element leaching behavior and slag weight

性,随着液固比升高,溶液 pH 值逐渐降低,从而促进硅酸钠水解生成硅酸。而硅酸为胶状物,析出并包裹在固体颗粒上会阻碍浸出的进行。而在工业生产中,较小的液固比有利于工艺操作,也能够减少浸出介质消耗。因此寻找液固比优化值,对该方法有较大的意义。

碱渣质量比 1.5∶1,熔炼温度 550 ℃,熔炼时间 1.5 h,浸出温度 50 ℃,浸出时间 20 min,液固比对多元素浸出率影响如图 5(c)所示。由图 5(c)可知,随着 $m(L)/m(S)$ 升高,Si、Al、Zn 元素的浸出率呈现先升高后降低的趋势,而 Fe 几乎不溶于水(最高为 0.56%)。在 $m(L)/m(S)=(4\sim 8)∶1$ 之间时,Si 浸出率提升较缓,仅仅由 35.26% 提升到

40.06%；在(8~15)∶1时，Si 浸出率提升显著，且在 $m(L)/m(S)$ 为 15∶1 时取得最大值 69.36%。$m(L)/m(S)$ 在(4~15)∶1时，Al、Zn 浸出率随着 $m(L)/m(S)$ 的升高而相应升高，浸出率分别从 22.36%、13.52%变化到 39.54%、27.07%。

液固比对渣重影响如图 5(d)所示。由图 5(d)可知，渣重随着 $m(L)/m(S)$ 明显呈现出先迅速降低后缓慢升高的趋势。在 $m(L)/m(S)$ 为(4~15)∶1时，随着 $m(L)/m(S)$ 升高，溶液体积增大，熔炼产物中可溶盐的溶解量也逐渐增大。而 $m(L)/m(S)$ 为(15~20)∶1时，由于各元素浸出率有一定的降低，故渣重也相应有所增加。综上所述，选取 $m(L)/m(S)=$ 15∶1 为最优条件。

2.6 浸出温度对多元素浸出率的影响

在水浸过程中，升高浸出温度有利于溶液中各粒子扩散，从而加快浸出过程，促进元素的有效分离和富集；但过高浸出温度也会导致能耗增长，增加了工艺成本，因此需找出合适的浸出温度。

碱渣质量比 1.5∶1、熔炼温度 550 ℃、熔炼时间 1.5 h、液固比 15∶1、浸出时间 20 min，浸出温度对多元素浸出率影响如图 5(e)所示。由图 5(e)可知，浸出温度对多元素浸出率影响不大。在 40 ℃时 Si 元素浸出率取得最大值 66.08%，而 Al、Zn 浸出率则在 70 ℃时分别达到最大值：39.37%与 27.41%。浸出温度对渣重影响如图 5(f)所示，浸出温度对渣重无较大影响。由于 Si 浸出率为主要考察目标，故浸出温度为 40 ℃为最优化条件。

2.7 熔炼浸出优化实验

通过以上单因素实验，确定优化实验条件为：铜尾渣 5 g、氢氧化钠 7.5 g、熔炼温度 550 ℃、熔炼时间 1.5 h、液固比 15∶1、浸出温度：40 ℃、浸出时间 20 min。优化条件下，渣成分含量如表 2 所示。由表 2 可知，Fe 品位由 47%上升至 61.82%，Si 质量分数由 17.44%下降至 1.78%，Si/Fe 由 0.371 下降至 0.029。实验过程中 Fe、Si 总回收率分别达 99.43%与 91.22%，Fe、Si 元素得到了的有效分离与富集。

表 2 综合实验铁精矿化学组成

Table 2 Chemical composition of iron concentrate in comprehensive experiment %

项目	Fe	Si	Si/Fe	O	Na	Zn	Cu	Al
原渣	47.00	17.44	0.371	24.54	0.91	3.37	0.33	2.25
实验1	62.15	1.74	0.028	26.86	1.56	1.34	0.25	0.26
实验2	61.48	1.83	0.030	26.54	1.43	1.42	0.23	0.31
均值	61.82	1.78	0.029	53.40	1.50	1.38	0.24	0.29

优化条件下，浸出液成分如表 3 所示。由表 3 可知，浸出液中主要存在 Si 元素。硅元素质量浓度为 8.41 g/L，两性金属 Zn、Al 含量较低，分别为 0.54 g/L、0.38 g/L，可通过后续工序进行脱除。同时对浸出液采用 1 mol/L 的盐酸，以酚酞为指示剂进行滴定，溶液碱度为 1.14 mol/L。因浸出液主要为 Si 元素，采用原位改性法制备白炭黑，实现 Fe、Si 资源高效利用[28]。

表 3　铜尾渣浸出液成分

Table 3　Element of copper tailings smelting lixivium　　　　g/L

成分	Fe	Si	Zn	Al	Ca	Mg	Pb	Cu
质量浓度	0	8.41	0.54	0.38	0	0	0	0

由图 6 可以看出，经过处理后的浸出渣颜色为赤红色，推断其主要物质为 Fe_2O_3。

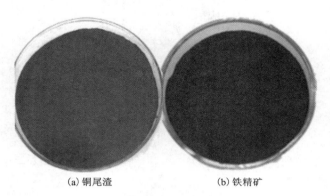

(a) 铜尾渣　　　　　　(b) 铁精矿

图 6　原料及产物对比图

Fig. 6　Comparison of raw materials and products

铜尾渣反应前后的 SEM 检测结果（如图 7 所示），原渣形貌不规则，呈块状，表面比较平滑，无孔状结构，通过熔炼-浸出后产物呈疏松多孔块状结构，也从侧面证明了其具有较好的吸附性能。与原铜尾渣相比，形貌发生了较大变化，表明铜尾渣在低温碱性熔炼-浸出过程中反应较剧烈。

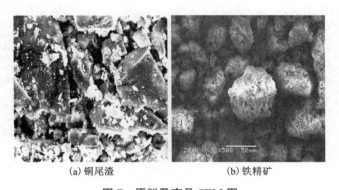

(a) 铜尾渣　　　　　　(b) 铁精矿

图 7　原料及产品 SEM 图

Fig. 7　SEM of raw materials and products

浸出渣 XRD 检测结果如图 8 所示，可以看出浸出渣中主要存在 Fe_2O_3、Fe_3O_4 和 $Fe(OH)_3$，并结合铁元素含量可推算出铁精矿中铁氧化物主要为 Fe_2O_3。

综上所述，通过低温碱性熔炼-水浸的方法处理铜浮选尾渣，可将其中硅铁元素进行有

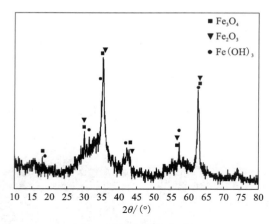

图 8 铁精矿 XRD 谱

Fig. 8 XRD patterns of the iron-rich concentrate

效分离,所制备的铁精矿品位高达 62.15%,铁品位达到 GB/T 2593—2010 二级铁精矿标准(见表 4),但各杂质元素中 Cu 有少量超标,故属高铁三级铁精矿。

表 4 铁精矿成分分析

Table 4 Composition analysis of the iron-rich concentrate %

品级	TFe 不小于	杂质含量,不大于					
		S	P	SiO_2	Al_2O_3	As	Cu
一级	65.00	0.60	0.04	6.00	1.00	0.05	0.10
二级	60.00	0.70	0.06	8.00	1.50	0.10	0.20
三级	55.00	0.80	0.08	10.00	2.00	0.15	0.30
原渣	47.00	0.19	—	36.42	4.25	0.15	0.33
实验结果	61.82	0.19	—	3.18	0.60	—	0.24

注:"—"指未测到。

3 结论

(1)低温碱性熔炼处理铜尾渣,温度对硅、铁分离效果影响最明显,其次是碱渣质量比,再次是熔炼时间。碱性熔炼优化条件为:熔炼时间 1.5 h,熔炼温度 550 ℃,碱渣质量比 1.5∶1。水浸处理低温碱性熔炼产物,液固比对硅、铁分离效果影响最明显,浸出时间与浸出温度对其影响较弱。浸出优化条件为:浸出时间 20 min,浸出温度 40 ℃,液固比 15∶1。

(2)Fe、Si 总回收率分别达 99.43% 与 91.22%,Fe、Si 元素得到了的有效分离与富集;所制得的铁精矿的铁品位为 61.82%,满足 GB/T 25953—2010 中三级铁精矿铁标准,且各杂质含量除铜外均低于一级标准中的限制值,可直接用于钢铁行业。

参考文献

[1] 赵凯,程相利,齐渊洪,等.配碳还原回收铜渣中铁、铜的影响因素探讨[J].环境工程,2012,30(2):76-78.

[2] 曹志成,孙体昌,吴道洪,等.转底炉直接还原铜渣回收铁、锌技术[J].材料与冶金学报,2017,16(1):38-41.

[3] GORAI B. Characteristics and utilisation of copper slag-a review[J]. Minerals Engineering, 2003, 39(4):299-313.

[4] 谭晓恒,郭少毓,喻相标,等.焙烧铜渣中磁铁矿的物性转变研究[J].有色金属科学与工程,2020,11(5):83-89.

[5] 刘金生,姜平国,肖义钰,等.从铜渣中回收铁的研究现状及其新方法的提出[J].有色金属科学与工程,2019,10(2):19-24.

[6] DHIR R, BRITO J, MANGABHAI R, et al. Sustainable Construction Materials-Copper Slag[M]. UK:Woodhead Publishing, 2016.

[7] NAJIMI M, POURKHORSHIDI A. Properties of concrete containing copper-slag-waste[J]. Magazine of Concrete Research, 2011, 63(8):605-615.

[8] LONG T, PALACIO J, SANCHES M, et al. Recovery of molybdenum from copper slag[J]. Tetsu-to-Hagane, 2012, 98(2):48-54.

[9] 王维,许向群,李杰,等.磷石膏与铜尾渣的高效耦合固定/稳定化处理[J].硅酸盐通报,2021,40(5):1601-1609.

[10] 侯霖杰,孟昕阳,王宏宇,等.铜渣改质、磁选及磁选尾渣制备陶瓷的基础研究[J].有色金属科学与工程,2021,12(2):23-29.

[11] HEO J, CHUNG Y, PARK J. Recovery of iron and removal of hazardous elements from waste copper slag via a novel aluminothermic smelting reduction (ASR) process[J]. Journal of Cleaner Production, 2016, 137:777-787.

[12] YANG Z, LIN Q, LU S, et al. Effect of CaO/SiO_2 ratio on the preparation and crystallization of glass-ceramics from copper slag[J]. Ceramics International, 2014, 40(5):7297-7305.

[13] HE R, ZHANG S, ZHANG X, et al. Copper slag: The leaching behavior of heavy metals and its applicability as a supplementary cementitious material[J]. Journal of Environmental Chemical Engineering, 2021, 9(1):105132-105143.

[14] HONG C W, LEE J I, RYU J H. Effect of copper slag as a fine aggregate on the properties of concrete[J]. Journal of Ceramic Processing Research, 2017, 18(4):324-328.

[15] ANJOS M, SALES A, ANDRADE N. Blasted copper slag as fine aggregate in Portland cement concrete[J]. Journal of Environmental Management, 2017, 196:607-613.

[16] EDWIN R S, SCHEPPER M D, GRUYAERT E, et al. Effect of secondary copper slag as cementitious material in ultra-high performance mortar[J]. Construction & Building Materials, 2016, 119:31-44.

[17] 鲁兴武,桑利,何国才,等.选矿后含铜尾渣选择性浸出的研究[J].有色金属(冶炼部分),2014(9):5-7.

[18] 王爽,倪文,王长龙,等.铜尾渣深度还原回收铁工艺研究[J].金属矿山,2014(3):156-160.

[19] 王建雄，张淑敏，李艳军，等.鞍山某铁矿石磁选——反浮选试验研究[J/OL].矿产保护与利用，2021(3)：150-154.

[20] 韩珍堂.中国钢铁工业竞争力提升战略研究[D].北京：中国社会科学院研究生院，2014.

[21] 侯永丰.浅谈中国进口铁矿石定价机制及变化趋势[J].现代营销（经营版），2021(7)：10-11.

[22] SHEN L, QIAO Y, GUO Y, et al. Preparation and formation mechanism of nano-iron oxide black pigment from blast furnace flue dust[J]. Ceramics International, 2013, 39(1)：737-744.

[23] LI D, GAO G, MENG F, et al. Preparation of nano-iron oxide red pigment powders by use of cyanided tailings[J]. Journal of Hazardous Materials, 2008, 155(1-2)：369-377.

[24] LEGODI M A, WAAL D D. The preparation of magnetite, goethite, hematite and maghemite of pigment quality from mill scale iron waste[J]. Dyes & Pigments, 2007, 74(1)：161-168.

[25] MADHESWARAN C K, AMBILY P S, DATTATREYA J K, et al. Studies on use of copper slag as replacement material for river sand in building constructions[J]. Journal of the Institution of Engineers (India)：Series A, 2014, 95(3)：169-177.

[26] GB/T 2593—2010.有色金属选矿回收铁精矿[S].北京：中国标准出版社，2011.

[27] 古国榜，李朴.无机化学[M].第二版.北京：化学工业出版社，2007.

[28] WANG Q, LI Z, LI D, et al. A method of high-quality silica preparation from copper smelting slag[J]. Journal of the Minerals Metals & Materials Society, 2020, 72(7)：2676-2685.

铜阳极泥典型处理工艺研究进展

摘　要：铜阳极泥含 Au、Ag、Se、Te、Cu 等多种有价金属，高效回收其中的稀贵金属不仅可实现资源综合利用，而且具有显著的社会经济效益。目前，铜阳极泥传统处理工艺存在环境污染大、回收率低等问题，卡尔多炉工艺是解决问题的有效途径。作为铜阳极泥典型处理工艺，本文系统介绍了卡尔多炉工艺的原理和方法，重点阐述了阳极泥预处理、卡尔多炉熔炼及金银合金电解的工艺特点。针对卡尔多炉熔炼周期长、稀贵金属分离提纯工艺复杂及回收率低等问题，在总结国内外先进技术及最新科研成果的基础上，总结出阳极泥预处理、熔炼效率提升及稀贵金属强化回收的工艺优化措施。本文旨在为卡尔多炉工艺高效回收铜阳极泥中稀贵金属提供指导。

关键词：卡尔多炉工艺；铜阳极泥；稀贵金属；工艺优化

Review on research progress for typical treatment process of copper anode slime

Abstract: Copper anode slime (CAS) contains valuable metals such as Au, Ag, Se, Te, Cu, therefore efficient recovery of CAS not only realizes the comprehensive utilization of resources, it also brings significant social and economic benefits. At present, there exists several problems in the CAS treatment process such as environmental pollution and low recovery rate. The Kaldo furnace process is an effective method for the treatment of CAS. As a typical treatment process of CAS, the principle and flow of the Kaldo furnace process are presented in this paper, including the pre-treatment, melting and Dore alloy electrolysis processes. To address the problems in the present process of long melting cycle of Kaldo furnace, complex separation and purification process of rare and precious metals, and low recovery rate, the optimazation measures are proposed including pre-treatment for CAS, enhancement in smelting effeicency and improvement for recyling rare and precious metals. This paper aims to provide guidance for the Kaldo furnace process to strengthen the recovery efficiency of rare and precious metals from CAS.

Keywords: Kaldo furnace process; copper anode slime; rare and precious metals; process optimization

---

本文发表在《中国有色金属学报》，网络首发。作者：郭学益，陈建儒，王松松，李明钢，王亲猛。

引言

随着电子信息、新能源等产业的蓬勃发展,稀贵金属需求量大幅增加[1-3]。目前,除 Au、Ag 等部分贵金属拥有独立开发的矿床外,Se、Te 等稀散金属尚未发现具有开采价值的矿床。因此,从二次资源中回收稀贵金属成为国内外冶金工作者的重要研究方向[4-5]。铜阳极泥是粗铜电解精炼时的不溶性产物,其产率通常为粗铜量的 0.2%~0.8%,含 Au、Ag、Se、Te、Cu 等多种有价金属,我国每年铜产量不断增加(见图 1),阳极泥产量也随之提高[6]。高效处理铜阳极泥不仅可实现资源综合利用,而且具有显著的社会经济效益[7]。

目前,铜阳极泥处理工艺主要有传统火法工艺、半湿法工艺、选冶联合工艺和全湿法工艺等[8]。传统火法工艺包括硫酸化焙烧蒸硒、酸浸分铜、贵铅炉还原熔炼、分银炉氧化精炼、金银合金电解及铂钯回收等工序,该工艺被大型冶炼厂广泛采用,但存在能耗高、环境污染大等问题。半湿法工艺以氯化分金、亚硫酸钠浸出银代替金银合金制备工序,避免了传统工艺带来的资金积压。选冶联合工艺为我国云南铜业铜阳极泥处理技术,浮选可提高炉料中贵金属含量,极大程度改善分银炉处理能力,但存在尾矿金银含量高、废水量大、经济效益差等缺点。全湿法工艺可综合回收 Au、Ag、Pb、Bi 等有价金属,但工艺流程复杂,生产成本高、连续性差等问题始终无法解决。作为铜阳极泥典型处理工艺,卡尔多炉工艺克服了传统火法工艺和湿法工艺的局限性,具有金属回收率高、原料适应性强、工艺流程短的技术优势,得以广泛应用,我国铜陵有色、阳谷祥光、紫金铜业等大型铜冶炼企业先后引入该工艺[9-11]。

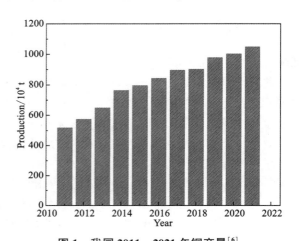

图 1 我国 2011—2021 年铜产量[6]
Fig. 1 China's copper production from 2011 to 2021

近年来,国内外学者对卡尔多炉工艺处理铜阳极泥展开了广泛研究,但对工艺原理的归纳总结较少,尚未形成系统有效的工艺优化措施。本文在总结国内外先进技术及最新成果的基础上,详细介绍了卡尔多炉工艺处理铜阳极泥的原理和流程,并针对稀贵金属分离回收工艺复杂等问题提出了工艺优化措施,为卡尔多炉工艺综合回收铜阳极泥提供指导。

1 典型铜阳极泥组成

国内外学者针对铜阳极泥综合利用开展了广泛的研究工作。表 1 总结了不同冶炼厂铜阳极泥的化学组成,由表 1 可知,铜阳极泥含 Au、Ag、Se、Te 等稀贵金属及 As、Sb、Bi 等杂质元素。不同冶炼厂的物料成分存在一定差异,这主要与矿石组成、冶炼工艺和电解制度等因素有关,此外,比利时 Umicroe 公司、芬兰 Outokumpu 公司、我国大冶有色等冶炼厂,利用铜捕集电子废弃物中的稀贵金属,铜阳极泥中稀贵金属含量将显著提高[12]。

表 1 不同冶炼厂铜阳极泥的元素含量
Table 1 Copper anode slime composition in different refineries

%

Smelter	Au	Ag	Cu	As	Ni	Se	Te	Sb	Pb	Pt	Pd	Bi	Ba
Yunnan Copper Co., Ltd.[17]	2.32	6.39	15.08	4.95	0.91	4.21	0.96	4.59	13.74	—	—	1.41	2.36
Jinchuan Group Co., Ltd.[18]	0.23	15.30	14.96	—	1.88	4.17	1.55	0.19	13.75	—	—	0.29	—
Zijin Mining Co., Ltd.[19]	0.27	13.80	16.80	3.90	0.13	5.33	1.09	1.41	10.78	—	—	3.60	—
Western Mining Co., Ltd[20]	1161.4*	70446.1*	18.01	0.58	—	—	—	—	25.43	—	—	—	—
Qinghai Copper Industry Co., Ltd[21]	0.13	5.73	12.54	2.39	1.39	4.17	—	2.62	27.02	—	—	4.61	2.34
Daye Non-ferrous Metals[22]	2.36#	83.21#	14.49	3.84	0.55	3.44	0.94	4.09	20.34	8.58*	97.86*	1.61	—
Jiangxi Copper Company[23]	0.19	6.29	21.80	6.22	—	4.52	2.86	—	—	—	—	—	—
Baiyin Nonferrous Metals[24]	0.27	8.33	34.53	—	—	13.24	0.62	—	0.27	—	—	—	—
Japan Toyama[25]	0.42	7.92	26.67	—	—	13.25	2.08	—	9.44	1.80#	9.00#	0.70	—
Japan Saganoseki[26]	2.10	19.00	25.20	3.30	0.40	12.10	3.90	1.80	4.00	—	—	1.50	—
Canada Montreal[26]	1.40	29.00	15.00	1.50	3.80	5.00	2.00	4.50	16.00	—	—	0.77	—
Canada Noranda[27]	0.18	19.50	18.70	1.14	—	10.00	1.20	1.68	8.00	—	—	0.14	—
America Inco[27]	0.12	6.37	21.00	0.50	17	8.40	1.80	0.09	1.70	—	—	2.62	—
America Phelps Dodge[27]	0.12	12.20	27.10	1.70	0.64	8.80	3.10	0.66	4.65	0.007	0.006	0.50	—
Finland Outokumpu[27]	0.50	9.38	11.20	0.21	54.21	4.23	—	0.04	—	—	—	0.15	0.68
Sweden Boliden[28]	0.65	21.40	9.50	0.90	5.80	3.30	1.60	2.60	7.80	—	—	0.11	5.87
Turkey Sakuysan[29]	0.23	2.80	25.80	3.93	0.29	4.68	0.90	0.99	12.93	—	—	0.08	—
Turkey Er-Bakir[30]	21.90*	2204.2*	23.10	—	0.82	413*	83*	0.24	15.42	—	—	—	—
Iran Sarcheshmeh[31]	0.08	7.07	7.00	0.42	0.03	13.66	—	0.10	4.42	—	—	—	—
India Ghatsila[32]	0.10	1.54	12.00	—	37.0	10.50	3.38	—	—	—	—	—	—

Remarks: *—g/t; #—kg/t

2 卡尔多炉处理工艺流程及原理

卡尔多炉（Kaldo furance），又称氧气顶吹旋转炉（top-blown rotary converter，TBRC），由瑞典波立登公司研发，该设备用于锡、铅等传统冶炼和阳极泥、废杂铜、电子废弃物等二次资源综合回收[13-15]。卡尔多炉将传统贵铅炉还原熔炼和分银炉氧化精炼两阶段合在一台炉子完成，同时替代了回转窑焙烧蒸硒工序。相对反射炉和转炉，卡尔多炉具有原料适应性强、能耗低、烟气排放量小、自动化程度高等特点。

卡尔多炉工艺处理铜阳极泥流程如图 2 所示，根据 Au、Ag、Cu 等金属的物质流向可将该工艺大致分为 5 个步骤[13, 16]：预处理脱铜碲、卡尔多炉熔炼、金银合金电解精炼、氯化分金和铂钯铑回收。相对传统工艺，该工艺通过氧压浸出工序回收铜、碲等有价金属，卡尔多炉熔炼工序完成贵铅到合金的一步转化，金银合金电解工序实现银与其他贵金属的高效分离。

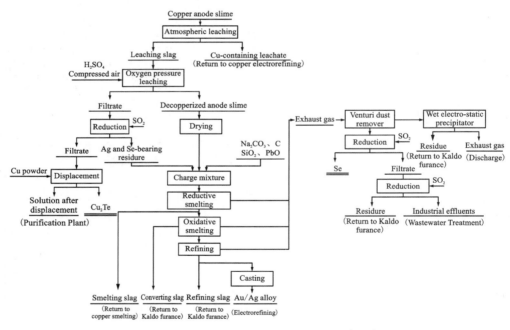

图 2 卡尔多炉工艺处理铜阳极泥流程图[13, 16]

Fig. 2　Flowchart of Kaldo furnace process for treating copper anode slime

2.1 预处理脱铜碲

卡尔多炉处理铜阳极泥时，铜在熔炼期形成冰铜溶解金银，精炼时难以氧化去除，致使生产周期延长和金银合金品位下降。可通过"常压浸出—加压酸浸"两段预处理工艺确保阳极泥含铜量小于 1%，以满足入炉要求[13]。常压浸出工序回收铜阳极泥中 CuO、Cu_2O 和

$CuSO_4$ 等,通常控制工艺条件为硫酸浓度 100~180 g/L,浸出温度 65~85 ℃和鼓风时间 5 h[33]。反应结束后,常压浸出液送铜电解精炼工序,滤渣泵入高压釜进一步脱除 Cu、As、Sb 等贱金属。加压酸浸工艺通过控制体系中氧分压(0.8~1.5 MPa)和反应温度(120~150 ℃)等条件,强化脱铜反应过程。此时 Cu_2S、Cu_2Se 和 Cu_2Te 等物相氧化进入加压浸出液(见表2),主要反应如式(1)~式(6)所示。除少量 Ag 外,其余贵金属保留在脱铜阳极泥中实现初步富集。

$$Cu_2Te + 2H_2SO_4 + 2O_2 =\!=\!= 2CuSO_4 + H_2TeO_3 + H_2O \quad (1)$$

$$Cu_2Se + 2H_2SO_4 + 2O_2 =\!=\!= 2CuSO_4 + H_2SeO_3 + H_2O \quad (2)$$

$$2Ag_2Se + 2H_2SO_4 + 3O_2 =\!=\!= 2Ag_2SO_4 + 2H_2SeO_3 \quad (3)$$

$$2Ag_2Te + 2H_2SO_4 + 3O_2 =\!=\!= 2Ag_2SO_4 + 2H_2TeO_3 \quad (4)$$

$$2CuS + 2H_2SO_4 + O_2 =\!=\!= 2CuSO_4 + 2H_2O + 2S \quad (5)$$

$$NiO + H_2SO_4 =\!=\!= NiSO_4 + H_2O \quad (6)$$

表 2 铜阳极泥氧压浸出工艺条件及 Cu,Se,Te 浸出率

Table 2 Reaction conditions for oxygen pressure leaching of copper anode slime and leaching efficiency of Cu, Se, Te

Pressurized medium	Reaction conditions	Leaching efficiency	Ref.
Industry pure oxygen	250 g/L H_2SO_4, T=150 ℃, t=2.5 h, L∶S=5∶1, OPP=0.9 MPa	93.60% Cu; 0.20% Se; 68.10% Te	[34]
Industry pure oxygen	120 g/L H_2SO_4, T=140 ℃, t=2 h, L∶S=5∶1, OPP=1 MPa	97.67% Cu*; 58.39% Te*	[35]
Compressed air	180 g/L H_2SO_4, 180 g/L NaCl, T=130 ℃, t=6 h, L∶S=6∶1, OPP=0.7~0.9 MPa, stirring rate=300 r/min	99.88% Cu; 69.28% Te	[36]
Air	100 g/L H_2SO_4, T=140 ℃, t=1.5 h, L∶S=5∶1, OPP=0.8 MPa, stirring rate=600 r/min	98.00% Cu; 49.00% Te; 13.00% Se	[37]
Oxygen	150 g/L H_2SO_4, T=160 ℃, t=8 h, L∶S=5∶1, OPP=1 MPa, stirring rate=275 r/min	99.90% Cu; 46.80% Se; 90.70% Te	[18]
Oxygen	50 g/L H_2SO_4, T=150 ℃, t=2 h, L∶S=10∶1, OPP=1.5 MPa, stirring rate=500 r/min	99.54% Cu; 96.95% Se; 68.54% Te	[38]
Oxygen	100 g/L H_2SO_4, T=150 ℃, t=1.5 h, L∶S=5∶1, OPP=0.8 MPa, stirring rate=600 r/min	Cu>98.00%; 52.90% Te#; 92.87% Se#	[39]
Oxygen	100 g/L H_2SO_4, T=120 ℃, t=1.5 h, L∶S=4∶1, OPP=0.8 MPa	99.80% Cu; 15.32% Se; 55.54% Te	[19]

Remarks: T=temperature, t=time, L∶S=liquid to solid, OPP=oxygen partial pressure, *-Average, #-Average retention rate.

冶炼厂通常采用还原法回收加压酸浸液中有价金属，以银硒摩尔比作为还原剂的选择依据[40]。当 Ag∶Se≥1 时，还原剂为铜粉；若该比值小于 1，采用 SO_2 作还原剂，反应原理如式(7)~式(9)所示。因酸性体系中 $E_{Ag^+/Ag}$、$E_{SeO_3^{2-}/Se}$、$E_{HTeO^{2+}/Te}$ 和 $E_{Cu^{2+}/Cu}$ 的标准电极电位分别为 0.80 V、0.74 V、0.57 V 和 0.34 V，SO_2 标准电极电位为 0.138 V，理论上可将 Ag、Se、Te 和 Cu 全部还原为金属单质[38]。然而，SO_2 在实际还原碲时动力学条件较弱，还原时间长且效率低。可加入卤素离子催化还原，提高碲粉收得率，如反应(10)~(11)所示。也有学者提出阳极泥酸浸液经蒸发结晶脱铜后，碲以 $TeOSO_4$ 形式存在。此时卤素离子(X^-)可作为碲还原反应过程的催化剂，反应机理见式(12)~式(15)，总反应如式(16)所示[41-43]。加压酸浸液还原工序通常包括两个阶段，控制反应温度 60 ℃时通入 SO_2，当 Ag 和 Se 质量浓度小于 0.01 g/L 时一段还原过程结束。固液分离后得到银硒渣和脱铜阳极泥送卡尔多炉配料系统，滤液中的碲以铜粉还原剂进行回收[13]。铜和碲的标准电位较为接近，还原驱动力较弱，但在硫酸体系中两者存在更强的相互化合作用，因此铜粉还原沉银硒后液得到碲化铜渣[44]。

$$Cu + 2Ag^+ \rightleftharpoons 2Ag + Cu^{2+} \tag{7}$$

$$2Cu + H_2SeO_3 + 2H_2SO_4 \rightleftharpoons 2CuSO_4 + 3H_2O + Se \tag{8}$$

$$2SO_2 + H_2SeO_3 + H_2O \rightleftharpoons 2H_2SO_4 + Se \tag{9}$$

$$TeO_3^{2-} + 6Cl^- + 6H^+ \rightleftharpoons TeCl_6^{2-} + 3H_2O \tag{10}$$

$$TeCl_6^{2-} + 2SO_3^{2-} + 2H_2O \rightleftharpoons Te + 2SO_4^{2-} + 4H^+ + 6Cl^- \tag{11}$$

$$TeOSO_4 + 2HX \rightleftharpoons TeOX_2 + H_2SO_4 \tag{12}$$

$$TeOX_2 + H_2O + SO_2 \rightleftharpoons TeX_2 + H_2SO_4 \tag{13}$$

$$TeX_2 + SO_2 \rightleftharpoons Te + SO_2X_2 \tag{14}$$

$$SO_2X_2 + 2H_2O \rightleftharpoons H_2SO_4 + 2HX \tag{15}$$

$$TeOSO_4 + 2SO_2 + 3H_2O \rightleftharpoons Te + 3H_2SO_4 \tag{16}$$

2.2 卡尔多炉熔炼

卡尔多炉作为整个熔炼工序的核心设备，其炉体如图 3 所示。该熔炼系统主要包括配料系统、加料系统、炉体、附属喷枪和液压装置等。炉体系统主要由炉子基础、拖轮和支撑臂、倾炉驱动装置等部件组成，底部配备支撑圈和旋转驱动装置，确保其以 1~20 r/min 绕自身旋转；同时在水平轴线安装倾斜系统，液压马达驱动炉子以 0.1~1 r/min 沿径向自由倾动。在不同冶炼阶段，卡尔多炉可通过调节炉体转速和倾斜角度强化动力学反应条件，确保熔炼反应充分进行[13]。

卡尔多炉熔炼目的是脱除脱铜阳极泥中 Pb、As、Sb 等杂质金属，生产符合电解要求的金银合金。该工序包括物料熔化、还原熔炼、氧化吹炼和精炼四个阶段。其中，还原熔炼和氧化吹炼是卡尔多炉熔炼工序的关键环节，其反应原理如图 4 所示。

还原熔炼是将金银化合物(Au_2Te、Ag_2Te 等)和部分 PbO 还原为金属单质。根据 Ellingham 图和选择性还原规律，元素还原顺序依次为：Ag、Cu、Pb、S、P。因此，只需控制还原剂用量确保 Ag、Cu 和少量 Pb 还原即可[45-46]。铅合金是贵金属良好的捕集剂，在还原期，Pb 吸收 Au、Ag 等贵金属沉于炉底形成贵铅，部分 As、Sb 和少量 Pb 与加入的苏打、石英砂生成炉渣(铅硅渣和钠硅渣)，实现贵金属与杂质元素的分离。

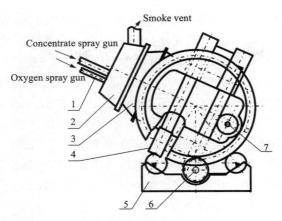

1—喷枪；2—活动烟罩；3—炉体装置；4—炉内旋转机构；
5—炉体倾翻支撑拖轮及其支架；6—炉体倾翻机构；7—止推托毂。

图 3 卡尔多炉设备图[47]

Fig. 3 Diagram of the Kaldo furnace

(1—Spray gun; 2—Adjustable hood; 3—Furnace device; 4—Rotating mechanism in furnace;
5—Furnace tilting support tugboat and support; 6—Furnace tilting mechanism; 7—Thrust hub)

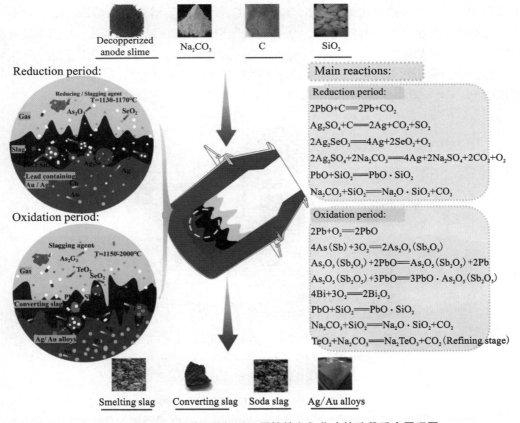

图 4 卡尔多炉处理脱铜阳极泥还原熔炼和氧化吹炼阶段反应原理图

Fig. 4 Diagram of reaction principle in reductive smelting and oxidative
converting phases of decopperized anode slime treated by Kaldo furnace

氧化期包括吹炼和精炼两个阶段：①吹炼阶段：贵铅中各金属的热力学氧化顺序为：Sb、As、Pb、Bi、Cu、Te、Se、Ag，而实际氧化过程与元素含量有关。在吹炼期，Pb氧化生成PbO与SiO_2造渣（铅硅渣），砷锑的一部分氧化为低价氧化物（Ⅲ）挥发进入烟气，一部分氧化为As_2O_5和Sb_2O_5与PbO生成砷锑酸铅进入炉渣。Pb、As、Sb氧化脱除后，Bi氧化形成Bi_2O_3与PbO形成低黏度的炉渣，整个吹炼过程伴随着Se和Te氧化。待贵铅中Pb、As等杂质质量分数<0.01%时，吹炼作业结束。②精炼阶段：该阶段的任务是深度脱除贵铅中的Cu、Se和Te。Cu氧化较为困难，需提高炉温至1200 ℃和增加氧气吹入量，将Cu氧化进入炉渣，Se氧化挥发进入烟气，Te氧化为TeO_2，与加入的Na_2CO_3反应生成苏打渣。待贵铅中$w_{(Au+Ag)}>97\%$，$w_{Cu}<2.5\%$，Pb、Bi等质量分数不大于0.01%时，卡尔多炉精炼工序完成，排渣后贵铅送至圆盘浇铸机得到金银合金板[8,13]。

2.3 金银合金电解精炼

电解精炼目的是制备纯度较高的电解银产品。电解过程以金银合金板为阳极，不锈钢板、银板或钛板为阴极，在HNO_3和$AgNO_3$电解液中生产纯度大于99.99%的电解银粉后，经模具熔铸得到银锭。图5为山东恒邦冶炼厂高效银电解设备示意图[48]。电解时，Pb、Cu等负电性金属（相对于Ag）溶解进入电解液；正电性金属理论上不发生化学溶解，但实际生产中部分铂、钯在阳极被氧化为$Pd(Pt)O_2·nH_2O$，经硝酸溶解进入电解液；化合物不发生反应，随阳极溶解脱落至电解槽底[49]。银在阴极沉积时以树枝晶方式生长，与阳极接触后造成短路，需用自动刮板机刮落至电解槽底部。电解液含有Pb、Cu、Sb等杂质元素，积累量一定时会在阴极表面析出，通常采用加热分解、铜粉置换或旋流电解等工艺净化[50-51]。针对废电解液中Pt、Pd等铂族金属，可采用沉淀法、椰壳碳吸附等工艺富集分离[52-53]。

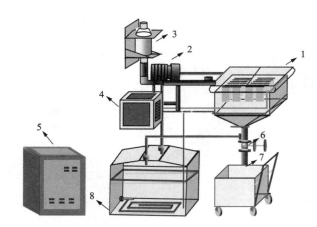

1—电解槽；2—自动刮银粉装置；3—侧抽风系统；4—配电柜；5—制冷机；
6—固液分离装置；7—银粉车；8—循环槽。

图5 高效银电解设备[48]

Fig. 5 High efficiency silver electrolysis equipment
(1—Electrolytic tanks; 2—Automatic silver powder scraper; 3—Side exhaust system; 4—Distribution box; 5—Refrigerator; 6—Solid liquid separator; 7—Silver powder car; 8—Circulating channel)

任同兴等[54]采用恒流电解法制备高纯银粉,并探究了银粉晶粒尺寸的影响因素。结果表明提高反应温度、使用添加剂(含酸性基团的烷烃基铵盐)、降低 HNO_3 浓度和沉积电位均可减小晶粒尺寸,$AgNO_3$ 浓度对其影响较小。Arif. T 研究结果表明,HNO_3 浓度对银溶解速率有显著影响,硝酸可分解生成 HNO_2,这种自催化反应改善了银溶解动力学[55]。为解决传统工艺生产效率低、劳动强度大等问题,铜陵有色贵溪冶炼厂研发了大阳极板银电解精炼新工艺。该工艺可大幅提高阴极电流密度和生产效率,实现电解工序的机械化生产[56]。针对电解车间酸雾治理难题,豫光金铅提出稳定温度、稳定酸度和设计新电解槽三项改进措施。电解槽与抽风系统一体化设计,可极大程度改善车间工作环境,这对企业的清洁化生产具有借鉴意义[57]。

电解产生的银阳极泥可由二次电解法进一步富集贵金属,或直接采用"氯化分金—黄嘌呤酸或 SO_2 还原沉淀—氯化铵沉淀工艺"回收 Au、Pt 和 Pd 等贵金属[58]。

3 卡尔多炉工艺优化

卡尔多炉工艺在实际生产中存在诸多问题,如预处理产品附加值低;卡尔多炉熔炼周期长、处理量小;贵金属分离提纯工艺复杂等。本章通过总结国内外大型冶炼厂改进措施及新工艺、新技术,得出了阳极泥预处理、熔炼工艺效率提升及稀贵金属强化回收的工艺优化措施。

3.1 阳极泥预处理工艺优化

3.1.1 硫酸钡脱除技术

铜冶炼行业通常以硫酸钡作为铜阳极板浇筑过程的脱模剂。钡盐经自动清刷装置清洗后去除率仅为55%,其余在电解精炼时沉入槽底。某些企业为提高阳极板合格率,增加脱模剂使用量,导致阳极泥中钡含量持续偏高[59]。卡尔多炉工艺处理高钡物料存在以下问题:①钡砂比重较大,预处理阶段沉积堵塞管道;②硫酸钡熔点为1580 ℃,延长物料熔化时间;③钡砂易于包覆物料,熔炼期加剧贵金属损失[60]。

紫金矿业利用钡砂与铜阳极泥的比重差异,对常压浸出槽底进行技术改造。改造后硫酸钡脱除率为70%以上,为脱铜渣的后续处理释放了产能[61]。具体改造措施如图6(a)和图6(b)所示:在浸出槽底部出料口位置,增加一个高约30 cm 的内伸套管,外部用法兰固定。铜阳极泥经常压浸出后,比重较大的钡砂先行沉降槽底。预浸槽泵将浸出渣送高压釜进一步脱铜时,套管对槽底的高钡泥浆起阻拦作用,达到定向脱除硫酸钡的目的。处理一定量铜阳极泥后,需拆除套管,清理槽底。钡砂沉降分离时,Au、Ag 等贵金属由于重力夹杂作用存在少量损失[图6(c)]。紫金矿业为此设计并研发了钡砂清洗槽,如图6(d)所示。该槽同样利用了钡砂与贵金属的密度差异,高钡泥浆水洗后由侧口回收含金银的上清液,硫酸钡从槽底排出。通过两次水洗工艺,铜阳极泥中 Au、Ag 总体回收率可提高0.3%,同时实现硫酸钡的高效回收。

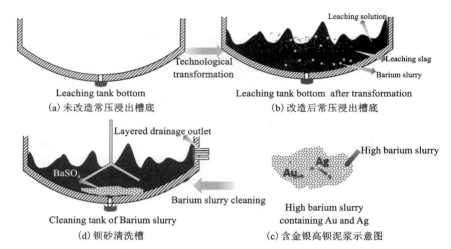

图 6 紫金矿业针对高钡阳极泥开展槽底结构改造及钡砂清洗槽示意图[60-61]

Fig. 6 Schematic diagram of leaching tank bottom structure transformation and barium sand cleaning tank for high barium anode slime in Zijin Mining Group (a) Leaching tank bottom without transformation; (b) Leaching tank bottom with transformation; (c) High barium anode slime containing Au and Ag; (d) cleaning tank for high barium anode slime

3.1.2 硒和碲强化浸出

卡尔多炉工艺处理铜阳极泥时,回收工序冗长致使 Te、Se 回收率仅为 60% 和 90%。为避免稀散金属分散,应尽可能在预处理阶段回收。为此,可通过加入氧化剂或改变浸出方式,提高铜阳极泥中 Se 和 Te 浸出率,其结果如表 3 所示。

表 3 铜阳极泥中 Se, Te 强化浸出工艺条件及其浸出率

Table 3 Reaction condition for enhanced leaching process of copper anode slime and leaching efficiency of Se and Te

Technology process	Oxidant	Reaction conditions	Leaching efficiency	Ref.
Atmospheric leaching	HNO_3	1.5 mol/L H_2SO_4, 0.5 mol/L HNO_3, T = 90 ℃, t = 3 h, L∶S = 5∶1, stirring rate = 300 r/min	98.30% Se	Li. et al.[62]
Atmospheric leaching	H_2O_2	150 g/L H_2SO_4, 2 mol/L H_2O_2, T = 60 ℃, t = 2 h, L∶S = 5∶1, stirring rate = 300 r/min	93.60% Se	Dong. et al.[21]
Atmospheric leaching	MnO_2 graphite	2 mol/L H_2SO_4, 0.8/0.8/1 MnO_2/graphite/CAS mass ratio, T = 90 ℃, t = 6 h, stirring rate = 500 r/min	81.90% Se 90.80% Te	K. K. et al.[63]

续表3

Technology process	Oxidant	Reaction conditions	Leaching efficiency	Ref.
Oxygen pressure leaching	O_2	first stage: 0.5 mol H_2SO_4, L:S = 10:1, secound stage: 1 mol/L H_2SO_4, L:S = 20:1; T = 150 ℃, t = 2 h, OPP = 1.5 MPa, stirring rate = 500 r/min	99.50% Se 98.60% Te	Rao. et al.[64]
Ultrasound-assisted leaching	H_2O_2/O_2	1.5 mol/L H_2SO_4, 0.2 mol/L H_2O_2, T = 70 ℃, t = 40 min, L:S = 5:1, aeration rate = 0.3 L/min	88.23% Te 33.98% Se	Ma Zhi-yuan[65]
Microwave-assisted leaching	H_2O_2	1.5 mol/L H_2SO_4, 0.8 mol/L H_2O_2, T = 110 ℃, t = 5 min, L:S = 10:1, stirring rate = 500 r/min, microwave power = 450 W	95.37% Se 95.97% Te	Yang. et al.[66]
Ultrasonic & microwave-assisted leaching	H_2O_2	1.5 mol/L H_2SO_4, 1 mol/L H_2O_2, T = 110 ℃, t = 403 s, L:S = 8:1, ultrasonic power = 638 W, microwave power = 834 W	98.57%±0.56 Se 98.96%±0.78 Te	Ma. et al.[67]

Remarks: T = temperature; t = time; L:S = liquid to solid; OPP = oxygen partial pressure

铜阳极泥常压酸性浸出时采用 HNO_3、MnO_2 或 H_2O_2 等做氧化剂, Se 和 Te 浸出率均大于 80%, 避免了稀散金属过度分散。传统浸出工艺以空气中 O_2 为氧化剂, 室温溶解度仅为 30 mg/L, 氧化能力较弱。表4 表明 H_2O_2 等具有更强的氧化能力, 促进浸出反应进行。为解决 Se 在酸性体系中回收率低的问题, Liu 等[68]提出氧压碱性分硒与酸浸分碲相结合的工艺优化措施。在 NaOH 物质的量浓度 2 mol/L、液固比 5:1、反应温度 200 ℃、浸出时间 2 h、氧分压 0.7 MPa 的最佳工艺条件下, Se 浸出率可达 99%。脱硒渣采用稀硫酸作浸出剂, Te 浸出率为 77.50%。该工艺实现了稀散金属的分步、高效提取。

表4 铜阳极泥浸出常用氧化剂的标准电位
Table 4 Standard potential of frequently-used oxidants in copper anode slime leaching process

Oxidant	Electrode reaction	Standard electrode potential $\varphi^{\ominus}$/V
O_2	$O_2 + 4H^+ + 4e^- = 2H_2O$	1.299
HNO_3	$NO_3^- + 3H^+ + 2e^- = HNO_2 + H_2O$	0.94
MnO_2	$MnO_2 + 4H^+ + 2e^- = Mn^{2+} + 2H_2O$	1.23
H_2O_2	$H_2O_2 + 2H^+ + 2e^- = 2H_2O$	1.8
OH^* [21]	—	2.8
O_3	$O_3 + 2H^+ + 2e^- = O_2 + H_2O$	2.07

超声波和微波辅助浸出工艺可大幅提高生产效率, 具有广阔的应用前景。研究表明: 超

声波作用于液体介质时，产生的"空化效应"可更新反应界面，提高扩散速率，显著降低反应活化能。微波加热是将电磁能转化为热能，在微观层面促进反应发生，避免了传统传导和对流传热的限制[69]。He 等[37]采用氧压酸浸工艺回收铜阳极泥，反应 90 min 后，Se 和 Te 浸出率仅为 13%和 49%；Rao 等[64]提出两段氧压浸出工艺高效回收阳极泥中稀散金属，反应时间共计 4 h，此时 Se 和 Te 浸出率达 99.3%和 94.4%；Yang 等[66]利用微波强化常压浸出过程，最佳工艺条件下 Se 和 Te 浸出率均大于 95%，而反应时间仅为 5 min。该工艺能够在提高浸出率的同时大幅缩短反应时间，提高生产效率 90%以上。

3.2 高温熔炼工艺优化

3.2.1 喷枪结构改进

卡尔多炉在不同冶炼阶段会调用不同类型的喷枪。燃料喷枪喷嘴可确保柴油与氧气充分混合，但喷料范围较为发散，会直接灼烧炉砖；同时柴油也存在燃尽度不够和含硫高的问题。吹炼喷枪向高温熔体中持续鼓入压缩空气，会降低炉膛温度，增加燃料消耗和作业时间。SHCHIKOV 等[70]针对卡尔多炉处理黄铜废料过程开展热量衡算，在 30%富氧条件下，废气热损失达总耗热量的 50%，据此推断卡尔多炉以压缩空气进行吹氧熔炼时热效率更低。实际生产中，可通过调控喷枪位移或优化燃料类型扩大卡尔多炉生产能力。

李庆纵等[71]通过调控油枪位移提高燃油效率；同时对燃料喷嘴进行改造，减少炉砖受热辐射时间。天然气价格仅为柴油的 30%，卡尔多炉采用天然气作还原剂和燃料时，可大幅缩减冶炼成本。铜陵有色稀贵冶炼厂[72]设计了以天然气为燃料的特殊喷枪结构，与氧气混合后火焰强度和燃烧区域稳定，避免了对耐火砖的直接灼烧。对比改造前后工艺参数，渣含银降低 0.02%，每年可增加阳极泥处理量 500 t。Wang 等[12]基于冶金理论计算和工艺分析，完成了卡尔多炉燃烧转换系统 30%富氧管道改造，平均每炉减少天然气消耗量 231～329 N·m³，缩短冶炼时间 1.5～3 h，节约运营和维护成本的 241 万元。

3.2.2 物料配比优化

卡尔多炉可通过优化炉料配比扩大生产能力。配料时除要求脱铜阳极泥粒度小于 50 mm、含水量小于 3%外，其余熔剂用量也需进行相应调整。焦粉入炉量过多时，燃烧产生过量还原性气体增加贵铅中杂质含量，吹炼阶段难以有效脱除；过少时则无法有效捕集贵金属。陈占飞等[73]以贵铅中 18%～20%的铅含量为配料依据，熔炼时增加 Na_2CO_3 用量并改变加料顺序，Ag 回收率提高 1.95%，作业时间缩短 2.3 h。郑伟忠等[74]通过控制石英砂用量减少冶炼时间。石英砂熔点高达 1650 ℃，而冶炼温度通常在 1150～1170 ℃，过量的石英砂会延长化料时间。优化后以贵铅中银、铅含量作为理论石英砂消耗量的判断依据，采用 3 批加料作业方式，可减少吹炼时间 2～3 h。

3.2.3 熔炼渣处理工艺

卡尔多炉炉渣属于高铅二次资源，含 Au、Ag、Se、Te、Pb 等多种有价金属，具有较高的回收价值[75]。熔炼渣产率约为入炉阳极泥量的 72%～80%，直接返炉将延长冶炼周期[13]。目前，高铅物料的处理工艺主要包括火法工艺、湿法工艺和选冶联合工艺。火法工艺发展最为成熟，但金属直收率低、烟气量大等问题仍未解决，始终制约着铅冶炼行业的绿色发展[76]。因此，本小节主要介绍选冶联合工艺和湿法工艺。

3.2.3.1 选冶联合工艺

选冶联合工艺为熔炼渣经选矿法富集贵金属得到精矿后,再利用卡尔多炉处理。精矿产率为原始炉料量的 1/3~1/2,可大幅提高工艺流程效率。我国铜陵有色、中原黄金等冶炼厂均采用此工艺综合回收卡尔多炉熔炼渣,其工艺流程如图 7 所示。浮选产出精矿送卡尔多炉熔炼工序,尾矿可生产铅合金[77]。

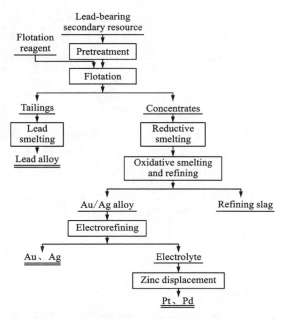

图 7 选冶联合工艺处理含铅二次资源的工艺流程图[77]

Fig. 7 Flowchart of processing lead containing secondary resources by combined and dressing and metallurgy process

浮选工艺通常以丁基黄药、PAC、BK-301 作捕集剂,通过固定的选矿制度,将熔炼渣中贵金属高效富集(见表 5)。代献仁等[78]通过浮选—重选工艺回收熔炼渣中贵金属。浮选时以 $FeSO_4$ 和 $N_2S_2O_3$ 为活化剂,丁基黄药和 PAC 为捕集剂,采用"两粗两精两扫"工艺流程,得到含 Au、Ag、Pt、Pd 品位分别为 0.05%、55.78%、9.36% 和 16.92% 的浮选精矿,尾矿采用重选法进一步回收 Au 和 Ag。熔炼渣经全流程处理后 Au、Ag 回收率可达 97.87% 和 93.37%,而浮选精矿产率仅为 11.50%,有效解决了返料的循环积累问题。

表 5 熔炼渣经浮选富集后 Au、Ag 回收率

Table 5 Recovery of Au and Ag from smelting slag after flotation process

Flotation system	Process parameters	Recovery	Ref.
Two roughing; two cleaning; two scavenging	Smelting slag of −74 μm account for 95%, Roughing 1: $FeSO_4$ 400 g/t, KBX 100 g/t, PAC 50 g/t, terpineol 40 g/t; Roughing 2: Na_2S 100 g/t, KBX 40 g/t, PAC 20 g/t; terpineol 20 g/t; Scavenging 1: KBX 40 g/t, PAC 20 g/t, terpineol 20 g/t; Scavenging 2: KBX 20 g/t, PAC 10 g/t, terpineol 10 g/t	95.60% Au 90.09% Ag	[78]

续表5

Flotation system	Process parameters	Recovery	Ref.
Two roughing; three scavenging	Smelting slag of -74 μm account for 90%, Roughing 1 and 2: KBX 160 g/t, terpineol 30 g/t; Scavenging 1 and 2: KBX 50 g/t, terpineol 15 g/t, Scavenging 3: KBX 20 g/t, terpineol 10 g/t	94.54% Au 86.69% Ag	[79]
Two roughing; one cleaning; one scavenging	Smelting slag of -74 μm account for 95%, terpineol 70 g/t, $FeSO_4$ 400 g/t, $Na_2S_2O_3$ 1000 g/t, KBX 90 g/t, PAC 60 g/t	91.62% Au 83.73% Ag	[80]

3.2.3.2 湿法工艺

湿法工艺是处理高铅物料的另一有效途径，其工艺流程如图 8 所示（以铅阳极泥为例）。铅阳极泥经预处理得到富铅渣，除返回卡尔多炉外，也可采用氯盐法进一步回收铅；分铅渣则通过氯化分金、亚硫酸钠分银等工序回收贵金属。相对传统火法工艺，该工艺具有生产周期短、能耗少、金属收得率高等特点[81]。

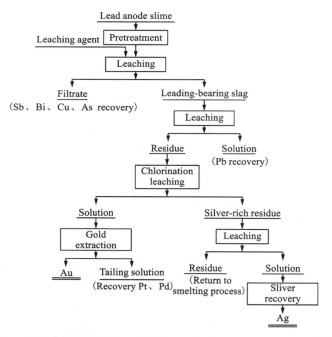

图 8 湿法工艺处理含铅二次资源（以铅阳极泥为例）的流程图[81]

Fig. 8 Flowchart of hydrometallurgy peocess for secondary resources containing lead (lead anode slime as an example)

预处理采用 HCl 或 H_2SO_4 和 Cl^- 组成的混合溶液脱除部分贱金属。含铅物料经烘料氧化、自然氧化或氧化剂氧化后，As、Bi 等转化为相应的氧化物。此时，物料易于与 HCl 反应，生成相应的氯化物进入溶液，反应如式（17）~式（19）所示。采用 Cl_2、$NaClO_3$ 等强氧化剂脱除贱金属时，应控制溶液体系电位为 350~500 mV，以避免贵金属分散。控电位氯化浸出相对传统火法工艺，金、银回收率可提高 10%左右[81]。

$$Sb_2O_3 + 6H^+ + 8Cl^- = 2SbCl_4^- + 3H_2O \quad (17)$$

$$As_2O_3 + 6H^+ + 8Cl^- = 2AsCl_4^- + 3H_2O \quad (18)$$

$$Bi_2O_3 + 6H^+ + 8Cl^- = 2BiCl_4^- + 3H_2O \quad (19)$$

吕建军[77]提出"两段 HCl 脱硅预处理—乙酸-乙酸钠浸出铅—Na_2CO_3 转化钡—H_2SO_4 沉淀钡"的全湿法工艺流程，综合回收卡尔多炉熔炼渣中 Ba、Pb、Bi、Sn 等有价金属。结果表明 Au 直收率接近 100%，Ag 直收率为 92.36%，Ba、Pb 回收率均大于 95%。该工艺实现了 Au、Ag、Pb 等有价金属的高效提取，同时具有富集比高、试剂消耗少、成本低等特点，这对熔炼渣的综合回收具有借鉴意义。张永锋等[82]采用盐酸浸出工艺回收熔炼渣中铅和铋，酸浸渣直接返回卡尔多炉熔炼，有效减少了铅、铋循环累积对冶炼产生的不利影响。徐磊等[83]依托铜陵有色冶炼厂，采用"氧化浸出—降温沉铅—水解沉锑铋"工艺综合回收铅阳极泥。铋渣经碳酸钠碱转化去除氯离子，与脱铜阳极泥混合后送卡尔多炉熔炼。该工艺整体生产平稳，为我国卡尔多炉一体化回收铜、铅阳极泥提供了依据。

3.3 稀贵金属强化回收

3.3.1 碲强化回收

卡尔多炉工艺处理铜阳极泥时，碲仅在氧压浸出工序分离，经铜粉置换后得到碲化铜渣。根据我国某冶炼厂稀贵车间总结的碲物质流分析，碲主要分布于铜碲渣、文丘里泥和卡尔多炉炉渣。熔炼渣产率可达入炉阳极泥量的 72%~80%，而碲质量分数仅为 0.5%~1%，回收成本高。因此，较为合适的提碲原料为碲化铜渣、文丘里泥和精炼渣(苏打渣)[87]。

碲化铜渣中碲主要以 Cu-Te 化合物形式存在。Cu_2Te 性质稳定，常压难溶于碱或非氧化性酸，可采用硫酸化焙烧或氧化焙烧—碱性浸出[84]、定向硫化—真空精馏[85]、加压碱性浸出[86]等工艺回收。文丘里泥中碲质量分数为 6%~12%，主要以 TeO_2 和碲酸盐形式存在，可采用两段碱浸工艺制备纯度>98%的 TeO_2 产品[87]。苏打渣中碲则以 Na_2TeO_3 和 Na_2TeO_4 为主，可采用"碱液浸出—中和沉碲—碱溶电解"工艺对其高效回收。含碲溶液除中和沉淀外，还可采用 SO_2 还原制备碲单质[88-89]。各物料回收碲的反应原理如式(20)~式(24)所示。

$$Cu_2Te + 6H_2SO_4 = 2CuSO_4 + TeO_2 + 4SO_2 + 6H_2O \quad (20)$$

$$Cu_2Te + 2O_2 = 2CuO + TeO_2 \quad (21)$$

$$TeO_2 + 2NaOH = Na_2TeO_3 + H_2O \quad (22)$$

$$Na_2TeO_3 + H_2SO_4 = TeO_2 + Na_2SO_4 + H_2O \quad (23)$$

$$Na_2TeO_4 + H_2SO_4 = H_2TeO_4 + Na_2SO_4 \quad (24)$$

Guo 等[90]针对高碲渣综合回收提出了 Na_2S 浸出—Na_2SO_3 还原新工艺，工艺流程如图 9 所示，反应如式(25)~式(28)所示。通过 XPS 表征，明确原料中的主要物相包括 Na_2TeO_4、$Na_2Sb(OH)_6$ 和 Pb_3O_4 等。最优工艺条件下，Te、Sb 浸出率分别为 91.24%和 92.51%。浸出液采用 Na_2SO_3 还原，反应 15 min 后，98.85%的 Te 被选择性沉淀，沉碲后液用于制备 $NaSb(OH)_6$。该工艺可为含碲物料的清洁、高效分离提供技术支撑。

$$Na_2TeO_4 + 4Na_2S + 4H_2O = Na_2TeS_4 + 8NaOH \quad (25)$$

$$NaSb(OH)_6 + 4Na_2S = Na_3SbS_4 + 6NaOH \quad (26)$$

$$Na_2TeS_4 + 3Na_2SO_3 = 3Na_2S_2O_3 + Na_2S + Te \qquad (27)$$
$$2Na_2SbS_4 + 4NaOH + 2H_2O_2 = 2NaSb(OH)_6 + 4Na_2SO_4 + 16H_2O \qquad (28)$$

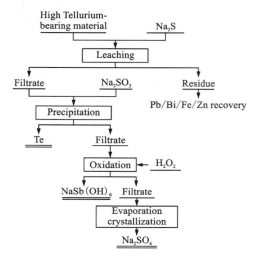

图9 含碲渣回收工艺流程图[90]

Fig. 9 The flowchart for recovery of high tellurium-bearing slag

3.3.2 铂钯铑强化回收

铂、钯和铑传统精炼工艺的原理,是利用不同价态铂族金属在氯化溶液中生成铵盐溶解度的不同,实现相互之间的分离提纯[91-92]。传统工艺如下:铂采用氯化铵沉淀法制备出粗氯铂酸铵[$(NH_4)_2PtCl_6$],经王水(或氯化造液工艺所得浸出液)溶解后,添加氯化铵进行反复沉淀、洗涤得到纯度较高的氯铂酸铵,最后经干燥、煅烧后产出纯度>99.95%的海绵铂。钯采用氯化铵沉淀法制备出粗氯钯酸铵[$(NH_4)_2PdCl_6$],经氨水溶解后添加盐酸,得到二氯二铵络亚钯黄色沉淀[$Pd(NH_3)_2Cl_2$],将其反复氨化、酸化提纯后采用高温煅烧产出氧化钯(PdO),最后用氢气还原制备出纯度大于99.95%的海绵钯。铑采用亚硝酸钠络合水解—硫化除杂质—亚硫酸铵除铱等工艺去除溶液中 Pd、Ir 等杂质后,经氯化铵沉淀得到六亚硝基络铑酸钠铵[$(NH_4)_2NaRh(NO_2)_6$]沉淀,随后采用氯化铵—盐酸溶液反复洗涤得到纯度较高的氯铑酸(H_2RhCl_6),再用甲醛或水合肼还原制备铑黑,最终氢气还原得到金属铑[91]。铂族金属传统精炼工艺存在工艺复杂、回收率低、产生 NO_x 气体污染环境等问题,亟须研发清洁高效的精炼工艺。萃取法和离子交换法因具有分离效果好、流程短、回收率高等特点,在铂族金属回收领域表现出巨大潜力[93]。

3.3.2.1 萃取法

萃取法(solvent extraction,SX)是利用金属离子在有机相和水相中分配系数存在差异,实现铂、钯和铑的定向分离回收。目前,国际大型铜冶炼企业如 Inco 的 Acton 公司、英国 Mathey-Rusterbury 冶炼厂、南非 Lonrho 冶炼厂等,国内金川公司、江西铜业和大冶有色冶炼厂等均采用萃取法回收阳极泥中 Pt、Pd 等铂族金属[91,94-96]。萃取剂包括硫醚类、亚枫类、肟类、喹啉类和螯合萃取剂等[97-99]。

萃取时应根据体系中的元素含量确定金属分离顺序,我国冶炼厂通常采用先萃取钯、再分离铂的工艺路线[100]。赖建林等[94]结合贵溪冶炼厂,提出"铂钯精矿预处理—氯化分金—S201萃取钯—N235萃取铂"的工艺路线,Pt和Pd回收率均大于90%,实现了铂族金属的清洁化提取。Shen等[95]针对金川公司贵金属分离工艺流程复杂、回收率低等问题,提出了以N530(活性成分为2-羟基-4-仲辛烷二苯酮肟)为萃取剂的多级逆流萃取工艺,高效分离盐酸介质中Au、Pt和Pd,工艺流程如图10所示。研究表明HCl浓度是分级萃取的关键因素,分别控制料液HCl浓度为0.2 mol/L和5 mol/L,将其多级逆流萃取、反萃后,Pd、Au回收率均大于99%。反萃液和萃金后液采用还原法或氯化铵沉淀法制备纯度大于99.99%的Au、Pt和Pd产品。该工艺实现了铂族金属的高效梯级分离,具有金属回收率高、试剂消耗少、流程短等特点。

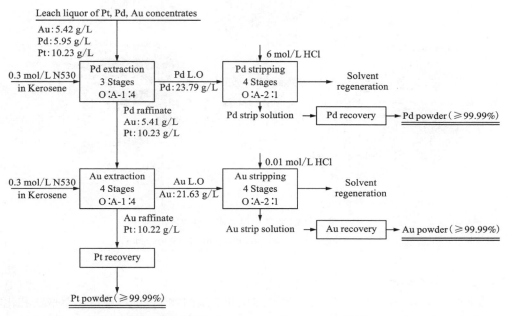

图10 从贵金属精矿的HCl浸出液中回收Pd(Ⅱ)、Au(Ⅲ)和Pt(Ⅳ)的工艺流程图[95]

Fig. 10 Flowsheet of the process for the recovery of Pd(Ⅱ), Au(Ⅲ) and Pt(Ⅳ) from HCl leach liquor of precious metals concentrate

国内外学者同样对离子液体开展了广泛深入的研究工作[101]。离子液体(ionic liquid, IL),又称室温离子液体、室温熔融盐或有机离子液体等,是指熔点低于某一温度的离子化合物。相对传统萃取剂,离子液体具有"零"蒸气压、溶解度高和化学性质稳定等优良特性[102]。孙旭[103]合成了[C_6C_1Pyr]Br离子液体萃取分离Au(Ⅲ)和Pt(Ⅳ),在最佳工艺条件下,Au和Pt萃取率分别为99.5%和95%,其萃取机理为萃取剂阴离子与溶液中[$AuCl_4$]$^-$和[$PtCl_6$]$^{2-}$进行离子交换,氮原子提供活性位点。

传统萃取工艺存在萃取级数多、自动化程度低、占地面积大等缺点,3D打印微反应器技术能够有效解决上述问题[104]。极小的腔室和通道使流体的物理量梯度急剧增加,因此具有更为优良的传质特能。刘晓玲等[105]以TBP为萃取剂,利用3D打印多通道微流体反应器萃取分离盐酸体系中Pt、Pd和Rh,其结构示意如图11所示。使用微反应器A时Pt、Pd萃取率分别为 84.26% 和 96%，将其增加流体混合室、延长通道长度并改变分配室结构得到微反应器 B，此时 Pt、Pd 萃取率分别提高 2% 和 1%。该技术相比常规单级萃取工艺能够有效提高铂族金属萃取率，减少萃取时间，且连续化生产更加符合工业大规模应用。

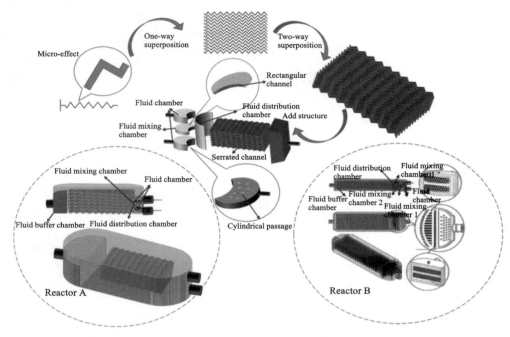

图 11　微反应器放大原理图[105]
Fig. 11　Enlarged schematic diagram of microreactor

3.3.2.2　离子交换法

离子交换法（ion exchange，IX）集分离、纯化和金属回收于一体，克服了萃取法和化学沉淀法无法处理低浓度物料的缺陷[106-107]。针对 Pt、Pd 等铂族金属分离回收，相关学者已研发性能优良的离子交换树脂，如 QPTU[108]、Amberlyst[109]、AG[110]、Diaion[111] 和 MSA[112] 系列等。表 6 总结了不同阴离子交换树脂从氯化物体系中分离铂、钯和铑的研究结果。

表 6　利用阴离子交换树脂从氯化物体系中分离铂、钯和铑的研究结果
Table 6　Research results of separation of Platinum Palladium and Rhodium from chlorinated materials by anion exchange resin

Resin	Functional group	Results	Eluent	Ref.
PS-GA	Glucosamine	The adsorption dynamics indicated that adsorption was controlled by the liquid film diffusion for Pt(IV) The Langmuirmodel was to describe the isothermal process	2% thiourea+ 0.1 mol/L HCl	[113]

续表6

Resin	Functional group	Results	Eluent	Ref.
S300	Sulfoether (—C—S—C—)	Pd(II) equilibrium adsorption rate exceeded 99% after equilibrium adsorption time about 30 min. The adsorption of Pd is proceeded following the mechanism of ion-dipole action between $[PdCl_4]^{2-}$ and the S300 resin. Kinetic modelling shows that the adsorption followe the pseudo-second order kinetic expression and the Langmuir model is to describe the isothermal process	1 mol/L $NH_3 \cdot H_2O$ +1 mol/L NH_4Cl	[114]
Diaion SA10A Diaion SA20A Diaion PA308	Tertiary amines	Diaion SA10A was the most effective to separate Pt from Rh at low acid concentrations. Decreasing the HCl concentration favored Pt adsorption, but suppressed that of Rh	0.1 mol/L thiourea+ 0.1 mol/L HCl	[111]
Lewatit M+ MP 600 Purolite S985 XUS43600.00	Quaternary ammonium type 2 Polyamine and Thiouronium	The Langmuir model was to describe the isothermal process for XUS 43600.0 resin, and pseudo-second order kinetic model was suitable for kinetic analysis Lewatit Mono Plus (M+) MP 600 and Purolite S985 the adsorption was best described by the Freundlich isotherm	1 mol/L thiourea+ 2 mol/L H_2SO_4	[115]
Amberlyst A21 Amberlyst A29	The dimethylamine and type 2 with a dimethyl-hydroxyethyl-quaternary amine group	Amberlyst A21 is more promising for adsorbing palladium(II) chloro-complexes from acidic chloride solutions than Amberlyst A29. Kinetic modelling of Amberlyst A21 was suitable for pseudo-second order linear kinetic model	—	[109]
Amberlite XAD-7	Monomer-type4-tert-butyl (dimethylthiocarbamoyloxy)benzene; dimer-type1, 1'-bis[(dimethyl-thiocarbamoyl) oxy]-2, 2'-thiobis[4-t-butylbenzene]	MTCA is more promising for adsorbing palladium (II) chloro-complexes than DTCA. The adsorption rate of MTCA and DTCA are 95% and 78%, respectively	1 mol thiourea+ HCl solution	[116]

续表6

Resin	Functional group	Results	Eluent	Ref.
TAPEHA	1,3,5-Triazine-Pentaethylenehexamine polymer	Polymer was used as an adsorbent for the recovery of palladium (Ⅱ) ions from chloride-containing solutions. Kinetic modelling of TAPEHA followe the pseudo-second order kinetic expression and the Langmuir model is to describe the isothermal process	—	[117]

阴离子交换树脂根据其官能团差异分为强阴离子交换树脂(strong anionic resin, SAR)和弱阴离子交换树脂(weak anionic resin, WAR)。强阴离子交换树脂对铂、钯吸附效果主要与溶液酸度和氯离子浓度有关。Sun等[110]研究了盐酸体系中AG1-x2、AG1-x8和AGMP-1树脂对Pt和Rh的负载行为,表明此类树脂对Pt均有一定的吸附能力,其中AG1-x8树脂对其选择性更强。采用AG1-x8树脂从混合溶液中分离Pt和Rh时,HCl浓度是将其分离的关键因素,低浓度时有利于吸附Pt,并在一定范围内遵循Langmuir等温吸附模型。李伟等[118]以沉金后液为原料,研究了QPTU树脂对Au、Pt和Pd的吸附效果,得到吸附后液中贵金属质量浓度均稳定在0.2 mg/L以下。相对于传统锌粉置换工艺,离子交换法具有工作环境好、选择性高等优点,但吸附时间较长,容易造成贵金属积压。

4 结论

铜阳极泥是回收稀贵金属的重要二次资源,卡尔多炉工艺是实现其清洁高效处理的有效方法。该工艺主要包括氧压浸出脱铜碲、卡尔多炉熔炼、电解精炼、氯化分金和铂钯铑回收五个步骤。本文针对当前工艺存在的稀贵金属回收率低、卡尔多炉熔炼周期长、铂族金属分离提纯工艺复杂等问题,在总结国内外先进技术及最新科研成果的基础上,得出了一系列工艺优化措施,具体包括:在阳极泥预处理阶段,采用硫酸钡脱除技术和硒碲强化浸出工艺提高稀贵金属回收率;在卡尔多炉熔炼阶段,通过喷枪结构改进、优化物料配比,减少熔炼渣返炉量以提高熔炼效率;在铂族金属精炼阶段,采用萃取法和离子交换法代替传统沉淀工艺。展望卡尔多炉工艺的未来发展,鉴于铅阳极泥与铜阳极泥化学成分相近的特点,可预先脱除Pb、As等贱金属后,将残渣与脱铜阳极泥混合送入卡尔多炉熔炼系统回收稀贵金属,实现二次资源的综合回收与高效利用。

参考文献

[1] LUO Yi-fan, ZHANG Chao, ZHENG Bing-bing, et al. Hydrogen sensors based on noble metal doped metal-oxide semiconductor: A review[J]. International Journal of Hydrogen Energy, 2017, 42(31): 20386-20397.

[2] WEI Xuan, LIU Chun-wei, CAO Hong-bin, et al. Understanding the features of pgms in spent ternary automobile catalysts for development of cleaner recovery technology[J]. Journal of Cleaner Production, 2019, 239: 118031.

[3] BOURGEOIS D, LACANAU V, MASTRETTA R, et al. A simple process for the recovery of palladium from wastes of printed circuit boards[J]. Hydrometallurgy, 2020, 191: 105241.

[4] KAVLAK G, GRAEDEL T E. Global anthropogenic tellurium cycles for 1940—2010[J]. Resources Conservation & Recycling, 2013, 76: 21-26.

[5] ZHANG Lin-gen, XU Zhen-ming. A critical review of material flow, recycling technologies, challenges and future strategy for scattered metals from minerals to wastes[J]. Journal of Cleaner Production, 2018, 202: 1001-1025.

[6] 中国国家统计局. 中华人民共和国2011—2021年国民经济和社会发展统计公报[R]. 2011-2021.

[7] LIU Gong-qi, WU Yu-feng, TANG Ai-jun, et al. Recovery of scattered and precious metals from copper anode slime by hydrometallurgy: A review[J]. Hydrometallurgy, 2020, 197: 105460.

[8] 王吉坤, 张博亚. 铜阳极泥现代综合利用技术[M]. 北京: 冶金工业出版社, 2008.

[9] 何云龙, 徐瑞东, 何世伟, 等. 铅阳极泥处理技术的研究进展[J]. 有色金属科学与工程, 2017, 8(5): 40-51.

[10] 花少杰, 胡鹏举, 布金峰. 卡尔多炉处理高杂铜阳极泥的工艺改进[J]. 有色金属(冶炼部分), 2020, 2: 45-48.

[11] 姬长征. 卡尔多炉工艺在铜阳极泥回收贵金属中大显身手[EB/OL]. [2014-05-23]. http://www.cnmn.com.cn/ShowNews1.aspx?id=291364.

[12] 翟保金, 王勇, 王亲猛, 等. 铜精矿富氧顶吹熔炼协同规模处置废电路板关键技术及应用[R]. 湖北省, 大冶有色金属集团控股有限公司, 2020-09-22.

[13] 衷水平, 王俊娥, 张焕然, 等. 铜阳极泥卡尔多炉法提取工艺[M]. 北京: 冶金工业出版社, 2019.

[14] GREGUREK D, REINHARTER K, MAJCENOVIC C, et al. Overview of wear phenomena in lead processing furnaces[J]. Journal of the European Ceramic Society, 2015, 35(6): 1683-1698.

[15] CUI Ji-rang, ZHANG Li-feng. Metallurgical recovery of metals from electronic waste: A review[J]. Journal of Hazardous Materials, 2008, 158(2-3): 228-256.

[16] 肖鹏, 王红军, 叶逢春, 等. 稀散金属硒、碲回收工艺现状与展望[J]. 金属矿山, 2020(4): 52-60.

[17] LIU Jian, WANG Shi-xing, LIU Chen-hui, et al. Decopperization mechanism of copper anode slime enhanced by ozone[J]. Journal of Materials Research and Technology, 2021, 15: 531-541.

[18] 钟清慎, 贺秀珍, 马玉天, 等. 铜阳极泥氧压酸浸预处理工艺研究[J]. 有色金属(冶炼部分), 2014, 7: 14-16.

[19] 蔡创开, 庄荣传, 林鸿汉. 从铜阳极泥中氧压浸出有价金属试验研究[J]. 湿法冶金, 2015, 34(5): 376-379.

[20] 张慧婷, 翁存建, 赖春华, 等. 青海某铜阳极泥的工艺矿物学[J]. 矿产综合利用, 2022, 2: 200-205.

[21] DONG Zhong-lin, JIANG Tao, XU Bin, et al. Comprehensive recoveries of selenium, copper, gold, silver and lead from a copper anode slime with a clean and economical hydrometallurgical process[J]. Chemical Engineering Journal, 2020, 393: 124762.

[22] 钟菊芽. 大冶铜阳极泥处理过程中有价金属元素物质流分析研究[D]. 长沙: 中南大学, 2010.

[23] 刘永平, 雷刚. 从铜阳极泥中浸出铜、砷、碲试验研究[J]. 湿法冶金, 2021, 40(4): 298-301.

[24] CHEN Ai-liang, PENG Zhi-wei, HWANG Jiann-yang, et al. Recovery of silver and gold from copper anode slimes[J]. JOM, 2014, 67: 493-502.

[25] SANUKI S, MINAMI N, ARAI K, et al. Oxidative leaching treatment of copper anode slime in a nitric acid solution containing sodium chloride[J]. Materials Transactions, JIM, 2007, 30(10): 781-788.

[26] MAHMOUDI A, SHAKIBANIA S, MOKMELI M, et al. Tellurium, from copper anode slime to high purity product: A review paper[J]. Metallurgical and Materials Transactions B, 2020, 51(6): 2555-2575.

[27] LEE J C, KURNIAWAN K, CHUNG K W, et al. Metallurgical process for total recovery of all constituent metals from copper anode slimes: A review of established technologies and current progress[J]. Metals and Materials International, 2020, 27: 2160-2187.

[28] 董凤书. 波立登隆斯卡尔冶炼厂阳极泥的处理[J]. 有色冶炼, 2003, 32(4): 25-27.

[29] KILIC Y, KARTAL G, TIMUR S. An investigation of copper and selenium recovery from copper anode slimes[J]. International Journal of Mineral Processing, 2013, 124: 75-82.

[30] AYDIN, NADERI M, GHAZITABAR A. Optimization of gold recovery from copper anode slime by acidic ionic liquid[J]. Korean Journal of Chemical Engineering, 2017, 34(11): 2958-2965.

[31] KHANLARIAN M, RASHCHI F, SABA M. A modified sulfation-roasting-leaching process for recovering Se, Cu, and Ag from copper anode slimes at a lower temperature[J]. Journal of Environmental Management, 2019, 235: 303-309.

[32] HAIT J, JANA R K, KUMAR V, et al. Some studies on sulfuric acid leaching of anode slime with additives[J]. Journal of Environmental Management, 2002, 41(25): 6593-6599.

[33] 王军正, 刘平, 赵景龙. 铜阳极泥常压氧化脱铜研究[J]. 大众科技, 2019, 21(5): 53-54+84.

[34] 陈志刚. 从铜阳极泥中加压浸出铜[J]. 湿法冶金, 2010, 29(3): 181-183.

[35] 熊家春, 王瑞祥, 衷水平, 等. 铜阳极泥加压酸浸脱除铜碲工艺研究[J]. 有色金属(冶炼部分), 2017, 02: 44-46.

[36] 夏彬, 邓成虎, 黄绍勇, 等. 高杂质铜阳极泥预处理的工艺研究[J]. 矿冶, 2013, 22(1): 69-71.

[37] HE Shan-ming, WANG Ji-kun, XU Zhi-feng, et al. Removal of copper and enrichment of precious metals by pressure leaching pretreatment of copper anode slime in sulfuric acid medium[J]. Precious Metals, 2014, 35(4): 48-53.

[38] 刘溢. 加压浸出废杂铜阳极泥高效提取稀散金属的实验研究[D]. 昆明: 昆明理工大学, 2020.

[39] 易超, 王吉坤, 李皓, 等. 铜阳极泥氧压酸浸脱铜试验研究[J]. 云南冶金, 2009, 38(3): 32-35.

[40] 王海荣. 铜阳极泥湿法预处理工艺研究[J]. 中国有色冶金, 2018, 47(4): 70-73.

[41] 马辉. 从碲渣中浸出和分离碲的工艺研究[D]. 长沙: 中南大学, 2009.

[42] 郑雅杰, 陈昆昆, 孙召明. SO_2 还原沉金后液回收硒碲及捕集铂钯[J]. 中国有色金属学报, 2011, 21(9): 2258-2264.

[43] 郑雅杰, 孙召明. 催化还原法从含碲硫酸铜母液中回收碲的工艺研究[J]. 中南大学学报(自然科学版), 2010, 41(6): 2109-2114.

[44] 刘兴芝, 宋玉林, 武荣成, 等. 碲化铜法回收碲的物理化学原理[J]. 材料研究与应用, 2002, 12: 55-58.

[45] ZHOU Yi, JIANG Wen-long, GUO Xin-yu, et al. A selective volatilization and condensation process for extracting precious metals from noble lead[J]. Journal of Cleaner Production, 2021, 294: 126330.

[46] 陈占飞. 铜阳极泥卡尔多炉处理工艺浅析[J]. 中国金属通报, 2020, 9: 7-8.

[47] 张乐如. 现代铅冶金[M]. 长沙: 中南大学出版社, 2013.

[48] 张腾, 张善辉, 王瑞强, 等. 高效银电解技术在恒邦股份的实践应用[J]. 世界有色金属, 2021, 22: 7-9.

[49] 王天丰. 铜阳极泥中贵金属的分离提纯[D]. 沈阳:东北大学, 2013.

[50] KUNTYI O I, BILAN O I, OKHREMCHUK E V. Morphology of silver electrolytically precipitated from acetonitrile solutions of silver nitrate[J]. Russian Journal of Applied Chemistry, 2011, 84(2):199-203.

[51] 衷水平, 王俊娥, 张焕然, 等. 银电解液水解净化新工艺及应用实践[J]. 中国有色冶金, 2018, 47(5):45-48.

[52] 杨德香, 周先辉, 杨继生, 等. 银电解液中有价金属综合回收生产实践[J]. 黄金, 2016, 37(11):62-65.

[53] 陈杭, 张永锋, 吴健辉, 等. 用椰壳炭从开路银电解液中富集铂、钯试验研究[J]. 湿法冶金, 2017, 36(3):222-226.

[54] 任同兴, 曹华珍, 王志伟, 等. 电解银粉的制备及其电结晶[J]. 粉末冶金材料科学与工程, 2010, 15(3):206-211.

[55] AJI A T, AROMAA J, WILSON B P, et al. Kinetic study and modelling of silver dissolution in synthetic industrial silver electrolyte as a function of electrolyte composition and temperature[J]. Corrosion Science, 2018, 138:163-169.

[56] 谢太李, 夏兴旺, 黄绍勇. 大阳极银电解工艺工业试验[J]. 有色金属(冶炼部分), 2017, 8:50-53.

[57] 刘超, 王光忠, 赵红浩. 银电解车间酸雾治理研究与槽体设计[J]. 中国有色冶金, 2015, 44(3):57-59.
LIU Chao, WANG Guang-zhong, ZHAO Hong-hao. Acid mist Treatment and cell body design of silver electrolysis workshop[J]. China Nonferrous Metallurgy, 2015, 44(03):57-59.

[58] JHA M K, LEE J C, KIM M S, et al. Hydrometallurgical recovery / recycling of platinum by the leaching of spent catalysts: A review[J]. Hydrometallurgy, 2013, 133:23-32.

[59] XIAO Li, WANG Yong-liang, YU Yang, et al. Enhanced selective recovery of selenium from anode slime using MnO_2 in dilute H_2SO_4 solution as oxidant[J]. Journal of Cleaner Production, 2018, 209:494-504.

[60] 林家永, 郑伟忠, 张永锋, 等. 硫酸钡含量对卡尔多炉处理铜阳极泥的影响[J]. 有色金属(冶炼部分), 2017, 11:29-31.

[61] 张焕然, 王俊娥, 陈杭, 等. 铜阳极泥预处理工艺改进生产实践[J]. 中国有色冶金, 2018, 47(5):20-23.

[62] YANG Hong-ying, LI Xue-jiao, TONG Lin-lin, et al. Leaching kinetics of selenium from copper anode slimes by nitric acid-sulfuric acid mixture[J]. Transactions of Nonferrous Metals Society of China, 2018, 28(1):186-192.

[63] KURNIAWAN K, LEE J C, KIM J C, et al. Augmenting metal leaching from copper anode slime by sulfuric acid in the presence of manganese (IV) oxide and graphite [J]. Hydrometallurgy, 2021, 205:105745.

[64] RAO Shuai, LIE Yi, WANG Dong-xing, et al. Pressure leaching of selenium and tellurium from scrap copper anode slimes in sulfuric acid-oxygen media[J]. Journal of Cleaner Production, 2020, 278:123989.

[65] 马致远. 铜阳极泥微波/超声波辅助浸出新工艺及理论研究[D]. 沈阳:东北大学, 2016.

[66] YANG Hong-ying, MA Zhi-yuan, HUANG Song-tao, et al. Intensification of pretreatment and pressure leaching of copper anode slime by microwave radiation[J]. Journal of Central South University, 2015, 22(12):4536-4544.

[67] MA Zhi-yuan, YANG Hong-ying, HUANG Song-tao, et al. Ultra fast microwave-assisted leaching for the recovery of copper and tellurium from copper anode slime [J]. International Journal of Minerals, Metallurgy, and Materials, 2015, 22(6):582-588.

[68] LIU Wei-feng, YANG Tian-zu, ZHANG Du-chao, et al. Pretreatment of copper anode slime with alkaline pressure oxidative leaching[J]. International Journal of Mineral Processing, 2014, 128: 48-54.

[69] 马致远, 杨洪英, 陈国宝, 等. 基于田口法的铜阳极泥微波浸出工艺[J]. 中国有色金属学报, 2014, 24(8): 2152-2157.

[70] MEN SHCHIKOV V A, AGEEV N G, KOLMACHIKHIN B V, et al. Features of the thermal performance of the TROF converter[J]. Metallurgist, 2017, 61: 597-601.

[71] 李庆枞, 张伟杰. 提升卡尔多炉炉龄指标的生产实践[J]. 铜业工程, 2014, 3: 32-35.

[72] 王海荣. 卡尔多炉喷枪油改气的实践[J]. 中国有色冶金, 2014, 43(1): 22-25.

[73] 陈占飞. 卡尔多炉处理铜阳极泥生产多尔合金配料工艺优化实践[J]. 中国金属通报, 2020, 5: 85-86.

[74] 郑伟忠, 赖寿华, 张永锋, 等. 卡尔多炉处理铜阳极泥生产工艺优化实践[J]. 有色金属(冶炼部分), 2017, 11: 18-20.

[75] LI Wei-feng, ZHAN Jing, FAN Yan-qing, et al. Research and industrial application of a process for direct reduction of molten high-lead smelting slag[J]. JOM, 2017, 69(4): 784-789.

[76] LU Sun-jun, LI Juan, CHEN Da-lin, et al. A novel process for silver enrichment from Kaldo smelting slag of copper anode slime by reduction smelting and vacuum metallurgy[J]. Journal of Cleaner Production, 2020, 261: 121214.

[77] 吕建军. 卡尔多炉渣全湿法综合利用研究[D]. 长沙: 中南大学, 2013.

[78] 代献仁. 铜陵有色卡尔多炉渣中稀贵元素的预富集试验[J]. 金属矿山, 2016, 5: 200-203.

[79] 杨静, 田静, 麻瑞苡, 等. 卡尔多炉熔炼渣浮选试验研究[J]. 黄金, 2018, 39(9): 61-65.

[80] 汪水红. 卡尔多炉熔炼渣选矿贫化的研究与实践[J]. 有色冶金节能, 2016, 32(3): 10-12.

[81] 何云龙. 高铋铅阳极泥中有价金属分离与富集的应用基础研究[D]. 昆明: 昆明理工大学, 2018.

[82] 张永锋, 张焕然, 衷水平, 等. 卡尔多炉熔炼渣提取铅铋工艺研究[J]. 有色金属(冶炼部分), 2017, 11: 21-24.

[83] 徐磊, 阮胜寿. 铅阳极泥湿法预处理的工业化试生产实践[J]. 中国有色冶金, 2016, 45(3): 16-19+38.

[84] 许志鹏, 郭学益, 田庆华, 等. 含碲物料分离提取理论及工艺研究[M]. 北京: 冶金工业出版社, 2020.

[85] LIU Wei-feng, JIA Rui, SUN Bai-qi, et al. A novel process for extracting tellurium from the calcine of copper anode slime via continuous enrichment[J]. Journal of Cleaner Production, 2020, 264: 121637.

[86] LI Zhi-chao, DENG Ju-hai, LIU Da-chun, et al. Waste-free separation and recovery of copper telluride slag by directional sulfidation-vacuum distillation[J]. Journal of Cleaner Production, 2022, 335: 130356.

[87] 熊家春. 铜阳极泥卡尔多炉处理工艺中碲化铜渣及文丘里泥综合提取碲的研究[D]. 赣州: 江西理工大学, 2017.

[88] 郑雅杰, 乐红春, 孙召明. 铜阳极泥处理过程中中和渣中碲的提取与制备[J]. 中国有色金属学报, 2012, 22(8): 2360-2365.

[89] ROBLES-VEGA A, SANCHEZ-CORRALES V M, Castillon-BARRAZA F. An improved hydrometallurgical route for tellurium production[J]. Mining, Metallurgy & Exploration, 2009, 26(3): 169-173.

[90] GUO Xue-yi, XU Zhi-peng, LI Dong, et al. Recovery of tellurium from high tellurium-bearing materials by alkaline sulfide leaching followed by sodium sulfite precipitation[J]. Hydrometallurgy, 2017: 355-361.

[91] 宾万达, 卢宜源. 贵金属冶金学[M]. 长沙: 中南大学出版社, 2011.

[92] NARITA H, KASUYA R, SUZUKI T, et al. USA: Precious Metal Separations[M]. 2020.

[93] SADEGHI N, ALAMDARI E K. Selective extraction of gold (Ⅲ) from hydrochloric acid-chlorine gas leach solutions of copper anode slime by tri-butyl phosphate (TBP) [J]. Transactions of Nonferrous Metals Society of China, 2016, 26(12): 3258-3265.

[94] 赖建林, 周宇飞, 饶红, 等. 从铂钯精矿中回收铂、钯和金[J]. 贵金属, 2015, 36(3): 10-13.

[95] SHEN Y F, XUE W Y. Recovery palladium, gold and platinum from hydrochloric acid solution using 2-hydroxy-4-sec-octanoyl diphenyl-ketoxime[J]. Separation and Purification Technology, 2007, 56(3): 278-283.

[96] 王立, 季婷, 李睿, 等. 从复杂铂钯精矿中高效提取钯的新工艺研究[J]. 有色金属(冶炼部分), 2019, 5: 28-32.

[97] WANG Jun-lian, LIU Lu, XU Guo-dong, et al. New compound N-methyl-N-isopropyl octanthioamide for palladium selective extraction and separation from chloride media[J]. Transactions of Nonferrous Metals Society of China: 1-18. https://kns.cnki.net/kcms/detail/43.1239.TG.20220512.1636.006.html

[98] 李荣, 张金燕, 陈慕涵, 等. 采用溴代十六烷基吡啶从碱性氰化液中萃取和富集钯、铂[J]. 中国有色金属学报, 2016, 26(11): 2449-2460.

[99] 李耀威, 古国榜. 盐酸介质中异戊基苯并噻唑亚砜萃取Pd(Ⅱ)的机理[J]. 中国有色金属学报, 2007, 99(6): 1014-1018.

[100] 张钦发. 从铜阳极泥分金钯后的铂精矿中提取分离铂钯金新工艺及萃取机理研究[D]. 长沙: 中南大学, 2007.

[101] WANG Yang-yang, CHEN Shu-wen, LIU Rong-hao, et al. Toward green and efficient recycling of Au(Ⅲ), Pd(Ⅱ) and Pt(Ⅳ) from acidic medium using UCST-type ionic liquid[J]. Separation and Purification Technology, 2022, 298: 121620.

[102] 马肃, 谢笑天, 张刚, 等. 离子液体萃取铂, 钯和铑离子的研究进展[J]. 化学通报, 2021, 84(6): 530-534.

[103] 孙旭. 吡咯烷离子液体对盐酸介质中金和铂的萃取分离研究[D]. 济南: 山东大学, 2019.

[104] ZHOU Ao, JU Shao-hua, KOPPALA S, et al. Extraction of In^{3+} and Fe^{3+} from sulfate solutions by using a 3D-printed "Y"-shaped microreactor[J]. Green Processing and Synthesis, 2018, 8: 163-171.

[105] 刘晓玲, 李熙腾, 巨少华, 等. 3D打印大流量微反应器在萃取分离铂钯铑中的小型化应用[J]. 贵金属, 2020, 41(S1): 98-106.

[106] TAVLARIDES L L, BAE J H, LEE C K. Solvent extraction, membranes, and ion exchange in hydrometallurgical dilute metals separation[J]. Separation Science and Technology, 1987, 22(2-3): 581-617.

[107] LEE J C, Kurniawan, Hong H J, et al. Separation of platinum, palladium and rhodium from aqueous solutions using ion exchange resin: A review[J]. Separation and Purification Technology, 2020, 246: 116896.

[108] 王爱萍. PDTU系列聚酯基硫脲树脂的合成及其对Pt(Ⅳ)、Pd(Ⅱ)吸附性能与机理的研究[D]. 长沙: 中南大学, 2008.

[109] HUBICKI Z, WLOWICZ A. Adsorption of palladium(Ⅱ) from chloride solutions on Amberlyst A29 and Amberlyst A21 resins[J]. Hydrometallurgy, 2009, 96(1-2): 159-165.

[110] SUN P P, LEE J Y, LEE M S. Separation of platinum(Ⅳ) and rhodium(Ⅲ) from acidic chloride solution by ion exchange with anion resins[J]. Hydrometallurgy, 2012, 113-114: 200-204.

[111] SUN P P, KIM T Y, MIN B J, et al. Separation of platinum(Ⅳ) and rhodium(Ⅲ) from hydrochloric acid solutions using diaion resins[J]. Materials Transactions, 2015, 56(11): 1863-1867.

[112] WOOWICZ A, HUBICKI Z. Adsorption characteristics of noble metals on the strongly basic anion exchanger Purolite A-400TL[J]. Journal of Materials Science, 2014, 49(18): 6191-6202.

[113] 黄海兰, 曲荣君, 孙昌梅, 等. 新型螯合树脂聚苯乙烯负载葡糖胺对Pt(Ⅳ)吸附性能[J]. 离子交换与吸附, 2010, 26(5): 385-392.

[114] 熊延杭, 侯雪, 程衎锟, 等. S300树脂对汽车失效催化剂浸出液中Pd(Ⅱ)的吸附行为[J]. 中国有色金属学报, 2021, 31(1): 151-160.

[115] NIKOLOSKI A N, ANG K L, Li Dan. Recovery of platinum, palladium and rhodium from acidic chloride leach solution using ion exchange resins[J]. Hydrometallurgy, 2015, 152: 20-32.

[116] HAGA K, SATO S, GANDHI M R, et al. selective recovery of palladium from PGM containing hydrochloric acid solution using Thiocarbamoyl-substituted adsorbents[J]. International Journal of the Society of Material Engineering for Resources, 2018, 23(2): 173-177.

[117] SAYIN M, CAN M, IMAMOGLU M, et al. 1, 3, 5-Triazine-pentaethylenehexamine polymer for the adsorption of palladium(Ⅱ) from chloride-containing solutions[J]. Reactive & Functional Polymers, 2015, 88: 31-38.

[118] 李伟, 余珊, 房孟钊, 等. 树脂吸附法从沉金后液中回收金铂钯的实践[J]. 湖南有色金属, 2019, 35(3): 29-32.

镓的分离提取及高纯化制备方法

摘　要：镓是一种重要的稀散金属，在众多新兴高精尖领域有着重要的应用。本文介绍了镓分离提取及高纯化制备的常规工艺。镓在地壳中的丰度较低，通常和铝、锌等金属伴生，主要作为冶炼过程的副产品回收。常采用酸或碱浸出矿石或二次资源中的镓，后采用离子交换法、分级沉淀法、溶剂萃取法富集浸出液中的镓，并最终通过电积制备粗镓(4N)。镓的高纯化制备通常以工业生产的粗镓为原料，借助镓低熔点、高沸点的特点，以及主金属和杂质元素在不同相分配比不同的性质，通过电解精炼法、部分结晶法、单晶生长法、区域熔炼法、真空蒸馏法和真空热解法等技术进一步提纯制备高纯镓(5N~8N)。

关键词：镓；分离提取；高纯化

Extraction and purification process of gallium

Abstract: Gallium, as a rare metal, occupies an important position in many emerging high-precision fields. This paper introduced several conventional processes of gallium extraction and purification. Gallium has a low abundance in the earth's crust and is usually associated with aluminum and zinc, which is usually recovered as a byproduct. Acid or alkali leaching is first used to extract gallium from raw materials. Subsequently, ion exchange, fractional precipitation, and solvent extraction are generally applied to enrich gallium from lixivium, and then crude gallium (4N) can be obtained by electrowinning, which is the raw material for further purification. Gallium has a low melting point and a high boiling point, and the distribution ratios of gallium and impurity elements are different, which is the theoretical basis for the purification. Finally, the high-purity gallium (5N-8N) can be prepared by electrorefining, partial crystallization, single crystal growth, zone smelting, vacuum distillation and vacuum pyrolysis processes.

Keywords: gallium; extraction; purification

本文发表在《有色金属科学与工程》，网络首发。作者：刘左伟，许志鹏，郭学益，田庆华。

引言

镓是一种重要的稀散金属,镓及其化合物具有优良的光电和化学性能,被广泛应用于半导体材料、太阳能电池、合金、化工、医疗等领域,是现代高科技发展的关键原料[1-4]。镓在地壳中的丰度仅为 17 mg/L,没有独立存在且值得开采的矿床[5, 6],因此镓主要作为冶炼过程的副产品回收。自然界中的镓主要与铝土矿、锌矿和煤矿等矿物伴生,铝冶炼过程中的拜耳母液和锌冶炼过程的浸出渣是镓提取的主要来源。拜耳法产出的镓产量约占全球镓总产量 90%[7],其中,约 70% 的镓被浸出到拜耳法母液中,而剩余的 30% 则通过赤泥回收[8, 9]。在湿法炼锌过程中,多数的镓会残留在浸出渣中,这部分镓约占全球镓总产量的 10%[10-13]。此外,粉煤灰、磷厂电炉烟尘、半导体加工废料等也是镓的重要资源。随着全球对镓的需求量不断上升,从其他资源中回收镓备受关注。尽管目前从二次资源中回收镓的占比不高,但随着通信、半导体等领域的快速发展,对镓的需求量将不断上升,实现镓二次资源的高效清洁回收可用于缓解未来镓金属的供需矛盾[14, 15]。此外,随着高精尖材料行业的需求,对原料镓的纯度要求也在不断提高。高纯镓不仅要求镓的纯度达到一定程度,还对主要有害杂质含量有着严格的要求[16, 17],因此粗镓的高纯化制备工艺也是未来的重要研究方向。

1 镓的分离提取概述

镓的生产主要包括分离提取与高纯化制备两大步骤。首先通过酸或碱浸出镓,浸出液中的镓离子可通过离子交换、分级沉淀或溶剂萃取法与大部分杂质离子分离,纯化后的母液通过电积可制得粗镓,粗镓再通过电解精炼、结晶法等进一步提纯制得高纯镓。基于镓的性质,分离提取与高纯化的方法种类繁多且各有优劣,镓的分离提取与高纯化制备流程图如图 1 所示。

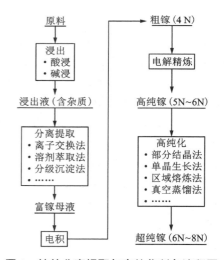

图 1 镓的分离提取与高纯化制备流程图

Fig. 1 Flow chart of separation, extraction and purification process of gallium

2 镓的分离提取

2.1 镓在水溶液中的行为

镓在水溶液中常以 Ga(Ⅲ) 的形式存在,如图 2 所示,在酸性条件下主要以 Ga^{3+} 的形式存在,而碱性条件下存在的主要形式为 $[Ga(OH)_4]^-$/GaO_2^-。基于这些镓的化学性质,可认为强酸或强碱条件和高温有利于镓的分离提取。在实际生产过程中,酸性和碱性两种体系都可以用于镓的浸出[9, 18, 19],反应方程式分别如式(1)

和式(2)所示。

$$Ga_2O_3 + 6H^+ \longrightarrow 2Ga^{3+} + 3H_2O \quad (1)$$

$$Ga_2O_3 + 2OH^- \longrightarrow 2GaO_2^- + H_2O \quad (2)$$

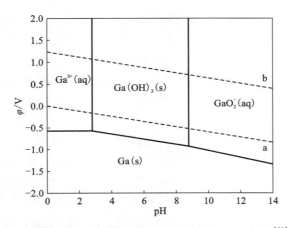

图 2 镓的 E_h–pH 图(25 ℃, 101325 Pa, a_{Ga} = 0.1)[20]

Fig. 2 Pourbaix diagram of gallium (at 25 ℃, 101325 Pa, a_{Ga} = 0.1)

由于原料中的镓含量较低,镓在浸出液中的浓度也较低,且浸出液中含有大量铝、锌、铁等杂质离子,难以进行电积回收,因此通常需要对浸出液中的镓进行进一步分离与富集。常用的分离富集方法包括离子交换法、分级沉淀法和溶剂萃取法。针对不同的主金属生产工艺,所采用的分离方法也往往不同。

2.2 离子交换法

离子交换法被认为是从拜耳液中回收 Ga 最有效的方法之一。1984 年,KATAONA 等发现某些螯合树脂含有活性基团,如=NOH,—NH₂,—OH,—SH 或=NH,对 Ga 具有优异的选择性萃取能力[21]。常见的树脂有 ES-346、DHG586、LSC600、Amberlite XAD-7 等。LSC600 吸附 Ga(Ⅲ)的机理如式(3)所示[22]。离子交换法由树脂吸附,酸洗解吸,中和沉淀除杂,电沉积四大步骤组成,其工艺流程图如图 3 所示。

$$RNH_2C=N[OH] + Ga(OH)_4^- \longrightarrow$$
$$RNH_2C=N[Ga(OH)_4]^- + H_2O \quad (3)$$

LU 等采用离子交换法处理赤泥酸性浸出液,LSD-396 树脂可完全去除浸出液中的铁,有效富集赤泥浸出液中少量的镓,镓的

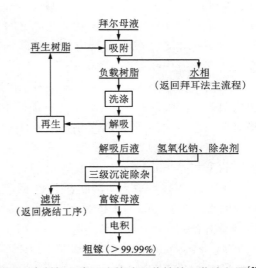

图 3 贵州铝业离子交换法回收镓的工艺流程图[23]

Fig. 3 Flow chart of the ion exchange method to recover gallium in Guizhou Aluminum Industry

吸附率和解吸率分别可达 59.84% 和 95.32%[24]。文朝璐等采用 LX-92 树脂吸附硫酸体系低浓度镓离子，在初始浓度为 260 mg/L，吸附温度为 55 ℃ 的条件下，树脂的最大动态平衡吸附容量为 56.65 mg/g；用硫酸进行洗脱，在最佳洗脱条件下，洗脱率达到 94.40%，经过吸附-脱附镓离子可富集 10 倍以上[25]。谢访友等采用离子交换法回收拜耳法种分母液中的镓，母液中含镓 190~240 mg/L，采用树脂吸附饱和后，使用碱性配合淋洗剂洗脱，后经蒸发浓缩、冷冻结晶、氧化等工序处理后再进行电解，可获得产品镓，最终镓的吸附率可达 65%，洗脱率大于 90%，电解回收率大于 90%[26]。路坊海等采用 LSC-700 树脂回收赤泥处理后液中的镓，在 LSC-700 树脂用量 0.6 g/L、温度 50 ℃、接触时间 24 h、振荡速率 120 r/min 的条件下，镓吸附效率可达 52.13%。对镓负载树脂采用酸法解吸，镓的平均解吸率为 92.29%，解吸液含镓浓度平均为 86.43 mg/L[27]。

离子交换法具有工艺简单，回收率高，操作简便，反应较快，对氧化铝加工影响不大等优点，广泛应用于拜尔母液提镓过程中，目前已被许多氧化铝厂应用。但目前工业生产常用的树脂存在易降解、易吸附钒等杂质等缺点，仍有待进一步开发新型树脂或优化反应体系来提升离子交换法的生产效率。此外，近年来研究者们也在探究和开发其他吸附剂，如生物吸附剂、高分子吸附剂等，这些吸附剂对镓具有良好的选择性，但距实现工业化应用尚有一段距离[28-30]。

2.3 分级沉淀法

铝镓分离是从拜耳液中回收镓的关键步骤，美国 Alcoa 铝厂于 1952 年首次提出了利用石灰沉淀分离的方法[31]。该方法首先使用 CO_2 气体对循环液进行碳酸化沉淀处理，以获得富镓沉淀。然后用石灰乳溶解沉淀，以分离镓和铝。随后，通过第二次碳酸化沉淀从 $NaGa(OH)_4$-$NaAl(OH)_4$ 溶液中回收镓，净化后用 NaOH 溶液重新溶解镓精矿，最终进行电积。石灰分级沉淀法的工艺流程图如图 4 所示。

WANG 等采用生物浸出-石灰沉淀分离的方法从铝渣中回收镓，在 2% 矿浆密度条件下镓的浸出率可达 100%，但树脂吸附法无法有效地从生物浸出液中回收镓，而采用石灰沉淀法回收镓回收率可达 60.6%，可达到镓铝的有效分离[32]。FONT 等采用碱浸-碳化沉淀法回收粉煤灰中的镓，在 1 mol/L NaOH，液固比 5:1 的条件下浸出 6 h，镓的浸出率可达 86%，后碳化沉淀回收镓，最终沉淀物的镓回收率可达 98.8%[33]。LI 等采用碱性浸出-分级沉淀法回收废弃太阳能薄膜电池中的铜铟镓硒（CIGS）材料，借助铟和镓的溶解度不同，将它们以氢氧化

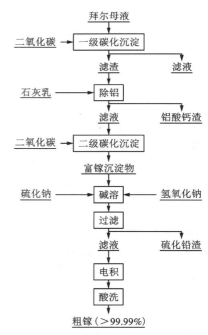

图 4 分级沉淀法从拜尔母液中回收镓的工艺流程图[7]

Fig. 4　Flow chart of recovery of gallium from Bayer pregnant by fractional precipitation

物的形态沉淀,经煅烧后分别可得到氧化铟和氧化镓,最终铟和镓的回收率分别可达 96.04%和 99.83%[34]。

分级沉淀法具有成本低、产品质量高等优势。传统的碳化沉淀法过程中需要大量的二氧化碳,需要配备专门的石灰窑来生产二氧化碳用于碳化沉淀过程,不符合"双碳"政策的趋势。在第一步碳化沉淀时会改变母液的组分,这会影响拜耳法主流程的正常运行。并且由于镓和铝溶解度较为相似,在生产镓的过程中会消耗部分铝,影响了拜耳法的产量。对于拜尔母液回收镓的过程,分级沉淀工艺已基本被离子交换法所替代。但在处理某些二次资源时,分级沉淀法仍是一种有效的从浸出液中分离镓的方法。

2.4 溶剂萃取法

溶剂萃取法最早应用于氯化物体系中,镓易从强酸性氯化物溶液中萃取[35]。溶剂萃取方法基于通过阳离子交换机制将镓萃取到有机相中,再利用镓和其他金属在有机相和水相中的分配系数不同进行分离。常见的萃取剂包括 Kelex 100、Cyanex 272 等,其工艺流程图如图 5 所示。

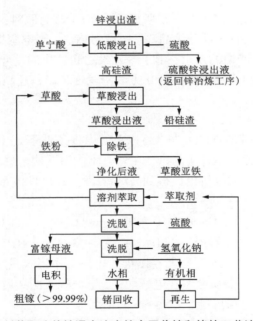

图 5 溶剂萃取法从锌浸出渣中综合回收镓和锗的工艺流程图[36]
Fig. 5 Flow chart for comprehensive recovery of gallium and germanium from zinc leach residue by solvent extraction

ZHANG 等采用溶剂萃取法提取锌浸出渣中的镓,先采用 N235 除铁,后采用 Cyanex 272 萃取镓,经过四级逆流萃取后,有机相负载镓可达 99.9%,酸洗后镓的总回收率可达 99.7%[37]。LIU 等采用 N235 和 TBP 萃取锌浸出渣中的镓和锗,并在不同 pH 值下进行反萃对镓和锗进行分离,最终镓和锗的回收率分别可达 99.0%和 99.8%[36]。HU 等采用氧化焙烧-浸出-萃取法处理回收废旧 CIGS,以 P204 为萃取剂选择性萃取镓和铟,并在不同酸度下洗涤，铜、铟、镓、硒四种元素均可得到有效地分离[38]。丘丽莉等采用 Cyanex 272 在 290 mg/L Ga^{3+}，pH=2.0 的硫酸镓溶液中萃取镓，在相比(有机相体积∶水相体积)1∶4，萃取温度25 ℃，时间 10 min，4 级逆流萃取后镓萃取率可达 99.50%，后用硫酸4级反萃后镓反萃率为 98.11%，镓的富集系数约 40 倍[39]。张魁芳等采用 P204 从硫酸体系萃取镓，在料液含 0.3 g/L Ga^{3+}，pH=1.2，相比 1∶3，温度 25 ℃下萃取 8 min，经过 3 级逆流萃取，镓萃取率可达 99.33%，负载有机相用硫酸反萃，经过 3 级逆流反萃后反萃率达 98.99%，镓浓度富集近 30 倍[40]。

溶剂萃取法是回收镓的有效方法，常用于提取锌浸出渣和拜尔母液中的镓[37]。该工艺流程简单，镓回收率高，且不会破坏循环溶液的成分。然而，溶剂萃取法尚未应用于拜尔母液回收镓的大规模生产中。主要因为其萃取动力学缓慢，且在高碱性介质中长期使用，萃取剂的降解和溶解导致其消耗量大、成本较高。

3 镓的高纯化制备

目前，高纯镓的制备往往以工业电积所得的粗镓(4N)为原料，通过各种精制方法精制得高纯金属镓，包括电解精炼法、部分结晶法、区域熔炼法、真空蒸馏法等[16]。高纯镓对于镓的纯度及杂质含量均有着严格的要求，由于各元素之间性质的差异，各工艺对杂质的选择性均有所不同，因此往往需要采用几种高纯化方法串联逐步制备超高纯镓。

3.1 电解精炼法

电解精炼法是在金属镓熔点(29.76 ℃)以上的温度条件下，以待提纯的粗镓为阳极，以高纯镓为阴极，用 NaOH 水溶液作电解液，在外电流作用下使金属粗镓在阳极溶解进入电解液后，通过迁移到达阴极并放电析出而得到高纯镓[41]。在碱性体系中，电解过程的主要反应如式(4)和式(5)所示。

$$阳极：Ga + 4OH^- - 3e^- \longrightarrow GaO_2^- + 2H_2O \tag{4}$$

$$阴极：GaO_2^- + 2H_2O + 3e^- \longrightarrow Ga + 4OH^- \tag{5}$$

在电解过程中 Ga 和 Al、Ca 等金属在阳极放电以离子形式进入电解液中；Ga 在阴极析出，而电位较负的 Al、Ca 等杂质会留在电解液中；电位较正的杂质如 Cu、Fe、Pb、Sn、Zn 不反应成为阳极泥。采用该方法可制备 5N~6N 的精制镓。

孙贤国等采用挥发熔炼-电解精炼法结合制备高纯镓，在真空度 1.3×10⁻³ Pa 以下，温度在 700 ℃左右，挥发熔炼时间 10~12 h 条件下真空蒸馏。随后 Ga 质量浓度 50~100 g/L、NaOH 质量浓度 100~180 g/L，电解液温度在 40~45 ℃，电流密度为 150~250 A/m²，槽电压 1~2 V 条件下电解精炼。最终粗镓一次电解完全能够达到 5N 高纯镓的要求，其中有 75% 以上的产品达到 6N 镓质量标准[42]。冯凤采用电解精炼法制备高纯镓，以 4N 粗镓为阳极，以 6N 高纯镓为阴极，在电流密度 0.02 A/m²，初始镓浓度 30 g/L 下电解，电流效率在 96% 以上，阴极镓纯度达 5N8[41]。

电解精炼法操作简便，对反应条件和设备的要求不高，但除杂效果较为一般，所制备镓的纯度最高仅可达到 6N，因此往往作为粗镓制备超纯镓的过渡工序。由于镓低熔点的特殊性质，电解过程中的镓往往以液态的形式存在于电解槽的底部，这会导致镓与电解液的接触面积较小，生产效率较低，有待对电解槽的结构进行进一步改进。此外，镓的标准还原电位较负，这使得电解过程中存在析氢反应严重、电流效率较低等问题。

3.2 结晶法

结晶法是通过使液态金属镓部分凝固，利用杂质元素在不同相态中的分布差异，使杂质在液态镓和固态镓中重新分布而得到较纯的金属镓的过程，常见的工艺包括部分结晶法、单晶生长法和直拉单晶法等。

HOU 等采用如图 6 所示部分结晶法的工艺制备高纯镓，其先将粗镓用盐酸和硝酸进行预处理，溶解其中的部分可溶性杂质，再将预处理的液态粗镓转移到洁净的结晶器中，当液态镓的温度降至结晶临界点时，加入 7N 镓作为晶种，并循环冷却水。当液态镓结晶到预定的结晶比例时，停止引入冷却水，将残留的液态镓排出结晶器外，再切换至加热，将结晶后的镓再次熔化。在冷却水温度 20 ℃，循环水流量 40 L/h，加入 6 个晶种的条件下重复结晶四次，最终制得 7N 的高纯镓[43]。DING 等采用单晶生长法制备高纯镓，首先将籽晶镓投入籽晶凹槽中，冷却至凝固。将粗镓液倒入所述结晶槽中，进行结晶，得到结晶镓。结晶率达到一定程度后，将剩余粗镓倒出，重新加热将镓熔化。重复上述步骤多次，在冷却液温度 15 ℃，重复 6 次的条件下制备出了 6N 级高纯镓[44]。YOON 等采用电解精炼-直拉单晶法制备超高纯镓，在 400 级洁净度的洁净室中于 -13 ℃下控制晶体生长，最终得到 8N 级高纯镓[45]。

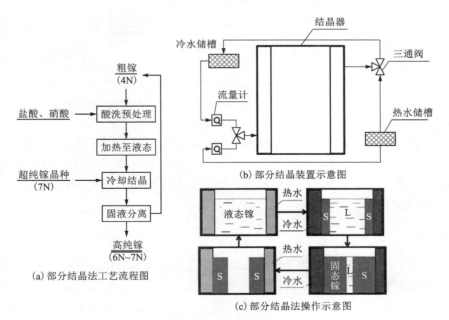

图 6 部分结晶法制备高纯镓的流程及设备示意图[43]

Fig. 6 Schematic diagram of the process and equipment for preparing high-purity gallium by partial crystallization

结晶法能够除去粗镓中的大多数杂质,可制备纯度较高的超纯镓。此外,由于镓的熔点较低,因此该过程无需消耗大量能量,在室温下即可进行操作。但存在对于部分杂质的脱除效果较差,过程繁琐,对设备和操作环境的要求较高,生产效率较低等问题。

3.3 区域熔炼法

区域熔炼法是通过采用互感加热线圈加热,使储于精炼设备内部的刚性管中的金属镓局部熔化,在精制过程中,加热线圈由刚性管的一端向另一端移动,熔区也随之移动,最后使金属镓中的杂质富集在刚性管一端的熔融镓中,分离富集了杂质的液态镓即可得到高纯金属镓。

RAMBABU 等采用如图 7 所示的区域熔炼设备制备高纯镓。将镓熔体置于聚四氟乙烯涂层的铜舟中,并向铜舟中通入 10 ℃的冷却液以凝固液态镓。全程通入高纯氢气防止镓被氧化。在熔区宽度为 44 mm,熔区移速 30 mm/h,加热温度为 40 ℃条件下,最终制得 6N 的高纯镓[46]。

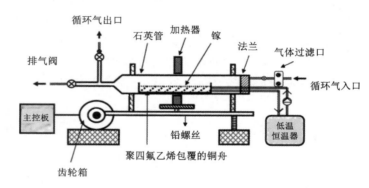

图 7 区域熔炼法制备高纯镓的设备示意图[46]

Fig. 7 Schematic diagram of zone smelting equipment for preparing high-purity gallium

区域熔炼法所制得的镓纯度较高,且由于镓的熔点较低,采用区域熔炼法制备高纯镓是在室温下进行的,从而避免了较高的能耗。但区域熔炼法存在处理量小,效率较低等问题。此外,为防止区域熔炼过程中液态镓的脱落,操作过程中镓应被置于容器内进行,这可能会导致镓的表面被容器所污染,从而降低产物的纯度。

3.4 真空法

真空法利用主金属和杂质间饱和蒸气压和挥发速度的差别,在挥发或冷凝的过程中将杂质去除,达到提纯的目的。金属镓具有低熔点和高沸点。因此,可以通过在真空下在较低温度下蒸馏除去具有高蒸气压的杂质元素。常见的工艺包括真空蒸馏法和真空热解法。

一种真空蒸馏法制备高纯镓的装置示意图如图 8 所示。该系统由一个三温区电子管式炉组成,采用扩散泵和旋转泵抽真空。冷阱装置于真空容器和泵之间,借由液态氮超低温使气体分子吸附于其壁上,以提高真空度。容器采用电子级高密度石墨舟以避免交叉污染。RAMBABU 等在 900~1050 ℃,4×10^{-3} Pa 下使用真空热处理,去除残留水分、溶解气体和高蒸气压金属杂质，最终制得 5N 的高纯镓[46]。

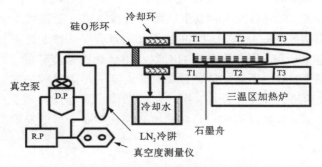

图 8 真空蒸馏法制备高纯镓的装置示意图[46]
Fig. 8　Schematic diagram of the vacuum distillation equipment for preparing high-purity gallium

真空法还可以直接以砷化镓等电子废弃物为原料，在高温真空下直接分离得到高纯镓。采用真空热分解法能有效回收砷化镓中的有价金属镓。胡亮等在系统压力为 3~8 Pa 下，GaAs 真空热分解最佳实验条件为蒸馏温度为 1000，恒温时间为 3 h，残留物 Ga 纯度在 4N 以上，挥发物中 As 质量分数为 87.97%，Ga 质量分数为 6.72%，可实现 Ga 和 As 的高效富集[47]。

真空法可以有效地除去饱和蒸气压较高的杂质元素，采用真空热解法可直接处理一些难处理物料。与传统的处理方法相比，真空热解法流程简单，且无废水、废气等污染物产生，过程中砷则以单质态回收，有效降低了污染治理的成本。但由于镓的沸点(2204 ℃)较高，因此相比于结晶法和区域熔炼法，真空法需要较高的能耗，且对于设备有很高的要求，对于沸点与镓相似的元素的脱除能力也比较有限，目前有关的研究还较少，因此仍有待进一步开发。

4　总结

(1)镓是一种重要的稀散金属，是现代高科技发展的关键原料。镓的主要产出仍来自于氧化铝或湿法炼锌过程，但随着对镓的需求量不断增加，从二次资源中回收镓正逐渐引起人们的关注。开发镓的高效分离提取和高纯化制备工艺是缓解镓资源匮乏，实现镓高值化利用的关键。

(2)镓的浸出可以在酸性和碱性体系下进行，浸出后的镓浓度较低，一般采用离子交换法、分级沉淀法、溶剂萃取法和电沉积法对浸出液中的镓进行富集，最终净化后的母液通过电沉积制备粗镓(4N)。开发高效的除杂工艺一直是镓分离提取领域的难题。分步沉淀法的成本较低，但流程较长，且会影响拜耳法的主流程；溶剂萃取法可有效地富集镓，但萃取动力学较缓慢；离子交换法可高效地回收镓，目前工业应用最为广泛，但存在树脂失活降解、成本较高等问题。研究者们也一直致力于开发和寻找选择性好、效率高、稳定性强的树脂或萃取剂。

（3）镓的高纯化制备通常以工业生产的粗镓（4N）为原料，借助镓低熔点、高沸点的特点，以及主金属和杂质元素在不同相分配比不同的性质，通过电解精炼法、部分结晶法、单晶生长法、区域熔炼法、真空蒸馏法和真空热解法等技术进一步提纯制备高纯镓（5N～8N）。不同提纯方法对各类杂质的脱除效果均不同，因此通常需要几种工艺进行串联。电解精炼法常作为镓高纯化的首个步骤，但其除杂能力有限。由于镓的熔点较低，因此结晶法和区域熔炼法提纯镓可在常温下操作，所需的能耗较低，而真空法则需要较高的温度。进一步探究杂质分布规律，开展理论研究和模拟计算，探寻缩短操作步骤和操作流程的方法，以及采用共沉积、共晶等方法一步制备高纯镓的化合物是镓高纯化制备今后的发展方向。

参考文献

[1] 李铁，刘人铭，曹家桢，等. 镓在医疗领域方面的应用现状及前景展望[J]. 西南医科大学学报，2022，45(1)：73-6+92.

[2] 刘延红，郭昭华，池君洲，等. 镓在新能源领域的应用[J]. 有色金属工程，2014，4(6)：78-80.

[3] 尹富强，赵玉辰，李赵春. 镓基液态金属应用的研究进展[J]. 现代化工，2022：1-6.

[4] 赵飞燕，张小东. 铝镓合金制氢技术研究进展[J]. 有色金属(冶炼部分)，2020(1)：60-5.

[5] REDLINGER M, EGGERT R, WOODHOUSE M. Evaluating the availability of gallium, indium, and tellurium from recycled photovoltaic modules[J]. Solar Energy Materials and Solar Cells, 2015, 138: 58-71.

[6] POŁEDNIOK J. Speciation of scandium and gallium in soil[J]. Chemosphere, 2008, 73(4): 572-9.

[7] ZHAO Z, YANG Y, XIAO Y, et al. Recovery of gallium from Bayer liquor: A review[J]. Hydrometallurgy, 2012, 125-126: 115-24.

[8] FU-JUN X, NUO-ZHEN X. Gallium production of three stage carbonization process[J]. Henan Chem Ind, 2002, 10(009).

[9] 张俊杰，李宏煦. 高铁铝土矿焙烧/高压碱浸提取镓的研究[J]. 有色金属(冶炼部分)，2017，(12)：36-9.

[10] LIU F, LIU Z, LI Y, et al. Extraction of gallium and germanium from zinc refinery residues by pressure acid leaching[J]. Hydrometallurgy, 2016, 164: 313-20.

[11] 邱伟明，奚长生，丘秀珍，等. 从冶锌工业废渣中提取镓和铟[J]. 有色金属(冶炼部分)，2017(5)：28-32.

[12] 马帅兵，刘付朋，陈飞雄. 改性花生壳吸附分离湿法炼锌溶液中的镓和锗[J]. 有色金属科学与工程，2022，13(1)：18-26.

[13] 张伟，宫晓丹，周科华，等. 锌粉置换镓锗渣加压氧化浸出的生产实践[J]. 有色金属科学与工程，2020，11(5)：142-7.

[14] 邹铭金，李栋，田庆华，等. 从二次资源中分离回收镓的研究进展[J]. 有色金属科学与工程，2020，11(5)：45-51.

[15] 黄蒙蒙，李宏煦，刘召波. 不同二次资源中镓提取方法的研究进展[J]. 有色金属科学与工程，2017，8(1)：21-8.

[16] 郭学益，田庆华. 高纯金属材料[M]. 北京：冶金工业出版社，2011：203-209.

[17] 中国国家标准化管理委员会. 高纯镓：GB/T 10118—2009[S]. 北京：中国标准出版社，2009.

[18] 蒋应平, 赵磊, 王海北, 等. 从浸锌渣中高压浸出镓锗的研究[J]. 有色金属(冶炼部分), 2012(8): 27-9.

[19] 王克勤, 李生虎, 朱国海, 等. 盐酸浸出氧化铝赤泥回收镓[J]. 有色金属(冶炼部分), 2012(8): 34-6.

[20] WOOD S A, SAMSON I M. The aqueous geochemistry of gallium, germanium, indium and scandium [J]. Ore Geology Reviews, 2006, 28(1): 57-102.

[21] YUSHIN KATAOKA M M, HIROSHI YOSHITAKE, YOSHIKAZU HIROSE. Method for recovery of gallium: US4468374[P]. 1984-08-28.

[22] ZHAO Z, LI X, CHAI Y, et al. Adsorption performances and mechanisms of amidoxime resin toward gallium (Ⅲ) and vanadium (Ⅴ) from Bayer liquor[J]. ACS Sustainable Chemistry & Engineering, 2016, 4(1): 53-9.

[23] LU F, XIAO T, LIN J, et al. Resources and extraction of gallium: A review[J]. Hydrometallurgy, 2017, 174: 105-115.

[24] LU F, XIAO T, LIN J, et al. Recovery of gallium from Bayer red mud through acidic-leaching-ion-exchange process under normal atmospheric pressure[J]. Hydrometallurgy, 2018, 175: 124-32.

[25] 文朝璐, 孙振华, 李少鹏, 等. LX-92树脂对硫酸体系低浓度镓离子的动态吸附[J]. 过程工程学报, 2021, 21(5): 567-78.

[26] 谢访友, 郭朋成, 王纪, 等. 用离子交换法从拜耳工艺溶液中提取镓的工业实践[J]. 湿法冶金, 2001(2): 66-71.

[27] 路坊海, 王芝成, 彭南丹, 等. 碱性溶液低浓度镓的回收[J]. 有色金属(冶炼部分), 2018, (11): 34-38.

[28] ZHAO Z, CUI L, GUO Y, et al. A stepwise separation process for selective recovery of gallium from hydrochloric acid leach liquor of coal fly ash [J]. Separation and Purification Technology, 2021, 265: 118455.

[29] SAIKIA S, COSTA R B, SINHAROY A, et al. Selective removal and recovery of gallium and germanium from synthetic zinc refinery residues using biosorption and bioprecipitation[J]. Journal of Environmental Management, 2022, 317: 115396.

[30] GUO W, ZHANG J, YANG F, et al. Highly efficient and selective recovery of gallium achieved on an amide-functionalized cellulose[J]. Separation and Purification Technology, 2020, 237: 116355.

[31] FRARY F C. Process of producing gallium: US2582376A[P]. 1952-01-15.

[32] WANG J, BAO Y, MA R, et al. Gallium recovery from aluminum smelting slag via a novel combined process of bioleaching and chemical methods[J]. Hydrometallurgy, 2018, 177: 140-5.

[33] FONT O, QUEROL X, JUAN R, et al. Recovery of gallium and vanadium from gasification fly ash[J]. Journal of Hazardous Materials, 2007, 139(3): 413-23.

[34] LI X, MA B, HU D, et al. Efficient separation and purification of indium and gallium in spent copper indium gallium diselenide (CIGS)[J]. Journal of Cleaner Production, 2022, 339: 130658.

[35] MIHAYLOV I, DISTIN P A. Gallium solvent extraction in hydrometallurgy: An overview [J]. Hydrometallurgy, 1992, 28(1): 13-27.

[36] LIU F, LIU Z, LI Y, et al. Recovery and separation of gallium(Ⅲ) and germanium(Ⅳ) from zinc refinery residues: Part Ⅱ: Solvent extraction[J]. Hydrometallurgy, 2017, 171: 149-56.

[37] ZHANG K, QIU L, TAO J, et al. Recovery of gallium from leach solutions of zinc refinery residues by stepwise solvent extraction with N235 and Cyanex 272[J]. Hydrometallurgy, 2021, 205: 105722.

[38] HU D, MA B, LI X, et al. Efficient separation and recovery of gallium and indium in spent CIGS materials[J]. Separation and Purification Technology, 2022, 282: 120087.

[39] 丘丽莉, 张魁芳, 刘志强, 等. 硫酸体系中 Cyanex272 的萃镓性能研究[J]. 有色金属(冶炼部分), 2021(5): 86-91.

[40] 张魁芳, 刘志强, 刘溢, 等. P204 从硫酸体系萃取镓性能研究[J]. 有色金属(冶炼部分), 2020(3): 50-54.

[41] 冯夙. 镓电解精炼工艺研究[D]. 沈阳: 东北大学, 2017.

[42] 孙贤国. 挥发熔炼-电解精炼联合法制备高纯镓的研究[J]. 湘南学院学报, 2016, 37(2): 22-5+39.

[43] HOU J, PAN K, TAN X. Preparation of 6N, 7N high-purity gallium by crystallization: Process optimization[J]. Materials, 2019, 12(16): 2549.

[44] DING Y-D, JIANG L, LI Z-S, et al. Numerical simulation and experimental verification of axial-directional crystallization purification process for high-purity gallium[J]. Transactions of Nonferrous Metals Society of China, 2020, 30(12): 3404-16.

[45] YOON J, YANG J. Study of refining and purification processing of gallium metal as a raw materials for compound semiconductor[J]. ECS Meeting Abstracts, 2020, MA2020-02(25): 1799.

[46] RAMBABU U, MUNIRATHNAM N R, PRAKASH T L. Purification of gallium from Indian raw material sources from 4N/5N to 6N5 purity[J]. Materials Chemistry and Physics, 2008, 112(2): 485-9.

[47] 胡亮, 刘大春, 陈秀敏, 等. 砷化镓真空热分解的理论计算与实验[J]. 中国有色金属学报, 2014, 24(9): 2410-2417.

从砷碱渣中回收锑、碱并固化砷的研究

摘　要：为实现砷碱渣中锑、碱的资源化和砷的固化，本文提出采用水浸分锑-水热石灰沉砷-浓缩结晶提碱工艺处理砷碱渣新思路，借助 ICP、XRD 等表征手段，考察了这一工艺的可行性。实验结果表明，在浸出温度为 90 ℃，浸出液固比为 3∶1，浸出时间为 5 h 的条件下，砷浸出率达到 97.96%，锑浸出率仅为 0.72%，所得浸出渣含砷量仅为 0.09%，由此实现了锑与砷、碱的有效分离；在钙砷物质的量比为 1.43，反应温度为 170 ℃ 的条件下，砷的沉淀率达到 99.68%，沉砷后液砷碱质量比仅为 1.65∶1000，砷碱分离效果优异；利用浓缩结晶法，可获得砷质量分数仅为 0.14‰ 的结晶碱，可满足锑精炼用碱的技术要求。

关键词：砷碱渣；水热；沉砷；回收；锑

Research on recovery of antimony and alkali from arsenic alkali slag and immobilization of arsenic

Abstract: In order to achieve the resource utilization of antimony and alkali in arsenic alkali slag, as well as the solidification of arsenic, a new process of the separation antimony by water leaching-hydrothermal precipitation arsenic by lime-extraction of alkali by evaporation was proposed to treat arsenic alkali slag. Using characterization methods such as ICP and XRD, the feasibility of this process was investigated. The experimental results showed that the leaching rate of arsenic reached 97.96%, while the leaching rate of antimony was only 0.72% under the conditions of leaching temperature of 90 ℃, liquid-solid ratio of 3∶1, and leaching time of 5 h, and the arsenic content in the obtained leaching residue was only 0.09%, which means that effective separation of antimony from arsenic and alkali was achieved; the precipitation rate of arsenic reached 99.68% under the conditions of a calcium arsenic molar ratio of 2.67 and a reaction temperature of 170 ℃, and the arsenic alkali mass ratio in final liquor was only 1.65‰, indicating excellent arsenic alkali separation effect achieve; A crystalline alkali with an arsenic content of only 0.14‰ can be obtained by using evaporation crystallization method, which can meet the technical requirements of alkali for antimony refining.

Keywords: arsenic alkali slag; hydrothermal; arsenic precipitation; recovery; antimony

---

本文发表在《矿冶工程》，网络首发。作者：康东升，金贵忠，徐志峰，杨裕东，李玉虎，金承永，罗燊。

引言

砷碱渣是锑精炼过程中产出的一种典型的副产品，其成分相对简单，含有较高含量的锑和碱，具有较高的经济价值，但砷含量较高，加之砷、锑性质相似，因而处理十分棘手[1]。目前，将砷稳定化-填埋处理是含砷固体废物最常用的方法，然而由于砷碱渣砷、碱含量较高，若直接稳定化-填埋处理，不仅砷固化效果不佳，易产生二次污染[2,3]，而且也会造成锑资源的浪费[4]。由于缺乏技术经济可行的处置技术，导致砷碱渣的资源化和无害化仍有待突破[5]。

目前，砷碱渣处置技术大致可分为火法和湿法两类。典型的火法处置技术是将砷碱渣作为配料直接返回锑冶炼工序，尽管这一处理可以回收一定量的锑，但易造成系统内砷的累积，导致锑产品品质恶化[6]。相较于火法工艺，湿法处置技术更具环境优势，因而受到广泛关注[7,8]。砷碱渣湿法处置技术通常是利用各类浸出剂选择性提取砷碱渣中的砷，然后再对浸出液进行处理，完成砷、碱分离[9,10]，典型湿法处置技术中包括砷酸钠混合盐工艺[11,12]、水浸-分步结晶工艺[13]、水浸-CO_2转化结晶工艺[14]等。利用这些工艺可将锑、砷分别以难溶性的锑盐、砷酸钠形式产出，而碱则以碳酸钠、碳酸氢钠或混合碱等结晶碱的形式回收。尽管这些工艺实现了锑、砷、碱的分离，但由于分离不彻底，使得锑渣、结晶碱砷含量较高，无法满足生产要求，导致锑、碱资源未能被有效利用。此外，由于砷酸钠产品市场需求量小，价值低且环境风险高，因而砷以砷酸钠形式产出已不是最佳方案[15]。因此，如何实现砷、锑、碱的深度分离，获得低砷的锑、碱产品，并实现砷的固化无疑是砷碱渣的资源化、无害化处置的关键。基于此，本文提出"水浸分锑-水热石灰沉砷-浓缩结晶提碱"处理砷碱渣的新工艺，利用水浸实现锑与砷、碱的分离，在此基础上，采用水热强化石灰对砷进行脱除，从而完成砷碱渣中锑、砷、碱的有效分离，实现锑、碱的资源化和砷的无害化。

1 实验部分

1.1 实验原料

实验所用砷碱渣取自湖南某锑冶炼厂。样品经球磨破碎、分级处理后，取粒度小于74 μm的筛下粉作为实验原料，其成分、物相和微观形貌如表1、图1所示。实验所用砷碱渣砷、锑质量分数分别为9.98%、16.58%，碳酸钠质量分数为21.63%，主要结晶物相为$NaSbO_2$、$Na_2(SbO_3)_2$、$Na_3AsO_4 \cdot 12H_2O$和Na_2CO_3，砷碱渣为粒径分布较宽的不规则颗粒，表面十分粗糙，活性较高，这对后续的浸出较为有利。

表 1 砷碱渣成分分析结果

元素	As	Sb	Se	Si	S	Mo	Na_2CO_3
质量分数/%	9.98	16.58	0.15	0.04	0.04	0.05	21.63

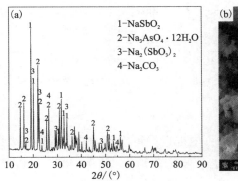

图 1 实验所用砷碱渣的 XRD 图谱(a)及 SEM 照片(b)

1.2 实验方法

1.2.1 砷碱渣浸出实验

砷碱渣资源化无害化工艺实验流程如图 2 所示,其反应原理如式(1)~式(4)所示。

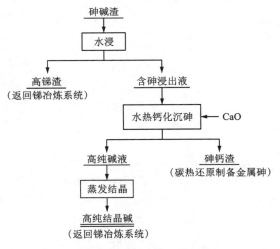

图 2 砷碱渣综合利用工艺流程图

$$Na_3AsO_4 = 3Na^+ + AsO_4^{3-} \tag{1}$$

$$2NaSbO_2 + O_2 + 6H_2O = 2NaSb(OH)_6 \tag{2}$$

$$3Na_3AsO_4 + 5CaO + 5H_2O = Ca_5OH(AsO_4)_3 + 9NaOH \tag{3}$$

$$Na_2CO_3 + CaO + H_2O = CaCO_3 + 2NaOH \tag{4}$$

浸出实验在 2 L 玻璃反应釜(DF-2L)中进行。每次称取 500 g 砷碱渣,在液固质量比为 3∶1 条件下进行浸出。浸出结束后,过滤、洗涤分别收集滤液、滤渣。滤液取样后测 As、Sb 浓度,计算砷、锑浸出率;滤渣烘干后,取样进行 X 射线衍射(XRD)、扫描电镜(SEM)分析。实验过程中砷、锑浸出率按式(5)、式(6)计算。

$$\alpha_{As} = \frac{V_1 \times C_{As}}{500 \times \eta_{As} \times 1000} \times 100\% \tag{5}$$

$$\alpha_{Sb} = \frac{V_1 \times C_{Sb}}{500 \times \eta_{Sb} \times 1000} \times 100\% \tag{6}$$

其中:α_{As} 为砷的浸出率,%;V_1 为滤液体积,L;C_{As} 和 C_{Sb} 分别为滤液中砷和锑的浓度,mg/L;η_{As} 和 η_{Sb} 分别为砷碱渣中砷和锑的质量分数,%。

1.2.2 水热钙化沉砷实验

水热钙化沉砷实验在不锈钢高压釜(XSF-2L)内进行。每次用 800 mL 浸出液,按照实验设计加入石灰,然后升温至目标温度并保温 2 h。实验结束后,过滤、洗涤分别收集滤液、滤渣。滤液取样测定 As 浓度,计算砷沉淀率;滤渣烘干后,取样进行 X 射线衍射(XRD)、扫描电镜(SEM)分析。实验过程中沉砷率按照式(7)进行计算。

$$\alpha = \left(1 - \frac{C_3 \times V_3}{C_2 \times V_2}\right) \times 100\% \tag{7}$$

其中,α 为沉砷率,%;C_2 和 C_3 分别为沉淀前后溶液的砷浓度,mg/L;V_2 和 V_3 分别为沉淀前后的溶液体积,L。

1.2.3 沉砷后液碱回收实验

每次量取 800 mL 沉砷后液于四氟烧杯中,置于 130 ℃ 油浴锅中,搅拌蒸发浓缩。待溶液体积达到所需浓缩比时,离心分离结晶碱和母液。取样测定结晶碱成分,计算砷含量。实验过程中蒸发浓度比按照式(8)进行计算。

$$\beta = \frac{V_5}{V_4} \times 100\% \tag{8}$$

其中,β 为蒸发浓度比,%;V_4 为沉淀后液的体积,L;V_5 为蒸发浓缩后溶液体积,L。

1.3 表征方法

采用电感耦合等离子体发射光谱仪(ICP, ICP-6300)分析溶液中砷、锑浓度;采用 X 射线衍射仪(XRD, Rigaku-TTRⅢ, Cu/K$_\alpha$, 波长 λ = 0.15406 nm)表征样品的结晶物相;采用扫描电子显微镜(SEM, Zeiss sigma 300)观测样品的微观形貌和粒度特征。

2 结果与讨论

2.1 水浸分锑

2.1.1 浸出温度的影响

在液固比为 3∶1,时间为 2 h 条件下,考察了浸出温度对砷、锑浸出的影响,结果如图 3

所示。由图3(a)可知,随着浸出温度的增加,砷的浸出率逐渐上升,而锑的浸出率则呈下降趋势;但反应温度由30 ℃上升至90 ℃时,砷的浸出率可由95.40%升至97.36%,而锑的浸出率则是由7.61%降低至2.71%。随着浸出温度增加,砷盐溶解度提高,这促进了砷的浸出,同时温度增加也有利于砷浸出速率的提高,此外高温促进了锑的水解析出,从而减少了锑的浸出。这点也可从浸出渣的XRD图谱[图3(b)]得到证实,随着浸出温度的提高,焦锑酸钠衍射峰逐渐增强,而砷盐的衍射峰逐渐消失。因此,为实现砷、锑的有效分离,浸出温度应控制在90 ℃为宜。

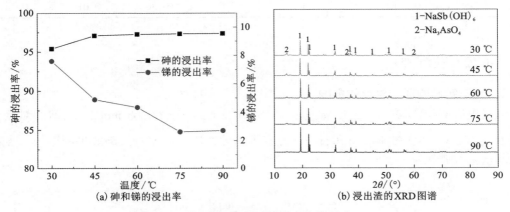

图3 浸出温度对砷锑浸出率的影响

2.1.2 浸出时间的影响

在液固比为3∶1,温度为90 ℃条件下,考察了浸出时间对砷、锑浸出的影响,结果如图3所示。由图4(a)可以看出,在所考察的时间范围内,浸出时间对砷的浸出影响不大,但对锑的影响较大。当反应时间由1 h增加至5 h时,砷的浸出率仅从94.85%增加至97.96%,而锑的浸出率则从10.19%降至0.71%;由于砷碱渣中砷主要以易溶性化合物形式存在,因而在较短的时间内即可完成浸出,而锑在起始阶段以Sb(Ⅲ)形式进入浸出液,随着反应时间

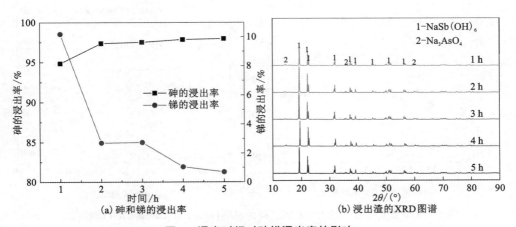

图4 浸出时间对砷锑浸出率的影响

的延长,Sb(Ⅲ)发生氧化水解,形成$NaSb(OH)_6$沉淀[图4(b)],从而可实现砷锑的良好分离。这点可从表2的浸出渣成分分析结果得到证实,经过水浸后浸出渣中砷含量大幅度降低,砷质量分数仅为0.09%,锑质量分数高达59.76%。因而为确保锑的回收率,浸出时间宜控制在5 h为宜。

表2 浸出渣成分分析结果

元素	Na	As	Sb	Se	Si	S	Mo
质量分数/%	10.40	0.09	59.70	0.02	0.36	0.01	0.14

2.2 水热石灰沉砷

2.2.1 钙砷物质的量比的影响

以浸出后液(c_{As} = 36.42 g/L,$c_{Na_2CO_3}$ = 0.68 mol/L,c_{NaOH} = 0.08 mol/L)为研究对象,在水热温度150 ℃,反应时间2 h条件下,考察了钙砷物质的量比对沉砷效果的影响,结果如图5所示。

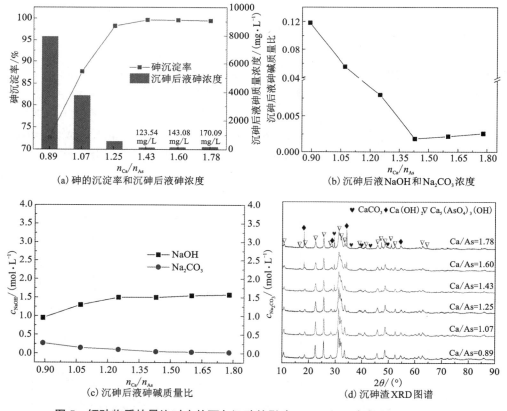

图5 钙砷物质的量比对水热石灰沉砷的影响(a~c)和沉砷渣XRD图谱(d)

如图 5 所示，钙砷物质的量比对沉砷效果影响较为显著。当钙砷物质的量比由 0.89 提高到 1.43 时，砷的沉淀率由 72.73% 上升至 99.63%，反应后液中砷浓度由 7492.13 mg/L 降低至 135.44 mg/L，沉砷后液砷碱质量比由 12∶1000 降低至 1.9∶1000。然而，当钙砷物质的量比超过 1.43 后，继续增加钙砷物质的量比，沉砷效果变化不大。可以发现，钙砷物质的量比明显高于理论用量（$n_{Ca}/n_{As}=0.89$），这是因为浸出液中的碱主要为碳酸钠，这使得除了砷的沉淀以外，碳酸钠的苛化过程也会消耗大量的 CaO，导致所需钙砷物质的量比较高；随着钙砷物质的量比的提高，反应后液中氢氧化钠浓度逐渐上升，同时碳酸钠浓度则逐渐下降，当钙砷物质的量比为 1.43 时，体系内 Na_2CO_3 物质的量浓度仅为 0.05 mol/L，而 NaOH 物质的量浓度上升至 1.55 mol/L。因此利用水热苛化可将碳酸钠转化为氢氧化钠，从而实现碱的纯化。当钙砷物质的量比低于 1.25 时，沉砷渣的主要结晶物相是 $Ca_5(AsO_4)_3(OH)$；当钙砷物质的量比超过 1.25 后，沉砷渣中 $CaCO_3$ 衍射峰逐渐增强；当钙砷物质的量比超过 1.60 后，沉砷渣中出现 $Ca(OH)_2$ 衍射峰，表明此时 CaO 已过量。沉砷渣的物相变化证实了砷沉淀反应优先进行。尽管不同钙砷物质的量比条件下得到的含砷物相均为 $Ca_5(AsO_4)_3(OH)$，但高钙砷物质的量比条件下，$Ca_5(AsO_4)_3(OH)$ 衍射峰强度降低，说明高钙砷物质的量比条件不利于得到高结晶度的 $Ca_5(AsO_4)_3(OH)$。因此，建议钙砷物质的量比以 1.43 为宜。

2.2.2 水热温度的影响

以浸出后液（$C_{As}=36.42$ g/L，$C_{Na_2CO_3}=0.68$ mol/L，$C_{NaOH}=0.08$ mol/L）为研究对象，在钙砷物质的量比为 1.43，反应时间 2 h 条件下，考察了水热温度对沉砷效果的影响，结果如图 6 所示。

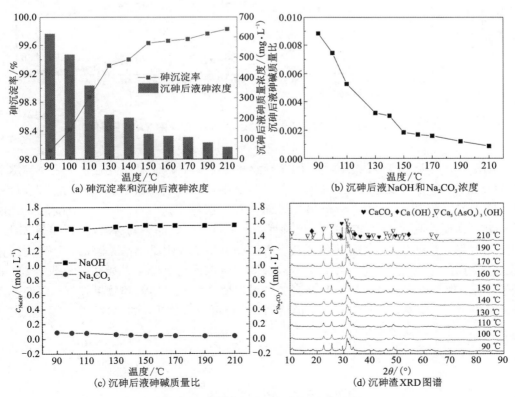

图 6 反应温度对水热石灰沉砷的影响

由图 6 可知,随着反应温度的上升,沉砷效果逐渐改善。当反应温度为 90 ℃ 时,沉砷率仅为 98.13%,沉砷后液砷、Na_2CO_3 分别为 616.57 mg/L、0.09 mol/L,砷碱质量比达到 8.9∶1000;当反应温度提高至 170 ℃ 时,沉砷率提高至 99.68%,沉砷后液砷浓度、Na_2CO_3 质量浓度分别降至 106.85 mg/L、0.05 mol/L,砷碱质量比为 1.65∶1000,砷碱质量比仅为 90 ℃ 体系下的 18%,可以达到较好的砷碱分离效果。值得注意的是,继续增加反应温度将进一步改善沉砷效果,并降低砷碱质量比。当反应温度较低时,沉砷渣中主要物相为 $Ca_5(AsO_4)_3(OH)$,而 $CaCO_3$ 的衍射峰并不明显;随着反应温度的提高,$Ca_5(AsO_4)_3(OH)$ 及 $CaCO_3$ 的衍射峰强度逐渐增强,表明水热条件下可以获得高结晶度的砷酸钙沉淀,同时有利于去除溶液中的碳酸盐将其转化为碳酸钙沉淀,从而实现碱液的纯化。为达到较好的沉砷效果,建议反应温度选择 170 ℃ 为宜。

2.3 浓缩结晶提碱

以水热沉砷后液(C_{NaOH} = 2.16 mol/L,$C_{Na_2CO_3}$ = 0.05 mol/L,C_{As} = 124 mg/L)为研究对象,考察了浓缩结晶过程中砷、锑的行为,实验结果如图 7 所示。

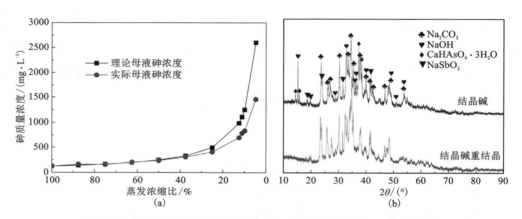

图 7 沉砷后液浓缩结晶过程中砷的走向(a)及结晶碱 XRD 图谱(b)

由图 7 可知,随着浓缩比的降低,砷在母液和结晶碱中的分配发生显著变化。当浓缩比高于 37.5% 时,砷不会发生结晶,主要保留在母液中。当浓缩比低于 37.5% 时,砷在母液中的分配降低,表明其已开始结晶析出。然而,由于水热沉砷后液砷质量浓度较低,仅为 124 mg/L,即使控制浓缩为 5% 时,仍可确保碱优先析出。结晶碱的 XRD 分析结果表明,在浓缩结晶过程中,砷主要以砷酸钙的形式进入结晶碱中。采用重结晶处理后,砷酸钙的衍射峰消失,表明利用重结晶可进一步提高碱的纯度。所得重结晶碱中砷质量分数仅为 0.14‰,远低于 5‰ 的现有技术标准。进一步的实验结果表明,由于水热沉砷效果较好,即使不采用分步结晶的方法,将碱液完全结晶,所得结晶碱中砷质量分数也不超过 1.5‰,明显低于 5‰ 的现有技术标准,可完全满足锑精炼要求。

3 结论

（1）采用水浸工艺可实现砷碱渣锑的有效分离和富集。在浸出温度为 90 ℃，浸出液固比为 3∶1，浸出时间为 5 h 条件下，砷浸出率达到 97.96%，所得浸出渣锑质量分数高达 59.76%，而砷含量仅为 0.09%。

（2）采用水热石灰沉砷不仅可实现浸出液中砷、碱的深度分离，还可完成碱的纯化。在钙砷物质的量比为 1.43，反应温度为 170 ℃ 的条件下，砷的沉淀率可达 99.68%，沉砷后液砷碱质量比仅为 1.65∶1000，砷碱分离效果优异。

（3）采用分步浓缩结晶法可进一步提升结晶碱的纯度。当控制浓缩比为 5% 时，可确保碱优先析出，所得结晶碱中砷含量仅为 0.14‰，远低于 5‰ 的现有技术标准。

参考文献

[1] Yang K, Qin W, Liu W. Extraction of metal arsenic from waste sodium arsenate by roasting with charcoal powder[J]. Metals, 2018, 8(7)：542.
[2] 郭文景, 张志勇, 符志友, 等. 锑的淡水水质基准及其对我国水质标准的启示[J]. 中国环境科学, 2020, 40(4)：1628-1636.
[3] 周亚明, 刘智勇, 刘志宏. 高砷锑烟尘浸出毒性研究[J]. 矿冶工程, 2018, 38(5)：107-110.
[4] 徐亚飞, 冯攀, 俞小花, 等. 典型含锑物料的湿法分离锑研究[J]. 矿产综合利用, 2017(6)：31-35.
[5] Wang T, Lin G, Gu L, et al. Role of organics on the purification process of zinc sulfate solution and inhibition mechanism[J]. Materials Research Express, 2019, 6(10)：106588.
[6] 邓卫华, 戴永俊. 我国锑火法冶金技术现状及发展方向[J]. 湖南有色金属, 2017, 33(4)：20-23.
[7] 王文祥, 王晓阳, 方红生, 等. 砷碱渣/高砷锑烟尘协同脱砷及有价金属回收[J]. 矿冶工程, 2022, 42(2)：102-105+109.
[8] 易宇, 叶逢春, 王红军. 含砷烟尘选择性浸出砷及其动力学[J]. 矿冶工程, 2020, 40(6)：99-102+107.
[9] 陈伟. 砷碱渣中砷锑分离并选择性回收锑的工艺研究[D]. 昆明：昆明理工大学, 2016.
[10] 邓卫华. 锑冶炼砷碱渣有价资源综合回收研究[D]. 长沙：中南大学, 2014.
[11] 单桃云, 金承永, 邓卫华, 等. 砷酸钠混合盐微波环保型干燥工艺探讨[J]. 湖南有色金属, 2012, 28(2)：35-37+57.
[12] 柯勇, 曹俊杰, 李童冰, 等. 砷碱渣中砷锑碱梯级分离及金属砷回收工艺[J]. 中南大学学报(自然科学版), 2023, 54(2)：495-505.
[13] 李志强, 陈文汨, 金承永. 分步结晶法分离砷碱的工艺研究[J]. 湖南有色金属, 2015, 31(1)：23-28.
[14] 张楠, 方紫薇, 龙华, 等. 砷碱渣稳定化处理合成臭葱石晶体固砷[J]. 中国有色金属学报, 2020, 30(1)：203-213.
[15] 王宝胜. 锑冶炼砷碱渣清洁化利用研究进展[J]. 中国资源综合利用, 2022, 40(8)：120-127+143.

碳热焙烧还原砷酸钙制备金属砷

摘　要：面对石灰沉淀法除砷工艺在含砷矿物冶炼中的广泛应用，砷酸钙的资源化被日益重视。通常砷酸盐的种类较多，高温特性不同，混合还原焙烧时渣型复杂，难以回收。而砷酸钙形态的砷酸盐是冶金工艺中较为常见且低廉的产物。因为不管是含砷废水、砷废渣、砷酸盐等，都可以通过低廉的石灰，在简单的冶金设备中沉淀或钙化转型生成砷酸钙而脱离体系。因此，本文致力于碳热焙烧还原砷酸钙制备具有商业价值的金属单质砷，为推进砷危废物无害化处理向砷资源化回收利用前进展开科学研究。其中热重分析表明，砷酸钙与碳粉混合热解的失重分为三个阶段，阶段一和阶段二为失水过程，阶段三为碳还原砷酸钙生成 CaO 和砷蒸气过程。且研究发现，可以利用相边界反应动力学模型解释阶段三反应机制。而单因素条件实验结果表明：在温度 1000 ℃、碳配入系数 1.4、恒温时长 60 min 条件下砷挥发率高达 99.94%。X 射线衍射仪（XRD）、扫描电镜能谱仪（SEM-EDS）对反应体系中有关产物表征表明，产品砷主要为片状金属砷和粉末不定型砷，焙烧残渣为 CaO。

关键词：砷酸钙；砷蒸气；金属砷；单质砷

Preparation of metallic arsenic from calcium arsenate by carbon thermal roasting reduction

Abstract: In face of the widespread application of lime precipitation process for arsenic removal in the smelting of arsenic-containing minerals, the resourcefulness of calcium arsenate has been paid increasing attention. In general, there are more types of arsenate with different high temperature characteristics, and the slag type is complicated when mixed reduction roasting, which is difficult to recover. And arsenate in the form of calcium arsenate is a more common and inexpensive product in the metallurgical process. Because whether it is arsenic-containing wastewater, arsenic slag, arsenate, and so on, all of them can be separated from the system by inexpensive lime, precipitation or calcification transformation in a simple metallurgical equipment to generate calcium arsenate. Therefore, this paper is devoted to the preparation of commercially valuable metallic monomers of arsenic by carbon thermal roasting reduction of calcium arsenate, and to start scientific research for advancing the harmless treatment of arsenic hazardous waste to

---

本文发表在《工程科学学报》，网络首发。作者：熊民，史冠勇，田磊，刘重伟，曹才放，张志辉，徐志峰。

arsenic resource recovery and utilization. Among them, thermogravimetric analysis shows that the weight loss of calcium arsenate mixed with carbon powder pyrolysis is divided into three stages, stage 1 and stage 2 are water loss process, stage 3 is carbon reduction of calcium arsenate to generate CaO and arsenic vapor process. And it was found that the phase Ⅲ reaction mechanism could be explained by using the phase boundary reaction kinetic model. The experimental results of single-factor conditions showed that the arsenic volatilization rate was as high as 99.94% at the temperature of 1000 ℃, carbon allotment factor of 1.4, and constant temperature of 60 min. The characterization of the relevant products in the reaction system by X-ray diffractometer (XRD) and scanning electron microscope energy spectrometer (SEM-EDS) showed that the product arsenic was mainly flake metallic arsenic and powder indefinite arsenic, and the roasted residue was CaO.

Keywords: calcium arsenate; arsenic vapor; metallic arsenic; elemental arsenic

砷对生物具有较强的毒害性，曾被广泛地用于生产农药、除草剂、杀虫剂以及木材防腐剂[1]。但基于砷对人体健康危害的长远考虑，这些使用被逐步淘汰。如今，砷常用于砷化镓(GaAs)半导体制备，汽车中铅电池的添加剂，轴承中的减摩剂，特殊玻璃的添加剂[2]。尽管砷的许多用途对社会发展的贡献不容小视，但砷的过度使用和不恰当处理会导致严重的环境问题[3]。为此，许多国家已开始对砷的使用加以限制。然而，砷废物的产生是不可避免的，因为它通常与自然界中其他所需的物质共存[4]。其中砷铜矿、砷铅矿、砷金矿等冶炼生产铜、铅、金等金属时，有关砷的处理是必不可少的。以砷铜矿为例，在火法炼铜过程中，铜精矿中53%~89%的砷挥发进入烟气中被电收尘收集进入烟灰中，还有7%~28%的砷进入熔炼渣，另一部分4%~19%的砷残留在铜锍，并由后续底吹、电解精炼进入阳极泥中[5]。因此，火法炼砷铜矿中的砷最终以高砷烟灰、熔炼渣和其他湿法含砷渣等砷物料形式离开系统。单铅铜冶炼而言，每年产生的高砷烟灰就大于150万吨[6]，导致大量含砷废渣堆积，不仅占用储存空间还会引起砷二次污染问题[7]。由于其日益严重，砷废渣进行无害化和资源化处置被广泛重视。

针对砷废渣的无害化相关研究人员提出了许多值得参考的治理方案。常见的有：稳定/固化含砷废渣方案。最早学者提出用低廉的水泥和石灰材料稳定/固化大量含砷的工业废渣[8-12]。后续随着各项研究的开展，各种不同黏合剂[13-15]的报道应运而生；砷玻璃化[16-17]，其特点在于用碎玻璃与砷酸钠混合加热到1200 ℃形成载砷玻璃，此砷玻璃稳定性高，对环境污染小。

不难看出含砷废渣处理方案较多[18-19]，但通过固定化砷来改善砷的稳定性似乎是解决相关环境问题的普遍方法。但基于白砷制备金属砷[20-22]的成功案例，直接从砷废物中提取金属砷的资源化处置，也是值得探索的方向。因为，一方面形成的金属砷具有一定的商业价值，可以作为产品出售。另一方面，可以大大减少要处置的砷废弃物总量和减轻砷二次污染的问题。

面对低廉的石灰沉淀法除砷[23-25]工艺在含砷矿物冶炼中的广泛应用，砷酸钙的资源化被日益重视。其中常见的操作是将砷废渣酸浸后再钙化沉淀处理，砷以砷酸钙的形式脱离体系，最后通过碳热还原制备稳定性高毒性小且具有商业价值的金属单质砷。因此，本文对碳热焙烧还原砷酸钙制备金属砷过程，展开热分析动力学和单因素条件实验研究。以期通过碳热还原砷酸钙制备金属砷来达到危废物资源化和高值化的目标。

1 实验原料及方法

1.1 原料的合成及表征

利用五氧化二砷溶液和氧化钙通过石灰沉淀法[26]合成砷酸钙，合成产物 X 射线荧光光谱（XRF）分析结果见表 1，X 射线衍射（XRD）分析如图 1 所示。结果表明合成产物主要物相为 $Ca_2As_2O_7$、$Ca_2As_2O_7 \cdot H_2O$、$Ca_5(AsO_4)_3OH$、$Ca(OH)_2$。

表 1 合成砷酸钙的主要成分

Table 1 Main components of synthetic calcium arsenate　　%

Ca	As	O	Other
19.81	31.43	38.41	10.35

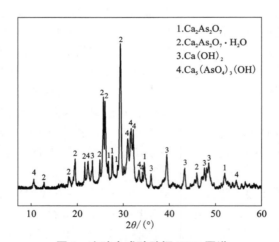

图 1 实验合成砷酸钙 XRD 图谱

Fig. 1 XRD pattern of experimentally synthesized calcium arsenate

1.2 实验流程

本文采用恒温管式炉进行碳热焙烧还原砷酸钙制备金属砷的实验。石英炉管二端用法兰密封连接，保护气氛采用高纯氩气，气体流速控制在 300 mL/min，仪器升温速率设置为 5 ℃/min。首先取砷酸钙物料 25 g 与活性炭粉（加入量与碳配入系数有关）混匀后加入刚玉方舟中，进行碳热焙烧热解实验。其中氩气将砷蒸气携带至低温区，冷凝至内管壁上，尾气则导入装有除砷剂的溶液中进行净化处理。实验装置如图 2 所示。实验完成后，收集还原渣和冷凝产物，并采用 X 射线衍射（XRD）和扫描电镜能谱仪（SEM-EDS）进行表征。

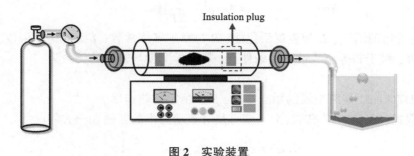

图 2　实验装置
Fig. 2　Experimental setup

1.3　相关计算公式及参数的检测方法

碳热焙烧还原砷酸钙中砷的挥发率计算公式如下：

$$\eta_{砷} = \frac{m_0 \times w_0 - m_1 \times w_1}{m_0 \times w} \times 100\% \tag{1}$$

式中：$\eta_{砷}$ 表示砷的挥发率；m_0 表示每次实验加入物料的质量；w_0 表示物料中砷的质量分数；m_1 表示每次焙烧后残渣的质量；w_1 表示每次焙烧后残渣中砷的质量分数。

公式（1）中物质质量 m_0、m_1 由百分位电子天平称量获取，而物相中砷的质量分数 w_0、w_1 则是采用电感耦合原子发射光谱法（ICP-AES）获得。其简要操作是用万分位天平称量少量物料或者焙烧残渣，将其加入一定酸度的盐酸中加热溶解后定容于 1000 mL 容量瓶中（砷浓度控制在 5 至 15 mg/L 之间）。溶液在检测前需优先获取砷标准曲线，曲线线性拟合程度必须大于 0.99 才可开始测量待测液。本文砷标准曲线是采用砷标液与蒸馏水配制 0 mg/L、5 mg/L、10 mg/L、15 mg/L、20 mg/L 获得。

2　热分析动力学

2.1　热分析动力学理论基础

在热解分析过程中，由于多相反应常存在多个失重阶段，把每个失重阶段类比是一个反应过程。因此，根据热重结果计算转化率 α 为：

$$\alpha = \frac{M_0 - M}{M_0 - M_f} \tag{2}$$

式中：M_0 为失重初始质量；M 为某时刻的质量；M_f 为失重阶段最终质量。

一般动力学方程为[27]20-21：

$$G(\alpha) = kt \tag{3}$$

式中：α 为转化率；t 为时间；$G(\alpha)$ 为积分形式反应机理函数；k 为动力学速率常数，且可用著名的 Arrhenius 方程[27]64-65 表示：

$$k = A\exp\left(-\frac{E}{RT}\right) \tag{4}$$

式中：A 为表观指前因子；E 为表观活化能；R 为摩尔气体常数；T 为热力学温度。

一般而言，对于非等温线性升温条件下，热力学温度与时间的关系为：

$$T = T_0 + \beta t \tag{5}$$

式中：T_0 为 DTG 曲线偏离基线的始点温度；β 为恒定加热速率。

联立方程式(3)、式(4)和式(5)，获得热分解动力学的普适积分方程：

$$G(\alpha) = \frac{A}{\beta}(T - T_0)\exp\left(-\frac{E}{RT}\right) \tag{6}$$

把方程式(6)改写成对数形式：

$$\ln(G(\alpha)) - \ln(T - T_0) = \ln\frac{A}{\beta} - \frac{E}{RT} \tag{7}$$

由一条 TG 曲线可以得到原始数据：T_i，$\alpha_i(i=1, 2, \cdots, n)$ 和 T_0，利用这些数据和线性最小二乘法处理方程式(7)，由斜率求 E，截距求 A。其中 $G(\alpha)$ 选取表 2 中所列机理函数[27]151-155。

表 2 动力学机理函数

Table 2 Kinetic mechanism function

Function number	Function name	Mechanism	Points form $G(\alpha)$
1	Ginstling-Brounshteine quation	Three-dimensional diffusion	$1 - \frac{2}{3}\alpha - (1-\alpha)^{\frac{2}{3}}$
2	Shrink globular	Phase boundary reaction	$1 - (1-\alpha)^{\frac{1}{3}}$

2.2 砷酸钙与碳粉混合热解特性

升温速率 2 ℃/min、氩气气氛下砷酸钙与碳粉混合热解的 TG-DTG 曲线如图 3(a)所示，恒温焙烧 30 min、氩气气氛下残渣 XRD 图如 3(b)所示。图 3(a)的 TG-DTG 曲线存在 3 个明显的波谷，说明热解过程存在三个阶段。而图 3(b)的残渣 XRD 图显示：物样（Ⅰ）到（Ⅱ）表观看出信号峰 2 减弱，信号峰 1 增强，第一个失重阶段为 $Ca_2As_2O_7 \cdot H_2O$ 脱水生成 $Ca_2As_2O_7$；物样（Ⅱ）到（Ⅲ）表观看出信号峰 1、3 减弱，峰 4 消失，而峰 5、6、7 出现，第二个失重阶段为 $Ca_2As_5O_7$ 到 $Ca_3(AsO_4)_2$ 的晶型转变，$Ca_5(AsO_4)_3OH$ 和 $Ca(OH)_2$ 到 $Ca_5(AsO_4)_3$、CaO 的失水过程；物样（Ⅲ）到（Ⅳ）表观看出信号峰 1、3、5、6 消失，只存在信号峰 7，第三个失重阶段为碳还原砷酸钙反应生成 CaO 和砷蒸气。

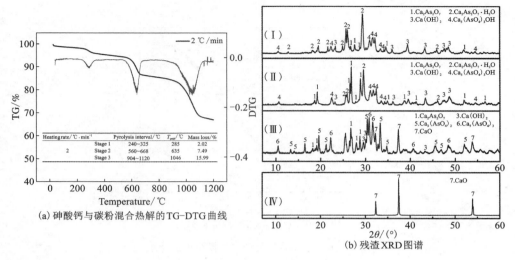

(a) 砷酸钙与碳粉混合热解的TG-DTG曲线

(b) 残渣XRD图谱

(b)：Ⅰ.原物料；Ⅱ.第一失重阶段，320 ℃；Ⅲ.第二失重阶段，700 ℃；Ⅳ.第三失重阶段，1000 ℃。

图3　砷酸钙与碳粉混合热解特性

[(a) TG-DTG curves of the pyrolysis of calcium arsenate mixed with carbon powder;
(b) XRD plots of the residue: Ⅰ. Raw material; Ⅱ. First weight loss stage, 320 ℃;
Ⅲ. Second weight loss stage, 700 ℃; Ⅳ. The third weight loss stage, 1000 ℃]

Fig. 3　Pyrolysis characteristics of calcium arsenate mixed with carbon powder

2.3　动力学模型拟合结果及分析

由上文分析发现前二个失重过程为脱 H 和 O 的失水过程，第三个失重阶段才是砷挥发热解过程。基于本文着力研究的是砷挥发热解过程，因此仅对热解失重第三阶段展开研究。筛选比较动力学模型后，文中展示两种动力学模型与图3(a)中第三个失重阶段数据拟合结果，结果如图4所示。可以看出：拟合结果符合相边界反应动力学模型，求得的表观活化能为 156.6 kJ/mol。且验算相边界反应模型计算的 $T-\alpha$ 曲线与实际测得 $T-\alpha$ 数据点之间吻合程度较高，说明可以利用相边界反应动力学模型解释砷酸钙与碳粉混合在第三个失重阶段温度范围的反应机制。

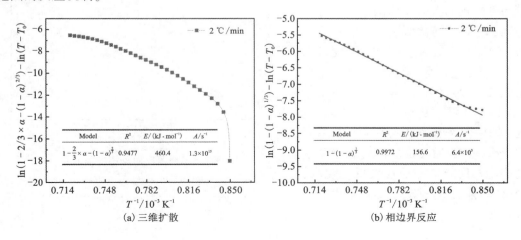

(a) 三维扩散

(b) 相边界反应

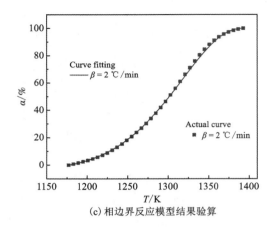

图 4 第三个失重阶段动力学模型拟合结果

[(a) 3D diffusion; (b) Phase boundary reaction; (c) Phase boundary reaction model result verification]

Fig. 4 Results of the third weightless phase kinetic model fitting

3 实验结果与讨论

3.1 焙烧温度对砷挥发率的影响

固定恒温时长 60 min，碳配入系数 1.8，氩气流速 300 mL/min，研究不同焙烧温度对砷酸钙中砷挥发率的影响，结果如图 5 所示。碳配入系数为：根据化学反应方程 $2Ca_3(AsO_4)_2 + 10C \Longrightarrow As_4(g) + 6CaO + 10CO(g)$，确定碳的加入量，并按理论计算用量的一定倍率配入。

由图 5 看出：焙烧温度对该反应体系砷挥发率有显著影响。随着焙烧温度的增加，砷挥发产率也随之上升。一般认为，温度影响分子运动速度，进而影响碳与砷酸钙的有效碰撞次数。焙烧温度越高，有效碰撞次数越多，体系反应越快，砷挥发率越高。因此，碳配入系数 1.8，恒定时长下，定量的砷酸钙，在特定温度 1000 ℃ 时，体系中砷挥发殆尽，砷挥发率可达 99%。继续增加反应温度，砷挥发率基本不再变化，其原因是反应率一定，增加反应温度影响分子运动速度，只能单一地加快反应速率，缩短反应时间，对反应率影响不大。因

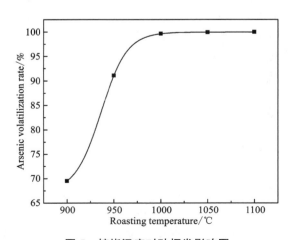

图 5 焙烧温度对砷挥发影响图

Fig. 5 Diagram showing the effect of roasting temperature on arsenic volatilisation

此，温度大于 1000 ℃ 后砷挥发率基本平稳于 99%。

3.2 碳配入系数对砷挥发的影响

固定焙烧温度 1000 ℃、恒温时长 60 min、氩气流速 300 mL/min 下，研究不同碳配入系数对砷酸钙中砷挥发率的影响，结果如图 6 所示。

由图 6 看出：碳配入系数从 1 增到 1.4 过程中，砷挥发率从 96.80% 增加到 99.5%，表明此阶段碳配入系数越大越有利于砷挥发。但当碳配入系数高于 1.4 时，砷挥发率变化幅度较小，稳定于 99.5%，后续再增加碳用量对体系影响甚微。可能的原因是定量的砷酸钙与碳满足实际反应方程计量比后，增加碳用量对反应影响不大。即砷酸钙中高价固态砷在被碳还原成低价气态砷过程，设定碳配入系数为 1.4 时，体系中高价砷刚好还原完全，砷挥发率不再随碳配入系数而增加，稳定在 99.5%。

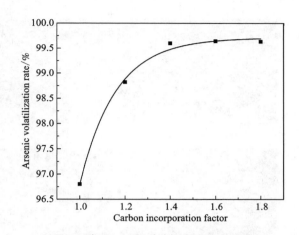

图 6 碳配入系数对砷挥发率的影响图

Fig. 6 Plot of the effect of carbon incorporation factor on the volatility of arsenic

3.3 恒温时长对砷挥发率的影响

固定焙烧温度 1000 ℃、碳配入系数 1.4、氩气流速 300 mL/min，研究不同保温时长对砷酸钙中砷挥发率的影响，结果如图 7 所示。

由图 7 看出：保温时间越长砷挥发率越高。时长从 30 min 增大到 60 min 时，砷挥发率从 99.7% 增加到 99.94%，后续再增加保温时长，砷挥发率稳定在 99.94%。这说明在 1000 ℃ 时碳热还原反应进行速度较快，在 30 min 时就已经达到很高的挥发率，60 min 后砷基本完全挥发，继续延长保温时间对还原渣中少量砷的挥发影响很小。

3.4 产物分析

由石英玻璃管后端冷凝区获得金属片状砷和粉末黑砷，金属片状砷在靠近高温热解区收集而得，粉末黑色砷在远离高温热解区收集而得。并采用 SEM、XRD 对冷凝产物砷进行分析，其中图 8(a)、图 8(b)、图 8(c)、图 8(d)、图 8(f) 为片状金属砷的电镜图，图 8(e) 为粉末不定型砷的电镜图。图 8

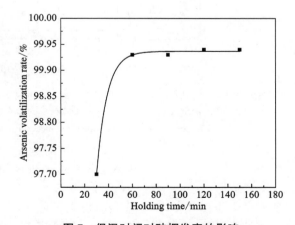

图 7 保温时间对砷挥发率的影响

Fig. 7 Effect of holding time on the volatility of arsenic

结果表明:SEM 图(c)、(f)发现金属片状砷存在明显的晶界,XRD 图显示尖锐的单质砷晶体特征峰;而粉末黑色砷 SEM 图(e)显示大小不一的颗粒球状砷,XRD 图显示的是馒头峰。因此,冷凝砷产物主要为片状金属砷和非晶体单质粉末球状砷。

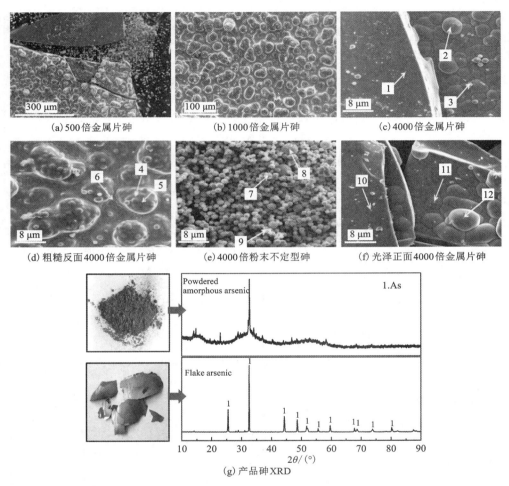

(a) 500 times the metal flakearsenic; (b) 1000 times the metal flake arsenic; (c) 4000 times the metal flake arsenic; (d) rough back side 4000 times the metal flake arsenic; (e) 4000 times powdered unshaped arsenic; (f) glossy front side 4000 times the metal flake arsenic; (g) Product arsenic XRD.

图 8 产品砷 SEM、XRD 图谱

Fig. 8 Product arsenic SEM, XRD patterns

实验选取了图 8 中 12 个代表的位点,通过 EDS 半定量分析砷和氧的含量,结果如表 3 所示。表明颗粒状砷冷凝物和片状砷冷凝物中砷含量都维持较高值,但颗粒状砷冷凝物中会出现氧含量较高的特例。如 4 号点位,砷质量分数 76.65%,氧质量分数 23.35%。分析颗粒状砷冷凝物中可能包裹或者夹杂着 As_2O_3,这可能是砷酸钙与碳升温到 1000 ℃之前,挥发出来的少量 As_2O_3(g)被氩气携带至低温区所致。但选取的 12 个位点中只出现一例氧含量高的情况,表明冷凝产物中 As_2O_3 的质量相对单质砷的质量较低。因此,说明冷凝物中主要产物是单质砷,并可能伴随部分氧化砷。

表3 金属砷的 EDS 分析结果
Table 3 Results of EDS analysis of arsenic metal   %

number	O	As
1	1.16	98.84
2	0.85	99.15
3	2.44	97.55
4	23.35	76.65
5	0.72	99.28
6	0.92	99.08
7	1.63	98.37
8	1.14	98.86
10	0.48	99.52
11	1.32	98.68
12	0.38	99.62

4 结论

本文针对碳热焙烧还原砷酸钙制备金属砷过程,展开热分析动力学和单因素条件实验研究,总结相关研究结论如下:

(1)热分析动力学表明:砷酸钙与碳粉混合热解的失重分为三个阶段,阶段一为 $Ca_2As_2O_7 \cdot H_2O$ 脱水生成 $Ca_2As_2O_7$;阶段二为 $Ca_2As_5O_7$ 到 $Ca_3(AsO_4)_2$ 的晶型转变, $Ca_5(AsO_4)_3OH$ 和 $Ca(OH)_2$ 到 $Ca_5(AsO_4)_3$、CaO 的失水过程;阶段三为碳还原砷酸钙生成 CaO 和砷蒸气。阶段三的反应机制可以用相边界反应动力学模型解释,求得的表观活化能为 156.6 kJ/mol。

(2)碳热焙烧还原砷酸钙单因素条件实验结果显示:在温度 1000 ℃、碳配入系数 1.4、恒温时长 60 min 条件下,砷挥发率可达 99.94%。

(3)采用 XRD、SEM-EDS 对砷蒸气冷凝物和还原焙烧残渣进行分析,结果表明:砷产物主要为晶型良好的片状金属砷和粉末球状不定性黑色砷,而还原残渣主要物相为 CaO。

参考文献

[1] Bothe J V, Brown P W. The stabilities of calcium arsenates at 23±1 ℃[J]. Journal of Hazardous Materials, 1999, 69(2): 197-207.

[2] Graham Long, Yong Jun Peng, Dee Bradshaw. A review of copper-arsenic mineral removal from copper concentrates[J]. Minerals Engineering, 2012, 36-38: 179-186.

[3] Lee P K, Yu S, Jeong Y J, et al. Source identification of arsenic contamination in agricultural soils surrounding a closed Cu smelter, South Korea[J]. Chemosphere, 2019, 217: 183-194.

[4] Leist M, Casey R J, Caridi D. The fixation and leaching of cement stabilized arsenic[J]. Waste Management, 2003, 23(4): 353-359.

[5] Yuan Yongfeng, Liu Suhong. Distribution and removal of arsenic in the process of bottom-blowing continuous copper smelting[J]. China Nonferrous Metallurgy, 2020, 49(2): 37-40.

[6] Jia Hai. Recycling and comprehensive utilization of metallurgical waste with high arsenic content[D]. Changsha: Central South University, 2013.

[7] A G L, B Y S, C G G, et al. Soil pollution characteristics and systemic environmental risk assessment of a large-scale arsenic slag contaminated site[J]. Journal of Cleaner Production, 2020, 251: 119721.

[8] V Dutré, C Vandecasteele. Solidification/stabilisation of arsenic-containing waste: Leach tests and behaviour of arsenic in the leachate[J]. Waste Management, 1995, 15(1): 55-62.

[9] V Dutré, C Vandecasteele. Solidification/stabilisation of hazardous arsenic containing waste from a copper refining process[J]. Journal of Hazardous Materials, 1995, 40(1): 55-68.

[10] Veronika Dutré, Vandecasteele C. An evaluation of the solidification/stabilisation of industrial arsenic containing waste using extraction and semi-dynamic leach tests[J]. Waste Management, 1996, 16(7): 625-631.

[11] Vandecasteele C, Veroniek Dutré, Geysen D, et al. Solidification/stabilisation of arsenic bearing fly ash from the metallurgical industry. Immobilisation mechanism of arsenic[J]. Waste Management, 2002, 22(2): 143-146.

[12] Choi W H, Lee S R, Park J Y. Cement based solidification/stabilization of arsenic-contaminated mine tailings[J]. Waste Management, 2009, 29(5): 1766-1771.

[13] Akhter H, Cartledge F K, Roy A, et al. Solidification/stabilization of arsenic salts: Effects of long cure times[J]. Journal of Hazardous Materials, 1997, 52(2-3): 247-264.

[14] Singh T S, Pant K K. Solidification/stabilization of arsenic containing solid wastes using portland cement, fly ash and polymeric materials[J]. Journal of Hazardous Materials, 2006, 131(1-3): 29-36.

[15] Yoon I H, Moon D H, Kim K W, et al. Mechanism for the stabilization/solidification of arsenic-contaminated soils with Portland cement and cement kiln dust[J]. Journal of Environmental Management, 2010, 91(11): 2322-2328.

[16] Zongwen Zhao, Yuxia, et al. XPS and FTIR studies of sodium arsenate vitrification by cullet[J]. Journal of Non-Crystalline Solids, 2016, 452: 238-244.

[17] Zhao Z W, Chai L Y, Peng B, et al. Arsenic vitrification by copper slag based glass: Mechanism and stability studies[J]. Journal of Non-Crystalline Solids, 2017, 466-467: 21-28.

[18] Xu Jianbing, Shen Qianghua, Chen Wen, et al. Current status and countermeasures of arsenic-containing

waste residue treatment(in Chinese)[J]. Mining and Metallurgy, 2017, 26(3): 82-86.

[19] Lu Xiaoyang. The status quo and progress of arsenic-containing waste residue treatment technology(in Chinese)[J]. Modern Salt and Chemical Industry, 2018, 45(5): 87-88.

[20] Lu Hongbo. The thermodynamic study on the preparation of crude metal arsenic by vacuum carbothermal reduction of As_2O_3(in Chinese)[J]. Nonferrous Metals(Extractive Metallurgy), 2012, 64(10): 55-59.

[21] Li Xuepeng. Experimental research on preparation of metal arsenic by DC arc furnace(in Chinese)[J]. Mining and Metallurgy, 2012, 21(3): 56-59.

[22] Pan Chongfa. An effective way to improve the quality of metallic arsenic(in Chinese)[J]. Nonferrous Metals(Extractive Metallurgy), 1994, 46(2): 32-38.

[23] Huang Zili, Liu Yuanyuan, Tao Qingying, et al. Influencing factors of arsenic removal by lime precipitation [J]. Chinese Journal of Environmental Engineering, 2012, 6(3): 734-738.

[24] Zhang Hua. Study on the solubility and stability of calcium arsenate[D]. Guilin: Guilin University of Technology; Guilin Institute of Technology, 2005.

[25] Liu Huili, Zhu Yinian. Thermodynamic analysis of the CO_2 effects on the stability of calcium arsenates[J]. Environmental Protection Science, 2006, 32(3): 7-18.

[26] Bothe J V, Brown P W. Arsenic immobilization by calcium arsenate formation[J]. Environmental Science and Technology, 1999, 33(21): 3806-3811.

[27] Hu Rongzu, Shi Qizhen. Thermal analysis kinetics (Second Edition)[M]. Beijing: Science Press, 2008: 20-21; 64-65; 151-155.

含砷固废碱性浸出脱砷工艺研究

摘 要：本文通过考察含砷固废氢氧化钠-硫化钠选择性浸出脱砷过程中各个影响因素对浸出过程各主要元素的浸出行为的影响，确定了含砷固废氢氧化钠-硫化钠选择性浸出脱砷过程的最佳工艺条件：碱料比为0.5、硫化钠用量为14 g/60 g 含砷固废、液固比（mL/g）为5∶1、浸出温度为90 ℃、浸出时间为2.0 h、搅拌速度为400 r/min。在最佳条件下，浸出液中锑、铅、锡、锌、铜、铁的平均质量浓度分别为2.41 g/L、0.31 g/L、1.18 g/L、0.24 g/L、0.002 g/L 和 0.005 g/L，浸出渣中砷的质量分数在0.84%至0.95%之间；砷、锑、铅、锡、锌、铜和铁的平均浸出率分别为92.74%、14.73%、0.35%、29.74%、3.68%、0.12% 和 0.15%。

关键词：含砷固废；碱性浸出；脱砷；湿法冶金

Arsenic removal from arsenic solid waste by alkaline leaching

Abstract: In this paper, the effects of various factors on the leaching behavior of the main elements in the leaching process were investigated by studying the effects of various factors on the leaching of the main elements during the leaching and arsenic removal process. The optimum conditions for the selective leaching of solid waste sodium hydroxide-sodium sulfide containing arsenic as follows: the ratio of alkali to material was 0.5, the amount of sodium sulfide was 14 g/60 g high arsenic dust, the liquid to solid ratio (mL/g) was 5∶1, the leaching temperature was 90 ℃, the leaching time was 2.0 h, the leaching speed was 400 r/min. The average mass concentration of antimony, lead, tin, zinc, copper and iron in the leaching solution were 2.41 g/L, 0.31 g/L, 1.18 g/L, 0.24 g/L, 0.002 g/L and 0.005, respectively. The average leaching rates of arsenic, antimony, lead, tin, zinc, copper and iron were 92.74%, 14.73%, 0.35%, 29.74%, 3.68%, 0.12% and 0.15% respectively.

Keywords: high arsenic dust, alkaline leaching, arsenic removal, wet metallurgy

作者：张磊，郭学益，田庆华。

引言

砷元素常伴生于重金属、贵金属矿中，而一般的选矿法不能将砷完全除去[1]，因此，会有部分的砷进入冶炼过程中，在冶炼各个工序以不同的形式输出冶炼系统，如冶炼废渣、冶炼烟尘、冶炼废水等[2-4]。含砷固废中常含有重金属和砷等有害元素，如果直接放在堆场或填埋，在雨水冲刷、溶浸、微生物等作用下，会进入水体或土壤，造成二次污染[5-7]。含砷固废不进行正确处理，将造成巨大的环境污染。由于含砷固废产生于有色冶炼火法冶炼过程[8-10]，往往包含了除砷以外的其他有价金属，将其直接进行无害化处理[11-14]，并不利于资源的有效利用。因此，将含砷固废中的砷进行选择性浸出[15]，使含砷固废从有害固废转变成有价资源，是实现我国可持续发展的必经之路[16]。

针对含砷物料中砷的脱除问题，国内外学者开展了一系列卓有成效的研究[17]，主要分为火法焙烧脱砷、湿法浸出脱砷和火法–湿法联合工艺。火法焙烧脱砷主要是在高温下使含砷物料中的砷以三氧化二砷的形式挥发[18]，使其与其他有价金属分离，再通过冷凝收尘得到粗制三氧化二砷产品[19]。湿法浸出脱砷主要是指使用合适的浸出剂搅拌浸取含砷物料，使砷从固相进入浸出液中，按照浸出剂的种类一般可以分为热水浸出、酸浸脱砷和碱浸脱砷[20-21]。与火法焙烧脱砷相比，湿法浸出脱砷具有脱砷率高、环境污染小、适用范围广、能耗较低等优点，且在浸出液的后续处理过程中还可以直接制备不同的砷系列产品[22]，但亦存在浸出液的处理流程较长、工序比较繁琐、工业废水处理困难等缺点。火法–湿法联合工艺[23]主要是指采用纯碱/烧碱焙烧然后水浸脱砷，该生产工艺生产能耗较高、纯碱/烧碱消耗量大、环境污染比较严重。因此，如何简单、环保地分离含砷固废中砷与其他有价金属是当前亟待解决的问题。

本研究涉及的含砷固废成分复杂，含砷固废中砷的分布比较分散，既有砷酸盐，又有砷的氧化物和硫化物。因此，本文作者采用氢氧化钠–硫化钠浸出体系对含砷固废进行选择性强化浸出脱砷，将铅锑抑制在浸出渣中，浸出渣中砷含量低，可以直接返回铅厂回收铅锑。

1 实验

1.1 原料

本研究所用含砷固废来自广西成源矿冶有限公司铅冶炼厂铜浮渣鼓风炉熔炼过程产生的烟尘。含砷固废为粒度细小、黑色的粉末状固体，暴露在空气中极易吸潮，因此从企业采集的含砷固废样品先置于鼓风干燥箱内，105 ℃烘干 24 h。干燥后的含砷固废经破碎后全部通过 100 目分样筛，混匀后用自封袋密封包装，作为后续实验的原料。含砷固废样品的定量化学成分分析和 XRD 谱分别见表 1 和图 1。

表 1　含砷固废的化学成分
Table 1　Chemical composition of high arsenic dust   %

元素	Pb	Sb	As	Sn	Zn	Cu	Fe	S	In
质量分数	49.13	9.55	6.86	2.80	2.40	1.10	1.80	5.50	0.26

从表 1 可以看出，含砷固废成分比较复杂，其中质量分数在 1% 以上的元素有铅、锑、砷、锡、锌、铜、铁和硫，且砷质量分数较高。在含砷固废中铅、锑的质量分数分别达到 49.13%、9.55% 和 6.86%，是非常重要的二次资源，进行回收处理，具有较高的经济效益和环境效益。由图 1 可知，含砷固废的主要物相是方铅矿 [PbS]、砷华 [As_2O_3]、方锑矿 [Sb_2O_3]、砷铅矿 [$Pb_5(AsO_4)_3OH$] 和砷酸铅 [$Pb_3As_2O_8$]。

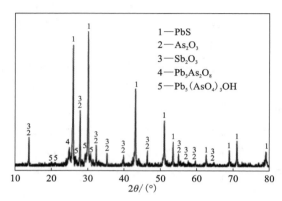

图 1　含砷固废的 XRD 谱
Fig. 1　XRD pattern of high arsentic dust

2　结果与讨论

2.1　碱料比对浸出率的影响

在含砷固废为 60 g、硫化钠用量为 10 g、反应温度为 80 ℃、反应时间为 2 h、液固体积质量比为 5∶1、搅拌速度为 400 r/min 的条件下，考察了碱料比分别为 0.1、0.2、0.3、0.4、0.5、0.6、0.8 和 1.0 时对浸出过程各金属浸出率 (质量分数) 的影响，实验结果如图 2 所示。

从图 2 可知，当碱料比从 0.1 增加至 0.3 时，锡的浸出率从 0 迅速增加至 25.02%，当碱料比进一步增加，锡的浸出率仅有少许增加；铅、锑、锌的浸出率随着碱料比的增加而呈线性增加，分别从 2.95%、0.03% 和 0.34% 增加至 20.16%、6.87% 和 17.01%；砷的浸出率随着碱料比的增加先增加后减小，当碱料比小于 0.6 时，砷浸出率随着碱料比的增加从 46.35% 迅速增加至 88.98%，当碱料比达到 1.0 时，砷浸出率反而降低至 74.55%。在实验过程中，当碱料比达到 0.6 以后，在抽滤过程中，滤渣表

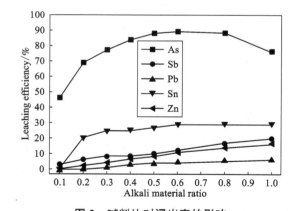

图 2　碱料比对浸出率的影响
Fig 2　Effect of alkail material ratio on leaching efficiency

面开始出现白色结晶物,且随着碱料比的增加逐渐增多,分析发现白色结晶的主要成分为砷酸钠;当碱料比达到 1.0 以后,浸出液开始变得黏稠,浸出液的过滤性能恶化。As_2O_5 在 NaOH 溶液中的溶解度随着 NaOH 浓度的增加而降低,且随着温度的降低而降低。铅锌的砷酸盐和亚砷酸盐的溶度积很小(如 $K_{sp[Pb_3(AsO_4)_2]} = 4.0×10^{-33}$,$K_{sp[Zn_3(AsO_4)_2]} = 2.8×10^{-28}$),当浸出液中存在大量的铅锌离子时就会抑制砷的浸出时。在确保较高的砷浸出率和较低的有价金属损失条件下,同时兼顾浸出液的过滤性能,选择碱料比为 0.5 比较合适。

2.2 硫化钠用量对浸出率的影响

在含砷固废为 60 g、碱料比为 0.5、反应温度为 80 ℃、反应时间为 2.0 h、液固质量比为 5∶1、搅拌速度为 400 r/min 的条件下,考察了硫化钠用量分别为 3.0 g、5.0 g、7.0 g、10.0 g、14.0 g 和 20.0 g 时对浸出过程各金属浸出率(质量分数)的影响,实验结果如图 3 所示。

从图 3 可知,在考察的硫化钠用量范围内,锡的浸出率没有明显的变化,锡浸出率在 28% 左右波动。铅和锌浸出率随着硫化钠用量的增加而降低;当硫化钠用量大于 14.0 g 后,浸出液中的铅和锌离子基本上被沉淀完全,含砷固废中的铅和锌基本上都被抑制在浸出渣中;随着硫化钠用量的增加,铅和锌浸出率分别由最高值的 5.96% 和 18.93% 降低至 0.11% 和 0.14%。砷的浸出率随着硫化钠用量的增加而增加,当硫化钠用量小于 14.0 g 时,砷浸出率从 72.17% 快速增加至 91.88%;进一步

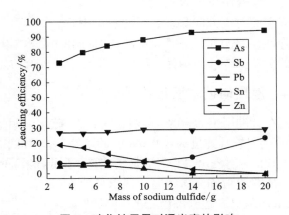

图 3 硫化钠用量对浸出率的影响

Fig. 3 Effect of Na_2S addition on leaching efficiency

增加硫化钠的用量,砷的浸出率基本上维持稳定。随着硫化钠用量的增加,浸出液中的可溶性铅逐渐转化为难溶的硫化铅沉淀,促进了砷酸铅溶解反应的向右进行;同时,浸出液中 Zn(Ⅱ)离子与 S^{2-} 反应生成硫化锌沉淀,导致浸出液中的锌离子浓度下降。锑的浸出率随着硫化钠用量的增加而增加,当硫化钠用量增加到 10.0 g 以后,锑浸出率增加的幅度越来越大。这是由于随着硫化钠用量的增加,浸出液中 Pb(Ⅱ)离子浓度逐渐降低,浸出液中游离的 S^{2-} 开始逐渐累积,含砷固废中的三氧化二锑在高浓度的硫化钠溶液中转化为可溶性的硫代亚锑酸钠,导致锑的浸出率急剧增加至 25.02%。综合考虑,选择硫化钠用量为 14.0 g 比较合适。

2.3 浸出温度对浸出率的影响

在含砷固废为 60 g、碱料比为 0.5、硫化钠用量为 14.0 g、反应时间为 2 h、液固体积质量比为 5∶1、搅拌速度为 400 r/min 的条件下,考察了反应温度分别为 30 ℃、40 ℃、50 ℃、60 ℃、70 ℃、80 ℃、90 ℃ 和 100 ℃ 时对浸出过程各金属浸出率(质量分数)的影响,实验结果如图 4 所示。

从图 4 可知,在考察的浸出温度范围内,锌的浸出率变化不明显,为 3.03% 左右;砷、锑

和锡的浸出率随着浸出温度的增加而增加；铅的浸出率随着浸出温度的增加而降低。在 90 ℃ 之前，砷的浸出率由 57.98%快速增加至 91.45%。一方面，砷在氢氧化钠溶液中的溶解度随着温度的增加而增加[20]，另一方面，砷酸铅的分解反应是吸热反应，因此，随着浸出温度的升高，反应平衡向正方向移动，砷浸出率逐渐增加。在 90 ℃ 之后，砷酸铅基本上分解完全，砷的浸出率增加趋于平缓。硫化铅的沉淀属于吸热反应，因此，随着浸出温度的增加，铅的沉淀反应平衡向正方向移动，铅的浸出率逐

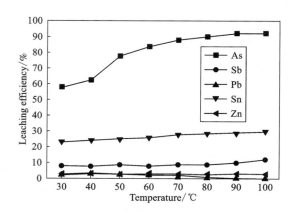

图 4 浸出温度对浸出率的影响
Fig. 4 Effect of temperature on leaching efficiency

渐降低，在 90 ℃ 之后，浸出液中的可溶性铅基本上沉淀完全，铅的浸出率由最高值的 3.18%降低至 0.22%。为确保较高的砷浸出率和较低的能耗，浸出温度选择 90 ℃ 比较合适。

2.4 浸出时间对浸出率的影响

在含砷固废为 60 g、碱料比为 0.5、硫化钠用量为 14.0 g、液固质量比为 5∶1、浸出温度为 90 ℃、搅拌速度为 400 r/min 的条件下，考察了浸出时间分别为 0.5 h、1.0 h、2.0 h、3.0 h 和 5.0 h 时对浸出过程各金属浸出率(质量分数)的影响，实验结果如图 5 所示。

从图 5 可知，锑和锡的浸出反应在 0.5 h 内就已经达到平衡，进一步延长浸出时间，锑和锡的浸出率基本上没有什么变化。砷的浸出率随着浸出时间的增加而增加，当浸出时间延长至 2.0 h 时，砷的浸出率增加至 92.77%；而在 2.0 h 以后，砷的浸出率增加的幅度可以忽略。铅和锌浸出率随着反应时间的增加而降低，浸出 2.0 h 之前，铅和锌的浸出率分别由 1.61%和 6.56%降低至 0.31%和 3.12%；浸出 2.0 h 之后，铅浸出率逐渐降低至 0.14%，锌的浸出率不再发生变化。在确保较高的砷浸出率和较低的有价金属损失条件下，

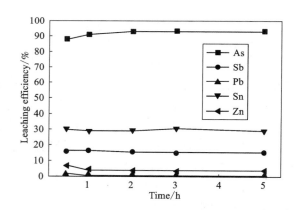

图 5 浸出时间对浸出率的影响
Fig. 5 Effect of time on leaching efficiency

综合考虑能耗、产能等因素，反应时间选择 2.0 h 比较合适。

2.5 液固比对浸出率的影响

在含砷固废为 60 g、碱料比为 0.5、硫化钠用量为 14.0 g、浸出时间为 2.0 h、浸出温度为 90 ℃、搅拌速度为 400 r/min 的条件下，考察了液固体积质量比分别为 3、4、5、7 和 10 时对浸出过程各金属浸出率(质量分数)的影响，实验结果如图 6 所示。

从图 6 可知，随着液固比的增加，铅的浸出率没有明显的变化，铅的浸出率在 1.0% 以下。砷浸出率随着液固比的增加先增加然后趋于稳定，当液固比小于 5∶1 时，砷浸出率由 80.78% 快速增加至 90.32%；当液固比大于 5∶1 时，砷的浸出反应趋于平衡。当硫化钠的用量一定时，随着液固比的增加，浸出体系中硫化钠的浓度逐渐降低，导致三氧化二锑转化为可溶性的硫代锑酸钠的转化率越来越低，甚至于不能够转化。虽然较高的液固比可以在保证较高的砷浸出率同时降低有价金属的损失，但是过高的液固比将导致生产能力的降低和能耗的增加。综合考虑，液固体积质量比选择 5∶1 比较合适。

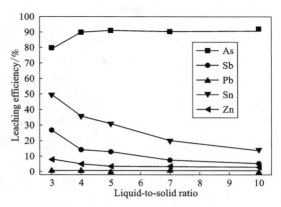

图 6　液固比对浸出率的影响

Fig.6　Effect of liquid-to-solid ratio on leaching efficiency

2.6　搅拌速度对浸出率的影响

在含砷固废为 60 g、碱料比为 0.5、硫化钠用量为 14.0 g、液固体积质量比为 5∶1、浸出温度为 90 ℃、浸出时间度为 2.0 h 的条件下，考察了搅拌速度分别为 200 r/min、250 r/min、300 r/min、350 r/min、400 r/min 和 500 r/min 时对浸出过程各金属浸出率（质量分数）的影响，实验结果如图 7 所示。

从图 7 可知，随着搅拌速度的增加，铅、锡和锌的浸出率没有明显的变化，锡的浸出率在 29.13% 至 31.44% 之间波动，铅的浸出率在 0.16% 至 0.31% 之间波动，锌的浸出率在 3.13% 至 3.52% 之间波动，铅和锌基本上都被抑制在浸出渣中。随着搅拌速度的增加，砷的浸出率逐渐增加，而锑的浸出率却逐渐降低。当搅拌速度从 200 r/min 增加至 500 r/min 时，砷的浸出率由 90.79% 缓慢增加至 93.23%；锑的浸出率由 18.33% 线性降低至 12.47%。随着搅拌速度的增加，浸出液与空气的接触增加，

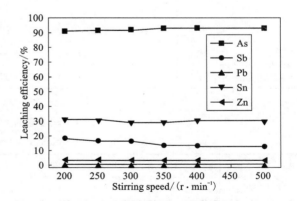

图 7　搅拌速度对浸出率的影响

Fig.7　Effect of stirring on leaching efficiency

更多的空气进入浸出液中，将浸出液中的可溶性锑氧化为不溶性的焦锑酸钠浸出渣中，导致锑浸出率逐渐降低。因此，需要控制合适的搅拌速度，确保含砷固废在浸出体系中呈悬浮状态均匀分散在浸出剂中，强化颗粒表层的传质传热过程。综合考虑，搅拌速度选择 400 r/min 比较合适。

2.7 综合实验

通过以上的系列单因素实验研究，可得出含砷固废氢氧化钠/硫化钠碱性浸出的优化工艺条件：碱料比为 0.5、硫化钠用量为 14 g/60 g 含砷固废、液固比(mL/g)为 5∶1、浸出温度为 90 ℃、浸出时间为 2.0 h、搅拌速度为 400 r/min。在此优化工艺条件进行了 3 次实验，其实验结果如表 2 和表 3 所示。由表 2 和表 3 可以看出，在综合实验中，浸出渣中砷的质量分数在 0.84% 至 0.95% 之间；砷、锑、铅、锡、锌、铜和铁的平均浸出率为 92.74%、14.73%、0.35%、29.74%、3.68%、0.12% 和 0.15%，含砷固废浸出渣的化学成分如表 4 所示。

表 2 优化实验结果
Table 2 Experiment results of optimal experiment

实验编号	浸出液体积/mL	浸出液中各元素质量浓度/(g·L^{-1})						浸出渣质量/g	浸出渣中砷的质量分数/%
		Sb	Pb	Sn	Zn	Cu	Fe		
1	355	2.360	0.297	1.119	0.229	0.002	0.005	48.3	0.95
2	350	2.337	0.257	1.169	0.249	0.003	0.005	49.6	0.84
3	340	2.536	0.383	1.250	0.246	0.002	0.004	48.2	0.87

表 3 优化实验中各元素的浸出率
Table 3 Leaching efficiencies of elements based on composition of leaching liquor %

实验编号	As	Sb	Pb	Sn	Zn	Cu	Fe
1	92.28	14.70	0.34	28.78	3.57	0.11	0.16
2	92.99	14.35	0.29	29.65	3.82	0.16	0.16
3	92.95	15.13	0.42	30.80	3.67	0.10	0.13
平均值	92.74	14.73	0.35	29.74	3.68	0.12	0.15

表 4 碱浸渣的化学成分
Table 4 Compositions of alkali leaching residue %

As	Sb	Pb	Sn	Zn	Cu	Fe	S	In	Na	Al	Ca
0.87	10.25	64.26	2.07	2.94	1.35	2.22	9.77	0.47	2.03	0.04	0.21

从表 4 中可以看出浸出渣中铅、锑和铟的质量分数分别为 64.26%、10.25% 和 0.47%，相对于含砷固废原料来说有一定程度的富集，浸出渣中的砷质量分数降低至 0.87%。

3 结论

通过考察含砷固废氢氧化钠-硫化钠选择性浸出脱砷过程中各影响因素对浸出过程各主要元素的浸出行为的影响，确定了含砷固废氢氧化钠-硫化钠选择性浸出脱砷过程的最佳工艺条件：碱料比为0.5、硫化钠用量为14 g/60 g 含砷固废、液固比(mL/g)为5∶1、浸出温度为90 ℃、浸出时间为2.0 h、搅拌速为400 r/min。在最佳条件下，浸出液中锑、铅、锡、锌、铜、铁的平均质量分数分别为2.41 g/L、0.31 g/L、1.18 g/L、0.24 g/L、0.002 g/L 和0.005 g/L，浸出渣中砷的质量分数在0.84%至0.95%之间；砷、锑、铅、锡、锌、铜和铁的平均浸出率为92.74%、14.73%、0.35%、29.74%、3.68%、0.12%和0.15%。

参考文献

[1] 伯英,罗立强.砷的地球化学特征与研究方向[J].岩矿测试,2009,28(6)：569-575.
[2] 王萍,王世亮,刘少卿,等.砷的发生、形态、污染源及地球化学循环[J].环境科学与技术,2010,33(7).无页码
[3] 马啸,王金生.砷的地球化学成因[J].地球科学进展,2012(S1)：388-389.
[4] 张鹏.次氧化锌中砷的固化与脱除研究[D].长沙：中南大学,2009.
[5] 项斯芬,严宣申,曹庭礼,等.无机化学丛书第四卷氮、磷、砷分族[M].北京：科学出版社,1995.
[6] 姜琪.重有色金属冶炼中砷的回收与利用[D].昆明：昆明理工大学,2002.
[7] 邓卫华.锑冶炼砷碱渣有价资源综合回收研究[D].长沙：中南大学,2014.
[8] 王勇.铜冶炼含砷废水处理新工艺及其基础理论研究[D].长沙：中南大学,2009.
[9] 彭春艳.三氧化二砷抗白血病的表观遗传学机制[D].汕头：汕头大学,2006.
[10] 李玉虎.有色冶金含砷固废中砷的脱与固化[D].长沙：中南大学,2011.
[11] 蒋学先,何贵香,李旭光,等.含砷固废脱砷试验研究[J].湿法冶金,2010(3)：199-202.
[12] 汤海波,秦庆伟,郭勇,等.含砷固废酸性氧化浸出砷和锌的试验研究[J].武汉科技大学学报,2014(5)：341-344.
[13] 陈维平,李仲英,边可君,等.湿式提取砷法在处理工业废水及废渣中的应用[J].无页码
[14] 徐养良,黎英,丁昆,等.艾萨炉含砷固废综合利用新工艺[J].中国有色冶金,2005(5)：25-27.
[15] 刘湛,成应向,曾晓冬.采用氢氧化钠溶液循环浸出法脱除高砷阳极泥中的砷[J].化工环保,2008(2)：141-144.
[16] Li Y, Liu Z, Li Q, et al. Removal of arsenic from Waelz zinc oxide using a mixed NaOH-Na_2S leach[J]. Hydrometallurgy, 2011, 108：165-170.
[17] Tongamp W, Takasaki Y, Shibayama A. Arsenic removal from copper ores and concentrates through alkaline leaching in NaHS media[J]. Hydrometallurgy, 2009(3-4)：213-218.
[18] Tongamp W, Takasaki Y, Shibayama A. Selective leaching of arsenic from enargite in NaHS-NaOH media[J]. Hydrometallurgy, 2010, 101：64-68.
[19] Chen Y, Liao T, Li G, et al. Recovery of bismuth and arsenic from copper smelter flue dusts after copper and zinc extraction[J]. Minerals Engineering, 2012, 12：23-28.

[20] Guolin Y, Zhang Y, Zheng S L, et al. Extraction of arsenic from arsenic-containing cobalt and nickel slag and preparation of arsenic-bearing compounds[J]. Transactions of Nonferrous Metals Society of China, 2014, 24: 1918-1927.

[21] Nishimura T, Tozawa K, Robins R G. The calcium-arsenic-water-air system[C]//CIM BULLETIN. 101 6TH AVE SW, STE 320, CALGARY AB TZP 3P4, CANADA: CANADIAN INST MINING METALLURGY PETROLEUM, 1985, 78(878): 75-75.

[22] Palfy P, Vircikova E, Molnar L. Processing of arsenic waste by precipitation and solidification[J]. Waste Management, 1999, 19(1): 55-59.

[23] Fujita T, Taguchi R, Shibata E, et al. Preparation of an As(V) solution for scorodite synthesis and a proposal for an integrated As fixation process in a Zn refinery[J]. Hydrometallurgy, 2009, 96: 300-312.

高镍锍浸出渣高效清洁利用工艺

摘　要：高镍锍是镍冶炼过程的重要中间产品，通过生产高镍锍可以使镍铁、电镍、硫酸镍等镍产品在市场中进行相互转化和维持平衡，对于镍产业的稳健发展具有十分重要的意义。高镍锍经湿法提镍后产生大量含镍、铜、钴及金、银、铂、钯等贵金属的浸出渣，具有巨大的综合利用价值。科研人员攻克了高镍锍浸出渣有价组分分离提取关键技术，实现了企业从最初的单一电解镍生产发展为镍、铜、钴和金、银、铂、钯等贵金属综合回收。近年来，在环保政策愈趋严格以及冶炼工艺低碳转型的形势驱动下，进一步提出了热压浸出降低渣含镍量、氧压浸出替代回转窑焙烧处理浸铜后渣、含钠高盐废水乏汽蒸发及资源化利用等技术对现有工艺进行优化升级和提质增效。本文主要介绍高镍锍浸出渣冶金工艺现状，剖析存在的问题，阐述改进工艺最新进展及其涉及的反应原理和工业实践情况，为相关企业提供参考。

关键词：高镍锍；浸出渣；热压浸出；氧压浸出；金属分离提取；硫酸钠；废水处理；资源化利用

High-efficiency clean utilization process of leaching residue of high nickel matte

Abstract: High nickel matte is an important intermediate product in the nickel metallurgical process. By using nickel matte, nickel products such as ferronickel, electrolytic nickel and nickel sulfate can be mutually transformed and balanced in the market, which is of great significance for the steady development of nickel industry. A large number of leaching residues containing nickel, copper, cobalt, gold, silver, platinum, palladium and other precious metals are produced from high matte nickel after nickel hydrometallurgical extraction, which has great comprehensive utilization value. Researchers have solved the key technologies of separation and extraction of valuable components of high nickel matte leaching residue, and realized the development of the enterprise from the original single electrolytic nickel production workshop to an industrial park for comprehensive recovery of nickel, copper, cobalt and precious metals such as gold, silver, platinum and palladium. In recent years, driven by the increasingly strict environmental protection policies and the low-carbon transformation of metallurgical process, technologies such as hot acid

---

本文发表在《中国有色冶金》，2023，52(5)：76-83。作者：谢铿，王海北，马育新，李渭鹏。

pressure leaching for nickel reduction in residues, oxygen pressure leaching instead of rotary kiln roasting to treat the residues after copper extraction, and resource utilization of sodium sulfate wastewater are further proposed to optimize and upgrade the existing process and improve the quality and efficiency. This paper mainly introduces the current situation of the metallurgical process of high nickel matte leaching residue, analyzes the existing problems, expounds the latest progress of the improved process, metallurgical principle and industrial practice, and provides reference for relevant enterprises.

Keywords: high nickel matte; leaching residue; hot acid pressure leaching; oxygen pressure leaching; extraction and separation of metals; sodium sulfate; wastewater treatment; resource utilization

高镍锍是镍、铜、钴、铁等金属的硫化物共熔体,既可用硫化镍矿生产,也可用红土镍矿生产[1-3]。硫化镍矿通常采用选矿富集得到镍精矿,然后经过熔炼、吹炼等火法造锍工艺除去镍精矿中的大量脉石、铁与一部分硫,产出高镍锍,该技术成熟且应用广泛,国内金川公司、新疆新鑫矿业阜康冶炼厂、吉恩镍业等都采用该传统工艺[4-5]。红土镍矿可以通过在还原熔炼过程中添加硫化剂(硫磺、石膏、黄铁矿或含硫的镍原料等)与还原得到镍、铜、铁等金属发生硫化反应,造就高镍锍的生产[6-7],采用该工艺的有印尼淡水河谷公司。红土镍矿还可以先通过回转窑焙烧-电炉熔炼工艺得到镍铁,再加入硫化剂进行转炉吹炼,制成高镍锍[8],该工艺被法国埃赫曼新喀里多尼亚的多尼安博厂所采用。高镍锍可进一步精炼产出电镍、硫酸镍等镍产品[9-11]。由此可见,高镍锍是一种重要的镍火法冶炼中间产品,能够作为硫化镍矿和红土镍矿的纽带和调节剂,通过生产高镍锍可以使电镍、镍铁、硫酸镍这几种镍产品在市场中进行相互转化和平衡,对于镍产业的稳健发展具有重要意义。

1 高镍锍多组分分离提取工艺现状

高镍锍多组分分离提取通常采用磨浮-硫化镍阳极电解精炼工艺和湿法冶金工艺处理[12-13]。磨浮-硫化镍阳极电解精炼工艺已被应用于加拿大国际镍公司汤普逊镍精炼厂、金川公司和成都电冶厂等,生产实践表明该工艺存在流程长、能耗高、贵金属分散、镍铜分离不彻底、金属回收率低等缺点。相比较磨浮-硫化镍阳极电解精炼工艺,湿法冶金工艺采用选择性浸出代替高镍锍磁选-浮选、熔炼、镍熔铸、酸性电解造液等工序,具有工艺流程短、原料适应性强、金属回收率高、介质便于循环利用、自动化程度高、环境友好等优点。因此,自20世纪50年代以来,湿法冶金工艺得到快速发展并实现了广泛应用,成为高镍锍精炼及镍产品生产的主要工艺。

矿冶科技集团有限公司(原北京矿冶研究总院)围绕我国新疆喀拉通克硫化镍矿冶炼-水淬高镍锍的处理,先后完成了高镍锍硫酸浸出试验研究和开发了高镍锍选择性浸出工艺,提出了硫酸常压浸出-加压浸出流程[12]。高镍锍经硫酸选择性浸出后,浸出液通过黑镍除钴和电积等工序生产电镍,黑镍除钴得到的钴渣经酸溶、萃取除杂和分离镍钴、电积等工序生产电钴;浸出渣含有部分镍,且富集了几乎全部的铜和金、银、铂、钯等贵金属,为了充分回收高镍锍浸出渣中的镍、铜和贵金属，采用氧化焙烧-浸出-电积提铜、浸铜后渣酸化焙烧-浸出提取镍铜和富集贵金属、贵金属渣电炉熔炼-焙烧-浸出-氯浸-置换-分步沉淀等组合工艺[13-14]，通过调控与优化技术参数，实现了目标有价金属的定向分离、精准富集和综合回收，得到了阴极铜、银粉、金粉、海绵铂、钯粉等产品，同时含镍、钴的溶液可返回主系统进一步提取镍和钴。该工艺在阜康冶炼厂进行生产实践，年产超12000 t 镍、11400 t 铜、110 t 钴。

运行过程中，该工艺存在浸出渣含镍高、硫酸化焙烧工序环境污染风险大、中和工序纯碱消耗大、硫酸钠废液难以利用等问题，通过大量的工艺试验和技术方案论证，提出了热压浸出降低高镍锍浸出渣中镍含量、氧压浸出替代酸化焙烧处理浸铜后渣、含钠高盐废水乏汽蒸发及资源化利用等技术升级和提质增效措施，致力于形成高镍锍浸出渣多组分深度分离提取成套技术体系，最终实现高镍锍浸出渣的高效绿色清洁综合利用。

2 高镍锍浸出渣冶金工艺

2.1 高镍锍浸出渣来源及性质

高镍锍浸出渣来源于镍提取车间提镍后滤渣，主要成分范围见表1，某代表性高镍锍浸出渣组成见表2。其中，渣中70%以上的镍和铜以硫化物存在，钴以 CoS 存在，铁以 Fe_3O_4 存在；主要矿物为 Cu_2S（包括辉铜矿、铜蓝）、Cu_2O、碱式硫酸铜、Ni_3S_2、NiS、CuS 和碱式硫酸镍。

表 1　高镍锍浸出渣主要成分

Table 1　Main compositions of leaching residue of high nickel matte　　%

成分	Ni	Cu	Co	Fe	S	Au*	Ag*	Pt*	Pd*
质量分数	5~9	43~53	<0.15	1.1~1.5	20~24	4.1~5.2	350~440	2~3	2~3

注：*单位为g/t。

表 2　代表性高镍锍浸出渣主要成分

Table 2　Main composition of leaching residue of representative high nickel matte　　%

成分	Ni	Cu	Fe	Co	S	Au*	Ag*	Pt*	Pd*	O	V	Ti
质量分数	7.64	53.12	1.25	0.12	21.86	4.64	372	2.51	2.43	22.3	0.02	0.02
成分	Ba	Ca	Na	Mg	Al	Si	Se	K	Pb	As	Cr	Sr
质量分数	4.60	0.17	1.23	0.22	0.20	0.56	0.03	0.05	0.02	0.08	0.01	0.02

注：*单位为g/t。

2.2 高镍锍浸出渣冶金工艺

高镍锍浸出渣冶金工艺流程如图1所示。高镍锍浸出渣首先送入铜提取车间，采用氧化焙烧-浸出工艺处理，浸出液通过电积生产阴极铜，浸铜后渣富集了几乎全部贵金属，主要成分范围见表3。

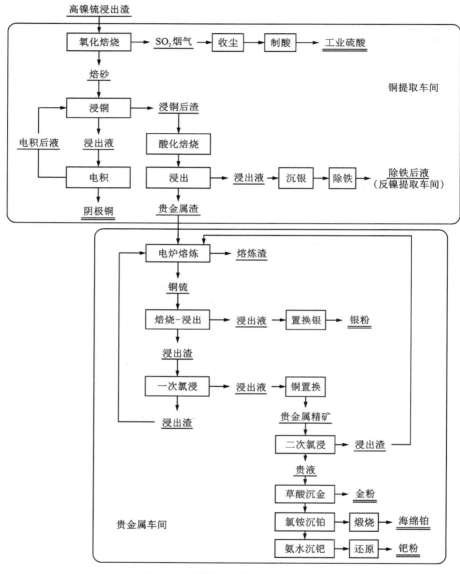

图 1　高镍锍浸出渣冶金工艺流程

Fig. 1　Flow sheet for metallurgical process of leaching residue of high nickel matte

表 3　浸铜后渣主要成分

Table 3　Main compositions of residue after copper leaching　　%

成分	Ni	Cu	Au*	Ag*	Pt*	Pd*
质量分数	18~23	7~11	16~21	700~1550	7.9~9.5	7.9~9.5

注：*单位为g/t。

采用酸化焙烧-浸出处理浸铜后渣，浸出液经过沉银-除铁工艺处理后送镍提取车间，浸出渣成分见表4。

表 4　贵金属渣主要成分
Table 4 Main compositions of noble metal residue　　　　　　　　%

成分	Cu	Ni	Au*	Ag*	Pt*	Pd*
质量分数	2.5~4	6~9	35~45	900~3000	15~18	14~16

注：* 单位为 g/t。

贵金属渣送入贵金属提取车间，经电炉熔炼得到铜锍，铜锍再通过多次回转窑硫酸化焙烧-浸出处理，浸出液采用沉银和除铁工序处理分别产生银渣和铁渣，银渣送贵金属车间，铁渣返回喀拉通克铜镍矿熔炼工序，除铁后液送镍提取车间；浸出渣经一次氯化浸出-铜板置换得到贵金属精矿，Au、Pt、Pd三者总质量分数约40%，再经二次氯浸得到贵液。贵液用草酸选择性还原金，沉金后液用氯化铵沉铂，铂盐经煅烧得到海绵铂。沉铂后液用氨水络合沉钯，钯盐用水合肼还原为钯粉，从而得到金、铂和钯产品。

2.3　现有高镍锍浸出渣工艺存在的问题

（1）铜镍选择性浸出效果不佳，高镍锍浸出渣中镍含量高，有时为8%以上。一方面，大量的镍进入铜提取系统后导致铜镍分离困难，对电铜质量产生影响；另一方面，为了回收镍，需将大量含镍的铜电积后液返回镍提取系统，增大了体系酸膨胀的压力。

（2）回转窑硫酸化焙烧处理浸铜后渣存在二氧化硫烟气污染、能耗高、酸耗大、工艺条件复杂、操作环境差等缺点，且焙砂常压浸出时绝大部分铁进入溶液，后续除铁压力大。

（3）为了维持体系酸平衡和钠平衡，使用纯碱中和一部分镍阳极液，沉淀产出碳酸镍并排出硫酸钠，为此消耗了大量纯碱，且产生大量TDS(Total Dissolved Solids，总溶解固体)质量分数160~200 g/L的废水，同时羟基氧化镍制备工序也产生大量TDS质量分数40~60 g/L的废水[15]，这些含钠高盐废水没有得到有效回收利用，不但造成资源浪费，而且增加了环保风险。

3　高镍锍浸出渣冶金工艺优化及工业实践

3.1　热压浸出降低高镍锍浸出渣中镍含量

在现有两段常压浸出—一段加压浸出工序后添加热压浸出工序，充分利用镍铜与硫之间亲和力差异及体系中镍铜交互反应特性，提高镍浸出的选择性，使镍最大限度被浸出进入溶液，从而降低浸出渣中镍含量。

高镍锍浸出渣热压浸出工艺流程如图2所示。高镍锍浸出渣与镍阳极液、热压浸出返液、浸铜后渣氧压浸出液、高镍锍浸出渣洗水等在加压配料槽中配料后用高压泵经矿浆加热器送入密闭加压釜中进行热压浸出,反应基本不需要氧气参与,主要完成渣中镍硫化物与料液中硫酸铜的反应,渣中的镍与溶液中的铜在高温条件下进行交互反应,渣中的镍转变为硫酸镍进入溶液,溶液中的铜离子生成硫化铜进入渣中,从而实现镍的选择性浸出;铁与镍的浸出行为类似,同时少量的硫酸也会参与镍离子的浸出反应,涉及的主要反应见式(1)~式(6)[16-17]。

$$4Ni_3S_4 + 9CuSO_4 + 8H_2O = Cu_9S_5 + 9NiS + 3NiSO_4 + 8H_2SO_4 \quad (1)$$

$$CuSO_4 + NiS = NiSO_4 + CuS\downarrow \quad (2)$$

$$NiS + H_2SO_4 = NiSO_4 + H_2S\uparrow \quad (3)$$

$$Ni_3S_2 + 3CuSO_4 = 3NiSO_4 + Cu_2S + CuS\downarrow \quad (4)$$

$$FeS + H_2SO_4 = FeSO_4 + H_2S\uparrow \quad (5)$$

$$H_2S + CuSO_4 = H_2SO_4 + CuS\downarrow \quad (6)$$

控制浸出工艺参数液固比(体积质量比)(5~6):1、浸出温度约160 ℃、浸出时间2.5~3 h,热压浸出矿浆经减压降温后,输送至浓密机液固分离,底流经离心脱水后得到热压浸出渣。镍和铁浸出率可分别达到91%和94%以上,渣中镍可由8%降至2%以下,热压浸出渣组成含 Cu 55%~68%、Ni 0.9%~2.0%、Fe 0.1%~0.3%、Au 3.3%~6.0 g/t、Ag 250~420 g/t、Pt 1.6~3.2 g/t、Pd 1.5~3.0 g/t。某代表性热压浸出渣和浸出液组成见表5。对比表1和表5的数据可看出,热压浸出后,镍含量降低,铜进入渣中进一步被富集。热压浸出渣送铜提取车间处理。热压浸出液溢流经压滤后送除铁工序,采用铁矾法除铁,经压滤得到除铁后液,除铁后液返回镍提取车间浸出工序配料。

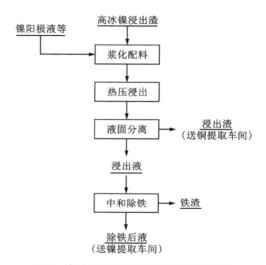

图2 高镍锍浸出渣热压浸出工艺流程

Fig. 2 Flow sheet for hot acid presussure leaching process of leaching residue of high nickel matte

表5 热压浸出渣和浸出液主要成分

Table 5 Main compositions of representative leaching residue and solution after hot acid pressure leaching

元素	Ni	Cu	Fe	Co	S	Au*	Ag*	Pt*	Pd*	H_2SO_4
浸出渣/%	0.95	65.41	0.14	0.10	22.47	4.21	345	2.21	2.01	—
浸出液/(g·L^{-1})	44.21	6.04	5.64	0.003	—	≤0.1	≤0.5	≤0.1	≤0.1	33.39

注:* 单位为 g/t 或 mg/L。

3.2 氧压浸出替代酸化焙烧-常压浸出处理浸铜后渣

采用氧压浸出工艺替代硫酸化焙烧-常压浸出工艺，反应在密闭加压釜中进行，从源头上消除二氧化硫污染风险，改善操作环境；通过氧压高温强化浸出，实现镍钴铜的一步完全溶出和贵金属在浸出渣中定向富集；浸铜后渣直接氧压酸浸，可与铜提取车间常压浸出工艺良好衔接，省去焙烧前所需的干燥和堆放等环节，简化了工艺流程，利于全流程连续化操作。

氧压浸出原料是铜提取车间日常生产产生的浸铜后渣及堆存的浸铜后渣。生产产出的浸铜后渣通过压滤机下料斗放入浆化槽，浆化后再泵送至调浆槽。堆存的浸铜后渣经破碎筛分后和新产生的浸铜后渣按既定的比例加入调浆槽。同时按比例往调浆槽内加入电积后液、新水、洗液，充分搅拌后泵送到加压供料槽。加压供料槽的矿浆通过加压泵泵入矿浆加热器后再进入加压釜，物料在氧压环境下进行反应。浸铜后渣中镍70%以上存在于铁矿物和硅酸盐中，另外还有少量以硫酸镍、氧化镍和硫化镍形式存在；铜主要存在于铁矿物中以水溶性硫酸铜形式存在。常压条件下，难以浸出的铁矿物中镍和铜等可在加压釜高温下分解，从而获得较高的浸出率。涉及的主要反应见式(7)~式(14)[9, 18]。

$$NiS + 2O_2 = NiSO_4 \tag{7}$$

$$CuS + 2O_2 = CuSO_4 \tag{8}$$

$$FeS + 2O_2 = FeSO_4 \tag{9}$$

$$2NiS + 2H_2SO_4 + O_2 = 2NiSO_4 + 2S + 2H_2O \tag{10}$$

$$NiFe_2O_4 + 4H_2SO_4 = NiSO_4 + Fe_2(SO_4)_3 + 4H_2O \tag{11}$$

$$CuFe_2O_4 + 4H_2SO_4 = CuSO_4 + Fe_2(SO_4)_3 + 4H_2O \tag{12}$$

$$Fe_2(SO_4)_3 + 6H_2O = 3H_2SO_4 + 2Fe(OH)_3 \downarrow \tag{13}$$

$$2S + 3O_2 + 2H_2O = 2H_2SO_4 \tag{14}$$

氧压浸出约3 h后通过闪蒸槽进行排料。闪蒸槽产生的蒸汽通过除沫装置除去夹带的矿浆后排空，矿浆则从底部自流进入室外两级串联的冷却槽，矿浆温度从100 ℃左右下降到80 ℃以下后用泵直接送加压浸出压滤机进行压滤，得到滤液含Cu 40~50 g/L、Ni 60~75 g/L、Fe 1.0~3.0 g/L、Co 0.01~0.02 g/L、$\rho(Au) \leqslant 0.2$ mg/L、Ag 0.5~1.5 mg/L、$\rho(Pt) \leqslant 0.1$ mg/L、$\rho(Pd) \leqslant 0.1$ mg/L、H_2SO_4 30~55 g/L，通过储槽收集后泵送至镍提取车间浸出工序配料；滤渣即为贵金属渣，组成含Cu 2.5%~4.5%、Ni 12%~16%、Fe 16%~30%、Au 40~50 g/t、Ag 2000~2800 g/t、Pt 16~25 g/t、Pd 16~25 g/t，送贵金属生产车间进一步处理。贵金属渣年产量超1500 t，某代表性贵金属渣组成见表6。

表6 贵金属渣主要成分

Table 6 Main composition of representative precious metals-bearing residue after oxygen pressure leaching %

成分	Ni	Cu	Fe	Au*	Ag*	Pt*	Pd*
质量分数	13.04	3.12	16.34	37	2200	18.93	18.40

注：*单位为g/t。

浸铜后渣氧压浸出工程建成投产后，年增加渣处理量约 3000 t，新增销售收入超 5000 万元，工艺改进后处理 1 t 渣的能耗可降低 370 kgce 以上[18]，铁的浸出率由大于 80%降至 20%以下，大幅缓解了浸出后液除铁的压力，实现了镍钴铜的一步浸出及贵金属在浸出渣中的定向富集，从根本上解决了二氧化硫烟气污染问题，实现了浸铜后渣的清洁提取。

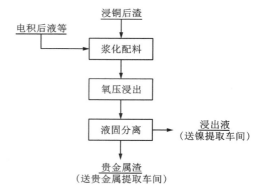

图 3　浸铜后渣氧压浸出工艺流程

Fig. 3　Flow sheet for oxygen presussure leaching process of leaching residue after copper extraction

3.3　含钠高盐废水乏汽蒸发及资源化利用

研发了高盐废水处理及纯碱再生利用技术，充分利用加压釜闪蒸乏汽热量对废水进行低温热法蒸发浓缩，然后采用碳铵复分解-碳铵回收-促进剂诱导蒸发结晶技术将废水中硫酸钠制备出工业级纯碱并联产硫酸铵，满足纯碱自给，同时废水处理后可回用，实现高盐废水清洁处置及资源化利用，有效减少污水排放量、固废产生量及用水总量。

镍阳极液沉镍后液与黑镍废水等含钠高盐废水在混合池中混合均匀，混合废水含 TDS 约 126 g/L，送至酸化反应槽中，加入浓硫酸调节 pH 值至 4.5，将混合液中的碳酸钠转化为硫酸钠，反应后液泵至管道混合器中，加入氢氧化钠溶液调 pH 值至 7.0，调节后液输送至低温热法蒸发浓缩。经三效蒸发，产生的二次蒸汽经冷凝得到产水，$\rho(TDS) \leqslant 100$ mg/L，可作为初纯水用于生产。

废水中 TDS 含量可浓缩至 2~3 倍以上，硫酸钠浓度达到 300 g/L，浓水由浓水泵送至蒸发结晶工序。蒸发结晶采用两效强制循环蒸发-稠厚结晶-离心分离工艺，蒸发结晶热源采用净化洗涤后的加压釜闪蒸乏汽，乏汽温度 92~97 ℃，蒸发出来的二次蒸汽温度约 70 ℃，通过管道输送至前述的低温热法蒸发浓缩工序用作热源。通过强制循环蒸发和汽水分离将废水增浓蒸发至过饱和状态，然后进入结晶稠厚器，稠厚器中晶浆通过离心机分离产出硫酸钠。离心滤液送至冷冻除杂工序，先用循环水冷却到 35 ℃，由冷冻滤液冷却至 30 ℃，然后由冷冻液冷却到 10 ℃，再进行离心分离产出芒硝，经溶解和预热后返回蒸发结晶工序制备硫酸钠。进一步以硫酸钠为原料，通过复分解反应、蒸汽煅烧、碳酸铵回收、蒸发结晶等工序制备得到碳酸钠和硫酸铵产品。

首先将硫酸钠和碳酸氢铵溶解浆化后输送至复分解反应系统进行混合，发生的反应见式（15）。

$$2NH_4HCO_3 + Na_2SO_4 = 2NaHCO_3 + (NH_4)_2SO_4 \tag{15}$$

反应生成碳酸氢钠和硫酸铵，由于碳酸氢钠溶解度小，所以容易以晶体形式从溶液中析出。经过滤得到碳酸氢钠滤饼，然后通过蒸汽煅烧炉煅烧制得纯碱产品，反应见式（16）。产品含 Na_2CO_3 大于 98.54%，可达到《工业碳酸钠及其试验方法》（GB/T 210.1—2004）Ⅱ类合格品（一般工业用碳酸钠）要求。

$$2NaHCO_3 = Na_2CO_3 + CO_2\uparrow + H_2O \tag{16}$$

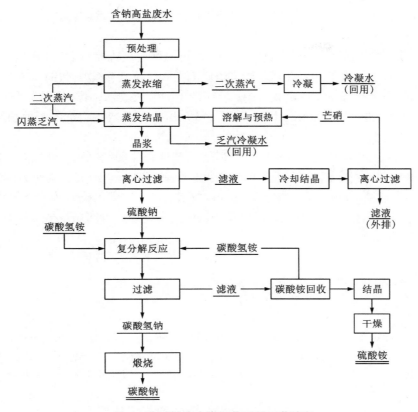

图 4　硫酸钠废水资源化利用工艺流程
Fig. 4　Flow sheet for resource utilization of sodium sulfate wastewater

复分解反应结束后，碳酸氢钠母液中未反应的碳酸氢铵，可采用微分式的碳酸氢铵分解与循环再生耦合技术进行回收，在低温下实现碳酸氢铵的再生，且不产生游离氨；回收的碳酸氢铵返回复分解反应工序。回收碳酸氢铵后的母液，通过引入促进剂推动氨水的解离，提高铵离子浓度，从而扩大硫酸铵的结晶区间，此时蒸发结晶可先进入硫酸铵结晶区，在未进入硫酸钠结晶区前停止蒸发，然后经离心分离、干燥，得到硫酸铵产品，产品中 NH_3-N 含量大于 20.5%，能够满足《肥料级硫酸铵》(GB/T 535—2020) 中 I 型产品要求。

4　结论

高镍锍作为一种镍冶炼中间产品，对于优化镍产业链结构以及保障铜镍钴和铂族金属等资源供给具有十分重要的意义。本文针对高镍锍多组分分离和精炼，提出了高镍锍浸出渣的高效和清洁利用工艺流程，并进行生产实践，取得了良好效益。

(1) 利用镍铜与硫之间亲和力差异及硫化镍与铜离子交互反应特性，可选择浸出渣中镍，降低渣中镍含量。在现有两段常压浸出—一段加压浸出工序后添加热压浸出工序，于液固比 (5~6):1、浸出温度约 160 ℃、浸出时间 2.5~3 h 的条件下，镍的浸出率为 91% 以上，渣中镍可由 8% 降至 2% 以下，可从源头上实现镍铜的高效分离，为后续铜提取及贵金属富集创造良好条件。

（2）氧压浸出替代回转窑焙烧处理浸铜后渣，可消除二氧化硫污染，改善操作环境，且利于与前后工序衔接，简化了工艺流程。通过氧压酸浸一步强化溶出镍钴铜，浸出率为 90% 以上，金银铂钯几乎全部富集在贵金属渣中；氧压浸出的节能效果显著，相比硫酸化焙烧工艺，每处理 1 t 渣可减少能耗 370 kgce 以上；采用连续加压浸出工艺，提高了自动化程度，有效降低了工人劳动强度。

（3）采用闪蒸乏汽蒸发-碳酸铵复分解-煅烧工艺处理硫酸钠废水，可短流程制备得到纯碱，实现硫酸钠废水资源化利用和纯碱的再生循环。采用高压釜闪蒸乏汽作为热源浓缩和结晶废水，利用了余热，有效降低了综合能耗和碳排放；年产纯碱和硫酸铵可分别达到 1.4 万 t 和 2 万 t，纯碱可以满足厂区自用，硫酸铵可外售，废水处理后可回用，有效解决了硫酸钠废水的利用难题，提高了工业用水循环利用水平。

参考文献

[1] ZHAO Kun, GAO Feng, YANG Qunying. Comprehensive review on metallurgical upgradation processes of nickel sulfide ores[J]. Journal of Sustainable Metallurgy, 2022, 8(1)：37-50.

[2] 张振芳, 陈秀法, 李仰春, 等."双碳"目标下镍资源的综合利用发展趋势[J]. 矿产综合利用, 2022(2)：31-39.

[3] 司俊起, 赵云, 王传强. 吉恩镍业转炉吹炼生产高冰镍生产实践[J]. 中国有色冶金, 2019, 48(6)：30-33.

[4] 贾玉斌, 张霜华. 喀拉通克金属化高冰镍吹炼及水淬工业化的回顾与探讨[J]. 新疆有色金属, 2003, 26(2)：23-25, 27.

[5] 刘燕庭, 许怀军. 富氧侧吹熔池熔炼铜镍矿[J]. 中国有色冶金, 2011, 40(5)：12-14.

[6] 武兵强, 齐渊洪, 周和敏, 等. 红土镍矿火法冶炼工艺现状及进展[J]. 矿产综合利用, 2020(3)：78-83, 93.

[7] WANG Hongyang, HOU Yongchang, CHANG Heqiang, et al. Preparation of Ni–Fe–S matte from nickeliferous laterite ore using CaS as the sulfurization agent[J]. Metallurgical and Materials Transactions B-Process Metallurgy and Materials Processing Science, 2022, 53(2)：1136-1147.

[8] 许欣. 红土镍矿火法冶炼制备高镍锍工艺及关键设备研发方向展望[J]. 有色设备, 2022, 36(5)：28-32.

[9] 钟清慎, 贺秀珍. 镍精炼工艺比较及发展方向[J]. 有色冶金节能, 2021, 37(3)：4-10.

[10] 李勇, 丁剑, 林洁媛, 等. 含镍硫化物湿法冶金技术应用及研究进展[J]. 中国有色冶金, 2020, 49(5)：9-15.

[11] 蒋开喜, 王玉芳, 郑朝振, 等. 硫化镍加压浸出研究进展与应用[J]. 矿冶, 2018(3)：45-50.

[12] 谢铿, 王海北, 马育新, 等. 高冰镍浸出渣冶金过程多金属走向行为探究[J]. 有色金属(冶炼部分), 2021(12)：20-27.

[13] 陈廷扬. 技术创新, 完善工艺, 提高企业经济效益——阜康冶炼厂铜渣处理工艺简介[J]. 新疆有色金属, 2002, 25(4)：27-29.

[14] 李春雷. 阜康冶炼厂高镍锍湿法精炼工业化回顾[J]. 中国有色冶金, 2007, 36(3)：20-24.

[15] 王春海.浅谈阜康冶炼厂体系降钠的摸索与实践[J].新疆有色金属,2014,37(4):75-76.
[16] 贺来荣,苏俊敏,赵秀丽,等.铜镍合金选择性两段加压浸出试验研究[J].中国有色冶金,2023,52(2):157-163.
[17] 王鹏伟.金属化高冰镍硫酸选择性浸出工艺中酸铜平衡及生产实践[J].新疆有色金属,1995(4):31-35.
[18] 蒋应平,马育新,苏立峰,等.红渣浸出工艺优化及工业试验验证研究[J].中国资源综合利用,2021,39(11):12-16.

图书在版编目(CIP)数据

典型冶金固废有价金属清洁提取与资源化利用 / 郭学益等著. —长沙：中南大学出版社，2023.12
ISBN 978-7-5487-5632-3

Ⅰ.①典… Ⅱ.①郭… Ⅲ.①冶金工业—固体废物处理—研究 Ⅳ.①X756.5

中国国家版本馆 CIP 数据核字(2023)第 227420 号

典型冶金固废有价金属清洁提取与资源化利用
DIANXING YEJIN GUFEI YOUJIA JINSHU QINGJIE TIQU YU ZIYUANHUA LIYONG

郭学益　王亲猛　田庆华　谢铿　管建红　徐志峰　著

□出版人	林绵优
□责任编辑	史海燕
□责任印制	李月腾
□出版发行	中南大学出版社
	社址：长沙市麓山南路　　邮编：410083
	发行科电话：0731-88876770　　传真：0731-88710482
□印　装	湖南省众鑫印务有限公司
□开　本	787 mm×1092 mm 1/16　□印张 48.75　□字数 1450 千字
□版　次	2023 年 12 月第 1 版　□印次 2023 年 12 月第 1 次印刷
□书　号	ISBN 978-7-5487-5632-3
□两册定价	328.00 元

图书出现印装问题，请与经销商调换

典型冶金固废有价金属清洁提取与资源化利用

Clean Extraction and Resource Utilization of
Valuable Metals from Typical Metallurgical Solid Waste

郭学益　王亲猛　田庆华　谢铿　管建红　徐志峰　著
Guo Xueyi　Wang Qinmeng　Tian Qinghua　Xie Keng　Guan Jianhong　Xu Zhifeng

下册

·长沙·

前言

冶金固废产生量大、资源化利用率低是制约有色金属行业可持续发展的瓶颈问题，冶金固废中有价金属清洁提取和资源化利用是实现固废"减量化、资源化、无害化"的必然要求和有效途径，是有色金属冶金重要发展方向。

镍、钴、钨、锑是支撑国防建设和经济发展的重要战略金属，但其矿产资源复杂，有价金属与有害元素伴生。由于缺乏固废源头减量及全过程控制技术，镍、钴、钨、锑金属提取过程易产生大量多源固废，如红土镍矿冶金渣、硫基镍钴渣、钨碱煮渣、锑砷碱渣等，导致冶炼过程有价/有害组元分散、存在潜在环境风险；同时，常规镍、钴、钨、锑冶金固废处理工艺侧重于有价金属提取，对有害组分迁移及演变规律认知不深，无法解决有害物质深度脱除与安全处置难题，导致处理过程存在流程长、金属回收率低、二次污染大等问题。因此，亟需开发典型冶金固废有价金属清洁提取与资源化利用关键技术，以实现固废源头减量、无害化处置和有价金属清洁高效提取，促进我国镍、钴、钨、锑战略金属产业可持续发展。

本书作者及其团队一直从事固废资源化领域人才培养、科学研究和工程实践工作，承担了国家重点研发计划项目"镍钴/钨/锑战略金属冶金固废清洁提取与无害化技术"，提出了固废源头减量-有价金属深度提取-有害组分安全处置-尾渣资源化利用系统性解决新思路，从科学理论-关键技术-集成示范全创新链开展研究。通过技术集成及工艺优化，建立工程示范，形成可推广应用技术体系。以该研究项目为主体，作者将最新研究成果归纳总结形成此书。

本书涵盖了典型冶金固废有价金属清洁提取与资源化利用，内容丰富，创新性强，反映了作者及其科研团队近年来在有色固废处置领域的学术成就和研究成果，是团队集体智慧的结晶。

全书由郭学益、王亲猛、田庆华、谢铿、管建红和徐志峰牵头，共同完成了本书研究内容。本书是作者研究团队集体智慧的结晶，研究过程中得到了中南大学、锡矿山闪星锑业有限责任公司、北京科技大学、江西理工大学、北京矿冶科技集团有限公司、中国恩菲工程技术有限公司、中冶瑞木新能源科技有限公司、江西钨业控股集团有限公司、新疆新鑫矿业股份有限公司和长沙矿冶研究院有限责任公司等单位的支持。对本书出版做出贡献的还有金贵忠、廖春发、黄凯、王雪亮、严康、李忠岐、马育新、李渭鹏、卢东昱、张金祥、朱红斌、黎敏、霍广生、雷湘、李玉虎、金承永、姚芾、田磊、苏华、李国、董波、张加美、王青鹜、王松松、陈远林、李中臣、田苗、黄柱、邓卫华、黄少波、杨必文、邓彤、谢岁、朱佳俊、刘召波、吕东等人，对他们的支持表示感谢。

由于作者水平有限，书中难免有疏漏和不妥之处，敬请批评指正。

Contents

Tungsten and Arsenic Substance Flow Analysis of a Hydrometallurgical
    Process for Tungsten Extracting from Wolframite .... (1)

Copper and Arsenic Substance Flow Analysis of Pyrometallurgical Process for
    Copper Production .... (22)

Physicochemical and Environmental Characteristics of Alkali Leaching Residue of
    Wolframite and Process for Valuable Metals Recovery .... (41)

Antimony and Arsenic Substance Flow Analysis in Antimony Pyrometallurgical Process
    .... (57)

A Sustainable Process for Tungsten Extraction from Wolframite Concentrate .... (79)

Sustainable Extraction of Tungsten from the Acid Digestion Product of Tungsten Concentrate
    by Leaching-solvent Extraction Together with Raffinate Recycling .... (100)

Wolframite Concentrate Conversion Kinetics and Mechanism in Hydrochloric Acid ...... (119)

Analysis of Antimony Sulfide Oxidation Mechanism in Oxygen-Enriched Smelting Furnace
    .... (136)

Comparative Atmospheric Leaching Characteristics of Scandium in Two Different Types
    of Laterite Nickel Ore from Indonesia .... (150)

Leaching Behavior of Scandium from Limonitic Laterite Ores Under Sulfation
    Roasting-water Leaching .... (174)

The Effect of Pre-roasting on Atmospheric Sulfuric Acid Leaching of Saprolitic Laterites
    .... (193)

Selective Recovery of Gold from Dilute Aqua Regia Leachate of Waste Printed Circuit
    Board by Thiol-modified Garlic Peel .... (209)

Thermodynamic and Experimental Analyses of the Carbothermic Reduction of
　　Tungsten Slag ··· (231)

Removal and Recovery of Arsenic(Ⅲ) from Hydrochloric Acid Leach Liquor of
　　Tungsten Slag by Solvent Extraction with 2-ethylhexanol ················ (242)

Separation of Molybdenum and Tungsten from Iron in Hydrochloric-phosphoric Acid
　　Solutions Using Solvent Extraction with TBP and P507 ···················· (255)

Process and Mechanism Investigation on Comprehensive Utilization of
　　Arsenic-Alkali Residue ·· (268)

The Catalytic Aerial Oxidation of As(Ⅲ) in Alkaline Solution by Mn-loaded Diatomite
　　·· (285)

Catalytic Oxidation Effect of $MnSO_4$ on As(Ⅲ) by Air in Alkaline Solution ··············· (301)

Enhanced Removal of Organic Matter from Oxygen-pressure Leaching Solution by
　　Modified Anode Slime ·· (320)

Thermodynamic Analysis and Process Optimization of Zinc and Lead Recovery from
　　Copper Smelting Slag with Chlorination Roasting ························· (339)

A Method of High-quality Silica Preparation from Copper Smelting Slag ················ (358)

Recovery of Zinc and Lead from Copper Smelting Slags by Chlorination Roasting ········· (373)

Recovery of Copper, Lead and Zinc from Copper Flash Converting Slag by the
　　Sulfurization-reduction Process ·· (387)

Mechanism and Kinetics for the Carbothermal Reduction of Arsenic-alkali Mixed Salt
　　Produced from Treatment of Arsenic-alkali Residue ······················· (404)

Valuable Metals Substance Flow Analysis in High Pressure Acid Leaching Process of Laterites
　　·· (421)

Tungsten and Arsenic Substance Flow Analysis of a Hydrometallurgical Process for Tungsten Extracting from Wolframite

Abstract: In this study, the metabolism of a hydrometallurgical process for tungsten extracting from wolframite was studied through substance flow analysis. The mass balance accounts, substance flow charts of tungsten and arsenic were established to evaluate the metabolism efficiency of the investigated system. The results showed that, the total tungsten resource efficiency of the system was 97.56%, and the tungsten recovery of unit process autoclaved alkali leaching, ion exchange, Mo removing, concentration and crystallization was 98.16%, 98.94%, 99.71%, 99.89%, respectively. Meanwhile, for extracting 1 ton of tungsten into the qualified ammonium paratungstate, 10.0414 kg of arsenic was carried into the system, with the generation of 7.2801 kg of arsenic in alkali leaching residue, 1.5067 kg of tungsten in arsenic waste residue, and 1.2312 kg of tungsten in Mo residue. Besides, 7.9 g of arsenic was discharged into nature environment with waste water, 15.5 g of arsenic was entrained into the final APT. The distribution and transformation behaviors of arsenic during production were analyzed through phases change analysis, and some recommendations for improving the resource efficiency of tungsten and pollution control during production were also proposed based on the substance flow analysis in this study.

Keywords: Tungsten; Arsenic; Substance flow analysis; Resource efficiency; Pollution control

1 Introduction

Tungsten is a kind of important strategic metal as it is widely used in many fields such as machinery manufacturing, chemical industry, aerospace industry and national defense industry[1].

Published in *Tungsten*, 2021, 3(3): 348–360. Authors: Chen Yuanlin, Guo Xueyi, Wang Qinmeng, Tian Qinghua, Huang Shaobo, Zhang Jinxiang.

With the continued exploitation of high-grade tungsten resource, the content of impurity elements in tungsten ores becomes higher and higher[2, 3]. For example, arsenic commonly occurs in wolframite concentrate, while it is strictly controlled in tungsten products[4]. Moreover, the emission of arsenic into nature environment during tungsten extraction process is a serious environmental issue[5, 6]. At present, scholars have done a great amount of researches on the efficient extraction of tungsten in a unit process, including tungsten minerals decomposition[7, 8], purification and transition of leaching solution[9, 10]. However, few researches focus on the metabolism of tungsten or other detrimental elements within a production process. Taking deep insight into the metabolism of tungsten and other key elements in a production system could provide a meaningful reference for process improvement and pollution control.

Substance flow analysis (SFA), a practical analytical tool, has been employed widely for studying the cycle and metabolism of a specific substance in a given system, tracking the released substances and then evaluating the effects of substance metabolism on economy and environment[11, 12]. In the past decades, SFA has been applied widely especially for analyzing the stock and consumption flows of metals, such as lead, zinc, copper, nickel, and other substances in a scope of local region[13, 14], country[15-19], or global[20, 21]. Recently, SFA was also applied as a supported tool in mineral resources management[22, 23], industry chain evaluation[24], waste management[25], environmental risk assessment and control[26, 27], and sustainable development assessment[28]. However, most of the previous studies mainly focused on the metals stock and potential waste generation in a system based on the statistics of production and consumption amount or analyzing the environmental impacts in the long life cycle of objective substance on a perspective of large system.

Applying SFA to trace the substance distribution and migrating in a production process could evaluate the metabolism efficiency of the process and reveal the pathways through which pollutants are generated, which is beneficial to the resource management and pollution control of a factory. Yoshida et al.[29] studied the fate of total organic carbon, 32 elements and four groups of organic pollutants in a conventional wastewater treatment plant through SFA, and provided an assessment on the treatment efficiency and environmental impact of the plant. Bai et al.[12] applied SFA to calculate lead mass balance in each stage of a lead smelting process, and the indicators such as waste circulation rate and resource efficiency were used to evaluate the metabolism efficiency of the system, through which some recommendations on improving emission control and pollution prevention for the lead smelting factory were put forward. To our knowledge, there is no report of the SFA on the metallurgical processes or life cycle of tungsten in current.

In this study, SFA was used as an analytical tool to investigate the metabolism of a hydrometallurgical process for producing ammonium paratungstate (APT) from wolframite. Tungsten and arsenic were selected as the objective substances, and then the mass balance accounts, substance flow charts of tungsten and arsenic in the production system were established. The metabolism efficiency of the system was evaluated by indicators including tungsten recovery of the main unit extraction processes, waste recycle ratio, resource efficiency, and the distributing and transforming behaviors of arsenic in production were also analyzed. This study focuses on the element metabolism efficiency from a microcosmic perspective, providing a reference for process improvement.

2 Calculation methodology

Compared with applying SFA on a large system with the scope of global or regional scale, applying SFA to a production process is more microcosmic and more specific. The SFA model of a production process always includes the substance flow of a unit process and the whole production system which is composed of several unit processes[7].

2.1 Definition of substance flows in a unit process

If we define one unit process in the production system as process j, for a continuous production system, the substance flow input to and output from process j are defined as follows:

(1) Input raw material flow, A_j.

(2) Input upstream product substance flow, P_{j-m}.

(3) Recycle substance flow from downstream process i, $R_{i,j}$. The substance flow which is recycled to upstream process k from process j is called $R_{j,k}$.

(4) Emission substance flow, E_j. It includes the by-products and pollutants which are discharged outside the production system from process j.

(5) Output product substance flow, P_j. The substance flow which is the by-product of process j and translated to process k for further treating is called $P_{j,k}$.

(6) If some products are stocked in warehouse temporarily during production, the stock substance flow, S_j, should be taken into consideration. In this study, no product is stored temporarily, and the stock substance flow S_j is not under calculation.

According to the conservation of mass, for the unit process j, the substance flows can be calculated as:

$$A_j + P_{j-m} + P_{i,j} + R_{i,j} = P_j + P_{j,k} + R_{j,k} + E_j \tag{1}$$

2.2 Substance flows of a whole system

If all the unit processes are combined, the whole system is formed, and the substance flows of the whole system are the sums of corresponding flows in all unit processes, as follows:

(1) Input raw material flow A is expressed as

$$A = \sum_{j=1}^{m} A_j \tag{2}$$

(2) Recycle substance flow R is expressed as

$$R = \sum_{j=1}^{m} \sum_{i=1}^{m} R_{i,j} \tag{3}$$

(3) Emission substance flow E is expressed as

$$E = \sum_{j=1}^{m} E_j \qquad (4)$$

(4) Output product substance flow P is expressed as

$$P = \sum_{j=1}^{m} P_j \qquad (5)$$

Correspondingly, according to the conservation of mass, for the whole system, it can be expressed as

$$\sum_{j=1}^{m} M_{j,\,\text{input}} = \sum_{j=1}^{m} M_{j,\,\text{output}} \qquad (6)$$

2.3 Mass balance calculation of a specific element

To analyze the substance flow of a specific element, two parts of data are required. One part is the flow quantity M_j, which is the amount of each material which contains the objective element, and this part of data is always collected from the daily production reports of plant. The other part is the content C_j of objective element in each material, and this part of data is obtained by sampling and analyzing for each material. Then the flow quantity of an objective element m_j is calculated as

$$m_j = M_j \times C_j \quad j = 1, 2, \cdots, m \qquad (7)$$

In this study, the mass balance calculation of unit processes and the whole system is based on one ton of tungsten output from the production system, and the flow ratio of the objective element in each material is expressed as f_j

$$f_j = m_j / m_m \qquad (8)$$

where m_j is the quantity of the objective element in the substance flow j, and m_m is the quantity of the objective element in final product output from the production system. The unit of f_j (t/t or kg/t) means the quantity of an objective element in each substance flow for per ton of tungsten in APT.

2.4 Evaluation indicators of SFA

To evaluate the production efficiency of the process and its influence on environment, three indicators are proposed in this study as follows.

(1) Tungsten recovery of a unit tungsten extraction process, γ, and the proportion of tungsten in product flow to the total output flows of a unit process, %. For a unit process j, γ_j is calculated as

$$\gamma_j = P_j / (P_j + P_{j,k} + R_{j,k} + E_j) \times 100\% \qquad (9)$$

(2) Waste recycle ratio of the whole process, α, and the proportion of recycle substance flows in all the substance flows not included in the final product of the whole process, %:

$$\alpha = R / (R + E) \times 100\% \qquad (10)$$

(3) Resource efficiency, ε, and the proportion of objective element in final product to the

total input raw material flows (wolframite concentrate in this study), %:
$$\varepsilon = P/A \times 100\% = (A - E)/A \times 100\% \tag{11}$$

3 Description of the system boundary

In this study, a hydrometallurgical process for producing APT from wolframite is chosen as system boundary. The process is now in operation in a tungsten products production enterprise in Jiangxi province, China, with a production scale of 5000 tons APT per year. And this kind of process accounts for a large proportion of APT production capacity in China. The simplified flowsheet of objective process is shown in Fig. 1.

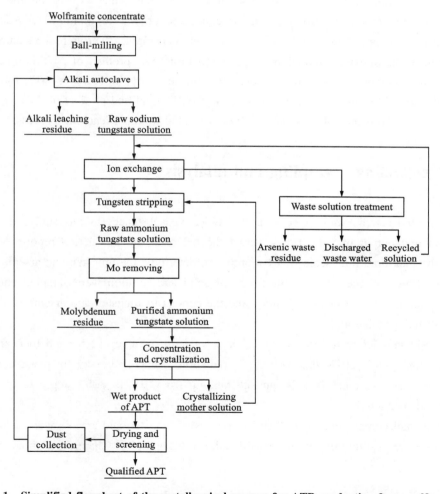

Fig. 1 Simplified flowsheet of the metallurgical process for ATP production from wolframite

The preprocessing for wolframite concentrate is first conducted through ball-milling, the slurry after grinding is transported into autoclaved alkali leaching process for tungsten minerals decomposition with NaOH solution, alkali leaching residue and raw sodium tungstate solution are separated from slurry. The alkali leaching residue is considered as hazardous waste according to *National Hazardous Waste Category (2016) of China*, and is always stockpiled in specialized warehouse. The raw sodium tungstate solution is diluted to a proper concentration for ion exchange process, the tungstate in prepared solution is extracted through ion exchange and is transformed into ammonium tungstate after stripping from tungsten loaded resin, and then raw ammonium tungstate solution is obtained. The waste solution from ion exchange process is further treated for recovering of residual tungsten and removal of arsenic before discharging, and the tungsten containing solution is recycled. Normally, there is a certain portion of Mo in the wolframite concentrate and is extracted into tungstate solution with tungsten due to their extremely similar chemical properties. For the purpose of reducing the effect of Mo on the quality of subsequent tungsten products, Mo removing is conducted on raw ammonium tungstate solution through sulfide precipitation, and Mo residue is generated. Through concentration and crystallization of the purified ammonium tungstate solution, wet product of APT is obtained, and the crystallizing mother solution are recycled to tungsten stripping process which follows ion exchange process. After drying and screening of wet APT, qualified product of APT is produced, the dust generated from screening is collected and recycled to alkali leaching process.

4 Data collecting, sampling and analysis

Tungsten and arsenic were selected as the objective substances in this study. The quantity data of input and output material flows were collected from the production reports of objective plant, and the measured time scale was 1 month. To determine the tungsten and arsenic content in each material flow, all the flows in the forms of solid and solution were sampled and analyzed every production day in the measured time, and the content of tungsten and arsenic were averaged for mass balance calculation.

Some solid samples were dried at 50 ℃ in a vacuum oven for 12 h, and then analyzed by X-ray diffraction (XRD, TTR Ⅲ, with Cu K_α radiation, Rigaku, Japan) for phase compositions analysis. The operating conditions is with 2θ ranging from 10° to 80°, step size of 0.02 and scanning speed of 2°/min.

The elemental compositions and chemical phases of tungsten and arsenic were determined by Changsha Research Institute of Mining and Metallurgy Co., Ltd in Changsha, China.

5 Results and discussion

5.1 Substance flow analysis of tungsten

5.1.1 Mass balance calculation

Based on the flows quantities and analysis data, the mass balance account of tungsten in all unit processes of the production system is built upon 1 ton of tungsten extracted into the qualified APT. The input and output flow ratios of tungsten in each substance flow are shown in Table 1.

Table 1　Input and output flow ratios of tungsten in each substance flow of the production system

No.	Process unit	Input			Output		
			Substance flow	Flow ratio /(t·t^{-1})		Substance flow	Flow ratio /(t·t^{-1})
1	Ball-milling	A_1	Wolframite concentrate	1.0250	P_1	Slurry	1.0250
		Total input		1.0250	Total output		1.0250
2	Autoclaved alkali leaching	P_1	Slurry	1.0250	E_2	Alkali leaching residue	0.0189
		$R_{9,2}$	Recycled dust	0.0014	P_2	Raw sodium tungstate solution	1.0075
		Total input		1.0264	Total output		1.0264
3	Ion exchange	P_2	Raw sodium tungstate solution	1.0075	P_3	Tungsten-loaded resin	1.0021
		$R_{4,3}$	Recycled solution	0.0053	$P_{3,4}$	Waste solution	0.0107
		Total input		1.0128	Total output		1.0128
4	Waste solution treatment	$P_{3,4}$	Waste solution	0.0107	$R_{4,3}$	Recycled solution	0.0053
					E_{4-1}	Arsenic waste residue	0.0038
					E_{4-2}	Discharged waste water	0.0016
		Total input		0.0107	Total output		0.0107
5	Tungsten stripping	P_3	Tungsten-loaded solution	1.0021	P_5	Raw ammonium tungstate solution	1.0036

Continue the Table 1

No.	Process unit	Input		Output		
		Substance flow	Flow ratio /(t·t^{-1})		Substance flow	Flow ratio /(t·t^{-1})
5	Tungsten stripping	$R_{7,5}$	Crystallizing mother solution	0.0015		
		Total input		1.0036	Total output	1.0036
6	Molybdenum removing	P_5	Raw ammonium tungstate solution	1.0036	E_6	Molybdenum residue	0.0007
					P_6	Purified ammonium tungstate solution	1.0029
		Total input		1.0036	Total output	1.0036
7	Concentration and crystallization	P_6	Purified ammonium tungstate solution	1.0029	P_7	Wet product of APT	1.0014
					$R_{7,5}$	Crystallizing mother solution	0.0015
		Total input		1.0029	Total output	1.0029
8	Drying and screening	P_7	Wet product of APT	1.0014	P_8	Qualified APT	1.0000
					$P_{8,9}$	Dust	0.0014
		Total input		1.0014	Total output	1.0014
9	Dust collection	$P_{8,9}$	Dust	0.0014	$R_{9,2}$	Recycled dust	0.0014
		Total input		0.0014	Total output	0.0014

5.1.2 Substance flow chart of tungsten

The substance flow chart of tungsten, shown in Fig. 2, is plotted based upon tungsten substance flow data and the tungsten production process. The production system includes 9 unit processes and there are 17 strands of substance flows within the whole system, and each of the substance flows has been identified with name and flow code as shown in Table 1.

5.1.3 Evaluation of the tungsten metallurgical process

The evaluation results of the production system with the defined indicators (Sect. 2.4) are presented in Table 2.

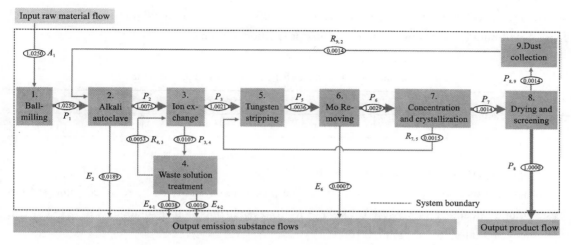

Fig. 2 Tungsten substance flow chart of the investigated production system (t/t)

Table 2 Evaluation results of the tungsten metallurgical system

No.	Evaluation indicator	Unit	Result
1	Input raw material flow for per ton of copper, A	t/t	1.0250
2	Cycle substance flow for per ton of copper, R	t/t	0.0082
3	Emission substance flow for per ton of copper, E	t/t	0.0250
4	Tungsten recovery of autoclaved alkali leaching process, γ_2	%	98.16
5	Tungsten recovery of ion exchange process, γ_4	%	98.94
6	Tungsten recovery of Mo removing process, γ_7	%	99.71
7	Tungsten recovery of concentration and crystallization process, γ_8	%	99.89
8	Resource efficiency, ε	%	97.56

The tungsten recovery of the main tungsten extraction processes including autoclaved alkali leaching, ion exchange, Mo removing, concentration and crystallization are evaluated. As can be seen, the tungsten recovery of these processes is 98.16%, 98.94%, 99.71%, and 99.89%, respectively. The high tungsten extracting efficiency of these unit processes contributes to a high resource efficiency of the whole system, 97.56%. The waste recycle ratio of the production system is 24.70%. In other words, 24.70% of the tungsten loss in rejected products during production is recycled to the system. The details of the defined substance flows are discussed in the following sections.

(1) Input substance flows of tungsten, A

As shown in Table 1, the total input substance flow is 1.0250 t/t, indicating that for extracting 1 ton of tungsten into the final product (qualified APT), 1.0250 tons of pure tungsten contained in wolframite concentrate is consumed. Table 3 shows the contents of main elements in wolframite concentrate. The tungsten content is 42.10% (the content of WO_3 is 53.08%)—2.4347 tons of wolframite concentrate is put into this production system.

Table 3 Contents of the main elements in wolframite concentrate

Element	W	Fe	Mn	Ca	As
Content/%	42.10	7.60	6.88	1.28	0.41

(2) Recycle substance flows of tungsten, R

There are three recycle substance flows in the production system: $R_{4,3}$, recycled solution from waste solution treatment process, (5.3 kg/t); $R_{7,5}$, crystallizing mother solution, (1.5 kg/t); and $R_{9,2}$, recycled dust, (1.4 kg/t).

As a whole, the recycle substance flow is (8.2 kg/t), accounts for 24.70% of the tungsten loss in rejected products which consist of recycle and emission substance flows. The high tungsten extracting efficiency of the unit processes is a main reason for the low recycle ratio. Among the recycle substance flows, recycled solution from the waste solution treatment process accounts for the largest portion, 64.63%.

(3) Emission substance flows of tungsten, E

The emission substance flows consist of the alkali leaching residue (18.9 kg/t), arsenic waste residue (3.8 kg/t), discharged waste water (1.6 kg/t), and Mo residue (0.7 kg/t). These flows are discharged outside the production system for further treatment, but not into the nature environment, except for the treated waste water that meets emissions standard.

The alkali leaching residue, accounting for 75.60% of all the emission substance flows, is the main emission substance. According to the main elemental compositions of alkali leaching residue (Table 4), the tungsten content is 1.82%. The result indicates that, for extracting 1 ton of tungsten into qualified APT, 1.0385 tons of residue is generated from autoclaved alkali leaching process. Table 5 shows the chemical phase analysis of tungsten in wolframite concentrate and alkali leaching residue. The result shows that, 90.83% of residual tungsten in leaching residue occurs as scheelite, and the leaching efficiency of scheelite occurring in concentrate should be improved for a higher resource efficiency. Moreover, the contents of some valuable metals, such as W, Sn, and Nb are higher than that of raw ore, and the resource value of the residue is fairly high.

Table 4 Main elemental compositions of alkali leaching residue

Element	W	Mn	Fe	Sn	Nb	Bi
Content/%	1.82	12.48	16.12	0.74	0.37	0.54
Element	Mo	As	Pb	Al	Ca	Si
Content/%	0.038	0.71	0.32	1.16	12.43	5.14

Table 5 Chemical phase analysis of tungsten in wolframite concentrate and alkali leaching residue

Phase	Mass fraction/%	
	Wolframite concentrate	Alkali leaching residue
Wolframite	76.87	3.49
Scheelite	21.78	90.83
Tungstite	1.36	5.68

However, the residue also contains some toxic elements (mainly As and Pb), and is classified as hazardous waste according to *National Hazardous Waste Category* (2016) *of China*. Therefore, the alkali leaching residue possesses dual properties of resource value and environmental pollution risk. Clean extraction of valuable metals from the residue and the safe disposal is in great necessity, which will also be presented in our later works.

The tungsten loss in arsenic waste residue and discharged waste water accounts for 50.47% of the tungsten in waste solution from ion exchange, in other words, only 49.53% of tungsten in waste solution is recovered. The low concentration of tungsten in waste solution after ion exchange is the main reason for the difficulty in recovering. Thus, improving the selectivity extraction for tungsten in ion exchange and waste solution treatment process is the key measure to reduce the loss of tungsten.

Mo residue accounts for a small portion of emission flows, 5.60%. The molybdenum content of this residue is high, and therefore it is used as raw material for molybdenum metallurgy.

5.2 Substance flow analysis of arsenic

As arsenic is considered as a hazardous element entrained into the system rather than valuable element, the evaluation indicators defined in Sect. 2.4 is not conducted for arsenic substance flows. However, for the purpose of giving a deep insight into the arsenic migrating and translating during production, the distribution and transformation behaviors of arsenic in the main tungsten extraction and impurity removing processes (autoclaved alkali leaching, ion exchange, waste solution treatment, Mo removing, concentration and crystallization) are analyzed based on arsenic substance flow analysis and the phase analysis of some products.

5.2.1 Mass balance calculation

The mass balance account of arsenic in all unit processes of the system is also built upon extracting 1 ton of tungsten output from the system. The flow ratio of each substance flow is shown in Table 6.

Table 6 Input and output flow ratios of arsenic in each substance flow of the production system

No.	Process unit	Input			Output		
		Substance flow		Flow ratio /(t·t^{-1})	Substance flow		Flow ratio /(t·t^{-1})
1	Ball-milling	A_1	Wolframite concentrate	10.0414	P_1	Slurry	10.0414
		Total input		10.0414	Total output		10.0414
2	Autoclaved alkali leaching	P_1	Slurry	10.0414	E_2	Alkali leaching residue	7.2801
		$R_{9,2}$	Recycled dust	0.0001	P_2	Raw sodium tungstate solution	2.7614
		Total input		10.0415	Total output		10.0415
3	Ion exchange	P_2	Raw sodium tungstate solution	2.7614	P_3	Tungsten-loaded resin	1.2468
		$R_{4,3}$	Recycled solution	0.0065	$P_{3,4}$	Waste solution	1.5211
		Total input		2.7679	Total output		2.7679
4	Waste solution treatment	$P_{3,4}$	Waste solution	1.5211	$R_{4,3}$	Recycled solution	0.0065
					E_{4-1}	Arsenic waste residue	1.5067
					E_{4-2}	Discharged waste water	0.0079
		Total input		1.5211	Total output		1.5211
5	Tungsten stripping	P_3	Tungsten-loaded resin	1.2468	P_5	Raw ammonium tungstate solution	1.2914
		$R_{7,5}$	Crystallizing mother solution	0.0446			
		Total input		1.2914	Total output		1.2914
6	Molybdenum removing	P_5	Raw ammonium tungstate solution	1.2914	E_6	Molybdenum residue	1.2312
					P_6	Purified ammonium tungstate solution	0.0602
		Total input		1.2914	Total output		1.2914
7	Concentration and crystallization	P_6	Purified ammonium tungstate solution	0.0602	P_7	Wet product of APT	0.0156
					$R_{7,5}$	Crystallizing mother solution	0.0446
		Total input		0.0602	Total output		0.0602

Continue the Table 6

No.	Process unit	Input		Flow ratio /(t · t^{-1})	Output		Flow ratio /(t · t^{-1})
		Substance flow			Substance flow		
8	Drying and screening	P_7	Wet product of APT	0.0156	P_8	Qualified APT	0.0155
					$P_{8,9}$	Dust	0.0001
		Total input		0.0156	Total output		0.0156
9	Dust collection	$P_{8,9}$	Dust	0.0001	$R_{9,2}$	Recycled dust	0.0001
		Total input		0.0001	Total output		0.0001

5.2.2 Substance flow chart of arsenic

The substance flow chart of arsenic based on substance flow data and the production process is shown in Fig. 3. All the arsenic substance flows are corresponding to the tungsten substance flows in Fig. 2. The total quantity of each arsenic substance flow in the system is calculated as shown in Table 7, and the details of each substance flow are discussed in the following sections.

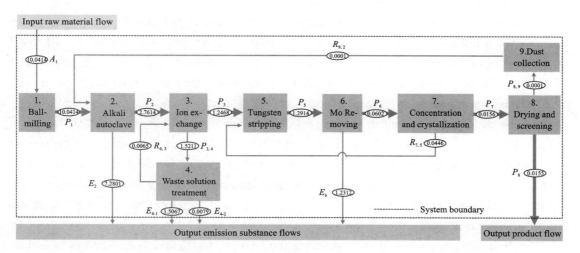

Fig. 3 Arsenic substance flow chart of the investigated production system (kg/t)

Table 7 The total quantity of arsenic in each substance flow of the production system

No.	Substance flow	Unit	Result
1	Input substance flows for per ton of tungsten, A	kg/t	10.0414
2	Recycle substance flows for per ton of tungsten, R	kg/t	0.0512
3	Emission substance flows for per ton of tungsten, E	kg/t	10.0259

(1) Input substance flows of arsenic

The total input substance flows of arsenic is 10.0414 kg/t tungsten, which indicates that extracting 1 ton of tungsten into the qualified APT (10.0414 kg) of arsenic is carried into the production system. All the input arsenic is carried by wolframite concentrate, and the arsenic content of wolframite concentrate is 0.41% (Table 3), which belongs to the kind of high arsenic wolframite concentrate. This result also reveals that, with the tungsten resources being complex globally, the amount of arsenic input to tungsten production system increases so that clarifying the evolution behavior of arsenic in tungsten extracting process is very necessary.

(2) Recycle substance flows of arsenic

Arsenic is recycled to the system with three flows: $R_{4,3}$, recycled solution; $R_{7,5}$, crystallizing mother solution; and $R_{9,2}$, recycled dust. The total recycle arsenic substance flow is 51.2 g/t tungsten, accounting for a very small portion (0.51%) of the arsenic distributes in rejected products (including alkali leaching residue 7.2801 kg/t), waste solution 1.5211 kg/t, Mo residue 1.2312 kg/t, crystallizing mother solution 44.6 g/t, and a small amout of recycled dust 0.1 g/t). The low recycle ratio of arsenic attributes to the high efficiency of impurity removing processes and helps to avoid the accumulation of arsenic in the production system and improve the quality of APT.

(3) Emission substance flows of arsenic

Most of arsenic carried into the production system is discharged outside the system in the forms of alkali leaching residue E_2 (7.2801 kg/t tungsten), arsenic waste residue E_{4-1} (1.5067 kg/t), and Mo residue E_6 (1.2312 kg/t). Besides, a small amount of arsenic is in waste water (7.9 g/t). These flows are discharged outside the production system for further treatment through other processes which are not included in the discussed production system.

Among these emissions, alkali leaching residue accounts for 72.61%, and the arsenic content in the residue is as high as 0.71% (Table 4), and Pb (0.32%) is also contained. As the residue is specified as hazardous waste, it can only be stored in special warehouse. After decades of production, the stock of this residue is in a very large amount. What is worse, the slow releasing of the toxic elements is a huge threat to the environment and therefore the safe disposal of this residue is in great necessity.

The arsenic waste residue from waste solution treatment, which is also specified as hazardous waste, accounts for 15.03% of the emission substance flows. The stabilization level of arsenic in this residue should be assessed and some safe disposal measures should be carried out for the purpose of minimizing the risk of environmental pollution.

Some arsenic is deposited with Mo in the form of Mo residue (12.28% of the emission substance of arsenic), and is transported to the Mo metallurgy process. The amount of arsenic discharged into nature environment with waste water is very small, 0.08%. However, the discharged arsenic with waste water is also a potential threat to environmental organisms in consideration of biological accumulation. Recycling the waste water to the leaching process as much as possible is a reasonable way.

5.3 The distribution and transformation behaviors of arsenic

The distribution of arsenic in the main tungsten extracting and impurity removing processes is shown in Table 8. The results indicate that, in the autoclaved alkali leaching process, 72.50% of the arsenic in wolframite concentrate stays in the alkali leaching residue, and the rest is extracted into raw sodium tungstate solution.

Table 8 Distribution of arsenic in the main tungsten extracting and impurity removing processes of the system

No.	Process	Product	Flow ratio /($kg \cdot t^{-1}$)	Distribution ratio/%
1	Autoclaved alkali leaching	Alkali leaching residue	7.2801	72.50
		Raw sodium tungstate solution	2.7614	27.50
2	Ion exchange	Tungsten-loaded resin	1.2468	45.04
		Waste solution	1.5211	54.96
3	Waste solution treatment	Arsenic waste residue	1.5067	99.05
		Discharged waste water	0.0079	0.52
		Recycled solution	0.0065	0.43
4	Mo removing	Molybdenum residue	1.2312	95.35
		Purified ammonium tungstate solution	0.0602	4.65
5	Concentration and crystallization	Wet product of APT	0.0156	25.91
		Crystallizing mother solution	0.0446	74.09

Slow scanning XRD was applied to analyzed the tungsten and arsenic phases in wolframite concentrate and alkali leaching residue, and the results were presented in Fig. 4. The main occurring forms of arsenic in the concentrate [Fig. 4(a)] are pararealgar (As_4S_4), arsenic sulfide (As_2S_3), arsenopyrite (FeAsS), lollingite ($FeAs_2$), iron arsenate ($FeAsO_4$), and arsenic oxide (As_2O_5). While the main occurring forms of arsenic in the leaching residue [Fig. 4(b)] are arsenic sulfide (As_2S_3), iron arsenate ($FeAsO_4$), ferrous arsenate ($Fe_3(AsO_4)_2$), and lollingite ($FeAs_2$). The chemical phase analysis for arsenic was also carried out, and the results were shown in Table 9. For the concentrate, the mass fraction of arsenic oxides, arsenic sulfides, arsenate, and insoluble arsenic compounds is 7.55%, 8.21%, 3.08%, and 81.16%, respectively. And for the leaching residue, the mass fraction of arsenic oxides, arsenic sulfides, arsenate, and insoluble arsenic compounds is 1.46%, 8.11%, 41.22%, and 49.21%, respectively. The insoluble arsenic compounds in chemical phase analysis may include arsenopyrite (FeAsS) and lollingite ($FeAs_2$). Compared to wolframite concentrate, the mass fraction of arsenate in residue increases significantly, while the mass fraction of arsenic oxides and insoluble arsenic compounds decrease sharply and the mass fraction of arsenic sulfides decrease slightly (considering that 27.50% of arsenic have been leached).

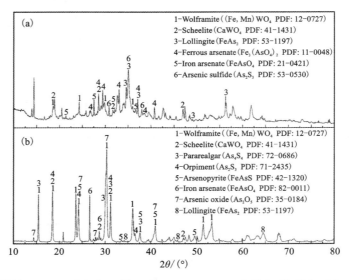

Fig. 4 XRD pattern of (a) wolframite concentrate and (b) alkali leaching residue

Table 9 Chemical phase analysis of the arsenic in wolframite concentrate and alkali leaching residue

Phase	Mass fraction/%	
	Wolframite concentrate	Alkali leaching residue
Arsenic oxides	7.55	1.46
Arsenic sulfides	8.21	8.11
Arsenate	3.08	41.22
Insoluble arsenic compounds	81.16	49.21
Total	100.00	100.00

During the leaching process with strong alkalinity, arsenic oxides and arsenate will be decomposed as follows:

$$2FeAsO_4 + 6NaOH = 2Na_3AsO_4 + Fe_2O_3 + 3H_2O \quad (12)$$

$$As_2O_5 + 6NaOH = 2Na_3AsO_4 + 3H_2O \quad (13)$$

Considering the presence of dissolved oxygen in the solution at the initial leaching stage, the following reaction may also occur:

$$As_2S_3 + 7O_2 + 12NaOH = 2Na_3AsO_4 + 3Na_2SO_4 + 6H_2O \quad (14)$$

$$As_4S_4 + 11O_2 + 20NaOH = 4Na_3AsO_4 + 4Na_2SO_4 + 10H_2O \quad (15)$$

$$2FeAsS + 7O_2 + 10NaOH = Fe_2O_3 + 2Na_3AsO_4 + 2Na_2SO_4 + 5H_2O \quad (16)$$

As a result, a part of arsenic is dissolved into raw sodium tungstate solution.

Due to the lack of oxygen and that some of arsenic minerals may be wrapped by other minerals, the reactions (12-16) may not proceed completely. In addition, the AsO_4^{3-} in solution may be deposited back into residue as follows (which may result in the increase of the mass fraction of arsenate in the leaching residue):

$$3Fe^{2+} + 2AsO_4^{3-} \rightleftharpoons Fe_3(AsO_4)_2 \qquad (17)$$

$$3Ca^{2+} + 2AsO_4^{3-} \rightleftharpoons Ca_3(AsO_4)_2 \qquad (18)$$

In the ion exchange process, 45.04% of the arsenic in sodium tungstate solution is entrained into tungsten-loaded resin, which leads to the dispersion of arsenic in molybdenum residue, and then increases the difficulty of arsenic treatment in subsequent process and also increase environmental pollution risk. Thus, improving the selectivity extraction of tungsten in ion exchange is necessary for reducing the entrainment of arsenic. The arsenic stays in waste solution after ion exchange is transported to waste solution treatment process, in which most of arsenic (99.05%) is removed into arsenic waste residue, 0.52% of arsenic is discharged into nature environment with water, and 0.43% of arsenic is recycled to the production system with recovered tungsten.

In Mo removing process, 95.35% of arsenic in raw ammonium tungstate solution is removed into Mo residue through sulfide precipitation. Fig. 5 shows the XRD pattern of molybdenum residue. The detected molybdenum phases incluing molybdenite (MoS_2) and copper molybdenum sulfide ($CuMo_2S_3$). The detected arsenic phases including arsenic sulfide (As_2S_3), enargite (Cu_3AsS_4), and sulfur (S). In this process, Mo has been proved to be precipitated through the following reactions[30, 31]:

$$MoO_4^{2-} + 4S^{2-} + 4H_2O \rightleftharpoons MoS_4^{2-} + 8OH^- \qquad (19)$$

$$MoS_4^{2-} + CuS \longrightarrow Cu_xMo_yS_z + S^{2-} \qquad (20)$$

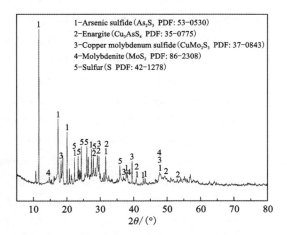

Fig. 5 XRD pattern of molybdenum residue

In addition to the reactions (19) and (20), as MoS_2 and S are also detected, another reaction may occur:

$$MoO_4^{2-} + 3S^{2-} + 4H_2O \longrightarrow MoS_2 + S + 8OH^- \qquad (21)$$

The peaks of molybdenum phases (including MoS_2 and $CuMo_2S_3$) in XRD pattern are fairly weak, which may be due to the low crystallinity of molybdenum phases.

According to the arsenic phases (As_2S_3 and Cu_3AsS_4) in Mo residue, it is speculated that, the AsO_4^{2-} in raw ammonium tungstate solution may be precipitated as follows:

$$AsO_4^{3-} + 4S^{2-} + 4H_2O \rightleftharpoons AsS_4^{3-} + 8OH^- \qquad (22)$$

$$AsS_4^{3-} + 3Cu^+ \longrightarrow Cu_3AsS_4 \qquad (23)$$

$$2AsO_4^{3-} + 5S^{2-} + 8H_2O \longrightarrow As_2S_3 + 2S + 16OH^- \qquad (24)$$

74.09% of arsenic in purified ammonium tungstate solution stays in mother solution during the concentration and crystallization process, and the rest small part is entrained into wet product of APT. The arsenic content in final APT product is 0.001%, and the product meets the APT-0 standard specified in GB/T 10116—2007 of China.

6 Conclusion and recommendations

(1) For the substance flow of tungsten, our results show that, the resource efficiency of tungsten for the discussed production system is 97.56%, and the tungsten recovery of autoclaved alkali leaching, ion exchange, Mo removing, concentration and crystallization is 98.16%, 98.94%, 99.71%, 99.89%, respectively.

For extracting 1 ton of tungsten into the qualified APT, 2.4347 tons of wolframite concentrate containing 42.10% W is consumed, 8.2 kg of extra tungsten recycles repeatedly within the system. Meanwhile, 18.9 kg of tungsten in alkali leaching residue, 3.8 kg of tungsten in arsenic waste residue, and 0.7 kg of tungsten in Mo residue are generated, and 1.6 kg of tungsten is discharged into nature environment with waste water.

(2) For the substance flow of arsenic, our results further reveal that, for extracting 1 ton of tungsten, 10.0414 kg of arsenic is carried into the system. 7.2801 kg of arsenic in alkali leaching residue, 1.5067 kg of arsenic in arsenic waste residue, and 1.2312 kg of arsenic in Mo residue are generated, and 7.9 g of arsenic is discharged into nature environment with waste water.

In autoclaved alkali leaching process, 27.50% of the arsenic is extracted into raw sodium tungstate solution, and most of the arsenic stays in the alkali leaching residue. In ion exchange process, 45.04% of the arsenic in sodium tungstate solution is entrained into tungsten-loaded resin; for the arsenic stays in waste solution after ion exchange, 99.05% of that is removed into arsenic waste residue by further treatment, 0.52% of that is dis-charged into nature environment with waste water, and 0.43% of that is recycled to the system with recovered tungsten. In Mo removing process, 95.35% of arsenic is removed into Mo residue through sulfide precipitation process, and only 4.65% of arsenic stays in purified ammonium tungstate solution. Finally, 25.91% of arsenic in purified ammonium tungstate solution is entrained into wet product of APT.

(3) Based on the substance flow analysis of the production system, some recommendations

for production improvement can be obtained:

The leaching efficiency of scheelite occurring in concentrate should be improved during autoclaved alkali leaching process to improve the resource efficiency of tungsten.

Alkali leaching residue is the main emission of the system which accounts for 75.60% of all the tungsten emissions and 72.61% of the arsenic emissions. On one hand, the resource value of this residue is fairly high for the contents of some valuable metals, such as W, Sn, and Nb. On the other hand, the residue contains some toxic elements (mainly As and Pb), and is classified as hazardous waste. Therefore, the safe disposal of this residue is in great necessity. The utilization and harmless disposal of the alkali leaching residue will also be investigated in our later works.

In the ion exchange process, 45.04% of the arsenic in sodium tungstate solution is entrained into tungsten-loaded resin. The result leads to the dispersion of arsenic in molybdenum residue, which means the difficulty of arsenic treatment in subsequent process is increased and a higher environmental pollution risk. In addition, as the content of arsenic is strictly controlled in tungsten products, the selectivity extraction of tungsten in ion exchange should be improved.

The stabilization level of arsenic waste residue from waste solution treatment should be assessed and some safe disposal measures should be carried out for the purpose of minimizing the risk of environmental pollution. In addition, the discharged arsenic with waste water is a potential threat to environmental organisms despite its small amount in consideration of biological accumulation, and recycling the waste water to the leaching process as much as possible is a reasonable way.

Acknowledgements

This paper was financially supported by the National Key R & D Program of China (Grant No. 2019YFC1907400) and the National Natural Science Foundation of China (Grant Nos. 51904351 and 51620105013).

References

[1] ZHAO Z W, XIAO L P, SUN F, et al. Study on removing Mo from tungstate solution by activated carbon loaded with copper[J]. Int J Refract Met Hard Mater, 2010, 28(4): 503.
[2] LI Y L, ZHAO Z W. Separation of molybdenum from acidic high-phosphorus tungsten solution by solvent extraction[J]. JOM, 2017, 69(10): 1920.
[3] YANG Y, XIE B Y, WANG R X, et al. Extraction and separation of tungsten from acidic high-phosphorus solution[J]. Hydrometallurgy, 2016, 164: 97.
[4] XIE H, ZHAO Z W, CAO C F, et al. Behavior of arsenic in process of removing molybdenum by sulfide method[J]. J Cent South Univ, 2012, 42(2): 435 (in Chinese).

[5] LI Y K, ZHU X, QI X J, et al. Removal and immobilization of arsenic from copper smelting wastewater using copper slag by in situ encapsulation with silica gel[J]. Chem Eng J, 2020, 394: 124833.

[6] OTONES V, ÁLVAREZ-AYUSO E, GARCÍA-SÁNCHEZ A, et al. Arsenic distribution in soils and plants of an arsenic impacted former mining area[J]. Environ Pollut, 2011, 159: 2637.

[7] LI J T, ZHAO Z W. Kinetics of scheelite concentrate digestion with sulfuric acid in the presence of phosphoric acid[J]. Hydrometallurgy, 2016, 163: 55.

[8] LI J T, MA Z L, LIU X H, et al. Sustainable and efficient recovery of tungsten from wolframite in a sulfuric acid and phosphoric acid mixed system[J]. ACS Sustain Chem Eng, 2020, 8: 13583.

[9] LIU X H, HU F, ZHAO Z W. Treatment of high concentration sodium tungstate with macroporous resin[J]. Chin J Nonferrous Met, 2014, 24(7): 1895 (in Chinese).

[10] YANG J L, CHEN X Y, LIU X H, et al. Separating W(Ⅵ) and Mo(Ⅵ) by two-step acid decomposition[J]. Hydrometallurgy, 2018, 179: 20.

[11] LOISEAU E, JUNQUA G, ROUX P, et al. Environmental assessment of a territory: An overview of existing tools and methods[J]. J Environ Manag, 2012, 112: 213.

[12] BAI L, QIAO Q, LI Y P. Substance flow analysis of production process: A case study of a lead smelting process[J]. J Clean Prod, 2015, 104: 502.

[13] LINDQVIST A, VON MALMBORG F. What can we learn from local substance flow analyses? The review of cadmium flows in Swedish municipalities[J]. J Clean Prod, 2004, 12(8-10): 909.

[14] HUANG C L, MA H W, YU C P. Substance flow analysis and assessment of environmental exposure potential for triclosan in mainland China[J]. Sci Total Environ, 2014, 499: 265.

[15] GUO X Y, ZHONG J Y, SONG Y, et al. Substance flow analysis of zinc in China[J]. Resour Conserv Recycl, 2010, 54: 171.

[16] LIN S H, MAO J S, CHEn W Q, et al. Indium in mainland China: Insights into use, trade, and efficiency from the substance flow analysis[J]. Resour Conserv Recycl, 2019, 149: 312.

[17] ZHANG L, CAI Z J, YANG J M, et al. The future of copper in China—a perspective based on analysis of copper flows and stocks[J]. Sci Total Environ, 2015, 536: 142.

[18] DAIGO I, MATSUNO Y, ADACHI Y. Substance flow analysis of chromium and nickel in the material flow of stainless steel in Japan[J]. Resour Conserv Recycl, 2010, 54(11): 851.

[19] ZENG X Y, ZHENG H X, GONG R Y, et al. Uncovering the evolution of substance flow analysis of nickel in China[J]. Resour Conserv Recycl, 2018, 135: 210.

[20] ELSHKAKI A, GRAEDEL T E. Dynamic analysis of the global metals flows and stocks in electricity generation technologies[J]. J Clean Prod, 2013, 59: 260.

[21] LI L, WANIA F. Tracking chemicals in products around the world: Introduction of a dynamic substance flow analysis model and application to PCBs[J]. Environ Int, 2016, 94: 674.

[22] SAKAMORNSNGUAN K, KRETSCHMANN J. Substance flow analysis and mineral policy: The case of potash in Thailand[J]. Extr Ind Soc, 2016, 3(2): 383.

[23] YELLISHETTY M, MUDD G M. Substance flow analysis of steel and long term sustainability of iron ore resources in Australia, Brazil, China and India[J]. J Clean Prod, 2014, 84: 400.

[24] LIU J, AN R, XIAO R G, et al. Implications from substance flow analysis, supply chain and supplier' risk evaluation in iron and steel industry in Mainland of China[J]. Resour Policy, 2017, 51: 272.

[25] ARENA U, GREGORIO F D. A waste management planning based on substance flow analysis[J]. Resour Conserv Recycl, 2014, 85: 54.

[26] CHÈVRE N, COUTU S, MARGOT J. Substance flow analysis as a tool for mitigating the impact of pharmaceuticals on the aquatic system[J]. Water Res, 2013, 47(9): 2995.

[27] CHU J W, YIN X B, HE M C, et al. Substance flow analysis and environmental release of antimony in the life cycle of polyethylene terephthalate products[J]. J Clean Prod, 2021, 291: 125252.

[28] HUANG C L, VAUSE J, MA H W. Using material/substance flow analysis to support sustainable development assessment: A literature review and outlook[J]. Resour Conserv Recycl, 2012, 68(9): 104.

[29] YOSHIDA H, CHRISTENSEN T H, GUILDAL T. A comprehensive substance flow analysis of a municipal wastewater and sludge treatment plant[J]. Chemosphere, 2015, 138: 874.

[30] ZENG L Q, ZHAO Z W, HUO G S, et al. Mechanism of selective precipitation of molybdenum from tungstate solution[J]. JOM, 2020, 72(2): 800.

[31] ZHAO Z W, ZHANG W G, CHEN X Y, et al. Study on removing Mo from tungstate solution using coprecipitation adsorption method based on novel Mo sulphidation process[J]. Can Metall Q, 2013, 52(4): 358.

Copper and Arsenic Substance Flow Analysis of Pyrometallurgical Process for Copper Production

Abstract: The metabolism of copper and arsenic in a copper pyrometallurgy process was studied through substance flow analysis method. The mass balance accounts and substance flow charts of copper and arsenic were established, indicators including direct recovery, waste recycle ratio, and resource efficiency were used to evaluate the metabolism efficiency of the system. The results showed that, the resource efficiency of copper was 97.58%, the direct recovery of copper in smelting, converting, and refining processes was 91.96%, 97.13% and 99.47%, respectively. Meanwhile, for producing 1 t of copper, 10 kg of arsenic was carried into the system, with the generation of 1.07 kg of arsenic in flotation tailing, 8.50 kg of arsenic in arsenic waste residue, and 0.05 kg of arsenic in waste water. The distribution and transformation behaviors of arsenic in the smelting, converting, and refining processes were also analyzed, and some recommendations for improving copper resource efficiency and pollution control were proposed based on substance flow analysis.

Keywords: copper smelting; substance flow analysis; copper; arsenic; distribution behavior

1 Introduction

Copper industry is a crucial industry in the global economy as copper is widely used in many fields and it is irreplaceable in some industries[1]. Currently, copper is mainly extracted from sulfide minerals and pyrometallurgical process is the most dominant technology for copper production[2-4]. However, the pyrometallurgical process for copper production is facing the challenge of pollution control, as copper resources become more and more complex. For example, arsenic commonly occurs in copper concentrates, and the emission of arsenic into atmosphere during the pyrometallurgical production causes a serious environmental issue[5, 6]. At present,

---

Published in *Transactions of Nonferrous Metals Society of China*, 2022, 32(1): 364-376. Authors: Guo Xueyi, Chen Yuanlin, Wang Qinmeng, Wang Songsong, Tian Qinghua.

many studies about improving production efficiency or reducing pollutants discharge have been conducted, but few of them have focused on the metabolism of copper or other detrimental elements flow in production process. Substance metabolism affects both energy and resource consumption of a factory, taking a deep insight into the metabolism of copper and other key elements is the foundation for process improvement.

Substance flow analysis (SFA) is an analytical tool that could be applied to investigating the metabolism of a specific substance flow in a given system and providing insight into the effects of the substance metabolism on economy and environment[7, 8]. In the past decades, SFA has experienced a growing tendency in many fields[9]. Specially, it is extensively used to analyze the stock and flows of metals (lead, zinc, copper, etc.) and other substances in local region[10], country[11-14], or global level[15, 16]. Moreover, SFA was exploited as a supported tool in mineral resources management[17, 18], industry chain evaluation[19], waste management[20], environmental risk assessment and pollution control during industrial production[21, 22]. However, most previous studies that analyzed a specified substance flow just concentrated on the stock and potential waste generation in a national or regional scope[10-16].

Applying SFA to trace the substances quantities and migration in a production process could determine the metabolism efficiency of the process and reveal the pathways through which pollutants are generated, which is beneficial to the resource management and pollution control of a factory. Yoshida et al.[23] studied the fate of total organic carbon, 32 elements and 4 groups of organic pollutants in a conventional wastewater treatment plant through SFA, and then an assessment on the treatment efficiency and potential environmental impact of the plant was given out. Bai et al.[8] applied SFA to calculating lead substance balance in a lead smelting process and evaluating the metabolism efficiency of the system, through which some recommendations on improving emission control and pollution prevention for the smelting factory were also proposed. Wang et al.[24] emphasized copper stocks and flows variation in production stage in the USA from 1974 to 2012, which was a meaningful reference for copper production industry of newly industrialized countries.

Most previous studies on copper metabolism were conducted at the state level, focusing mainly on the manufacturing and consumption stages, and there is rare report of SFA on the metallurgical processes of copper[25-28]. Especially, as an important associated element in copper concentrate, arsenic receives little attention in industrial metabolism analysis[29]. SFA of copper and arsenic in copper metallurgical process helps to find the measures for improving the utilization efficiency of copper concentrate and reveal the generation of arsenic-containing contaminant flows. In this study, SFA was used as an analytical tool to study the metabolism of a pyrometallurgical process for copper production. The mass balance accounts and substance flow charts of copper and arsenic in the production system were established. The metabolism efficiency of the system was evaluated, and the distribution and transformation behaviors of arsenic were analyzed. This study focused on the element metabolism efficiency with a microcosmic perspective, providing a reference for the process improvement.

2 Methodology

Compared to applying SFA to a large system with the scope of global or regional scale, applying SFA to a production process is more microcosmic and more specific. The SFA model of a production process generally includes the substance flow of a single process and the whole production system composed of several single processes[8].

2.1 Definition of substance flows in single process

If a single process in the production system is defined as process j, the substance flows input to and output from process j include the following stages.

(1) Input raw material flow, A_j.

(2) Input upstream product substance flow, $P_{m,j}$.

(3) Recycle substance flow from downstream process i, $R_{i,j}$. The substance flow which is recycled to upstream process k from process j is called $R_{j,k}$.

(4) Stock substance flow, S_j. This kind of product is temporarily stocked in the warehouse and will be put into the production process when needed.

(5) Emission substance flow, E_j. This flow includes the by-products and pollutants which are discharged outside of the objective system from process j.

(6) Output product substance flow, P_j. The substance flow which is the by-product of process j and is transported to process k for further treating is called $P_{j,k}$.

According to the conservation of mass, for the single process j, it can be expressed as

$$A_j + P_{m,j} + R_{i,j} = P_j + P_{j,k} + R_{j,k} + S_j + E_j \tag{1}$$

2.2 Substance flows of the whole system

The combination of all the single processes makes up the whole system, and the substance flows of the whole system are the sums of corresponding flows in all single processes, which are described as follows.

(1) Input raw material flow A

$$A = \sum_{j=1}^{m} A_j \tag{2}$$

(2) Recycle substance flow R

$$R = \sum_{j=1}^{m} \sum_{i=1}^{m} R_{i,j} \tag{3}$$

(3) Stock substance flow S

$$S = \sum_{j=1}^{m} S_j \tag{4}$$

(4) Emission substance flow E

$$E = \sum_{j=1}^{m} E_j \tag{5}$$

(5) Output product substance flow P

$$P = \sum_{j=1}^{m} P_j \tag{6}$$

2.3 Mass balance calculation

To analyze the substance flow of a specific element, two parts of data are required. One part is the flow quantity M_j, which is the amount of each material that contains the objective element, and this part of data is collected from the daily production reports. The other part is the content C_j of an objective element in each material, and it is obtained by sampling and analyzing for each material. Then, the flow quantity of the objective element is calculated by Eq. (7):

$$m_j = M_j \times C_j \tag{7}$$

where m_j is the quantity of the objective element in the substance flow j.

In this study, the mass balance calculation of all single processes and the whole system is based on 1 t copper output from the production system, the flow ratio of objective element in each substance is expressed as f_j:

$$f_j = m_j/m_m \tag{8}$$

where m_m is the quantity of the objective element in the final product output from the production system.

2.4 Evaluation indicators of SFA

In order to evaluate the production efficiency of a process and its influence on environment, three indicators are proposed in this study as follows.

(1) Direct recovery of the primary single processes, D, the proportion of objective element in qualified products to the total output flows of a single process, %. For a single process j, D_j is calculated by Eq. (9):

$$D_j = P_j/(P_j + S_j + R_{j,k} + P_{j,k} + E_j) \times 100\% \tag{9}$$

A high direct recovery of a process means low resource consumption for per ton of qualified product from the process.

(2) Waste recycle ratio of the process, α, the proportion of recycle substance flow in all the substance flows which are not included in the final product, %:

$$\alpha = R/(S + R + E) \times 100\% \tag{10}$$

Increasing the waste recycle ratio helps to reduce the environmental load caused by discharging of rejected substance flows and to improve resource efficiency.

(3) Resource efficiency, ε, the proportion of objective element in final product to the total input flows, including both natural and secondary resources, %:

$$\varepsilon = P/A = (A - S - E)/A \times 100\% \tag{11}$$

3　System boundary description and data collecting

In this study, a pyrometallurgical process of copper production, including the primary processes of smelting, converting, refining, and the assistant processes for slags, flue dust, and tail gas treatment, is chosen as system boundary. The process is exploited by a copper production enterprise in Shandong Province, China, which produces copper of 3×10^5 t/a. And this kind of process is with great development potential. The simplified flowsheet of the objective process is shown in Fig. 1.

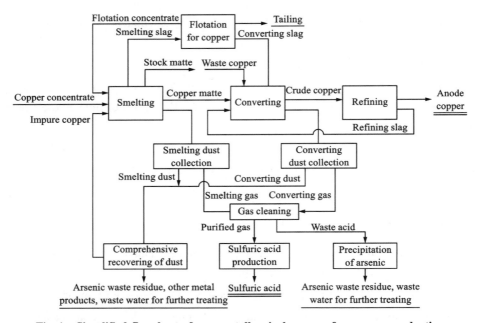

Fig. 1　Simplified flowsheet of pyrometallurgical process for copper production

The oxygen-enriched smelting is conducted in a bottom blown furnace, in which the oxidation reactions occur. Copper concentrate with a grade of 20%-25% and some recycled products are fed into the bottom blown furnace along with fuel and flux. Copper matte with a grade of 65%-75% and some stock matte as a by-product in reserve are produced from smelting. The copper matte is converted into crude copper with a grade of 98.5%-99% by converting in a PS converter, and some waste copper is also put into the converting process. The crude copper is upgraded to about 99.5% through refining to meet the impurity content standard of subsequent electrolysis process. During the production process, some slags and flue dust are also generated. The smelting slag and converting slag are cooled slowly and then transported to a flotation system for recovering residual copper. Then, the flotation concentrate is recycled to smelting process, and the refining slag is recycled to converting process. For the purposes of reducing arsenic accumulation and maximizing economic profit, the flue dust from smelting and converting is gathered to remove arsenic and recover copper after separating from tail gas, and the impure copper product from dust treatment is recycled to smelting process. The tail gas from smelting and converting, which contains a large amount of sulfur dioxide, is gathered for sulfuric acid production after cleaning. Arsenic precipitation is carried out for the waste acid produced from the gas cleaning process. The arsenic waste and waste water generated from flue dust and tail gas treating systems are considered as emission substance flows and then discharged outside the production system for further treatment. The by-product sulfuric acid is also considered as emission substance flow.

The quantities of input and output material flows were measured for the time scale of 30 d, and the data were collected from the production reports of objective plant. In order to determine the copper and arsenic contents in each material flow, all the flows were sampled and analyzed each production day. The sampling period was 30 d corresponding to the production reports, and the copper and arsenic contents of samples were averaged for mass balance calculation.

Some solid samples were dried at 50 ℃ in a vacuum oven for 12 h, and then characterized through X-ray diffraction (XRD, TTR Ⅲ, with Cu K_α radiation, Rigaku, Japan) for phase composition analysis by slow scanning, at the 2θ range from 10° to 80°, a step size of 0.02° and a scanning speed of 2(°)/min.

4 Results and discussion

4.1 Substance flow analysis of copper

4.1.1 Mass balance calculation

Based on the production data, the mass balance account of copper in all single processes of the system was built upon 1 t of copper output from the system. The flow ratio of copper in each substance flow is shown in Table 1.

Table 1 Input and output flow ratios of copper in each substance flow of production system

No.	Process unit	Input		Output	
		Substance flow	Flow ratio, f_j	Substance flow	Flow ratio, f_j
1	Smelting	Copper concentrate (A_1)	1.0201	Copper matte (P_1)	1.0253
		Flotation concentrate ($R_{3,1}$)	0.0833	Stock matte (S_1)	0.0172
		Impure copper ($R_{6,1}$)	0.0116	Smelting slag ($P_{1,3}$)	0.0657
				Smelting dust and gas ($P_{1,4}$)	0.0068
		Total input	1.115	Total output	1.115

Continue the Table 1

No.	Process unit	Input		Output	
		Substance flow	Flow ratio, f_j	Substance flow	Flow ratio, f_j
2	Converting	Copper matte (P_1)	1.0253	Crude copper (P_2)	1.0053
		Waste copper (A_2)	0.0047	Converting slag ($P_{2,3}$)	0.025
		Refining slag ($R_{10,2}$)	0.0053	Converting dust and gas ($P_{2,5}$)	0.005
		Total input	1.0353	Total output	1.0353
3	Flotation system	Smelting slag ($P_{1,3}$)	0.0657	Flotation concentrate ($R_{3,1}$)	0.0833
		Converting slag ($P_{2,3}$)	0.025	Tailing (E_3)	0.0074
		Total input	0.0907	Total output	0.0907
4	Smelting dust collection	Smelting dust and gas ($P_{1,4}$)	0.0068	Smelting dust ($P_{4,6}$)	0.0068
				Smelting gas ($P_{4,7}$)	0
		Total input	0.0068	Total output	0.0068
5	Converting dust collection	Converting dust and gas ($P_{2,5}$)	0.005	Converting dust ($P_{5,6}$)	0.005
				Converting gas ($P_{5,7}$)	0
		Total input	0.005	Total output	0.005
6	Comprehensive recovering of dust	Smelting dust ($P_{4,6}$)	0.0068	Impure copper ($R_{6,1}$)	0.0116
		Converting dust ($P_{5,6}$)	0.005	Other metal products (E_{6-1})	0
				Arsenic waste residue (E_{6-2})	0
				Waste water (E_{6-3})	0.0002
		Total input	0.0118	Total output	0.0118
7	Gas cleaning	Smelting gas ($P_{4,7}$)	0	Purified gas ($P_{7,8}$)	0
		Converting gas ($P_{5,7}$)	0	Waste acid ($P_{7,9}$)	0
		Total input	0	Total output	0
8	Sulfuric acid production	Purified gas ($P_{7,8}$)	0	Sulfuric acid (E_8)	0
		Total input	0	Total output	0
9	Precipitation of arsenic	Waste acid ($P_{7,9}$)	0	Arsenic waste residue (E_{9-1})	0
				Waste water (E_{9-2})	0
		Total input	0	Total output	0
10	Refining	Crude copper (P_2)	1.0053	Anode copper (P_{10})	1
				Refining slag ($R_{10,2}$)	0.0053
		Total input	1.0053	Total output	1.0053

4.1.2 Substance flow chart of copper

Fig. 2 shows the substance flow chart of copper. There are 26 strands of substance flows within the whole system, and each of them has been identified with name and flow code, as shown in Table 1.

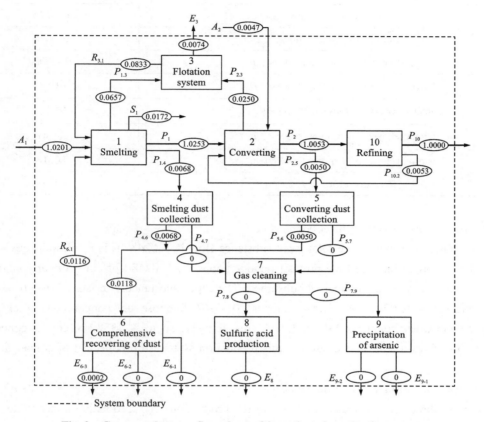

Fig. 2 Copper substance flow chart of investigated production system

4.1.3 Evaluation of copper production process

Table 2 shows the evaluation results based on the defined indicators. As can be seen, the direct recovery of copper in smelting, converting, and refining processes is 91.96%, 97.13%, and 99.47%, respectively. According to the definition of direct recovery, the proportion of by-product (for example, stock matte) is not included in direct recovery. The waste recycle ratio of the production system is 80.16%, which means that 80.16% of the rejected products produced during production are recycled to the system for further copper recovering. Thus, a high resource efficiency of the whole system, 97.58%, is observed due to such a high waste recycle ratio. The details of the defined substance flows of the whole production system are discussed in the following sections.

Table 2 Results of evaluation indicators for copper production system

Evaluation indicator	Value
Input substance flow, A	1.0248
Recycle substance flow, R	0.1002
Stock substance flow, S	0.0172
Emission substance flow, E	0.0076
Output product substance flow, P	1.0000
Direct recovery of melting process, $D_1/\%$	91.96
Direct recovery of converting process, $D_2/\%$	97.13
Direct recovery of refining process, $D_8/\%$	99.47
Waste recycle ratio of production system, $\alpha/\%$	80.16
Resource efficiency, $\varepsilon/\%$	97.58

(1) Input substance flow of copper

As is shown in Table 2, the total input substance flow is 1.0248. This result indicates that, to produce 1 t of copper into the final product (anode copper), 1.0248 t of pure copper contained in copper concentrate or waste copper is consumed. The input substance consists of two flows: A_1, copper concentrate 1.0201; A_2, waste copper, 0.0047. Suppose the copper content of copper concentrate and waste copper is 23% and 96%, respectively. Then, 4.4352 t of copper concentrate and 0.0050 t of waste copper are put into this production system to produce 1 t of copper in anode copper.

(2) Recycle substance flow of copper

There are three recycle substance flows in the production: $R_{3,1}$, flotation concentrate; $R_{6,1}$, impure copper; $R_{10,2}$, refining slag.

Among the recycle substance flows, flotation concentrate accounts for the largest portion, 83.13%. The flotation recovery of copper is up to 91.84%. The slag generated in refining process is recycled to converting process directly, and this strand accounts for a small portion of all the recycle substance flows, 5.29%.

As a whole, the recycle substance flow is 0.1002, accounts for 80.16% of the rejected products (including recycle substance flow of 0.1002, stock substance flow of 0.0172, and emission substance flow of 0.0076), and that is the guarantee of high resource efficiency.

(3) Emission substance flow of copper

The emission substance flows consist of the flotation tailing, arsenic waste and waste water from comprehensive recovering of dust, sulfuric acid, arsenic waste and waste water during tail gas treatment. In fact, almost no copper is carried in the tail gas from the separation of flue dust for the low volatility of copper. As we can see, no copper is discharged outside the system with sulfuric acid, arsenate waste and waste water generated from tail gas treatment processes. The flotation tailing, 0.0074, accounts for 97.37% of emission substance flows, and the copper content in tailing is 0.3%. Thus, improving flotation efficiency may contribute to a higher resource efficiency. Another emission substance flow is the waste water from dust treating process, 0.0002, which contains a certain amount of copper (about 500 mg/L), accounting for 2.63% of emission substance flows. Before the waste water is discharged into environment, further copper removal should be carried out.

(4) Stock substance flow of copper

Only one stock substance flow is generated during stable production. The main portion of stock substance flow is the extra copper matte from smelting process. When the production quantity of smelting process is larger than that of converting process, the product is not digested completely. Thus, some of the copper matte is stocked in the warehouse. The other source of stock substance flow is the low-grade matte produced from the settling of slags. Usually, the stock substance flow is relatively unstable compared to other substance flows, as it depends on the production quantity of smelting and converting to a large extent. The stock substance flow will be put into the production process when the production capacity is sufficient.

4.2 Substance flow analysis of arsenic

As arsenic is a hazardous element that flows into the system associated with copper rather than valuable element, the evaluation defined in Section 2.4 is not conducted for arsenic substance flow. However, for the purpose of giving an insight into the arsenic migration and translation in the production process, the distribution and transformation behaviors of arsenic in the primary single processes (smelting, converting, and refining) were analyzed based on arsenic substance flow analysis and phase analysis of some products.

4.2.1 Mass balance calculation

The mass balance account of arsenic in all single processes of the system was also built upon producing 1 t of copper output from the system, and the results are shown in Table 3.

Table 3 Input and output flow ratios of arsenic in each substance flow of production system

No.	Process unit	Input		Output	
		Substance flow	Flow ratio, f_j	Substance flow	Flow ratio, f_j
1	Smelting	Copper concentrate, A_1	0.01000	Copper matte, P_1	0.0034
		Flotation concentrate, $R_{3,1}$	0.00120	Stock matte, S_1	0.00007
		Impure copper, $R_{6,1}$	0	Smelting slag, $P_{1,3}$	0.00183
				Smelting dust and gas, $P_{1,4}$	0.0059
		Total input	0.0112	Total output	0.0112

Continue the Table 3

No.	Process unit	Input		Output	
		Substance flow	Flow ratio, f_j	Substance flow	Flow ratio, f_j
2	Converting	Copper matte, P_1	0.0034	Crude copper, P_2	0.00032
		Waste copper, A_2	0	Converting slag, $P_{2,3}$	0.00044
		Refining slag, $R_{10,2}$	0.00001	Converting dust and gas, $P_{2,5}$	0.00265
		Total input	0.00341	Total output	0.00341
3	Flotation system	Smelting slag, $P_{1,3}$	0.00183	Flotation concentrate, $R_{3,1}$	0.0012
		Converting slag, $P_{2,3}$	0.00044	Tailing, E_3	0.00107
		Total input	0.00227	Total output	0.00227
4	Smelting dust collection	Smelting dust and gas, $P_{1,4}$	0.0059	Smelting dust, $P_{4,6}$	0.00432
				Smelting gas, $P_{4,7}$	0.00158
		Total input	0.0059	Total output	0.0059
5	Converting dust collection	Converting dust and gas, $P_{2,5}$	0.00265	Converting dust, $P_{5,6}$	0.00194
				Converting gas, $P_{5,7}$	0.00071
		Total input	0.00265	Total output	0.00265
6	Comprehensive recovering of dust	Smelting dust, $P_{4,6}$	0.00432	Impure copper, $R_{6,1}$	0
		Converting dust, $P_{5,6}$	0.00194	Arsenic waste residue, E_{6-1}	0.00622
				Other metal products, E_{6-2}	0
				Waste water, E_{6-3}	0.00004
		Total input	0.00626	Total output	0.00626
7	Gas cleaning	Smelting gas, $P_{4,7}$	0.00158	Purified gas, $P_{7,8}$	0
		Converting gas, $P_{5,7}$	0.00071	Waste acid, $P_{7,9}$	0.00229
		Total input	0.00229	Total output	0.00229
8	Sulfuric acid production	Purified gas, $P_{7,8}$	0	Sulfuric acid, E_8	0
		Total input	0	Total output	0
9	Precipitation of arsenic	Waste acid, $P_{7,9}$	0.00229	Arsenic waste residue, E_{9-1}	0.00228
				Waste water, E_{9-2}	0.00001
		Total input	0.00229	Total output	0.00229
10	Refining	Crude copper, P_2	0.00032	Anode copper, P_{10}	0.00031
				Refining slag, $R_{10,2}$	0.00001
		Total input	0.00032	Total output	0.00032

4.2.2 Substance flow chart of arsenic

Fig. 3 shows the substance flow chart of arsenic. All the arsenic substance flows are corresponding to the copper substance flows in Fig. 2. The total quantity of arsenic in each kind of substance flow is shown in Table 4. The details of each substance flow are discussed in the following sections.

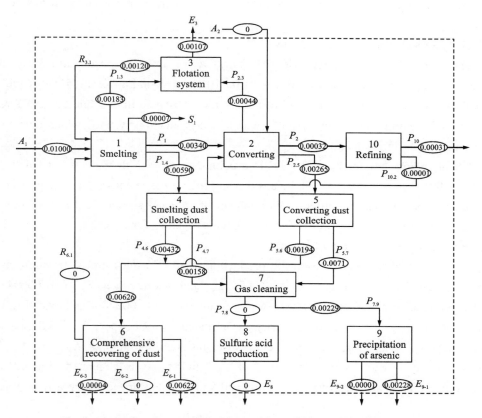

Fig. 3　Arsenic substance flow chart of investigated production system

Table 4　Flow ratio of arsenic in each substance flow of production system

Evaluation indicator	Value
Input substance flow, A	0.01000
Stock substance flow, S	0.00007
Recycle substance flow, R	0.00121
Emission substance flow, E	0.00962

(1) Input substance flow of arsenic

The total input substance flow of arsenic is 0.01000, which indicates that to produce 1 t of copper into the final product (anode copper), 10 kg of arsenic is carried into the system. All the input arsenic is carried by copper concentrate, and the arsenic content of copper concentrate is 0.23%, which belongs to the kind of high arsenic copper concentrate. This result also reveals that, with the copper resources being complex globally, the amount of arsenic input to copper production system increases. It is necessary to clarify the transformation behavior of arsenic in the copper production process. Another input substance flow of copper, i.e. waste copper, does not carry arsenic into the system, as the arsenic content of copper products conforms to quality standard.

(2) Recycle substance flow of arsenic

Arsenic is recycled to the system with two flows: $R_{3,1}$, flotation concentrate; $R_{10,2}$, refining slag. The total recycled arsenic substance flow is 0.00121, accounting for 11.10% of the arsenic distributed in rejected products. Almost all of the arsenic distributed in dust and tail gas is removed in the separated treating process and is not cycled to the production system, which helps to avoid arsenic accumulation in production system.

The flotation concentrate ($R_{3,1}$ = 0.00120) accounts for most of the recycled arsenic substance flows (R = 0.00121), 99.17%, which means that improving the selective depressing of arsenic in flotation process could contribute to a smaller amount of recycled arsenic. And for this purpose, it is necessary to do sufficient liberation of arsenic minerals before flotation.

(3) Emission substance flow of arsenic

Some arsenic is discharged outside the production system in the form of flotation tailing E_3 (0.00107), arsenic waste residue E_{6-1} (0.00622) and waste water E_{6-3} (0.00004) from comprehensive recovering of dust, arsenic waste residue E_{9-1} (0.00228) and waste water E_{9-2} (0.00001) from tail gas treating. In summary, to produce 1 t of copper, 1.07 kg of arsenic in flotation tailing, 8.50 kg of arsenic in arsenic waste residue (the sum of E_{6-1} and E_{9-1}), and 0.05 kg of arsenic in waste water (the sum of E_{6-3} and E_{9-2}) is generated. The other metal products from comprehensive recovering of dust and sulfuric acid do not carry arsenic out of the system.

Among these emissions, arsenic waste residue from dust and tail gas treatment accounts for 88.36%, and is considered as hazardous waste, thus the stabilization level of arsenic in waste residue should be assessed for the purpose of minimizing environmental pollution risk. And the waste water should be treated further before discharging into the natural environment, as the arsenic content of waste water (1.4 mg/L) exceeds emission standard in China (*Emission standard of pollutants for copper, nickel, cobalt industry*, GB 25467-2010). Reusing the waste water to the gas cleaning process is a reasonable way.

(4) Stock substance flow of arsenic

The amount of arsenic in stock flow, 0.00007, is very small compared to the other substance flows. And this substance flow will be put into the production process along with copper.

4.2.3 Distribution and transformation behaviors of arsenic

With the copper resources being complex globally, the amount of arsenic entering copper production system increases. It is necessary to clarify the evolution behaviors of arsenic during copper production. The content of arsenic in copper concentrate and some products from the production system is shown in Table 5. The distribution and transformation behaviors of arsenic are analyzed in detail in the following sections.

Table 5 Contents of arsenic in copper concentrate and some products   %

Copper concentrate	Copper matte	Smelting dust	Smelting slag	Crude copper
0.23	0.22	7.01	0.10	0.04
Converting dust	Converting slag	Anode copper		Refining slag
5.93	0.08	0.03		0.04

(1) Smelting process

Table 6 shows the distribution of arsenic in the primary production processes of system. The results indicate that, in smelting process, a large portion of arsenic flows with flue gas (including 38.56% in smelting dust, and 14.11% in tail gas) for its low volatile point. Then, most of the volatile arsenic is collected in smelting dust by dust collection process, which results in the high arsenic content of smelting dust, 7.01%. Besides, there are still large portion of arsenic (30.36%) distributed in copper matte followed by 16.34% in smelting slag and 0.63% in stock matte.

Table 6 Distribution of arsenic in the primary production processes of system

Process unit	Product	Flow ratio, f_j	Distribution rate/%
Smelting	Copper matte	0.00340	30.36
	Stock matte	0.00007	0.63
	Smelting slag	0.00183	16.34
	Smelting dust	0.00432	38.56
	Smelting gas	0.00158	14.11
Converting	Crude copper	0.00032	9.38
	Converting slag	0.00044	12.91
	Converting dust	0.00194	56.89
	Converting gas	0.00071	20.82
Refining	Anode copper	0.00031	96.88
	Refining slag	0.00001	3.12

The main occurring forms of arsenic in raw material (copper concentrate) are Cu_3AsS_4 and CuAsS [shown in Fig. 4(a)], and these complex compounds are decomposed rapidly into As_2S_3 at high temperatures. During copper smelting, arsenic sulfides are oxidized and decomposed as follows:

$$As_2S_3 + O_2(g) \Longrightarrow 2AsS(g) + SO_2(g) \quad (12)$$
$$2AsS + 2O_2(g) \Longrightarrow As_2(g) + 2SO_2(g) \quad (13)$$
$$As_2(g) + O_2(g) \Longrightarrow 2AsO(g) \quad (14)$$
$$2As_2S_3 \Longrightarrow 4AsS(g) + S_2(g) \quad (15)$$
$$2AsS \Longrightarrow As_2(g) + S_2(g) \quad (16)$$

Therefore, arsenic flows into flue gas in the forms of $As_2(g)$, $AsS(g)$, and $AsO(g)$, some arsenic is further oxidized to As_2O_3 during flowing, and part of it is collected into dust. The form of arsenic in smelting dust is shown in Fig. 4(c).

Some arsenic is also oxidized in the molten bath and enters smelting slag in the form of As_2O_3 which is detected by slow scanning XRD, as shown in Fig. 4(d), and the following reactions occur:

$$4AsO(g) + O_2(g) \Longrightarrow 2As_2O_3(l) \quad (17)$$
$$As_2(g) + 1.5O_2(g) \Longrightarrow As_2O_3(l) \quad (18)$$

$FeAsO_4$ is also detected as shown in Fig. 4(d). This result indicates that in the slag, $As_2O_3(l)$ further reacts with iron oxides to form iron arsenate.

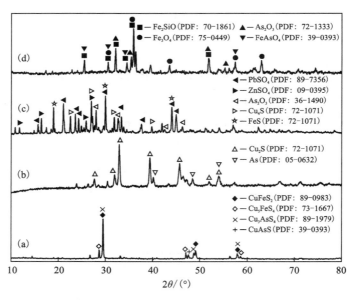

Fig. 4 Slow scanning XRD patterns of copper concentrate (a), copper matte (b), smelting dust (c), and smelting slag (d)

During the flowing of the melt, As_2O_3 is reduced to $As(l)$ that is the existing form in copper matte [shown in Fig. 4(b)] by FeS as follows:

$$As_2O_3(l) + FeS(l) \Longrightarrow 2As(l) + FeO(l) + SO_2(g) \quad (19)$$

Meanwhile, a part of $As(l)$ in copper matte can also be oxidized to $As_2O_3(l)$ by Cu_2O and then enters the smelting slag:

$$2As(l) + 2Cu_2O(l) + 2FeS(l) + 1.5O_2(g) \Longleftrightarrow As_2O_3(l) + 2FeO + 2Cu_2S(l) \quad (20)$$
$$2As(l) + 1.5O_2(g) \Longleftrightarrow As_2O_3(l) \quad (21)$$

Besides, a part of As_2O_3 is also reduced to $As_2(g)$, and flows back into flue dust:
$$As_2O_3(l) + FeS(l) \Longleftrightarrow As_2(g) + FeO(l) + SO_2(g) \quad (22)$$

In fact, all of the above reactions form a balanced cycle in the smelting process.

(2) Converting process

As shown in Table 6, in converting process, the arsenic collected in converting dust accounts for the largest portion, 56.89%, resulting in high arsenic content of the dust, 5.93%. 20.82% of arsenic flows into converting gas, 12.91% of arsenic is distributed in converting slag, and 9.38% of arsenic is distributed in crude copper.

During this process in PS converter, the arsenic in copper smelt is oxidized to As_2O_3 by O_2, and Eqs. (20) and (21) occur. A large amount of As_2O_3 flows into flue gas and part of it is collected into dust [shown in Fig. 5(a)] in converting dust collection process. A part of As_2O_3 enters converting slag and can be detected by XRD as shown in Fig. 5(b). Similar to the smelting process, some As_2O_3 further reacts with iron oxides to form iron arsenate, and $FeAsO_4$ can be detected [shown in Fig. 5(b)]. A small amount of arsenic in copper matte is not oxidized and stays in crude copper.

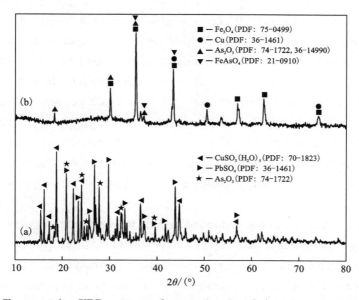

Fig. 5 Slow scanning XRD patterns of converting dust (a) and converting slag (b)

(3) Refining process

In refining process, 96.88% of arsenic stays in anode copper and the rest in slag is recycled to converting process. During this process, the arsenic in crude copper is difficult to be oxidized due to its low content (0.04%). Thus, most of arsenic is distributed in anode copper in the form of $As(l)$, and a small part of arsenic is oxidized to As_2O_3 (shown in Fig. 6) and enters refining slag. Besides, some As_2O_3 also reacts with iron oxides, and thus iron arsenate $FeAsO_4$ is formed as shown in Fig. 6.

It is worth noting that, in consideration of the high arsenic content in dust from smelting and converting, centralized treatment of the two dusts is a reasonable way to avoid arsenic accumulation in the production system instead of recycling it to smelting and converting processes directly.

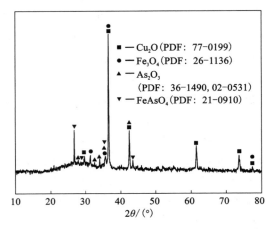

Fig. 6 Slow scanning XRD pattern of refining slag

5 Conclusions and recommendations

(1) For the copper substance flows, the resource efficiency of copper for the production process is 97.58%. The direct recovery of copper in smelting, converting, and refining processes is 91.96%, 97.13%, and 99.47%, respectively.

(2) For the arsenic substance flows, to produce 1 t of copper, 10 kg of arsenic is carried into the system, with the generation of 1.07 kg of arsenic in flotation tailing, 8.50 kg of arsenic in arsenic waste residue, and 0.05 kg of arsenic in waste water. In smelting process, the distribution rate of arsenic in smelting dust, tail gas, copper matte, smelting slag, and stock matte is 38.56%, 14.11%, 30.36%, 16.34%, and 0.63%, respectively. In converting process, the distribution rate of arsenic in converting dust, tail gas, crude copper, and converting slag is 56.89%, 20.82%, 9.38%, and 12.91%, respectively. In refining process, the distribution ratio of arsenic in anode copper and slag is 96.88% and 3.12%, respectively. For the transformation behaviors, arsenic occurs in dust from smelting and converting as As_2O_3, in slag from smelting, converting and refining as As_2O_3 and arsenate, in copper matte, crude copper and anode copper as As.

(3) Some recommendations for improving copper resource efficiency and pollution control were proposed based on substance flow analysis: The centralized treatment of dust from smelting and converting with high arsenic content is a reasonable way for avoiding arsenic accumulation in the production system. The selective depressing of arsenic in flotation process should be improved for a smaller amount of recycled arsenic. Meanwhile, improving flotation efficiency may contribute to a higher copper resource efficiency. The stabilization level of arsenic in waste residue should be assessed to minimize the risk of environmental pollution. The waste water from dust and tail gas treatment should be treated further before discharging into natural environment. Since the arsenic content is high, it is with good environmental and economic benefits to reuse the waste water to the gas cleaning process.

Acknowledgments

The authors are grateful for the financial supports from the National Key R&D Program of China (No. 2019YFC1907400), and the National Natural Science Foundation of China (Nos. 51904351, 51620105013).

References

[1] NIKOLIC I, MILOSEVIC I, MILIJIC N, et al. Cleaner production and technical effectiveness: Multi-criteria analysis of copper smelting facilities[J]. Journal of Cleaner Production, 2019, 215: 423-432.

[2] LIU Z H, XIA L G. The practice of copper matte converting in China[J]. Mineral Processing and Extractive Metallurgy, 2019, 128: 117-124.

[3] LIU W F, FU X X, YANG T Z, et al. Oxidation leaching of copper smelting dust by controlling potential [J]. Transactions of Nonferrous Metals Society of China, 2018, 28(9): 1854-1861.

[4] ZHENG Y X, LV J F, LAI Z N, et al. Innovative methodology for separating copper and iron from Fe-Cu alloy residues by selective oxidation smelting[J]. Journal of Cleaner Production, 2019, 231: 110-120.

[5] LI Y K, ZHU X, QI X J, et al. Removal and immobilization of arsenic from copper smelting wastewater using copper slag by in situ encapsulation with silica gel[J]. Chemical Engineering Journal, 2020, 394: 124833.

[6] YAO L W, MIN X B, XU H, et al. Physicochemical and environmental properties of arsenic sulfide sludge from copper and lead-zinc smelter[J]. Transactions of Nonferrous Metals Society of China, 2020, 30(7): 1943-1955.

[7] LOISEAU E, JUNQUA G, ROUX P, et al. Environmental assessment of a territory: An overview of existing tools and methods[J]. Journal of Environmental Management, 2012, 112: 213-225.

[8] BAI L, QIAO Q, LI Y P. Substance flow analysis of production process: A case study of a lead smelting process[J]. Journal of Cleaner Production, 2015, 104: 502-512.

[9] HUANG C L, MA H W, YU C P. Substance flow analysis and assessment of environmental exposure potential for triclosan in Mainland of China[J]. Science of the Total Environment, 2014, 499: 265-275.

[10] LINDQVIST A, von MALMBORG F. What can we learn from local substance flow analyses? The review of cadmium flows in Swedish municipalities[J]. Journal of Cleaner Production, 2004, 12(8-10): 909-918.

[11] GUO X Y, ZHONG J Y, SONG Y, et al. Substance flow analysis of zinc in China[J]. Resources, Conservation and Recycling, 2010, 54(3): 171-177.

[12] LIN S H, MAO J S, CHEN W Q, et al. Indium in Mainland of China: Insights into use, trade, and efficiency from the substance flow analysis [J]. Resources, Conservation and Recycling, 2019, 149: 312-321.

[13] ZHANG L, CAI Z J, YANG J M, et al. The future of copper in China-a perspective based on analysis of copper flows and stocks[J]. Science of the Total Environment, 2015, 536: 142-149.

[14] DAIGO I, MATSUNO Y, ADACHI Y. Substance flow analysis of chromium and nickel in the material flow

of stainless steel in Japan[J]. Resources, Conservation and Recycling, 2010, 54(11): 851-863.

[15] ELSHKAKI A, GRAEDEL T. Dynamic analysis of the global metals flows and stocks in electricity generation technologies[J]. Journal of Cleaner Production, 2013, 59: 260-273.

[16] LI L, WANIA F. Tracking chemicals in products around the world: Introduction of a dynamic substance flow analysis model and application to PCBs[J]. Environment International, 2016, 94: 674-686.

[17] SAKAMORNSNGUAN K, KRETSCHMANN J. Substance flow analysis and mineral policy: The case of potash in Thailand[J]. The Extractive Industries and Society, 2016, 3(2): 383-394.

[18] YELLISHETTY M, MUDD G. Substance flow analysis of steel and long term sustainability of iron ore resources in Australia, Brazil, China and India[J]. Journal of Cleaner Production, 2014, 84: 400-410.

[19] LIU J, AN R, XIAO R G, et al. Implications from substance flow analysis, supply chain and supplier' risk evaluation in iron and steel industry in Mainland of China[J]. Resources Policy, 2017, 51: 272-282.

[20] ARENU U, GREGORIO F. A waste management planning based on substance flow analysis[J]. Resources, Conservation and Recycling, 2014, 85: 54-66.

[21] CHEVRE N, COUTU S, MARGOT J, et al. Substance flow analysis as a tool for mitigating the impact of pharmaceuticals on the aquatic system[J]. Water Research, 2013, 47(9): 2995-3005.

[22] CHU J W, YIN X B, HE M C, et al. Substance flow analysis and environmental release of antimony in the life cycle of polyethylene terephthalate products[J]. Journal of Cleaner Production, 2021, 291: 125252.

[23] YOSHIDA H, CHRISTENSEN T, GUILDAL T. A comprehensive substance flow analysis of a municipal wastewater and sludge treatment plant[J]. Chemosphere, 2015, 138: 874-882.

[24] WANG M X, CHEN W, LI X. Substance flow analysis of copper in production stage in the U.S. from 1974 to 2012[J]. Resources, Conservation and Recycling, 2015, 105: 36-48.

[25] GUO X Y, SONG Y. Substance flow analysis of copper in China[J]. Resources, Conservation and Recycling, 2008, 52: 874-882.

[26] HAN F, YU F, CUI Z J. Industrial metabolism of copper and sulfur in a copper-specific eco-industrial park in China[J]. Resources, Conservation and Recycling, 2016, 133: 459-466.

[27] ZHANG L, CHEN T M, YANG J M, et al. Characterizing copper flows in international trade of China, 1975-2015[J]. Science of the Total Environment, 2017, 601/602: 1238-1246.

[28] PFAFF M, GLOSER-CHAHOUD S, CHRUBASIK L, et al. Resource efficiency in the German copper cycle: Analysis of stock and flow dynamics resulting from different efficiency measures[J]. Resources, Conservation and Recycling, 2018, 139: 205-218.

[29] HU Z Q, QU J, GUO L, et al. Selective leaching of arsenic and valuable metals in copper smelting soot strengthened by ball milling pretreatment[J]. The Chinese Journal of Nonferrous Metals, 2020, 30(8): 1915-1924. (in Chinese)

Physicochemical and Environmental Characteristics of Alkali Leaching Residue of Wolframite and Process for Valuable Metals Recovery

Abstract: Physicochemical and mineralogical characteristics of an alkali leaching residue of wolframite were studied by XRD, SEM-EDS, chemical phase analysis, mineral liberation analyzer (MLA), and TG-DSC methods. Batch leaching tests, toxicity characteristic leaching procedure (TCLP) tests and Chinese standard leaching tests (CSLT) were conducted to determine the environmental mobility of toxic elements. The results show that, due to the high contents of W, Fe, Mn, Sn, and Nb, the residue is with high resource value, but the content of a toxic element, As, is also high. The existing minerals of the investigated elements mainly occur as monomer particles, but it is difficult to extract these valuable metals by conventional acid leaching due to their mineral properties. The release of As increases over time in acidic environment. The leaching concentration of all investigated harmful elements through TCLP is within the limiting value, while the leaching concentration of As through CSLT exceeds the limiting value by more than 4 times, so the residue is classified as hazardous solid waste based on the Chinese standard. A process for valuable metals recovery from this residue was proposed. Preliminary experimental results indicated that the main valuable metals could be extracted effectively.

Keywords: Alkali leaching residue; Physicochemical characteristics; Environmental characteristics; Valuable metals recovery

1 Introduction

Tungsten is an important strategic metal that is widely used in many fields such as machinery manufacturing, chemical industry, aerospace industry and national defense industry[1, 2]. The main

---

Published in *Transactions of Nonferrous Metals Society of China*, 2022, 32(5): 1638−1649. Authors: Chen Yuanlin, Guo Xueyi, Wang Qinmeng, Tian Qinghua, Zhang Jinxiang, Huang Shaobo.

economical tungsten mineral resources are wolframite $[(Fe, Mn)WO_4]$ and scheelite ($CaWO_4$). The decomposition of tungsten minerals is one of the key processes for tungsten extraction[3]. After a long period of development, an efficient process for tungsten minerals decomposition-autoclaved alkali leaching has been proposed and has been widely applied in China[4]. However, large amount of leaching residue is inevitably generated during minerals decomposition process. According to statistics, the cumulative amount of tungsten residue has reached 1×10^6 t in China, and it increases by more than 70000 t per year[5, 6]. Most of the alkali leaching residue is mainly stockpiled so far.

Generally, the alkali leaching residue contains a fair amount of valuable metals, such as W, Mo, Mn, Sn, Ta and Nb. In some cases, the contents of some metals, such as W (above 1%) and Sn (above 0.4%), are higher than that of raw ore[7]. Therefore, this type of residue can be regarded as a high valued secondary resource for some valuable metals. On the other hand, some toxic elements, such as As and Pb, are also enriched in the residue, which poses a potential threat to the local ecologic system[6, 8]. What is worse, the alkali leaching tungsten residue has been classified as a hazardous solid waste in China according to the latest environmental law. As a result, the tungsten production industry is required to pay an environmental tax, about RMB ￥1000 per ton of residue[9, 10]. In recent decades, many processes have been proposed for the recovery of valuable metals and the harmless disposal of the residue: (1) beneficiation of W, Mo, Bi, and Sn by flotation and gravity separation[10-12]; (2) extraction of W, Sc, Ta, and Nb by hydrometallurgical methods[10-13]; (3) recovery of W, Mn, Fe, Ta and Nb by pyrometallurgical methods[10-12]; (4) solidification/stabilization of the toxic elements[10, 14]. However, the choice of recovery and stabilization processes for the residue strongly depends on the physicochemical and mineralogical characteristics, resource value, and the potential environmental characteristics of the residue[15]. Therefore, in order to develop an efficient technology for the tungsten residue treatment, it is essential to take a systematic analysis on the characteristics of the residue. In fact, a detailed research on the characteristics of the alkali leaching residue was rarely reported, especially for the mineralogical state of valuable and toxic elements, the releasing behaviors of toxic elements, and the evaluation of environmental characteristics.

To clarify the occurrence state of valuable and harmful elements, evaluate the potential of valuable metals recovery and environmental risk of toxic elements, and select an appropriate treatment process for the residue, this study is devoted to systematically investigate the characteristics of an alkali leaching residue of wolframite, including chemical compositions, microstructure features, mineralogical characteristics of the main valuable and toxic elements, thermal properties, and releasing behaviors and environmental mobility of toxic elements. The residue was collected from an ammonium paratungstate (APT) production enterprise in Jiangxi province, China, one of the largest tungsten producers in China. Basing on the physicochemical characteristic investigation, this work also proposes a process for the comprehensive recovery of valuable metals and the harmless treatment of this residue, which provides a reference for the technology development of tungsten residue treatment.

2 Experimental

2.1 Material

A tungsten residue, which was generated from autoclaved alkali leaching process of wolframite under the conditions of 160 ℃ reaction temperature, (6.06-7.07) MPa reaction pressure, and NaOH solution as leaching agent, was chosen as material in the tests, since wolframite has been used as the main raw material in tungsten industry for a long time and autoclaved alkali leaching process is widely used for wolframite decomposition in China. The agents used in this work are analytically pure.

2.2 Determination of physicochemical characteristics

The contents of the main elements such as W, Mo, Mn, Fe, Sn, Pb, Nb, and As in the residue and the phase composition of some elements were determined by Changsha Research Institute of Mining and Metallurgy, China.

The microtopography features of the residue were observed by a scanning electron microscope (SEM-EDS, JSM-7900F, JEOL, Japan).

The crystallographic composition was analyzed by X-ray diffraction (XRD, TTR Ⅲ, with Cu K_α radiation, Rigaku, Japan), and the analysis was operated with a 2θ range from 10° to 80°, a step size of 0.02°, and a scanning speed of 2 (°)/min. The XRD data were analyzed by Jade 6.5 software. The occurrence states of the main valuable metals and toxic elements were analyzed by mineral liberation analyzer (MLA) (MLA 650, configured with a scanning electron microscope FEI Quanta 650, an energy spectrometer Bruker Quantax 200 with Dual XFlash 5010, and a software MLA 3.1 for automatic analysis of mineral parameters).

Thermal stability of the residue was tested by TG-DSC test (Model: NETZSCH STA 449 F3) under flowing air with a flow velocity of 50 mL/min at a heating rate of 10 ℃/min.

2.3 Determination of environmental mobility of toxic elements

The toxic metals releasing behaviors of the residue were determined through batch leaching experiments. The effects of leaching time in the range of 2 to 24 h and initial pH in the range of 2 to 12 on the releasing behaviors of toxic metals were investigated. In each run, moisture content of the residue ($w(H_2O)$, %), was determined firstly. A sample with a dry mass of 10 g (the mass of the sample was calculated by the formula of $10/(100\%-w(H_2O))$), was transferred to a 250 mL Erlenmeyer flask, and leaching agent with specified pH was then added into the flask (pH was adjusted by diluted mixture of sulfuric acid and nitric acid with a mass ratio of 2∶1). The flask was shaken at the vibrational frequency of 150 r/min using a shaker for specified time during leaching experiments. The leaching solution was acidified and analyzed.

The leaching toxicity was evaluated through the toxicity characteristic leaching procedure (TCLP)[16, 17] and Chinese standard leaching test (CSLT)[15, 18]. In the TCLP, a sample with a dry mass of 50 g was transferred to a 2 L extraction bottle, then the acetic acid solution with a pH of 2.88±0.05 was added into the bottle as leaching agent under a liquid to solid ratio of 20 : 1. The leaching test was conducted in a flip oscillator running at (30±2) r/min and was maintained for 24 h. The leachate was separated by a pressure filter using a filter membrane with 0.45 μm aperture.

The CSLT was conducted according to the standard method (GB 5085.3-2007 and HJ/T 299-2007) of China. The leaching agent was the diluted mixture of sulfuric acid and nitric acid with a mass ratio of 2 : 1, the pH was adjusted to 3.20±0.05, and the liquid to solid ratio was 10 : 1. The leaching experiment was maintained for 18 h in a flip oscillator. The leachate was separated by a pressure filter using a filter membrane with 0.45 μm aperture.

The concentrations of metals in the leachate obtained from the above experiments were determined by ICP-OES (Avio-500, PerkinElmer) and ICP-MS (iCAP-RQ, Thermo Scientific). The calibration standards used for ICP analysis were purchased from the National Nonferrous Metal and Electronic Material Analysis and Testing Center of China. The concentration of fluorine was determined by fluorinion selective electrode (PHSJ-4F). The experiments and tests were conducted three times, and the average was taken as the final result.

2.4 Experiments on valuable metals recovery

Carbothermal reduction experiments were conducted in a box type furnace (SQFL-1700C, Shanghai Jujing Precision Instrument Manufacturing Co., LTD, China) with the atmosphere of nitrogen. In each run, 200 g of dried alkali leaching residue was mixed with a certain amount of carbon (reductant) in a crucible, then the crucible was placed in the furnace and nitrogen was injected. After 0.5 h, the system was heated up to the designed temperature and maintained for a certain period of time. The multicomponent alloy obtained from carbothermal reduction experiments was crushed to <74 μm.

The alkaline smelting of multicomponent alloy was conducted in a box type furnace (SQFL-1700C, Shanghai Jujing Precision Instrument Manufacturing Co., LTD, China) with argon atmosphere. In each run, 5 g of multicomponent alloy powder was mixed with a certain amount of alkali medium in a crucible, and the crucible was then placed in the furnace. After that, argon flow was injected, and the system was heated up to the designed temperature and smelted for 90 min.

The alkaline slag obtained from alkaline smelting was leached for 90 min in hot water of 60 ℃ with a stirrer, and the leachate and leaching residue were separated by a pressure filter. The leachate was then transferred to a beaker and stirred with a magnetic stirrer, and the pH of the leachate was adjusted to 8 under stirring. After pH adjustment, the system was kept stirring for 30 min, and then the precipitate was separated from the solution using a pressure filter.

3 Results and discussion

3.1 Chemical composition of residue

The main elemental composition of the alkali leaching residue is presented in Table 1. The residue contains W(1.22%), Fe(18.24%), Mn(17.25%), Ca(4.74%), and Si(7.03%) as the major elemental compositions, some other valuable metals including Sn(0.74%), Ta(0.16%), Nb(0.61%), Mo(0.27%), and Bi(0.54%) also exist in the residue, which contribute to a high resource value of the residue.

Table 1 Main elemental composition of alkali leaching residue   %

W	Mn	Fe	Sn	Nb	Ta	Mo
1.22	17.25	18.24	0.74	0.61	0.16	0.27
Bi	Pb	As	Al	Ca	Si	Cr
0.54	0.32	0.75	1.96	4.74	7.03	0.019
Sc	Na	Ti	S	C	F	Others
0.03	3.68	0.23	1.84	1.09	0.20	39.08

With the continued exploitation of high-grade mineral resources around the world, the raw ore grades of many metals have become extremely low. It was reported that the grades of WO_3, Sn, and Nb_2O_5 in the raw ore of mineral processing industry were as low as 0.30%, 0.17%, and 0.12%, respectively[19, 20, 21]. Obviously, the content of W, Sn, and Nb in this residue is higher than that of corresponding raw ore. Therefore, this residue can be considered as a high value secondary resource. Besides, the main toxic elements in the residue are Pb(0.32%), As(0.75%), and trace amount of Cr(0.019%). The As content is fairly high, so the occurrence state and releasing behavior of As should be the main concern regarding the environmental characteristics of the residue.

3.2 Occurrence state of main valuable and toxic elements

The microtopography of the residue was observed with a scanning electron microscope, as microtopography features of a solid waste play an important role in the environmental characteristics and the selection of treatment process. Fig. 1(a) clearly shows that, the residue consists of irregular particles with different sizes, in which the majority of these particles are ultra-fine, even smaller than 10 μm. There are many cracks on the surface of relatively large particles, which may be due to the grinding pretreatment before tungsten extraction. Some large particles are covered with many fine particles. The fine particles and cracks on the large particles may cause the release of toxic elements when the residue is exposed to natural environment. Table 2 shows the EDS analysis results of large particles (Area 1) and fine particles (Area 2). The results reveal that, the large particles (Area 1) contain high content of Fe and relatively low content of Mn, while the fine particles (Area 2) contain high content of Mn and relatively low content of Fe. Besides, Bi and Sn are not detected on the surface of fine particles (Area 2), and the content of Pb and As on the surface of large particles is higher than that on fine particles.

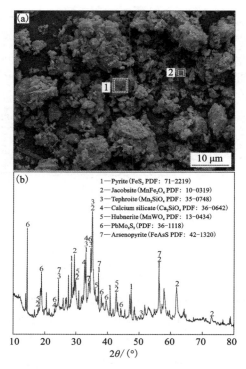

Fig. 1 SEM image (a) and XRD pattern (b) of residue

Table 2 EDS analysis results of large (Area 1 in Fig. 1(a)) and fine (Area 2 in Fig. 1(a)) particles %

Area	Fe	Mn	W	Si	Al	Ca	Mo
1	37.67	7.77	7.38	18.88	2.26	4.35	1.95
2	27.05	26.33	3.68	15.29	3.90	6.33	1.59

Area	Bi	Pb	Sn	As	Cr	Na
1	3.28	3.57	0.52	1.19	0.15	8.65
2	0.00	0.81	0.00	0.50	0.00	12.60

The XRD pattern [Fig. 1(b)] obtained by slow scanning shows that, the main crystalline phases in the residue are fairly complicated. Pyrite, jacobsite, silicate of manganese and calcium are the main phases in the XRD pattern. Meanwhile, lead molybdenum sulfide, hubnerite (one of the sources of wolframite), and arsenopyrite are also detected. The phases of some other elements, such as Bi and Nb, are not detected, which may be due to their low content or low crystallinity.

The elemental composition results show that, tungsten is a main valuable metal with relatively high content, while arsenic is a toxic element with high content. The phase compositions of tungsten and arsenic were analyzed through chemical analysis method. Table 3 indicates that, the main phases of tungsten are scheelite and wolframite, and the mass fractions of scheelite and wolframite are 44.31% and 37.97%. A small amount of tungsten exists in the form of tungstite (17.72%). For the phase composition of arsenic (Table 4), the mass fractions of arsenate, arsenic sulfide, and arsenic oxides are 41.33%, 8.13%, and 1.47%, respectively. The other 49.07% of arsenic exists in insoluble arsenic compound. It is inferred that this part of insoluble compound may be arsenopyrite. Although the arsenic oxides account for a very small proportion, the oxides cannot be ignored as their impacts on environment are critical[15].

Table 3 Phase composition of tungsten in residue   %

Wolframite	Scheelite	Tungstite	Total
37.97	44.31	17.72	100.00

Table 4 Phase composition of arsenic in residue   %

Arsenic oxides	Arsenate	Arsenic sulfide	Insoluble arsenic compound	Total
1.47	41.33	8.13	49.07	100.00

To further understand the occurrence states of various valuable metals and toxic elements, mineralogical characteristics analysis was conducted by MLA. Fig. 2 shows the SEM images of the minerals of the main valuable and toxic elements. The occurrence of these elements is summarized as follows.

(1) The detected tungsten minerals are scheelite and wolframite, and these two minerals mainly exist in the form of monomer particles [Figs. 2(a) and (c)]; a small portion of scheelite is associated with iron manganese oxide colloid [Fig. 2(b)]; some wolframite is associated with calcium-rich iron manganese oxide colloid [Fig. 2(d)].

(2) The detected iron minerals are pyrite, hematite and oxide colloids, and both the pyrite and hematite exist in the form of monomer particles [Figs. 2(e) and (f)]; some Fe forms oxide colloids with Mn, Si, and Ca, and is associated with various minerals of W, Nb Mo, Bi, and As.

(3) The detected manganese minerals are vernadite and oxide colloids, and the vernadite mainly exists in the form of monomer particles [Fig. 2(g)]; the majority of Mn forms oxide colloids with Fe, Si, and Ca.

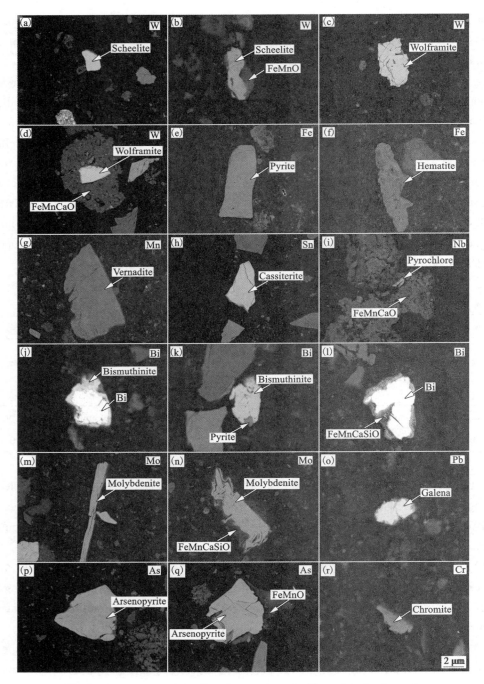

Fig. 2　Occurrence state of various main valuable and toxic elements

(4) The detected tin mineral is cassiterite which exists in the form of monomer particles [Fig. 2(h)].

(5) The detected niobium mineral is pyrochlore which mainly exists in the form of ultrafine

particles associated with calcium-rich iron manganese oxide colloid [Fig. 2(i)].

(6) The detected bismuth minerals are metal Bi and bismuthinite, some Bi is associated with bismuthinite which grows on the edges of Bi particles [Fig. 2(j)], and some Bi is associated with iron manganese calcium silicon oxide colloid [Fig. 2(l)]; a part of bismuthinite is associated with ultrafine pyrite [Fig. 2(k)] except the interlocked particles with Bi.

(7) The detected molybdenum mineral is molybdenite, the majority of molybdenite exists in the form of needle-like monomer particles [Fig. 2(m)], and some molybdenite is associated with iron manganese calcium silicon oxide colloid [Fig. 2(n)].

(8) The detected lead mineral is galena which exists in the form of monomer particles [Fig. 2(o)].

(9) The detected arsenic mineral is arsenopyrite, the majority of arsenopyrite exists in the form of monomer particles [Fig. 2(p)], and some arsenopyrite is associated with iron manganese oxide colloid (Fig. 2(q)).

(10) The detected chromium mineral is chromite which mainly exists in the form of fine monomer particles [Fig. 2(r)].

The MLA analysis results reveal that, the detected minerals are mainly monomer particles, which is the result of sufficient liberation of most minerals in wolframite concentrate caused by grinding pretreatment before leaching. The tungstite, arsenic oxides, and arsenate detected in chemical phase analysis are not found by MLA, which may be due to the low crystallinity degree of these minerals or that the ultra fine particles are wrapped by oxide clays. The lead molybdenum sulfide ($PbMo_6S_8$) detected by XRD may be the interlocked particles of undeveloped molybdenite and galena. Based on the presented mineral properties of the main valuable metals, it can be concluded that, it is difficult to extract these valuable metals from the existing minerals, such as pyrite, cassiterite, and pyrochlore, by conventional acid leaching. Pretreatment (for example, high-temperature roasting) is necessary for decomposing the minerals of valuable metals into leachable compounds.

3.3 Thermal characteristics of residue

TG-DSC test was conducted to evaluate the thermal stability of the residue. Fig. 3 shows the TG-DSC curves of the residue. As presented in Fig. 3, two mass loss stages and a mass gain stage are observed. A mass loss of 5.55% is observed from room temperature to 469.9 ℃, which is due to the volatilization of adsorbed and bonded water. The result also illustrates that the residue has a good thermal stability when temperature is lower than 469.9 ℃. It is worth noting that, an abnormal phenomenon during the test is that a slight mass gain of 1.45% occurs from 469.9 to 557.6 ℃, and a strong exothermic peak is observed on the DSC curve. Considering that a certain amount of pyrite exists in the residue according to the XRD pattern (Fig. 1) and MLA analysis [Fig. 2(e)], it can be concluded that, the mass increase may be attributed to the oxidation of pyrite (an exothermic reaction) and the formation of sulfate under air flow and high temperature. HU et al.[22] have reported that, during the oxidation process of pyrite, sulfates including ferrous sulfate (FeSO$_4$) and ferric sulfate [Fe$_2$(SO$_4$)$_3$] can be formed as minor products from 803 to 873 K (530 to 600 ℃, which is similar to the temperature range in this work) and will result in mass gain. Wang et al.[23] also stated that, for the oxidation of pyrite, Fe-S-O was first formed under certain conditions, which might cause mass gain. The additional mass loss above 557.6 ℃ is mainly caused by the decomposition of sulfides and arsenic compounds. In this temperature range, the molybdenite and bismuthinite contained in the residue (see Fig. 2) can be oxidized into MoO$_3$ and Bi$_2$O$_3$[24-26]. Besides, some residual arsenopyrite may ultimately convert to As$_2$O$_3$ at about 650 ℃[27].

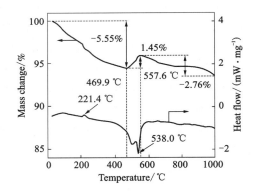

Fig. 3 TG-DSC curves of residue

3.4 Environmental characteristics of residue

3.4.1 Leaching behaviors of toxic elements

The release of toxic elements in the residue can be affected by natural conditions when the residue is disposed in a segregated landfill. The effects of contacting time and pH value of solution on the dissolution of As and Pb were investigated. The results are shown in Fig. 4. Fig. 4(a) indicates that, under the initial pH of 3.00±0.05 and liquid to solid ratio of 10∶1, the leaching concentration of As increases over time within 12 h and remains stable over 12 h, while the leaching concentration of Pb is fairly low (<1 mg/L). Fig. 4(b) shows that, under the liquid to solid ratio of 10∶1 and leaching time of 20 h, the leaching concentration of As increases with the increase of initial pH from 2 to 8 and then decreases. The result is similar to the releasing behavior of As in a sludge from copper and lead-zinc smelter reported by Yao et al.[15]. Some researchers

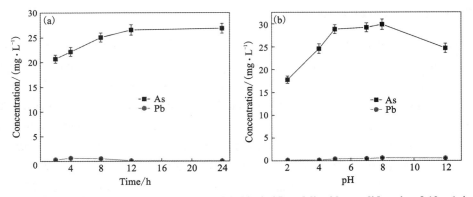

Fig. 4 Effects of leaching time under initial pH of 3.00±0.05 and liquid to solid ratio of 10∶1 (a), and initial pH under liquid to solid ratio of 10∶1 and leaching time of 20 h (b) on leaching concentrations of As and Pb

stated that, under alkaline condition, the formation of iron-oxyhydroxide phases and arsenic oxide passivation films on the arsenopyrite surface (which was the main arsenic mineral in this residue) reduced the diffusion of dissolved oxygen in the solution and the oxidation dissolution of arsenopyrite[28]. The leaching concentration of Pb is fairly low in the whole investigated pH range. Therefore, there is a high risk of As release during the disposal of the leaching residue even in the neutral environment.

3.4.2 Leaching toxicity assessment

To determine the mobility of harmful elements and assess the potential risk of the residue when disposed in a segregated landfill, TCLP and CSLT were conducted on the residue.

Environmental risk coefficient is put forward to assess the environmental risk caused by the harmful elements leached from the residue. The coefficient γ is defined as Eq. (1):

$$\gamma = c_i / C_i \tag{1}$$

where c_i is the concentration of an element in leachate, and C_i is the standard limiting concentration. The higher the coefficient, the greater the environmental risk, and vice versa.

Table 5 lists the leaching concentrations and corresponding environmental risk coefficients of the main harmful elements. As can be seen in Table 5, the leaching concentrations of all the investigated elements obtained by TCLP do not exceed the limiting values of US Environmental Protection Agency (EPA)[15], and the coefficients of these elements are fairly low. However, for the results obtained by CSLT, the leaching concentration of As exceeds the limiting value of GB 5085.3−2007 of China[29], the environmental risk coefficient of As is as high as 4.524, while the concentrations of other elements do not exceed the limiting values similar to those of TCLP tests. The results also indicate that, the arsenic compounds in this residue are dissolved easily in oxidizing strong acid medium (sulfuric acid and nitric acid).

Table 5 Leaching concentrations and environmental risk coefficients of the main harmful elements

Method	Index	As	Pb	Zn	Cr	Cu	F
TCLP	Leaching concentration/(mg·L^{-1})	0.18	0.0078	0.0042	0.0028	0.45	6.88
	Limiting value#/(mg·L^{-1})	5.0	5.0	N	5.0	15.0	N
	γ	0.036	0.0016	N	0.0006	0.03	N
CSLT	Leaching concentration/(mg·L^{-1})	22.62	0.052	0.0057	0.0185	0.11	9.54
	Limiting value*/(mg·L^{-1})	5	5	100	15	100	100
	γ	4.524	0.0104	0.0001	0.0012	0.0011	0.0954

#The limiting value of US; * The limiting value of China; N: Not limited in the standard

4 Process for valuable metals recovery

As shown by above studies, it is clear that the resource value of this residue is fairly high for the contents of some valuable metals, such as W, Sn, Nb, and Bi. On the other hand, the residue contains some toxic elements (mainly As and Pb), and is classified as hazardous waste. Therefore, the clean utilization and safe disposal of this residue is in great necessity. The occurrence state analysis of the main valuable metals by MLA indicates that it is difficult to extract these valuable metals from the existing minerals by conventional acid leaching. In addition, the treatment process will be complicated by doing arsenic removal from leaching solution since arsenic is dissolved easily in sulfuric acid solution (see Table 5). According to the above analysis, a potential process for valuable metals recovery from this residue has been designed, and the simplified flowsheet is shown in Fig. 5. The characteristics of this proposed route are the combination of carbothermal reduction smelting, low temperature alkaline smelting, and hydrometallurgical process to realize the comprehensive utilization of valuable metals and low value components. One outstanding advantage of the process is that toxic metals As and Pb flow into the flue dust during the carbothermal reduction smelting, the small amount of flue dust simplifies the subsequent treatment of As and Pb.

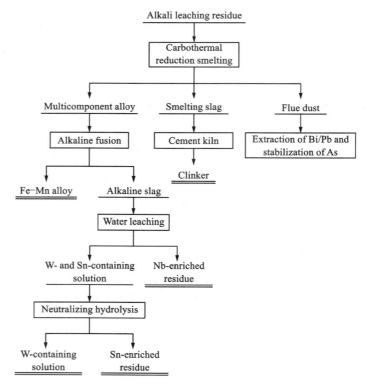

Fig. 5 Simplified flowsheet of designed process for alkali leaching residue treatment

Preliminary experiments have been carried out to treat the residue with the proposed process, and the results are listed in Table 6. Carbothermal reduction smelting is firstly conducted with the reduction temperature of 1500 ℃, reaction time of 60 min, and carbon dosage of 16% (mass fraction of residue). A multicomponent alloy can be obtained, and the recoveries of W, Fe, Mn, Sn, and Nb in the alloy are 91.84%, 90.64%, 56.95%, 65.09%, and 82.98%, respectively. The As, Pb, and Bi flow into flue dust during smelting. As mentioned before, the small amount of flue dust containing As and Pb simplifies the subsequent extraction of Bi and Pb, and the solidification/stabilization of As. The smelting slag can be utilized as supplementary raw material of cement, as the main components of smelting slag (SiO_2, CaO, Al_2O_3, and a certain amount of iron oxide) are similar to those of cement raw material. This is an appropriate way for the comprehensive utilization of the low value component. Alkaline fusion with smelting temperature of 800 ℃, reaction time of 90 min, NaOH and $NaNO_3$ as alkali medium, is then conducted on crushed multicomponent alloy. W, Sn and Nb are transformed into fused salt (Na_2WO_4, Na_2SnO_3, Na_5NbO_5) during alkaline fusion, and a Fe-Mn alloy with 69.28% Fe and 20.82% Mn can be obtained, the final recoveries of Fe and Mn are 90.63% and 56.95%, respectively. After water leaching of salt slag, the insoluble Na_5NbO_5 is separated with water-soluble Na_2WO_4 and Na_2SnO_3, Nb-enriched residue with 47.92% Nb and a recovery of 73.31% is obtained. Then through further neutralizing hydrolysis of W- and Sn- containing solution under pH value of 8, Sn is separated and the Sn-enriched residue with 60.02% Sn and a recovery of 56.29% is obtained. The final Na_2WO_4 solution can be transported to ammonium paratungstate (APT) production system for tungsten extraction. The preliminary results show that, the valuable metals can be extracted effectively from the residue by the proposed technological process. Further investigations will be carried out systematically to realize comprehensive utilization of the residue.

Table 6 Preliminary results of valuable metals recovering experiments

Product	Index	W	Fe	Mn	Sn	Nb
Multicomponent alloy	Content/%	5.31	62.36	18.73	3.47	1.63
	Recovery/%	91.84	90.64	56.95	65.09	82.98
Fe-Mn alloy	Content/%	0.08	69.28	20.82	0.05	0.06
	Recovery/%	0.93	90.63	56.95	0.84	2.75
Nb-enriched residue	Content/%	0.58	—	—	0.96	47.92
	Recovery/%	0.30	—	—	0.54	73.31
Sn-enriched residue	Content/%	0.12	—	—	60.02	0.08
	Recovery/%	0.10	—	—	56.29	0.20

All the indexes are calculated based on the raw material (alkali leaching residue)

5 Conclusions

(1) The residue is with fairly high resource value for the high contents of some valuable metals (W 1.22%, Fe 18.24%, Mn 17.25%, Sn 0.74%, Nb 0.61%), while the content of the toxic element As is also high (0.75%). Mineralogical analysis shows that, W exists in both wolframite and scheelite; Fe exists as pyrite, hematite and oxide colloids; Mn exists as vernadite and oxide colloids; Sn mainly exists as cassiterite; Nb mainly exists as pyrochlore; As mainly exists as arsenopyrite. The particle size of the residue is ultrafine. Although the existing minerals of the investigated elements mainly occur as monomer particles, it is difficult to extract these valuable metals by conventional acid leaching due to their mineral properties. The residue is in good thermal stability below 469 ℃.

(2) The release of As from the residue increases over time even in neutral environment. The leaching concentrations of all the investigated harmful elements through TCLP are within the limiting values of US, while the leaching concentration of As through CSLT exceeds the limiting value of China by more than 4 times, and the residue is classified as hazardous solid waste by the Chinese standard.

(3) A potential process for valuable metals recovery from this residue was proposed. The preliminary experiment results indicated that the main valuable metals could be extracted effectively: the Fe-Mn alloy with 69.28% Fe, 20.82% Mn, Fe and Mn recovery of 90.63% and 56.95% was obtained; the Nb-enriched residue with 47.92% Nb and a recovery of 73.31% was obtained; the Sn-enriched residue with 60.02% Sn and a recovery of 56.29% was obtained; the Na_2WO_4 solution could be used for tungsten extraction.

Acknowledgments

The authors are grateful for the financial supports from the National Key R & D Program of China (No. 2019YFC1907400), and the National Natural Science Foundation of China (Nos. 51904351, 51620105013)

References

[1] ZHAO Z W, XIAO L P, SUN F, et al. Study on removing Mo from tungstate solution by activated carbon loaded with copper [J]. International Journal of Refractory Metals and Hard Materials, 2010, 28(4): 503-507.

[2] LI Y L, YANG J H, ZHAO Z W. Recovery of tungsten and molybdenum from low-grade scheelite [J].

JOM, 2017, 69(10): 1958-1962.
[3] CAO C F, QIU X C, LI Y H, et al. Study on leaching behaviour of tungstates in acid solution containing phosphoric acid[J]. Hydrometallurgy, 2020, 197: 105392.
[4] LI J T, MA Z L, LIU X H, et al. Sustainable and efficient recovery of tungsten from wolframite in a sulfuric acid and phosphoric acid mixed system[J]. ACS Sustainable Chemistry & Engineering, 2020, 8(36): 13583-13592.
[5] LI J J, HE D W, ZHOU K G, et al. Research status of comprehensive utilization of tungsten slag[J]. Conservation and Utilization of Mineral Resources, 2019, 39(3): 125-132. (in Chinese)
[6] YANG J Z, GAO H F, WANG N, et al. Research on polluting characteristic of tungsten residue from ammonium paratungstate (APT)[J]. Journal of Environmental Engineering Technology, 2015, 5(6): 525-530. (in Chinese)
[7] YANG J Y, QI S, LIU H, et al. Progress of research related to the comprehensive recovery and utilization of tungsten smelting slag[J]. Chinese Journal of Engineering, 2018, 40(12): 1468-1475. (in Chinese)
[8] WANG S F, JIAO B B, ZHANG M M, et al. Arsenic release and speciation during the oxidative dissolution of arsenopyrite by O_2 in the absence and presence of EDTA[J]. Journal of Hazardous Materials, 2018, 346: 184-190.
[9] FAN Z K, HUANG C, XU G Z, et al. Experimental study on decomposition of tungsten smelting slag for emission reduction and resource recycling[J]. Rare Metals and Cemented Carbides, 2020, 48(2): 1-4. (in Chinese)
[10] LIU H, LIU H L, NIU C X, et al. Comprehensive treatments of tungsten slags in China: A critical review [J]. Journal of Environmental Management, 2020, 270: 110927.
[11] ZHU H L, DENG H B, WU C H, et al. The current comprehensive recovery technology of tungsten slag [J]. China Tungsten Industry, 2010, 25(4): 15-18. (in Chinese)
[12] XIE X X, ZHANG X X. The research status and development trend of tungsten residue recycling in China [J]. Shanghai Chemical Industry, 2014, 39(5): 26-29. (in Chinese)
[13] NIE H P, WANG Y B, WANG Y L, et al. Recovery of scandium from leaching solutions of tungsten residue using solvent extraction with Cyanex 572[J]. Hydrometallurgy, 2018, 175: 117-123.
[14] JING Q X, WANG Y Y, CHAI L Y, et al. Adsorption of copper ions on porous ceramsite prepared by diatomite and tungsten residue[J]. Transactions of Nonferrous Metals Society of China, 2018, 28(5): 1053-1060.
[15] YAO L W, MIN X B, XU H, et al. Physicochemical and environmental properties of arsenic sulfide sludge from copper and lead-zinc smelter[J]. Transactions of Nonferrous Metals Society of China, 2020, 30(7): 1943-1955.
[16] KE Y, CHAI L Y, MIN X B, et al. Sulfidation of heavy-metal-containing neutralization sludge using zinc leaching residue as the sulfur source for metal recovery and stabilization[J]. Minerals Engineering, 2014, 61: 105-112.
[17] PENG B LEI J, MIN X B, et al. Physicochemical properties of arsenic- bearing lime-ferrate sludge and its leaching behaviors[J]. Transactions of Nonferrous Metals Society of China, 2017, 27(5): 1188-1198.
[18] YAO L W, MIN X B, KE Y, et al. Release behaviors of arsenic and heavy metals from arsenic sulfide sludge during simulated storage[J]. Minerals, 2019, 9(2): 130.
[19] ZHANG H, ZHANG F M, JIANG H Y, et al. Beneficiation test of a scheelite mine in Hunan[J]. China Tungsten Industry, 2020, 35(1): 23-28. (in Chinese)

[20] WANG P R, WANG J. Beneficiation tests of a low-grade high slime tin oxide ore from Yunnan[J]. Metal Mine, 2020(7): 83-88. (in Chinese)

[21] LI J, XIE X, LV J F, et al. Overview of niobium resources and research progress in mineral processing technology[J]. Metal Mine, 2021(2): 120-126. (in Chinese)

[22] HU G L, DAM-JOHANSEN K, WEDEL S, et al. Decomposition and oxidation of pyrite[J]. Progress in Energy and Combustion Science, 2006, 32(3): 295-314.

[23] WANG T, ZHANG H, YANG H R, et al. Oxidation mechanism of pyrite concentrates (PCs) under typical circulating fluidized bed (CFB) roasting conditions and design principles of PCs' CFB roaster[J]. Chemical Engineering and Processing-Process Intensification, 2020, 153: 107944.

[24] LI X B, WU T, ZHOU Q S, et al. Kinetics of oxidation roasting of molybdenite with different particle sizes [J]. Transactions of Nonferrous Metals Society of China, 2021, 31(3): 842-852.

[25] UTIGARD T. Oxidation mechanism of molybdenite concentrate[J]. Metallurgical and Materials Transactions B, 2009, 40: 490-496.

[26] ZHAN J, WANG Z J, ZHANG C F, et al. Separation and extraction of bismuth and manganese from roasted low-grade bismuthinite and pyrolusite: Thermodynamic analysis and sulfur fixing[J]. JOM, 2015, 67(5): 1114-1122.

[27] HAFFERT L, CRAW D. Mineralogical controls on environmental mobility of arsenic from historic mine processing residues, New Zealand[J]. Applied Geochemistry, 2008, 23(6): 1467-1483.

[28] ASTA M P, CAMA J, AYORA C, et al. Arsenopyrite dissolution rates in O_2-bearing solutions[J]. Chemical Geology, 2010, 273(3/4): 272-285.

[29] GB 5085.3-2007. Identification standards for hazardous wastes—Identification for extraction toxicity[S]. Beijing: China Environmental Science Press, 2007. (in Chinese)

Antimony and Arsenic Substance Flow Analysis in Antimony Pyrometallurgical Process

Abstract: Substance flow analysis was applied to an antimony pyrometallurgical system. By taking antimony and arsenic as the objective elements, the mass balance and substance flow charts based on the production system were established, and evaluating indicators such as the direct recovery rate, waste recovery rate, and resource efficiency were set up. The results show that the resource efficiency of antimony is 89.21%, and the recovery rates of antimony in volatilization smelting, reduction smelting, and refining are 78.79%, 91.00%, 96.06%, respectively. At the same time, for 1 t of antimony produced, 11.94 kg of arsenic is carried into the smelting system. Arsenic is a major impurity element in the smelting process. The distribution behavior of arsenic in the main process was analyzed. Based on the substance flow analysis, some recommendations for improving the resource efficiency of antimony and cleaner production were proposed.

Keywords: Antimony metallurgy; Substance flow analysis; Antimony resource efficiency; Arsenic distribution behavior

1 Introduction

Antimony is an important strategic metal. Antimony metal and compounds are mainly used in producing semiconductors, far-infrared materials, lead-antimony alloys, flame retardants, catalysts, and other products. They are widely used in the military industry, electronics, aerospace, and other fields[1-3]. According to the survey data of the U.S. Geological Survey in 2021, the global antimony reserves in 2020 are 1.9 million tones, and the reserves in China are 480000 tones, accounting for 25% of the total reserves[4]. China is the largest producer of antimony products in the world[5]. With the increasing complexity of antimony resources and the extensive use of antimony-gold concentrate with high arsenic concentration, the concentrations of arsenic and antimony in arsenic-alkali residue produced in the production process have increased significantly (up to 20% arsenic and 20%-30% antimony)[6,7]. With the improvement of environmental protection requirements, it is urgent to solve the problems of efficient separation of high arsenic and antimony and final disposal of arsenic. There are many studies on smelting treatment technology[8-10] and arsenic-containing solid waste treatment[11,12]. Still, few people pay attention to the flow and distribution of antimony and arsenic in the production system. Substance flow profoundly impacts the resource efficiency and environmental load of the production system. Therefore, an in-depth analysis of the distribution behavior of antimony and impurity elements is of great significance in improving the utilization efficiency of antimony resources and reducing the emission level of pollutants.

As an analytical tool, substance flow analysis (SFA) is an important method to study substance flow state of a given system (production, economy, society, etc.) in a certain range. In the past few decades, SFA has been widely used to analyze the stock and flow of various metals (copper[13-15], lead[16], aluminum[17], zinc[18], indium[19], cobalt[20], etc.) In addition, SFA was applied to waste management[21,22], pollution prevention and control[23], resource recycling, and whole industry chain analysis[24,25]. In the past, most studies were focused on material flow and waste management at the regional or national level. Using SFA to track the quantity and migration of substances in the production process can determine the production efficiency of the process and reveal the ways of pollutant generation, which is conducive to the resource management and pollution control of the plant. It has important guiding significance for the efficient utilization of resources, pollutant prevention and control, and the formulation of environmental protection policies for enterprises and industries. Bai et al.[26] took SFA as the research tool, established the substance flow model of a lead-smelting system, evaluated the resource utilization, circulation, and emission level of the system, and put forward some suggestions on emission control and pollution prevention of lead production enterprises. Chen et al.[27] established the mass balance and material flow chart of the tungsten hydrometallurgy system and analyzed the tungsten resource efficiency of the system and the distribution behavior of arsenic in the production process.

The researches on substance flow analysis of antimony have been mainly focused on the national stock, consumption, and circulation of antimony[28] and the environmental impact caused by the use of antimony-containing products[29,30]. There was less research on substance flow analysis of the antimony production process. In this study, SFA was applied to an antimony pyrometallurgical production system to study the substance flow of antimony and arsenic in the whole system. By setting evaluation indicators such as the metal recovery rate, waste recovery rate, resource efficiency and studying the utilization level of antimony resources, the emission characteristics of arsenic pollutants in the production process were revealed, which provided theoretical support for improving the utilization rate of resources in the smelting process and reducing the emission level of contaminants.

2 Methodology

Compared with applying SFA in global or regional large-scale systems, applying SFA to a production process is more specific. It can be used to analyze the impact of various logistics changes on environmental load and resource efficiency and then determine the key links and main factors to put forward corresponding improvement measures. The SFA model of the production process generally includes the substance flow of a unit process and the substance flow of the whole production system composed of several unit processes.

2.1 Definition of substance flows in unit process

If we define a unit process in the production system as Process j, the substance flow model of Process j can be decomposed into six substance flows. Six substance flows are explained below and shown in Fig. 1.

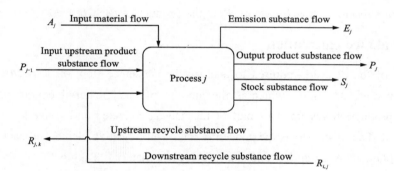

Fig. 1 Decomposition of substance flow chart of a unit process

(1) Input material flow, A_j.

(2) Input upstream product substance flow, P_{j-1}.

(3) Recycle substance flow: upstream recycle substance flow, $R_{j,k}$, meaning that the upstream processes reuse the substance flow; downstream recycle substance flow, $R_{i,j}$, meaning that the substance flow is reused from downstream back to Process j.

(4) Emission substance flow, E_j: This flow includes the by-products and pollutants that are discharged outside of the objective system from Process j.

(5) Output product substance flow, P_j.

(6) Stock substance flow, S_j: This kind of product is temporarily stocked in the warehouse and will be put into production when needed.

According to the mass-balance principle of steady-state process, a unit process j can be expressed as

$$A_j + P_{j-1} + R_{i,j} = P_j + R_{j,k} + E_j + S_j \tag{1}$$

2.2 Substance flows of the whole system

By connecting all unit processes in a particular order, the substance flow model of the whole system can be obtained. The composition of each substance flow is as follows:

(1) Input material flow A:

$$A = \sum_{j=1}^{m} A_j \tag{2}$$

(2) Recycle substance flow R:

$$R = \sum_{i=1}^{m} \sum_{j=1}^{m} R_{i,j} \tag{3}$$

(3) Stock substance flow S:

$$S = \sum_{j=1}^{m} S_j \tag{4}$$

(4) Emission substance flow E:

$$E = \sum_{j=1}^{m} E_j \tag{5}$$

(5) Output substance flow P, where P_j is the product of each process.

2.3 Mass balance calculation

Two parts of the data are required to analyze the substance flow of a specific element. One part is the flow rate, M_i, t/d, which is the amount of each material containing an objective element in the production system. This part of the data is collected and converted from enterprise production report. The other part is the concentration of an objective element in each material, C_i, %, which is obtained by sampling and analyzing each material. Taking Sb as an example, the concentration of antimony in each substance flow, m_i, can be obtained by

$$m_i = M_i \cdot C_i, \ i = 1, 2, \cdots, m \tag{6}$$

In this study, the mass balance calculation of unit processes and the whole system was based on 1 t of antimony output from the production system. We expressed this as flow ratio, the mass of substance i for each tonne of antimony produced within the balance area. The flow ratio of substance flow i, f_i, is calculated as

$$f_i = m_i / m_m \tag{7}$$

where m_m is the quantity of the objective element in the final product output from the production system, t(Sb)/d.

2.4 Unsuspected losses

Balancing mass flow is a difficult task for substance flow analysis. Even if the mass-balance principle is applied to each unit process in the production system, 100% mass balance cannot always be accurately obtained. Usually, when the output material of a process is less than the input material, some material loss occurs in the process. In most metallurgical enterprises, the general measurement unit of raw materials and products is tonne. Still, some pollutant emissions, such as tail gas, need to be accurate to kilogram or even gram. Therefore, measurement error is one of the main reasons for the failure of mass balance. In addition, fugitive emission-pollutant generated in the production process that cannot be effectively collected is also an important influencing factor. For example, in the process of volatilization smelting, feeding, slag discharge, and other operations can cause some dust to be discharged into the environment, resulting in mass balance failure.

To obtain 100% mass balance of each process and the whole system, the unsuspected losses caused by (1) measurement errors and (2) fugitive emissions are regarded as a virtual substance flow in this study. Considering that this part of mass loss does not enter any product in the system, the unsuspected loss is regarded as an exceptional emission substance flow[26]. The input and output of each unit process have been measured and calculated. Taking antimony as an example, the failure degree of mass balance (γ) is calculated by the following formula:

$$\gamma = \frac{I_{antimony} - O_{antimony}}{I_{antimony}} \times 100\% \tag{8}$$

where I is input and O is output.

Previous studies have shown that 10% balance difference between input and output is acceptable and insignificant for conclusions[31, 32].

2.5 Evaluation indicators of SFA

To explain the relationship between different substance flows and their resource efficiency and environmental load, three indicators are proposed in this study as follows.

(1) Direct recovery of the primary processes, α, the proportion of objective element in qualified products to the total output flows of the process, %. For the Process j, α_j is calculated as

$$\alpha_j = [P_j/(P_j + R_j + S_j + E_j)] \times 100\% \tag{9}$$

(2) Waste recovery of the process, ω, the proportion of recycle substance flow in all the substance flows not included in the final product, %:

$$\omega = [R/(R + S + E)] \times 100\% \tag{10}$$

(3) Resource efficiency, ε, the proportion of objective element in the final product to the total input flows, %:

$$\varepsilon = (A - S - E)/A = (P/A) \times 100\% \tag{11}$$

3 System definition and data collecting

This study takes a pyrometallurgical process for producing antimony ingot from antimony-gold concentrate as the system boundary. The process includes primary processes of volatilization smelting, reduction smelting, refining, and assistant processes for slags, flue dust, and tail gas treatment. The process is developed and operated by an antimony production enterprise with an annual output of 20000 t of antimony ingot in Hunan province, China. Many domestic enterprises adopt this representative treatment technology. The simplified flow of the target process is shown in Fig. 2.

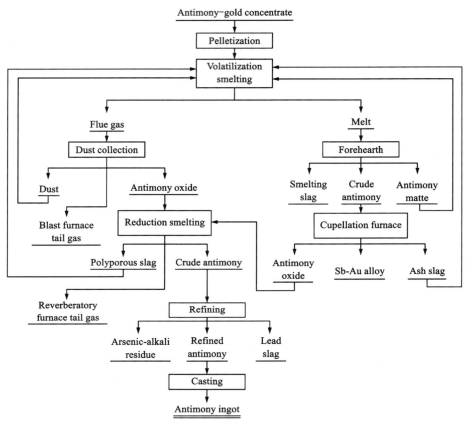

Fig. 2　Simplified flowsheet of pyrometallurgical process for antimony production

The antimony-gold concentrate is granulated with approximately 10% lime to obtain granular ore with appropriate particle size. Granular ore with an antimony grade of 40%-50% is sent into the blast furnace with the recycled material, fuel, and fluxing agent of the system. Smelting slag, antimony matte, and crude antimony are obtained from the melt produced in the volatilization smelting process through forehearth treatment. The crude antimony is then treated in a cupellation furnace to produces Sb-Au alloy, antimony oxide, and ash slag. The Sb-Au alloy is sent to the gold production system. Antimony oxide is used as the raw material for the production of antimony. At the same time, the dust collecting system treats the antimony-containing dust produced in the volatilization smelting process.

The antimony oxide produced in the above process is used as the raw material for reduction smelting, combined with an appropriate amount of reduction coal, so that the antimony oxide is reduced to antimony. The produced slag contains high antimony and returns to the blast furnace; crude antimony contains arsenic, lead, copper, iron, and other impurity elements, which need to be further refined. In the process of refining, through the application of arsenic removal agent (sodium carbonate) and lead removal agent (phosphoric acid), the impurity elements, arsenic and lead, in crude antimony are reduced to national standard. Arsenic and lead are enriched in arsenic-alkali residue and lead slag, respectively. The refined antimony is then cast to obtain antimony ingot.

In this study, antimony and arsenic were selected as target elements to study their flow and distribution in the above production systems. The input, recycle, emission, stock, and output data of each substance flow in the system were obtained from the production report of the enterprise. To determine the content of antimony and arsenic in each logistics, all materials in the system were sampled and analyzed on the stable production day. The sampling period was three consecutive days. The antimony and arsenic contents of samples were averaged and compared with the production data of the enterprise. The mass balance was calculated in combination with the production flow data.

The collected solid samples were dried in an oven at 60 ℃ for 24 h. The concentration of the main elements in the sample was determined by using an ICP-OES (Optima 7300 V, Perkin Elmer, USA). An XRD analyzer (D8 Discover 2500) using a PANalytical X'Pert X-ray diffractometer (Cu K_α radiation) and a scanning electron microscope (SEM, JSMIT500LV, JEOL, Japan) equipped with energy dispersive spectrometer (EDS) were used to determine the phase composition of some solid samples. Before scanning electron microscope (SEM) analysis, antimony-gold concentrate particles were dispersed in epoxy resin, then polished with fine diamond spray, and the sample was sputtered with gold (10-20 nm gold film). HSC chemistry 6.0 was a thermochemical software often used for chemical reaction and equilibrium calculation[33, 34]. The reaction equation module in the software was used to calculate the Gibbs free energy of reactions.

4 Results and discussion

4.1 Substance flow analysis of antimony

4.1.1 Mass balance calculation

Based on the enterprise production data and analysis of all materials streams, the antimony mass balance for producing 1 t antimony-containing products was established according to the method in Section 2.3. The antimony flow of input, recycle, emission, stock, and output in each unit process is listed in Table 1.

4.1.2 Substance flow chart of antimony

Fig. 3 shows the substance flow chart of antimony. The whole system includes 27 substance flows of five types: input, recycle, stock, emission, and output. Each substance flow is identified with both name and flow code, as shown in Table 1.

Table 1　Input and output flow ratios of antimony in each substance flow of production system　　t/t Sb

No.	Process unit	Input		Output	
		Substance flow	Flow ratio, f_i	Substance flow	Flow ratio, f_i
1	Pelletization	Antimony-gold concentrate, A_1	1.121	Granular ore, P_1	1.121
		Total input	1.121	Total output	1.121
2	Volatilization smelting	Granular ore, P_1	1.121	Melt, $P_{2,3}$	0.189
		Antimony matte, $R_{3,2}$	0.071	Flue gas, $P_{2,5}$	1.15
		Ash slag, $R_{4,2}$	0.01	Loss from volatilization smelting, E_2	0.028
		Dust, $R_{5,2}$	0.064		
		Polyporous slag, $R_{6,2}$	0.101		
		Total input	1.367	Total output	1.367
3	Forehearth	Melt, $P_{2,3}$	0.189	Crude antimony, P_3	0.097
				Antimony matte, $R_{3,2}$	0.071
				Smelting slag, E_{3-1}	0.018
				Loss from forehearth, E_{3-2}	0.003
		Total input	0.189	Total output	0.189
4	Cupellation furnace	Crude antimony, P_3	0.097	Sb-Au alloy, E_{4-1}	0.004
				Antimony oxide, P_4	0.082
				Ash slag, $R_{4,2}$	0.01
				Loss from cupellation furnace, E_{4-2}	0.001
		Total input	0.097	Total output	0.097
5	Dust collection	Flue gas, $P_{2,5}$	1.15	Dust, $R_{5,2}$	0.064
				Antimony oxide, P_5	1.062
				Antimony oxide, S_5	0.015
				Blast furnace tail gas, E_{5-1}	0.0005*
				Loss from dust collection, E_{5-2}	0.009
		Total input	1.15	Total output	1.15
6	Reduction smelting	Antimony oxide, P_4	0.082	Crude antimony, P_6	1.041
		Antimony oxide, P_5	1.062	Polyporous slag, $R_{6,2}$	0.101
				Reverberatory furnace tail gas, E_{6-1}	0.0003*
				Loss from reduction smelting, E_{6-2}	0.002
		Total input	1.144	Total output	1.144

Continue the Table 1

No.	Process unit	Input		Output	
		Substance flow	Flow ratio, f_i	Substance flow	Flow ratio, f_i
7	Refining	Crude antimony, P_6	1.041	Refined antimony, P_7	1
				Arsenic-alkali residue, E_{7-1}	0.022
				Lead slag, E_{7-2}	0.016
				Loss from refining, E_{7-3}	0.003
		Total input	1.041	Total output	1.041
8	Casting	Refined antimony, P_7	1	Antimony ingot, P_8	1
		Total input	1	Total output	1

* The unit is kg/t(Sb)

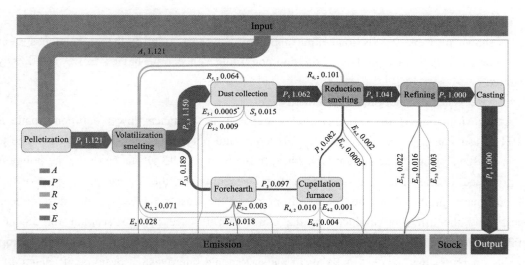

Fig. 3 Antimony substance flow chart of antimony pyrometallurgical production system (t/t(Sb))

4.1.3 Evaluation of antimony production process

Table 2 gives the evaluation results based on the defined indicators. The antimony recovery rates of volatilization smelting, reduction smelting, and refining processes are 78.79%, 91.00%, and 96.06%, respectively. The resource efficiency of antimony production process reaches 89.21%. The reason for this high resource efficiency is a high waste recovery rate, 67.03%, which means that 67.03% of antimony "lost" in the production process is returned to the production system.

Table 2 Results of evaluation indicators for antimony production process

Evaluation indicator	Value
Input raw substance flow, $A/(\text{t}\cdot\text{t}^{-1}\text{ Sb})$	1.121
Recycle substance flow, $R/(\text{t}\cdot\text{t}^{-1}\text{ Sb})$	0.246
Stock substance flow, $S/(\text{t}\cdot\text{t}^{-1}\text{ Sb})$	0.015
Emission substance flow, $E/(\text{t}\cdot\text{t}^{-1}\text{ Sb})$	0.106
Direct recovery of volatilization smelting, $\alpha_2/\%$	78.79
Direct recovery of reduction smelting, $\alpha_6/\%$	91.00
Direct recovery of refining, $\alpha_7/\%$	96.06
Waste recovery rate of process, $\omega/\%$	67.03
Resource efficiency, $\varepsilon/\%$	89.21

(1) Input raw substance flow of antimony, A

As shown in Table 2, the total input antimony substance flow of the production system is 1.121. For 1 t of antimony produced, the antimony-gold concentrate having 1.121 t antimony is consumed.

(2) Recycle substance flow of antimony, R

As shown in Fig. 3, the system includes four recycle substance flows: antimony matte $R_{3,2}$ (0.071), ash slag $R_{4,2}$ (0.01), dust $R_{5,2}$ (0.064), and polyporous slag $R_{6,2}$ (0.101). The total recycle substance flow of the system is 0.246 t. In the total recycle substance flow, the polyporous slag produced in the reduction smelting process accounts for the largest part, 41.06%; the antimony matte produced in the forehearth process, the dust produced in the dust collection process, and the ash slag produced in the cupellation furnace process account for 28.86%, 26.02%, and 4.06%, respectively. Blast furnace has the characteristics of handling complex materials: it reuses the recycled materials of the system and improves the resource efficiency of the system.

(3) Stock substance flow of antimony, S

One stock substance flow in the production process is S_5, antimony oxide (0.015). The antimony oxide obtained from the flue gas produced in the volatilization smelting after condensation and separation is the raw material for the subsequent reduction smelting process. The antimony oxide is stored in the tank through the conveying device. According to production plan, the enterprise allocates the corresponding mass of antimony oxide to the reduction smelting process. In this production period, the antimony oxide required for the reduction smelting process is less than that produced by the volatilization smelting process, so part of the antimony oxide is stored in the tank.

(4) Emission substance flow of antimony, E

There are 12 emission substance flows in the system, which can be divided into four categories: 1) by-product: Sb-Au alloy E_{4-1} (0.004); 2) solid waste: smelting slag E_{3-1} (0.018), arsenic-alkali residue E_{7-1} (0.022), lead slag E_{7-2} (0.016); 3) tail gas: blast furnace tail gas E_5 (0.0005*), reverberatory furnace tail gas E_6 (0.0003*); 4) unsuspected losses. The total emission substance flow of the system is 0.106 t. The gold resources in the antimony-gold concentrate processed in the production system are mainly enriched in the by-product: the antimony in Sb-Au alloy can be recycled in the gold production system. Arsenic-alkali residue and lead slag produced in the refining process account for 20.75% and 15.09% of the emission substance flow, respectively. Blast furnace tail gas and reverberatory furnace tail gas are purified by a gas collection device (including a surface cooler, bag dust collection chamber, etc.).

Unsuspected losses caused by measurement errors mainly exist in the following unit processes: 1) volatilization smelting (E_2): the blast furnace used in this process consumes about 150 t antimonial materials per day, and the weighing system in the factory cannot accurately measure the quality of each material; 2) dust collection (E_{5-2}): part of antimony oxide remains in the conveying equipment and pipelines, and the factory cannot clean it every day, resulting in inaccurate measurements of the mass of antimony oxide; 3) reduction smelting and refining (E_{6-2} and E_{7-3}): the two processes are carried out in the same reverberatory furnace, and the mass of crude antimony cannot be measured, but is indirectly obtained through the calculation and measurement of other input and output materials in the two processes. Fugitive emissions mainly exist in the following unit processes: volatilization smelting (E_2), forehearth (E_{3-2}), cupellation furnace (E_{4-2}), and reduction smelting (E_{6-2}). The equipment used in the above processes is not fully enclosed, and some dust escaped during feeding, slag discharge, and other operations, resulting in unsuspected losses.

Therefore, the unsuspected losses are caused by measurement errors and fugitive emissions, but the contribution of the two factors to each lost substance flow is different. The balance difference of antimony in the production process is calculated using the criteria in Section 2.4, and the results are acceptable, as shown in Table 3.

Table 3 Ratio of antimony loss from each process

No.	Process	Antimony loss ratio/%
1	Pelletization	<0.01
2	Volatilization smelting	2.05
3	Forehearth	1.59
4	Cupellation furnace	1.03
5	Dust collection	0.78
6	Reduction smelting	0.17
7	Refining	0.29
8	Casting	<0.01

According to the mass-balance principle, the total input of the system is equal to the total output, and the corresponding total output substance flow includes emission, stock, and output substance flow, as shown in Fig. 3. The output substance flow accounts for 89.21%, which is the system antimony resource efficiency. The proportion of other substance flows is shown in Fig. 4.

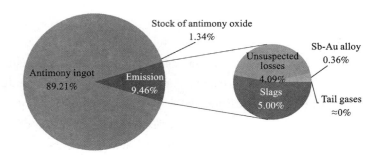

Fig. 4　Distribution of antimony in total output substance flows

4.2　Substance flow analysis of arsenic

Arsenic is an associated impurity element of antimony-gold concentrate and is a harmful component that is difficult to remove during antimony production. Therefore, relevant evaluation based on the definition in Section 2.5 is not conducted. The distribution behavior of arsenic in the primary process of antimony smelting is clarified through the substance flow analysis of arsenic and the phase analysis of related products.

4.2.1　Mass balance calculation

The mass balance of arsenic based on the qualified products containing 1 t antimony produced by the system is established. The substance flow ratio of arsenic in each unit process is given in Table 4.

Table 4　Input and output flow ratios of arsenic in each substance flow of production system　　kg/t

No.	Process unit	Input		Output	
		Substance flow	Flow ratio, f_i	Substance flow	Flow ratio, f_i
1	Pelletization	Antimony-gold concentrate, A_1	11.94	Granular ore, P_1	11.94
		Total input	11.94	Total output	11.94

Continue the Table 4

No.	Process unit	Input		Output	
		Substance flow	Flow ratio, f_i	Substance flow	Flow ratio, f_i
2	Volatilization smelting	Granular ore, P_1	11.94	Melt, $P_{2,3}$	0.97
		Antimony matte, $R_{3,2}$	0.09	Flue gas, $P_{2,5}$	13.80
		Ash slag, $R_{4,2}$	0.08	Loss from volatilization smelting, E_2	0.05
		Dust, $R_{5,2}$	0.45		
		Polyporous slag, $R_{6,2}$	2.26		
		Total input	14.82	Total output	14.82
3	Forehearth	Melt, $P_{2,3}$	0.97	Crude antimony, P_3	0.79
				Antimony matte, $R_{3,2}$	0.09
				Smelting slag, E_{3-1}	0.09
		Total input	0.97	Total output	0.97
4	Cupellation furnace	Crude antimony, P_3	0.79	Sb-Au alloy, E_{4-1}	0.01
				Antimony oxide, P_4	0.69
				Ash slag, $R_{4,2}$	0.08
				Loss from cupellation furnace, E_{4-2}	0.01
		Total input	0.79	Total output	0.79
5	Dust collection	Flue gas, $P_{2,5}$	13.80	Dust, $R_{5,2}$	0.45
				Antimony oxide, P_5	13.11
				Antimony oxide, S_5	0.19
				Blast furnace tail gas, E_{5-1}	0.0004
				Loss from dust collection, E_{5-2}	0.05
		Total input	13.80	Total output	13.80
6	Reduction smelting	Antimony oxide, P_4	0.69	Crude antimony, P_6	11.48
		Antimony oxide, P_5	13.11	Polyporous slag, $R_{6,2}$	2.26
				Reverberatory furnace tail gas, E_{6-1}	0.0003
				Loss from reduction smelting, E_{6-2}	0.06
		Total input	13.80	Total output	13.80

Continue the Table 4

No.	Process unit	Input		Output	
		Substance flow	Flow ratio, f_i	Substance flow	Flow ratio, f_i
7	Refining	Crude antimony, P_6	11.48	Refined antimony, P_7	0.31
				Arsenic-alkali residue, E_{7-1}	10.84
				Lead slag, E_{7-2}	0.26
				Loss from refining, E_{7-3}	0.07
		Total input	11.48	Total output	11.48
8	Casting	Refined antimony, P_7	0.31	Antimony ingot, P_8	0.31
		Total input	0.31	Total output	0.31

4.2.2 Substance flow chart of arsenic

Based on the production process and arsenic substance flow data, a substance flow chart of arsenic is established. Each arsenic substance flow corresponds to the antimony substance flow in Fig. 5. The total amount of each type of arsenic substance flow in the system is given in Table 5. The arsenic concentration in some products in the antimony smelting process is given in Table 6.

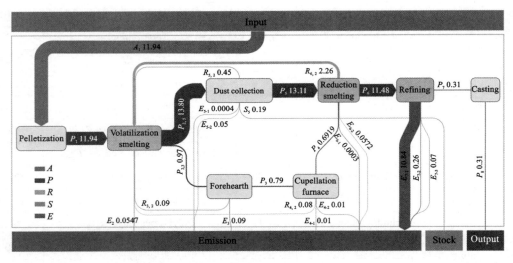

Fig. 5 Arsenic substance flow chart of antimony pyrometallurgical production system (kg/t)

Table 5 Total quantity of arsenic in each substance flow of production system   kg/t

Evaluation indicator	Value
Input substance flow, A	11.94
Recycle substance flow, R	2.88

Continue the Table 5

Evaluation indicator	Value
Stock substance flow, S	0.19
Emission substance flow, E	11.44

Table 6 Arsenic mass concentration in some products   %

Product	Antimony-gold concentrate	Antimony matte	Smelting slag	Dust	Antimony oxide
Arsenic concentration	0.513	0.009	0.007	0.327	0.924
Product	Crude antimony (for refining)	Polyporous slag	Arsenic-alkali residue	Lead slag	Antimony ingot
Arsenic concentration	1.46	1.261	15.693	0.040	0.031

(1) Input substance flow of arsenic

The total input substance flow ratio of arsenic is 11.94 kg/t, indicating that 11.94 kg of arsenic enters the production system for 1 t of antimony produced. The antimony-gold concentrate is the only input source in the system, with an arsenic concentration of 0.513%. It is complex antimony-gold resource with a high arsenic concentration. Therefore, clarifying the distribution behavior of arsenic in the smelting process plays a vital role in controlling the emission of arsenic-containing pollutants.

(2) Recycle substance flow of arsenic

Part of the arsenic circulates in the system with four recycle substance flows. The total recycle flow ratio of arsenic is 2.88 kg/t. Among them, polyporous slag $R_{6,2}$ (2.26) and dust $R_{5,2}$ (0.45) account for 78.47% and 15.63%, respectively. Some of the intermediate products are reused to obtain a high recovery of antimony in production system. However, the arsenic in the intermediate products re-enters the system, resulting in the continuous potential accumulation of arsenic in the system. Therefore, improving the reduction smelting process and reducing the output of polyporous slag is beneficial to both improving direct antimony recovery and reducing the risk of arsenic accumulation in the system.

(3) Stock substance flow of arsenic

The system has only one stock substance flow of antimony oxide, S_5. The total flow ratio of stock is 0.19 kg/t. The arsenic content of antimony oxide is 0.924%. When the antimony oxide required for reduction smelting is greater than the output of volatilization smelting in a subsequent production cycle, the stock of antimony oxide can be used.

(4) Emission substance flow of arsenic

The total arsenic emission substance flow ratio of the system is 11.44 kg/t. The emission substance flow consists of 11 streams: smelting slag E_3 (0.09), arsenic-alkali residue E_{7-1} (10.84), lead slag E_{7-2} (0.26), blast furnace tail gas E_{5-1} (0.0004), reverberatory furnace tail gas E_{6-1} (0.0003), and Sb-Au alloy E_{4-1} (0.01). Arsenic entering Sb-Au alloy can be further treated in the gold production system. The arsenic content in the two tail gases is very low, which can be treated later by desulfurization. The total unsuspected loss of arsenic (0.24) accounts for 2.10% of the total emission substance flow of arsenic. Using the criterion in Section 2.4, the calculation results of arsenic balance difference are acceptable, as given in Table 7.

Table 7　Ratio of arsenic loss from each process

No.	Process	Arsenic loss ratio/%
1	Pelletization	<0.01
2	Volatilization smelting	0.34
3	Forehearth	<0.01
4	Cupellation furnace	1.27
5	Dust collection	0.36
6	Reduction smelting	0.43
7	Refining	0.61
8	Casting	<0.01

Among these emissions, arsenic-alkali residue accounts for 94.76%, and arsenic concentration is 15.693%, which is a hazardous waste. The high arsenic content makes this solid waste need to be stored in a special warehouse. After decades of production, the stockpile of arsenic-alkali residue has been massive. Some enterprises return the arsenic-alkali residue to the smelting system for antimony recovery treatment, resulting in secondary arsenic-alkali residue with low antimony content. Past nonstandard treatment and long-term storage have caused severe harm to the environment[35]. Therefore, it is necessary to safely and adequately deal with these solid wastes.

4.3　Distribution behavior of arsenic

Arsenic associated with antimony-gold concentrate is a harmful element in the smelting process. It brings technological challenges to the antimony smelting and purification process. Moreover, the generation of arsenic-containing pollutants in this process is a major environmental problem. Therefore, it is essential to clarify the distribution behavior of arsenic in the smelting process to realize the clean production of antimony.

(1) Volatilization smelting

Table 8 gives the distribution of arsenic in the main process. In the process of volatilization smelting, 93.12% of raw arsenic enters the flue gas, and 6.55% of arsenic enters the melt due to the volatile characteristics of arsenic compounds. The flue gas is treated by the dust collection system, 3.28% of arsenic enters the dust, and 96.38% of arsenic enters antimony oxide, which is the input source of arsenic in subsequent smelting process. The melt is separated in forehearth. A total of 9.28% arsenic enters the smelting slag, 9.28% arsenic enters the antimony matte and 81.44% arsenic enters the crude antimony.

Table 8 Distribution of arsenic in primary production processes of system

Unit process	Intermediate product	Flow ratio/(kg · t^{-1})	Distribution ratio/%
Volatilization smelting	Melt	0.97	6.55
	Flue gas	13.80	93.12
	Unsuspected loss	0.05	0.37
Forehearth	Smelting slag	0.09	9.28
	Antimony matte	0.09	9.28
	Crude antimony	0.79	81.44
Dust collection	Dust	0.45	3.28
	Antimony oxide	13.30	96.38
	Tail gas	0.0004	~0
	Unsuspected loss	0.05	0.36
Reduction smelting	Crude antimony	11.48	83.19
	Polyporous slag	2.26	16.38
	Tail gas	0.0003	~0
	Unsuspected loss	0.06	0.43
Refining	Refined antimony	0.31	2.70
	Arsenic-alkali residue	10.84	94.43
	Lead slag	0.26	2.26
	Unsuspected loss	0.07	0.61

According to results of SEM-EDS analysis, arsenic in antimony-gold concentrate mainly exists in the form of arsenopyrite (FeAsS), as shown in Fig. 6 and Table 9. In the blast furnace, arsenopyrite decomposes into AsS(g), As$_2$(g), AsO(g)[36]. Part of arsenic is further oxidized to As$_2$O$_3$ and collected into flue gas. Reaction between arsenopyrite and sulfur dioxide is expressed as Reaction (14)[37]. The form of arsenic in the antimony oxide is shown in Fig. 7(b).

$$FeAsS + FeS_2 \Longrightarrow 2FeS(l) + AsS(g) \qquad (12)$$
$$FeAsS \Longrightarrow FeS(l) + 0.5As_2(g) \qquad (13)$$
$$As_2(g) + O_2(g) \Longrightarrow 2AsO(g) \qquad (14)$$
$$FeAsS + 3/4SO_2(g) \Longrightarrow FeS(l) + 1/4As_4O_6(g) + 3/8S_2(g) \qquad (15)$$

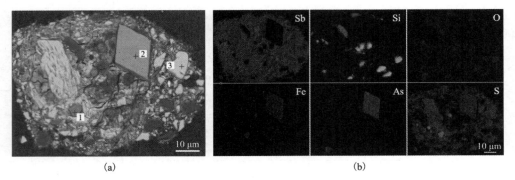

Fig. 6 SEM image of antimony-gold concentrate (a) and EDS mappings of antimony-gold concentrate (b)

Table 9 Analytical results of antimony-gold concentrate (molar fraction) %

Point in Fig. 7(a)	Sb	S	As	Fe	Si
1	—	67.6	—	32.4	—
2	—	35.9	33.6	30.5	—
3	41.4	57.7	—	—	0.9

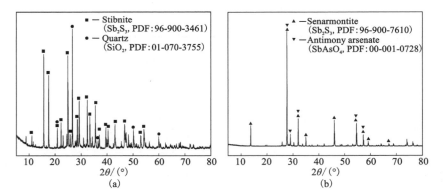

Fig. 7 XRD pattern of antimony-gold concentrate (a) and antimony oxide (b)

(2) Reduction smelting and refining

In the process of reduction smelting, 83.19% arsenic enters crude antimony, and remaining 16.38% arsenic enters polyporous slag. As_2O_3 in antimony oxide undergoes the following reduction, and the generated As(l) enters crude antimony.

$$As_2O_3(l) + 3C = 2As(l) + 3CO(g) \tag{16}$$

$$As_2O_3(l) + 3CO(g) = 2As(l) + 3CO_2(g) \tag{17}$$

$$As_2O_3(l) + 3/2C = 2As(l) + 3/2CO_2(g) \tag{18}$$

$$As_2O_3(l) + 2Sb(l) = 2As(l) + Sb_2O_3(l) \tag{19}$$

In refining, 94.43% arsenic enters the arsenic-alkali residue, the remaining 2.70% and

2.26% arsenic enter the refined antimony and lead slag, respectively. Through the oxygen blowing operation and using the arsenic removing agent (sodium carbonate), the elemental arsenic is oxidized to arsenic trioxide. It then reacts with sodium carbonate to form sodium arsenate and sodium arsenite into arsenic-alkali residue[38].

$$4/5As(l) + O_2(g) + 6/5Na_2CO_3 = 4/5Na_3AsO_4 + 6/5CO_2(g) \quad (20)$$
$$4/3As(l) + O_2(g) + 2/3Na_2CO_3 = 4/3NaAsO_2 + 2/3CO_2(g) \quad (21)$$

The curves of the above reaction standard Gibbs free energy change with temperature are drawn, as shown in Fig. 9. In the production process, the following complex reactions of arsenic-containing compounds occur during volatilization smelting, reduction smelting, and refining, which can be carried out spontaneously under normal production conditions.

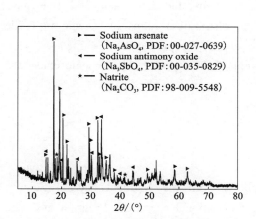

Fig. 8 XRD pattern of arsenic-alkali residue

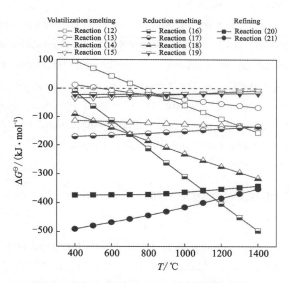

Fig. 9 Relationship between standard free energy change and temperature of multiphase arsenic reaction in antimony smelting process
(Data source: HSC Chemistry 6.0)

Based on the systematic substance flow analysis, the following improvement measures are proposed: In the volatilization smelting process, due to the relatively high vapour pressures of Sb_2O_3 and As_2O_3, antimony and arsenic are enriched in the flue gas, which brings certain difficulties to the subsequent separation of antimony and arsenic. Therefore, if the controllable separation of antimony and arsenic is achieved in the dust collection stage of the flue gas to obtain relatively pure antimony oxide, the output of arsenic-alkali residue in the refining process can be reduced correspondingly, and the goal of reducing the source of solid waste can be achieved. The waste recycling rate of the whole system reaches 67.03%, which makes the resource efficiency of antimony reach 89.21%. Still, this also causes the circulation and accumulation of arsenic in the system, which has potential environmental risks. Using natural gas as a heat source in the reduction smelting stage can correspondingly reduce the output of polyporous slag and SO_2, reduce the antimony arsenic cycle rate, and improve the clean production level of the enterprise.

5 Conclusions

(1) The antimony resource efficiency of the smelting system is 89.19%, and the antimony direct recovery of volatilization smelting, reduction smelting, and refining processes are 78.78%, 91.00%, and 96.03%, respectively. The antimony loss caused by measurement error and fugitive emission is collectively referred to as unsuspected loss. The difference in mass balance in each unit process is within 5%.

(2) For the arsenic substance flow in the smelting process, 11.94 kg of arsenic is carried into the system with antimony-gold concentrate for 1 t of antimony produced. In the process of volatilization smelting, 6.55% and 93.12% of arsenic are fed into melt and flue gas, respectively. In the process of reduction smelting, the distribution ratios of arsenic in crude antimony and slag are 83.19% and 16.38%, respectively. In the process of refining, 94.43% arsenic enters the arsenic-alkali residue, and the remaining 2.70% and 2.26% arsenic enters the refined antimony and lead slag, respectively.

Acknowledgments

The authors are grateful for the financial supports from the National Natural Science Foundation of China (Nos. 51904351, U20A20273), the National Key R&D Program of China (No. 2019YFC1907400), the Science and Technology Innovation Program of Hunan Province, China (No. 2021RC3005), and the Innovation Driven Project of Central South University, China (No. 2020CX028).

References

[1] KRENEV V A, DERGACHEVA N P, FOMICHEV S V. Antimony: Resources, application fields, and world market[J]. Theoretical Foundations of Chemical Engineering, 2015, 49(5): 769-772.
[2] ANDERSON C G. The metallurgy of antimony[J]. Geochemistry, 2012, 72: 3-8.
[3] WANG Q, LAI Y Q, LIU F Y, et al. Sb_2S_3 nanorods/ porous-carbon composite from natural stibnite ore as high-performance anode for lithium-ion batteries[J]. Transactions of Nonferrous Metals Society of China, 2021, 31(7): 2051-2061.
[4] USGS. Mineral commodity summaries[M]. Reston, VA: US Geological Survey, 2021.
[5] Editorial Committee of Yearbook of Nonferrous Metals Industry of China. The yearbook of nonferrous metals

industry of China [M]. Beijing: China Non-ferrous Metals Industry Association, 2020. (in Chinese)

[6] WANG X, DING J Q, WANG L L, et al. Stabilization treatment of arsenic-alkali residue (AAR): Effect of the coexisting soluble carbonate on arsenic stabilization [J]. Environment International, 2020, 135: 105406.

[7] LIU W F, YANG T Z, CHEN L, et al. Development of antimony smelting technology in China [C]// JIANG T, HWANG J Y, MACKEY P J, et al. 4th International Symposium on High-Temperature Metallurgical Processing 2013. San Antonio, TX: TMS, 2013: 341-351.

[8] ZHANG Z T, DAI X. Effect of Fe/SiO_2 and CaO/SiO_2 mass ratios on metal recovery rate and metal content in slag in oxygen-enriched direct smelting of jamesonite concentrate[J]. Transactions of Nonferrous Metals Society of China, 2020, 30(2): 501-508.

[9] ZHANG Z T, DAI X, ZHANG W H. Thermodynamic analysis of oxygen-enriched direct smelting of Jamesonite concentrate[J]. JOM, 2017, 69(12): 2671-2676.

[10] LIU W F, HUANG K H, YANG T Z, et al. Selective leaching of antimony from high-arsenic antimony-gold concentrate[J]. The Chinese Journal of Nonferrous Metals, 2018, 28(1): 205-211. (in Chinese)

[11] LONG H, HUANG X Z, ZHENG Y J, et al. Purification of crude As_2O_3 recovered from antimony smelting arsenic-alkali residue[J]. Process Safety and Environmental Protection, 2020, 139: 201-209.

[12] SU R, MA X, LIN J R, et al. An alternative method for the treatment of metallurgical arsenic-alkali residue and recovery of high-purity sodium bicarbonate[J]. Hydrometallurgy, 2021, 202: 105590.

[13] GUO X Y, SONG Y. Substance flow analysis of copper in China[J]. Resources, Conservation and Recycling, 2008, 52(6): 874-882.

[14] HAO M, WANG P, SONG L L, et al. Spatial distribution of copper in-use stocks and flows in China: 1978-2016[J]. Journal of Cleaner Production, 2020, 261: 121260.

[15] WANG M X, CHEN W, LI X. Substance flow analysis of copper in production stage in the U. S. from 1974 to 2012[J]. Resources, Conservation and Recycling, 2015, 105: 36-48.

[16] LIU W, CUI Z J, TIAN J P, et al. Dynamic analysis of lead stocks and flows in China from 1990 to 2015 [J]. Journal of Cleaner Production, 2018, 205: 86-94.

[17] YUE Q, WANG H M, LU Z W, et al. Analysis of anthropogenic aluminum cycle in China [J]. Transactions of Nonferrous Metals Society of China, 2014, 24(4): 1134-1144.

[18] GUO X Y, ZHONG J Y, SONG Y, et al. Substance flow analysis of zinc in China[J]. Resources, Conservation and Recycling, 2010, 54(3): 171-177.

[19] LIN S H, MAO J S, CHEN W Q, et al. Indium in mainland of China: Insights into use, trade, and efficiency from the substance flow analysis [J]. Resources, Conservation and Recycling, 2019, 149: 312-321.

[20] SUN X, HAO H, LIU Z W, et al. Tracing global cobalt flow: 1995-2015[J]. Resources, Conservation and Recycling, 2019, 149: 45-55.

[21] CAMPITELLI A, SCHEBEK L. How is the performance of waste management systems assessed globally? A systematic review[J]. Journal of Cleaner Production, 2020, 272: 122986.

[22] QU C S, LI B, WANG S, et al. Substance flow analysis of lead for sustainable resource management and pollution control[J]. Advanced Materials Research, 2014, 878: 30-36.

[23] HUANG C L, VAUSE J, MA H W, et al. Using material/substance flow analysis to support sustainable development assessment: A literature review and outlook[J]. Resources, Conservation and Recycling, 2012, 68: 104-116.

[24] LI Q F, DAI T, GAO T M, et al. Aluminum material flow analysis for production, consumption, and trade in China from 2008 to 2017[J]. Journal of Cleaner Production, 2021, 296: 126444.

[25] HAN F, LI W F, YU F, et al. Industrial metabolism of chlorine: A case study of a chlor-alkali industrial chain[J]. Environmental Science and Pollution Research International, 2014, 21(9): 5810-5817.

[26] BAI L, QIAO Q, LI Y P, et al. Substance flow analysis of production process: A case study of a lead smelting process[J]. Journal of Cleaner Production, 2015, 104: 502-512.

[27] CHEN Y L, GUO X Y, WANG Q M, et al. Tungsten and arsenic substance flow analysis of a hydrometallurgical process for tungsten extracting from wolframite[J]. Tungsten, 2021, 3(3): 348-360.

[28] CHU J W, MAO J W, HE M C. Anthropogenic antimony flow analysis and evaluation in China[J]. Science of the Total Environment, 2019, 683: 659-667.

[29] CHU J W, HU X Y, KONG L H, et al. Dynamic flow and pollution of antimony from polyethylene terephthalate (PET) fibers in China[J]. Science of the Total Environment, 2021, 771: 144643.

[30] CHU J W, CAI Y P, LI C H, et al. Dynamic flows of polyethylene terephthalate (PET) plastic in China [J]. Waste Management, 2021, 124: 273-282.

[31] BAI L, QIAO Q, LI Y P, et al. Statistical entropy analysis of substance flows in a lead smelting process [J]. Resources, Conservation and Recycling, 2015, 94: 118-128.

[32] BRUNNER P H, RECHBERGER H. Handbook of material flow analysis: For environmental, resource, and waste engineers[M]. 2nd ed. Boca Raton: Taylor & Francis, CRC Press, 2017.

[33] ROINE A. HSC Chemistry, vers. 6.0 [EB/OL]. Pori, Finland: Outotec Research, 2006. https://www.mogroup.com/portfolio/hsc-chemistry/.

[34] LI P L, QIU X Y, HU Z, et al. Mechanism of leaching on complex Mn-Ag-Cu bound ore and industrial optimizations[J]. The Chinese Journal of Nonferrous Metals, 2022, 32(2): 563-573. (in Chinese)

[35] GUO X J, WANG K P, HE M C, et al. Antimony smelting process generating solid wastes and dust: Characterization and leaching behaviors[J]. Journal of Environmental Sciences, 2014, 26(7): 1549-1556.

[36] LUGANOV V A, ANDERSON C G. Sulfidization of Arsenopyrite [C]//REDDY R G, RAMACHANDRA V. Pro Arsenic Metallurgy 2005. San Francisco, CA: TMS, 2005: 305-316.

[37] DUNN J G, IBRADO A S, GRAHAM J. Pyrolysis of arsenopyrite for gold recovery by cyanidation[J]. Minerals Engineering, 1995, 8(4): 459-471.

[38] CHAI L Y, WANG J Q, WANG Y Y, et al. Preparation of colloidal Sb_2O_5 from arsenic-alkali residue[J]. Transactions of Nonferrous Metals Society of China, 2005, 15(6): 1401-1406.

A Sustainable Process for Tungsten Extraction from Wolframite Concentrate

Abstract: Aiming to explore a sustainable process for tungsten extraction from wolframite concentrate, the digestion behavior and mechanism of wolframite in HCl solution at atmospheric pressure were firstly investigated in this work. The results showed that the digestion efficiency of wolframite reached 99.3% at the optimum condition: particle size of $D(95) = 20$ μm, stirring intensity of 250 r/min, reaction temperature of 90 ℃, HCl stoichiometric ratio of 3.0, liquid to solid ratio of 3 : 1, and duration of 4 h. In the initial stage of digestion, tungsten was mainly dissolved into HCl solution, when the tungsten dissolution reached balance, solid H_2WO_4 layers were formed on the particle surface. The accumulation of Fe^{2+} and Mn^{2+} in HCl solution reduced the digestion efficiency of wolframite. The process of atomization-oxidative thermal decomposition is feasible for the recovery of Fe and Mn and recycling of HCl from the mother solution of digestion. Additionally, it was showed that the digested product had an excellent leachability of tungsten in $NH_3 \cdot H_2O$ solution with WO_3 leaching efficiency above 99.5%, and the $(NH_4)_2WO_4$ solution was qualified for ammonium paratungstate (APT) production. The wrapping of wolframite by quartz, as well as the solid H_2WO_4 layers covering on wolframite particles, is a crucial factor for the incomplete digestion. This work has the potential to develop a novel technique for tungsten extraction from wolframite concentrate by passing the alkali leaching process.

Keywords: Wolframite; Tungsten extraction; HCl digestion; HCl recycling; Sustainable process

---

Published in *International Journal of Refractory Metals and Hard Materials*, 2022, 107: 105903. Authors: Chen Yuanlin, Huo Guangsheng, Guo Xueyi, Wang Qinmeng.

1 Introduction

Tungsten is a strategic metal with high melting point, good mechanical properties, and excellent corrosion resistance. Tungsten and tungsten compounds are widely used in many fields such as machinery manufacturing, aerospace, national defense, and chemical industry[1-3]. Wolframite [(Fe, Mn)WO$_4$] and scheelite (CaWO$_4$) are the main economical tungsten mineral resources[4, 5]. In the past five years, China's tungsten concentrate output accounted for more than 80% of the world's tungsten concentrate output[6]. Although there are fewer wolframite reserves than scheelite in the world, wolframite still accounts for a considerable part of China's tungsten reserves[7], such as 230000 tons tungsten reserves of Shizhuyuan mine in Hunan and 150000 tons tungsten reserves of Xingluokeng mine in Fujian[8]. Moreover, wolframite is easier to be exploited, and therefore has long been an important raw material for ammonium paratungstate (APT, an important intermediate product in tungsten industry) production. Thus, there is an urgent need to develop sustainable processes for extracting tungsten from wolframite.

Autoclave-based alkali leaching is still a conventional process applied to wolframite digestion to produce APT today, and it shows excellent performance in treating high-grade tungsten resources[9, 10]. However, this method requires high reaction temperature, which consumes a large amount of energy[11, 12]. In addition, the alkali leaching process is faced with several problems: the large excess amount of alkali consumed to achieve high WO$_3$ recovery and the difficulty of alkali recycling raise the problems of cost and environment. And the subsequent ion exchange or solvent extraction process converting Na$_2$WO$_4$ to (NH$_4$)$_2$WO$_4$ consumes a large amount of water to dilute the leaching solution or large quantities of acid to acidify the crude Na$_2$WO$_4$ solution, which result in the discharge of a large amount of sodium salt wastewater (20-100 tons wastewater for one ton of APT)[13-15]. What is worse, the alkali leaching residue has been classified as a hazardous solid waste in China according to the latest environmental law, the tungsten production enterprises are required to pay an environmental tax, about 1000 Yuan (140 Dollars or 130 Euro) per ton of residue, which further increases the production cost[6].

Considering of the above problems of alkali leaching process, acid digestion method has attracted more and more attention in recent years. Acid digestion method can avoid discharging large amount of sodium salt wastewater and reduce the production cost owe to its high thermodynamic driving force and short processing flow, through which (NH$_4$)$_2$WO$_4$ can be directly produced from digested product[16-19]. However, in acid digestion process, dense solid H$_2$WO$_4$ are formed and cover the surface of tungsten mineral particles, which hinders the mass transfer and leads to incomplete digestion[20]. Some researchers have applied strengthening methods to enhance the digestion efficiency, such as reducing the particle size[21], ultrasonic enhancement[22], mechanical activation[23], and using of complex reagents (a typical complex leaching system is H$_2$SO$_4$-H$_3$PO$_4$ mixture) to promote water soluble tungsten compounds[17, 24].

However, most of the previous studies are on scheelite digestion, yet a considerable amount of tungsten exists in the forms of wolframite and wolframite-scheelite mixed ores, it is unclear that whether acid digestion is efficient for the digestion of wolframite. Moreover, some studies found that under the same conditions as scheelite digestion, the acid digestion extent of wolframite was much lower than that of scheelite, and the digestion rate was also slower[7]. Shen et al.[20] clarified the digestion behavior of wolframite in treating a mixed concentrate with H_2SO_4 solution, which showed that it was more difficult to decompose wolframite than scheelite in H_2SO_4 acid solution. It was also found that the wolframite digestion efficiency was only 92% and 93% through wet-grinding the concentrate to obtain a particle size of $D(90) = 13.3$ μm and grinding while digesting, the low digestion efficiency means that the product layer was not the decisive factor although it had a negative influence on the digestion. Li et al.[25] found that, in a H_2SO_4 and H_3PO_4 mixed system, the leaching efficiency of wolframite could be improved through mechanical activation with the addition of $Ca(OH)_2$ into mechanical milling process, in which the wolframite was transformed to scheelite. Undoubtedly, the transformation of wolframite by mechanical activation is with high energy consumption and increases the number of treatment steps. The use of $Ca(OH)_2$ also increases the production of solid waste. Luo et al.[7] extracted tungsten from wolframite through pressure leaching in H_2SO_4-H_3PO_4 mixture, the wolframite digestion efficiency under the optimum condition reached 98.8% which was significantly higher than that of atmospheric pressure leaching. Nevertheless, this method requires a temperature of more than 140 ℃, which thereby leads to high energy consumption. In addition, the phosphorus removal in the subsequent step complicates the process. Therefore, it is urgent to develop a sustainable technology to digest wolframite in single acid medium with mild conditions. Martins[9] calculated the leaching thermodynamics and equilibrium constants of scheelite and wolframite in acid medium and suggested that all the commercial tungsten minerals can be digested in acidic medium (HCl, HNO_3, and H_2SO_4), but HCl is desirable. The standard Gibbs free energies for the digestion of $FeWO_4$ and $MnWO_4$ in HCl were -33.96 kJ/mol and -62.85 kJ/mol, and the calculated equilibrium constant of $FeWO_4$ reached around 1000. Martins[9] also believed that in the acid digestion of tungsten minerals, it is not necessary to use high temperatures and large excesses of acidic medium on the stoichiometric amount, the high consumption observed in practice resulted from the reaction kinetics. Though the HCl can theoretically digest wolframite at low temperature, there is limited literature available and the digestion mechanism is still unclear. Moreover, there is no research on the recycling of HCl solution during digestion process.

Therefore, this paper aims to clarify the digestion behavior and mechanism of wolframite in HCl solution at atmospheric pressure. Based on HCl digestion of wolframite, the leachability of tungsten from the digested product was tested in $NH_3 \cdot H_2O$ solution to prepare a qualified $(NH_4)_2WO_4$ solution for APT production, and a process of atomization oxidative thermal decomposition for the recycling of HCl solution was proposed. Finally, a sustainable process for extracting tungsten from wolframite concentrate was developed.

2 Material and methods

2.1 Materials and reagents

The wolframite concentrate was provided by Jiangxi Tungsten Group Co., Ltd., China. The concentrate contains WO_3 69.50%, Fe 11.12%, and Mn 6.56% as the main compositions as presented in Table 1. The X-ray diffraction (XRD) pattern of the concentrate (Fig. 1) shows that it is with high content of wolframite [(Fe, Mn)WO_4] and trace amount of scheelite ($CaWO_4$) and phlogopite [$KMg_3(Si_3Al)O_{10}(OH)_2$]. The HCl was purchased from Chron Chemicals, the NaOH and $NH_3 \cdot H_2O$ solution was purchased from Sinopharm Chemical Reagent Co., Ltd., China. All the reagents used in this work were analytically pure, and all solutions were prepared with deionized water.

Table 1 The elemental composition of wolframite concentrate   %

Element	WO_3	Fe	Mn	SiO_2	CaO	Al	As
Content	69.50	11.12	6.56	3.34	2.18	0.72	0.01
Element	S	Bi	Mg	Ti	Nb	K	Na
Content	0.43	0.23	0.20	0.14	0.11	0.23	0.05
Element	Mo	Sn	Ta	P	F	Others	Total
Content	0.02	0.06	0.05	0.05	0.17	4.88	100.00

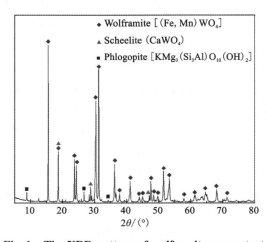

Fig. 1 The XRD pattern of wolframite concentrate

2.2 Experimental procedure

2.2.1 Digestion of wolframite using HCl solution

Prior to the digestion experiments, the concentrate was ground to a certain particle size. The digestion experiments were conducted in a 1 L three-neck round-bottom flask which was heated thermostatically by a water bath. In each run, a certain volume of HCl solution with a certain concentration was added into the flask and heated to the required temperature, then ground slurry containing 200 g concentrate (or 200 g of ground sample) was added into the solution. The slurry was stirred at a set speed using a polytetrafluoroethylene coated agitator. After a certain duration, the slurry was filtered and washed, the filtrate (mother solution) was collected for analysis, and the digested product was then transferred to a beaker containing 800 mL of NaOH solution (2 mol/L). Subsequently, the digested product was leached at room temperature for 2 h, and the leaching residue was obtained by filtering and washing. NaOH is used to ensure that all H_2WO_4 in the digested product is leached. Some digested products at optimum condition were dried in an oven at 90 ℃ for 12 h, then used for testing its leachability in $NH_3 \cdot H_2O$ solution.

The digestion of wolframite in HCl solution is accompanied by the formation of H_2WO_4 and the dissolution of Fe^{2+} and Mn^{2+} as follows:

$$[(Fe, Mn)WO_4](s) + 2HCl(aq) \longrightarrow (Fe, Mn)Cl_2(aq) + H_2WO_4(s) \qquad (1)$$

During the digestion, a trace amount of tungsten will be dissolved into solution. The digestion efficiency of wolframite based on H_2WO_4 was calculated using Formula (2).

$$\eta = [1 - (w_{residue} \cdot m_1)/(w_{concentrate} \cdot m_0) - (c \times V)/(w_{concentrate} \cdot m_0)] \times 100\% \qquad (2)$$

where η is the digestion efficiency (%) of wolframite based on H_2WO_4, $w_{residue}$ and $w_{concentrate}$ is the WO_3 content (mass fraction) of NaOH leaching residue and wolframite concentrate, m_1 and m_0 is the weight of NaOH leaching residue and wolframite concentrate sample used in each run. $c(g/L)$ and $V(L)$ are the WO_3 concentration and volume of the mother solution in each run.

In addition to a group of control experiments, wolframite concentrate samples with a particle size of 50-75 μm were digested in a 500 mL three-neck round-bottom flask. After a certain duration of digestion, the slurry was filtered, the solution was analyzed for the concentration of WO_3. The digested product was dry and analyzed by scanning electron microscope (SEM) and energy dispersive spectrometer (EDS) to clarify the acid digestion mechanism of wolframite. The samples for observation were prepared by mixing the digested product with epoxy and triethanolamine to form grain mounts, followed by sectioning and polishing.

2.2.2 Leaching of digested products by $NH_3 \cdot H_2O$ solution

The leaching experiments of digested products of wolframite concentrate were carried out in a 250 mL conical flask which was heated thermostatically by a water bath and stirred by a magnetic stirrer. The digested products were washed with low concentration HCl solution prior to leaching experiments. In each run, a certain volume of $NH_3 \cdot H_2O$ solution was added into the flask and heated to the required temperature, then 50 g digested product sample was added into the solution. After a certain duration, the slurry was filtered and washed. A leaching experiment using NaOH solution was also carried out as a reference of complete leaching to calculate the leaching efficiency by $NH_3 \cdot H_2O$ solution. The leaching efficiency of tungsten from digested product by $NH_3 \cdot H_2O$ solution was calculated by Formula (3).

$$L = [1 - (w_{residue'} \cdot m')/(w_{product} \cdot M')]/[1 - (w_{residue''} \cdot m'')/(w_{product} \cdot M')] \times 100\% \quad (3)$$

Where L is the leaching efficiency by $NH_3 \cdot H_2O$ solution (%), $w_{residue'}$ and $w_{residue''}$ is the WO_3 content (mass fraction) of leaching residue by $NH_3 \cdot H_2O$ and NaOH solution, m' and m'' is the weight of leaching residue by $NH_3 \cdot H_2O$ and NaOH solution, $w_{product}$ and M' is the WO_3 content (mass fraction, 87.67%) and weight of digested product (50 g) used in each run.

2.3 Analysis methods

The elemental composition of wolframite concentrate and the leaching residue of digestion product were determined by chemical analysis by Changsha Research Institute of Mining and Metallurgy in Changsha, China. Phase analysis of the wolframite concentrate was performed by X-ray diffraction (XRD, TTR III, with Cu K_α radiation, Rigaku, Japan), the analysis was operated with a step size of 0.02°, and a scanning speed of 2°/min. The concentrations of WO_3, Fe, Mn, P, and As in solutions were measured by ICP-OES (ICAP7400, Thermo Scientific). The concentrations of K and Na in $(NH_4)_2WO_4$ solution were measured by atomic absorption spectrometer (iCE3500, Thermo Scientific). The particle size of grinding samples was tested by laser particle analyzer (Mastersizer 2000, Malvern Instruments Ltd.). The morphology and the element distribution of the digested product particles were determined by a scanning electron microscope (SEM-EDS, JSM-IT500LV, JEOL, Japan). The mineralogical characteristic of $NH_3 \cdot H_2O$ solution leaching residue was analyzed by mineral liberation analyzer (MLA 650, configured with a scanning electron microscope FEI Quanta 650, an energy spectrometer Bruker Quantax 200 with Dual XFlash 5010, and a software MLA 3.1 for automatic analysis of mineral parameters).

3 Results and discussion

3.1 Wolframite digestion behavior in HCl solution

To explore the optimum condition for the digestion of wolframite in HCl solution at atmospheric pressure, a group of conditional experiments were carried out to clarify the effects of particle size, stirring intensity, the stoichiometric ratio of HCl, reaction temperature, and reaction time on the digestion efficiency.

3.1.1 Effect of particle size

Generally, mineral particle size affects the liquid-solid reaction by affecting the specific surface area of the particles. As the particle size decreases, the surface area of particles increases, and the contact area with leaching agent also increases, therefore, the reaction rate is accelerated.

In order to determine an appropriate particle size, the wolframite concentrate samples with different particle sizes were obtained by wet-grinding in ball mill and dry grinding in vibration mill for different time. Fig. 2(a) shows the dependence of digestion efficiency on particle size at reaction temperature of 95 ℃, stirring intensity of 250 r/min, HCl stoichiometric ratio of 3.0, liquid to solid ratio of 3 : 1, duration of 4 h, which indicates that the digestion efficiency rises significantly with the particle size ($D(95)$) reduction from 220 μm to 20 μm. When the particle size reaches 20 μm by ball milling, wolframite can be almost completely digested (a digestion efficiency of 99.3%), and this particle is appropriate for the digestion of wolframite. It is interesting that at the same particle size, the digestion efficiency obtained by ball milling is higher than obtained by vibration milling. Fig. 2(b)-(e) present the morphology of ground concentrate samples with particle size of 60 μm and 20 μm obtained by vibration milling and ball milling. Apparently, at the same particle size, the particle surface of vibration milling samples is very flat, while the ball milling samples have many defects on the particle surface, and therefore milling samples are more conducive to the liquid-solid reaction. The reason for the difference of

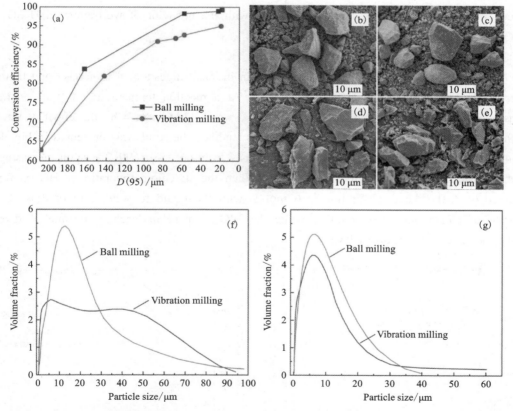

Fig. 2 (a) effect of particle sizes on the digestion efficiency of wolframite; (b, c) vibration milling samples with particle size of $D(95)=60$ μm and $D(95)=20$ μm; (d, e) ball milling samples with particle size of $D(95)=60$ μm and $D(95)=20$ μm; (f, g) particle size distribution of the samples with $D(95)=60$ μm and $D(95)=20$ μm obtained by ball milling and vibration milling

morphology of samples obtained by different grinding methods may be that the vibration grinding process is mainly about impact, while the ball milling process has the dual action of impact and grinding. Fig. 2(f) and (g) show the particle size distribution of the samples with $D(95) = 60$ μm and $D(95) = 20$ μm obtained by ball milling and vibration milling. The results show that for the samples with $D(95) = 60$ μm, the proportion of coarse particle size of vibration grinding sample is greater than that of ball milling sample, while for the sample with $D(95) = 20$ μm, the particle size distribution of the samples obtained by the two grinding methods tends to be similar.

3.1.2 Effect of stirring intensity

The stirring intensity plays a crucial role on mass transfer in liquid solid reaction. The effect of stirring intensity on the digestion efficiency of wolframite was studied with reaction temperature of 95 ℃, HCl stoichiometric ratio of 3.0, liquid to solid ratio of 3 : 1, and duration of 4 h. Fig. 3 shows that the digestion efficiency is almost independent of stirring intensity in the stirring speed range of 150-300 r/min, when the stirring intensity reached 250 r/min, the WO_3 content in leaching residue of digested product decreases to the lowest level (7.8%). The slight decrease of the digestion efficiency at 350 r/min might be caused by the adhesion of wolframite concentrate to the container wall. The result here is consistent with the behavior of wolframite in H_2SO_4 and H_2SO_4-H_3PO_4 mixture leaching system[8, 20].

3.1.3 Effect of the stoichiometric ratio of HCl

Acid concentration is the main driving force in the acid digestion of minerals, and therefore influence the digestion efficiency remarkably[26]. At a reaction temperature of 95 ℃, stirring intensity of 250 r/min, liquid to solid ratio of 3 : 1, and duration of 4 h, the effect of the HCl stoichiometric ratio on the digestion efficiency was studied, the result was presented in Fig. 4. In Fig. 4, the digestion efficiency increases continuously from 66.1% to 99.2% upon increasing the stoichiometric ratio of HCl from 1.0 to 2.5. When the stoichiometric ratio of HCl is further increased to 3.0 (the concentration is 6 mol/L with the liquid to solid ratio of 3 : 1), the digestion efficiency increases slightly, while the WO_3 content in leaching residue of digested product decreases to the lowest level.

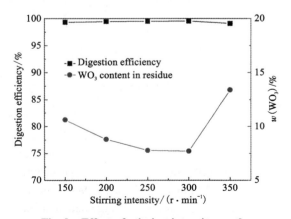

Fig. 3　Effect of stirring intensity on the digestion efficiency of wolframite

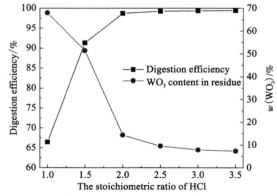

Fig. 4　Effect of HCl stoichiometric ratio on the digestion efficiency of wolframite

3.1.4 Effect of reaction temperature

Previous studies have shown that temperature affects wolframite digestion in acid solution[8, 27]. From the perspective of industrial production, increasing reaction temperature increases the cost. Therefore, determining the optimum reaction temperature is crucial for the industrialization of this method. Under the stirring intensity of 250 r/min, HCl stoichiometric ratio of 3.0, liquid to solid ratio of 3 : 1, and duration of 4 h, the effect of reaction temperature on the digestion efficiency of wolframite was studied, the result was as shown in Fig. 5. The results indicates that the digestion efficiency is positively correlated with the temperature. When the temperature increases from 60 ℃ to 90 ℃, the digestion efficiency rises greatly from 75.0% to 99.3%. As the temperature continues to rise, the digestion efficiency is almost constant. Therefore, the most suitable reaction time is determined to be 90 ℃, which is significantly lower than the pressure leaching temperature using a H_2SO_4-H_3PO_4 mixture[7], and a high digestion efficiency is obtained with single HCl solution under atmospheric pressure.

3.1.5 Effect of reaction time

When the digestion reaches equilibrium, extending reaction time will increase energy consumption to maintained reaction temperature during the digestion process. Thus, reaction time is also a key factor for the industrialization of a technology. Under the reaction temperature of 90 ℃, stirring intensity of 250 r/min, HCl stoichiometric ratio of 3.0, and liquid to solid ratio of 3 : 1, the effect of reaction time on the wolframite digestion efficiency was investigated with the results presented in Fig. 6. Fig. 6 shows that the digestion efficiency of wolframite gradually increased upon prolonging the reaction time. When the reaction time was prolonged from 1 h to 4 h, the digestion efficiency increased from 89.9% to 99.3%. After 4 h, the increase of digestion efficiency becomes very slight, although the WO_3 content in leaching residue of digested product decreases to a lower value. The obvious decrease of WO_3 content in the residue is mainly due to the very low yield of leaching residue. Considering both digestion efficiency and production efficiency, the most suitable reaction time is determined to be 4 h.

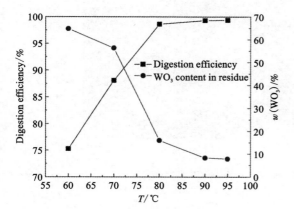

Fig. 5 Effect of reaction temperature on the digestion efficiency of wolframite

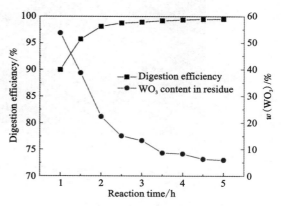

Fig. 6 Effect of reaction time on the digestion efficiency of wolframite

3.2 Digestion mechanism of wolframite in HCl solution

As discussed above, the wolframite concentrate was almost completely digested in HCl solution at atmospheric pressure. Nevertheless, some conditions were much more rigorous than theoretical conditions, such as a very fine particle size ($D(95) = 20$ μm) and high stoichiometric ratio of HCl (3.0). In order to optimize the digestion process, it is necessary to clarify the digestion mechanism.

Fig. 7 shows the SEM images of digested products after a series of durations under reaction temperature of 90 ℃, stirring intensity of 250 r/min, the HCl concentration of 3 mol/L, liquid to solid ratio of 100∶1. As can be seen in Fig. 7(a) and Fig. 7(b), no product layer was formed on the surface of product particles in the first 20 min of the digestion process, while solid layers can be clearly observed coating the particles at 30 min as shown in Fig. 7(c), and the solid layers became thicker with the shrinking of the encapsulated core after 60 min of digestion [Fig. 7(d)]. The EDS analysis (Fig. 8) of the solid layer and encapsulated core shows that, the layer (point B) is principally H_2WO_4 and the core (point A) is the unreacted $(Fe, Mn)WO_4$. The above analysis means that the digestion of wolframite is a shrinking core reaction advancing layer by layer with the continuous formation of H_2WO_4.

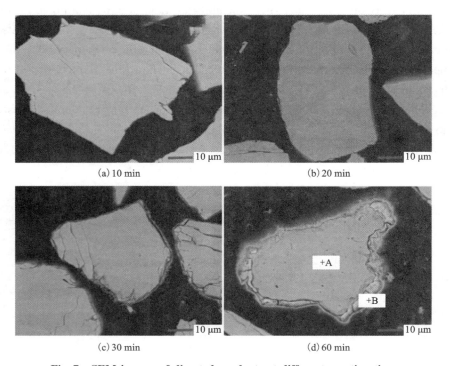

Fig. 7 SEM images of digested products at different reaction time

In order to further understand the digestion mechanism of wolframite in HCl solution, the dissolution amount of tungsten in HCl solution at different reaction times was measured, and the

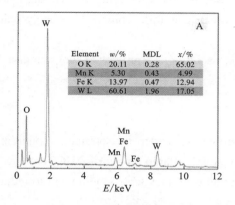
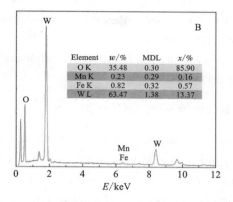

Fig. 8　EDS analysis of digested product at the reaction time of 60 min

results were presented in Fig. 9. Obviously, the WO_3 concentration in HCl solution gradually increases in the first 30 min of digestion process, and the concentration remains at about 330 mg/L after 30 min. Li et al.[28] reported that H_2WO_4 had a certain solubility in water. Combined with the analysis of particle morphology under different duration, it is reasonable to infer that at the initial stage of wolframite digestion, tungsten is mainly dissolved into HCl solution, when the tungsten dissolution reaches a balance, solid H_2WO_4 is deposited on the particle surface and the product layers begin to form. According to the digestion of scheelite by HCl solution reported by Martins et al.[21], the dissolution of tungsten and the formation of solid H_2WO_4 during wolframite digestion can be explained by the following aspects:

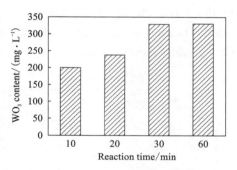

Fig. 9　WO_3 concentration in the HCl solution at different reaction time

(1) The HCl attacks the solid wolframite particles with the formation of the structural lattice of H_2WO_4 on the particles:

$$\text{Solid} + H^+ \rightleftharpoons \text{Solid} \cdot H^+_{(ads)} \text{ (ads represents adsorption state)} \quad (4)$$

$$\text{Solid} \cdot H^+_{(ads)} \rightleftharpoons \text{Solid} \cdot HWO^-_{4(ads)} + Me^{2+}_{(aq)} \text{ (Me represents Fe and Mn)} \quad (5)$$

$$\text{Solid} \cdot HWO^-_{4(ads)} + H^+ \rightleftharpoons \text{Solid} \cdot H_2WO_{4(ads)} \quad (6)$$

$$\text{Solid} \cdot H_2WO_{4(ads)} \rightleftharpoons \text{Solid} \cdot H_2WO_{4(s)} \quad (7)$$

(2) Outside the formation of solid H_2WO_4, the formation of isopolytungstate ions, which is the main reason for the dissolution of tungsten in HCl solution, would be regarded as an accompanied process according to the following scheme[21]:

$$\text{Solid} \cdot H_2WO_{4(s)} + 5H^+ \rightleftharpoons \text{Solid} + [H_2W_6O_{21} \cdot x]^{5-}_{(aq)} \quad (8)$$

$$[H_2W_6O_{21} \cdot x]^{5-}_{(aq)} \xrightarrow[2H^+]{\text{Fast (some minutes)}} [H_2W_{12}O_{40} \cdot x]^{6-}_{(aq)} \quad (9)$$

On the other hand, the wolframite concentrate used in this work contains a certain amount of phosphorus and arsenic (as shown in Table 1), which can form soluble heteropoly acid with tungsten (for example $[PW_{12}O_{40}]^{3-}$)[7, 29-31].

In conclusion, the digestion of wolframite in HCl solution is a process of tungsten dissolution-H_2WO_4 deposition.

3.3 Recycling of HCl solution

According to the results of conditional experiments, under the optimum condition, the required stoichiometric ratio of HCl reaches 3.0, that is, a large amount of HCl remains in the digestion mother solution. The recycling of HCl solution is an inevitable requirement to reduce HCl consumption and waste water discharge.

3.3.1 Effect of cations accumulation in HCl solution on wolframite digestion efficiency

The digestion of wolframite is accompanied by the continuous dissolution of Fe^{2+} and Mn^{2+} into HCl solution, as shown in reaction (1). The accumulation of these cations in solution may influence the wolframite digestion, as the high concentration of Fe^{2+} and Mn^{2+} will resist reaction (1). The influence of Fe^{2+} and Mn^{2+} on the digestion efficiency is confirmed in Fig. 10 which shows that when the initial concentrations of Fe^{2+} and Mn^{2+} in the solution reach 40 g/L, the digestion efficiency is remarkably reduced, and the residual WO_3 content in the leaching residue of digested product increases sharply.

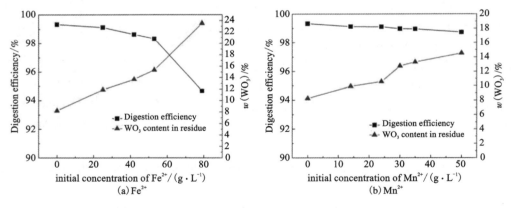

(reaction temperature of 90 ℃, stirring intensity of 250 r/min,
HCl stoichiometric ratio of 3.0, liquid to solid ratio of 3 : 1, duration of 4 h).

Fig. 10 Effect of initial cation concentration on the digestion efficiency of wolframite

Table 2 shows that in addition to a large amount of H^+, the concentrations of Fe^{2+} and Mn^{2+} in the mother solution of digestion reach 33.74 g/L and 21.48 g/L. As discussed above, the accumulation of Fe^{2+} and Mn^{2+} in the solution will reduce the wolframite digestion efficiency when the acid mother solution is directly reused. Therefore, the recovery of Fe and Mn from mother solution is a route not only to eliminate the influence of Fe^{2+} and Mn^{2+} on the recycling of mother solution but also to achieve comprehensive utilization of resources.

Table 2 The analysis result of the main ions concentration in digestion mother solution   g/L

Ion	H^+	TFe	Fe^{2+}	Mn
Concentration	3.6 mol/L	37.01	33.74	21.48

3.3.2 A potential process of HCl solution recycling

In order to recover Fe and Mn from mother solution while recycling HCl, a process of atomization-oxidative thermal decomposition for the mother solution was proposed. In this process, the following reactions will occur:

$$4FeCl_2(aq) + O_2(g) + 4H_2O(g) \longrightarrow 2Fe_2O_3(s) + 8HCl(g) \qquad (10)$$
$$4MnCl_2(aq) + O_2(g) + 4H_2O(g) \longrightarrow 2Mn_2O_3(s) + 8HCl(g) \qquad (11)$$

Through this process, the newly generated and remaining HCl can be recycled, and Fe_2O_3 and Mn_2O_3 powder can be obtained. The Gibbs free energy and equilibrium constants of the two reactions as a function of reaction temperature were calculated by Factsage 7.1 to predict the reaction trend, the results were presented in Fig. 11. Theoretically, reaction (10) is easy to occur, while reaction (11) begins to occur above 400 ℃ (at this temperature, the Gibbs free energy becomes below 0). The reaction trend of these two reactions increases with increasing temperature. Moreover, the equilibrium constants of both reactions reach the order of 1010 ($Log(K) > 10$) at 500 ℃. Xie et al.[32] have reported that high quality Fe_2O_3 powder can be prepared from $FeCl_2$ at 400 to 500 ℃.

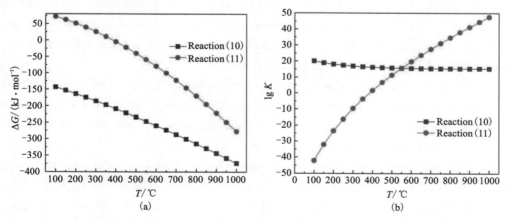

Fig. 11 Gibbs free energy (a) and reaction rate constants (b) of reaction (10, 11) at various temperatures

We have designed a device for the atomization and thermal decomposition of nickel salt solution and prepared nickel metal powder through this process[33, 34], and we have improved it for the decomposition of the mother solution in this work. Firstly, the mother solution is atomized by an ultrasonic atomizer. Then, the atomized solution and a certain volume of air are sprayed into a heated vertical tubular furnace through pressure with a flow rate of 300-500 mL/min. After oxidative thermal decomposition in the tubular furnace, oxide powder and HCl will be generated. The oxide powder will be separated through a cyclone separator connected with the tubular furnace, and the soluble HCl will be recovered by an eluting column. Detailed investigation will be presented in our future work.

3.4 Tungsten extraction from digested product using $NH_3 \cdot H_2O$ solution

The digested product is with high content of H_2WO_4, as most of Fe and Mn in wolframite are dissolved into mother solution. Direct conversion of H_2WO_4 to $(NH_4)_2WO_4$ solution is a convenient way to produce APT. In view of the strong trend of the reaction between H_2WO_3 and alkaline solutions and the convenient recycle of NH_3, the investigation on tungsten extraction from digested product was conducted by using $NH_3 \cdot H_2O$ as leaching agent.

Fig. 12(a) shows the tungsten leaching efficiency as a function of reaction time under different $NH_3 \cdot H_2O$ concentrations and reaction temperature of 25 ℃, liquid to solid ratio of 5 : 1. When the $NH_3 \cdot H_2O$ concentration is increased from 2.0 mol/L to 2.7 mol/L, the leaching efficiency and leaching rate of tungsten from digested product are significantly improved. With the further increasing of $NH_3 \cdot H_2O$ concentration to 3.5 mol/L, the improvement of tungsten leaching efficiency is very weak. Especially when the reaction time reaches 2 h, the leaching efficiency is almost 100% at the $NH_3 \cdot H_2O$ concentrations above 2.7 mol/L.

The experiments were carried out to study the effect of reaction temperature on the tungsten leaching efficiency with $NH_3 \cdot H_2O$ concentration of 2.7 mol/L and liquid to solid ratio of 5 : 1. As illustrated in Fig. 12(b), although increasing the temperature can promote the reaction between H_2WO_4 and $NH_3 \cdot H_2O$, the promotion is weak when the temperature is above 35 ℃ and the reaction time reaches 1.5 h, which confirms the strong trend of the reaction between H_2WO_3 and alkaline solutions.

The influence of the liquid to solid ratio on the leaching efficiency was investigated using a $NH_3 \cdot H_2O$ concentration of 2.7 mol/L and reaction temperature of 35 ℃. The Fig. 12(c) shows that when the reaction time is less than 2 h, the leaching efficiency of tungsten with liquid to solid ratio of 3 : 1 is considerably lower than that with liquid to solid ratio of 5 : 1, while the leaching efficiency is not improved as the liquid to solid ratio is increased to 7 : 1. The main reason may be that when the liquid to solid ratio is 3 : 1, the solution viscosity is high, which affects the mass transfer. In addition, the stoichiometric ratio of $NH_3 \cdot H_2O$ is only 1.2 with the liquid-solid ratio of 3 : 1. As a result, the chemical driving force of the reaction is not strong enough.

In conclusion, the leaching efficiency of tungsten from the digested product of wolframite was above 99.5% in $NH_3 \cdot H_2O$ solution with $NH_3 \cdot H_2O$ concentration of 2.7 mol/L, reaction temperature of 35 ℃, liquid to solid ratio of 5 : 1, and reaction time of 1.5 h. In the process of preparing APT by crystallization of $(NH_4)_2WO_4$ solution, the precipitation of some impurities will affect the quality of APT. In particular, when the concentrations of K and Na in $(NH_4)_2WO_4$

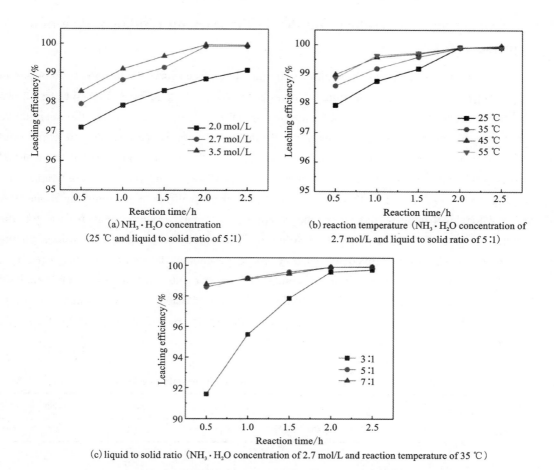

Fig. 12 Effect of the following conditions on the leaching efficiency of tungsten from digested product in $NH_3 \cdot H_2O$ solution

solution is high, it is not feasible for direct conversion of H_2WO_4 to $(NH_4)_2WO_4$ by $NH_3 \cdot H_2O$ solution, as the removal of K and Na from $(NH_4)_2WO_4$ solution requires complex processes such as ion exchange or extraction. Therefor, the concentrations of WO_3 and the main impurity elements in the $(NH_4)_2WO_4$ solution obtained at the optimum conditions were analyzed. Table 3 shows that the $(NH_4)_2WO_4$ solution is with low concentrations of impurity elements, which is suitable for producing APT.

Table 3 The analysis result of the $(NH_4)_2WO_4$ solution obtained under the optimized conditions   g/L

Element	WO_3	P	As	Si	K	Na
Mass concentration	172.91	0.016	0.012	0.11	0.005	0.001

3.5 Existence state of residual tungsten in the leaching residue of digestion product

The wolframite concentrate can be almost completely digested in HCl solution at the optimum condition. However, the leaching residue of digested product still contains 8.22% WO_3 as shown in Table 4. In order to clarify occurrence state of residual tungsten in the residue, the quantitative analysis of mineral composition and observation of tungsten mineral in the residue were carried out by MLA, with the result presented in Table 5 and Fig. 13. The results indicate that the residual tungsten is wolframite, including monomer particles [Fig. 13(a)] and fine particles wrapped by quartz [Fig. 13(b)]. Table 5 further shows that the wrapped wolframite accounts for about 50% of total residual wolframite, indicating the wrapping of quartz is a crucial reason for the incomplete digestion of wolframite by HCl solution. The hindrance of quartz inclusions on the digestion efficiency has been proved by the digestion experiments of leaching residue using HCl solution. Under the optimum condition of conditional experiments, the digestion efficiency of residual wolframite in leaching residue is only 20%. The existence of wolframite monomer particles is mainly due to the covering of unreacted particles by H_2WO_4 layers during digestion, which is also confirmed by Fig. 7.

Table 4 The content of the main elements in leaching residue   %

Element	WO_3	Sn	Ta	Nb	SiO_2
Mass fraction	8.22	1.37	0.41	1.15	67.33

Table 5 The mineral composition of leaching residue   %

Mineral	Wolframite	Fine wolframite wrapped by quartz	Fine wolframite and cassiterite wrapped by quartz	Cassiterite
Content	6.67	4.08	3.96	3.18
Mineral	Pyrite	Quartz	Rutile containing Ta and Nb	Zircon
Content	5.51	57.13	2.90	3.91
Mineral	Mica	Tourmaline	Others	Total
Content	2.55	3.20	6.91	100.00

It is also obviously presented by Table 4 that the leaching residue contains high contents of Sn (1.37%), Ta, and Nb (Ta + Nb 1.56%) in addition to W. As Fe and Mn do not enter the solid product during digestion, the leaching residue yield is greatly reduced, which contributes to the effective enrichment of these trace valuable metals in wolframite concentrate. It is worth noting that this is a prominent advantage of acid digestion process. Apparently, the content of these valuable metals is much higher than that of the corresponding raw ore in mineral processing

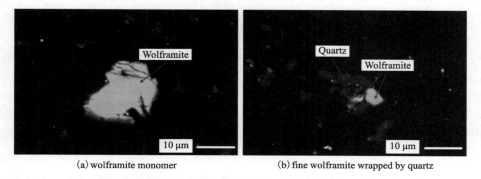

(a) wolframite monomer (b) fine wolframite wrapped by quartz

Fig. 13 The occurrence state of residual tungsten in leaching residue

industry according to the literatures[35-37]. It is worth recovering these valuable metals to make further comprehensive utilization of wolframite concentrate.

3.6 A sustainable process for extracting tungsten from wolframite concentrate

Based on the above studies, a sustainable process of HCl solution digestion-$NH_3 \cdot H_2O$ solution leaching-atomization and oxidative thermal decomposition of mother solution-HCl recycling for extracting tungsten from wolframite is developed in this work. The flowsheet is presented in Fig. 14. Compared with the conventional process of autoclave-based NaOH leaching-leaching solution conversion through extraction or ion exchange, the process of this study has the following advantages:

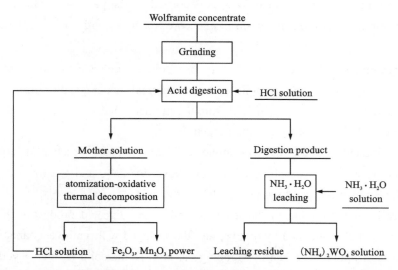

Fig. 14 Flowsheet of the sustainable process for extracting tungsten from wolframite concentrate

(1) The digestion medium HCl is recycled. In the autoclave-based NaOH leaching process, the excess NaOH is difficult to be recycled, which results in the discharge of a large amount of

sodium salt wastewater in subsequent ion exchange or solvent extraction process.

(2) The Fe and Mn in wolframite concentrate are recovered. In the autoclave-based NaOH leaching process, the Fe and Mn enter the leaching residue and are discarded.

(3) No hazardous solid waste is generated, and the leaching residue of digested product has the value of extracting W, Sn, Ta and Nb.

4 Conclusions

HCl solution was used to digest wolframite at atmospheric pressure and the digestion behavior and mechanism were studied in this work. Based on the HCl digestion of wolframite, the leachability of tungsten from the digested product by $NH_3 \cdot H_2O$ solution was tested to prepare a qualified $(NH_4)_2WO_4$ solution for APT production, and a process of atomization-oxidative thermal decomposition for the recycling of HCl solution was proposed. The following conclusions were obtained:

(1) The digestion efficiency of wolframite reached 99.3% in HCl solution under atmospheric pressure at the optimum condition: particle size of $D(95) = 20$ μm, stirring intensity of 250 r/min, reaction temperature of 90 ℃, HCl stoichiometric ratio of 3.0, liquid to solid ratio of 3:1, and duration of 4 h.

(2) The digestion of wolframite in HCl solution is a process of tungsten dissolution-H_2WO_4 deposition. In the initial stage of digestion, tungsten is mainly dissolved into HCl solution, when the dissolution reaches balance, solid H_2WO_4 is deposited on the particle surface. The subsequent process of wolframite digestion is a shrinking core reaction with the continuous formation of H_2WO_4.

(3) The accumulation of Fe^{2+} and Mn^{2+} in HCl solution reduced the digestion efficiency of wolframite. The process of atomization oxidative thermal decomposition is feasible for the recovery of Fe and Mn and recycling of HCl from the mother solution of digestion according to the thermodynamic calculation of decomposition reactions.

(4) The digested product of wolframite had an excellent leachability of tungsten in $NH_3 \cdot H_2O$ solution with WO_3 leaching efficiency above 99.5%, and the $(NH_4)_2WO_4$ solution was qualified for ammonium paratungstate (APT) production.

(5) The residual wolframite in the leaching residue of digested product includes monomer particles and fine particles wrapped by quartz, and the wrapped wolframite accounts for about 50% of total residual wolframite. The quartz wrapping, as well as the solid H_2WO_4 layers covering the unreacted wolframite particles, is a crucial factor for the difficulty of complete wolframite digestion. The leaching residue contains 1.37% Sn, Ta, and Nb (Ta+Nb 1.56%) in addition to WO_3 (8.22%), the recovery of these valuable metals will make a further comprehensive utilization of wolframite concentrate.

Acknowledgment

This work was supported by the National Key R&D Program of China (grant number 2019YFC1907400) and the National Natural Science Foundation of China (grant numbers 51620105013).

References

[1] LI T, MIAO J, GUO E, et al. Tungsten containing high-entropy alloys: A focused review of manufacturing routes, phase selection, mechanical properties, and irradiation resistance properties[J]. Tungsten, 2021(3): 181-196. https://doi.org/10.1007/s42864-021-00081-x.

[2] WANG X, HUANG H, SHI J, et al. Recent progress of tungsten-based high entropy alloys in nuclear fusion[J]. Tungsten, 2021(3): 143-160. https://doi.org/10.1007/s42864-021-00092-8.

[3] LI Y, YANG J, ZHAO Z. Recovery of tungsten and molybdenum from low-grade scheelite[J]. JOM, 2017(69): 1958-1962. https://doi.org/10.1007/s11837-017-2440-5.

[4] LIU C, ZHANG W, SONG S, et al. Study on the activation mechanism of lead ions in wolframite flotation using benzyl hydroxamic acid as the collector[J]. Miner. Eng, 2019(141): 105859. https://doi.org/10.1016/j.mineng.2019.105859.

[5] CHEN W, FENG Q, ZHANG G, et al. Effect of energy input on flocculation process and flotation performance of fine scheelite using sodium oleate[J]. Miner. Eng, 2017(112): 27-35. https://doi.org/10.1016/j.mineng.2017.07.002.

[6] LIU H, LIU H, NIE C, et al. Comprehensive treatments of tungsten slags in China: A critical review[J]. Environ. Manag, 2020(270): 110927. https://doi.org/10.1016/j.jenvman.2020.110927.

[7] LUO Y, CHEN X, ZHAO Z, et al. Pressure leaching of wolframite using a sulfuric-phosphoric acid mixture[J]. Miner. Eng, 2021(169): 106941. https://doi.org/10.1016/j.mineng.2021.106941.

[8] YANG K, ZHANG W, HE L, et al. Leaching kinetics of wolframite with sulfuric-phosphoric acid[J]. Chinese J. Nonferrous Metals, 2018(28): 175-182. https://doi.org/10.19476/j.ysxb.1004.0609.2018.01.21. (In Chinese)

[9] MARTINS J. Leaching systems of wolframite and scheelite: A thermodynamic approach[J]. Miner. Process. Extr. Metall. Rev., 2014(35): 23-43. https://doi.org/10.1080/08827508.2012.757095.

[10] ZHAO Z. Tungsten metallurgy: Fundamentals and applications[M]. Beijing: Tsinghua University Press, 2013. (In Chinese)

[11] LI X, SHEN L, ZHOU Q, et al. Scheelite conversion in sulfuric acid together with tungsten extraction by ammonium carbonate solution[J]. Hydrometallurgy, 2017(171): 106-115. https://doi.org/10.1016/j.hydromet.2017.05.005.

[12] KALPAKLI A, ILHAN S, KAHRUMAN C, et al. Dissolution behavior of calcium tungstate in oxalic acid solutions[J]. Hydrometallurgy, 2012(121): 7-15. https://doi.org/10.1016/j.hydromet.2012.04.014.

[13] LI J, ZHAO Z. Kinetics of scheelite concentrate digestion with sulfuric acid in the presence of phosphoric

acid[J]. Hydrometallurgy, 2016(163): 55-60. https://doi.org/10.1016/j.hydromet.2016.03.009.

[14] SHEN L, LI X, ZHOU Q, et al. Kinetics of scheelite conversion in sulfuric acid[J]. JOM, 2018(70): 2499-2504. https://doi.org/10.1007/s11837-018-2787-2.

[15] REN H, LI J, TANG Z, et al. Sustainable and efficient extracting of tin and tungsten from wolframite-scheelite mixed ore with high tin content[J]. J. Clean. Prod, 2020(269): 122282. https://doi.org/10.1016/j.jclepro.2020.122282.

[16] LIU L, XUE J, LIU K, et al. Complex leaching process of scheelite in hydrochloric and phosphoric solutions[J]. JOM, 2016(68): 2455-2462. https://doi.org/10.1007/s11837-016-1979-x.

[17] SHEN L, LI X, ZHOU Q, et al. Sustainable and efficient leaching of tungsten in ammoniacal ammonium carbonate solution from the sulfuric acid converted product of scheelite[J]. J. Clean. Prod, 2018(197): 690-698. https://doi.org/10.1016/j.jclepro.2018.06.256.

[18] SHEN L, LI X, LINDBERG D, et al. Tungsten extractive metallurgy: A review of processes and their challenges for sustainability[J]. Miner. Eng., 2019(142): 105934. https://doi.org/10.1016/j.mineng.2019.105934.

[19] YIN C, JI L, CHEN X, et al. Efficient leaching of scheelite in sulfuric acid and hydrogen peroxide solution[J]. Hydrometallurgy, 2020(192): 105292. https://doi.org/10.1016/j.hydromet.2020.105292.

[20] SHEN L, LI X, ZHOU Q, et al. Wolframite conversion in treating a mixed wolframite-scheelite concentrate by sulfuric acid[J]. JOM, 2018(70): 161-167. https://doi.org/10.1007/s11837-017-2691-1.

[21] MARTINS J, MOREIRA A, COSTA S. Leaching of synthetic scheelite by hydrochloric acid without the formation of tungstic acid[J]. Hydrometallurgy, 2003(70): 131-141. https://doi.org/10.1016/S0304-386X(03)00053-7.

[22] ZHAO Z, DING W, LIU X, et al. Effect of ultrasound on kinetics of scheelite leaching in sodium hydroxide[J]. Can. Metall. Q., 2013(52): 138-145. https://doi.org/10.1179/1879139512Y.0000000050.

[23] ZHAO Z, LI H, LIU M, et al. Soda decomposition of low-grade tungsten ore through mechanical activation[J]. J. Cent. South Univ., 1996(3): 181-184. https://doi.org/10.1007/BF02652201.

[24] ZHANG W, LI J, ZHAO Z. Leaching kinetics of scheelite with nitric acid and phosphoric acid[J]. Int. J. Refract. Met. Hard Mater., 2015(52): 78-84. https://doi.org/10.1016/j.ijrmhm.2015.05.017.

[25] LI J, MA Z, LIU X, et al. Sustainable and efficient recovery of tungsten from wolframite in a sulfuric acid and phosphoric acid mixed system[J]. ACS Sustain. Chem. Eng., 2020(8): 13583-13592. https://doi.org/10.1021/acssuschemeng.0c04216.

[26] XIE H. Study on new technology of acid extraction from wolframite[D]. Changsha: Central South University, 2011. (In Chinese)

[27] YANG K. Technical study on treating wolframite and wolframite-scheelite mixed ore by sulfuric-phosphoric mixed acid method[D]. Changsha: Central South University, 2017. (In Chinese)

[28] LI H, YANG J, LI K. Tungsten metallurgy[M]. Changsha: Central South University Press, 2010. (In Chinese)

[29] ZHANG W, CHEN Y, CHE J, et al. Green leaching of tungsten from synthetic scheelite with sulfuric acid-hydrogen peroxide solution to prepare tungstic acid[J]. Sep. Purif. Technol, 2020(241): 116752. https://doi.org/10.1016/j.seppur.2020.116752.

[30] ZHONG L. Research on the decomposition of scheelite by hydrochloric acid in the present of phosphorus and fluorin[J]. Changsha: Central South University, 2018. (In Chinese)

[31] WANG W. Hydrothermal synthesis, structure characterization and properties of arsenotungstates compounds

[D]. Zhengzhou: Henan University, 2004. (In Chinese)

[32] XIE X, HU Y, ZHANG W, et al. Study on manufacture technology of high quality red iron oxide from ferrous chloride[J]. Inorganic Chemicals Industry, 2015(47): 41-44. (In Chinese)

[33] GUO X, YI Y, TIAN Q. A method for preparing nickel and cobalt metal or alloy powder by solution atomization[P]. China Patent CN201010577129.0, 2010.

[34] GUO X, TIAN Q, FENG Q, et al. A special device for the atomization and oxidation decomposition of solution[P]. China Patent ZL200920063526.9, 2009.

[35] ZHANG H, ZHANG F, JIANG H, et al. Beneficiation test of a scheelite mine in Hunan[J]. China Tungsten Industry, 2020(35): 23-28. https://doi.org/10.3969/j.issn.1009-0622.2020.01.005. (In Chinese)

[36] WANG P, WANG J. Beneficiation tests of a low-grade high slime tin oxide ore from Yunnan[J]. Metal Mine, 2020(7): 83-88. https://doi.org/10.19614/j.cnki.jsks.202007012. (In Chinese)

[37] LI J, XIE X, LV J, et al. Overview of niobium resources and research progress in mineral processing technology[J]. Metal Mine, 2021(2): 120-126. https://doi.org/10.19614/j.cnki.jsks.202102019. (In Chinese)

Sustainable Extraction of Tungsten from the Acid Digestion Product of Tungsten Concentrate by Leaching-solvent Extraction Together with Raffinate Recycling

Abstract: Acid digestion of tungsten concentrates has attracted more and more attention in recent years, sustainable extraction of tungsten from acid digestion product of tungsten concentrates to obtain $(NH_4)_2WO_4$ solution with low impurity content is the key for ammonium paratungstate production. In this work, a process of extracting tungsten from acid digestion product by leaching-solvent extraction together with raffinate circulation was introduced. The acid digestion product was firstly leached by Na_2CO_3, and the leaching efficiency was 99.47% with circulation of Na_2CO_3 solution converted from the raffinate of solvent extraction. The tungsten in pure Na_2WO_4 solution with 121.26 g/L WO_3 could be almost completely extracted by solvents containing 15% N235 or 15% Alamine 308 in one stage. The extraction mechanism of N235 and Alamine 308 was the same, while Alamine 308 showed better performance and higher saturated loading capacities for tungsten extraction. The loaded tungsten was stripped from organic phases using $NH_3 \cdot H_2O$ with stripping efficiency of 99.98%, obtaining $(NH_4)_2WO_4$ solutions with WO_3 concentration above 260 g/L and low impurity content. The raffinate containing high Na_2SO_4 concentration was converted to Na_2CO_3 solution by adding $BaCO_3$ and recycled for the leaching of digestion product, which could significantly reduce the discharge of sodium salt wastewater. The presented work provides a cleaner production route for preparing pure $(NH_4)_2WO_4$ solution from acid digestion product of tungsten concentrates.

Keywords: Tungsten; Leaching; Solvent extraction; Raffinate recycling

Published in *Journal of Cleaner Production*, 2022, 375: 133924. Authors: Chen Yuanlin, Huo Guangsheng, Guo Xueyi, Wang Qinmeng.

1 Introduction

Tungsten is an important rare metal with high melting point, high modulus of elasticity, good thermal and electrical conductivity and excellent corrosion resistance, and it is an essential component for many products in a wide range of fields (Wang et al., 2021). Particularly, the use of tungsten in the production of cemented carbide represents its most critical application, and cemented carbide is widely applied in machinery manufacturing, aerospace, and national defense (Shen et al., 2019; Furberg et al., 2019; Ma et al., 2017). The main tungsten mineral resources with economic value are wolframite [(Fe,Mn)WO_4] and scheelite ($CaWO_4$) (Liu et al., 2019; Chen et al., 2017). For a long time, China's tungsten concentrate output has accounted for more than 80% of the world's tungsten concentrate output (USGS, 2021). Alkali digestion is a conventional process for tungsten concentrates digestion in China. However, according to the latest environmental law of China, the alkali leaching residue has been classified as hazardous solid waste. Tungsten production enterprises are required to pay an environmental tax, about 1000 Yuan (140 Dollars or 130 Euros) for per ton of residue discharged, and this environmental tax raises production cost (Liu et al., 2020).

In view of the environmental problems faced by alkali digestion process, acid digestion is considered a cleaner method and has been paid more and more attention (He et al., 2014; Shen et al., 2018a, 2018b; Li and Zhao, 2016). Digesting scheelite with HCl is a classical and efficient method, in which tungsten is converted to solid H_2WO_4 and most impurities are dissolved into solution (Martins, 2014; Huo et al., 2020). Some previous works confirmed that acid digestion was also effective for wolframite (Luo et al., 2021; Chen et al., 2022). The digestion of scheelite and wolframite in acid medium can be expressed by Equations (1)-(5).

The digestion of scheelite (Martins, 2014; Lei et al., 2022):

$$CaWO_4(s) + 2H^+(aq) \Longleftrightarrow Ca^{2+}(aq) + H_2WO_4(s) \text{ (Using HCl or HNO}_3\text{)} \quad (1)$$

$$CaWO_4(s) + 2H^+(aq) + SO_4^{2-} \Longleftrightarrow CaSO_4(s) + H_2WO_4(s) \text{ (Using H}_2SO_4\text{)} \quad (2)$$

The digestion of wolframite (Martins, 2014; Lei et al., 2022):

$$FeWO_4(s) + 2H^+(aq) \Longleftrightarrow Fe^{2+}(aq) + H_2WO_4(s) \text{ (Using HCl or H}_2SO_4\text{)} \quad (3)$$

$$3FeWO_4(s) + 10H^+(aq) + NO_3^-(aq) \Longleftrightarrow$$
$$3Fe^{3+}(aq) + 3H_2WO_4(s) + NO(g) + 2H_2O \text{ (Using HNO}_3\text{)} \quad (4)$$

$$MnWO_4(s) + 2H^+(aq) \Longleftrightarrow Mn^{2+}(aq) + H_2WO_4(s) \quad (5)$$

The digestion products mainly containing H_2WO_4 can be leaching directly with $NH_3 \cdot H_2O$ to produce $(NH_4)_2WO_4$ solution. For the products containing H_2WO_4 and $CaSO_4 \cdot nH_2O$, the expected $(NH_4)_2WO_4$ solution can not be obtained, because the leached WO_4^{2-} can react with $CaSO_4 \cdot nH_2O$ to form $CaWO_4$ precipitation (secondary scheelite) as shown in Equation (6) (Lei et al., 2022).

$$H_2WO_4(s) + 2NH_3 \cdot H_2O(aq) + CaSO_4 \cdot nH_2O(s) =\!=\!= $$
$$CaWO_4(s) + (NH_4)_2SO_4(aq) + (n+2)H_2O \tag{6}$$

Li et al. (2017) and Shen et al. (2018c) reported that $NH_3 \cdot H_2O$ and $(NH_4)_2CO_3$ mixed solution could be used to leached H_2WO_4 selectively, where CO_3^{2-} promoted the conversion of $CaSO_4 \cdot nH_2O$ to $CaCO_3$ instead of $CaWO_4$ through Equation (7).

$$CaSO_4 \cdot nH_2O(s) + (NH_4)_2CO_3(aq) =\!=\!= (NH_4)_2SO_4(aq) + CaCO_3(s) + nH_2O \tag{7}$$

It seems that, leaching tungsten from digestion products with $NH_3 \cdot H_2O$ or $(NH_4)_2CO_3$ is an economical method due to the short process. However, these studies ignored the impurities (especially K) in the $(NH_4)_2WO_4$ solution obtained by leaching acid digestion products with $NH_3 \cdot H_2O$ or $(NH_4)_2CO_3$. K is a strictly controlled element in ammonium paratungstate (APT) (Li et al., 2010). The high concentration of K in $(NH_4)_2WO_4$ solution will seriously affect the quality of APT, as K will precipitate during the crystallization of APT by evaporation. In our work, we found that when $NH_3 \cdot H_2O$ solution was used to leach tungsten from the acid digestion product of complex tungsten concentrate, the K concentration in $(NH_4)_2WO_4$ solution was as high as 12 mg/L with the WO_3 concentration of 180 g/L. Liu (2022) reported that during the acid digestion of tungsten concentrate, some K bearing minerals (for example potassium feldspar) would be wrapped by solid H_2WO_4, this resulted in the dissolution of K into $(NH_4)_2WO_4$ solution during $NH_3 \cdot H_2O$ leaching process. It is not feasible to directly convert crude H_2WO_4 to $(NH_4)_2WO_4$ by $NH_3 \cdot H_2O$ leaching in this case. However, this problem has not attracted attention in previous studies.

To obtain $(NH_4)_2WO_4$ solution with low K and Na concentration, solvent extraction by amines is commonly used in tungsten extractive metallurgy due to its high efficiency (Liu et al., 2021; Ning et al., 2009; Yang et al., 2016). Yang et al. (2016) used the organic phase containing 10% Aliquat 336 to extract tungsten from acidic high-phosphorus solution, the extraction efficiency reached 99.8% by one extraction stage and high quality $(NH_4)_2WO_4$ solution was obtained. While the treatment of the raffinate produced in tungsten extraction process was not solved in the previous studies. In the traditional tungsten solvent extraction, 20-100 tons of raffinate with high concentration of sodium salt is produced for 1 ton of APT (Shen et al., 2018c; Li et al., 2020). Recently, most countries have strengthened environmental management, requiring zero discharge of high salt wastewater from industrial production (Liu et al., 2021). At present, the desalination technologies of high salt wastewater mainly include ultrafiltration by membrane and evaporative crystallization (Shi et al., 2020; Cichy et al., 2017), these treatment processes require high cost. Thus, to achieve sustainable extraction of tungsten, it is essential to recycle the raffinate while preparing pure $(NH_4)_2WO_4$ solution.

This work aims to develop a sustainable process for extracting tungsten from acid digestion product (crude tungstic acid) of tungsten concentrate. The crude tungstic acid was firstly leached by Na_2CO_3 solution, then the P, As, and Si removal and acidification of crude Na_2WO_4 solution were conducted to prepare purified Na_2WO_4 solution for solvent extraction. After that, the extraction behavior and mechanism of tungsten with different extractants and the subsequent stripping process were systematically studied. The qualified $(NH_4)_2WO_4$ solution with high WO_3 concentration was obtain. The raffinate containing Na_2SO_4 was converted to Na_2CO_3 solution by adding $BaCO_3$ and was recycled for crude tungstic acid leaching, this significantly reduced the discharge of sodium salt wastewater and leaching cost. In addition, the crude Ba_SO_4 generated from raffinate conversion could be purified by HCl and H_2SO_4 to prepare pure $BaSO_4$, which increased the value-added output of the process. Based on the above studies, a sustainable process of Na_2CO_3 leaching-solvent extraction together with raffinate circulation for preparing pure $(NH_4)_2WO_4$ solution from crude tungstic acid was developed with the flowsheet shown in Fig. 1.

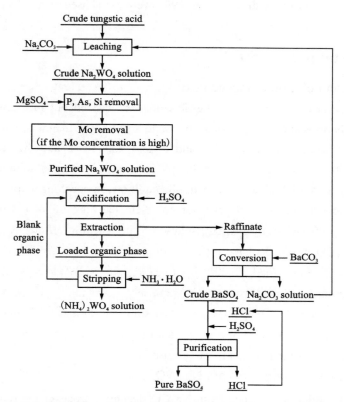

Fig. 1 Flow diagram of the process for sustainable extraction of tungsten from crude tungstic acid

2 Materials and methods

2.1 Materials and reagents

The crude tungstic acid was obtained by HCl digestion of a wolframite concentrate with reaction temperature of 90 ℃, HCl stoichiometric ratio of 3.0, liquid to solid ratio of 3 : 1, and duration of 4 h. The digestion of wolframite concentrate by HCl can be expressed as Equations

(3) and (5), and the digestion mechanism and kinetics have been reported in our previous studies (Chen et al., 2022). The crude tungstic acid contains 85.07% WO_3 and 3.30% SiO_2 as the main compositions.

The analytical-grade Na_2CO_3, $MgSO_4$ and $NH_3 \cdot H_2O$ were purchased from Aladdin; the H_2SO_4 and HCl were also analytical-grade and purchased from Sinopharm Chemical Reagent Co., Ltd, China; the extractant tri(octyl/decyl) alkyl tertiary amine (N235) and solvent oil were purchased from Hecheng New Material Technology Co., Ltd, Zhengzhou, China; the extractant tri-n-octylamine (Alamine 308) was purchased from BASF; the phase modifier 2-octanol with purity above 99% and $BaCO_3$ with purity of 99% were purchased from Aladdin. All solutions were prepared with deionized water.

2.2 Experimental procedures

2.2.1 Leaching of crude tungstic acid

The leaching experiments of crude tungstic acid were carried out in a 250 mL conical flask which was heated thermostatically by a water bath and stirred by a magnetic stirrer. In each run, a desired volume of Na_2CO_3 solution was added into the flask and heated to the required temperature, then 50 g crude tungstic acid sample was added into the solution. After a certain duration, the slurry was filtered and washed. In addition, a leaching experiment using NaOH solution was also carried out as a reference of complete leaching of WO_3 from tungstic acid to calculate the leaching efficiency by Na_2CO_3 solution. The leaching efficiency of WO_3 from crude tungstic acid by Na_2CO_3 was calculated by Equation (8).

$$L = [1 - (w_{residue} \cdot m)/(w_{product} \cdot M)]/[1 - (w_{residue'} \cdot m')/(w_{product} \cdot M)] \times 100\% \quad (8)$$

where L is the leaching efficiency by Na_2CO_3 (%), $w_{residue}$ and $w_{residue'}$ is the WO_3 content (mass fraction) of leaching residue by Na_2CO_3 and NaOH solution, m and m' is the weight of leaching residue by Na_2CO_3 and NaOH solution, $w_{product}$ and M is the WO_3 content (mass fraction, 85.07%) and weight of crude tungstic acid (50 g) used in each run.

2.2.2 Extraction of tungsten from Na_2WO_4 solution

For the Na_2WO_4 solution obtained from Na_2CO_3 leaching process, the removal of P, As and Si was firstly conducted with $MgSO_4$. The dosage of $MgSO_4$ was 1.2 times of the theoretical amount required for the complete precipitation of P, As and Si in Na_2WO_4 solution. If the concentration ratio of Mo/WO_3 in the Na_2WO_4 solution exceeds 1×10^{-4}, Mo removal is also carried out with selective precipitation (Zeng et al., 2020). The purified Na_2WO_4 solution was then acidified with concentrated H_2SO_4 to a designed pH value as the feeding solution of solvent extraction.

The organic phase acidification, extraction and stripping experiments were carried out in separating funnels which were settled in a temperature-controlled water bath shaker. In each run, the organic phase was prepared by dissolving known amount of extractant in solvent oil along with 15% of 2-octanol as phase modifier. The acidification of organic phase was conducted with

0.5 mol/L H_2SO_4 and O/A phase ratio of 1 : 4. After solvent extraction, the concentration of WO_3 in raffinate was determined to calculate the extraction efficiency. The organic phase after tungsten stripping by $NH_3 \cdot H_2O$ was subject to secondary stripping by NaOH solution (2 mol/L), and the WO_3 concentration in the secondary stripping liquor was determined to calculate the stripping efficiency. The extraction efficiency and stripping efficiency were calculated by Equations (9) and (10).

$$E = [1 - (C_1 \cdot V_1)/(C_0 \cdot V_0)] \times 100\% \qquad (9)$$

$$S = [1 - (C_2 \cdot V_2)/(C_0 \cdot V_0 \cdot E)] \times 100\% \qquad (10)$$

where E is the extraction efficiency, C_0(g/L) and V_0(L) are the WO_3 concentration and volume of feeding solution, C_1(g/L) and V_1(L) are the WO_3 concentration and volume of raffinate, S is the stripping efficiency, C_2(g/L) and V_2(L) are the WO_3 concentration and volume of secondary stripping liquor.

2.2.3 Raffinate conversion with $BaCO_3$

In each run of raffinate conversion experiments, 50 mL of raffinate was transferred to a beaker, then the beaker was placed in a water bath set at 50 ℃ and with magnetic stirrer. After that, the $BaCO_3$ powder with a desired stoichiometric ratio was slowly added to the raffinate, and the slurry was stirred for a certain duration. The slurry after reaction was filtered, and 10 mL of filtrate was taken to analyze the SO_4^{2-} concentration. The conversion efficiency of raffinate was calculated via Equation (11).

$$E_r = (1 - C_s/C_{s0}) \times 100\% \qquad (11)$$

where E_r is the conversion efficiency of raffinate, C_s(mol/L) and C_{s0}(mol/L) are the SO_4^{2-} concentration of filtrate and feeding raffinate.

2.2.4 Purification of crude $BaSO_4$

The crude $BaSO_4$ obtained from raffinate conversion was mixed with water with a liquid to solid ratio of 1.5 : 1. The crude $BaSO_4$ slurry was stirred with magnetic stirrer, and HCl solution was slowly added to the slurry until no bubble was generated. Then, a certain amount of concentrated H_2SO_4 was added to the slurry to precipitate the remaining Ba^{2+}. Finally, purified $BaSO_4$ was obtained after filtration and washing.

2.3 Analysis methods

The elemental composition of crude tungstic acid and the leaching residue of digestion product were determined by chemical analysis by Changsha Research Institute of Mining and Metallurgy in Changsha, China. The concentrations of WO_3, P, As and Si in solutions were measured by ICP-OES (ICAP7400, Thermo Scientific). The concentration of K in $(NH_4)_2WO_4$ solutions were measured by atomic absorption spectrometer (WFX-220B, Beijing Beifen-Ruili Analytical Instrument (Group) Co., Ltd). The Fourier transform infrared spectroscopy (FTIR) test was carried out by Is-50, Thermo Scientific.

The concentration of SO_4^{2-} in the solutions were measured by gravimetric method. A certain volume of solution was firstly mixed with excessive HCl, then the solution was heated to almost boiling. After that, $BaCl_2$ solution with 0.5 mol/L was slowly added to the tested solution under stirring followed by standing the solution to settle the sediment, then a small amount of $BaCl_2$ solution was added to check whether the precipitation was complete. After precipitation, the tested solution was heated for 1 h on the heating plate and then filtered through a pressure filter with 0.15 μm filter membrane. The precipitate was dried and then weighed. The SO_4^{2-} concentration of the tested solution was calculated via Equation (12).

$$C_t = [m_s/(233 \cdot V_t)] \times 1000 \quad (12)$$

where C_t(mol/L) is SO_4^{2-} concentration of the tested solution, m_s(g) is the weight of precipitate, V_t(mL) is the volume of tested solution.

3 Results and discussion

3.1 Leaching of crude tungstic acid by Na_2CO_3 solution

The crude tungstic acid was with high content of WO_3, as most of Ca, Fe and Mn in tungsten concentrate were dissolved into solution during acid digestion. Na_2CO_3 was selected as the crude tungstic acid leaching agent to provide conditions for subsequent recycling of converted raffinate, so as to form a complete sustainable tungsten extraction process.

Fig. 2(a) showed the tungsten leaching efficiency as a function of reaction time under different Na_2CO_3 concentrations and reaction temperature of 50 ℃, liquid to solid ratio of 3 : 1. The leaching rate increased slightly as the Na_2CO_3 concentration was increased from 1.5 mol/L to 2.0 mol/L. When the leaching time reached 60 min, further increasing of Na_2CO_3 concentration barely affected the leaching efficiency. The leaching efficiency of 99.6% could be obtained under the Na_2CO_3 concentration of 1.5 mol/L (the stoichiometric ratio of Na_2CO_3 was 1.2), which confirmed the strong trend of the reaction between H_2WO_3 and Na_2CO_3 solution.

Fig. 2(b) presented the effect of reaction temperature on the tungsten leaching efficiency with Na_2CO_3 concentration of 1.5 mol/L and liquid to solid ratio of 3 : 1. Although increasing the temperature could promote the reaction between H_2WO_4 and Na_2CO_3, the promotion was weak when the temperature was above 40 ℃.

The influence of the liquid to solid ratio on the leaching efficiency was also investigated with a Na_2CO_3 stoichiometric ratio of 1 : 2 and reaction temperature of 40 ℃. Fig. 2(c) revealed that the leaching efficiency with liquid to solid ratio of 2 : 1 was considerably lower than that with liquid to solid ratio of 3 : 1, and the leaching efficiency was slightly improved as the liquid to solid ratio is increased to 4 : 1. The main reason may be that when the liquid to solid ratio was 2 : 1, the solution viscosity was high, which affected the mass transfer.

In summary, the tungsten leaching efficiency of 99.6% from the crude tungstic acid was obtained using Na_2CO_3 solution with Na_2CO_3 stoichiometric ratio of 1 : 2, reaction temperature of

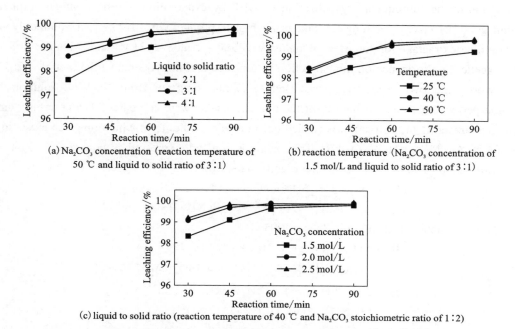

Fig. 2 Effect of the following conditions on the leaching efficiency of tungsten from crude tungstic acid in Na_2CO_3 solution

40 ℃, liquid to solid ratio of 4 : 1, and reaction time of 60 min. Table 1 shows the concentrations of the main impurity elements in the crude Na_2WO_4 solution. The removal of P, As and Si of crude Na_2WO_4 solution with $MgSO_4$ was conducted, the purified Na_2WO_4 solution was diluted to about 120 g/L to avoid the influence of excessive SO_4^{2-} in raffinate on the subsequent conversion efficiency. The concentrations of the main impurity elements in the diluted purified Na_2WO_4 solution are showed in Table 1.

Table 1 The concentrations of the main impurity elements in Na_2WO_4 solutions g/L

Element	WO_3	P	As	Si	K
Crude Na_2WO_4 solution	212.91	0.19	0.15	1.90	0.013
Purified Na_2WO_4 solution	121.26	0.012	0.009	0.045	0.008

3.2 Solvent extraction of tungsten from Na_2WO_4 solution

Zhang et al. (2016) have reported the direct solvent extraction of tungsten from alkaline medium with quaternary ammonium salt extractant. However, several problems limited the industrial application of this process. The phase separation during extraction was poor, and the low WO_3 concentration of the obtained stripping liquor (about 100 g/L) leaded to high energy consumption in the subsequent crystallization of APT by evaporation. Tertiary amine extractants are with good selectivity for tungsten and bring higher productivity. N235 is a commonly used tertiary amine extractant for tungsten in industry (Long et al., 2021), and Alamine 308 also shows excellent performance on the extraction of tungsten and molybdenum (Sahu et al., 2013). In this work, the two extractants were used to extract tungsten from Na_2WO_4 solution, the extraction performance and mechanism of the two extractants were compared. Using tertiary amine sulfate as extractant, the extraction is carried out in acidic medium, in which tungsten exists in the form of $[W_{12}O_{39}]^{6-}$ or $[H_2W_{12}O_{40}]^{6-}$. The main reactions are expressed as Equations (13)-(17).

Acidification of organic phase (Zhang and Zhao, 2005):

$$2R_3N(org) + H_2SO_4(aq) \longrightarrow (R_3NH)_2SO_4(org) \tag{13}$$

$$R_3N(org) + H_2SO_4(aq) \longrightarrow (R_3NH)HSO_4(org) \tag{14}$$

Solvent extraction (Zhang and Zhao, 2005):

$$4(R_3NH)HSO_4(org) + 2H^+ + [W_{12}O_{39}]^{6-}(aq) \longrightarrow$$
$$(R_3NH)_4H_2W_{12}O_{39}(org) + 4HSO_4^-(aq) \tag{15}$$

$$3(R_3NH)_2SO_4(org) + [H_2W_{12}O_{40}]^{6-}(aq) \longrightarrow$$
$$(R_3NH)_6H_2W_{12}O_{40}(org) + 3SO_4^{2-}(aq) \tag{16}$$

$$5(R_3NH)_2SO_4(org) + 2H^+ + 2[H_2W_{12}O_{40}]^{6-}(aq) \longrightarrow$$
$$2(R_3NH)_5H(H_2W_{12}O_{40})(org) + 5SO_4^{2-}(aq) \tag{17}$$

3.2.1 Effect of extractant concentration on tungsten extraction

To optimize the concentrations of the extractants for tungsten extraction, the volume concentrations of N235 and Alamine 308 were varied from 7% to 15%. The initial pH of the Na_2WO_4 solution and O/A phase ratio were maintained at 3.2 and 1. The extraction efficiencies of tungsten were found to increase with the increase of extractant concentrations from 7% to 15%, and Alamine 308 showed better extraction performance than N235 (Fig. 3). Using Alamine 308 as extractant, the WO_3 concentration lower than 100 mg/L in raffinate was obtained at the concentration of 12%, while it needed 15% when using N235 as extractant.

The axis pointed by the arrow is the coordinate axis corresponding to the curve.

For feeding Na_2WO_4 solution with 121.26 g/L WO_3, tungsten was completely extracted by N235 or Alamine 308 in one stage at O/A of 1 here. Compared with industrial feeding concentration of 90 g/L (Zhang and Zhao, 2005), this extraction process significantly reduced the generation of sodium salt wastewater and production cost.

3.2.2 Effect of feeding solution pH on tungsten extraction

The extraction efficiency of tungsten was found to be critically dependent on the pH of feeding solution, as tungsten formed different anionic species at low pH (Moris et al., 1999). The effect of initial pH on the extraction efficiency of tungsten was studied with the pH ranging from 3.2 to 6.0. The extraction data and equilibrium pH of raffinate as a function of initial solution pH were represented in Fig. 4. The results indicated that when the initial pH was lower than 4.5, the extraction of tungsten was hardly affected. In fact, within this pH range of feeding solution, the equilibrium pH of raffinate was still lower than 3. The reason was that when the organic phase was acidified under high acidity (0.5 mol/L in this work), the extractant was loaded with H^+ through reaction (14), which had a certain buffer effect on the pH rising of the solution.

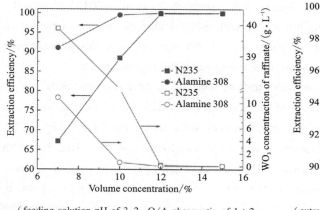

(feeding solution pH of 3.2, O/A phase ratio of 1 : 2, contact time of 4 min)

Fig. 3 Effect of extractant concentration on the extraction of tungsten

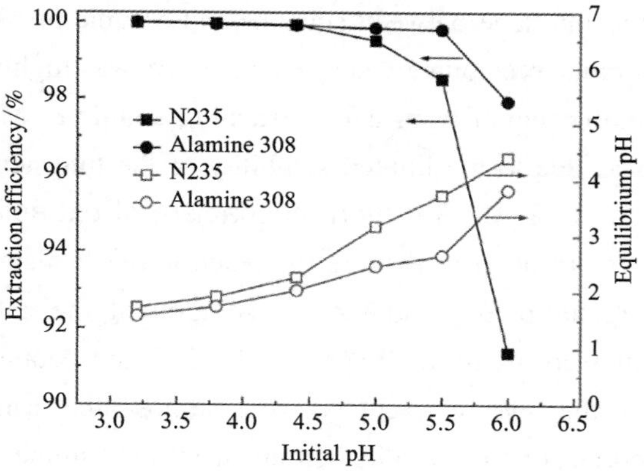

(extractant concentration of 15%, O/A phase ratio of 1 : 2, contact time of 4 min)

Fig. 4 Effect of feeding solution pH on the extraction of tungsten

The axis pointed by the arrow is the coordinate axis corresponding to the curve.

It was also apparent in Fig. 4 that when the equilibrium pH of raffinate was higher than 3, the tungsten uptake decreased sharply with both extractants. Therefore, in order to obtain satisfactory tungsten extraction efficiency, it is necessary to keep the pH of the extraction system lower than 3.

3.2.3 Effect of O/A phase ratio on the extraction of tungsten with N235 and Alamine 308

The organic phases containing 15% N235 and 15% Alamine 308 were contacted with Na_2WO_4 solution with initial pH of 3.2 at different O/A ratios to establish extraction isotherms, the results were presented in Fig. 5. This helps to determine the saturation loading capacity of the solvents. The loading of tungsten by both the two extractants were found to increase with O/A ratio decreasing. At the O/A ratio of 0.6 for N235 and 0.5 for Alamine 308, the extraction of tungsten was almost saturated, the solvents containing 15% N235 and 15% Alamine 308 were

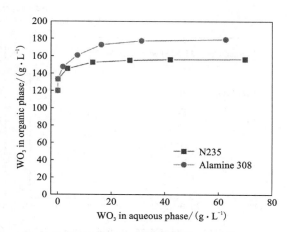

Fig. 5 The extraction isotherms of organic phase containing 15% N235 and 15% Alamine 308

loaded with 156.4 g/L and 179.0 g/L of WO_3 which was considered as the saturated loading capacity. This result also indicated that Alamine 308 showed better performance on tungsten extraction than N235.

It was observed that from the O/A ratio of 0.6 onwards, the extracted tungsten precipitated at the interface between aqueous and organic in Alamine 308 system. Sahu et al. (2013) observed a similar precipitate when Alamine 308 was highly loaded with molybdenum during the extraction of molybdenum from acidic leaching solution. Moris et al. (1999) believed that this phenomenon was due to the limited solubility of the tungsten-amine complex in the organic phase.

3.2.4　Reaction mechanism of extraction process with N235 and Alamine 308

In order to explore the reaction mechanism of extraction process with the two extractants, the organic phases and feeding solution of the extraction system were tested by FTIR. Fig. 6(a) showed the infrared spectra of N235 and Alamine 308, and it was found that the infrared spectra of the two extractants were almost the same. The characteristic peaks in region I were corresponding to alkyl group in tertiary amine (R_3N), and the characteristic peaks in region II were corresponding to amino group (C-N). Although there was a characteristic peak representing

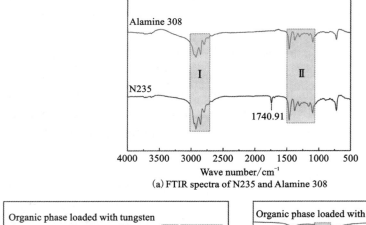

(a) FTIR spectra of N235 and Alamine 308

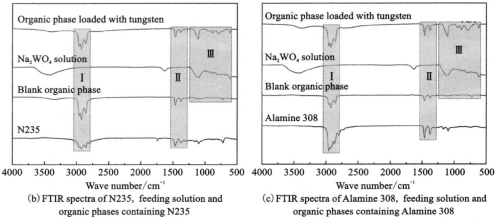

(b) FTIR spectra of N235, feeding solution and organic phases containing N235

(c) FTIR spectra of Alamine 308, feeding solution and organic phases containing Alamine 308

Fig. 6　FTIR spectra of extractants, feeding solution and organic phases

C=O bond in the infrared spectrum of N235 at 1740.91 cm^{-1}, this peak almost disappeared in the blank organic phase. Cao (2015) believed that this was due to the formation of hydrogen bond after preparing the organic phase. Therefore, this group had no effect on the extraction process.

Fig. 6 (b) and (c) revealed that the characteristic peaks of region I and region II barely changed after tungsten was loaded on the organic phases. This finding confirmed that in the solvent extraction process, the polymeric anions of tungsten replaced the SO_4^{2-} on the acidified extractant molecules as shown in reactions (15)-(17) rather than reacting with the N atoms on N235 and Alamine 308. It was also observed that new characteristic peaks appeared in region III of the two organic phases loaded with tungsten. According to the infrared spectrum of purified Na_2WO_4 solution, the characteristic peaks in region III represented the polymeric anions of tungsten associated with extractant molecules.

3.3 Stripping of loaded tungsten from organic phase

In order to obtain $(NH_4)_2WO_4$ solution directly from tungsten stripping process, $NH_3 \cdot H_2O$ was used as stripping agent in the presented study. The stripping of tungsten with $NH_3 \cdot H_2O$ are as Equations (18)-(20), the organic phase after stripping can be reused to the extraction process after acidification (Zhang and Zhao, 2005).

$$(R_3NH)_4H_2W_{12}O_{39}(org) + 24NH_4OH(aq) \longrightarrow$$
$$4R_3N(org) + 12(NH_4)_2WO_4(aq) + 15H_2O \quad (18)$$
$$(R_3NH)_6H_2W_{12}O_{40}(org) + 24NH_4OH(aq) \longrightarrow$$
$$6R_3N(org) + 12(NH_4)_2WO_4(aq) + 16H_2O \quad (19)$$
$$(R_3NH)_5H(H_2W_{12}O_{40})(org) + 24NH_4OH(aq) \longrightarrow$$
$$5R_3N(org) + 12(NH_4)_2WO_4(aq) + 16H_2O \quad (20)$$

The solvents containing 15% N235 and 15% Alamine 308 were loaded with 133.3 g/L WO_3 at the O/A phase ratio of 0.9 : 1 in extraction, and the solvents were washed with deionized water at an O/A phase ratio of 1 : 1 to remove impurity ions entrained into the organic phases. Since the extraction system was acidic and the stripping was carried out under alkaline condition, part of tungsten would form APT and precipitate in the stripping process due to the change of acidity. The effect of NH_3 concentration on tungsten stripping from N235 and Alamine 308 at the O/A phase ratio of 2 : 1 was as presented in Fig. 7. It was observed that the stripping efficiencies from loaded N235 and Alamine 308 reached 99.3% and 99.4% with the NH_3 concentration of 3 mol/L (the stoichiometric ratio of 1.2), and the precipitation ratio of tungsten was about 2%. The stripping efficiencies increased to almost 100% as the NH_3 concentrations were increased to 4 mol/L and the tungsten precipitation ratio were also reduced to about 1% for N235 and 0.5% for Alamine 308. However, when the NH_3 concentrations were further increased, the tungsten precipitation ratios increased dramatically. The main reason was that under the high NH_3 concentration, the stripping rate of tungsten from extractants was greater than the conversion rate of polymeric anions to WO_4^{2-}, which resulted in a sharply increase of the tungsten precipitation ratio. Moreover, when

Alamine 308 was used, the tungsten precipitation ratio during stripping was lower than that of N235.

Fig. 8 showed the variation of tungsten stripping efficiency and tungsten precipitation ratio with stripping time. The results indicated that when the stripping time was 5 min, the stripping efficiencies of tungsten from two extractants was more than 99%. As the stripping time was extended to 10 min, both the stripping efficiencies gradually increased to 99.98%, and the $(NH_4)_2WO_4$ solution with WO_3 concentration of above 260 g/L was obtained. For the tungsten precipitation, the precipitation ratio increased rapidly in the first 5 min, while decreased gradually with the further extension of stripping time, which indicated that there was a dissolution process after the precipitation of APT. The K concentration of the $(NH_4)_2WO_4$ solution obtained from N235 and Alamine 308 was 2 mg/L and 3 mg/L, indicating that the $(NH_4)_2WO_4$ solution was qualified for APT production.

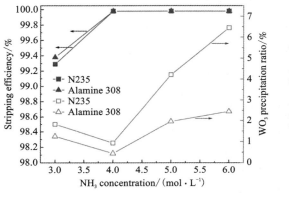 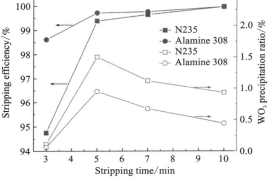

(O/A phase ratio of 2, stripping time of 10 min)

Fig. 7 Effect of NH_3 concentration on tungsten stripping from loaded N235 and Alamine 308

(O/A phase ratio of 2, NH_3 concentration of 4 mol/L)

Fig. 8 Variation of tungsten stripping efficiency and tungsten precipitation ratio with stripping time

The axis pointed by the arrow is the coordinate axis corresponding to the curve.

3.4 Conversion and circulation of raffinate

The above findings confirm that tungsten can be completely extracted from Na_2WO_4 solution by the two tertiary amine extractants in one stage, and high concentration $(NH_4)_2WO_4$ solution can be obtained by $NH_3 \cdot H_2O$ stripping. However, as shown by reactions (15)-(17), the raffinate contains Na_2SO_4, and the SO_4^{2-} concentration reaches 1.03 mol/L according to analysis. The improper disposal of sodium salt wastewater can lead to environmental pollution (Lu et al., 2017). The recycling of the raffinate into industrial process can not only realize cleaner production, but also reduce the consumption of water resources (Santos et al., 2020). Previous studies mainly focused on the extraction efficiency and selectivity of tungsten in the solvent extraction process, while little attention was paid to the recycling of raffinate (Moris et al., 1999;

Li et al., 2018; Zeng et al., 2020). In addition, in industry, the raffinate is mainly discharged after desalination treatment, which increases the production cost. Since Na_2CO_3 is needed for the leaching of crude tungstic acid in this work, it is an economic and cleaner production route to convert Na_2SO_4 into Na_2CO_3 to realize the circulation of raffinate. In this study, a route of converting Na_2SO_4 into Na_2CO_3 by $BaCO_3$ was proposed based on Equation (21).

$$Na_2SO_4(aq) + BaCO_3(s) = BaSO_4(s) + Na_2CO_3(aq) \qquad (21)$$

Fig. 9(a) showed the effect of stoichiometric ratio of $BaCO_3$ on the conversion of raffinate with different SO_4^{2-} concentrations. It was found that for the raffinate with 1.03 mol/L SO_4^{2-}, as the stoichiometric ratio of $BaCO_3$ increased from 1.2 to 1.8, conversion efficiency increased from 83.42% to 88.36%. Higher conversion efficiency could be obtained by diluting the raffinate to SO_4^{2-} concentration of 0.8 mol/L, the conversion efficiency was 89.80% at the $BaCO_3$ stoichiometric ratio of 1.8. However, further diluting the raffinate would reduce the conversion efficiency. And when the reaction time reached 6 h [as shown in Fig. 9(b)], the conversion reaction was in equilibrium.

After purifying of the crude $BaSO_4$, the $BaSO_4$ product with purity of 97.8% and whiteness of 95.4 was obtained, which meet the quality requirements of first-grade product in GB/T 2899 - 2008 standard of China. The preparing of pure $BaSO_4$ increases the value-added output of the process and avoids the generation of waste residue. In addition, the recycling of acid medium in the purification process contributes to zero discharge of wastewater.

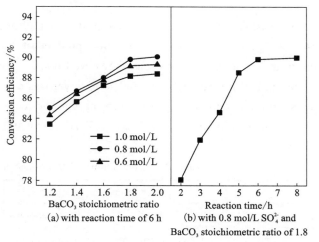

Fig. 9 Effect of the stoichiometric ratio of $BaCO_3$ (a) and reaction time (b) on the raffinate conversion

In order to validate the feasibility of the conversion-circulation route of raffinate, the converted raffinate was supplemented with Na_2CO_3 to reach the optimized concentration and then reused for a new round of crude tungstic acid leaching. The repeated experiments of crude tungstic acid leaching, solvent extraction with reusing of organic phase (Alamine 308) after stripping, raffinate conversion and purifying of crude $BaSO_4$ were carried out at the optimized conditions for 5 cycles. The results of 5 cycles were presented in Fig. 10.

According to Fig. 10(a), the efficiencies of crude tungstic acid leaching, solvent extraction of tungsten and raffinate conversion processes were slightly affected, the average efficiencies were 99.47%, 99.97% and 89.34% in 5 cycles of converted raffinate and organic phases circulation. Besides, Fig. 10(b) showed that the quality of $BaSO_4$ product also remained stable. The results confirm that the sustainable process for extracting tungsten from crude tungstic acid presented in

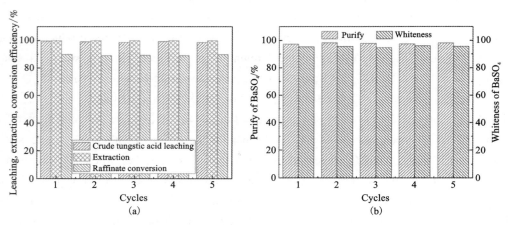

Fig. 10 Leaching, extraction and conversion efficiency (a) and BaSO$_4$ product quality (b) in repeated experiments

Fig. 1 is feasible and efficient, which provides a cleaner production route for preparing pure (NH$_4$)$_2$WO$_4$ solution from acid digestion product of tungsten concentrates. The mass balance based on the cycle experiments was shown in Table 2. The tungsten entrained by the precipitation generated in the process of P, As and Si removal was calculated as 0.31% of the total tungsten in crude Na$_2$WO$_4$ solution. The results showed that the recovery of tungsten from crude tungstic acid to (NH$_4$)$_2$WO$_4$ solution was 99.13%, indicating that this process is efficient.

Table 2 Mass balance of the tungsten extraction process

Process	Input		Output	
	Material	Mass ratio/%	Material	Mass ratio/%
Leaching	Crude tungstic acid	100	Crude Na$_2$WO$_4$ solution	99.47
	Na$_2$CO$_3$ solution	0	Leaching residue	0.53
P, As, Si removal	Crude Na$_2$WO$_4$ solution	99.47	Purified Na$_2$WO$_4$ solution	99.16
			Precipitation	0.31
Solvent extraction	Purified Na$_2$WO$_4$ solution	99.16	Tungsten loaded organic phase	99.15
	Blank organic phase	0.02	Raffinate	0.03
Stripping	Tungsten loaded organic phase	99.15	Blank organic phase	0.02
	NH$_3$·H$_2$O solution	0	(NH$_4$)$_2$WO$_4$ solution	99.13

4 Conclusions

This work presented a cleaner production route for preparing pure $(NH_4)_2WO_4$ solution from acid digestion product of tungsten concentrates, the leaching of acid digestion product, efficient solvent extraction of tungsten, conversion and recycling of raffinate were carried out, the following conclusions were obtained:

(1) The tungsten in crude tungstic acid could be leached efficiently by Na_2CO_3. Under the Na_2CO_3 stoichiometric ratio of 1.2, reaction temperature of 40 ℃, liquid to solid ratio of 4:1 and reaction time of 60 min, the average leaching efficiency of tungsten was 99.47% with circulation of Na_2CO_3 solution converted from raffinate.

(2) For Na_2WO_4 solution with 121.26 g/L WO_3, tungsten could be completely extracted by organic phases containing 15% N235 or 15% Alamine 308 in one stage at O/A of 1 and the pH of extraction system lower than 3. The extraction mechanism of tungsten by N235 and Alamine 308 was the same, while Alamine 308 showed better performance and higher saturated loading capacity for tungsten extraction. The stripping efficiencies of loaded tungsten from N235 and Alamine 308 reached 99.98% with $NH_3 \cdot H_2O$ at the NH_3 concentration of 4 mol/L and stripping time of 10 min. The $(NH_4)_2WO_4$ solutions with WO_3 concentration above 260 g/L and low K concentration was obtained. The recycling of organic phase after stripping hardly affected the tungsten extraction.

(3) The raffinate containing Na_2SO_4 was converted to Na_2CO_3 solution by adding $BaCO_3$ with conversion efficiency above 89%. The converted raffinate was recycled for the leaching of crude tungstic acid, and the leaching efficiency remained stable. The recycling of raffinate indicates that the process is sustainable.

(4) Although the process established in this work is clean and sustainable, it is complicated to a certain extent due to the additional impurity removal process. Adding impurity precipitant in the crude tungstic acid leaching process to obtain Na_2WO_4 solution with low impurity content in one step can simplify the process. But the leaching regulation of impurities needs further study.

Acknowledgment

This work was supported by the National Key R&D Program of China [grant number 2019YFC1907400].

References

[1] CAO W. Solvent Extraction of Vanadium from the Leachate of Stone Coal with N235[D]. Wuhan: Wuhan University of Technology, 2015. (In Chinese).

[2] CHEN W, FENG Q, ZHANG G, et al. Effect of energy input on flocculation process and flotation performance of fine scheelite using sodium oleate[J]. Miner. Eng. 2017, 112: 27-35. https://doi.org/10.1016/j.mineng.2017.07.002.

[3] CHEN Y, HUO G, GUO X, et al. Wolframite concentrate conversion kinetics and mechanism in hydrochloric acid[J]. Miner. Eng. 2022, 179: 107422 https://doi.org/10.1016/j.mineng.2022.107422.

[4] CICHY B, JAROSZEK H, MIKOLAJCZAK W, et al. Application of nanofiltration for concentration of sodium sulfate waste solution[J]. Desalination Water Treat. 2017, 76: 349-357. https://doi.org/10.5004/dwt.2017.20594.

[5] FURBERG A, ADVIDSSON R, MOLANDER S. Environmental life cycle assessment of cemented carbide (WC-Co) production[J]. J. Clean. Prod. 2019, 209: 1126-1138. https://doi.org/10.1016/j.jclepro.2018.10.272.

[6] HE G, ZHAO Z, WANG X, et al. Leaching kinetics of scheelite in hydrochloric acid solution containing hydrogen peroxide as complexing agent[J]. Hydrometallurgy, 2014, 144-145: 140-147. https://doi.org/10.1016/j.hydromet.2014.02.006.

[7] HUO G, LIU Z, ZENG L, et al. A method for digesting scheelite with hydrochloric acid based on mother liquor circulation[J]. Chinese Patent CN108342597B. (In Chinese)

[8] LEI Y, SUN F, LIU X, et al. Understanding the wet decomposition processes of tungsten ore: Phase, thermodynamics and kinetics[J]. Hydrometallurgy, 2022, 213: 105928. https://doi.org/10.1016/j.hydromet.2022.105928.

[9] LI H, YANG J, LI K. Tungsten metallurgy[M]. Changsha: Central South University Press. (In Chinese)

[10] LI J, MA Z, LIU X, et al. Sustainable and efficient recovery of tungsten from wolframite in a sulfuric acid and phosphoric acid mixed system[J]. ACS Sustainable Chem. Eng. 2020, 8: 13583-13592. https://doi.org/10.1021/acssuschemeng.0c04216.

[11] LI J, ZHAO Z. Kinetics of scheelite concentrate digestion with sulfuric acid in the presence of phosphoric acid[J]. Hydrometallurgy, 2016, 163: 55-60. https://doi.org/10.1016/j.hydromet.2016.03.009.

[12] LI X, SHEN L, ZHOU Q, et al. Scheelite conversion in sulfuric acid together with tungsten extraction by ammonium carbonate solution[J]. Hydrometallurgy, 2017, 171: 106-115. https://doi.org/10.1016/j.hydromet.2017.05.005.

[13] LI Z, ZHANG G, GUAN W, et al. Separation of tungsten from molybdate using solvent extraction with primary amine N1923[J]. Hydrometallurgy, 2018, 175: 203-207. https://doi.org/10.1016/j.hydromet.2017.10.018.

[14] LIU C, ZHANG W, SONG S, et al. Study on the activation mechanism of lead ions in wolframite flotation using benzyl hydroxamic acid as the collector[J]. Miner. Eng. 2019, 141: 105859. https://doi.org/10.1016/j.mineng.2019.105859.

[15] LIU H, LIU H, NIE C, et al. Comprehensive treatments of tungsten slags in China: A critical review[J]. J. Environ. Manag. 2020, 270: 110927 https://doi.org/10.1016/j.jenvman.2020.110927.

[16] LIU J. Extraction of tungsten and enrichment of tin from high arsenic tungsten-tin ore[M]. Changsha: Central South University. (In Chinese)

[17] LIU X, DENG L, CHEN X, et al. Recovery of tungsten from acidic solutions rich in calcium and iron[J]. Hydrometallurgy, 2021, 204: 105719. https://doi.org/10.1016/j.hydromet.2021.105719.

[18] LIU X, ZHANG Z, ZHANG L, et al. Separation of sodium sulfate from high-salt wastewater of lead-acid batteries[J]. Chem. Eng. Res. Des. 2021, 176: 194-201. https://doi.org/10.1016/j.cherd.2021.10.008.

[19] LONG Y, ZHANG G, ZENG L, et al. Study on tungsten extraction from crystal mother liquor of ammonium paratungstate in N235/isooctanol system[J]. Rare Met. Cemented Carbides, 2021, 49(1): 1-6. (In Chinese)

[20] LU H, WANG J, WANG T, et al. Crystallization techniques in wastewater treatment: An overview of applications[J]. Chemosphere, 2017, 173: 474-484. https://doi.org/10.1016/j.chemosphere.2017.01.070.

[21] LUO Y, CHEN X, ZHAO Z, et al. Pressure leaching of wolframite using a sulfuric-phosphoric acid mixture[J]. Miner. Eng. 2021, 169: 106941. https://doi.org/10.1016/j.mineng.2021.106941.

[22] MA X, QI C, YE L, et al. Life cycle assessment of tungsten carbide powder production: a case study in China[J]. J. Clean. Prod, 2017, 149: 936-944. https://doi.org/10.1016/j.jclepro.2017.02.184.

[23] MARTINS J. Leaching systems of wolframite and scheelite: A thermodynamic approach[J]. Miner. Process. Extr. Metall. Rev, 2014, 35: 23-43. https://doi.org/10.1080/08827508.2012.757095.

[24] MORIS M, DIEZ F, COCA J. Solvent extraction of molybdenum and tungsten by Alamine 336 and DEHPA in a rotating disc contactor[J]. Separ. Purif. Technol. 1999, 17: 173-179. https://doi.org/10.1016/S1383-5866(99)00022-2.

[25] NING P, CAO H, ZHANG Y. Selective extraction and deep removal of tungsten from sodium molybdate solution by primary amine N1923[J]. Separ. Purif. Technol. 2009, 70: 27-33. https://doi.org/10.1016/j.seppur.2009.08.006.

[26] SAHU K, AGRAWAL A, MISHRA D. Hazardous waste to materials: Recovery of molybdenum and vanadium from acidic leach liquor of spent hydroprocessing catalyst using alamine 308[J]. J. Environ. Manag. 2013, 125: 68-73. https://doi.org/10.1016/j.jenvman.2013.03.032.

[27] SANTOS P G, SCHERER C M, FISCH A G, et al. Petrochemical wastewater treatment: water recovery using membrane distillation[J]. J. Clean. Prod. 2020, 267: 121985 https://doi.org/10.1016/j.jclepro.2020.121985.

[28] SHEN L, LI X, LINDBERG D, et al. Tungsten extractive metallurgy: A review of processes and their challenges for sustainability[J]. Miner. Eng. 2019, 142: 105934. https://doi.org/10.1016/j.mineng.2019.105934.

[29] SHEN L, LI X, ZHOU Q, et al. Kinetics of scheelite conversion in sulfuric acid[J]. JOM (J. Occup. Med.) 2018, 70(11): 2499-2504. https://doi.org/10.1007/s11837-018-2787-2.

[30] SHEN L, LI X, ZHOU Q, et al. Wolframite conversion in treating a mixed wolframite-scheelite concentrate by sulfuric acid[J]. JOM (J. Occup. Med.) 2018, 70(2): 161-167. https://doi.org/10.1007/s11837-017-2691-1.

[31] SHEN L, LI X, ZHOU Q, et al. Sustainable and efficient leaching of tungsten in ammoniacal ammonium carbonate solution from the sulfuric acid converted product of scheelite[J]. J. Clean. Prod. 2018, 197: 690-698. https://doi.org/10.1016/j.jclepro.2018.06.256.

[32] SHI J, HUANG W, HAN H, et al. Review on treatment technology of salt wastewater in coal chemical industry of China[J]. Desalination, 2020, 493: 114640. https://doi.org/10.1016/j.desal.2020.114640.

[33] USGS, 2021. Tungsten, U.S. Geological Survey, Mineral Commodity Summaries 2021.

[34] WANG X, HUANG H, SHI J, et al. Recent progress of tungsten-based highentropy alloys in nuclear fusion [J]. Tungsten, 2021, 3(2): 143-160. https://doi.org/10.1007/s42864-021-00092-8.

[35] YANG Y, XIE B, WANG R, et al. Extraction and separation of tungsten from acidic high-phosphorus solution[J]. Hydrometallurgy, 2016, 164: 97-102. https://doi.org/10.1016/j.hydromet.2016.05.018.

[36] ZENG L, ZHAO Z, HUO G, et al. Mechanism of selective precipitation of molybdenum from tungstate solution[J]. JOM (J. Occup. Med.) 2020, 72: 800-805. https://doi.org/10.1007/s11837-019-03915-9.

[37] ZENG L, YANG T, YI X, et al. Separation of molybdenum and tungsten from iron in hydrochloric-phosphoric acid solutions using solvent extraction with TBP and P507[J]. Hydrometallurgy, 2020, 198: 105500. https://doi.org/10.1016/j.hydromet.2020.105500.

[38] ZHANG Q, ZHAO Q. Tungsten and molybdenum metallurgy [M]. Beijing: Metallurgical Industry Press, 2005.

[39] ZHANG G, GUAN W, XIAO L, et al. A novel process for tungsten hydrometallurgy based on direct solvent extraction in alkaline medium[J]. Hydrometallurgy, 2016, 165: 233-237. https://doi.org/10.1016/j.hydromet.2016.04.001.

Wolframite Concentrate Conversion Kinetics and Mechanism in Hydrochloric Acid

Abstract: Efficient conversion of wolframite in acid solution plays a crucial role in exploration of sustainable technology for ammonium paratungstate (APT) production. In this work, the factors influencing wolframite concentration conversion in HCl solution and the conversion mechanism were investigated to establish a kinetics model for the decomposition process. The results indicated that the conversion rate increased significantly with increasing temperature and HCl concentration and reducing particle size, while it was independent of stirring intensity. Scanning electron microscope (SEM) analysis indicated that layered H_2WO_3 films were formed on the surface of converted particles, and the experimental data was consistent with the shrinking core model. The conversion process was divided into two stages with a transition point of 30 min, the first stage was a surface reaction controlling process while the second stage was a hybrid controlling process. The kinetic rate equations were established as: $1-(1-x)^{1/3} = 0.19 \times C^{1.60} \cdot D^{-0.81} \cdot \exp(-40570/RT) \cdot t$ for the first stage, and $4 \times [1-(1-x)^{1/3}] + [1-2x/3-(1-x)^{2/3}] = 1.41 \times C^{0.75} \cdot D^{-0.31} \cdot \exp(-38160/RT) \cdot t$ for the second stage. This work contributes to better understanding of wolframite conversion in HCl solution and the development of a cleaner route for APT production.

Keywords: Kinetics; Mechanism; Wolframite; Acid decomposition; Shrinking core model

1 Introduction

Tungsten is a strategic rare metal. Owing to its high melting point, hardness, and excellent corrosion resistance, tungsten is widely used in many fields such as machinery manufacturing, chemical industry, aerospace industry and national defense industry (Li et al., 2021; Wang et

---

Published in *Minerals Engineering*, 2022, 179: 107422. Authors: Chen Yuanlin, Huo Guangsheng, Guo Xueyi, Wang Qinmeng.

al., 2021; Li et al., 2017a, 2017b). The main economical tungsten mineral resources are wolframite [(Fe,Mn)WO$_4$] and scheelite (CaWO$_4$) (Liu et al., 2019; Chen et al., 2017). There are fewer wolframite reserves than scheelite in the world, but wolframite has long been an important raw material for ammonium paratungstate (APT, an important intermediate product in tungsten industry) production and accounts for a considerable part of China's tungsten reserves (Luo et al., 2021), such as 230000 tons tungsten in Shizhuyuan mine of Hunan and 150000 tons tungsten in Xingluokeng mine of Fujian (Yang et al., 2018). Thus, there is an urgent need to develop sustainable processes for wolframite decomposition.

Autoclave-based alkali leaching is widely applied to decompose tungsten ores to produce APT and shows excellent performance in treating high-grade tungsten resources (Martins, 2014; Zhao, 2013). However, this method requires high reaction temperature (the typical temperature condition is 433-498 K) and consumes a large amount of energy (Li et al., 2017a, 2017b; Kalpakli et al., 2012). In addition, the alkali leaching process is faced with environmental problems: the excess alkali consumption (for example, the stoichiometric ratio of Na$_2$O to WO$_3$ is as high as 4.5 to achieve high WO$_3$ recovery) and the difficulty of recycling alkali result in the discharge of 20-100 tons sodium salt wastewater for 1 ton of APT produced (Li and Zhao, 2016; Shen et al., 2018a; Ren et al., 2020). What is worse, the alkali leaching residue is dangerous, because of this it is classified as a hazardous solid waste in China according to the latest environmental law (Liu et al., 2020).

Compared with the alkali leaching process, acid decomposition method has the potential to remarkably reduce the production cost and the discharge of high-salinity wastewater owe to its high thermodynamic driving force and short processing flow (Liu et al., 2016; Yin et al., 2020). However, in acid decomposition process, dense solid layers of H$_2$WO$_3$ are formed and then cover the surface of tungsten mineral particles, which hinders the mass transfer and leads to incomplete conversion. Many works have been carried out to enhance the acid decomposition process, such as reducing the particle size (Martins et al., 2003), using of complex reagents (oxalic, citric, and phosphoric acids) during the acid decomposition to promote water soluble tungsten salts (Zhang et al., 2015; Shen et al., 2018b), mechanical activation (Zhao et al., 1996), and ultrasonic enhancement (Zhao et al., 2013). However, most of the previous works were on scheelite decomposition. It is unclear whether this method is effective for wolframite decomposition, which is still an important raw material for tungsten extraction. Moreover, some preliminary experiments found that under the same conditions as scheelite decomposition, the conversion extent of wolframite was much lower than that of scheelite, and the conversion rate was also slower (Luo et al., 2021). Shen et al. (2018c) clarified the conversion behavior of wolframite in treating a mixed concentrate with sulfuric acid, which showed that it was more difficult to decompose wolframite than scheelite in sulfuric acid solution due to the accumulation of Fe^{2+} (and/or Mn^{2+}) as well as the formation of solid H$_2$WO$_4$ layer. Li et al. (2020) found that, in a sulfuric acid and phosphoric acid mixed system, the leaching of FeWO$_4$ was more difficult than the leaching of MnWO$_4$ in wolframite, as the Fe-O bond energy was stronger than Mn-O. The leaching efficiency of wolframite with high ferberite content can be improved after mechanical activation with the addition of Ca(OH)$_2$ into mechanical milling process.

Undoubtedly, the further optimization of acid decomposition process for wolframite is a prerequisite to achieve high tungsten recovery and assists in exploring sustainable technology for APT production. And it is essential to better understand the wolframite conversion kinetics and mechanism in mineral acid solution. Therefore, the effects of stirring intensity, temperature, acid concentration, and particle size on the wolframite conversion rate with HCl solution were studied in this work, and a kinetic model for the process was established. The results of this study are expected to provide a support for developing a sustainable process for producing APT from wolframite concentrate via acid decomposition.

2 Materials and methods

2.1 Materials

The HCl used in this work was analytically pure, purchased from Sinopharm Chemical Reagent Co., Ltd, China. The wolframite concentrate was provided by Jiangxi Tungsten Group Co., Ltd., China, and contained WO$_3$ 69.50%, Fe 11.12%, and Mn 6.56% as the main compositions as presented in Table 1. The X-ray diffraction (XRD) pattern of the concentrate (Fig. 1) showed that it was with high content of wolframite [(Fe, Mn)WO$_4$)] and trace mount of scheelite (CaWO$_4$) and phlogopite [KMg$_3$(Si$_3$Al)O$_{10}$(OH)$_2$]. The concentrate was sieved to different size fractions ($-100+75$ μm, $-75+50$ μm, $-50+38$ μm, $-38+25$ μm). The WO$_3$, Fe, and Mn mass fractions of different size fractions of the wolframite concentrate were presented in Table 2. The contents of WO$_3$, Fe, and Mn in each size fraction are about 70%, 11%, and 6%, respectively.

Table 1 The elemental composition of wolframite concentrate   %

Element	WO$_3$	Fe	Mn	SiO$_2$	CaO	Al	Bi
Mass fraction	69.11	11.12	6.56	4.76	2.10	0.65	0.22
Element	Mo	S	Mg	Sn	Nb	Ta	Zn
Mass fraction	0.02	0.32	0.17	0.06	0.11	0.05	0.05
Element	K	Na	Ti	P	F	Others	Total
Mass fraction	0.18	0.03	0.13	0.06	0.17	4.13	100.00

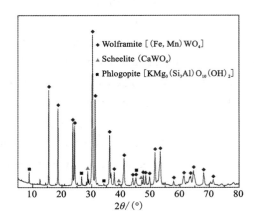

Fig. 1 The XRD pattern of wolframite concentrate

Table 2 Mass fractions of WO_3, Fe, and Mn in different size fractions

Size fraction /μm	−100+75	−75+50	−50+38	−38+25
Mass fraction of WO_3/%	69.73	68.56	68.50	69.56
Mass fraction of Fe/%	11.08	10.88	10.99	11.19
Mass fraction of Mn/%	6.66	6.37	6.44	5.96

2.2 Methods

The decomposition experiments were conducted in a 1 L three-neck round-bottom flask, which was heated thermostatically by a water bath. In each run, 1 L of HCl solution with a certain concentration was added into the flask and heated to the required temperature, then 10 g of sample was added into the solution. The slurry was stirred at a set speed using a polytetrafluoroethylene coated agitator. After a certain duration, 5 mL of the solution was sampled to measure the conversion efficiency. A total of three replicate decomposition experiments were carried out for each condition, and the average values were used for kinetic model fitting. The large liquid-solid ratio of 100 mL/g and large solution volume could ensure the stability and reproducibility of the experiments.

The analysis results of composition (Table 2 and Fig. 1) show that, the concentrate is with high content of WO_3, and the main tungsten mineral is wolframite [(Fe, Mn)WO_4]. The conversion of wolframite in HCl solution is accompanied by the dissolution of Fe^{2+} and Mn^{2+} as follows:

$$(Fe, Mn)WO_4(s) + 2HCl(aq) \longrightarrow (Fe, Mn)Cl_2(aq) + H_2WO_4(s) \qquad (1)$$

Therefore, the total leaching efficiency of Fe and Mn is considered as the conversion efficiency of wolframite concentrate in this study. The concentrations of Fe, Mn and W of sampled solutions in each run were analyzed by inductively coupled plasma optical emission spectrometer (ICP-OES, ICAP7400, Thermo Scientific). The leaching efficiency of Fe and Mn during the decomposition process was calculated by Eq. (2).

$$\eta(Fe + Mn) = [(C_{Fe} + C_{Mn}) \times V]/[(w_{Fe} + w_{Mn}) \times m_0] \cdot 100\% \quad (2)$$

where $\eta(Fe + Mn)$ is the total leaching efficiency (%) of Fe and Mn from the wolframite concentrate, C_{Fe} and C_{Mn} are the Fe and Mn concentrations (g/L) of the leaching solution, V is the volume of leaching solution (1 L), w_{Fe} and w_{Mn} are the mass fractions of Fe and Mn in the wolframite concentrate sample, m_0 is the weight of wolframite concentrate sample used in each run (10 g). For each condition, three replicate experiments were carried out, the average concentrations of Fe, Mn, and W in the leaching solutions of three replicate experiments were used to calculate the leaching efficiency and the concentration (mol/L) ratio of (Fe+Mn)/W. The standard deviation of leaching efficiency and concentration ratio under each condition was presented as an error bar.

Phase analysis of the wolframite concentrate was performed by X-ray diffraction (XRD, TTR Ⅲ, with Cu K_α radiation, Rigaku, Japan), the analysis was operated with a 2θ range from 10° to 80°, a step size of 0.02°, and a scanning speed of 2°/min. The microtopography feature of the wolframite concentrate and converted product particles were observed through a scanning electron microscope (SEM-EDS, JSMIT500LV, JEOL, Japan). The elemental distributions of product particles were analyzed by energy dispersive spectrometer (EDS).

3 Results and discussion

3.1 Decomposition experiments

The effect of stirring intensity on the leaching efficiency of Fe and Mn from −50+38 μm size fraction in HCl solution was firstly studied with 7 mol/L HCl and a temperature of 363 K. As shown in Fig. 2(a), the leaching efficiency is almost independent of stirring intensity in the studied speed range 200-300 r/min, especially when the stirring intensity is above 250 r/min. This result is consistent with all previous studies on the conversion behavior of tungsten minerals under different stirring intensity in acid solutions (Kahruman and Yusufoglu, 2006; He et al., 2014; Shen et al., 2018a; Yang et al., 2018). It is reasonable to conclude that the external diffusion through the bulk fluid does not act as rate-controlling step of the decomposition process. A stirring intensity of 250 r/min was selected in the subsequent experiments.

Fig. 2(b) shows the leaching efficiency of Fe and Mn from −50+38 μm size fraction under different temperatures (333 K-368 K), 7 mol/L HCl, and stirring intensity of 250 r/min. It is observed that the leaching efficiency is significantly affected by temperature, increasing continuously with the increase of temperature from 333 K to 363 K. The leaching efficiency of Fe and Mn is only 17% after 120 min at 333 K. However, the leaching efficiency achieves 51% after 120 min at 363 K. When the reaction temperature reached 368 K, the increase of leaching

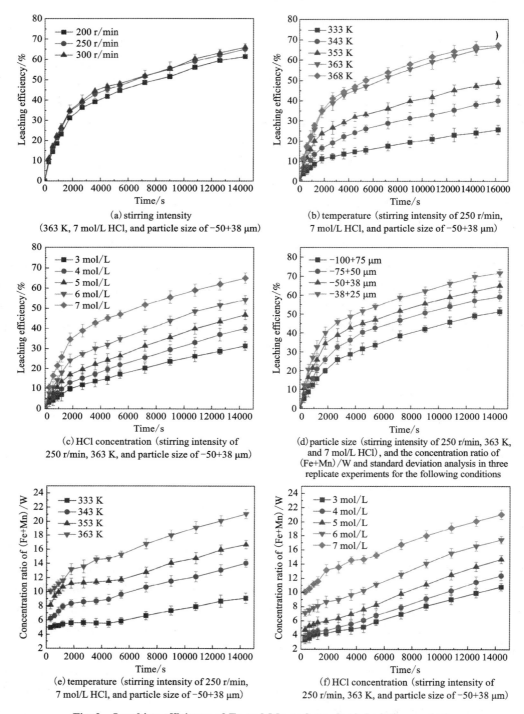

Fig. 2　Leaching efficiency of Fe and Mn and standard deviation analysis in three replicate experiments for the following conditions

efficiency was not obvious.

The effect of HCl concentration on the leaching efficiency of Fe and Mn from $-50+38$ μm size fraction was studied with a temperature of 363 K and stirring intensity of 250 r/min. The leaching efficiency curves [Fig. 2(c)] as a function of time show that the leaching efficiency increases significantly with the increase of HCl concentration in the range of 3 mol/L to 7 mol/L.

The effect of particle size on the leaching efficiency of Fe and Mn was also investigated using the four sieved samples of wolframite concentrate with a temperature of 363 K, 7 mol/L HCl, and stirring intensity of 250 r/min. The results are presented in Fig. 2(d). As expected usually, reducing the wolframite concentrate particle size accelerates the leaching rate. The leaching efficiency of Fe and Mn reached 59% after 120 min for the $-38+25$ μm size fraction, while it was only 38% for the $-100+75$ μm size fraction.

Fig. 2(e) and Fig. 2(f) show the concentration (mol/L) ratio of (Fe+Mn)/W in HCl solution under various temperatures and HCl concentrations. The results indicate that the concentration ratio of (Fe+Mn)/W increases continuously during the decomposition of wolframite, as Fe and Mn were continuously dissolved into the solution, while W was slightly dissolved in the initial stage and then reached balance. Li et al. (2017a, 2017b) reported that H_2WO_4 was almost insoluble ($K_{sp} = 10^{-5.98}$). In this study, it was found that the concentration of WO_3 in HCl solution did not exceed 230 mg/L.

3.2 Wolframite conversion mechanism analysis

To identify the conversion mechanism of wolframite during decomposition process, the electron microscope observation on the wolframite concentrate and incompletely converted products with different reaction time was performed, and the elemental distributions of product particles were analyzed by EDS mapping. As shown in Fig. 3(a) and Fig. 3(b), compared with wolframite concentrate, the particle size of the converted product changes little, whereas new layered products on the surface of converted particles are observed. Moreover, it is observed from Fig. 3(c) and Fig. 3(e) that the layer thickened as the reaction time increases from 30 min to 60 min. The elemental distribution analysis results of product particles [presented in Fig. 3(d) and Fig. 3(f)] show that the layered products are with high content of W and trace amounts of Fe and Mn, which indicates that the conversion process of wolframite concentrate is a shrinking core model with partial solid product of H_2WO_4. Shen et al. (2018c) studied the conversion mechanism of mixed wolframite-scheelite concentrate in H_2SO_4 solution and also proved that the product layers coating the particles were principally made up of solid H_2WO_4.

3.3 Kinetics model selection

Yang et al. (2018) studied the leaching kinetics of wolframite concentrate in mixed acid solution of H_2SO_4 and H_3PO_4, and it was found that the experimental data was inconsistent with the shrinking core model. The observation of the surface morphology of product particles showed that the wolframite decomposition is a process in which the leaching agent eroded wolframite

Fig. 3　The SEM image of wolframite concentrate (a, −75+50 μm) and incompletely converted product (b, e with 7 mol/L HCl, temperature of 363 K, and reaction time of 60 min; c with 7 mol/L HCl, temperature of 363 K, and reaction time of 30 min), and elemental distributions of the product particles (d, the EDS map of particle c; f, the EDS map of particle e).

particle surface irregularly rather than advancing layer by layer. Nevertheless, the leaching of wolframite in H_2SO_4-H_3PO_4 mixed acid system is a liquid-solid reaction without solid product layers, which is distinctly different from the conversion process with solid product layers in this work. Shen et al. (2018a) suggested that the scheelite conversion in H_2SO_4 solution (which was similar to this study, and solid product layers were formed on the particles surface) was inconsistent with shrinking core model and the rate control step was surface reaction.

According to the analysis of wolframite conversion mechanism above, the conversion is a core shrinking process advancing layer by layer. Therefore, the shrinking core model is selected to describe the kinetics of wolframite conversion in HCl solution in this work. In shrinking core model, the rate-controlling step may be surface reaction, internal diffusion through the product layer, or hybrid control (external diffusion is not the rate-controlling step in this work). The kinetic equations of surface reaction control, internal diffusion control, and hybrid control in the shrinking core model can be expressed as Eqs. (3), (4), and (5) (Guo et al., 2017; Demirkıran and Künkül, 2007):

$$1 - (1 - x)^{1/3} = K_1 \times t \tag{3}$$

$$1 - 2x/3 - (1 - x)^{2/3} = K_2 \times t \tag{4}$$

$$\alpha[1 - (1 - x)^{1/3}] + \beta[1 - 2x/3 - (1 - x)^{2/3}] = K_3 \times t \tag{5}$$

where x represents the leaching efficiency of Fe and Mn from wolframite concentrate at a point in time, K_1, K_2, and K_3 are the overall reaction rate constants (s^{-1}), and t is the reaction time (s), α and β are weight coefficients which represent the weights of surface reaction and internal diffusion in the hybrid control process.

According to the well-known apparent activation energy equation, the dependence of rate constant K on the leaching agent concentration and particle size is described as Eqs. (6) and (7) (Guo et al., 2017; Demirkıran and Künkül, 2007).

$$K = A \times \exp(-E_a/RT) \text{ with } A = k \times C^m \times D^n \times W^i \tag{6}$$

$$\ln K = \ln A - (E_a/RT) = \ln k + m\ln C + n\ln D + i\ln W - (E_a/RT) \tag{7}$$

where k is the frequency factor, E_a is the apparent activation energy (J/mol), R is the universal gas constant (8.314 J/(mol·K)), T is the reaction temperature (K), C is the concentration of leaching agent (HCl, mol/L), m is the reaction order with respect to HCl concentration, D is the initial equivalent particle radius of wolframite particles (mm), n is the reaction order with respect to equivalent initial particle radius, and W is the stirring speed. Based on the analysis above, external diffusion is not the rate-controlling step of conversion process. Thus, $i=0$.

According to Eqs. (3)-(7), the overall reaction rate constant K under a certain condition can be determined by the slope of the curve $1-(1-x)^{1/3}$, $1-2x/3-(1-x)^{2/3}$, or $\alpha[1 - (1 - x)^{1/3}] + \beta[1 - 2x/3 - (1 - x)^{2/3}]$ versus reaction time based on the experimental data. Then E_a is calculated from the slope ($-E_a/R$) of the fitting line $\ln K$ versus $1/T$ using the series of kinetics data at various temperatures, and $\ln A$ is the intercept of the fitting line. Furthery, m and n can be calculated through plotting the fitting lines of $\ln K$ versus $\ln C$ and $\ln K$ versus $\ln D$ based on the experimental data at various HCl concentrations and particle sizes. Finally, the frequency factor k in the kinetic rate equation is determined through the equation: $A = k \cdot C^m \cdot D^n$.

3.4 Kinetics model verification

3.4.1 Effect of temperature

To verify the kinetics model, the curves of $1-(1-x)^{1/3}$ versus reaction time and $1-2x/3-(1-x)^{2/3}$ versus reaction time were plotted in Fig. 4(a) and Fig. 4(b) based on the experimental data in Fig. 2(b). It can be seen that the kinetic data acceptably fits the equation of internal diffusion control, while the data of the entire reaction time range does not fit the equation of surface reaction control.

According to Eqs. (6) and (7), the plot of $\ln K_2$ versus $1/T$ was depicted in Fig. 5 based on

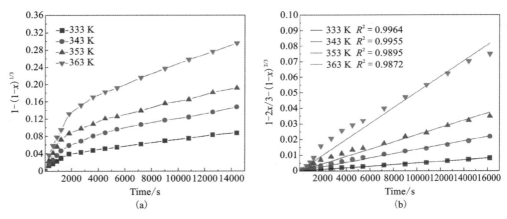

Fig. 4 Relationships of $1-(1-x)^{1/3}$ (a) and $1-2x/3-(1-x)^{2/3}$ (b) versus reaction time at various temperatures (stirring intensity of 250 r/min, 7 mol/L HCl, and particle size of $-50+38$ μm)

the slopes of the fitting lines in Fig. 4(b). The apparent activation energy E_a, which is calculated from the slope of the fitting line in Fig. 5, is 74.58 kJ/mol. However, this value does not conform to the apparent activation energy range of internal diffusion control. In normal metallurgical reaction, the activation energy is lower than 10 kJ/mol for an internal diffusion control process, the activation energy is higher than 40 kJ/mol for a surface reaction control process, and the activation energy is between 10 and 40 kJ/mol for a hybrid control process (Guo et al., 2017).

Although the data in Fig. 2(b) does not fit the equation of reaction control in the entire reaction time range, it is observed that there are transition points for the linearity of the curves in Fig. 4(a), and the transition points are evident at about 30 min at high reaction temperature (higher than 353 K). It is reasonable to infer that in the initial stage the control step is surface reaction as the wolframite concentrate particles are not covered with product films. After a period of reaction time, solid H_2WO_3 films are formed on

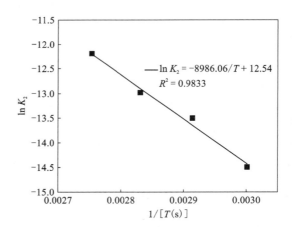

Fig. 5 Relationship between $\ln K_2$ and $1/T$

the surface of wolframite concentrate particles (this phenomenon is confirmed by Fig. 3), which changes the control step to hybrid control. Therefore, taking 30 min as the transition point, the kinetics of the conversion process is divided into two stages for analysis.

(1) The kinetics model for the first stage

The plots of $1-(1-x)^{1/3}$ versus reaction time were plotted in Fig. 6(a) based on the experimental data of the first 30 min in Fig. 2(b). Obviously, the kinetic data is consistent with reaction control model with the correlation coefficients over 0.97. The plot of $\ln K_1$ versus $1/T$ was plotted in Fig. 6(b) based on the slopes of the fitting lines in Fig. 6(a). The activation energy E_a of this stage is calculated as 40.57 kJ/mol. The activation energy value clearly indicates that the first stage of the wolframite conversion is most likely controlled by surface reaction. According to Eqs. (6) and (7), the intercept (3.99) of the fitting line in Fig. 6(b) is the value of $\ln A$ for the first stage.

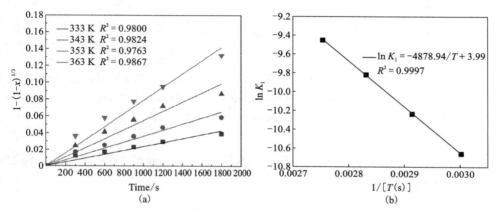

Fig. 6 Relationships between $1-(1-x)^{1/3}$ and the first stage reaction time (a) at various temperatures (stirring intensity of 250 r/min, 7 mol/L HCl, and particle size of $-50+38$ μm) and relationship between $\ln K_1$ and $1/T$ (b)

(2) The kinetics model for the second stage

By comparing the correlation coefficients of the plots of $\alpha[1-(1-x)^{1/3}]+\beta[1-2x/3-(1-x)^{2/3}]$ versus the second stage reaction time (after 30 min) and the corresponding activation energy with different values of α and β, it is determined that the values $\alpha = 4$ and $\beta = 1$ are appropriate. The plots of $4\times[1-(1-x)^{1/3}]+[1-2x/3-(1-x)^{2/3}]$ versus the second stage reaction time were plotted based on the kinetics data in Fig. 2(b), the results were presented in Fig. 7(a). The plot of $\ln K_3$ versus $1/T$ was also plotted in Fig. 7(b). The activation energy E_a of this stage is calculated as 38.16 kJ/mol, the value is in the range of 10 to 40 kJ/mol, which indicates that the second stage kinetics data is consistent with the equation of hybrid control. The value of $\ln A$ for the second stage is 2.77 which is presented by Fig. 7(b).

3.4.2 Effect of HCl concentration

As shown by the studies above, the kinetics of the conversion process is divided into two stages with a transition point of 30 min. The first stage is under surface reaction control, and the second stage is under hybrid control. Therefore, Eqs. (3) and (5) were used to fit the first and second stage of the kinetic data under different HCl concentrations. Fig. 8(a) shows the plots of $1-(1-x)^{1/3}$ versus reaction time using the kinetics data of the first 30 min in Fig. 2(c). Based on Eq. (7) and the slopes of the fitting lines in Fig. 8(a), the relationship between $\ln K_1$ and $\ln C$

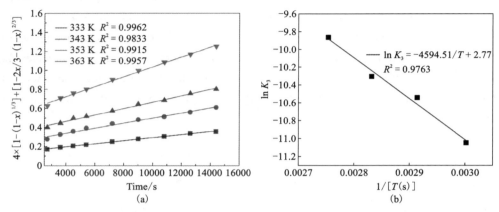

Fig. 7 Relationships between $4\times[1-(1-x)^{1/3}]+[1-2x/3-(1-x)^{2/3}]$ and the second stage reaction time (a) at various temperatures (stirring intensity of 250 r/min, 7 mol/L HCl, and particle size of $-50+38$ μm) and relationship between $\ln K_3$ and $1/T$ (b)

was plotted in Fig. 8(b). The reaction order of HCl concentration is 1.60, which is the slope of the fitting line in Fig. 8(b).

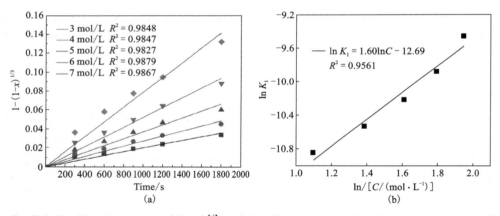

Fig. 8 Relationships between $1-(1-x)^{1/3}$ and the first stage reaction time (a) at various HCl concentrations (stirring intensity of 250 r/min, 363 K, and particle size of $-50+38$ μm) and relationship between $\ln K_1$ and $\ln C$ (b)

Fig. 9(a) shows the plots of $4\times[1-(1-x)^{1/3}]+[1-2x/3-(1-x)^{2/3}]$ versus the second stage reaction time based on the kinetic data under various HCl concentrations, and Fig. 9(b) shows the relationship between $\ln K_3$ and $\ln C$ in the second stage. As can be seen in Fig. 9(b), the reaction order of HCl concentration in the second stage is 0.75.

3.4.3 Effect of particle size

Similar to the analysis on the effects of temperature and HCl concentration, the equations of surface reaction control and hybrid control were used to fit the kinetic data of different size

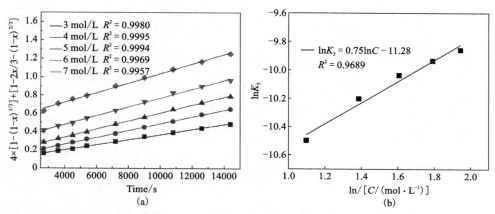

Fig. 9 Relationships between $4\times[1-(1-x)^{1/3}]+[1-2x/3-(1-x)^{2/3}]$ and the second stage reaction time (a) at various HCl concentrations (stirring intensity of 250 r/min, 363 K, and particle size of −50+38 μm) and relationship between $\ln K_3$ and $\ln C$ (b).

fractions. Fig. 10(a) shows the plots of $1-(1-x)^{1/3}$ versus the first stage reaction time based on the kinetic data of various size fractions, and Fig. 10(b) shows the relationship between $\ln K_1$ and $\ln D$ in the first stage. The average particle sizes of −100+75 μm, −75+50 μm, and −50+38 μm size fraction were taken as 87.5 μm, 62.5 μm, and 44.0 μm. The reaction order of particle size in the first stage is −0.81, which is shown in Fig. 10(b).

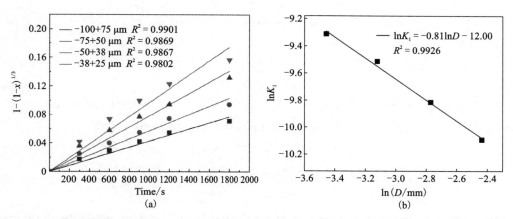

Fig. 10 Relationships between $1-(1-x)^{1/3}$ and the first stage reaction time (a) at various particle sizes (stirring intensity of 250 r/min, 363 K, and 7 mol/L HCl) and relationship between $\ln K_1$ and $\ln D$ (b)

Fig. 11(a) shows the plots of $4\times[1-(1-x)^{1/3}]+[1-2x/3-(1-x)^{2/3}]$ versus the second stage reaction time based on the kinetic data of various size fractions, and Fig. 11(b) shows the relationship between $\ln K_3$ and $\ln D$ in the second stage. As can be seen in Fig. 11(b), the reaction order of particle size in the second stage is −0.31.

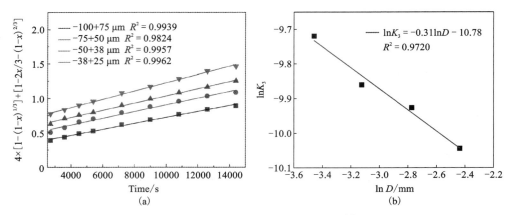

Fig. 11 Relationships between $5\times[1-(1-x)^{1/3}]+[1-2x/3-(1-x)^{2/3}]$ and the second stage reaction time (a) at various particle sizes (stirring intensity of 250 r/min, 363 K, and 7 mol/L HCl) and relationship between $\ln K_3$ and $\ln C$ (b)

3.5 Kinetics model establishment

According to the discussion above, the kinetic rate equations for the two stages of wolframite conversion process can be expressed as follows:

$$1-(1-x)^{1/3} = k_1 \times C^{1.60} \times D^{-0.81} \times \exp(-40570/RT) \times t \tag{8}$$

$$4[1-(1-x)^{1/3}] + [1-2x/3-(1-x)^{2/3}]$$
$$= k_2 \times C^{0.75} \times D^{-0.31} \times \exp(-38160/RT) \times t \tag{9}$$

Furtherly,

$$\ln(k_1 \times C^{1.60} \times D^{-0.81}) = 3.99 \tag{10}$$

$$\ln(k_2 \times C^{0.75} \times D^{-0.31}) = 2.77 \tag{11}$$

where $C = 7$ mol/L and $D = 0.044$ mm. Therefore, $k_1 = 0.19$, $k_2 = 1.41$ can be obtained. The concrete kinetic rate equations of the conversion process are expressed as follows:

$$1-(1-x)^{1/3} = 0.19 C^{1.60} \times D^{-0.81} \times \exp(-40570/RT) \times t \tag{12}$$

$$4[1-(1-x)^{1/3}] + [1-2x/3-(1-x)^{2/3}]$$
$$= 1.41 C^{0.75} \times D^{-0.31} \times \exp(-38160/RT) \times t \tag{13}$$

The kinetic rate equations for wolframite conversion in HCl solution established in this work is obviously different from the previous studies on the conversion kinetics of tungsten minerals in acid solutions. For the process of converting tungsten minerals into soluble tungsten salts by using mixed acid solutions rather than producing solid products (Kahruman and Yusufoglu, 2006; He et al., 2014; Li and Zhao, 2016; Yang et al., 2018), the mechanisms of tungsten minerals conversion are completely different from that in this work. Moreover, in some studies, the tungsten minerals conversion was inconsistent with the shrinking core model, other models such as Avrami-Erofeev and Mampel equation were applied to fit the kinetic data (Yang et al., 2018).

For the scheelite conversion in H_2SO_4 solution with the formation of solid product layers presented by Shen et al., (2018a), the entire process was under surface reaction control, which was different from the conclusion of this work. The main reason might be that the low concentration H_2SO_4 solution (lower than 1.5 mol/L) used in the process resulted in insufficient driving force of surface reaction. As a result, when the solid product layers were formed, the proportion of surface reaction in the control step was still much greater than that of internal diffusion. However, in this work, the HCl concentration used in the wolframite conversion at various temperatures was high enough (7 mol/L).

4 Conclusions

The effects of stirring intensity, temperature, HCl concentration, and particle size on the conversion performance of wolframite concentrate have been investigated. The results indicate that the conversion rate of wolframite increases significantly with increasing temperature and HCl concentration and the reduction of particle size, while the stirring intensity has little influence on the conversion rate.

The analysis on wolframite conversion mechanism indicates that layered H_2WO_3 films are formed on the surface of converted particles, the conversion is a core shrinking process advancing layer by layer. And the kinetic data is consistent with the shrinking core model.

The conversion process is divided into two stages with a transition point of 30 min, the first stage is a surface reaction controlling process while the second stage is a hybrid controlling stage. The following kinetic equations can be used to describe the conversion process:
$1-(1-x)^{1/3} = 0.19 \times C^{1.60} \cdot D^{-0.81} \cdot \exp(-40570/RT) \cdot t$ for the first stage, and $4 \times [1-(1-x)^{1/3}] + [1-2x/3-(1-x)^{2/3}] = 1.41 \times C^{0.75} \cdot D^{-0.31} \cdot \exp(-38160/RT) \cdot t$ for the second stage. The activation energies of the two stages are 40.57 kJ/mol and 38.16 kJ/mol.

Acknowledgment

This work was supported by the National Key R&D Program of China [grant number 2019YFC1907400] and the National Natural Science Foundation of China [grant numbers 51620105013].

References

[1] CHEN W, FENG Q, ZHANG G, et al. Effect of energy input on flocculation process and flotation performance of fine scheelite using sodium oleate[J]. Miner. Eng., 2017: 112: 27-35. https://doi.org/

[2] DEMIRKıRAN N, KÜNKÜL A. Dissolution kinetics of ulexite in perchloric acid solutions[J]. Int. J. Miner. Process., 2007, 83(1-2): 76-80. https://doi.org/10.1016/j.minpro.2007.04.007.

[3] GUO X, XIN Y, WANG H, et al. Leaching kinetics of antimony-bearing complex sulfides ore in hydrochloric acid solution with ozone[J]. Trans. Nonferr. Met. Soc. China. 2017, 27(9): 2073-2081. https://doi.org/10.1016/S1003-6326(17)60232-2.

[4] HE G, ZHAO Z, WANG X, et al. Leaching kinetics of scheelite in hydrochloric acid solution containing hydrogen peroxide as complexing agent[J]. Hydrometallurgy, 2014, 144-145: 140-147. https://doi.org/10.1016/j.hydromet.2014.02.006.

[5] KAHRUMAN C, YUSUFOGLU I. Leaching kinetics of synthetic $CaWO_4$ in HCl solutions containing H_3PO_4 as chelating agent[J]. Hydrometallurgy, 2006, 81(3-4): 182-189. https://doi.org/10.1016/j.hydromet.2005.12.003.

[6] KALPAKLI A, ILHAN S, KAHRUMAN C, et al. Dissolution behavior of calcium tungstate in oxalic acid solutions[J]. Hydrometallurgy, 2012, 121: 7-15. https://doi.org/10.1016/j.hydromet.2012.04.014.

[7] LI J, ZHAO Z. Kinetics of scheelite concentrate digestion with sulfuric acid in the presence of phosphoric acid[J]. Hydrometallurgy, 2016, 163: 55-60. https://doi.org/10.1016/j.hydromet.2016.03.009.

[8] LI J, MA Z, LIU X, et al. Sustainable and efficient recovery of tungsten from wolframite in a sulfuric acid and phosphoric acid mixed system[J]. ACS Sustain. Chem. Eng., 2020, 8(36): 13583-13592. https://doi.org/10.1021/acssuschemeng.0c04216.

[9] LI T, MIAO J, GUO E, et al. Tungsten-containing high-entropy alloys: A focused review of manufacturing routes, phase selection, mechanical properties, and irradiation resistance properties[J]. Tungsten, 2021, 3(2): 181-196. https://doi.org/10.1007/s42864-021-00081-x.

[10] LI X, SHEN L, ZHOU Q, et al. Scheelite conversion in sulfuric acid together with tungsten extraction by ammonium carbonate solution[J]. Hydrometallurgy, 2017a, 171: 106-115. https://doi.org/10.1016/j.hydromet.2017.05.005.

[11] LI Y, YANG J, ZHAO Z. Recovery of tungsten and molybdenum from low-grade scheelite[J]. JOM. 2017b, 69(10): 1958-1962. https://doi.org/10.1007/s11837-017-2440-5.

[12] LIU C, ZHANG W, SONG S, et al. Study on the activation mechanism of lead ions in wolframite flotation using benzyl hydroxamic acid as the collector[J]. Miner. Eng., 2019, 141: 105859. https://doi.org/10.1016/j.mineng.2019.105859.

[13] LIU H U, LIU H, NIE C, et al. Comprehensive treatments of tungsten slags in China: A critical review[J]. J. Environ. Manage., 2020, 270: 110927. https://doi.org/10.1016/j.jenvman.2020.110927.

[14] LIU L, XUE J, LIU K, et al. Complex leaching process of scheelite in hydrochloric and phosphoric solutions[J]. JOM, 2016, 68(9): 2455-2462. https://doi.org/10.1007/s11837-016-1979-x.

[15] LUO Y, CHEN X, ZHAO Z, et al. Pressure leaching of wolframite using a sulfuric-phosphoric acid mixture[J]. Miner. Eng., 2021, 169: 106941. https://doi.org/10.1016/j.mineng.2021.106941.

[16] MARTINS J I, MOREIRA A, COSTA S C. Leaching of synthetic scheelite by hydrochloric acid without the formation of tungstic acid[J]. Hydrometallurgy, 2003, 70(1-3): 131-141. https://doi.org/10.1016/S0304-386X(03)00053-7.

[17] MARTINS J I. Leaching systems of wolframite and scheelite: a thermodynamic approach[J]. Miner. Process. Extr. Metall. Rev., 2014, 35(1): 23-43. https://doi.org/10.1080/08827508.2012.757095.

[18] REN H, LI J, TANG Z, et al. Sustainable and efficient extracting of tin and tungsten from wolframite-

scheelite mixed ore with high tin content[J]. J. Clean. Prod. , 2020, 269: 122282. https://doi. org/10. 1016/j. jclepro. 2020. 122282.

[19] SHEN L, LI X, ZHOU Q, et al. Kinetics of scheelite conversion in sulfuric acid[J]. JOM, 2018a, 70(11): 2499-2504. https://doi. org/10. 1007/s11837-018-2787-2.

[20] SHEN L, LI X, ZHOU Q, et al. Sustainable and efficient leaching of tungsten in ammoniacal ammonium carbonate solution from the sulfuric acid converted product of scheelite[J]. J. Clean. Prod. , 2018b, 197: 690-698. https://doi. org/10. 1016/j. jclepro. 2018. 06. 256.

[21] SHEN L, LI X, ZHOU Q, et al. Wolframite conversion in treating a mixed wolframite-scheelite concentrate by sulfuric acid[J]. JOM, 2018c, 70(2): 161-167. https://doi. org/10. 1007/s11837-017-2691-1.

[22] WANG X, HUANG H E, SHI J, et al. Recent progress of tungstenbased high-entropy alloys in nuclear fusion[J]. Tungsten, 2021, 3(2): 143-160. https://doi. org/10. 1007/s42864-021-00092-8.

[23] YANG K, ZHANG W, HE L, et al. Leaching kinetics of wolframite with sulfuric-phosphoric acid[J]. Chin. J. Nonferr. Met. , 2018, 28(1): 175-182. https://doi. org/10. 19476/j. ysxb. 1004. 0609. 2018. 01. 21 (in Chinese).

[24] YIN C, JI L, CHEN X, et al. Efficient leaching of scheelite in sulfuric acid and hydrogen peroxide solution [J]. Hydrometallurgy, 2020, 192: 105292. https://doi. org/10. 1016/j. hydromet. 2020. 105292.

[25] ZHANG W J, LI J T, ZHAO Z W. Leaching kinetics of scheelite with nitric acid and phosphoric acid[J]. Int. J. Refract. Metals Hard Mater. , 2015, 52: 78-84. https://doi. org/10. 1016/j. ijrmhm. 2015. 05. 017.

[26] ZHAO Z, LI H, LIU M, et al. Soda decomposition of low-grade tungsten ore through mechanical activation[J]. J. Cent. South Univ. , 1996, 3(2): 181-184. https://doi. org/10. 1007/BF02652201.

[27] ZHAO Z. Tungsten metallurgy: fundamentals and applications[M]. Beijing: Tsinghua University Press, 2013. (in Chinese).

[28] ZHAO Z, DING W, LIU X, et al. Effect of ultrasound on kinetics of scheelite leaching in sodium hydroxide [J]. Can. Metall. Quart. , 2013, 52 (2): 138-145. https://doi. org/10. 1179/1879139512Y. 0000000050.

Analysis of Antimony Sulfide Oxidation Mechanism in Oxygen-Enriched Smelting Furnace

The phase transformation mechanism of antimony sulfide in the oxygen-enriched smelting system was investigated. The standard Gibbs free energy change ($\Delta G^\ominus$) of reactions in the smelting system at 100-1400 ℃ were calculated, indicating that the reaction between Sb_2S_3 and oxygen was liable to occur first, and that the high-valent antimony oxide decomposed at high temperatures. The thermodynamic analysis of the Sb-S-O system at different temperatures indicated that the phase transformation of antimony was feasible, and that the increase in temperature was beneficial to the existence of Sb_2O_3. Equilibrium calculation of Sb_2S_3 and oxygen showed that the antimony mainly volatilized into the gas phase, and that a small amount of antimony formed non-volatile antimony oxides. Thermogravimetric analysis of antimony oxide verified the accuracy of the thermodynamic analysis. A conceptual oxidation model in the oxygen-enriched smelting process is proposed, combined with the thermodynamic analysis, equilibrium calculation, and thermogravimetry-differential scanning calorimetry (TG-DSC) analysis of the oxygen-enriched smelting system.

1 Introduction

Antimony is considered an important metal, and various antimony compounds are widely used in electronics, flame retardants, chemicals, alloys, and semiconductor materials[1]. Antimony exists mainly in the form of antimony sulfide, which is usually smelted by volatilization in the blast furnace. However, the process has some problems, such as a complex process flow, low-concentration SO_2 emissions, high energy consumption, and occupational disease due to the volatile arsenic and antimony compounds[2-4]. Therefore, the efficient and clean utilization of antimony concentrate has been a concern in recent years. Oxygen-enriched smelting is a

---

Published in *JOM*, 2023, 75(2): 506-514. Authors: Huang Mingxing, Wang Qinmeng, Wang Songsong, Guo Xueyi.

strengthening technology widely used in the smelting field of non-ferrous metals, such as copper and lead[5,6], but references for pyrometallurgical smelting antimony are few[7]. The principle of the antimony oxygen-enriched smelting process is similar to that of the blast furnace, which is based on antimony sulfide being easily oxidized to Sb_2O_3 and volatilized into the gas phase[8-10]. The oxygenenriched smelting process solved the problem of SO_2 pollution in the traditional method and reduced energy consumption. Nevertheless, non-volatile antimony compounds and other impurity metallic elements were generated and entered the slag in industrial practice, which had a negative effect on the recovery rate of antimony[1,11-13]. However, whether antimony can be effectively recovered in a smelting process depends on the content of Sb_2O_3. Therefore, to solve the above industrial practice problems and to realize the application of the oxygen-enriched smelting process, it is necessary to study the phase transformation of antimony compounds in the oxygen-enriched smelting process.

Liu et al.[8] investigated the direct oxidation of stibnite (Sb_2S_3) using a nitrogen-oxygen gas mixture and realized autogenous oxygen smelting. The results showed that the recovery of Sb and Au through direct reduction of the slag from stibnite oxygen-enriched bath smelting is feasible. Dai et al.[9,14] studied the phase transformation, element migration mechanism, and thermodynamic analysis. The results clarified the occurrence state of metallic lead, metallic antimony, and other impurity metallic elements in oxygen-enriched direct smelting jamesonite concentrate. Zhang et al.[15] simulated the volatilization behavior of stibnite in the oxygen-enriched top-blowing process, analyzed the phase and distribution of Sb in the slag, and found that Sb existed as micro-elemental particles in the slag. Hua et al.[16] studied the volatilization of Sb_2S_3 in a steam atmosphere from 650 ℃ to 850 ℃, and proposed a complex gas phase reaction mechanism that Sb_2S_3 could be oxidized/decomposed by water vapor to Sb_2O_3 and metallic Sb at high temperature. Qin et al.[17] investigated the volatilization of stibnite (Sb_2S_3) in nitrogen from 700 ℃ to 1000 ℃. The results indicated that stibnite could be volatilized most efficiently in a nitrogen atmosphere, and quickly oxidized into antimony oxides, such as Sb_2O_3 and SbO_2, in oxygen. Padilla[18,19] was concerned with the oxidation/volatilization of stibnite (Sb_2S_3) in mixtures of nitrogen-oxygen of various compositions, and determined the kinetics of oxidation stibnite at temperatures below the melting point of Sb_2S_3. Cody[20] observed the thermal behaviors of all the antimony oxides, which were characterized by combining thermogravimetric analysis and Raman spectroscopy. The results revealed that Sb_2O_3 exhibited simultaneous volatilization and oxidation in the heating process. Aracena et al.[21] analyzed the volatilization kinetics of senarmontite (Sb_2O_3) in a neutral atmosphere in two temperature ranges, 550-615 ℃ (roasting temperature) and 660-1100 ℃ (melting temperature), by using a thermogravimetric analysis method under various gas flow rates. Zhao et al.[22] assessed stibnite concentrate's heating characteristics, thermodynamic behavior, and pure antimony compounds. The results indicated that the phase changes in the microwave heating process appeared in the roasting stage. Moreover, they investigated the volatilization/oxidation kinetics of Sb_2S_3 and Sb_2O_3 and the behavior of impurity elements in oxygen-enriched smelting.

The existing references have focused on the volatilization behavior of Sb_2S_3 and Sb_2O_3, and conducted kinetic analyses on them. There are few studies on the phase transformation mechanism of antimony in the smelting process of antimony gold resources, and the relevant mechanism studies were aimed at a certain temperature range. The research methods were experimental, which were not combined with industrial production. In addition, the oxygen-enriched smelting mechanism of antimony sulfide has been clarified, but researches on the formation and disappearance mechanism of nonvolatile antimony oxides are few. Given the complex phase evolution of antimony during the oxygenenriched smelting process, it is necessary to study the evolution mechanism of the antimony phase.

Therefore, the thermodynamic behaviors of Sb_2S_3 from 25 ℃ to 1400 ℃ have been analyzed in this paper. Based on the study of thermodynamic analysis, the phase occurrence state and transformation mechanism were clarified with the help of analysis and detection methods. This work revealed the oxidation mechanism of antimony sulfide in the oxygen-enriched smelting process, giving theoretical guidance and promoting the application of oxygen-enriched smelting.

2 Materials and methods

2.1 Materials

The antimony sulfide concentrates were obtained from a domestic antimony smelter in China. The main mineralogical phases present in the antimony sulfide concentrates were Sb_2S_3 and SiO_2, as shown in Fig. 1. As listed in Tables 1 and 2, the antimony sulfide concentrates contained considerable antimony (44.52%). The main phase compositions of antimony in raw materials were antimony oxide 9.48%, antimony sulfide 33.77%, and antimonate 1.27%. The analytical reagent, Sb_2S_3 (98.00%), was purchased from Shanghai Macklin Biochemical. The material of Sb_2O_3 was obtained from a domestic antimony smelter in China.

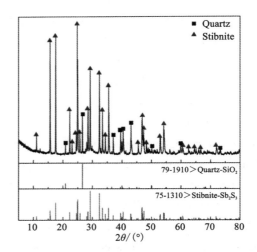

Fig. 1 XRD pattern of antimony sulfide concentrates

Table 1 Major components of antimony sulfide concentrate   %

Element	Sb	S	Fe	Ca	Al	Mg	Cu	Pb	SiO$_2$	Au/(g·t^{-1})
Mass fraction	44.52	22.6	3.27	5.75	0.56	0.45	0.15	0.17	9.71	71.61

Table 2 Phase analysis of antimony element in antimony sulfide concentrate   %

Phase	Antimony oxide	Antimony sulfide	Antimonate	Total antimony
Mass fraction	9.48	33.77	1.27	44.52

2.2 Oxygen side-blown smelting process

A domestic antimony smelting plant adopts a side-blown furnace for oxygen-enriched smelting, and the schematic of the oxygen side-blown antimony smelting process is shown in Fig. 2. The molten pool depth is 1100-2500 mm, and the oxygen lances are installed at the side of the furnace to blow oxygen and natural gas into the molten bath.

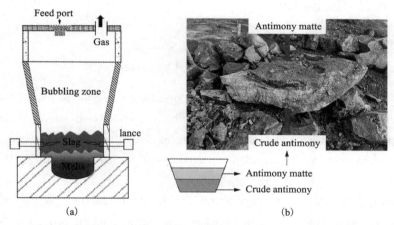

Fig. 2 (a) Schematic of antimony side-blown smelting furnace,
(b) Solidified antimony matte and crude antimony

Minerals are continuously charged into the furnace through a feed port with flux and coke, as depicted in Fig. 2(a). Oxygen enrichment of 70%-95% is injected by lances to create turbulent agitation in the smelting furnace, promoting rapid reactions between the minerals, flux, and oxygen, and providing heat to maintain a smelting temperature of 1250-1350 ℃. The reactions of the furnace include melting, dissociation, desulfurization, oxidation, volatilization, and slagging, which generate high concentration antimony-containing flue gas and high-temperature melt.

The smelting slag floats on top of the molten bath because of its low density, and is tapped off from a slag tap hole. The mixture of melts is released by siphon, and later settles into two layers, i.e., antimony matte and crude antimony [from top to bottom, Fig. 2(b)]. The high-concentration antimony containing flue gas is oxidized and desulfurized, and settles in the condensation dust collection system to obtain crude antimony oxide products.

2.3 Sampling and characterization

Weight changes of these samples were obtained using thermogravimetry-differential scanning calorimetry (TG-DSC; NETZSCH STA 449F3). It consisted of a high-temperature vertical tube furnace, an electronic balance connected to a personal computer for continuously recording the mass variation, and a gas delivery system. The powder sample was placed in a graphite crucible and the heating rate was 10 K/min. The chemical composition of these samples was analyzed by X-ray fluorescence spectroscopy, while elements with low content were detected by inductively coupled plasma atomic emission spectroscopy (IRIS Intrepid 3 XRS; Thermo Electron, USA). X-ray diffraction (XRD) patterns of these samples were obtained using a JEOL TTRAX-3 X-ray diffractometer operated at 40 kV, 250 mA, and 10 ℃/min. Thermodynamic modeling was carried out with the help of FactSage™ 7.1 software.

3 Results and discussion

3.1 Thermodynamic analysis in the Sb-S-O system

The Gibbs free energy ($\Delta G^\ominus$) values of the antimony compounds were obtained from the HSC Chemistry database. The main component of the raw material is antimony sulfide, and the essence of antimony oxygen-enriched smelting is the oxidation of antimony sulfide. This process involved multiple processes such as oxidation, decomposition, and interaction, which are shown below.

The dependency of $\Delta G^\ominus$ with temperatures for the main reactions in the antimony sulfide smelting process is shown in Fig. 3, which presents the thermodynamic process of Sb_2S_3 oxidation in the oxygen-enriched system. It can be concluded from Fig. 3 that the Sb_2S_3 can be transformed into Sb, Sb_2O_3, Sb_2O_4, and Sb_2O_5 spontaneously in theory, because the $\Delta G^\ominus$ of the Eqs. (1)-(5) was less than 0. The $\Delta G^\ominus$ of Eq. 3 is the most negative, indicating that the reaction trend of producing Sb_2O_3 is the largest. Sb_2O_4 and Sb_2O_5 are non-volatile, which increases the antimony content of the slag, and should be avoided during the smelting process.

Based on the above analysis, it is known that the conversion from Sb_2S_3 to Sb, Sb_2O_3, Sb_2O_4, and Sb_2O_5 is thermodynamically feasible. It can be seen from Fig. 3 that, when the temperature is more than 526.1 ℃, the $\Delta G^\ominus$ of Eq. 9 is less than 0, and the Sb_2O_5 converts to Sb_2O_4. However, converting Sb_2O_4 to Sb_2O_3 is difficult, and the temperature should be greater than 1137.5 ℃. Moreover, the $\Delta G^\ominus$ of Eq. 7 is less than 0 and decreases with the increase of temperature when the temperature is higher than 269.5 ℃, indicating that Sb_2O_4 reacts with Sb_2S_3

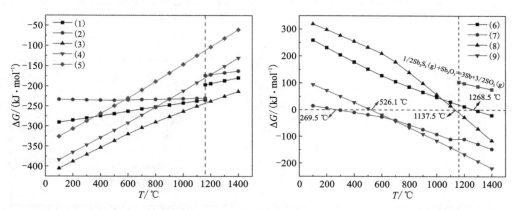

Fig. 3 Dependency of $\Delta G^{\ominus}$ on temperature for the main reactions in the antimony trisulfide smelting process

to form Sb_2O_3. When the temperature is higher than 1268.5 ℃, it can be concluded from the $\Delta G^{\ominus}$ of Eq. 6 that the non-gas phase Sb_2S_3 and Sb_2O_3 interact to form Sb. Although the $\Delta G^{\ominus}$ of Eq. 2 is negative, the $\Delta G^{\ominus}$ of Eq. 3 is more negative than Eq. 2, so that Eq. 3 is more likely to occur. The antimony gold resources smelting process is an oxygen-rich system, and the antimony sulfide in the system is converted to Sb, which continues to react with the residual oxygen to generate Sb_2O_3. In addition, when the Sb_2S_3 is a gas, the reaction cannot occur. Therefore, to avoid the formation of by-products Sb, Sb_2O_4, and Sb_2O_5, the temperature should be controlled to be over 1268.5 ℃.

$$2/9Sb_2S_3 + O_2(g) = 2/9Sb_2O_3 + 2/3SO_2(g) \tag{1}$$
$$1/3Sb_2S_3 + O_2(g) = 2/3Sb + SO_2(g) \tag{2}$$
$$4/3Sb + O_2(g) = 2/3Sb_2O_3 \tag{3}$$
$$Sb + O_2(g) + 1/2Sb_2O_4 \tag{4}$$
$$4/5Sb + O_2(g) = 2/5Sb_2O_5 \tag{5}$$
$$Sb_2O_3 + 1/2Sb_2S_3 = 3Sb + 3/2SO_2(g) \tag{6}$$
$$Sb_2O_4 + 1/9Sb_2S_3 = 10/9Sb_2O_3 + 1/3SO_2(g) \tag{7}$$
$$2Sb_2O_4 = 2Sb_2O_3 + O_2(g) \tag{8}$$
$$2Sb_2O_5 = 2Sb_2O_4 + O_2(g) \tag{9}$$

The predom module of FactSage ™ 7.1 software was used to calculate the equilibrium stability zone of the Sb-S-O system at 1100-1400 ℃, as shown in Fig. 4(a), where predominance areas are shown for phases in equilibrium with the vapor phase at 1100-1400 ℃ as a function of the partial pressure of SO_2 and O_2 (marked as $p(SO_2)$ and $p(O_2)$, respectively). This equilibrium diagram indicates that the stable $Sb_2O_3(g)$ regions were presented with a high concentration range of p_{O_2} 10^{-4} to 10^6 and p_{SO_2} 10^{-16} to 10^8. The stable zone of $Sb_2S_3(g)$ moved directly in the direction of decreasing sulfur partial pressure and increasing oxygen partial pressure with the increase of temperature. As the temperature increased, the stable region of $Sb_2O_3(g)$ and Sb moved in the direction of increasing oxygen partial pressure, and the stable zone of $Sb_2O_4(s)$

decreased, indicating that the increase in temperature is beneficial to the oxidation of Sb_2S_3 and the decomposition of Sb_2O_4.

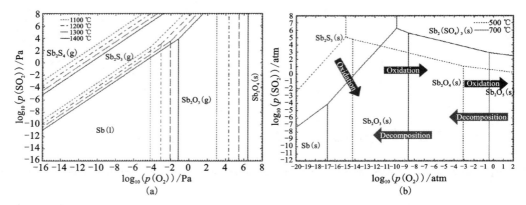

Fig. 4 (a) Equilibrium stability zone of the Sb-S-O system at 1100-1400 ℃, (b) equilibrium stability zone of the Sb-S-O system at 500 and 700 ℃

To further clarify the stable region's conditions of non-volatile antimony oxides, the equilibrium stability zone of the Sb-S-O system at 500 and 700 ℃ was drawn. As seen in Fig. 4(b), there were some stable zones in the Sb-S-O system, where $Sb_2O_3(s)$, $Sb_2O_4(s)$, $Sb_2O_5(s)$, $Sb_2S_3(s)$, and $Sb_2(SO_4)_3(s)$ are at 500 ℃. When the temperature increased to 700 ℃, the stable regions of $Sb_2O_4(s)$ and $Sb_2O_5(s)$ decreased, the stable zone of $Sb_2O_3(s)$ increased, and Sb(s) was generated at a low oxygen partial potential. The results indicated that the low temperature was favorable for forming $Sb_2O_4(s)$ and $Sb_2O_5(s)$, and their stable zone decreased with the increase in temperature. This equilibrium diagram indicates that Sb_2S_3 in the presence of oxygen was first oxidized to Sb_2O_3 as an intermediate. Then, it was oxidized to Sb_2O_4, and Sb_2O_4 was further oxidized to Sb_2O_5 under higher $p(O_2)$ and low $p(SO_2)$ in the vapor. Generally, in traditional antimony smelting, the stibnite concentrate is first oxidized to the volatile trioxide in the "Oxidation" area in Fig. 5, and the Sb_2O_4 and Sb_2O_5 are decomposed to Sb_2O_3 in the "Decomposition" area.

Figure 5 shows the lg P_{O_2}-T diagram of the Sb-S-O system. It can be seen that Sb existed in the form of $Sb_2O_3(s)$ and $Sb_2O_4(s)$ without $Sb_2O_5(s)$ at low temperatures. Figure 4(b) indicates that the stable existence conditions of $Sb_2O_5(s)$ were low temperature and high oxygen potential. However, the range of $p(O_2)$ was low, and $Sb_2O_5(s)$ did not appear in Fig. 5. When the temperature was 1200-1300 ℃, and the oxygen concentration of the system was 30%-80% (lg $p(O_2, atm) = -0.52$-0.097), the Sb-S-O system was all in the gas phase, indicating that antimony entered the gas phase under the condition.

The thermodynamic feasibility of oxygen-enriched enhanced smelting of Sb_2S_3 is determined

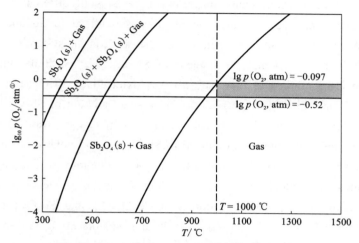

Fig. 5 $\lg p_{O_2}$-T diagram of the Sb-S-O system

by analyzing the equilibrium diagram and the dependency of $\Delta G^{\ominus}$ with the reaction temperatures. In the production process, the smelting temperature and the oxygen potential are important factors that affect production efficiency, and theoretical calculation research was carried out in this paper. However, the space in the paper was limited, and other influencing factors will be studied in follow-up research work. The equilibrium composition of the reaction between Sb_2S_3 and oxygen at different temperatures was calculated using the equilibrium composition module of HSC 6.0. This module is based on the Gibbs free energy minimization principle for isothermal, isobaric, and fixed mass conditions. The amount of Sb_2S_3 was fixed at 1 kmol, and the temperature was set at 1300 ℃. The calculated results under different oxygen amounts are shown in Fig. 6.

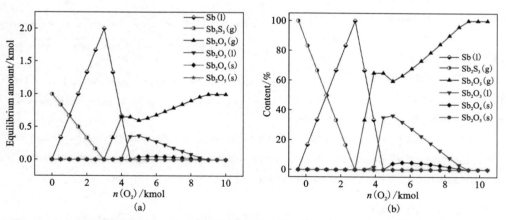

Fig. 6 (a) Equilibrium amount of 1 kmol Sb_2S_3 with different oxygen amounts at 1300 ℃,
(b) phase content of 1 kmol Sb_2S_3 with different oxygen amounts at 1300 ℃

① 1 atm = 1.01×10⁵ Pa.

Fig. 6 shows the equilibrium phase diagrams of Sb_2S_3 with different oxygen amounts at 1300 ℃, from which it can be seen that the Sb_2S_3 was oxidized to Sb and then oxidized to antimony oxide. When the amount of oxygen was 0, the Sb_2S_3 in the system is in a gaseous form. The content of Sb_2S_3 decreased and the content of Sb increased with the increase of oxygen amount. When the O_2 continued to increase, Sb was oxidized to $Sb_2O_3(l)$, $Sb_2O_3(g)$, and $Sb_2O_4(s)$. $Sb_2O_3(l)$ and $Sb_2O_4(s)$ decreased, and a small amount of $Sb_2O_5(s)$ was generated with the amount of oxygen increasing. When the amount of oxygen was 9 kmol, the reaction reached equilibrium, and antimony was oxidized to $Sb_2O_3(g)$ and volatilized into the gas phase.

The oxygen was fixed at 9 kmol, and equilibrium phase diagrams of 1 kmol Sb_2O_5 with O_2 at different temperatures were calculated, as shown in Fig. 7. $Sb_2O_5(s)$ decomposed to $Sb_2O_4(s)$, $Sb_2O_4(s)$ converts to Sb_2O_3, and the content of $Sb_2O_3(l)$ and $Sb_2O_3(g)$ increased with increasing temperature. These results indicate that the non-volatile $Sb_2O_4(s)$ and $Sb_2O_5(s)$ were generated at low temperatures, but that they started to decompose with the increase of temperature. When the temperature was 1300 ℃, the non-volatile oxides were converted into volatile Sb_2O_3. As the temperature increased, a small amount of $Sb_2O_3(l)$ entered the melt, resulting in a slight decrease in the content of $Sb_2O_3(g)$.

Thermodynamic analysis showed that the antimony loss in the oxygen-enriched smelting process of antimony gold resources was divided into three ways: oxidation into slag, metal phase loss, and volatilization escape. When the smelting temperature was low and the oxygen potential was high, antimony was oxidized to generate non-volatile antimony oxides in the slag. When the oxygen potential was low, the antimony loss was the metal phase loss and volatilization escape. The metal phase loss is that antimony generates elemental antimony into antimony matte or crude antimony because of the insufficient oxidation reaction. The volatilization escape means that the antimony sulfide oxidation reaction does not occur and that it is directly volatilized into the gas phase in the form of antimony sulfide.

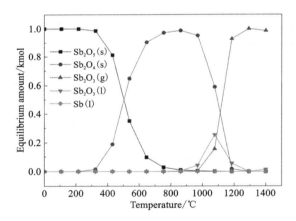

Fig. 7 Equilibrium phase diagrams of 1 kmol Sb_2O_5 with 9 kmol O_2 at different temperatures

3.2 TG-DSC Analysis

Padilla[19] and Zhao[22] et al. studied the thermogravimetric analysis of antimony sulfide in an oxygen atmosphere. The results showed that the weight loss of the antimony sulfide oxidization process was divided into four stages. The first stage (322.6-525.0 ℃) can be attributed to the oxidization of Sb_2S_3, wherein the SO_2 and Sb_2O_3 were produced, which contributed to the mass loss. The second stage (525.0-601.3 ℃) was interpreted as the synergistic effect between the volatilization of Sb_2O_3 and the formation of Sb_2O_4. The third stage (601.3-658.1 ℃) can be attributed to the formation of Sb_2O_4, leading to weight gain. The sample weight changed less in the first three stages, and we observed that the most considerable weight loss of the sample occurred at 1023-1100 ℃, which was attributed to the volatilization of the Sb_2O_3.

In this work, TG-DSC analysis of the Sb_2O_3 was conducted to investigate the mass loss and thermal effects. The sample was produced by a domestic antimony smelter. The specific composition is shown in Table 3. The thermogravimetric conditions are that the heating rate is 10 ℃/min and the atmosphere is nitrogen or oxygen. The results are presented in Fig. 8.

Table 3   Composition of antimony oxide samples   %

Composition	Sb_2O_3	As_2O_3	SO_3	PbO	Others
Mass fraction	94.34	1.43	1.22	0.75	2.26

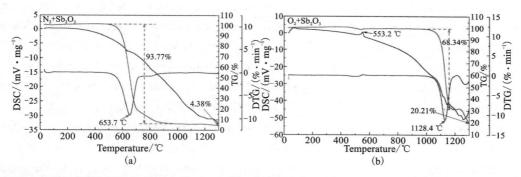

Fig. 8 (a) TG-DSC analysis of the Sb_2O_3 under nitrogen atmosphere with a heating rate of 10 ℃/min, (b) TG-DSC analysis of the Sb_2O_3 under the oxygen atmosphere with a heating rate of 10 ℃/min

There was no obvious heat shock or release peak under the nitrogen atmosphere. The sample was lost at 653.7 ℃, and the weight loss rate was 93.77%. The melting point of Sb_2O_3 was 656 ℃ and occurred in Sb_2O_3 melting volatilization, resulting in the reduced sample quality. According to Table 3, the residual rate of the sample was 4.38% at the end temperature, from which it was considered that the sample residue is not easy to melt.

There was an exothermic peak at 553.2 ℃ under the oxygen atmosphere. It can be seen from the TG curve that the weight of the sample slowly decreased and then slowly increased at this stage, and that a small amount of the sample melted and volatilized. When the temperature was 1128.4 ℃, a large amount of the sample melted and volatilized, and the sample weight loss rate was 68.34%. The third stage occurred in the weight loss, and the weight loss rate was 11.29% when the temperature was 1250 ℃. The residual rate of the sample was 20.21% at the end temperature. These results showed that the weight loss in the first stage is the chemical reaction between the sample and oxygen, and the weight loss in the second and third stages is the melting and volatilization of the samples.

Compared with the samples' TG-DSC curves under the nitrogen and oxygen atmosphere, it could be seen that the sample melted and volatilized under the nitrogen atmosphere. However, the weight loss temperature of the sample increased, and the volatilization of the sample was inhibited under the oxygen atmosphere, which resulted in the residual rate of the sample under the oxygen atmosphere being higher than that in the nitrogen atmosphere. Combined with the literature research[19], the reasons can be divided into two types. The first is that the oxygen atmosphere inhibited the melting and volatilization of Sb_2O_3, resulting in the increase of the sample residual rate at the end temperature. The second is that Sb_2O_3 occurred in the reverse reactions of Eqs. 8 and 9 under the oxygen atmosphere, which inhibited the normal volatilization and weight loss of the sample and increased the sample residual rate.

3.3 Oxidation mechanism of antimony sulfide

The main component of antimony sulfide concentrate is antimony sulfide, and understanding the antimony sulfide smelting process's oxidation mechanism is the key to obtain antimony oxide and metallic antimony from antimony sulfide concentrate. According to thermodynamic calculation results, current research results, and analysis results, the oxidation mechanism of antimony sulfide in the smelting process has been studied. The entire smelting process can be divided into five stages, according to the types of antimony phase. As shown in Table 4, from room temperature to 200 ℃, the main components are Sb_2S_3, O_2, and a slight amount of oxidation products Sb_2O_3 and Sb. This is because, although the reaction thermodynamics is feasible, the reaction kinetics is poor, and the content of oxidation products in the system is less. The reaction kinetics get better, and oxidation reactions occur with the increase of temperature. However, owing to the oxygen-rich system, the phenomenon of peroxidation occurs and Sb_2O_4 and Sb_2O_5 are produced. At this stage, the system components are mainly Sb_2O_4 and Sb_2O_5, accompanied by small amounts of Sb, Sb_2S_3, and Sb_2O_3. However, Sb_2O_5 begins to decompose and forms Sb_2O_4 when the temperature is higher than 400 ℃, which leads to an increase in Sb_2O_4 and a decrease in Sb_2O_5. When the temperature is from 1000 to 1200 ℃, Sb_2O_5 is completely decomposed, and the decomposition reaction of Sb_2O_4 occurs. At this time, the system is mainly Sb_2O_3, Sb_2O_4, and a small amount of Sb. When the temperature is higher than 1200 ℃, Eq. 8 occurs, and Sb_2O_4 is converted into Sb_2O_3. Finally, the system components are Sb_2O_3, SO_2, and O_2. The oxidation mechanism of the smelting process greatly enhances our understanding of the mobilization, oxidation, and transformation of Sb in antimony sulfides, which are the main Sb-bearing ore minerals.

Table 4 Analysis of reaction mechanism in the smelting process

Stage	Composition	Reaction	Schematic diagram
Raw materials	Sb_2S_3, O_2	—	
25~200 ℃	Sb_2S_3, O_2, Sb, SO_2, Sb_2O_3	$2Sb_2S_3+9O_2(g)=2Sb_2O_3+6SO_2(g)$ $Sb_2S_3+3O_2(g)=2Sb+3SO_2(g)$ $2Sb_2O_3+Sb_2S_3=6Sb+3SO_2(g)$ $4Sb+3O_2(g)=2Sb_2O_3$	
200~1000 ℃	Sb_2O_3, O_2, SO_2, Sb, Sb_2O_4, Sb_2O_5	$2Sb_2S_3+9O_2(g)=2Sb_2O_3+6SO_2(g)$ $2Sb_2O_3+O_2(g)=2Sb_2O_4$ $2Sb_2O_4+O_2(g)=2Sb_2O_5$ $2Sb+2O_2(g)=Sb_2O_4$ $4Sb+5O_2(g)=2Sb_2O_5$	
1000~1200 ℃	Sb_2O_3, O_2, SO_2, Sb, Sb_2O_4	$2Sb_2S_3+9O_2(g)=2Sb_2O_3+6SO_2(g)$ $2Sb_2O_5=2Sb_2O_4+O_2(g)$	
>1200 ℃	Sb_2O_3, O_2, SO_2	$2Sb_2S_3+9O_2(g)=2Sb_2O_3+6SO_2(g)$ $2Sb_2O_4=2Sb_2O_3+O_2(g)$ $9Sb_2O_4+Sb_2S_3=10Sb_2O_3+SO_2(g)$	

● O_2　● SO_2　● Sb　● Sb_2S_3　● Sb_2O_3　● Sb_2O_4　● Sb_2O_5

4 Conclusion

Thermodynamic analysis showed that the conversion reaction between different antimony oxides is feasible, and that the temperature has a significant effect on the stable existence of the antimony phase. The occurrence state of antimony in oxygen-enriched smelting of antimony sulfide concentrate was investigated by constructing the predominance area diagrams at different temperatures. It was found that non-volatile antimony oxides can be formed at low temperatures, gradually decompose into volatile antimony oxides at high temperatures, and volatilize into the gas phase.

The equilibrium content of the antimony phase in oxygen-enriched smelting of antimony sulfide was calculated. It can be concluded that antimony sulfide is oxidized to elemental antimony and Sb_2O_3 at low temperatures. As the temperature increased, non-volatile high-valent antimony oxide was formed and decomposed, and the final system mainly contained Sb_2O_3.

Thermodynamic analysis and industrial production results showed that the antimony loss in the oxygen-enriched smelting process of antimony gold resources was divided into three ways: oxidation into slag, metal phase loss, and volatilization escape. The results were elaborated on the oxidation mechanism of antimony sulfide and provided theoretical guidance for the oxygen-enriched smelting process of antimony sulfide resources.

Acknowledgements

The authors acknowledge financial support from the National Natural Science Foundation of China (No. U20A20273 and No. 51904351), National Key R&D Program of China (No. 2019YFC1907401), Natural Science Foundation for Distinguished Young Scholar of Hunan Province (No. 2022JJ10078), Science and Technology Innovation Program of Hunan Province (No. 2021RC3005), Innovation Driven Projects of Central South University (No. 2020CX028).

References

[1] ANDERSON C G. Chem. Erde-Geochem. 2012, 72: 3.
[2] LIU T, QIU K, HAZARD J. Mater. 2018, 347: 334.
[3] REN B, ZHOU Y, MA H, et al. Mater. Cycles Waste Manage. 2018, 20: 193.
[4] YE L, OUYANG Z, CHEN Y, et al. Miner. Eng. 2019, 144: 106049.
[5] GUO X, TIAN M, WANG S, et al. JOM. 2019, 71: 3941.
[6] GAO W, WANG C, YIN F, et al. Adv. Mater. Res. 2012, 2044: 904.
[7] LUO H, LIU W, QIN W, et al. Rare Met. 2019, 38: 800.
[8] LIU W, LUO H, QING W, et al. Metall. Mater. Trans. B. 2014, 45: 1281.
[9] CHEN M, DAI X. JOM. 2018, 70: 41.
[10] LAGER T, FORSSBERG K S E. Miner. Eng. 1989, 2: 543.
[11] OROSELA D, BALOGA P, LIU H, et al. Solid State Chem. 2005, 178: 2602.
[12] ZHANG Z, DAI X. Trans. Nonferrous Met. Soc. China. 2020, 30: 501.
[13] ZHOU K. Cent South Univ. 2014. (in Chinese)
[14] ZHANG Z, DAI X, ZHANG W. JOM. 2017, 69: 2671.
[15] ZHANG J, YANG X, DENG W, et al. Non. Metall. (Extra Metall). 2019, 3: 1. (in Chinese)

[16] HUA Y, YANG Y, ZHU F. J. Mater. Sci. Technol. 2003, 19: 619.
[17] QIN W, LUO H, LIU W, et al. South Univ. 2015, 22: 868.
[18] PADILLA R, ARACENA A, RUIZ M C. Min J. Metall. Sect. 2014, B 50: 127.
[19] PADILLA R, GUSTAVO R, MARIA C R. Metall. Mater. Trans. B. 2010, 41B: 1284.
[20] CODY C A, DICARLO L, DARLINGTON R K. Inorg. Chem. 1979, 18: 1572.
[21] ARACENA A, JEREZ O, ANTONUCCI C. Trans. Nonferrous Met. Soc. China. 2016, 26: 294.
[22] ZHAO P, LIU C, CHANDRASEKAR S, et al. Powder Technol. 2021, 379: 630.

Comparative Atmospheric Leaching Characteristics of Scandium in Two Different Types of Laterite Nickel Ore from Indonesia

Abstract: Atmospheric acid leaching behaviour of scandium (Sc) in two different types of laterite nickel ore from Indonesia was investigated. Ore and leaching residue characterization was performed by XRD, FTIR, XPS, and SEM-EDX. Ore characterisation showed that the major minerals in limonitic laterite were goethite, magnetite, hematite, and saprolitic laterite mainly consisted of goethite, magnetite, lizardite, clinochlore. Sc in two different types of laterite nickel ore are distributed widely among minerals, but it mainly hosts in Al-bearing goethite and silicate minerals. Sc host minerals in limonitic and saprolitic laterite nickel ore are different. 84.27% and 59.86% of Sc in limonitic and saprolitic laterite could be leached under the experimental conditions of 3 mol/L H_2SO_4, 80 ℃ reaction temperature, leaching duration 3 h and liquid to solid ratio 6∶1, respectively. The results show that Sc and Mn, Mg in limonitic laterite have similar dissolution characteristics because the extractions of Sc and Mn, Mg are linearly correlated. Sc in limonitic laterite is susceptible to acid attack and easier to be extracted than other metals except for Mn. Sc and Ni in saprolitic laterite have similar dissolution characteristics because Sc is not strongly related to metals other than Ni. Sc in saprolitic laterite is more difficult to extract than Mg and Ni, but it is easier to be leached than other metals. The dissolution kinetics was found to fit well to the shrinking core model with the diffusion through the product layer as the rate controlling step. Results of this research may assist in the development of a more efficient process for exacting Sc from laterite nickel ores.

Keywords: Laterite nickel ore; Scandium; Atmospheric sulfuric acid leaching; Comparative leaching behavior

---

Published in *Minerals Engineering*, 2021, 173: 107212. Authors: Tian Qinghua, Dong Bo, Guo Xueyi, Xu Zhipeng, Wang Qingao, Li Dong, Yu Dawei.

1 Introduction

Scandium (Sc), a member of rare earth elements (RE), is not scarce but highly dispersed in the earth's crust. Its average crustal abundance of 22 g/t, ranked the 34th most abundant element in the earth (Le et al., 2018; Qing et al., 2018; Ramasamy et al., 2018; Wang and Cheng, 2011; Wang et al., 2011). It has the characteristics of high activity, lightweight, softness, and high melting point (Chakhmouradian et al., 2015; Hu et al., 2020; Liu et al., 2019), and has been widely used in the fields of national defense and military industry, metallurgy and chemical industry, light high-temperature resistant alloy, and new electric light source material, etc. (Davris et al., 2016; Kerkove et al., 2014; Wang et al., 2011; Yin et al., 2011). At present, the main application of Sc is Al-Sc alloy and Zr-based solid oxide fuel cells (A et al., 2019; Kaya et al., 2017; Wang et al., 2011). The global supply of Sc is about 15 tons per year (Kim and Azimi, 2020). The ores with a Sc content range of 0.002%-0.005% can be used as Sc resources, which is worthy of deserving exploitation and utilization (Shaoquan and Suqing, 1996; Zhou et al., 2018). The Sc minerals containing appreciable quantities of Sc such as euxenite, thortveitite, and gadolinite are scarce and hard to meet the requirements of industrial exploitation in scale (Qing et al., 2018). However, a trace amount of Sc frequently coexists in the ores of aluminum, titanium, tungsten, nickel. Generally, it is obtained as a by-product in the production of other metals or recovered from the residues or waste liquid, such as wolframite residue, bauxite residue, waste liquor of titanium pigment, and so on (Borra et al., 2016; Fujinaga et al., 2013; Li et al., 2018; Liu and Li, 2015; Ochsenkuhnpetropulu et al., 1995; Onal and Topkaya, 2014; Shaoquan and Suqing, 1996; Wang et al., 2011). The absence of reliable and long-term production coupled with the high price of Sc has limited the commercial applications of Sc. In short, industrial applications are waiting for a sufficient, reliable, and reasonably priced Sc supply.

The laterite nickel ores containing from 50 g/t up to 600 g/t of Sc are proposed as the most promising Sc resources for its production shortly (Chasse et al., 2017; Guo et al., 2021; Kim and Azimi, 2020; Luo et al., 2015; Makuza et al., 2021; Meshram et al., 2019; Van der Ent et al., 2013; Yan et al., 2021). Laterites can be classified into limonites or saprolites, depending on the iron and magnesium content (Garces-Granda et al., 2018). High pressure acid leaching (HPAL) and atmospheric acid leaching (AL) are the two prevailing technologies for hydrometallurgical processing of laterite nickel ores (Luo et al., 2021). In recent years, AL for processing laterite nickel ores has become a research hotspot in hydrometallurgy because of the method's use of small equipment, mild reaction conditions, and low technical risk (Guo et al., 2015). Usually, extractions of valuable metals such as nickel and cobalt through AL rely on the complete dissolution of nickeliferous minerals. Hence, a proper understanding of mineral dissolution behavior in acidic solutions is helpful for leaching valuable metals from laterite nickel ores. Previous studies have provided some information on the dissolution behavior of laterite minerals at atmospheric pressure. Overall, these studies suggest that leaching behavior strongly depends on ore mineralogy and chemical composition, and process conditions. The sulphuric acid leachability of metal values associated with different minerals follows the order: lizardite > goethite > maghemite > magnetite ≈ hematite > chromite ≈ ringwoodite (Luo et al., 2015; Senanayake et al., 2011). It has been thought from extensive studies that metal cations exist in nickeliferous laterites in two modes, (a) weakly adsorbed to the mineral surface and (b) as a substitute in the mineral structure (Liu et al., 2009). The extent of substitution has a significant impact on the dissolution behavior of laterite minerals. In laterite nickel ores where nickel and cobalt are disseminated in different associated/interlocked minerals, the ore mineralogy type can dramatically impact H_2SO_4 leachability and consumption rate (Luo et al., 2015). Until recently, Sc, a potential by-product not considered by previous for its dissolution behavior during AL. Sc in laterite nickel ores is distributed widely among minerals but it may be especially associated with goethite, clay minerals, or manganese oxides, in which it substitutes for Fe^{3+} and Al^{3+} because of the similarities in ionic radius (Ferizoglu et al., 2018; Kaya et al., 2017). Kaya (Kaya et al., 2017) also speculated that Sc and nickel should have similar dissolution characteristics because Sc occurs together with nickel in the same minerals of laterite nickel ores.

The laterite nickel ores from Indonesia are a typical tropical laterite deposit, about 12% of world nickel resources (Luo et al., 2021). The aims of the present work were to investigate the leaching characteristics of Sc in two different types of laterite nickel ore from Indonesia during atmospheric acid leaching. Based on the leaching results, the kinetics and mechanism of Sc dissolution from limonitic and saprolitic laterite material, especially the relationship and interaction of Sc and other metals were studied. Results of this research may assist in the development of a more efficient process for exacting Sc from laterite nickel ores.

2 Experimental

2.1 Materials

Limonitic and saprolitic laterite ores used in this study were obtained from Sulawesi, Indonesia. Ore samples were initially dried at 105 ℃ overnight and ground to d_{90} = 43.449 μm (limonitic) and d_{90} = 49.407 μm (saprolitic) by sequential step-by-step crushing and grinding, respectively. The detailed applied mineralogical studies of the limonitic and saprolitic laterite material were carried out with a combination of X-ray fluorescence (XRF), inductively coupled plasma spectrometer (ICP), X-ray diffraction (XRD), Fourier transform infrared (FTIR) spectroscopy, X-ray photoelectron spectroscopy (XPS), scanning electron microscope (SEM) and energy dispersive X-ray analysis (EDX).

2.2 Methods

Chemical leaching experiments were conducted in a 500 mL three-neck round bottom flask fitted with an overhead mechanical stirrer. Processing parameters studied are summarized in Table 1, and only one of the above parameters was allowed to vary while all others were fixed. For each leaching experiment, the quantity of raw materials was 50.00 g. According to the predetermined liquid to solid ratio, sulfuric acid at a predetermined concentration was transferred into the reactor, which was then heated to the desired temperature through the digital homoeothermic water bath before adding the ore samples. Finally, the speed of the mechanical stirrer was set to 300 r/min and timing started. After a certain period of time, the mechanical stirrer and the digital homoeothermic water bath were both switched off. Subsequently, liquid/solid separation was performed with a Büchner funnel and the residues were washed thrice with deionized water. All solutions were collected for subsequent analysis. In the meantime, the leaching residues were dried for at least a day in a drying oven at 105 ℃, weighed, re-ground in a mortar prior to XRD and SEM-EDX analysis. The pregnant leach solution produced by each experiment was analyzed using an atomic emission spectrometer (ICP-AES, IRIS Intrepid II XSP, Thermo Electron Corporation, USA). For the mineralogical characterization of raw ore and the leaching residue, X-ray diffraction (XRD, Rigaku D/max-2550 X-ray diffractometer with Cu K_α radiation), under the conditions of tube current and voltage: 40 mA, 45 kV, sample size: 320 mesh, irradiated area: 10 mm×10 mm (fixed grating), scanning range: 5°-80° (2θ), step size: 0.02° (2θ) and scanning speed: 1°/min was utilized. Then the minerals identification was performed with the software of Jade 6.5. For detailed examinations of ore samples and leach residue, a TESCAN MIRA3 field emission scanning electron microscope coupled with energy-dispersive X-ray spectroscopy was used. Chemicals used in this study were all of the reagent grades and all the aqueous solutions were prepared by using deionized water. The leaching efficiency of metal was calculated according to Eq. (1).

$$\eta_i = \frac{C_i \times V}{m \times w_i} \times 100\% \tag{1}$$

where η_i is the leaching efficiency of metal i (%), C_i is the concentrations of metal i in the leaching solution (g/L), V is the volume of leaching solution (L), m is the mass of raw material (g), w_i is the mass fraction of metal i in raw material (%).

Table 1 List of process parameters studied during AL experiments

Experiments	Fixed parameters	Studied parameters
Sulfuric acid concentration /(mol·L^{-1})	80 ℃, 2 h, 5∶1 liquid/solid ratio	0.5, 1.0, 1.5, 2.0, 2.5, 3.0
Leaching duration/h	80 ℃, 3 mol/L, 5∶1 liquid/solid ratio	1.0, 2.0, 3.0, 4.0, 5.0
Liquid to solid ratio	80 ℃, 3 mol/L, 2 h	3∶1, 4∶1, 5∶1, 6∶1, 7∶1, 8∶1
Leaching temperature /℃	2 h, 3 mol/L, 6∶1 liquid/solid ratio	50, 60, 70, 80, 90

3 Results and discussion

3.1 Mineralogical analyses

The XRF analysis results of the chemical composition of laterite nickel ore for the experiment are listed in Table 2, and the ICP analysis results of essential elements are listed in Table 3. The limonitic laterite is characterized by high iron content (45.30%), low nickel content (0.69%). Compared with limonitic, the saprolitic laterite is featured by the high nickel (1.82%) and silicon dioxide (30.38%), lower iron content (26.10%). The X-ray diffraction (XRD) analysis results of the laterite sample are shown in Fig. 1(a, b). The main limonitic minerals, which are evident from the X-ray diffraction pattern shown in Fig. 1(a), are goethite (PDF#99-0055), magnetite (PDF#99-0073), hematite (PDF#99-0060). According to the XRD profile [Fig. 1(b)], the saprolitic sample mainly consisted of goethite (PDF#99-0055), magnetite (PDF#99-0073), lizardite (PDF#89-6275), clinochlore (PDF#79-1270). FTIR spectrum of the sample is shown in Fig. 2. In the IR pattern [Fig. 2(a)] of the limonitic laterite, the sharp band centered around 1628 cm^{-1} is attributed to the —OH bending vibration (Luo et al., 2010; Luo et al., 2009). The bands at 3164, 916, and 803 cm^{-1} are due to the bands between H and O (Girgin et al., 2011; Panda et al., 2014). Absorptions at 458 cm^{-1} are related to goethite resulting from FeO_5 hexagon (Prasad et al., 2006). From the FTIR spectrum [Fig. 2(b)] of saprolitic laterite material, the band at 3690 cm^{-1} can be assigned to the outer —OH stretching vibrations coordinated to three magnesium of the octahedral layer (Luo et al., 2010; Luo et al., 2009). Water in laterite gives broadband at 3418 cm^{-1} correspondings to the —OH stretching vibrations (Girgin et al., 2011; Kursunoglu et al., 2018; Luo et al., 2010; Luo et al., 2009; Panda et al., 2014). Those at 791 and 1011 cm^{-1} are related to quartz because of Si and O bonds (Kursunoglu et al., 2018); and this at 1640 cm^{-1} is due to the bonds between H and O (Kursunoglu et al., 2018; Panda et al., 2014). The intense band at 446 cm^{-1} is associated with the Si-O bending mode (Luo et al., 2010; Luo et al., 2009).

Table 2 The XRF analysis results of the chemical composition of limonitic and saprolitic laterite nickel ore   %

Composition	MgO	Al_2O_3	SiO_2	CaO	Sc_2O_3	TiO_2	Cr_2O_3	MnO	Fe_2O_3	CoO	NiO	CuO	ZnO
Limonitic	0.84	8.34	4.72	0.07	0.01	0.30	3.00	0.87	74.10	0.08	1.05	0.02	0.04
Saprolitic	14.19	3.44	30.38	0.54	0.01	0.04	1.51	0.67	38.29	0.08	2.93	0.02	0.04

Table 3 The ICP analysis results of essential elements %

Element	Ni	Co	Mn	Sc	Fe	Mg	Al	Cr
Limonitic	0.69	0.058	0.51	0.0054	45.30	0.37	3.64	1.16
Saprolitic	1.82	0.061	0.45	0.0043	26.10	6.35	1.61	0.57

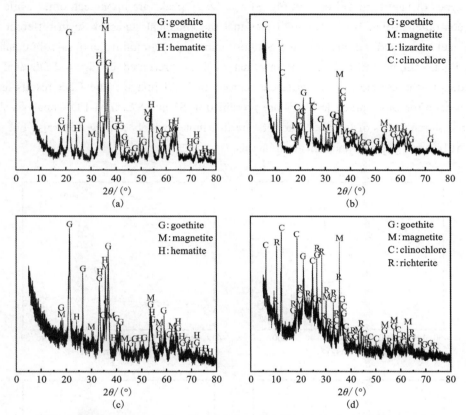

Fig. 1 XRD pattern of laterite nickel ore (a. limonitic, b. saprolitic) and leach residue (c. limonitic, d. saprolitic)

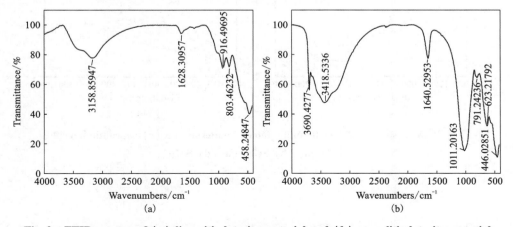

Fig. 2 FTIR spectra of (a) limonitic laterite material and (b) saprolitic laterite material

To search for possible Sc host minerals, XPS analysis and SEM-EDX analysis were conducted for raw materials. From the XPS spectrum [Fig. 3(a, c)] of laterite material, there are weak Sc2p peaks on the surface of limonitic and saprolitic laterite. It can be proved that some Sc hosts on the surface of minerals in the form of surface adsorption, which is much more easily dissolved by acid solution. Fig. 3(b, d) shows that although Sc hosts in the trivalent form in two different types of laterite nickel ore, in Fig. 3(b), Sc 2p peaks are upon each other while they are next to each other in Fig. 3(d). This indicates that the chemical states of Sc in different types of laterite nickel ore are different, that is, Sc host minerals in limonitic and saprolitic laterite are different. Backscattered SEM images of raw materials are presented in Fig. 4. EDS analysis was used to identify the composition of different regions in SEM images. The EDS results associated with the 6 main regions marked in Fig. 4 are presented in S1 and S2. SEM-EDX analysis show that Sc in two different types of laterite nickel ore are distributed widely among minerals, but it mainly hosts in Al-bearing goethite and silicate minerals.

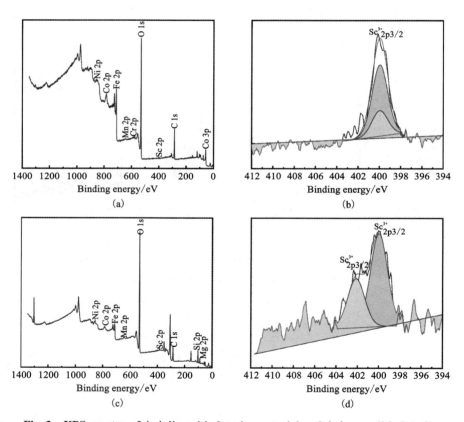

Fig. 3 XPS spectra of (a) limonitic laterite material and (c) saprolitic laterite material, (b) Sc2p and (d) Sc2p

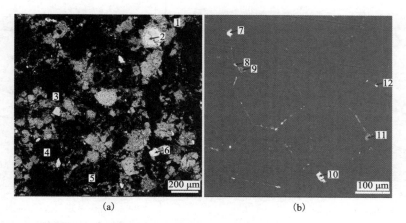

Fig. 4 SEM images of (a) limonitic laterite material and (b) saprolitic laterite material

3.2 Effect of sulfuric acid concentration

The complete leaching of minerals that contain Sc requires a sufficient acid addition to destroy the mineral inclusion. The common acid-consuming reaction (2) is given below where M = Ni, Co, Mn, Sc, $Fe^{2+, 3+}$, Al, Mg, Cr, and so on. The specific chemical reactions are relevant to leaching and thermodynamic data are summarized in Table 4. Based on chemical analysis, the theoretically calculated sulfuric acid requirement for limonitic and saprolitic laterite were 3.5 mol/L and 2.5 mol/L, respectively. Hence, The effect of the 0.5 mol/L to 3.0 mol/L sulfuric acid concentration at leaching temperature of 80 ℃, liquid to solid ratio of 5 ml/g, and leaching duration of 2 h was investigated as shown in Fig. 5.

$$M_xO_y + yH_2SO_4 \rightleftharpoons M_x(SO_4)_y + yH_2O \tag{2}$$

Table 4 The chemical reactions are relevant to leaching and thermodynamic data (Kaya et al., 2017)

Chemical reaction	$\Delta G_T^\ominus /(kJ \cdot mol^{-1})$	Chemical reaction	$\Delta G_T^\ominus /(kJ \cdot mol^{-1})$
$Sc_2O_3 + 6H^+ \rightleftharpoons 2Sc^{3+} + 3H_2O$	$-201.56 + 0.35T$	$Cr_2O_3 + 6H^+ \rightleftharpoons 2Cr^{3+} + 3H_2O$	$-92.49 + 0.14147T$
$NiO + 2H^+ \rightleftharpoons Ni^{2+} + H_2O$	$-99.25 + 0.09702T$	$Fe_2O_3 + 6H^+ \rightleftharpoons 2Fe^{3+} + 3H_2O$	$-42.997 + 0.1698T$
$CoO + 2H^+ \rightleftharpoons Co^{2+} + H_2O$	$-105.12 + 0.09598T$	$Fe_3O_4 + 8H^+ \rightleftharpoons 2Fe^{3+} + Fe^{2+} + 4H_2O$	$-52.76 + 0.15904T$
$MnO + 2H^+ \rightleftharpoons Mn^{2+} + H_2O$	$-121.65 + 0.06348T$	$2/3FeOOH + 2H^+ \rightleftharpoons 2/3Fe^{3+} + 4/3H_2O$	$-40.79 + 0.15694T$
$Al_2O_3 + 6H^+ \rightleftharpoons 2Al^{3+} + 3H_2O$	$-86.34 + 0.1637T$	$Mg_3Si_2O_5(OH)_4 + 6H^+ \rightleftharpoons 3Mg^{2+} + 2SiO_2(s) + 5H_2O$	$-111.57 + 0.07051T$

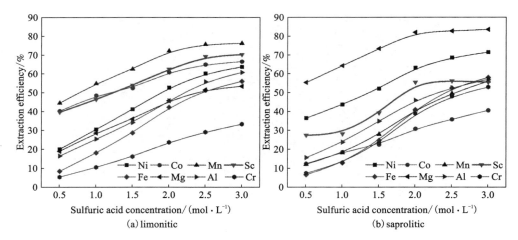

Fig. 5　Effect of sulfuric acid concentration on metal extractions

According to Fig. 5(a) and (b), the influence of sulfuric acid concentration on the leaching efficiency of Sc in limonitic is entirely different from that in saprolitic. As shown in Fig. 5(a), the leaching of Sc increased considerably with the increase in H_2SO_4 solution concentration from 0.5 to 2.5 mol/L and was then maintained steadily. It reached 70.30% when the sulfuric acid concentration was 3 mol/L, which indicates that the majority of Sc can be extracted from the limonitic laterite as long as the sulfuric acid concentration is high enough. Sc extraction was more elevated than nickel, cobalt, iron, aluminum, magnesium, and chromium, but lower than that of manganese from 0.5 to 3.0 mol/L sulphuric acid, indicating that Sc is susceptible to acid attack and easier to be extracted than other metals except for manganese. However, it can be seen from Fig. 5(b) that the leaching efficiency of Sc in saprolitic remains unchanged when the sulfuric acid concentration increases from 0.5 to 1.0 mol/L, and then increased gradually with sulfuric acid concentration up to 2 mol/L beyond which further increases have a marginal effect on the extractions. Sc extraction was much lower than magnesium and nickel, but slightly higher than other metals from 0.5 to 3.0 mol/L sulphuric acid, indicating that Sc is more difficult to extract than magnesium and nickel, but it is easier to be leached than other metals. To compare the leaching behavior of Sc in two different types of laterite nickel ore, 3 mol/L sulfuric acids were selected for the next leaching experiment.

3.3　Congruency of dissolution

An examination of the congruency of metal dissolution can provide information regarding the distribution of metal substituents within crystals. If Sc and associated metals dissolve at identical rates (i.e. unit slope for plot of% total metal dissolved versus% total Sc dissolved = 1), we can postulate that the Sc occurs together with other metals in the same minerals of laterite nickel ores. If non-congruent dissolution is evident (slope ≠ 1) then the distribution of the metal is not uniform or Sc and metal are not present in a single mineral (Liu et al., 2009; Luo et al., 2015; Matthew et al., 2007). It can be seen from Fig. 6(a) and Table 5 that Sc and Mn (slope = 1.02), Mg (slope = 1.09) were extracted simultaneously. Furthermore, it can be speculated that Sc occurs together with Mn and Mg in the same minerals of limonitic laterites because Sc and Mn, Mg have similar dissolution characteristics. As shown in Fig. 6(b) and Table 5, Sc in saprolitic laterite is not strongly related to metals other than Ni (slope = 0.99). Therefore, it can be inferred that Sc occurs together with Ni in the same minerals of saprolitic laterites.

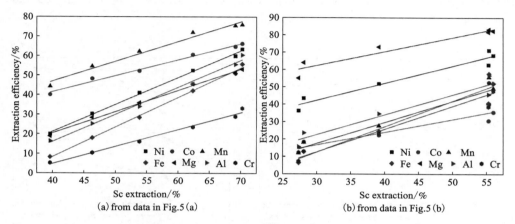

Fig. 6 Sc extractions versus other metals extractions

Table 5 The slopes of extraction correlation at different sulfuric acid concentrations

Element	Limonitic		Saprolitic	
	Slope	R^2	Slope	R^2
Ni	1.38	0.997	0.99	0.972
Co	0.82	0.990	0.73	0.943
Mn	1.02	0.979	1.21	0.958
Fe	1.51	0.997	1.47	0.963
Mg	1.09	0.996	0.81	0.962
Al	1.39	0.992	1.14	0.966
Cr	0.87	0.987	1.31	0.968

3.4 Effect of leaching duration

In order to investigate the effect of leaching duration on Sc extraction from this laterite, a series of leaching experiments was performed with the leaching duration varied from 1 to 5 h under the experimental condition of 3 mol/L H_2SO_4, reaction temperature 80 ℃, liquid to solid ratio 5 mL/g. Compared with sulfuric acid concentration, leaching duration has little effect on Sc leaching efficiency. The results in Fig. 7(a) showed that the leaching efficiency of Sc in the limonitic increased gradually within 3.0 h, and then keep almost constant at ~75% with a further extension of leaching. This is probably due to the remaining metals to be extracted being associated with refractory minerals (Kursunoglu and Kaya, 2016). As is shown in Fig. 7(b), the leaching efficiency of Sc in saprolitic increased gradually, and then decreased as the leaching duration increased. This may be caused by the adsorption of silicate (Alkan et al., 2018; McDonald and Whittington, 2008). Compared with Fig. 7(a) and (b), it can be seen that Sc in the limonitic is easier to be leached in a short time than that in saprolitic. Therefore, the leaching duration was determined as 3 h.

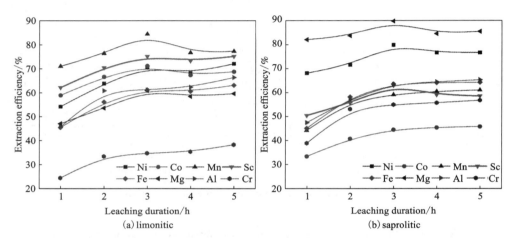

Fig. 7 Effect of leaching duration on metal extractions

3.5 Effect of liquid to solid ratio

The effect of liquid to solid ratio was measured at different liquid to solid ratios and other conditions include 3 mol/L H_2SO_4, reaction temperature 80 ℃, leaching duration 3 h. The results are shown in Fig. 8. It can be seen from Fig. 8(a) that the leaching efficiency of Sc increased gradually with a liquid to solid ratio up to 6∶1, beyond which they tended to be stable with a further increase in the liquid to solid ratio. A similar trend appears in Fig. 8(b), which is the leaching efficiency of Sc is greatly improved up to a liquid to solid ratio of 5∶1. After that, the leaching efficiency fluctuated slightly. In summary, the leaching efficiency of Sc in different types of laterite has the same changing trend with the liquid to solid ratio, and the liquid to solid ratio required to reach the optimal value of the leaching efficiency is not much different. Therefore, the selected liquid to solid ratio for the next leaching experiment was 6∶1.

3.6 Effect of reaction temperature

To investigate the effect of leaching temperature on leaching metals, a series of experiments were performed in the following conditions: 3 mol/L H_2SO_4, leaching duration 3 h, liquid to solid ratio 6 mL/g. The results are shown in Fig. 9. Fig. 9 demonstrated that the leaching efficiency

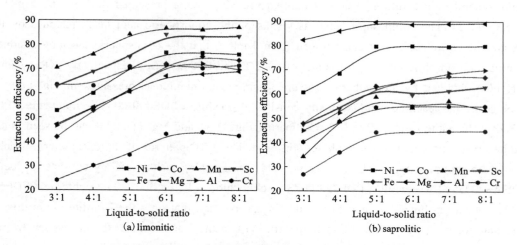

Fig. 8 Effect of liquid to solid ratio on metal extractions

of Sc in different types of laterite has the same changing trend with the leaching temperature, but the temperature has a more significant influence on the leaching efficiency of Sc in limonitic laterite. The leaching efficiency of Sc in saprolitic is higher than that of limonitic before the temperature reaches 60 ℃. The above phenomenon is due to serpentine minerals are easy to react with acid at low temperature (McDonald and Whittington, 2008). Therefore, the leaching temperature was determined as 80 ℃.

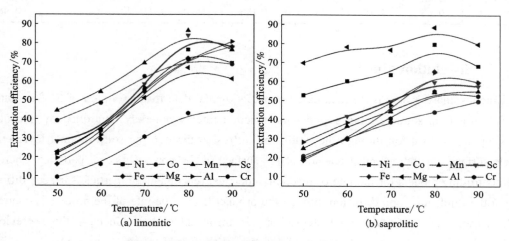

Fig. 9 Effect of temperature on metal extractions

3.7 Leach residue characterization

Analysis and characterization for the residue obtained by atmospheric pressure leaching of laterite using the most suitable conditions (H_2SO_4 concentration 3 mol/L, leaching duration 3 h,

Liquid to solid ratio 6 mL/g, leaching temperature 80 ℃) are presented in Table 6, Fig. 1(c, d), S3 and S4. The composition analysis results of the atmospheric acid leach residues by ICP-AES are shown in Table 6. It can be seen from Table 6 that the atmospheric leaching residue of limonitic and saprolitic laterite still contain 0.002% and 0.003% Sc, respectively. XRD data for the residue obtained by atmospheric pressure leaching of limonitic is presented in Fig. 1(c). As can be seen, the leaching residue mainly consists of goethite (PDF#99-0055), magnetite (PDF#99-0073), hematite (PDF#99-0060). Comparing Fig. 1(a) and Fig. 1(c), it can be seen that the goethite, magnetite, and hematite phases still exist. The XRD patterns of saprolitic laterite leaching residues are shown in Fig. 1(d). As can be seen, the leaching residue mainly consists of goethite (PDF#99-0055), magnetite (PDF#99-0073), clinochlore (PDF#79-1270), richterite (PDF#81-0724). Comparing Fig. 1(b) and Fig. 1(d), it can be seen that only the lizardite phase does not exist, and goethite, magnetite, clinochlore still exist in the leach residue. It can be seen from S3 that there are holes of different sizes on the surface of minerals, but the minerals structure is not completely destroyed. As shown in S4, this kind of minerals is not dissolved by atmospheric pressure acid leaching and still maintains a relatively intact structure.

Table 6 Chemical composition analysis results of the atmospheric acid leach residues   %

	Ni	Co	Mn	Sc	Fe	Mg	Al	Cr
Limonitic	0.34	0.04	0.14	0.002	26.90	0.26	2.19	1.40
	Ni	Co	Mn	Sc	Fe	Mg	Al	Cr
Saprolitic	0.68	0.06	0.36	0.003	16.85	1.32	1.05	0.48

3.8 Sc dissolution kinetics

Kinetics analysis was performed based on experimental data for dissolution of Sc from laterite material. In a fluid-solid reaction system, the reaction rate is generally controlled by one of the following steps: diffusion through the fluid film, diffusion through the solid product layer on the particle surface, or the chemical reaction at the surface of the core of reacted particles. There are three controlling models for the rate of reaction: chemical reaction at the particle surface, diffusion through the fluid film, and diffusion through the product layer. The rate of the process is controlled by the slowest of these sequential steps. For the metal dissolution kinetics, three previously established shrinking core model were used, expressed by Eq. (3), (4), and (5):

$$1 - (1 - x)^{\frac{1}{3}} = k_c t \tag{3}$$

$$1 - (1 - x)^{\frac{2}{3}} = k_{d1} t \tag{4}$$

$$1 - \frac{2}{3}x - (1 - x)^{2/3} = k_{d2} t \tag{5}$$

where x refers to the degree of extent of leaching. t is the time in minutes, k_c, k_{d1}, and k_{d2} are the overall rate constants. Eq. (3) assumes that the step controlling the leach rate is the chemical reaction taking place on the surface of the mineral and Eq. (4) and Eq. (5) assumes that the controlling step is the diffusion through the fluid film and diffusion through the product layer, respectively.

The apparent rate constant can be used for determining the temperature dependency by Arrhenius' law (6):

$$k = k_0 e^{-E/RT} \qquad (6)$$

By plotting the apparent rate constants for each experiment in an Arrhenius plot, the activation energy was determined.

Laterite nickel ores were leached at different times under the experimental conditions of 3 mol/L H_2SO_4, liquid to solid ratio 6 mL/g and reaction temperature 80 ℃. Examination of plots of the above kinetics equations as functions of time showed that only Eq. (5) gives excellent straight line fits for Sc leaching from laterite material (Fig. 10), which indicates that the reaction rate is controlled by the diffusion through the product layer. The apparent activation energy of Sc dissolution in limonitic and saprolitic laterite calculated from the Arrhenius plot of k_{d2} (Fig. 11) are 37.23 kJ/mol and 25.02 kJ/mol, respectively.

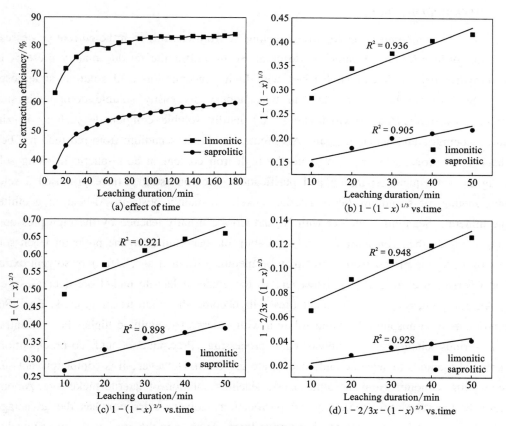

Fig. 10 Analyses of the kinetics of Sc leaching

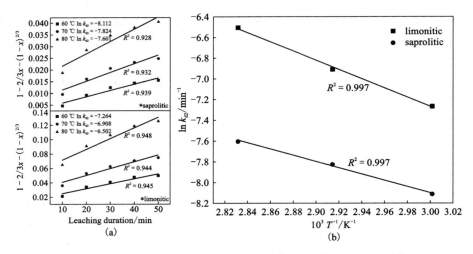

Fig. 11 Arrhenius plot for leaching of Sc from laterite material

3.9 Comparison of acid leaching (AL) for extracting Sc from laterites with current processes

Table 7 gives a comparison AL for extracting Sc from laterites with the current processes. The Sc mineralogy in different resources is different. Sc in wolframite residue mainly presents in the form of hydroxide, so it can be leached with high concentration acid solution at atmospheric pressure. Sc hosts in bauxite residues in two modes: a) easily soluble compounds such as carbonates, hydroxides and oxyhydroxides, b) hardly soluble compounds such as in zircon, ilmenite, goethite, and hematite. Using AL process to extract scandium from red mud has become the mainstream process. In order to avoid the high iron content in atmospheric leaching solution affecting the subsequent separation and purification, previous studies have proposed a selective leaching method of scandium. Sc in laterite nickel ore mainly hosts in Al-bearing goethite and silicate minerals. Scandium together with Ni and Co are usually leached by the high pressure acid leaching method, then concentration and separation of scandium from the pregnant leach solution (PLS) by using the pH-controlled precipitation method (Kaya et al., 2017) or solvent extraction method (Ferizoglu et al., 2018). However, as the grade of laterite nickel ore decreases year by year, AL for processing laterite nickel ores will become the mainstream process. But, Sc low concentrations in atmosphere leaching solution with orders of magnitude higher Fe concentration, raise a serious challenge for downstream processing. Precipitation and co-precipitation are conventional methods to separate and concentrate scandium. However, it is completely not suitable for recovering scandium from the atmospheric leaching solution of laterite nickel ore. Among the technologies for scandium separation and purification, solvent extraction has the advantages of high extraction capacity and ease of operating at large scales, and therefore is the most widely used technology for recovery of trace scandium from solutions with large amounts of impurity elements.

Table 8 gives a summary of extraction of scandium from various aqueous solutions by solvent extraction. Reference Table 8, P204 + N1923 extractant (Zou et al., 2020) or Cyanex 572 extractant (Nie et al., 2017) can be used for selective extraction of scandium from atmospheric leaching solution of laterite nickel ore.

Table 7 Comparison of AL for extracting Sc from laterites with current processes

Raw materials	Sc mineralogy	Extraction method	Conditions	Leaching efficiency	Ref(s).
In this work	Sc mainly hosts in Al-bearing goethite and silicate minerals	Acid leaching	3 mol/L H_2SO_4, 6 mL/g, 80 ℃, 3 h	Sc (limonitic): 84.27%, Sc (saprolitic): 59.86%	
Red mud	Sc_2O_3: 60-120 ppm, Sc hosts in bauxite residues in two modes: a) easily soluble compounds such as carbonates, hydroxides and oxyhydroxides, b) hardly soluble compounds such as in zircon, ilmenite, goethite, and hematite (Gentzmann et al., 2021)	Acid leaching	9.36 mol/L HCl, 4 mL/g, 75 ℃, 3 h	Sc: 93.3%, Fe: 95.9%	(Zhang et al., 2019)
			HCl (40% of theoretical value): red mud: EDTA = 40 mL : 10 g : 2 g, 70 ℃, 4 h	Sc: 79.6%, Fe: 6.12%	(Zhou et al., 2018)
			Microwave, 1.5 M H_2SO_4, 15 mL/g, 90 ℃, 30 min	Sc: 64.2%	(Reid et al., 2017)
			6-8 mol/L H_3PO_4, 10-12 mL/g, 120-140 ℃, 60-90 min	Sc: >90%	(Li et al., 2018)
			0.5 mol/L HNO_3, 50 mL/g, 25 ℃, 24 h	Sc: 80%, Fe: 3%	(Ochsenkühn-Petropulu et al., 1996)
		Bioleaching	10% (v/v) Gluconobacter oxydans (DSMZ 46616), 10% (w/v), 37 ℃, 18-20 d	Sc (Indian red mud): 83%, Sc (German red mud): 94%	(A et al., 2021)
			pyrite/RM mass ratio 2 : 1, 22 d	Sc: 78.6%	(Drz et al.)
		Sulphurizing roasting-water leaching	Roasting: H_2SO_4 (mass ratio of 1 : 1), 700 ℃, 1 h, Leaching: 2 d	Sc: 60%, Fe: <1%	(A et al., 2016)
		Acid baking-water leaching	Baking: 0.6-1.3 mL/g H_2SO_4, 400 ℃, 2 h, Leaching: >45 min	Sc: 80%	(John et al., 2019)

Continue the Table 7

Raw materials	Sc mineralogy	Extraction method	Conditions	Leaching efficiency	Ref(s).
Laterite nickel ores	Sc: 50 g/t-600 g/t, Sc mainly hosts in Al-bearing goethite and kaolinite by adsorption (Chasse et al., 2017)	High pressure acid leaching	260 kg/t H_2SO_4, 255 ℃, 6 h	Sc: 85%-90%	(Kaya and Topkaya, 2016)
		Carbothermic reduction-NaOH cracking	Reduction: lignite (mass fraction of 0.1-0.5), 1400-1600 ℃, 1.5 h, Cracking: 5 mL/g NaOH, 4 mL/g, 400 ℃, 90 min	Sc: 88.4%	(Kim and Azimi, 2020)
Wolframite residue	Sc: 0.04%-0.06%, Sc mainly presents in the form of hydroxide in the residue	Acid leaching	concentrated HCl, 100 ℃	Sc: 95.3%	(Wang and Cheng, 2011)
			18 mol/L H_2SO_4, 100-140 ℃, 6 h	Sc: 100%	(Wang and Cheng, 2011)
Rare earth ores	Sc: 20-50 ppm, Sc hosts in bastnasite	Sulphurizing roasting-water leaching	Roasting: H_2SO_4, 250-300 ℃	—	(Wang and Cheng, 2011)
Ilmenite slag	Sc: 132 ppm, Sc is concentrated principally in the residue as $ScCl_3$	Alkaline roasting-acid leaching	Roasting: Na_2CO_3, 900-1000 ℃, Leaching: 30% HCl, 2 mL/g, 80 ℃	Sc: 80%-90%	(Wang and Cheng, 2011)

Table 8 Summary of extraction of scandium from various aqueous solutions by solvent extraction

Solution composition	Extraction conditions	Extraction efficiency	Stripping conditions	Stripping efficiency	Ref(s).
Hydrochloric acid system (Sc: 8 mg/L, Na: 7.0 g/L, Fe: 7.0 g/L, Ca: 12.4 g/L, Al: 14.8 g/L, Ti: 2.8 g/L)	6.25 g/L activated carbon modified by TBP, 308K, 40 min	Sc: >90%, Fe: <10%	—	—	(Hualei et al., 2008)
Waste liquor from TiO production (Sc: 5.49×10^{-4} mol/L, Zr: 3.07×10^{-4} mol/L, Ti: 4.89×10^{-4} mol/L, Fe: 6.00×10^{-4} mol/L, Lu: 2.66×10^{-4} mol/L)	5%-7%v/v Cyanex 923, 2.0 mol/L H_2SO_4, O/A=1/5, 25 ℃, 30 min	Sc: 96%, Fe: 10%	4% ammonium oxalate solution, O/A=2/1	Sc: 75%	(Wang, 1998)

Continue the Table 8

Solution composition	Extraction conditions	Extraction efficiency	Stripping conditions	Stripping efficiency	Ref(s).
Spent sulfuric acid solution in titanium dioxide production (acidity: 230 g/L H_2SO_4, Sc: 23 mg/L, Fe: 28.36 g/L, Ti: 2.40 g/L, Mn: 2.40 g/L, Al: 1.03 g/L)	15%P204+N1923 (v/v =1), O/A=1/10, 15 min, 3 stages	Sc: 90%, Fe: 1.11%	6 mol/L HNO_3, O/A=1, 30 min	Sc: 89.31%	(Zou et al., 2020)
Leachate of wolframite residue (Mn: 5.6 g/L, Fe: 3 g/L, Sc: 14 mg/L)	HTTA in toluene, pH =1.8-2.0	Sc: ~95%	3 mol/L HCl	Sc: 84%	(Wang and Cheng, 2011)
Leaching solutions of tungsten residue (acidity: 1.38 mol/L H^+, Sc: 9.90 mg/L, Fe: 13.10 g/L, Mn: 9.53 g/L, Ca: 5.59 g/L)	0.28 mol/L Cyanex 572, O/A=1/4	Sc: 93%, Fe: ~1%	3 mol/L HCl, O/A=1, 12 stages	Sc: 90%	(Nie et al., 2017)
Sulfuric acid leaching of red mud (Sc: 3.57 mg/L, Fe: 497 mg/L)	5%Cextrant 230, O/A=1/5	Sc: 90%, Fe: 20%	Scrubbing: 0.2 mol/L H_2SO_4+3% (v/v) H_2O_2, Stripping: 0.010 mol/L EDTA, pH=4-5	Scrubbing: Sc: 14%, Fe: >95% Stripping: Sc: ~90%	(Le et al., 2018)
Simulated red mud leaching liquor(Sc: 0.142 g/L, Fe: 1.724 g/L)	0.157 mol/L P507 + 15% isooctanol, [H]+ = 1 mol/L, O/A = 1:1, T= 25 ℃, t = 30 min	Sc: 95.3%, Fe: 1.5%	H_2SO_4	—	(Liu et al., 2019)
MHP leaching solution (Sc: 46 mg/L, Fe: 2.36 g/L, Ni: 5.60 g/L, Co: 294 mg/L)	8% Primene JMT	Sc: ~93%, Fe: ~4%	—	—	(Ferizoglu et al., 2018)
	6% Cyanex 923	Sc: ~30%, Fe: ~1.5%	—	—	
	7% Ionquest 290	Sc: ~27%, Fe: ~2.5%	—	—	
	6% Cyanex 272	Sc: ~20%, Fe: ~3.5%	—	—	
	6% DEHPA	Sc: ~100%, Fe: ~1.5%	—	—	

Continue the Table 8

Solution composition	Extraction conditions	Extraction efficiency	Stripping conditions	Stripping efficiency	Ref(s).
Sulfuric acid liquor of laterite nickel ores (Fe: 0.253 mol/L, Sc: 0.003 mol/L, Ni: 0.01 mol/L, Co: 0.004 mol/L)	Cyanex272 (molar fraction: 0.6) + Cyanex923, 0.1 mol/L H_2SO_4	Sc: ~95%	10% $H_2C_2O_4$	Sc: 98.79%	(Hu et al., 2020)

4 Conclusions

Atmospheric acid leaching behaviour of Sc in two different types of laterite nickel ore from Indonesia was investigated. The mineralogical analysis showed that the major minerals in limonitic laterite were goethite, magnetite, hematite, and saprolitic laterite mainly consisted of goethite, magnetite, lizardite, clinochlore. Sc in two different types of laterite nickel ore are distributed widely among minerals, but it mainly hosts in Al-bearing goethite and silicate minerals. Sc host minerals in limonitic and saprolitic laterite nickel ore are different. 84.27% and 59.86% of Sc in limonitic and saprolitic laterite could be leached under the experimental conditions of 3 mol/L H_2SO_4, 80 ℃ reaction temperature, leaching duration 3 h and liquid to solid ratio 6 : 1, respectively. Sc in two different types of laterite nickel ore exhibit significantly different dissolution behaviour during sulphuric acid leaching at atmospheric pressure. The results show that Sc and Mn, Mg in limonitic laterite have similar dissolution characteristics because the extractions of Sc and Mn (slope = 1.02), Mg (slope = 1.09) are linearly correlated. Sc in limonitic laterite is susceptible to acid attack and easier to be extracted than other metals except for Mn. Sc and Ni in saprolitic laterite have similar dissolution characteristics because Sc is not strongly related to metals other than Ni (slope = 0.99). Sc in saprolitic laterite is more difficult to extract than Mg and Ni, but it is easier to be leached than other metals. The dissolution kinetics was found to fit well to the shrinking core model with the diffusion through the product layer as the rate controlling step. Results of this research may assist in the development of a more efficient process for exacting Sc from laterite nickel ores.

Acknowledgments

The authors gratefully acknowledge the financial support from National Key R&D Program of China (No. 2019YFC1907402), National Natural Science Foundation of China (No. 51922108 and No. 52074363), Hunan Natural Science Foundation (No. 2019JJ20031) and Hunan Key Research and Development Program (NO. 2019SK2061).

References

[1] A A Sh, B As B. Distribution of scandium in red mud and extraction using Gluconobacter oxydans[J]. Hydrometallurgy, 2021, 202(5): 105621. https://doi.org/10.1016/j.hydromet.2021.105621.

[2] A C R B, A J M, B B B, et al. Selective recovery of rare earths from bauxite residue by combination of sulfation, roasting and leaching[J]. Minerals Engineering, 2016, 92: 151-159. https://doi.org/10.1016/j.mineng.2016.03.002.

[3] A G A, B B Y A, C L G, et al. Selective silica gel free scandium extraction from iron-depleted red mud slags by dry digestion[J]. Hydrometallurgy, 2019, 185: 266-272. https://doi.org/10.1016/j.hydromet.2019.03.008.

[4] ALKAN G, YAGMURLU B, CAKMAKOGLU S, et al. Novel approach for enhanced scandium and titanium leaching efficiency from bauxite residue with suppressed silica gel formation[J]. Sci Rep, 2018, 8(1): 5676. https://doi.org/10.1038/s41598-018-24077-9.

[5] BORRA C R, BLANPAIN B, PONTIKES Y, et al. Recovery of rare earths and other valuable metals from bauxite residue (red mud): A review[J]. Journal of Sustainable Metallurgy, 2016, 2(4): 365-386. https://doi.org/10.1007/s40831-016-0068-2.

[6] CHAKHMOURADIAN A R, SMITH M P, KYNICKY J. From 'strategic' tungsten to 'green' neodymium: A century of critical metals at a glance[J]. Ore Geology Reviews, 2015, 64: 455-458. https://doi.org/10.1016/j.oregeorev.2014.06.008.

[7] CHASSE M, GRIFFIN W L, O'REILLY S Y., et al. Scandium speciation in a world-class lateritic deposit [J]. Geochemical Perspectives Letters, 2017, 3(2): 105-113. https://doi.org/10.7185/geochemlet.1711.

[8] DAVRIS P, BALOMENOS E, PANIAS D, et al. Selective leaching of rare earth elements from bauxite residue (red mud), using a functionalized hydrophobic ionic liquid[J]. Hydrometallurgy, 2016, 164: 125-135. https://doi.org/10.1016/j.hydromet.2016.06.012.

[9] DRZ A, HRC A, ZYN A, et al. Extraction of Al and rare earths (Ce, Gd, Sc, Y) from red mud by aerobic and anaerobic bi-stage bioleaching[J]. Chemical Engineering Journal, 2020, 401: 125914. https://doi.org/10.1016/j.cej.2020.125914.

[10] FERIZOGLU E, KAYA S, TOPKAYA Y A. Solvent extraction behaviour of scandium from lateritic nickel-cobalt ores using different organic reagents[J]. Physicochemical Problems of Mineral Processing, 2018, 54(2): 538-545. https://doi.org/10.5277/ppmp1855.

[11] FUJINAGA K, YOSHIMORI M, NAKAJIMA Y, et al. Separation of Sc(III) from ZrO(II) by solvent extraction using oxidized Phoslex DT-8[J]. Hydrometallurgy, 2013, 133: 33-36. https:////10.1016/j.hydromet.2012.11.014.

[12] GARCES-GRANDA A, LAPIDUS G T, RESTREPO-BAENA O J. The effect of calcination as pre treatment to enhance the nickel extraction from low-grade laterites[J]. Minerals Engineering, 2018, 120: 127-131. https://doi.org/10.1016/j.mineng.2018.02.019.

[13] GENTZMANN M C, SCHRAUT K, VOGEL C, et al. Investigation of Scandium in bauxite residues of different origin[J]. Applied Geochemistry, 2021(3): 104898. https://doi.org/10.1016/j.apgeochem.2021.104898.

[14] GIRGIN I, OBUT A, ÜÇYILDIZ A. Dissolution behaviour of a Turkish lateritic nickel ore[J]. Minerals Engineering, 2011, 24(7): 603-609. https://doi.org/10.1016/j.mineng.2010.10.009.

[15] GUO Q, QU J, HAN B, et al. Innovative technology for processing saprolitic laterite ores by hydrochloric acid atmospheric pressure leaching[J]. Minerals Engineering, 2015, 71: 1-6. https://doi.org/10.1016/j.mineng.2014.08.010.

[16] GUO X, ZHANG C, TIAN Q, et al. Liquid metals dealloying as a general approach for the selective extraction of metals and the fabrication of nanoporous metals: A review[J]. Materials Today Communications, 2021, 26(4): 102007. https://doi.org/10.1016/j.mtcomm.2020.102007.

[17] HU J S, ZOU D, CHEN J, et al. A novel synergistic extraction system for the recovery of scandium(III) by Cyanex272 and Cyanex923 in sulfuric acid medium[J]. Separation and Purification Technology, 2020, 233: 115977. https://doi.org/10.1016/j.seppur.2019.115977.

[18] ZHOU H L, LI D Y, TIAN Y J, et al. Extraction of scandium from red mud by modified activated carbon and kinetics study[J]. Rare Metals, 2008, 27(3): 223-227. https://doi.org/10.1016/S1001-0521(08)60119-9.

[19] JOHN, ANAWATI, GISELE, et al. Recovery of scandium from Canadian bauxite residue utilizing acid baking followed by water leaching[J]. Waste Management, 2019, 95: 549-559. https://doi.org/10.1016/j.wasman.2019.06.044.

[20] KAYA S, DITTRICH C, STOPIC S, et al. Concentration and separation of scandium from Ni laterite ore processing streams[J]. Metals, 2017, 7(12): 1-7. https://doi.org/10.3390/met7120557.

[21] KERKOVE M A, WOOD T D, SANDERS P G, et al. The diffusion coefficient of scandium in dilute aluminum-scandium alloys[J]. Metallurgical and Materials Transactions A-Physical Metallurgy and Materials Science, 2014, 45A(9): 3800-3805. https://doi.org/10.1007/s11661-014-2275-4.

[22] KIM J, AZIMI G. An innovative process for extracting scandium from nickeliferous laterite ore: Carbothermic reduction followed by NaOH cracking[J]. Hydrometallurgy, 2020, 191: 1-11. https://doi.org/10.1016/j.hydromet.2019.105194.

[23] KURSUNOGLU S, ICHLAS Z T, KAYA M. Dissolution of lateritic nickel ore using ascorbic acid as synergistic reagent in sulphuric acid solution[J]. Transactions of Nonferrous Metals Society of China, 2018, 28(8): 1652-1659. https://doi.org/10.1016/S1003-6326(18)64808-3.

[24] KURSUNOGLU S, KAYA M. Atmospheric pressure acid leaching of Caldag lateritic nickel ore[J]. International Journal of Mineral Processing, 2016, 150: 1-8. https://doi.org/10.1016/j.minpro.2016.03.001.

[25] LE W, KUANG S, ZHANG Z, et al. Selective extraction and recovery of scandium from sulfate medium by Cextrant 230[J]. Hydrometallurgy, 2018, 178: 54-59. https://doi.org/10.1016/j.hydromet.2018.04.005.

[26] LI G, YE Q, DENG B, et al. Extraction of scandium from scandium-rich material derived from bauxite ore residues[J]. Hydrometallurgy, 2018, 176: 62-68. https://doi.org/10.1016/j.hydromet.2018.01.007.

[27] LIU C, CHEN L, CHEN J, et al. Application of P507 and isooctanol extraction system in recovery of scandium from simulated red mud leach solution[J]. Journal of Rare Earths, 2019, 37(9): 1002-1008. https://doi.org/10.1016/j.jre.2018.12.004.

[28] LIU K, CHEN Q, HU H. Comparative leaching of minerals by sulphuric acid in a Chinese ferruginous nickel laterite ore[J]. Hydrometallurgy, 2009, 98(3-4): 281-286. https://doi.org/10.1016/j.hydromet.2009.05.015.

[29] LIU Z B, LI H X. Metallurgical process for valuable elements recovery from red mud: A review[J]. Hydrometallurgy, 2015, 155: 29-43. https://doi.org/10.1016/j.hydromet.2015.03.018.

[30] LUO J, LI G, RAO M, et al. Atmospheric leaching characteristics of nickel and iron in limonitic laterite with sulfuric acid in the presence of sodium sulfite[J]. Minerals Engineering, 2015, 78: 38-44. https://doi.org/10.1016/j.mineng.2015.03.030.

[31] LUO J, RAO M, LI G, et al. Self-driven and efficient leaching of limonitic laterite with phosphoric acid [J]. Minerals Engineering, 2021, 169: 106979. https://doi.org/10.1016/j.mineng.2021.106979.

[32] LUO W, FENG Q, OU L, et al. Kinetics of saprolitic laterite leaching by sulphuric acid at atmospheric pressure[J]. Minerals Engineering, 2010, 23(6): 458-462. https://doi.org/10.1016/j.mineng.2009.10.006.

[33] LUO W, FENG Q, OU L, et al. Fast dissolution of nickel from a lizardite-rich saprolitic laterite by sulphuric acid at atmospheric pressure[J]. Hydrometallurgy, 2009, 96(1-2): 171-175. https://doi.org/10.1016/j.hydromet.2008.08.001.

[34] MAKUZA B, TIAN Q, GUO X, et al. Pyrometallurgical options for recycling spent lithium-ion batteries: A comprehensive review[J]. Journal of Power Sources, 2021, 491. https://doi.org/10.1016/j.jpowsour.2021.229622.

[35] MATTHEW, LANDERS, ROBERT J, et al. Dehydroxylation and dissolution of nickeliferous goethite in New Caledonian lateritic Ni ore[J]. ScienceDirect. Applied Clay Science, 2007, 35(3-4): 162-172. https://doi.org/10.1016/j.clay.2006.08.012.

[36] MCDONALD R G, WHITTINGTON B I. Atmospheric acid leaching of nickel laterites review Part I. Sulphuric acid technologies[J]. Hydrometallurgy, 2008, 91(1-4): 35-55. https://doi.org/10.1016/j.hydromet.2007.11.009.

[37] MESHRAM P, ABHILASH, PANDEY B D. Advanced review on extraction of nickel from primary and secondary sources[J]. Mineral Processing and Extractive Metallurgy Review, 2019, 40(3): 157-193. https://doi.org/10.1080/08827508.2018.1514300.

[38] NIE H, WANG Y, WANG Y, et al. Recovery of scandium from leaching solutions of tungsten residue using solvent extraction with Cyanex 572[J]. Hydrometallurgy, 2017, 175: 117-123. https://doi.org/10.1016/j.hydromet.2017.10.026.

[39] OCHSENKÜHN-PETROPULU M, LYBEROPULU T, OCHSENKÜHN K M, et al. Recovery of lanthanides and yttrium from red mud by selective leaching[J]. Analytica Chimica Acta, 1996, 319(1-2): 249-254. https://doi.org/10.1016/0003-2670(95)00486-6.

[40] OCHSENKUHNPETROPULU M, LYBEROPULU T, PARISSAKIS G. Selective separation and determination of scandium from yttrium and lanthanides in red mud by a combined ion exchange/solvent extraction method[J]. Analytica Chimica Acta, 1995, 315: 231-237. https://doi.org/10.1016/0003-2670(95)00309-N.

[41] ONAL M A R, TOPKAYA Y A. Pressure acid leaching of Caldag lateritic nickel ore: An alternative to heap leaching[J]. Hydrometallurgy, 2014, 142: 98-107. https://doi.org/10.1016/j.hydromet.2013.11.011.

[42] PANDA L, RAO D S, MISHRA B K, et al. Characterization and dissolution of low-grade ferruginous nickel lateritic ore by sulfuric acid[J]. Minerals & Metallurgical Processing, 2014, 31(1): 57-65. https://doi.org/10.1007/BF03402349.

[43] PRASAD P, PRASAD K S, CHAITANYA V K, et al. In situ FTIR study on the dehydration of natural

goethite[J]. Journal of Asian Earth Sciences, 2006, 27(4): 503-511. https://doi.org/10.1016/j.jseaes.2005.05.005.

[44] QING Y, SHUNYAN N, WEI Z, et al. Recovery of scandium from sulfuric acid solution with a macro porous TRPO/SiO$_2$-P adsorbent[J]. Hydrometallurgy, 2018, 181: 74-81. https://doi.org/10.1016/j.hydromet.2018.07.025.

[45] RAMASAMY D L, PUHAKKA V, REPO E, et al. Selective separation of scandium from iron, aluminium and gold rich wastewater using various amino and non-amino functionalized silica gels-A comparative study [J]. Journal of Cleaner Production, 2018, 170: 890-901. https://doi.org/10.1016/j.jclepro.2017.09.199.

[46] REID S, TAM J, YANG M, et al. Technospheric mining of rare earth elements from bauxite residue (red mud): Process optimization, kinetic investigation, and microwave pretreatment[J]. Sci Rep, 2017, 7(1): 15252. https://doi.org/10.1038/s41598-017-15457-8.

[47] SENANAYAKE G, CHILDS J, AKERSTROM B D, et al. Reductive acid leaching of laterite and metal oxides: A review with new data for Fe(Ni, Co)OOH and a limonitic ore[J]. Hydrometallurgy, 2011, 110 (1-4): 13-32. https://doi.org/10.1016/j.hydromet.2011.07.011.

[48] SHAOQUAN X, SUQING L. Review of the extractive metallurgy of scandium in China (1978−1991)[J]. Hydrometallurgy, 1996, 42(3): 337-343. https://doi.org/10.1016/0304-386X(95)00086-V.

[49] VAN DER ENT A, BAKER A J M, VAN BALGOOY M M J, et al. Ultramafic nickel laterites in Indonesia (Sulawesi, Halmahera): Mining, nickel hyperaccumulators and opportunities for phytomining [J]. Journal of Geochemical Exploration, 2013, 128: 72-79. https://doi.org/10.1016/j.gexplo.2013.01.009.

[50] WANG L C. Solvent extraction of Scandium(Ⅲ) by Cyanex 923 and Cyanex 925[J]. Hydrometallurgy, 1998, 48(3): 301-312. https://doi.org/10.1016/S0304-386X(97)00080-7.

[51] WANG W, CHENG C Y. Separation and purification of scandium by solvent extraction and related technologies: A review[J]. Journal of Chemical Technology and Biotechnology, 2011, 86(10): 1237-1246. https://doi.org/10.1002/jctb.2655.

[52] WANG W, PRANOLO Y, CHENG C Y. Metallurgical processes for scandium recovery from various resources: A review [J]. Hydrometallurgy, 2011, 108(1-2): 100-108. https://doi.org/10.1016/j.hydromet.2011.03.001.

[53] YAN K, LIU L, ZHAO H, et al. Study on extraction separation of thioarsenite acid in alkaline solution by CO_3^{2-}-type tri-n-octylmethyl-ammonium chloride[J]. Frontiers in Chemistry, 2021, 8: 1-13. https://doi.org/10.3389/fchem.2020.592837.

[54] YIN Y M, XIONG M W, YANG N T, et al. Investigation on thermal, electrical, and electrochemical properties of scandium-doped $Pr_{0.6}Sr_{0.4}(Co_{0.2}Fe_{0.8})(1-x)Sc_xO_3$-delta as cathode for IT-SOFC[J]. International Journal of Hydrogen Energy, 2011, 36(6): 3989-3996. https://doi.org/10.1016/j.ijhydene.2010.12.113.

[55] ZHANG X K, ZHOU K G, CHEN W, et al. Recovery of iron and rare earth elements from red mud through an acid leaching-stepwise extraction approach[J]. Journal of Central South University, 2019, 26(2): 458-466. https://doi.org/10.1007/s11771-019-4018-6.

[56] ZHOU K G, TENG C Y, ZHANG X K, et al. Enhanced selective leaching of scandium from red mud[J]. Hydrometallurgy, 2018, 182: 57-63. https://doi.org/10.1016/j.hydromet.2018.10.011.

[57] ZOU D, LI H, CHEN J, et al. Recovery of scandium from spent sulfuric acid solution in titanium dioxide production using synergistic solvent extraction with D2EHPA and primary amine N1923 [J]. Hydrometallurgy, 2020, 197: 105463. https://doi.org/10.1016/j.hydromet.2020.105463.

Leaching Behavior of Scandium from Limonitic Laterite Ores Under Sulfation Roasting-water Leaching

Abstract: In the present study, the leaching behavior of scandium from limonitic laterites under sulfation roasting-water leaching (SAL) was explored. The mineralogical analysis of limonitic laterites showed that scandium was associated with the iron phase. The roasting temperature played an important role in the iron phase conversion during the sulfation roasting process. With an increase in the roasting temperature from 100 to 800 ℃, the iron phase (goethite, magnetite, and hematite) gradually converted to monoclinic $Fe_2(SO_4)_3$ (400 ℃), then to rhombohedral $Fe_2(SO_4)_3$ (600 ℃), and finally to hematite (800 ℃). Due to the iron phase conversion mechanism, the leaching efficiency of scandium gradually increased first and then decreased with the increase of roasting temperature from 100 to 800 ℃. When limonitic laterites roasted at temperature of 600 ℃ for 2 h with sulfuric acid/laterite ratio of 1/4 (mL/g), a total of 81.15% of scandium and only 0.37% of iron were extracted into leaching solution at 30 ℃ with the liquid-to-solid ratio of 4∶1 (mL/g) for 1 h with agitation. The SAL process enables the selective and efficient enrichment of scandium from low-grade limonitic laterites with lower costs for equipment and operation compared to high pressure acid leaching (HPAL), lower acid consumption, and lower dissolution of iron compared to atmospheric pressure acid leaching (AL).

Published in *Journal of Sustainable Metallurgy*, 2022, 8(3): 1078-1089. Authors: Dong Bo, Tian Qinghua, Guo Xueyi, Wang Qingao, Xu Zhipeng, Li Dong.

Leaching Behavior of Scandium from Limonitic Laterite Ores Under Sulfation Roasting-water Leaching

Graphical Abstract

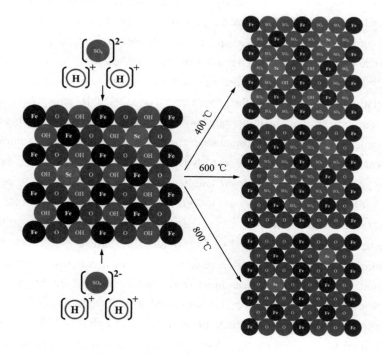

Keywords: Scandium; Limonitic laterite; Sulfation roasting-water leaching; Leaching characteristic; Roasting temperature

1 Introduction

Laterite nickel ore is a promising scandium resource with economically interesting concentrations and large amounts[1-6]. It can be simply divided into two types, limonitic and saprolitic laterite[7, 8]. Typical chemical composition and metallurgical process of two different types of laterite nickel ore as shown in Table 1[1-6, 9, 10]. The limonitic laterite is rich in iron in the form of goethite, while the content of valuable metals nickel, cobalt and scandium is relatively low[11, 12]. Goethite is the main Sc-bearing mineral in limonitic laterite[13, 14], and scandium is mainly presented in the following two states: (a) weakly adsorbed to the crystalline goethite surface, and (b) as lattice substituents[15, 16]. These results indicate that the extraction of scandium is highly dependent upon the degree of the decomposition of the Sc-bearing minerals[17]. Scandium is usually extracted by the high-pressure acid leaching (HPAL) process, followed by the concentration and separation of scandium from the pregnant leach solution by using the pH-controlled precipitation[18] or solvent extraction[19]. The HPAL process provides high recoveries of scandium, allows acceptable acid consumption, and produces low residual iron in solution, but requires expensive autoclaves and has high maintenance costs and more neutralization of acid[20, 21]. Processes with a lower-cost alternative to the HPAL process especially for lower grade ores have been widely studied due to growing interest from the industry. Among them, hydro-pyro integration in the processing of laterites could potentially be the way to process laterite with the complexion of their own merits[21-23]. The sulfation roasting-leaching route is shown to be quite promising[24]. Although corrosive gases are released during the above process, and the roasting temperature is relatively high, the whole procedure is conducted in atmosphere air, and so the requirement for equipment and facilities are not high. Conventional equipment with cheaper materials of construction and lower maintenance costs are acceptable. For example, rotary kilns can be used for roasting and non-pressure vessels in water leaching[25]. SAL has been investigated for the extraction of nickel and cobalt from laterites in which iron rejection is performed by the thermal decomposition of ferric sulfate at ~ 700 ℃, while nickel and cobalt are selectively converted into water leachable sulfates[17, 20, 21]. A summary of previous studies on the extraction of nickel and cobalt from laterites utilizing SAL is provided in Table 2(a). However, the leaching behavior of scandium in the SAL process is seldom studied in literatures. Anawati et al.[26] have previously explored the recovery of scandium from laterite nickel ore by acid roasting-water leaching. Nevertheless, the acid ore-water mixtures need to be heated at 50 ℃ for 24 h to ensure sulfation. Moreover, the leaching efficiency of iron is very high (~ 15%), and the selective extraction of scandium is not good enough. Most of the research in the selective sulfation roasting-water leaching field focused on the extraction of scandium from bauxite residue[27-31]. A summary of previous studies on the extraction of valuable metals from bauxite residue utilizing acid roasting water leaching is provided in Table 2(b). However, the properties of bauxite residue and laterites are quite different[32-35]. Bauxite residue is the hydrometallurgical slag produced in the production process of the Bayer process, which is less difficult to deal with than the primary mineral resources[36-44]. Scandium is usually hosted in bauxite residue in the form of surface adsorption and is easy to extract[36, 45], while scandium in laterites usually exists in refractory minerals such as goethite and silicate in the form of isomorphism[1, 13]. As the demand for Al-Sc alloys and solid fuel cells increases year by year, there is an urgent need for a scandium supply chain with sufficient resources, stable processes, and reasonable prices.

Table 1 Typical composition and metallurgical process of two different types of laterite nickel ore

Different types	Chemical composition (mass fraction)/%						Metallurgical process
	Ni	Co	Fe	Cr	MgO	Sc	
Limonitic laterite	0.8-1.5	0.1-0.2	40-50	0.8-3.0	0.5-5	0.005~0.06	Hydrometallurgy
Saprolitic laterite	1.8-3.0	0.02-0.1	10-25	0.8-1.6	15-35	0.005~0.06	Pyrometallurgy

Table 2 An overview of previous studies on acid roasting-water leaching of valuable metals from (a) laterites, and (b) bauxite residue

Reference	Source	Roasting conditions	Leaching conditions	Extraction
(a) Laterites				
[21]	China	6 g $(NH_4)_2SO_4$/8 g laterites, 400 ℃, 1.5 h	100 ℃, 1 h	Ni: 90.8%, Co: 85.4%, Mn: 86.7%, Fe: 9.98%
[25]	Papua New Guinea	0.45 g H_2SO_4/g laterites, 650 ℃, 4 h	80 ℃, L/S=7, 4 h	Ni: 100%, Fe: 2%
[20]	Philippines	40% H_2SO_4, 4% Na_2SO_4, 700 ℃, 60 min	80 ℃, 30 min	Ni: 88%, Co: 93%, Fe: <4%
[46]	India	25% H_2SO_4, 700 ℃, 15 min	90-95 ℃, L/S=10, 30 min	Ni: 85%, Fe: 2.5%
[26]	New Caledonia	Sulfation: 0.5-2 g H_2SO_4/g LO, 0-1 g H_2O/g LO, 50 ℃, 24 h Roasted: 600-800 ℃, 30-90 min	4-16 mL H_2O/g ARLO, 300-500 r/min	Sc: 80%, Fe: 15%
(b) Bauxite residue				
[47]	Canada	0.95 mL H_2SO_4/g BR, 400 ℃, 2 h	25 ℃, 9.5 mL H_2O/g ABBR, 2 h, 300 r/min	Sc: 80%
[30]	Jamaica	1.47 mL H_2SO_4/g BR, 700 ℃, 1 h	65 ℃, sonication: 5 h, ball-mill: 30 min	Sc: 89%, REE: 88%
[29]	China	1 mL H_2SO_4/g BR, 750 ℃, 40 min	65 ℃, 30 min	Sc: 60%, Na: >95%
[44]	Greece	1.84 mL H_2SO_4/g BR, 700 ℃, 1 h	25 ℃, non-agitated: 7 days, agitated: 2 days	Sc: 60%, REE: >80%, Fe and Ti<1%
[33]	Greece	1.84 mL H_2SO_4/g BR, 700 ℃, 2 h	25 ℃, 24 h	Sc: 60%, Fe and Ti: <1%

In this work, the feasibility of selective extracting scandium from limonitic laterite by traditional sulfation roasting-water leaching was evaluated. The effects of various parameters on the leaching behavior of metals in the SAL process have been systematically investigated. Although the high-temperature transformation mechanism and water leaching behavior of nickel and cobalt have been mentioned in this work, the extraction of these metals from limonitic laterite by SAL has been reported in the literature, so it is not the focus of this study. This work focuses on the migration and transformation mechanism of scandium in this process and optimizes the process parameters for efficient and selective recovery of scandium, to meet the growing demand for this rare and expensive metal in the world.

2　Experimental

2.1　Materials

Limonitic laterite used in the experiments was obtained from Sulawesi, Indonesia. The results for particle size, XRF, ICP, SEM-EDX, FTIR, XPS, and XRD analysis of the limonitic laterite material have already been reported elsewhere[13]. The composition of the raw material is listed in Table 3. Thermal gravimetric analysis and differential scanning calorimetry (TG-DSC, STA 449 C, Netzsch, German) were conducted on the samples from 30 to 800 ℃ with a linear heating rate of 10 ℃/min using an air atmosphere and a flow rate of 20 mL/min. The concentrated sulfuric acid (98%, Sinopharm) used in this study were all the reagent grades, and all aqueous solutions were prepared by using deionized water.

Table 3　The mass fraction of the raw material　　%

Ni	Co	Mn	Sc	Fe	Mg	Al	Cr
0.69	0.058	0.51	0.0054	45.30	0.37	3.64	1.16

2.2　Methods

Sulfation roasting of the limonitic laterite was carried out in a tubular atmosphere furnace (Brand: Hefei Kejing Material Technology Co., Ltd., Model: OTF-1200X). As an experimental procedure, 10.00 g ore was intensively mixed with 98% H_2SO_4 with a predetermined acid to ore ratio in a porcelain crucible, and then roasted in the furnace for a preset time at a desired temperature. The gas produced during the sulfation roasting process was absorbed with 2 mol/L NaOH solution. The acid-roasted samples were cooled and then leached with deionized water using a glass reaction vessel equipped with a magnetic stirrer. Subsequently, liquid/solid separation was performed with a Büchner funnel. The pregnant leach solution produced by each experiment was analyzed using an atomic emission spectrometer (ICP-AES, IRIS Intrepid Ⅱ XSP, Thermo Electron Corporation, USA). The leaching efficiency of the metal was calculated according to Eq. (1).

$$\eta_i = \frac{C_i \times V}{m \times w_i} \times 100\% \tag{1}$$

where η_i is the leaching efficiency of metal i(%), C_i is the concentration of metal i in the leaching solution (g/L), V is the volume of leaching solution (L), m is the mass of raw material (g), and w_i is the mass fraction of metal i in raw material (%).

3 Results and discussion

3.1 Effect of roasting temperature on leaching

The TG-DSC curves of the limonitic laterite (Fig. 1) show that there are two obvious endothermic peaks at 72.4 ℃ and 329.6 ℃. The endothermic peak was at 72.4 ℃ with a weight loss of 0.85% due to free water removal. Another major endothermic peak at 329.6 ℃ was due to the dehydroxylation of goethite (FeO(OH)) to hematite (Fe_2O_3)[16, 46]. For different laterites, the dehydroxylation temperature depended on the properties of the ore, such as crystallinity and the degree of cation substitution in the ore. The dehydroxylation temperature of goethite with poor crystallinity is lower than that with good crystallinity[16, 46]. Trivalent ions can substitute directly for iron, and due to differing bonding energies with the hydroxyl group, they can inhibit the dehydroxylation process[48].

In the sulfation roasting process, the roasting temperature is crucial for evaluating the sulfation effect, leaching efficiency, and energy consumption. The main driving mechanism for this selective process is the differences in thermal decomposition temperatures of different water-soluble metal sulfates[26]. The onset of $Fe_2(SO_4)_3$ decomposition is reported to be 545 ℃, and the onset of $Sc_2(SO_4)_3$ decomposition is 700 ℃[26, 44]. Some

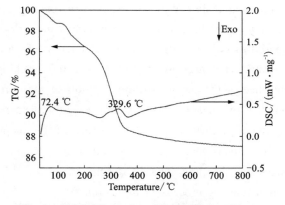

Fig. 1 TG-DSC results of limonitic laterite ore

reports indicate that the initial decomposition temperatures of pure nickel, and cobalt sulfates are in the ranges of 640 ℃ to 676 ℃, and 644 ℃ to 690 ℃, respectively, depending upon the composition of the atmosphere[20, 49, 50]. In this study, HSC Chemistry software (version 6.0) was utilized to establish the diagram of Gibbs free energy for the chemical reactions are relevant to sulfation roasting (Fig. 2). Firstly, the ore is mixed with concentrated H_2SO_4 to convert all the iron and scandium species to the corresponding sulfate forms (Eq. 2, 3, 4, and 5). Finally, the sulfated solids are roasted at 600 ~ 800 ℃ to convert the soluble contaminant sulfates ($Fe_2(SO_4)_3$), which have lower thermal stability to water-insoluble oxide forms, while driving off the sulfate groups as SO_3 vapors (Eq. 6). However, when the temperature is higher than 338 ℃, the concentrated sulfuric acid will decompose into sulfur trioxide (Eq. 7), which will inhibit the conversion of metal to the corresponding sulfate.

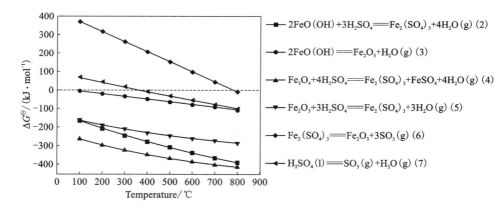

Fig. 2 The chemical reactions are relevant to sulfation roasting and thermodynamic data

Experiments to investigate the sulfation roasting temperature effect were carried out in the temperature range of 100~800 ℃. Samples mixed with 98% H_2SO_4 with an acid to ore ratio of 1 ∶ 1 (mL/g) were roasted for 1 h and then water leached at 80 ℃ and liquid-to-solid ratio of 2 ∶ 1 for 1 h. The results, shown in Fig. 3, show that the roasting temperature has a significant effect on metal conversion. With an increase in roasting temperature from 100 to 400 ℃, the leaching efficiency of nickel, cobalt, scandium, and iron gradually increased. When the temperature increased to 400 ℃, the leaching efficiency of valuable metals reached a maximum. When the temperature increases to 400~800 ℃, the leaching of iron dramatically decreases due to the occurrence of the decomposition reaction of the corresponding sulfates. Meanwhile, the leaching efficiency of nickel, cobalt, and scandium decreased slightly and then decreased greatly. Consequently, further tests were carried out by keeping the roasting temperature fixed at 600 ℃.

To investigate the mineralogical changes of the iron phases during the sulfation roasting process, samples roasted and water leaching residue at 400, 600, and 800 ℃ were characterized with XRD (Fig. 4 and Fig. 5) and SEM-EDX (Fig. 6 and Table 4). It was observed from the XRD results (Fig. 4) that the major iron phase in the laterite was goethite, magnetite, and hematite, indicating the iron phase was gradually digested and replaced with sulfate phases at 400 and 600 ℃. However, the crystal structures of the obtained sulfate phases are different. It is monoclinic $Fe_2(SO_4)_3$

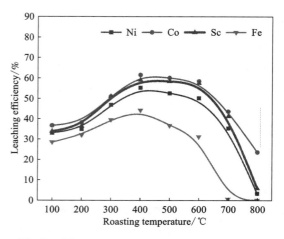

Fig. 3 Effect of sulfation roasting temperature on metal extractions

(iron(Ⅲ) sulfate: PDF#75-1767)[51, 52] at 400 ℃, and becomes rhombohedral $Fe_2(SO_4)_3$ (iron sulfate: PDF#42-0229)[53] as the temperature increases to 600 ℃. As the temperature further increased to 800 ℃, the hematite peaks became stronger due to the thermal decomposition of the iron sulfate. In addition, it can also be found from Fig. 4 that there is obvious quartz phase in the roasted sample. As shown in Fig. 5, the major phase of the water leaching residue at different roasting temperatures is hematite and the minor phase is quartz. Because scandium is associated with iron, the leaching of scandium is determined by the phase transformation from iron phases (goethite, magnetite, and hematite) to monoclinic $Fe_2(SO_4)_3$ (400 ℃), then to rhombohedral $Fe_2(SO_4)_3$ (600 ℃), and finally to hematite (800 ℃). Backscattered SEM images of samples roasted and water leaching residue are presented in Fig. 6. EDX analysis was used to identify the composition of different regions in SEM images. The EDX results associated with the 12 main regions marked in Fig. 6 are presented in Table 4. The results of SEM-EDX analysis corroborate the observed XRD spectra, confirming that the resulting iron phase after sulfating roasting depends on the roasting temperature.

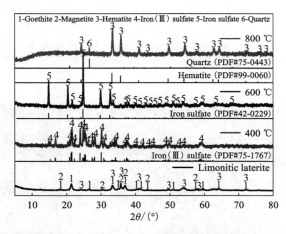

Fig. 4 XRD patterns of samples roasted at different temperatures

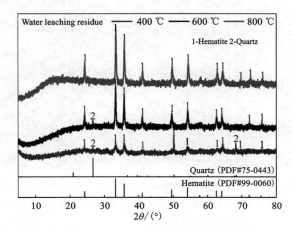

Fig. 5 XRD patterns of water leaching residue at different temperatures

Table 4 Point analysis results from samples roasted in Fig. 6 (a-c) and water leaching residue in Fig. 6 (d-f)

Point No	O	S	Fe	Ni	Co	Sc	Si
a-1	58.80	21.79	18.99	0.22	0.18	0.02	—
a-2	47.33	1.37	0.52	0.02	—	0.04	50.71
b-1	47.70	20.22	30.20	1.36	0.49	0.03	—
b-2	30.36	30.86	38.45	0.31	—	0.02	—
c-1	35.58	0.61	61.89	1.36	0.55	0.02	—
c-2	48.30	1.35	49.38	0.50	0.47	—	—
d-1	39.59	1.05	57.62	1.24	0.42	0.08	—
d-2	35.19	9.15	54.35	0.66	0.57	0.07	—
e-1	37.70	4.05	55.26	2.70	0.26	0.04	—
e-2	46.13	3.95	49.27	0.15	0.43	0.07	—
f-1	28.00	0.24	70.27	1.16	0.25	0.09	—
f-2	37.70	0.16	59.72	1.77	0.55	0.10	—

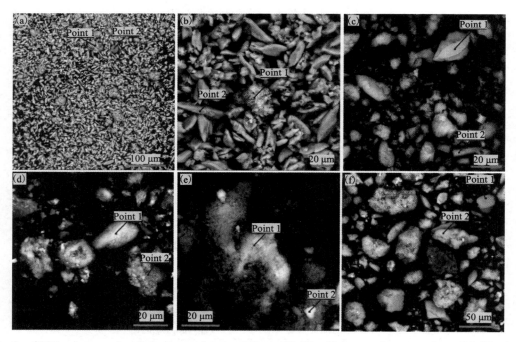

Fig. 6 SEM micrographs of (a) samples roasted at 400 ℃, (b) samples roasted at 600 ℃, (c) samples roasted at 800 ℃, (d) water leaching residue at 400 ℃, (e) water leaching residue at 600 ℃, (f) water leaching residue at 800 ℃

3.2 Effect of sulfuric acid amount on leaching

A series of experiments were carried out by varying the ratio of sulfuric acid to ore from 0.25 (1 : 4 ratio) to 1.25 (5 : 4 ratio). The samples were roasted for 1 h at a fixed temperature of 600 ℃. After roasting, the samples were leached with water at 80 ℃ and a liquid-to-solid ratio of 2 : 1 (mL/g) for 1 h with agitation. Fig. 7 shows the extraction of the scandium and major elements in the leachate with an increasing initial acid to ore ratio. As seen in Fig. 7, the scandium, nickel, and cobalt extraction gradually decreases with increasing acid to ore ratio. However, iron extraction dramatically increases with increasing acid to ore ratio. Probably, the excess acid leads to more iron sulfate formation, which to some extent hinders the subsequent gas-solid sulfation reaction of nickel, cobalt and scandium in the iron phase[54]. Furthermore, the excess acid can decompose into SO_3 with time (Eq. 7), hindering the decomposition of soluble iron sulfate into insoluble compounds (Eq. 6)[25].

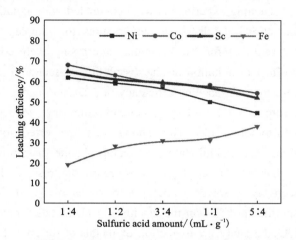

Fig. 7 Effect of sulfuric acid amount on metal extractions

3.3 Effect of roasting duration on leaching

Roasting experiments were conducted for different durations to understand the effect of roasting time on the dissolution of scandium and major elements. In these experiments, samples were prepared with a constant sulfuric acid to ore ratio of 1 : 4 (mL/g). Samples were roasted at 600 ℃ for different durations. Finally, the samples were leached with water at 80 ℃ and a liquid-to-solid ratio of 2 : 1 (mL/g) for 1 h with agitation. Fig. 8 shows the effect of roasting duration on the leaching of scandium and major elements. The scandium extraction slightly decreases with roasting time up to 2 h. A further increase in roasting time has a negative effect on scandium dissolution due to the increasing amount of sulfates that are decomposed to oxides that are insoluble in water. The extraction of iron decreases with an increase in roasting time, again

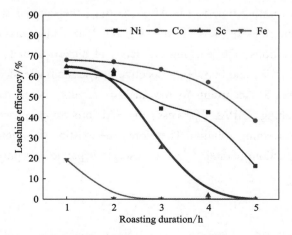

Fig. 8 Effect of roasting duration on metal extractions

because of the increased amount of sulfates that are decomposed. Therefore, roasting is performed at a higher temperature, and the roasting time needed for selective leaching will be shorter. The optimal roasting time is determined considering two main factors: scandium dissolution and iron dissolution. Based on the results, the optimal roasting time at 600 ℃ is 2 h. At this temperature, scandium dissolution is still high, while iron dissolution is low.

3.4 Effect of water immersion parameters on leaching

Leaching experiments were conducted with agitation at different leaching durations, liquid-to-solid ratios, and leaching temperatures for samples roasted at 600 ℃ and acid-to-ore ratios of 1:4 (mL/g) for 2 h. Leaching experiments were conducted at different L/S ratios to determine the effect of agitation on the dissolution of scandium and major elements [Fig. 9(a)]. Fig. 9(a) shows that the leaching efficiency of scandium and cobalt increased gradually with a liquid-to-solid ratio up to 4:1 (mL/g), beyond which they tended to be stable with a further increase in the liquid-to-solid ratio. This shows that the effect of the L/S ratio on nickel dissolution is insignificant. Additionally, no change in the dissolution of iron was observed. Iron dissolution is less than 1% at all L/S ratios. From the results, it can be concluded that scandium can be selectively dissolved even at L/S ratios of 4:1 (mL/g). The leaching temperature has an insignificant effect on the leaching of all metals. Therefore, the optimal leaching temperature was 30 ℃. To study the extraction of metals at different leaching times, a series of experiments were conducted while other conditions were kept constant, including 4:1 (mL/g) L/S ratios and 30 ℃ of leaching temperature. The results are shown in Fig. 9(c). Fig. 9(c) indicates that leaching time has a similar effect on scandium, nickel, and cobalt. When the leaching time is 1 h, the leaching efficiency of scandium, nickel, cobalt, and iron reaches 80.47%, 61.37%, 80.06%, and 0.35%, respectively. After 1 h, the leaching of scandium, nickel, and cobalt remained constant. Therefore, the optimal leaching time was 1 h.

Through the above single factor experimental results, determine the following suitable process conditions: the roasting temperature is 600 ℃, the roasting acid to ore ratio is 1/4 (mL/g), and the roasting time is 2 h. After roasting, the samples were leached with water at 30 ℃ and liquid to solid ratio of 4:1 (mL/g) for 1 h with agitation. As shown in Fig. 9(d), under these conditions, the leaching efficiency of nickel, cobalt, scandium, and iron are 62.53%, 80.83%, 81.15%, and 0.37%, respectively. The main chemical analysis of the leaching solution is given in Table 5. According to the analysis results, compared with the conventional high pressure acid leaching (HPAL) process, the SAL process proposed in this research has a lower iron content in the leaching solution. Therefore, the existing pH-controlled precipitation method[18, 55-57] or solvent extraction method[19] can be used to separate and purify scandium in the leaching solution.

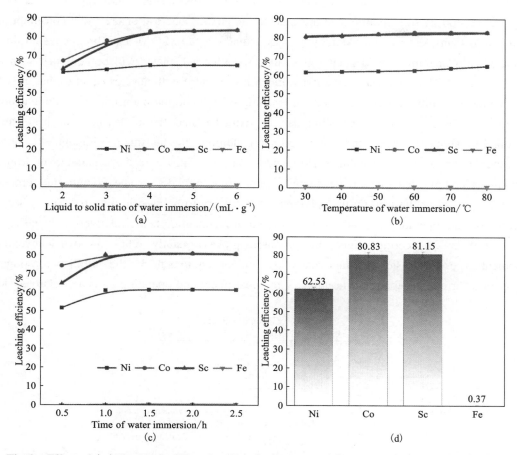

Fig. 9 Effect of (a) liquid to solid ratio, (b) temperature, (c) duration on leaching efficiency, and (d) optimum conditons for leaching

Table 5 Chemical composition analysis results of the leaching solution   mg/L

Element	Ni	Co	Sc	Fe
Concentration	1182.51	129.26	12.03	367.30

3.5 Preliminary economic analysis

Based on the mechanism described above and the optimized experimental results, SAL process is proposed (Fig. 10). Rather than HPAL, SAL process for laterites has not been commercially proven. Therefore, a preliminary economic analysis of the process is shown in Table 6, including the estimated costs of consumed reagents, values of the generated products, equipment and operating costs, energy consumption and gas exhaust treatments. While the calculations are based on the 10.00 g scale experiments, the results are given for 1 ton of laterites. Comparing the SAL process with AL and HPAL process, it can be seen that the proposed process has some advantages. First, the reagents consumption is reasonable. Based on the data of the study, about 460 kg of sulphuric acid are used for sulphating one ton of ore. Acid consumption is much lower than AL process. Moreover, gas exhaust (SO_2/SO_3) can be recovered in the associated sulphur burning acid plant (generally, metallurgical plants will be equipped with acid making system). Thus, the acid depletion is much lower and it will close or equal to that in HPAL under an ideal condition. Second, scandium leaching is high and the main impurity iron leaching is very low. About 81.15% of scandium can be extracted and the co-leaching of iron can be controlled to about 0.37%, under the optimal conditions in the study. Although AL process providing similar or better leaching performances than SAL process, the separation of iron and scandium is complicated due to the chemical similarity. Actually, some complicated processes including removal of the iron by precipitation are necessary to be used in industry. Thus, both the material cost and operating cost are high. Third, the investment costs are relatively low and easy to be achieved in engineering. Since the proposed process mainly relies on two conventional equipment (i. e., corrosion-resistant rotary kiln/mixing equipment for the sulfation roasting, an atmospheric stirred tank for water leaching), the investment and operating costs are relatively low.

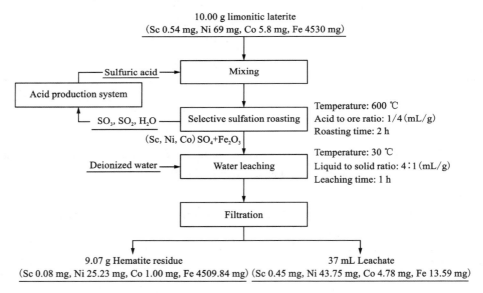

Fig. 10 Flowsheet for the SAL process

Table 6 A preliminary economic analysis of the process was carried out in this study

Item	SAL	AL	HPAL	Price reference standard
Raw material costs	$ 20	$ 86	$ 15	0.0428 $/kg H_2SO_4 [47]
Sc values	$ 312	$ 320	$ 342	4600.00 $/kg Sc_2O_3 [47]

Continue the Table 6

Item	SAL	AL	HPAL	Price reference standard
Equipment costs	$ 4200000	$ 1800000	$ 10000000	60 m^3 rotary kiln: $ 2400000[47] 120 m^3 stirred tank+mixer: $ 1800000[47] 60 m^3 titanium autoclave: $ 10000000
Energy consumption	High	Low	General	—
SO$_2$ disposal costs	$ 203-261	—	—	Levelized costs: $ 203- $ 261[59]

However, the SAL process also has some shortcomings. The energy consumption of this process is higher, mainly including the energy consumption of bituminous coal, electricity, water, and gas, among which electricity consumption accounts for the largest proportion[22, 58]. Moreover, there are still numerous low content SO$_2$ streams needing recovery and disposal, which will also increase the technical and production costs. Similarly, the AL process also produces a large amount of acid-containing wastewater that needs to be disposal. In addition, the physical characteristics of residue obtained by AL are unstable and difficult to handle. From the economic perspective, the SAL process is feasible as long as low content SO$_2$ and waste heat can be reasonably utilized.

4 Conclusions

In the present study, the leaching behavior of scandium from limonitic laterite ores under SAL was explored. The mineralogical analysis of limonitic laterites showed that scandium was associated with the iron phase. The roasting temperature played an important role in the iron phase conversion during the sulfation roasting process. With an increase in the roasting temperature from 100 to 800 ℃, the iron phase (goethite, magnetite, and hematite) gradually converted to monoclinic Fe$_2$(SO$_4$)$_3$ (400 ℃), then to rhombohedral Fe$_2$(SO$_4$)$_3$ (600 ℃), and finally to hematite (800 ℃). The iron phase conversion mechanism could determine the leaching of scandium. An increase in the sulfuric acid amount decreases nickel, cobalt, and scandium dissolution and increases the leaching efficiency of iron. Most likely, the excess acid can decompose into SO$_3$, which will hinder the decomposition of soluble iron sulfate into insoluble compounds and the conversion of scandium, nickel, and cobalt in the iron phase to the corresponding sulfate. At a roasting temperature of 600 ℃ for a roasting duration of 2 h and with a sulfuric acid to ore ratio of 1/4 (mL/g). A total of 81.15% of scandium and only 0.37% of iron could be extracted after leaching with water at 30 ℃ and a liquid-to-solid ratio of 4 : 1 (mL/g) for 1 h with agitation. The existing pH-controlled precipitation method or solvent extraction method can be used to separate and purify scandium in the leaching solution. The SAL process enables the selective and efficient enrichment of scandium from low-grade limonitic laterite with lower costs for equipment and operation compared to HPAL, lower acid consumption, and lower dissolution of iron compared to AL.

Acknowledgments

The authors gratefully acknowledge the financial support from the National Key R&D Program of China (Grant No. 2019YFC1907402 and 2018YFC1902501), National Natural Science Foundation of China (Grant No. 51922108, 52074363, and 52104355).

References

[1] CHASSE M, GRIFFIN W L, O'REILLY S Y, et al. Scandium speciation in a world-class lateritic deposit [J]. Geochemical Perspectives Letters, 2017, 3(2): 105-113. https://doi.org/10.7185/geochemlet.1711

[2] KIM J, AZIMI G. An innovative process for extracting scandium from nickeliferous laterite ore: Carbothermic reduction followed by NaOH cracking[J]. Hydrometallurgy, 2020, 191: 1-11. https://doi.org/10.1016/j.hydromet.2019.105194

[3] LUO J, LI G, RAO M, et al. Atmospheric leaching characteristics of nickel and iron in limonitic laterite with sulfuric acid in the presence of sodium sulfite[J]. Minerals Engineering, 2015, 78: 38-44. https://doi.org/10.1016/j.mineng.2015.03.030

[4] MESHRAM P, ABHILASH, PANDEY B D. Advanced review on extraction of nickel from primary and secondary sources[J]. Mineral Processing and Extractive Metallurgy Review, 2019, 40(3): 157-193. https://doi.org/10.1080/08827508.2018.1514300

[5] VAN DER ENT A, BAKER A J M, VAN BALGOOY M M J, et al. Ultramafic nickel laterites in Indonesia (Sulawesi, Halmahera): Mining, nickel hyperaccumulators and opportunities for phytomining [J]. Journal of Geochemical Exploration, 2013, 128: 72-79. https://doi.org/10.1016/j.gexplo.2013.01.009

[6] YAN K, LIU L, ZHAO H, et al. Study on extraction separation of thioarsenite acid in alkaline solution by CO_3^{2-}-type tri-n-octylmethyl-ammonium chloride[J]. Frontiers in Chemistry, 2021, 8: 1-13. https://doi.org/10.3389/fchem.2020.592837

[7] GARCES-GRANDA A, LAPIDUS G T, RESTREPO-BAENA O J. The effect of calcination as pre treatment to enhance the nickel extraction from low-grade laterites[J]. Minerals Engineering, 2018, 120: 127-131. https://doi.org/10.1016/j.mineng.2018.02.019

[8] XIAO J, XIONG W, ZOU K, et al. Extraction of nickel from magnesia-nickel silicate ore[J]. Journal of Sustainable Metallurgy, 2021, 7: 642-652. https://doi.org/10.1007/s40831-021-00364-0

[9] GUO X, ZHANG C, TIAN Q, et al. Liquid metals dealloying as a general approach for the selective extraction of metals and the fabrication of nanoporous metals: A review [J]. Materials Today Communications, 2021, 26(4): 102007. https://doi:10.1016/j.mtcomm.2020.102007

[10] MAKUZA B, TIAN Q, GUO X, et al. Pyrometallurgical options for recycling spent lithium-ion batteries: A comprehensive review[J]. Journal of Power Sources, 2021, 491: 229622. https://doi:10.1016/j.

jpowsour. 2021. 229622

[11] CHANG Y, ZHAI X, LI B, et al. Removal of iron from acidic leach liquor of lateritic nickel ore by goethite precipitate[J]. Hydrometallurgy, 2010, 101(1-2): 84-87. https://doi.org/10.1016/j.hydromet.2009.11.014

[12] ZHAO D, MA B, SHI B, et al. Mineralogical characterization of limonitic laterite from Africa and its proposed processing route[J]. Journal of Sustainable Metallurgy, 2020, 6(3): 491-503. https://doi.org/10.1007/s40831-020-00290-7

[13] TIAN Q, DONG B, GUO X, et al. Comparative atmospheric leaching characteristics of scandium in two different types of laterite nickel ore from Indonesia[J]. Minerals Engineering, 2021, 173: 107212. https://doi.org/10.1016/j.mineng.2021.107212

[14] PINTOWANTORO S, WIDYARTHA A B, SETIYORINI Y, et al. Sodium thiosulfate and natural sulfur: Novel potential additives for selective reduction of limonitic laterite ore[J]. Journal of Sustainable Metallurgy, 2021, 7(2): 481-494. https://doi.org/10.1007/s40831-021-00352-4

[15] LIU K, CHEN Q, HU H. Comparative leaching of minerals by sulphuric acid in a Chinese ferruginous nickel laterite ore[J]. Hydrometallurgy, 2009, 98(3-4): 281-286. https://doi.org/10.1016/j.hydromet.2009.05.015

[16] SCHWERTMANN U, SCHULZE D G, MURAD E. Identification of ferrihydrite in soils by dissolution kinetics, differential X-ray diffraction, and Mossbauer spectroscopy[J]. Journal of the Soil Science Society of America, 1982, 46(4): 869-875. https://doi.org/10.2136/sssaj1982.03615995004600040040x

[17] FAN C, ZHAI X, YAN F, et al. Extraction of nickel and cobalt from reduced limonitic laterite using a selective chlorination-water leaching process[J]. Hydrometallurgy, 2010, 105(1-2): 191-194. https://doi.org/10.1007/s40831-021-00352-4

[18] KAYA S, DITTRICH C, STOPIC S, et al. Concentration and separation of Sc from Ni laterite ore processing streams[J]. Metals, 2017, 7(12): 557. https://doi.org/10.3390/met7120557

[19] FERIZOGLU E, KAYA S, TOPKAYA Y A. Solvent extraction behaviour of scandium from lateritic nickel-cobalt ores using different organic reagents[J]. Physicochemical Problems of Mineral Processing, 2018, 54(2): 538-545. https://doi.org/10.5277/ppmp1855

[20] GUO X, LI D, PARK K H, et al. Leaching behavior of metals from a limonitic nickel laterite using a sulfation-roasting-leaching process[J]. Hydrometallurgy, 2009, 99(3-4): 144-150. https://doi.org/10.1016/j.hydromet.2009.07.012

[21] LI J, CHEN Z, SHEN B, et al. The extraction of valuable metals and phase transformation and formation mechanism in roasting-water leaching process of laterite with ammonium sulfate[J]. Journal of Cleaner Production, 2017, 140: 1148-1155. https://doi.org/10.1016/j.jclepro.2016.10.050

[22] OXLEY A, BARCZA N. Hydro-pyro integration in the processing of nickel laterites[J]. Minerals Engineering, 2013, 54: 2-13. https://doi.org/10.1016/j.mineng.2013.02.012

[23] ZHANG L, GUO X, TIAN Q, et al. Improved thiourea leaching of gold with additives from calcine by mechanical activation and its mechanism[J]. Minerals Engineering, 2022, 178: 107403. https://doi.org/10.1016/j.mineng.2022.107403

[24] RIBEIRO P, NEUMANN R, SANTOS I, et al. Nickel carriers in laterite ores and their influence on the mechanism of nickel extraction by sulfation-roasting-leaching process[J]. Minerals Engineering, 2019, 131: 90-97. https://doi.org/10.1016/j.mineng.2018.10.022

[25] WANG W, DU S, GUO L, et al. Extraction of nickel from Ramu laterite by sulphation roasting-water

leaching. In: 3RD INTERNATIONAL CONFERENCE ON CHEMICAL MATERIALS AND PROCESS (ICCMP 2017), 2017, 1879(1): 050004. https://doi.org/10.1063/1.5000474

[26] ANAWATI J, YUAN R, KIM J, et al. Selective recovery of scandium from nickel laterite ore by acid roasting-water leaching. In: Symposium on Rare Metal Extraction and Processing held during the 149th TMS Annual Meeting and Exhibition[J]. Minerals Metals & Materials Series, 2020: 77-90. https://doi.org/10.1007/978-3-030-36758-9_8

[27] REID S, TAM J, YANG M, et al. Technospheric mining of rare earth elements from bauxite residue (red mud): Process optimization, kinetic investigation, and microwave pretreatment[J]. Sci Rep, 2017, 7(1): 15252. https://doi.org/10.1038/s41598-017-15457-8

[28] RIVERA R M, ULENAERS B, OUNOUGHENE G, et al. Extraction of rare earths from bauxite residue (red mud) by dry digestion followed by water leaching[J]. Minerals Engineering, 2018, 119: 82-92. https://doi.org/10.1016/j.mineng.2018.01.023

[29] LIU Z, ZONG Y, HONGXU L I, et al. Selectively recovering scandium from high alkali bayer red mud without impurities of iron, titanium and gallium[J]. Journal of Rare Earths, 2017, 9(35): 844-941. https://doi.org/10.1016/S1002-0721(17)60992-X

[30] NARAYANAN R P N, KAZANTZIS N K, EMMERT M H. Selective process steps for the recovery of scandium from jamaican bauxite residue (red mud)[J]. ACS Sustainable Chemistry & Engineering, 2018, 6(1): 1478-1488. https://doi.org/10.1021/acssuschemeng.7b03968

[31] ONGHENA B, BORRA C R, GERVEN T V, et al. Recovery of scandium from sulfation-roasted leachates of bauxite residue by solvent extraction with the ionic liquid betainium bis(trifluoromethylsulfonyl) imide[J]. Separation and Purification Technology, 2017, 176: 208-219. https://doi.org/10.1016/j.seppur.2016.12.009

[32] GAMALETSOS P N, GODELITSAS A, FILIPPIDIS A, et al. The rare earth elements potential of greek bauxite active mines in the light of a sustainable REE demand[J]. Journal of Sustainable Metallurgy, 2018, 5: 20-47. https://10.1007/s40831-018-0192-2

[33] BORRA C R, BLANPAIN B, PONTIKES Y, et al. Recovery of rare earths and other valuable metals from bauxite residue (red mud): A review[J]. Journal of Sustainable Metallurgy, 2016, 2: 365-386. https://10.1007/s40831-016-0068-2

[34] BORRA C R, BLANPAIN B, PONTIKES Y, et al. Smelting of bauxite residue (red mud) in view of iron and selective rare earths recovery[J]. Journal of Sustainable Metallurgy, 2016, 2(1): 28-37. https://doi.org/10.1007/s40831-015-0026-4

[35] BORRA C R, BLANPAIN B, PONTIKES Y, et al. Recovery of rare earths and major metals from bauxite residue (red mud) by alkali roasting, smelting, and leaching[J]. Journal of Sustainable Metallurgy, 2016, 3(2): 393-404. https://doi.org/10.1007/s40831-016-0103-3

[36] GENTZMANN M C, SCHRAUT K, VOGEL C, et al. Investigation of scandium in bauxite residues of different origin[J]. Applied Geochemistry, 2021, 126: 104898. https://doi.org/10.1016/j.apgeochem.2021.104898

[37] ZHANG X K, ZHOU K G, CHEN W, et al. Recovery of iron and rare earth elements from red mud through an acid leaching-stepwise extraction approach[J]. Journal of Central South University, 2019, 26(2): 458-466. https://doi.org/10.1007/s11771-019-4018-6

[38] ZHOU K G, TENG C Y, ZHANG X K, et al. Enhanced selective leaching of scandium from red mud[J]. Hydrometallurgy, 2018, 182: 57-63. https://doi.org/10.1016/j.hydromet.2018.10.011

[39] DING W, BAO S X, ZHANG Y M, et al. Efficient selective extraction of scandium from red mud[J]. Mineral Processing and Extractive Metallurgy Review, 2022: 1-9. https://doi.org/10.1080/08827508.2022.2047044

[40] LI G, YE Q, DENG B, et al. Extraction of scandium from sc-rich material derived from bauxite ore residues[J]. Hydrometallurgy, 2018, 176: 62-68. https://doi.org/10.1016/j.hydromet.2018.01.007

[41] OCHSENKUHNPETROPULU M, LYBEROPULU T, PARISSAKIS G. Selective separation and determination of scandium from yttrium and lanthanides in red mud by a combined ion exchange/solvent extraction method[J]. Analytica Chimica Acta, 1995, 315: 231-237. https://doi.org/10.1016/0003-2670(95)00309-N

[42] ABHILASH, HEDRICH S, SCHIPPERS A. Distribution of scandium in red mud and extraction using Gluconobacter oxydans[J]. Hydrometallurgy, 2021, 202: 105621. https://doi.org/10.1016/j.hydromet.2021.105621

[43] ZHANG D, CHEN H, NIE Z, et al. Extraction of Al and rare earths (Ce, Gd, Sc, Y) from red mud by aerobic and anaerobic bi-stage bioleaching[J]. Chemical Engineering Journal, 2020, 401: 125914. https://doi.org/10.1016/j.cej.2020.125914

[44] BORRA C R, MERMANS J, BLANPAIN B, et al. Selective recovery of rare earths from bauxite residue by combination of sulfation, roasting and leaching[J]. Minerals Engineering, 2016, 92: 151-159. https://doi.org/10.1016/j.mineng.2016.03.002

[45] BOTELHO JUNIOR A B, ROMANO ESPINOSA D C, SOARES TENORIO J A. Extraction of scandium from critical elements-bearing mining waste: Silica gel avoiding in leaching reaction of bauxite residue[J]. Journal of Sustainable Metallurgy, 2021, 7: 1627-1642. https://doi.org/10.1007/s40831-021-00434-3

[46] SWAMY Y V, KAR B B, MOHANTY J K. Physico-chemical characterization and sulphatization roasting of low-grade nickeliferous laterites[J]. Hydrometallurgy, 2003, 69(1): 89-98 https://doi.org/10.1016/S0304-386X(03)00027-6

[47] ANAWATI J, AZIMI G. Recovery of scandium from Canadian bauxite residue utilizing acid baking followed by water leaching[J]. Waste Management, 2019, 95: 549-559. https://doi.org/10.1016/j.wasman.2019.06.044

[48] HARRIS C T, PEACEY J G, PICKLES C A. Selective sulphidation of a nickeliferous lateritic ore[J]. Minerals Engineering, 2011, 24(7): 651-660. https://doi.org/10.1016/j.mineng.2010.10.008

[49] TAGAWA H. Thermal decomposition temperatures of metal sulfates[J]. Thermochimica Acta, 1984, 80(1): 23-33. https://doi.org/10.1016/0040-6031(84)87181-6

[50] SIRIWARDANE R V, JR J, FISHER E P, et al. Decomposition of the sulfates of copper, iron (Ⅱ), iron (Ⅲ), nickel, and zinc: XPS, SEM, DRIFTS, XRD, and TGA study[J]. Applied Surface Science, 1999, 152(3-4): 219-236. https://doi.org/10.1016/S0169-4332(99)00319-0

[51] LONG G J, LONGWORTH G, BATTLE P, et al. A study of anhydrous iron (Ⅲ) sulfate by magnetic susceptibility, Moessbauer, and neutron diffraction techniques[J]. Inorganic Chemistry, 1979, 18(3): 624-632. https://doi.org/10.1021/ic50193a021

[52] COOMBS P, MUNIR Z. The decomposition of iron (Ⅲ) sulfate in air[J]. Journal of thermal analysis, 1989, 35(3): 967-976. https://doi.org/10.1007/BF02057253

[53] MASON C W, GOCHEVA I, HOSTER H E, et al. Iron (Ⅲ) sulfate: A stable, cost effective electrode material for sodium ion batteries[J]. Chemical Communications, 2014, 50(18): 2249-2251. https://doi.org/10.1039/c3cc47557c

[54] YU D, UTIGARD T A, BARATI M. Fluidized bed selective oxidation-sulfation roasting of nickel sulfide concentrate: Part II. sulfation roasting[J]. Metallurgical and Materials Transactions, 2014, B 45(2): 662-674. https://doi.org/10.1007/s11663-013-9959-9

[55] YAGMURLU B, DITTRICH C, FRIEDRICH B. Precipitation trends of scandium in synthetic red mud solutions with different precipitation agents[J]. Journal of Sustainable Metallurgy, 2016, 3(1): 90-98. https://doi.org/10.1007/s40831-016-0098-9

[56] SADRI F, KIM R, GHAHREMAN A. Behavior of light and heavy rare-earth elements in a two-step Fe and Al removal process from rare-earth pregnant leach solutions[J]. Journal of Sustainable Metallurgy, 2021, 7(3): 1327-1342. https://doi.org/10.1007/s40831-021-00423-6

[57] PETERS E M, KAYA S, DITTRICH C, et al. Recovery of scandium by crystallization techniques[J]. Journal of Sustainable Metallurgy, 2019, 5(1): 48-56. https://doi.org/10.1007/s40831-019-00210-4

[58] MCDONALD R G, WHITTINGTON B I. Atmospheric acid leaching of nickel laterites review Part I. Sulphuric acid technologies[J]. Hydrometallurgy, 2008, 91(1-4): 35-55. https://doi:10.1016/j.hydromet.2007.11.009

[59] WAN X, TASKINEN P, SHI J, et al. A potential industrial waste-waste co-treatment process of utilizing waste SO_2 gas and residue heat to recover Co, Ni, and Cu from copper smelting slag[J]. Journal of Hazardous Materials, 2021, 414: 125541. https://doi.org/10.1016/j.jhazmat.2021.125541

The Effect of Pre-roasting on Atmospheric Sulfuric Acid Leaching of Saprolitic Laterites

Abstract: The effect of pre-roasting on the atmospheric sulfuric acid leaching behavior of saprolitic laterites, obtained from Indonesia, was investigated. The phase transformation of laterite minerals roasted at different temperatures was investigated with X-ray powder diffraction (XRD). The TG-DTA analysis showed that there were two phase transformation processes for that dehydroxylation of goethite (FeO(OH)) and lizardite ($Mg_3(Si_2O_5)(OH)$) at roasting temperatures of 260.8 ℃ and 602.2 ℃, respectively. Roasting at temperatures up to 600 ℃ improved the metal extraction, because roasting at that temperature transforms the main mineral structures of lizardite and goethite, allowing a rapid interaction between the leachant and the entrapped nickel species during leaching. The transformation of the mineral structures caused by roasting is the decisive factor influencing the metal extraction, followed by changes to the specific surface area. These results contribute to the development of more efficient atmospheric acid leaching process for laterites.

Keywords: Pre-roasting; Saprolitic laterites; Leaching; Phase transformation

1 Introduction

China's new energy strategy is a low-carbon and circular development movement to achieve "carbon peaking" and "carbon neutrality" by 2030 through adopting new electricity initiatives (Ma et al., 2020; Zhang et al., 2019). Nickel and cobalt strategic metals in this plan are considered to be "white oil in the 21st century", as they are the key ingredients for the continued growth of the new energy industry and determines the lifeline of the development of the new energy industry chain (Li et al., 2018a; Li et al., 2018b). At present, there is increased interest

---

Published in *Hydrometallurgy*, 2023, 218: 106063. Authors: Dong Bo, Tian Qinghua, Xu Zhipeng, Guo Xueyi, Wang Qingao, Li Dong.

in extracting nickel and cobalt from laterites, and there is encouragement to urgently develop new green technologies to treat laterites (Khoo et al., 2017). Laterites can be classified into limonites or saprolites, depending on the iron and magnesium contents (Oxley et al., 2016; Thubakgale et al., 2013; Wanderley et al., 2020). Limonites refer to laterites with high iron contents (at least 40% by weight) and low magnesium contents (0.5%-5%) (Garces-Granda et al., 2018; Landers et al., 2009). The nickel in these ores is mainly hosted in goethite (α-FeOOH), showing contents of up to 1.5% (Chasse et al., 2017; Landers et al., 2009). Saprolites refer to laterites with high MgO contents (15%-35%) and lower iron contents (10%-25%) (Dong et al., 2022). Nickel content is usually in the range of 1% to 4%, and it mainly exists in the form of lattice substitution inside the magnesium silicate minerals (Tian et al., 2021; Zhang et al., 2016). The properties of different types of laterites determine the different extraction processes (Li et al., 2017; Zhang et al., 2022). Currently, nickel is recovered from limonites by hydrometallurgical processing, due to the low magnesium and high iron content of limonites (Oxley et al., 2016; Thubakgale et al., 2013; Wanderley et al., 2020). While pyrometallurgy is generally used to enable the extraction of nickel from saprolites, due to the high magnesium and low iron content of saprolites (Oxley et al., 2016; Thubakgale et al., 2013; Wanderley et al., 2020). The smelting process also limits the application fields of its products (Thompson and Senanayake, 2018). The ferronickel produced by pyrometallurgy is mainly used in the production of stainless steel, while the products of hydrometallurgy are the main raw materials for the new energy industry (Wang et al., 2019). In summary, the hydrometallurgical treatment of saprolites is seldom reported in the literature. Saprolites with higher grades have not yet become a source of raw materials for the new energy industry.

High pressure acid leaching (HPAL) and atmospheric acid leaching (AL) are the two main hydrometallurgical processes used to treat laterites (Luo et al., 2021). The AL process which can treat both limonite and saprolite ores, is often discussed in terms of lower capital investment and simpler process requirements when compared to the HPAL process. Despite the drawbacks of high acid consumption and prolonged leaching period, AL, if applied properly, can be the most cost-effective method for the extraction of nickel from laterites. However, the leaching efficiency of nickel by the AL process is relatively low. A few research papers and several patents have described significant enhancement of nickel extraction by AL, after simply roasting limonites at various temperatures (Cornell, 1993; Febriana et al., 2017; Landers and Gilkes, 2007; Landers et al., 2009; Landers et al., 2011; Li et al., 2013; Li et al., 2009; Lim-Nunez and Gilkes, 1985; O'Connor et al., 2006; Rhamdhani et al., 2009). Dehydroxylation of goethite (the dominant Ni host mineral in limonites) due to roasting causes a phase transformation to hematite which occurs via modification in the goethite structure (Gualtieri and Venturelli, 1999; Ruan et al., 2002; Watari et al., 1983; Wells et al., 2006). During progressive dehydroxylation of goethite some of the associated metals may not be compatible with the various hematite-like structures and may be ejected from the lattice into the abundant voids or onto the crystal surface (Landers and Gilkes, 2007; Landers et al., 2009). In these accessible locations, they will be much more easily dissolved by acid solutions. Moreover, due to the removal of free and combined moisture and to the partial collapse of the phase structure, a roasting step could alter the mineralogical composition of the ore and increase the surface and porosity of the raw ore, thus making it more amenable to leaching (Landers and Gilkes, 2007; Olanipekun, 2000). Different roasting conditions have been applied to treat different limonites (Landers et al., 2009; Li et al., 2009). These differences indicate that optimum thermal treatment conditions are ore-dependent and are consistent with the previous findings that the species and level of metal substitution in goethite can influence the dehydroxylation temperature of goethite (Schulze, 1984; Wells et al., 2006). Most of these previous findings have focused on limonites, and thermal pre-treatment of saprolites followed by the recovery of nickel under AL conditions is seldom studied in the literature. The effect of pre-roasting on hydrochloric acid leaching of the gamierite laterite ore, obtained from Yunnan province, China, was investigated by Li et al. (Li et al., 2009). Although the leaching was efficient, sulfuric acid is still the first choice for hydrometallurgical treatment of laterites due to economic and environmental reasons (McDonald and Whittington, 2008). The mechanisms of dehydroxylation and phase transformation of lizardite within laterite were investigated by Zhou and Wei (Zhou et al., 2017). However, the dominant nickel host minerals in saprolites from Indonesia are goethite and lizardite, so the dehydroxylation and phase transformation are more complex, and should be studied systematically.

This study investigated the effect of pre-roasting at different temperatures on the leachability of nickel from Indonesia saprolites. An attempt has been made to identify the mechanism of pre-roasting on the leaching of saprolite in sulfuric acid and efforts have been made to extract nickel selectively over iron by using a roast-leach process. These results contribute to the development of more efficient atmospheric acid leaching process for laterites.

2 Experimental

2.1 Materials

A saprolite sample from Sulawesi, Indonesia was used in this study. Chemical and mineralogical composition was determined by Inductively coupled plasma spectrometry (ICP) and X-ray diffraction (XRD). The chemical analysis result for the sample is presented in Table 1, where it determined that the saprolite contained 1.82% Ni, 0.061% Co, 6.35% Mg and 26.10% Fe.

Table 1 The chemical composition of the saprolite ore

Element	Ni	Co	Mn	Fe	Mg	Al	Cr	Si
Composition/%	1.82	0.061	0.45	26.10	6.35	1.61	0.57	13.58

2.2 Roasting and leaching

All roasting experiments comprised of, approximately 10 g of sample, passing 50 μm, which was placed in ceramic crucibles. The crucibles were placed in a furnace, and pre-roasted to the selected temperature (200 ℃, 400 ℃, 600 ℃, 800 ℃, 1000 ℃), for a preset time. The air was metered into the furnace using rotameters prior to heating. After roasting, the samples were removed from the furnace and allowed to cool to room temperature. Thermogravimetric and differential thermal analysis (TG-DSC, STA 449 C, Netzsch, German) was conducted on the ore sample in air. The temperature was scanned between 30 and 1000 ℃ with a heating ramp of 10 ℃/min. In addition, a Brunauer-Emmett-Teller (BET) analysis was performed to determine the specific surface area for all calcined and uncalcined samples.

All leaching tests were conducted at the same liquid to solid ratio, and the same sulfuric acid concentration where the acid was heated to the desired temperature through a digital homoeothermic water bath before adding 5 g of calcined or uncalcined sample. After a predetermined time, liquid/solid separation was performed with a Büchner funnel, and the residues were washed thrice with deionized water. The pregnant leach solution produced was analyzed using an atomic emission spectrometer (ICP-AES, IRIS Intrepid II XSP, Thermo Electron Corporation, USA). The leaching efficiency of metal was calculated according to Eq. (1):

$$\eta_i = \frac{C_i \times V}{m \times w_i} \times 100\% \tag{1}$$

where η_i is the leaching efficiency of metal i (%), C_i is the concentration (g/L) of metal i in the pregnant leach solution (PLS), V is the volume (L) of PLS, m is the mass of raw material (g), and w_i is the mass fraction of metal i in raw material (%).

3 Results and discussions

3.1 Mineralogical analysis

The TG-DTA curve of the raw ore is shown in Fig. 1. The effect of roasting on the mineral transformation of the saprolitic laterite sample under atmospheric pressure at different temperatures is shown in Fig. 2. The TG-DTA analysis showed three endothermic peaks at about 71.8 ℃, 260.8 ℃, and 602.2 ℃ due to the release of adsorbed water, dehydroxylation of goethite and lizardite, respectively (Li et al., 2009; Tartaj et al., 2000). According to the XRD patterns generated at the various roasting temperatures shown in Fig. 2, there are some reactions and structural changes that occurred. At 260.8 ℃, goethite (FeO(OH)) transforms to hematite (Fe_2O_3) as shown in Eq. (2) (Valix and Cheung, 2002; Demol et al., 2022), which can be observed from the difference between traces at 200 ℃ and 400 ℃. These decomposition temperatures are lower compared to the decomposition temperatures of 385 ℃ (Schwertmann et al., 1982) and 337 ℃ (Swamy et al., 2003), which have been reported for highly crystalline goethite. The lower decomposition temperature of laterite in this study is indicative of a poorly crystalline goethite structure. At a roasting temperature of 602.2 ℃, lizardite ($Mg_3Si_2(OH)_4O_5$) decomposes and forms an amorphous magnesium silicate phase. This is evident by the formation of broad peaks and eventual loss of X-ray diffraction signals for the lizardite minerals at 2θ angles of 12.1°, 18.64°, and 24.38°. At higher temperatures, above 800 ℃, the amorphous magnesium silicate phase appears to have recrystallized as forsterite (Mg_2SiO_4) and enstatite ($MgSiO_3$), as shown in Eq. (3) (Luo et al., 2009) and Fig. 2.

$$2FeO(OH) \rightleftharpoons Fe_2O_3 + H_2O(g) \quad \Delta G_T^\ominus = 11.33 - 0.145T, \text{ (kJ/mol)} \quad (2)$$

$$Mg_3Si_2(OH)_4O_5 \rightleftharpoons Mg_2SiO_4 + MgSiO_3 + 2H_2O, (g)$$
$$\Delta G_T^\ominus = 61.23 - 0.294T, \text{ (kJ/mol)} \quad (3)$$

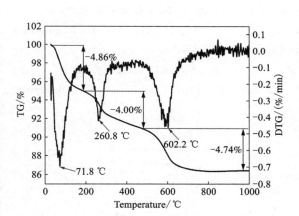

Fig. 1　TG-DTA curves for the saprolite ore

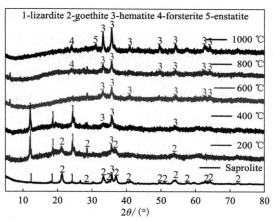

(roasting conditions: 10 g of raw ore, passing 50 μm, air flowrate 0 L/min, roasting time 1 h)

Fig. 2　XRD patterns of the raw ore and saprolite roasted ore at different temperatures

3.2　Effect of roasting temperature on metal dissolution

Experiments to investigate the effect of roasting temperature were carried out in the temperature range of 200~1000 ℃. Samples were roasted for 1 h and then leached with 2 mol/L sulfuric acid solution at 30 ℃ and a liquid-to-solid ratio of 6 : 1 for 2 h with agitation. Fig. 3 shows that the roasting temperature had a significant effect on metal extraction. Nickel became liberated after dehydroxylation that enables the nickel to be extracted (Kukura et al., 1979). Roasting at temperatures up to 600 ℃ appeared to provide the optimum metal extraction. It is very probable that the roasting transforms the main mineral structures of lizardite and goethite, allowing a rapid interaction between the leachant and the contained nickel species during leaching. The experimental results show that excessive temperature may lead to the incorporation or entrapment of nickel into the hematite and magnesium silicate phases during recrystallization (Park et al., 2015), which renders it inert and hinders further release of nickel trapped in this phase (O'Connor et al., 2006; Valix and Cheung, 2002). The results in Fig. 3 also show that the specific surface area of the roasted product reaches a maximum when the calcination temperature is 400 ℃, and the specific surface area decreases significantly when the calcination temperature exceeded 800 ℃. However, Fig. 2 shows that lizardite still exists when heated to 400 ℃. Therefore, the transformation of the mineral structure caused by heating is the decisive factor affecting metal extraction, followed by the specific surface area. Consequently, further tests were carried out by keeping the roasting temperature fixed at 600 ℃.

3.3 Effect of airflow rate during roasting on metal dissolution

A series of experiments were carried out at various airflow rates from 0.0 to 0.5 L/min during roasting. The sample had been roasted for 1 h at a fixed temperature of 600 ℃. After roasting, the samples were leached in 2 mol/L sulfuric acid solution at 30 ℃ and a liquid-to-solid ratio of 6 : 1 for 2 h with agitation. As shown in Fig. 4, the nickel, cobalt, iron and magnesium extractions improved slightly with increasing airflow rate. This increase can be attributed to the increase in the surface area. According to (Li et al., 2009), the leaching efficiency can be related to the specific surface area. For this reason, the optimal airflow rate used in subsequent experiments was 0.5 L/min.

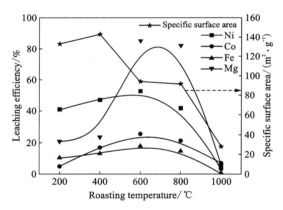

(roasting conditions: 10 g of raw ore, passing 50 μm, air flowrate 0 L/min, roasting time 1 h; leaching conditions: 5 g of calcined sample, sulfuric acid concentration 2 mol/L, S/L ratio 6 : 1 mL/g, leaching temperature 30 ℃, leaching time 2 h, stirring rate 400 r/min)

Fig. 3 Effect of roasting temperature on metal extractions

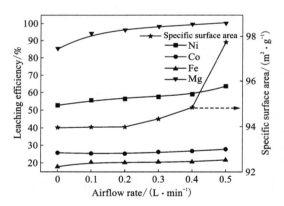

(roasting conditions: 10 g of raw ore, passing 50 μm, roasting temperature 600 ℃, roasting time 1 h; leaching conditions: 5 g of calcined sample, sulfuric acid concentration 2 mol/L, S/L ratio 6 : 1 mL/g, leaching temperature 30 ℃, leaching time 2 h, stirring rate 400 r/min)

Fig. 4 Effect of airflow rate during roasting on metal extractions

3.4 Effect of roasting time on metal dissolution

The duration of the roasting process was also evaluated to analyze its effect on metal extraction during the leaching stage. Samples were roasted at 600 ℃ in an airflow rate of 0.5 L/min for different durations. Finally, the samples were leached with 2 mol/L sulfuric acid solution at 30 ℃ and a liquid-to-solid ratio of 6 : 1 for 2 h with agitation. As shown in Fig. 5, the nickel, cobalt, iron, and magnesium extraction slightly decrease with increasing roasting time. These results may be related to the change in specific surface area of the particles. Fig. 5 shows a slight decrease in the surface area when the calcination time is increased. The dissolution occurs because acid attack occurs preferentially along cracks and areas of weakness (e.g. micropores and dislocations) (Cornell et al., 1976). This difference can also be attributed to the stability of the phase, longer roasting times allow the structure to stabilize, increasing the probability that nickel is included again into the iron oxide during recrystallization (O'Connor et al., 2006).

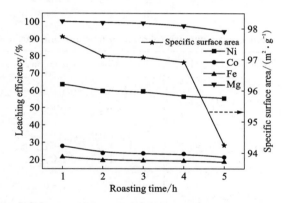

(roasting conditions: 10 g of raw ore, passing 50 μm, roasting temperature 600 ℃, air flowrate 0.5 L/min; leaching conditions: 5 g of calcined sample, sulfuric acid concentration 2 mol/L, S/L ratio 6 : 1 mL/g, leaching temperature 30 ℃, leaching time 2 h, stirring rate 400 r/min)

Fig. 5 Effect of roasting duration on metal extractions

3.5 Effects of leaching parameters

Leaching experiments were conducted with agitation at different leaching temperatures, liquid-to-solid ratios, sulfuric acid concentrations, and leaching durations for samples roasted at 600 ℃ with an airflow rate of 0.5 L/min for 1 h. The specific chemical reactions are relevant to leaching and thermodynamic data are summarized in Table 2. To investigate the effect of leaching temperature, a series of experiments were performed under the following conditions: 2 mol/L sulfuric acid, leaching duration 0.5 h, liquid to solid ratio 6 mL/g. The results are shown in Fig. 6(a) which demonstrates that the leaching efficiencies of nickel and cobalt increased gradually with increasing temperature, and the leaching efficiencies of iron and magnesium increased slightly. The effect of temperature is restricted to 70 ℃ because our previous studies showed that leaching efficiencies of nickel and iron increase with the increase of temperature when the uncalcined saprolite ore was leached at atmospheric pressure, reaching the maximum at 80 ℃ and then decreasing (Tian et al., 2021). However, the leaching efficiency of iron increases rapidly, especially in the range of 70-80 ℃. To 80 ℃, the leaching efficiency of iron has reached approximately 60% (Tian et al., 2021). Although the increase of temperature can increase the nickel leaching efficiency slightly, it will increase the iron content in the PLS, which will affect the subsequent impurity removal process. From the perspective of process integrity, 70 ℃ was selected as the optimal parameter under atmospheric pressure acid leaching conditions. The effect of the liquid to solid ratio was measured at different liquid to solid ratios, and other conditions included 2 mol/L H_2SO_4, at a leaching temperature of 70 ℃, and a leaching duration of 0.5 h. The results are shown in Fig. 6(b). With the increase in the liquid-solid ratio, the leaching efficiencies of nickel and cobalt increased gradually, the leaching efficiency of iron increased slightly, and the leaching efficiency of magnesium remained unchanged. Therefore, the optimal liquid to solid ratio for the next leaching experiment was 7 : 1 mL/g.

Table 2 The chemical reactions are relevant to leaching and thermodynamic data

Chemical reaction	$\Delta G_T^\ominus$ /(kJ·mol^{-1})	Chemical reaction	$\Delta G_T^\ominus$ /(kJ·mol^{-1})
$NiO_{(S)} + 2H^+_{(aq)} = Ni^{2+}_{(aq)} + H_2O_{(l)}$	$-99.25 + 0.097T$	$Fe_2O_{3(S)} + 6H^+_{(aq)} = 2Fe^{3+}_{(aq)} + 3H_2O_{(l)}$	$-43.00 + 0.170T$
$CoO_{(S)} + 2H^+_{(aq)} = Co^{2+}_{(aq)} + H_2O_{(l)}$	$-105.12 + 0.096T$	$FeOOH_{(S)} + 3H^+_{(aq)} = Fe^{3+}_{(aq)} + 2H_2O_{(l)}$	$-40.79 + 0.157T$
$MnO_{(S)} + 2H^+_{(aq)} = Mn^{2+}_{(aq)} + H_2O_{(l)}$	$-121.65 + 0.063T$	$Mg_3Si_2O_5(OH)_{4(S)} + 6H^+_{(aq)} = 3Mg^{2+}_{(aq)} + 2SiO_2(s) + 5H_2O_{(l)}$	$-111.57 + 0.071T$
$Al_2O_{3(S)} + 6H^+_{(aq)} = 2Al^{3+}_{(aq)} + 3H_2O_{(l)}$	$-86.34 + 0.164T$	$Mg_2SiO_{4(S)} + 4H^+_{(aq)} = 2Mg^{2+}_{(aq)} + SiO_2(s) + 2H_2O_{(l)}$	$-189.10 + 0.045T$
$Cr_2O_{3(S)} + 6H^+_{(aq)} = 2Cr^{3+}_{(aq)} + 3H_2O_{(l)}$	$-92.49 + 0.141T$	$MgSiO_{3(S)} + 2H^+_{(aq)} = Mg^{2+}_{(aq)} + SiO_2(s) + H_2O_{(l)}$	$-90.11 + 0.024T$

From Tian et al., 2021.

The effect of the 0.5 mol/L to 2.5 mol/L sulfuric acid concentration at a leaching temperature of 70 ℃, liquid to solid ratio of 7 : 1 mL/g, and leaching duration of 0.5 h was investigated as shown in Fig. 6(c). Here, the leaching of magnesium increased with the increase in H_2SO_4 solution concentration from 0.5 to 1.0 mol/L and to near completion. Fig. 6(c) also showed that the leaching efficiencies of nickel, cobalt and iron increased with increasing sulfuric acid concentration. The effect of acid concentration was restricted to 2.5 mol/L for two reasons. First of all, according to the results of chemical analysis, the theoretically calculated sulfuric acid requirement for saprolitic laterite was 2.5 mol/L. Secondly, our previous studies showed that

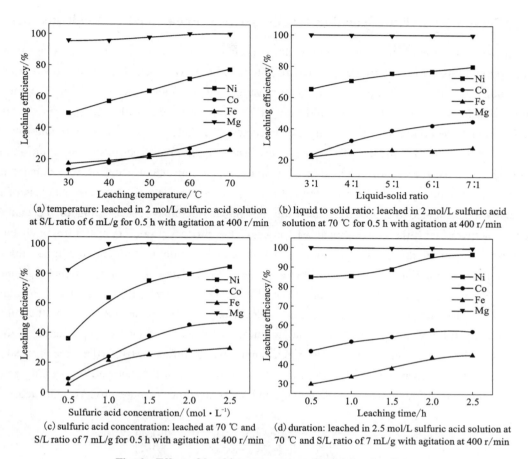

Fig. 6 Effect of leaching parameters on metal extractions

although the increase of acid concentration can increase the nickel leaching efficiency, it will increase the iron, aluminum and chromium content in the leaching solution and the concentration of residual acid, which will affect the subsequent impurity removal and pH adjustment process (Tian et al., 2021). Therefore, 2.5 mol/L sulfuric acid was selected for the proceeding leaching experiments. To investigate the effect of leaching duration on metal extraction, a series of leaching experiments was performed with the leaching duration varying from 0.5 to 2.5 h under the experimental conditions of 2.5 mol/L H_2SO_4, reaction temperature 70 ℃, and liquid to solid ratio 7 : 1 mL/g. As shown in Fig. 6(d), the leaching of nickel, cobalt, and iron increased gradually with increasing leaching duration from 0.5 to 2.0 h and then remained steady. The leaching efficiency of magnesium remained constant for all leaching times. Therefore, the optimal leaching duration was determined to be 2 h. In summary, the optimum leaching conditions were as follows: leaching temperature 70 ℃, liquid-solid ratio 7 : 1 mL/g, 2.5 mol/L sulfuric acid, and leaching duration 2 h.

3.6 Comparison of the leaching results between uncalcined and calcined ore

At the conclusion of the above single factor experimental series of tests, the following suitable process conditions were determined: a roasting temperature was 600 ℃, an airflow rate was 0.5 L/min, and a roasting time of 1 h. After roasting, the samples were leached in 2.5 mol/L sulfuric acid at 70 ℃ and a liquid to solid ratio of 7∶1 mL/g for 2 h with agitation. As shown in Fig. 7, under these conditions, the leaching efficiencies of nickel, cobalt, magnesium and iron are 96.4%, 58.0%, 100%, and 44.1%, respectively. The comparative experiments on an uncalcined sample showed that the use of calcination as a pretreatment can enhance the nickel, cobalt, and magnesium extractions, while decrease the extraction of iron. The leaching solution analysis of the uncalcined and calcined ore are given in Table 3 (based on Fig. 7). According to the analysis results, compared with the conventional AL process, the process proposed in this research has a lower iron content in the leaching solution, but its concentration is still about 7 times that of nickel. Solvent extraction is the more suitable process for selective separation of nickel directly over other metals, especially over iron. In recent years, a new and commercially available extractant of HBL110 was developed by Zeng and Zhang (Zeng et al., 2016) for the highly selective extraction of nickel over iron, aluminum, manganese, magnesium, calcium, zinc, chromium and silicon from sulfuric acid leach solutions. Therefore, HBL110 extractant can be used for selective extraction of nickel from the leaching solution of the calcined ore as long as the process parameters in the extraction, scrubbing and stripping processes are properly adjusted.

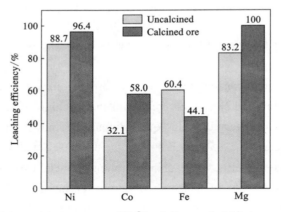

(roasting conditions: roasting temperature 600 ℃, air flowrate 0.5 L/min, roasting time 1 h; leaching conditions: 5 g of calcined or uncalcined sample, temperature of 70 ℃, liquid to solid ratio of 7∶1 mL/g, 2.5 mol/L sulfuric acid, and leaching duration of 2 h)

Fig. 7 Comparison of the leaching results between uncalcined and calcined ore

Table 3 The leach solution analysis results of the uncalcined ore and calcined ore

Element	Ni	Co	Fe	Mg	Remarks (volume of leaching solution)
Mass concentration of PLS of uncalcined ore/(mg · L^{-1})	1681	20	16413	5502	48.0 mL
Mass concentration of PLS of calcined ore/(mg · L^{-1})	1871	38	12279	6774	54.5 mL

4 Preliminary economic analysis

Based on the mechanism described above and the optimized experimental results, a roast-leach process is proposed (Fig. 8). In addition, the material balance of the proposed process is also provided. In this preliminary economic analysis, the objective is to provide a first look at the economic feasibility of using a roast-leach process to extract nickel and cobalt from Indonesia saprolites. Using approximate market values for the constituent metals (whichever appropriate) and data collected during the experiment, the values of the generated products, energy and reagent consumption were estimated. While the calculations are based on the 10 g scale experiments, the results are given for 1 ton of saprolitic laterites. A preliminary economic analysis of the process is shown in Table 4. Because the leaching efficiencies of nickel and cobalt have improved by the roasting-leaching process, the values of the generated products are slightly higher than that of the atmospheric pressure process. The energy consumption is relatively high, mainly due to the high energy consumption of the roasting process. However, if the waste heat can be used reasonably, the energy consumption of this process will be slightly reduced. In addition, the proposed process mainly relies on two conventional equipment (i.e., rotary kiln/mixing equipment for the roasting, an atmospheric stirred tank for leaching), the investment is relatively high but acceptable. On this basis, it may be economically and/or technologically feasible as long as waste heat can be reasonably utilized.

Table 4 A preliminary economic analysis of the process was carried out in this study

Item	Roast-leach	AL	Price reference standard
Ni and Co values	$456.83	$413.57	~25000 $/t Ni; ~52000 $/t Co.
Energy and reagent consumption (per ton of ore)	$91.74	$74.91	Formula: $Q = m^* C_p^{laterites *} (600\text{-}25\ ℃)$; The average heat capacity of the laterites: $C_p = 1.2$ kJ/kg laterites · K; Assuming 80% efficiency, the energy cost for heating from natural gas combustion: 19.51 $/GJ; ~0.0428 $ US/kg sulfuric acid; ~1 $ US/1000 kg process water (Anawati and Azimi, 2019).

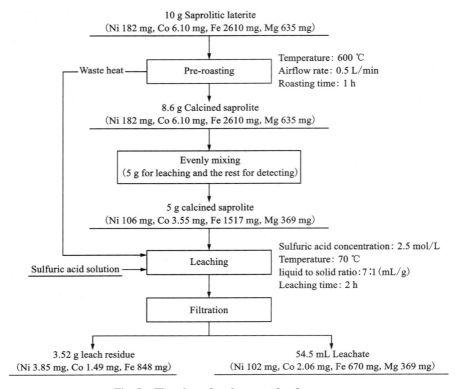

Fig. 8　Flowsheet for the roast-leach process

5　Conclusions

A saprolitic laterite from an Indonesian deposit was chosen to investigate the effect of pre-roasting on the recovery of metals during a subsequent leaching stage. The results determined that roasting temperature can have a significant effect on nickel metal recovery as the nickel is intricately associated with the goethite and lizardite phases in the saprolite. The transformation of the mineral structure caused by roasting is the decisive factor influencing the metal extraction, followed by the specific surface area. Increasing the roasting temperature of 600 ℃ gave the optimum nickel recovery, and further heating was detrimental to the nickel recovery. The results also showed that airflow rate during roasting and roasting time had little effect on the extraction of metal. The increase in the airflow rate during roasting will cause an increase in the specific surface area, thus leading to an increase in metal extraction. However, the extension of roasting time will cause a decrease in the specific surface area, thus leading to a decrease in metal extraction. The use of roasting as a pretreatment can enhance nickel extraction while decreasing the extraction of iron for the atmospheric acid leaching step, proving to have great potential for the extraction of nickel from saprolitic laterites. The results of this research also assist in the development of more efficient processes for atmospheric acid leaching laterites.

Acknowledgments

The authors gratefully acknowledge the financial support from the National Key R&D Program of China (Grant No. 2019YFC1907402 and 2018YFC1902501), National Natural Science Foundation of China (Grant No. 51922108, 52074363, and 52104355).

References

[1] ANAWATI J, AZIMI G. Recovery of scandium from Canadian bauxite residue utilizing acid baking followed by water leaching[J]. Waste Management, 2019, 95: 549-559. https://doi.org/10.1016/j.wasman.2019.06.044.

[2] CHASSE M, GRIFFIN W L, O'REILLY S Y, et al. Scandium speciation in a world-class lateritic deposit [J]. Geochemical Perspectives Letters, 2017, 3(2): 105-113. https://doi.org/10.7185/geochemlet.1711.

[3] CORNELL R M. Acid dissolution of hematites of different morphologies[J]. Clay Minerals, 1993, 28(2): 223-232. https://doi.org/10.1180/claymin.1993.028.2.04.

[4] CORNELL R M, POSNER A M, QUIRK J P. Kinetics and mechanisms of the acid dissolution of goethite (α-FeOOH)[J]. Journal of Inorganic & Nuclear Chemistry, 1976, 38(3): 563-567. https://doi.org/10.1016/0022-1902(76)80305-3.

[5] DONG B, TIAN Q, GUO X, et al. Leaching behavior of scandium from limonitic laterite ores under sulfation roasting-water leaching[J]. Journal of Sustainable Metallurgy, 2022, 8: 1078-1089. https://doi.org/10.1007/s40831-022-00551-7.

[6] DEMOL J, HO E, SOLDENHOFF K, et al. Beneficial effect of iron oxide/hydroxide minerals on sulfuric acid baking and leaching of monazite[J]. Hydrometallurgy, 2022, 211: 105864. https://doi.org/10.1016/j.hydromet.2022.105864.

[7] FEBRIANA E, MANAF A, PRASETYO A L, et al. Thermal characteristic of limonite ore upon calcination and reduction[C]//1st International Seminar on Metallurgy and Materials (ISMM), Jakarta, INDONESIA, 2017, 1964(1): 020026. https://doi.org/10.1063/1.5038308.

[8] GARCES-GRANDA A, LAPIDUS G T, RESTREPO-BAENA O J. The effect of calcination as pre treatment to enhance the nickel extraction from low-grade laterites[J]. Minerals Engineering, 2018, 120: 127-131. https://doi.org/10.1016/j.mineng.2018.02.019.

[9] GUALTIERI A F, VENTURELLI P. In situ study of the goethite-hematite phase transformation by real time synchrotron powder diffraction[J]. American Mineralogist, 1999, 84(5-6): 895-904. https://doi.org/doi.org/10.2138/am-1999-5-624.

[10] KHOO J Z, HAQUE N, WOODBRIDGE G, et al. A life cycle assessment of a new laterite processing technology[J]. Journal of Cleaner Production, 2017, 142: 1765-1777. https://doi.org/doi.org/10.1016/j.jclepro.2016.11.111.

[11] KUKURA M E, STEVENS L G, AUCK Y T. Development of the UOP process for oxide silicate ores of nickel and cobalt[C]//EVANS D J I, SHOEMAKER R S, VELTMAN H. International Laterite Symposium. SME-AIME, New York, USA, 1979: 527-552.

[12] LANDERS M, GILKES R J. Dehydroxylation and dissolution of nickeliferous goethite in New Caledonian lateritic Ni ore[J]. Applied Clay Science, 2007, 35(3-4): 162-172. https://doi.org/10.1016/j.clay.2006.08.012.

[13] LANDERS M, GILKES R J, WELLS M. Dissolution kinetics of dehydroxylated nickeliferous goethite from limonitic lateritic nickel ore[J]. Applied Clay Science, 2009, 42(3-4): 615-624. https://doi.org/10.1016/j.clay.2008.05.002.

[14] LANDERS M, GRAEFE M, GILKES R J, et al. Nickel distribution and speciation in rapidly dehydroxylated goethite in oxide-type lateritic nickel ores: XAS and TEM spectroscopic (EELS and EFTEM) investigation[J]. Australian Journal of Earth Sciences, 2011, 58(7): 745-765. https://doi.org/10.1080/08120099.2011.602985.

[15] LI G, ZHOU Q, ZHU Z, et al. Selective leaching of nickel and cobalt from limonitic laterite using phosphoric acid: An alternative for value-added processing of laterite[J]. Journal of Cleaner Production, 2018, 189: 620-626. https://doi.org/10.1016/j.jclepro.2018.04.083.

[16] LI J, BUNNEY K, WATLING H R, et al. Thermal pre-treatment of refractory limonite ores to enhance the extraction of nickel and cobalt under heap leaching conditions[J]. Minerals Engineering, 2013, 41: 71-78. https://doi.org/10.1016/j.mineng.2012.11.002.

[17] LI J, CHEN Z, SHEN B, et al. The extraction of valuable metals and phase transformation and formation mechanism in roasting-water leaching process of laterite with ammonium sulfate[J]. Journal of Cleaner Production, 2017, 140: 1148-1155. https://doi.org/10.1016/j.jclepro.2016.10.050.

[18] LI J, LI D, XU Z, et al. Selective leaching of valuable metals from laterite nickel ore with ammonium chloride-hydrochloric acid solution[J]. Journal of Cleaner Production, 2018, 179: 24-30. https://doi.org/10.1016/j.jclepro.2018.01.085.

[19] LI J, LI X, HU Q, et al. Effect of pre-roasting on leaching of laterite[J]. Hydrometallurgy, 2009, 99(1-2): 84-88. https://doi.org/10.1016/j.hydromet.2009.07.006.

[20] LIM-NUNEZ R, GILKES R J. Acid dissolution of synthetic metal-containing goethites and hematites[C]//Proceedings of the international clay conference. Clay Minerals Society. Am., 1985: 197-204. https://doi.org/10.1346/CMS-ICC-1.

[21] LUO J, RAO M, LI G, et al. Self-driven and efficient leaching of limonitic laterite with phosphoric acid[J]. Minerals Engineering, 2021, 169: 106979. https://doi.org/10.1016/j.mineng.2021.106979.

[22] LUO W, FENG Q, OU L, et al. Fast dissolution of nickel from a lizardite-rich saprolitic laterite by sulphuric acid at atmospheric pressure[J]. Hydrometallurgy, 2009, 96(1-2): 171-175. https://doi.org/10.1016/j.hydromet.2008.08.001.

[23] MA B, LI X, YANG W, et al. Nonmolten state metalized reduction of saprolitic laterite ores: Effective extraction and process optimization of nickel and iron[J]. Journal of Cleaner Production, 2020, 256: 120415. https://doi.org/10.1016/j.jclepro.2020.120415.

[24] MCDONALD R G, WHITTINGTON B I. Atmospheric acid leaching of nickel laterites review Part I. Sulphuric acid technologies[J]. Hydrometallurgy, 2008, 91(1-4): 35-55. https://doi.org/10.1016/j.hydromet.2007.11.009.

[25] O'CONNOR F, CHEUNG W H, VALIX M. Reduction roasting of limonite ores: Effect of dehydroxylation

[J]. International Journal of Mineral Processing, 2006, 80(2-4): 88-99. https://doi.org/10.1016/j.minpro.2004.05.003.

[26] OLANIPEKUN E O. Kinetics of leaching laterite[J]. International Journal of Mineral Processing, 2000, 60(1): 9-14. https://doi.org/10.1016/s0301-7516(99)00067-8.

[27] OXLEY A, SMITH M E, CACERES O. Why heap leach nickel laterites? [J]. Minerals Engineering, 2016, 88: 53-60. https://doi.org/10.1016/j.mineng.2015.09.018.

[28] PARK J O, KIM H S, JUNG S M. Use of oxidation roasting to control NiO reduction in Ni-bearing limonitic laterite [J]. Minerals Engineering, 2015, 71: 205-215. https://doi.org/10.1016/j.mineng.2014.11.011.

[29] RHAMDHANI M A, HAYES P C, JAK E. Nickel laterite Part 2- Thermodynamic analysis of phase transformations occurring during reduction roasting[J]. Mineral Processing & Extractive Metallurgy, 2009, 118(3): 146-155. https://doi.org/10.1179/174328509X431409.

[30] RUAN H D, FROST R L, KLOPROGGE J T, et al. Infrared spectroscopy of goethite dehydroxylation. II. Effect of aluminium substitution on the behaviour of hydroxyl units [J]. Spectrochimica Acta Part a—Molecular and Biomolecular Spectroscopy, 2002, 58(3): 479-491. https://doi.org/10.1016/s1386-1425(01)00556-x.

[31] SCHULZE D G. The influence of aluminium on iron oxides: X. properties of Al-substituted goethites[J]. Clay Minerals, 1984, 19(4): 521-539. https://doi.org/10.1180/claymin.1984.019.4.02.

[32] SCHWERTMANN U, SCHULZE D G, MURAD E. Identification of ferrihydrite in soils by dissolution kinetics, differential X-ray diffraction, and Mossbauer spectroscopy[J]. Journal of the Soil Science Society of America, 1982, 46(4): 869-875. https://doi.org/10.2136/sssaj1982.03615995004600040040x.

[33] SWAMY Y V, KAR B B, MOHANTY J K. Physico-chemical characterization and sulphatization roasting of low-grade nickeliferous laterites[J]. Hydrometallurgy, 2003, 69(1-3): 89-98. https://doi.org/10.1016/s0304-386x(03)00027-6.

[34] TARTAJ P, CERPA A, GARCIA-GONZALEZ M T, et al. Surface instability of serpentine in aqueous suspensions[J]. Journal of Colloid and Interface Science, 2000, 231(1): 176-181. https://doi.org/10.1006/jcis.2000.7109.

[35] THUBAKGALE C K, MBAYA R K K, KABONGO K. A study of atmospheric acid leaching of a South African nickel laterite[J]. Minerals Engineering, 2013, 54: 79-81. https://doi.org/10.1016/j.mineng.2013.04.006.

[36] THOMPSON G, SENANAYAKE G. Effect of iron (II) and manganese (II) on oxidation and co-precipitation of cobalt (II) in ammonia/ammonium carbonate solutions during aeration: An update and insight to cobalt losses in the Caron process for laterite ores[J]. Hydrometallurgy, 2018, 181: 53-63. https://doi.org/10.1016/j.hydromet.2018.07.017.

[37] TIAN Q, DONG B, GUO X, et al. Comparative atmospheric leaching characteristics of scandium in two different types of laterite nickel ore from Indonesia[J]. Minerals Engineering, 2021, 173: 107212. https://doi.org/10.1016/j.mineng.2021.107212.

[38] VALIX M, CHEUNG W H. Study of phase transformation of laterite ores at high temperature[J]. Minerals Engineering, 2002, 15(8): 607-612. https://doi.org/10.1016/s0892-6875(02)00068-7.

[39] WANDERLEY K B, BOTELHO JUNIOR A B, ESPINOSA D C R, et al. Kinetic and thermodynamic study of magnesium obtaining as sulfate monohydrate from nickel laterite leach waste by crystallization[J]. Journal of Cleaner Production, 2020, 272: 122735. https://doi.org/10.1016/j.jclepro.2020.122735.

[40] WANG X, SUNS T, WU S, et al. A novel utilization of Bayer red mud through co-reduction with a limonitic laterite ore to prepare ferronickel[J]. Journal of Cleaner Production, 2019, 216: 33-41. https://doi.org/10.1016/j.jclepro.2019.01.176.

[41] WATARI F, DELAVIGNETTE P, LANDUTY J V, et al. Electron microscopic study of dehydration transformations. Part Ⅲ: High resolution observation of the reaction process FeOOH → Fe_2O_3[J]. Journal of Solid State Chemistry, 1983, 48(1): 49-64. https://doi.org/10.1016/0022-4596(83)90058-0.

[42] WELLS M A, FITZPATRICK R W, GILKES R J. Thermal and mineral properties of Al-, Cr-, Mn-, Ni- and Ti-substituted goethite[J]. Clays and Clay Minerals, 2006, 54(2): 176-194. https://doi.org/10.1346/ccmn.2006.0540204.

[43] ZENG L, ZHANG G, XIAO L, et al. Direct solvent extraction of nickel from sulfuric acid leach solutions of low grade and complicated nickel resources using a novel extractant of HBL110[C]//Symposium on Rare Metal Extraction and Processing held during 145th The-Minerals-Metals-and-Materials-Society Annual Meeting and Exhibition, Nashville, TN, 2016: 47-53. https://doi.org/10.1002/9781119274834.ch5.

[44] ZHANG P, GUO Q, WEI G, et al. Leaching metals from saprolitic laterite ore using a ferric chloride solution[J]. Journal of Cleaner Production, 2016, 112: 3531-3539. https://doi.org/10.1016/j.jclepro.2015.10.134.

[45] ZHANG P, SUN L, WANG H, et al. Surfactant-assistant atmospheric acid leaching of laterite ore for the improvement of leaching efficiency of nickel and cobalt[J]. Journal of Cleaner Production, 2019, 228: 1-7. https://doi.org/10.1016/j.jclepro.2019.04.305.

[46] ZHANG L, GUO X, TIAN Q, et al. Improved thiourea leaching of gold with additives from calcine by mechanical activation and its mechanism[J]. Minerals Engineering, 2022, 178: 107403. https://doi.org/10.1016/j.mineng.2022.107403.

[47] ZHOU S, WEI Y, LI B, et al. Kinetics study on the dehydroxylation and phase transformation of $Mg_3Si_2O_5(OH)_4$[J]. Journal of Alloys and Compounds, 2017, 713: 180-186. https://doi.org/10.1016/j.jallcom.2017.04.162

Selective Recovery of Gold from Dilute Aqua Regia Leachate of Waste Printed Circuit Board by Thiol-modified Garlic Peel

Abstract: Garlic peel (GP) was chemically modified by using thiourea under hydrothermal conditions, which could selectively adsorb gold ions from the diluted aqua regia media directly without needing the dangerous evaparation operation. The synthetic chloroauric solution and practical leach liquor of the waste PCB powders in aqua regia were employed to assess the adsorption efficiency on the thiol-GP and the commercial anion resin of D201, respectively. It was confirmed that, the adsorption efficiency of gold onto the thiol-GP and D201 resin both reached 100%, and the maximum adsorption capacity of thiol-GP gel was evaluated as 42.59 mg Au/g that was much larger than that of D201 resin (3.33 mg Au/g). The Thio-GP gel adsorption efficiency of other coexisting base metal ions like Cu^{2+}, Ni^{2+}, Al^{3+} and Fe^{3+} from dilute aqua regia leach liquor of the waste PCB powder was near zero, and only gold could be enriched by selective adsorption onto the thiol-GP gel. At least 3 cycles of adsorption/elution could be obtained without decreasing the adsorption efficiency drastically for gold adsorption onto the thiol-GP gel. The adsorbed gold on the thiol-GP was able to be eluted effectively by using the mixture solution of 0.1 mol/L thiourea and 0.1 mol/L hydrochloric acid, and finally the soild gold could be recovered completely by sodium borohydride through a reduction process. This study demonstrated a green, environmentally friendly, low-cost, and efficient method for selective recovery of gold from the dilute leach liquor (aqua regia) of waste circuit boards.

Keywords: Aqua regia; Biosorption; Waste PCB; Gold; Thiol-garlic peel

1 Introduction

Gold, as one of the typical precious metals, has excellent corrosion resistance and ductility, as well as good electrical and thermal conductivity. It is widely used in the fields of electrics

---

Published in *Environmental Science and Pollution Research*, 2022, 29(37): 55990-56003. Authors: Zhu Jiajun, Huang Kai.

(Watanabe et al. 2020), catalysis (Huang et al., 2020), jewelry (Biswas et al., 2021), electronics (Hashem et al., 2020), medical treatment (Xue et al., 2019), aerospace, communications (Wang et al., 2019), and so on. Due to the incresingly depleted deposit in nature, the recovery of gold from various scraps like E-waste has caused a lot of concern of people, and its mining from the city ore has become a profitable industry.

The waste PCB is one of the typical E-waste that contained quite a few valuable metals like copper, nickel, zinc, gold and silver, etc. Usually, a circuit board is mainly, composed of three parts, including brominated epoxy resin, glass fiber and metals (Pietrellil et al., 2019; Zhang et al., 2021; Sodha et al., 2019). The composition of the glass fiber is silicon dioxide which is the bone of the circuit board that makes it have a certain mechanical strength. The brominated epoxy resin of the circuit board can help to lower the temperature of computing, and ensures that the electronic devices will not burn out at a relatively high temperature. The metals in the circuit board include copper, iron, nickel, tin and others (Seabra et al., 2020; Touze et al., 2020; Korf et al., 2019), in which a very small of gold will be used as the conducting parts in the circuit board to keep its excellent contacting performance. With the increasing of the number of electronic products year by year, the amount fo waste PCB correspondingly rose quickly (Ismail et al., 2019; Isildar et al., 2019; Tipre et al., 2021; Goleva et al., 2019; Einasr et al., 2020; Thakur et al., 2020). Comparing to the traditional crude ore in nature, the grade of the valuable metals contained in the waste PCB is much higher, which was the great value of research. Many technologies had been developed to recover the valuable metals from the waste PCB, including pyrometallurgical and hydrometallurgical processes, which were usually converted and improved from the conventional smelting processes. Before the smelting processes, there was a very important procedure for waste PCB treatment, i.e., disassembly and crushing the circuit boards. Disassembly to remove small electronic components from waste circuit boards, which usually took a long time and was tedious in current industry (Kumar et al., 2020). Copper pyrometallurgical processes were typically applied in the e-waste recovery in the inustrial practice, and in which the copper or matte was the excellent capture to enrich gold and other metals, and separate them from the valueless substances as the slag species. By this technique, the produced matte or crude copper contained a lot of valuable metals including gold and silver that could be further treated by roasting and leaching, then the gold could be extracted into the solution in the solubale species and condensed into the crude metal again from the aqueous solution. In above pyrometallurgical process of waste PCB, it occurred in a very short time and had high efficiency, but large pollution might be produced. For example, brominated epoxy resin might produce dioxins at high temperature, and as well known the toxicity of dioxins is ten times that of arsenic (Liu et al., 2019; Srivastava et al., 2020). So for the consideration of environmetnal safe, the more advanced technologies were in developing for avoidance of the dioxins production in current pyrometallurgical process of waste PCB. Another technical choice was by complete hydrometallurgical process for the waste PCB smelting, which ususally included the leaching of the waste PCB powder, enriching the gold from the leaching solution by cementation (Das Graças et al., 2019; Shen et al., 2019; Johnson et al. , 2019; Páez-Vélez et al. , 2019; Fan et al. , 2020) or solvent extraction(Perez et al. , 2019; Hsu et al. , 2019; Sodha et al. , 2019; Sodha et al. , 2020), and finally recovering the gold by reduction and precipitation. Hydrometallurgical recycling of gold from waste PCB did not produce dioxins and dust, but the leaching time was long, the extraction efficiency of the gold might be low. Due to the above reasons, the industrial plants preferred to use pyrometallurgy to recycle metals from waste circuit boards rather than hydrometallurgy. But if taking it into account of the eco-friendliness of the smelting process, the hydrometallurgical process showed much better competitive advantages than the pyrometallurgical one.

In a typical hydrometallurgical process, the leaching and cementation were extremely important steps for the recycling of the gold from waste PCB. Aqua regia was the frquently used lixiviant in the industry for gold recovery, but the evaporation of nitric acid by heating up to boiling point would cause sereve pollution of the workplace that was a quite headache problem. Considering the very dilute concentration of gold ions in the leachate, so the traditonal cementation process was not so efficient and economical by zinc, iron or aluminium powder. The best choice was by adsorption, and the resins had been extensively used in the industry, like the resin of D201 containing the ligand of quaternary ammonia that had strong affinity to the gold ions. But this resin was expensive, and small adsorption capacity from the dilute aqua regia media. Moreover, this resin was easy to be decayed when contacted the dilute aqua regia due to the oxidizing and strong acidity of aqua regia. Murakami had ever proposed recovery of gold from waste LED using ion exchange resin in the diluted aqua regia (Murakami et al. , 2015), but the resin had poor adsorption selectivity. So it was quite necessary to develop better adsorbents for gold recovery from the dilute aqua regia.

In this work, a thiol-GP gel was proposed that could absorb gold under a 10-fold dilution of aqua regia system, which had better adsorption selectivity, lower price, and higher gold adsorption capacity. It was worth to mentioning that biosorption technology could be applied to recovering gold in aqua regia leachate of the waste circuit board powder directly, without evaporation of the nitric acid before sorption. This method was verified with high recovery rate of gold and low cost.

2 Experimental

2.1 Chemicals

The chemicals used in present study were all of analytical grade and did not require further purification. Nitric acid (HNO_3, 63.01, ≥68.0%-70.0%), hydrochloric acid (HCl, 36.46, 36%-38%), ferric chloride ($FeCl_3$, 162.21, 98%), copper chloride ($CuCl_2$, 134.45, ≥98.5%), nickel chloride ($NiCl_2$, 129.60, ≥98%), and aluminum chloride ($AlCl_3$, 133.34, 99%) were purchased from Beijing Chemical Plant. Sodium borohydride ($NaBH_4$, 37.83,

98%), thiourea (CH$_4$N$_2$S, 76.12, ⩾99%) and chloroauric acid (HAuCl$_4$, 339.786, ⩾98%) were purchased from Tianjin Guangfu Fine Chemical Research Institute. The waste PCB samples were collected from the waste computers, crushed into fine powder and used for the leaching experiments. By the ICP (Inductive Coupled Plasma Emission Spectrometer, ICPS-7510, Shimadzu, Japan) analysis of the leach liquor of the PCB powder in aqua regia, the main compoents of metals in this waste PCB were determined as following, i.e., Au 0.2 mg/g and Al 20.8 mg/g, Fe 20.2 mg/g, Ni 19.4 mg/g, and Cu 338.4 mg/g, as shown in Table 1. D201 resin was purchased from Jiangsu Suqing resin company, which contained the quaternary ammonium ligand as the primary functional group for anions exchange.

Table 1　Main components in the waste PCB powder

Metal	Au	Al	Fe	Ni	Cu
Content/(mg·g^{-1})	0.2	20.8	20.2	19.4	338.4

2.2　Characterization

An IR Tracer-100 Fourier transform infrared spectrometer (FTIR) was used to identify the functional groups in the adsorbents before and after adsorption. Scanning electron microscopy (SEM) and energy dispersive X-ray spectroscopy (EDX) manufactured by ZEISS (EVO18) were used to observe the morphology and elemental compositions of the adsorbent. X-ray photoelectron spectroscopy (XPS) was used to analyze the elemental valent stated loaded on the adsorbent. The N$_2$ adsorption measurement of the Brunauer-Emmett-Teller (BET, ASAP 2020HD88, USA) surface area of the garlic peel and pore size distribution was performed by the Barrett-Joyner-Halenda (BJH) method from the desorption branch of the isotherm curves.

2.3　Leaching of waste PCB

Waste PCB were crushed and sieved into the size below 100 mesh and used for leaching in the aqua regia. In order to save the consumption of aqua regia, the copper in the PCB was removed in advance by leaching in 1 M FeCl$_3$ solution for 24 h, and then the residual powder was washed with tap water and dried at 80 ℃ for overnight. The oxidation-reduction potential (ORP) of the lixiviant was measured by ORP meter (HORIBA, 9300-10D, Japan). The obtained PCB powder was weighted and mixed with a certain volume of regia and stirred for a certain period of time. The suspension was further filtered and the filtrate solution was collected for sequential analysis or diluted by 10 factors for adsorption experiments.

2.4　Adsorption gold with thiol-GP gel

In present study, the thiol-GP gel, was prepared by a hydrothermally treatment of the mixture of raw garlic peel particle with the thiourea according to the mass ratio of 1 : 1 under 150 ℃ for

2 h. Then the above-mentioned gel was washed by distilled water by 3 times, and then dried at 80 ℃ for 12 h.

A series of batch and continuous adsorption experiments were conducted to evaluate the adsorption efficiency of thiol-GP by using the synthetic gold solution and real leach liquor of waste PCB, respectively. The main parameters including aqua regia concentration, dosage, and contact time were studied systematically. Aqua regia concentration of the solution was adjusted to 0.65 mol/L (ORP = 765 mV) to 5.2 mol/L (ORP = 926 mV), and the amount of adsorbent was chosen from 10 to 200 mg. The adsorption time period was set from 15 to 1440 min.

The kinetic data of adsorption of Gp for gold were then represented by the pseudo-second-order model, the model of Weber-Morris kinetic equations and the model of Elovich kinetic equations based on the following three equations:

$$\frac{t}{q_t} = \frac{1}{k_2 q_e^2} + \frac{t}{q_e} \quad (1)$$

where q_t is the amount adsorbed at time t, k_2 is the second order adsorption kinetic constant. By plotting t/q_t and t, the adsorption kinetic models of the pseudo-second-order model can help to confirm the value of q_e and t from the intercept and the slope.

$$q_t = k_{id} t^{1/2} + C \quad (2)$$

where q_t is the amount adsorbed at time t, k_{id} and C is the Weber-Morris adsorption kinetic constant. By plotting q_t and $t^{1/2}$, the adsorption kinetic models of the Weber-Morris can help to confirm the value of k_{id} and C from the intercept and the slope.

$$q_t = a + b \ln t \quad (3)$$

where q_t is the amount adsorbed at time t, a and b is the Elovich constant. By plotting q_t and $\ln t$, the adsorption kinetic models of the Elovich model can help to confirm the value of a and b from the intercept and the slope.

In order to examine the effect of impurity metal ions, cation of Al^{3+}, Ni^{2+}, Cu^{2+}, and Fe^{3+} was individually mixed in the gold solutions at an initial concentration of 1 g/L, respectively. After the adsorption, the solution was filtered, and the gold ion concentration after adsorption was measured by using ICP (inductively coupled plasma emission spectrometer, ICPS-7510, Shimadzu, Japan). The adsorption of gold efficiency and adsorption capacity were expressed by calculating the difference of gold concentration before and after adsorption based on the following two equations:

$$\%A = \frac{C_i - C_e}{C_i} \times 100\% \quad (4)$$

$$Q_e = \frac{C_i - C_e}{W} \times V \quad (5)$$

where C_i (mg/L) and C_e (mg/L) denote initial and equilibrium concentration of the gold, respectively. Q_e (mg/g) is adsorption capacity of the adsorbent, W (mg) is the dry weight of the adsorbent, and V (mL) is the volume of the solution for adsorption.

The maximum adsorption capacity of thiol-GP adsorbent was tested by varying the initial concentration of gold, and the adsorption isotherm models of Langmuir, D-R (Dubinin-Radushkevich), and Temkin was used to describe the adsorption behavior based on the following two equations:

$$\frac{C_e}{q_e} = \frac{1}{K_L q_m} + \frac{C_e}{q_m} \tag{6}$$

where C_e is the equilibrium concentration of gold ions in solution (mg/L); q_e is the amount adsorbed at equilibrium concentration (mg/g); q_m (mg/g) is the maximal adsorption capacity; K_L, is a binding constant. By plotting C_e/q_e and C_e, the adsorption isotherm models of Langmuir can help to confirm the value of q_m and K_L from the intercept and the slope can be obtained correspondingly.

$$\ln q_e = \ln a - b \ln^2 C_e \tag{7}$$

where C_e is the equilibrium concentration of gold ions in solution (mg/L); q_e is the amount adsorbed at equilibrium concentration (mg/g); a and b are a binding constant of D-R. By plotting $\ln q_e$ and $\ln^2 C_e$, the value of a and b from the intercept and the slope can be assessed correctly.

$$q_e = (RT/b)\ln A + (RT/b)\ln C_e \tag{8}$$

where C_e is the equilibrium concentration of gold ions in solution (mg/L); q_e is the amount adsorbed at equilibrium concentration (mg/g); $T(K)$ is absolute temperature; $R(J/(mol \cdot K))$ and A (min^{-1}) are molar gas constant and Arrhenius constant respectively; A and b are binding constants of Temkin model. By plotting q_e and $\ln C_e$, the value of A and b from the intercept and the slope can be estimated.

In order to check the feasibility to apply in the real leach liquor of waste PCB in aqua regia, a continuous operation was established and carried out in a glass column with an inner diameter of 1.0 cm and a height of 25 cm, respectively. The gold concentration in the dilute aqua regia for column adsorption experiment was set at 10 mg/L in synthetic gold solution and 18.3 mg/L in leach liquor of waste PCB. The initial aqua regia concentration was adjusted at 1.3 mol/L (ORP = 816 mV) for synthetic and real leach liquor, and the flow rate was 1.0 mL/min, and the amount of adsorbent packed in the column bed was 150 mg. C_i is defined as the initial concentration of gold ions in the solution when entering the u-tube, and C_t is defined as the concentration of gold ions after being adsorbed by garlic skin at t. The column adsorption experiment was performed until the gold concentration of the effluent approached the initial level. Then the fixed bed column was flown through deionized water to remove the possible physically adsorbed impurities and gold ions, and then eluted by using the previously prepared mixture of 0.1 mol/L HCl and 0.1 mol/L CH_4N_2S. The condensed elute was reduced by adding sodium borohydride solid to obtain gold particles, and the amount of sodium borohydride was chosen from 1 g/L to 20 g/L. The EDS was used to identify the elemental components of the reduced product.

3 Results and discussion

3.1 FTIR

Fig. 1 showed the results tested by Fourier transform infrared spectrometry before and after adsorption in 1.3 M aqua regia. It could be seen that the FTIR patterns for thiol-GP and thiol-GP-gold were quite different in their fingerprint wave numbers range, i.e., peaks at 3546.94 cm^{-1}, 3328.99 cm^{-1} and 601.77 cm^{-1} could be ascribed to the tensile and bending vibrations of hydroxyl groups of crystalline water molecules on the thiol-GP, those at 2910.46 cm^{-1}, 2057.96 cm^{-1}, 1623.99 cm^{-1}, 1317.32 cm^{-1}, 1103.23 cm^{-1} could be assigned to the —CH$_2$— bond, the —C≡C≡N— bond, the —NH$_2$— bond, the —C—O— bond and the —C—S— bond, respectively. The peaks at 775.35 cm^{-1} and 663.49 cm^{-1} could be ascribed to the —S—S— bond. By comparing the peaks before and after adsorption, it was easy to confirm that the —S—S— bond and NH$_2$— bonds on the thiol-GP surface were critical to the effective adsorption of gold.

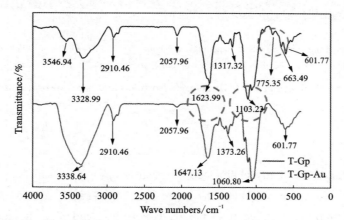

(initial gold concentration = 10 mg/L, aqua regia concentration = 1.3 mol/L, ORP = 816 mV, temperature = 23 ℃, solution volume = 10 mL, contact time = 60 min, weight of adsorbents = 100 mg)

Fig. 1 FTIR results of thiol-GP before adsorption and after adsorption for gold

3.2 SEM

Fig. 2 demonstrated the morphology and elemental compositions of the thiol-GP gel before and after adsorption of gold in the dilute aqua regia (1.3 mol/L, ORP = 816 mV). SEM photos showed that its surface was slightly compact and rough, and after adsorption of gold, the surface almost kept its original appearance. It could be clearly seen from the energy dispersive X-ray spectrum that the original composite of the thiol-GP gel mainly contained C, O and S elements

(Nitrogen not detected), and after adsorption the characteristic peaks of gold appeared that indicated the successful adsorption of gold. By combining with the FTIR results shown in Fig. 1, it could be confirmed that the surface of the thiol-GP was covered by gold element.

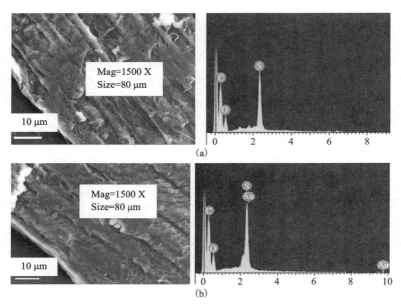

(initial gold concentration = 10 mg/L, aqua regia concentration = 1.3 M, ORP = 816 mV, temperature = 23 ℃, solution volume = 10 mL, contact time = 60 min, weight of adsorbents = 100 mg)

Fig. 2　SEM photos of thiol-GP before adsorption and after adsorption for gold

3.3　XPS

Fig. 3 demonstrated the X-ray photoelectron spectroscopy (XPS) analysis results of thiol-GP gel before and after adsorption of gold in the dilute aqua regia. Fig. 3(a)-(d) showed the binding energy peaks of the individual elements O 1 s, N 1 s, C 1 s, and S 2p. It showed that the broad peaks of the thiol-GP before adsorption indicated that O 1 s, N 1 s, C 1 s, and S 2p, were present in the adsorbent, and their binding energies were 531.17 eV, 398.44 eV, 283.44 eV, 162.45 eV and 168.01 eV, respectively. While it showed that the broad peaks of the thiol-GP after adsorption indicated that O 1 s, N 1 s, C 1 s, and S 2p, were present in the adsorbent, and their binding energies were 531.27 eV, 398.45 eV, 283.38 eV and 162.5 eV, respectively. Above peak variation fully confirmed the adsorption of gold on the adsorbent. The binding energy of N 1 s, was converted from 398.44 eV to 398.45 eV, which might be due to the adsorption of gold by $-NH_2-$ bond. And the binding energy of S 2p, was converted from 162.45 eV and 168.01 eV to 162.5 eV, which might be due to the adsorption of gold by the $-S-S-$ bond and the $-S-H-$ bond. After the adsorption of gold, a broad peak of Au 4f with a binding energy of 83.66 eV and 87.30 eV appeared. Owing to the spin-orbit splitting, the Au 4f spectrum exhibited double characteristic peaks. By combining with the FTIR, SEM and EDS results shown in Fig. 1 and Fig. 2, it could be confirmed that the surface of the thiol-GP was covered by gold element, and the bonds of $-S-S-$, $-S-H-$, and $-NH_2$ on the thiol-GP surface were critical to the effective adsorption of gold.

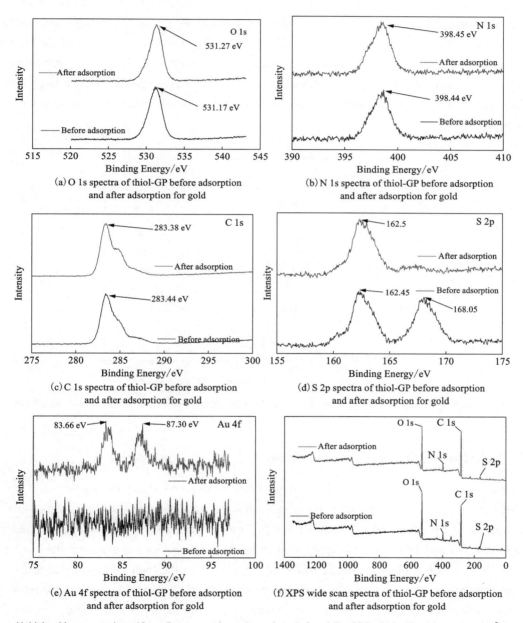

(initial gold concentration = 10 mg/L, aqua regia concentration = 1.3 mol/L, ORP = 816 mV, temperature = 23 ℃, solution volume = 10 mL, contact time = 60 min, weight of adsorbents = 100 mg)

Fig. 3　XPS results of thiol-GP before adsorption and after adsorption for gold

3.4　BET N_2 adsorption-desorption and pore size distribution

Fig. 4 showed the adsorption-desorption isotherm of the thiol-GP gel and the inserted small image showed the presence of a mesoporous structure in the adsorbent. The adsorption-desorption isotherm of the thiol-GP gel showed a type-Ⅲ isotherm. The multi-point BET specific surface area of the adsorbent was measured by static capacity method to be 2.236 m^2/g, and the total adsorption pore volume was 0.0221 cm^3/g. The inset of Fig. 4 showed that most of the pore size is distributed around 5.57 nm and the average pore diameter was 3.95 nm, indicating the existence of micropores. So many tiny mesopores and micropores suggested that the large specific surface area of thiol-GP gel would be beneficial to the effective adsorption of gold from the dilute aqua regia.

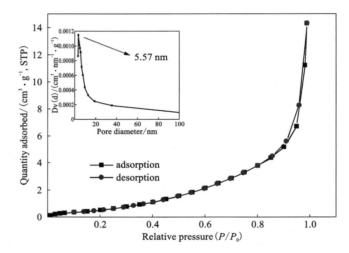

Fig. 4　Bet N_2 adsorption-desorption and pore size distribution(inset) of the thiol-GP adsorbent

3.5　Effect of aqua regia concentration

Fig. 5 indicated the effect of aqua regia concentration on the gold adsorption onto the modified garlic peel gel and D201 resin respectively, and it could be seen that when the concentration of aqua regia varied from 0.65 to 5.2 mol/L, the adsorption efficiency of gold onto the thiol-GP gel and D201 had the quite similar adsorption behavior, and almost decresed from 92% to 79%, which could be assigned to the inhabitation from the high ORP potential in the case of high concentration of aqua regia that might lead to destruction of sorbent to some extent. It was well known that the aqua regia was the stong oxidation reagent due to the nitrci acid, and the ORP potential values of its dilution would increase from 765 mV at 0.65 mol/L to 926 mV at 5.2 mol/L. The small decrease of adsorption efficiency of gold onto the thiol-GP gel made it more convenient for biosorption recovery after leaching of gold from the waste PCB.

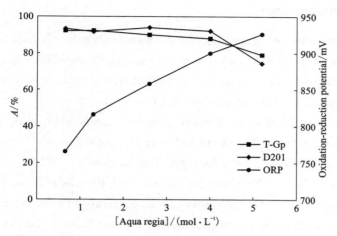

(initial gold concentration = 10 mg/L, temperature = 23 ℃, solution volume = 10 mL, contact time = 60 min, weight of adsorbents = 100 mg)

Fig. 5 Effect of aqua regia concentration on gold adsorption on thiol-GP and D201 resin

3.6 Effect of dosage

Fig. 6 showed the effect of dosage of GP on the adsorption gold efficicency, and it could be seen that with the increase of adsorbent amount in the solution, the adsorption percentage of gold increased correspondinlgy, and dosage of 10 g/L was quite enough for the effective adsorption gold in the diluted aqua regia (1.3 mol/L). Actually, in order to save the amount of GP sobrent, in the real application cases, 1 g/L dosage was recommended because the the redisual gold in the solution could be extracted completely by sequetial adsorption process in the dilute aqua regia (1.3 mol/L).

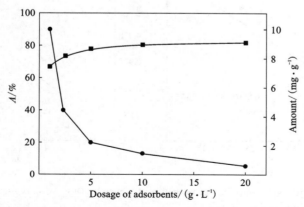

(initial gold concentration = 10 mg/L, aqua regia concentration = 1.3 mol/L, ORP = 816 mV, temperature = 25 ℃, solution volume = 10 mL, contact time = 60 min)

Fig. 6 Effect of adsorbents dosage on equilibrium time at different S/L ratio of adsorbents

3.7 Effect of contact time

Fig. 7 showed the effect of contact time on the adsorption gold effciency of thiol-GP gel, and it could be seen that only 10 min could reach a high adsorption efficiency in 1.3 mol/L aqua regia. With the prolonging of contact time, the adsorption gold efficiency would gradually approach to a platform. So fast adsorption meant that this thiol-GP sorbent had very good prospective in the real application for industry at a high rate, which made it quite attactive to shorten the gold recovery period. Murakami had ever proposed recovery of gold from waste LED using ion exchange resin in the diluted aqua regia (Murakami et al., 2015), but the resin had poor adsorption selectivity. Based on the adsorption results, with the model of the pesudo-second-order kinetic equation, the model of Weber-Morris kinetic equations and the model of Elovich kinetic equations, the related calculation parameters was listed in the Table 2 and Fig. 7(b), (c), (d). By the comparison of theoretical calculation of 0.98 mg/g and experimental data of 0.98, it was found that the result obtained by the model of pesudo-second-order kinetic equations was more reliable and the experimental process of adsorption could be better described by the model in the

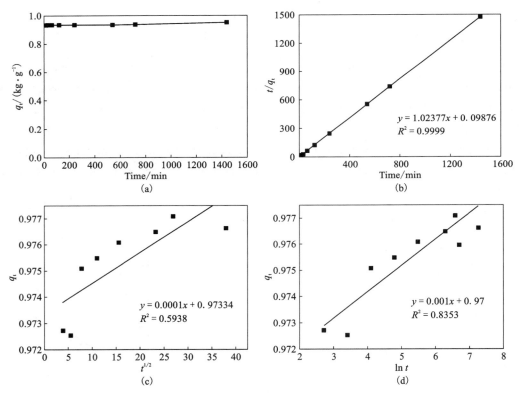

(initial gold concentration = 10 mg/L, aqua regia concentration = 1.3 mol/L, ORP = 816 mV, temperature = 25 ℃, solution volume = 10 mL, mass of adsorbents = 100 mg)

Fig. 7 Effect of contact time on gold adsorption on thiol-GP (a); the model of pesudo-second-order kinetic equations (b); the model of Weber-Morris kinetic equations (c); the model of Elovich kinetic equations (d)

dilute aqua regia. It could be seen that the adsorption of gold ions onto the thiol-GP gel was very fast, and the rate constant was determianed to be 10.59 g/(mg·min), which was quite beneficial to the quick recovery of gold from the dilute aqua regia solution directly.

Table 2 Pseudo-second-order model kinetic parameters of the gold adsorption onto the T-GP particles

$q_{e,exp}$ /(mg·g^{-1})	Pseudo-first-order model				Pseudo-second-order mode		
	q_e/(mg·g^{-1})	k_1/min^{-1}		R^2	$q_{e,cal}$/(mg·g^{-1})	k_2/[g/(mg·min)]	R^2
0.98	0.005	0.0004		0.423	0.98	10.59	0.999

3.8 Adsorption isotherm

Fig. 8 (a), (b), (c) and (d) showed the maximum capacity for gold adsorption on the thiol-GP gel, and it could be found that the maximum value was evaluated to be 35.61 mg/g in 1.3 mol/L aqua regia. Langmuir, D-R models, and Temkin model were used to fit the isotherm adsorption data, and it was found that the Langmuir model could better describe the adsorption process as shown in Fig. 8 (b) and Table 3. Based on this model, the theoretical maximum adsorption capacity can be obtained to be 42.59 mg/g, close to the experimental result. Most reseach work was reported about the gold sorption from the chloroauric solution, but seldom about the direct recovery from aqua regia media.

Fig. 8 (e) and (f) showed the maximum capacity for gold adsorption on the D201 resin, and it could be found that the maximum value was evaluated to be only 3.33 mg/g in 1.3 mol/L (ORP = 816 mV) aqua regia. Langmuir and Freundlich models were used to fit the isotherm adsorption data, and it was found that the Langmuir model could better describe the adsorption process as shown in Fig. 8(f) and Table 3, and the theoretical maximum adsorption capacity was obtained to be 3.33 mg/g, which was quite close to the experimental result.

Table 3 Parameters of the Langmuir and Freundlich isotherm models for gold adsorption

Sample	Langmuir model			Freundlich model		
	q_{max}/(mg·g^{-1})	k_1/(L·mg^{-1})	R^2	K_F/(mg·g^{-1})	n	R^2
T-GP	42.59	0.020	0.9259	0.646	1.3	0.78
D201	3.33	0.288	0.97	0.936	2.5	0.98

3.9 Selective adsorption of gold

Fig. 9 showed the selective adsorption behavior of gold ions onto the thiol-GP gel and D201 resin in 1.3 mol/L (ORP = 816 mV) aqua regia, respectively. It could be found that the adsorption of coexisting base metal ions released from the waste PCB board like Cu^{2+}, Ni^{2+}, Al^{3+}

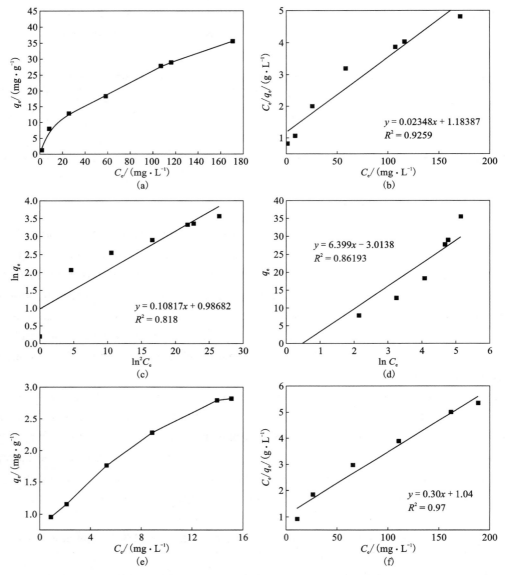

Fig. 8 Adsorption isotherms of gold on thiol-GP (a); corresponding Langmuir plot (b); corresponding D-R plot (c); corresponding Temkin plot (d) (temperature = 25 ℃, aqua regia concentration = 1.3 mol/L, ORP = 816 mV, solution volume = 10 mL, contact time = 60 min, weight of thiol-GP = 100 mg); (e) adsorption isotherm of gold on D201; (f) corresponding Langmuir plot (temperature = 25 ℃, aqua regia concentration = 1.3 M, ORP = 816 mV, solution volume = 10 mL, contact time = 60 min, weight of D201 = 100 mg)

and Fe^{3+} onto the thiol-GP gel was close to zero, while the gold adsorption was reached 100%. However, compared with the thiol-GP gel, D201 resin also had quite excellent selective adsorption for gold, and the adsorption efficiency for gold was almost the same as that of thiol-GP gel. It meant that both the thiol-GP gel and D201 resin had the quite similar selective adsoprtion behavior from the dilute aqua regia of 1.3 mol/L directly.

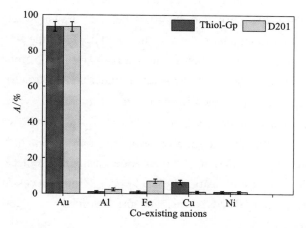

(initial gold concentration = 10 mg/L, individual Cu^{2+}, Ni^{2+}, Al^{3+} and Fe^{3+} concentration = 1 g/L temperature = 25 ℃, aqua regia concentration = 1.3 mol/L, ORP = 816 mV, solution volume = 10 mL, contact time = 60 min, weight of adsorbent = 100 mg)

Fig. 9 Selective adsorption of gold onto thiol-GP and D201 resin

3.10 Reusability

Fig. 10 showed the repeating adsorption/elution results, and it could be seen that the thiol-GP gel could maintain the adsorption efficiency for gold well even after 3 cycles of operations in dilute aqua regia of 1.3 mol/L (ORP= 816 mV). Based on above research, it could concluded that the thiol-GP gel prepared in present study could be reused quite well for the recover of gold from the leach liquor of PCB powder.

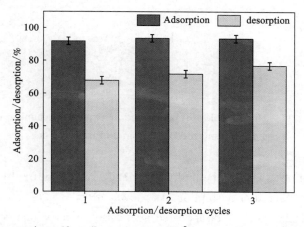

(initial gold concentration = 10 mg/L, temperature = 25 ℃, aqua regia concentration = 1.3 mol/L, ORP = 816 mV, solution volume = 10 mL, contact time = 60 min, weight of adsorbent = 100 mg, hydrochloric acid concentration = 0.1 mol/L, thiourea concentration = 0.1 mol/L)

Fig. 10 Adsorption/desorption cycle test of thiol-GP

3.11 Gold adsorption/desorption mechanism

On the basis of the above study, it was able to deduce the possible gold adsorption and desorption mechanism of D201 resin and thiol-GP gel, as drawn in Fig. 11. The D201 resin contains a quaternary amino functional group, with $-N^+(CH_3)_3$ as the fixed group and Cl^- as the mobile ion. When the solution was acidic, the D201 resin fixed group would be positively charged, attracting the anions ($AuCl_4^-$) in the solution under the electroattraction effect to form the complex of $-N^+(CH_3)_3AuCl_4^-$, as shown in the reaction formula (9). The gold on the surface of D201 resin dominantly existed in the form of $-N^+(CH_3)_3AuCl_4^-$, on which the adsorbed gold could be eluted off by using the mixture solution of hydrochloric acid and thiourea. Due to the high concentration of chloride ions and thiourea, the chloride ions would replace the adsorbed gold ions to form the more stable complex of $Au[SC(NH_2)_2]_2Cl_3$, thus successfully desorbing the gold ions on the surface of the thiol-GP adsorbent, and the desorbed resin/gel could be recycled, as shown in reaction formula (13).

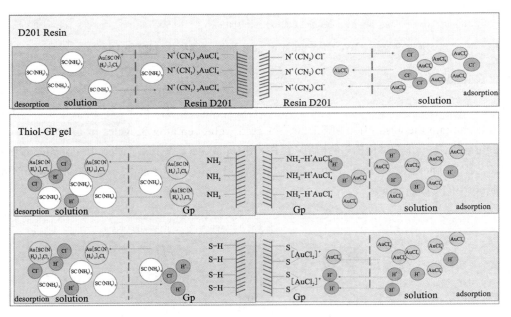

Fig. 11 The possible adsorption and desorption mechanism of gold ions onto the D201 resin and thiol-GP gel

According to SEM observation, the morphology and structure of thiol-GP gel was found to be almost no change before and after adsorption. As shown in the Fig. 1, it was easy to see that the peaks at wavenumber 1623.99 cm^{-1} disappeared after adsorption, which could be ascribed to the substitution of the amino on the thiol-GP surface by the gold in the adsorption process. And the peaks at wavenumber 775.35 cm^{-1} and 663.49 cm^{-1} disappeared after adsorption, which could be ascribed to the substitution of $-S-S-$ on the thiol-GP surface by the gold in the adsorption process. As shown in the Fig. 3, the binding energy of N 1s was converted from 398.44 eV to 398.45 eV, which may be due to the adsorption of gold by $-NH_2-$ bond. And the binding energy of S 2p was converted from 162.45 eV and 168.01 eV to 162.5 eV, which may be due to the adsorption of gold by the $-S-S-$ bond broken and the $-S-H-$ bond. According to above analysis, there were a large number of the $-S-S-$ bond, the $-S-H-$ bond, and the $-NH_2-$ bond on the surface of the tihol-modified garlic peel adsorbent. Hence, under acidic conditions, the adsorbent would be protonated causing the adsorbent showing a positive charge on its surface, which adsorbed the negatively complex of $AuCl_4^-$, as shown in formula (10). And under acidic conditions, the $-S-S-$ bond would be broken and the $-S-H-$ bond would be produced, as shown in formula (11). The $-S-H-$ bond of the Gp adsorbed the gold by the ion exchange, as shown in formula (12). The gold on the surface of the adsorbent mainly existed in the form of $[GP-NH_3]AuCl_4$ and $Gp-S-[AuCl_2]^+-S-Gp$, and the gold on the surface of the adsorbent was desorbed under the conditions of hydrochloric acid and thiourea. Due to the stronger coordination effect of thiourea, the adsorbed gold ions would be eluted off by thiourea to form a more stable $Au[SC(NH_2)_2]_2Cl_3$, as shown in reaction formula (14) and (15).

Adsorption process:

$$D201 - N^+(CH_3)_3Cl^- + AuCl_4^- = D201 - N^+(CH_3)_3AuCl_4^- + Cl^- \quad (9)$$

$$GP - NH_2 + H^+ + AuCl_4^- = [GP - NH_3]^+ + AuCl_4^- = [GP - NH_3]AuCl_4 \quad (10)$$

$$GP - S - S - GP + 2H^+ = 2GP - S - H \quad (11)$$

$$2GP - S - H + AuCl_4^- = GP - S - [AuCl_2]^+ - S - GP + 2Cl^- + 2H^+ \quad (12)$$

Elution process:

$$D201 - N^+(CH_3)_3AuCl_4^- + 2SC(NH_2)_2 =$$
$$Au[SC(NH_2)_2]_2Cl_3 + D201 - N^+(CH_3)_3Cl \quad (13)$$

$$[GP - NH_3]AuCl_4 + 2SC(NH_2)_2 = Au[SC(NH_2)_2]_2Cl_3 + HCl + GP - NH_2 \quad (14)$$

$$GP - S - [AuCl_2]^+ - S - GP + 2SC(NH_2)_2 + 2H^+ + Cl^- =$$
$$Au[SC(NH_2)_2]_2Cl_3 + 2GP - S - H \quad (15)$$

3.12 Column adsorption

Fig. 12 showed the breakthrough profiles of the column adsorption by flowing with the synthetic gold solution in 1.3 mol/L auqa regia (ORP = 816 mV). It could be found that the gold in the synthetic solution was effectively adsorbed onto the thiol-GP gel column, and even after more than 200 min percolating the gold cocentration in the flow-out solution was still kept almost zero, verifying the very strong adsorption capability of thiol-GP gel for gold ion capture. After adsorption, the mixing solution of 0.1 mol/L HCl and 0.1 mol/L CH_4N_2S was used to recover the adsorbed gold from the column bed, and it could be seen that the elution was effective and the solution could be condensed at least 12 factors of the initial concentration of gold.

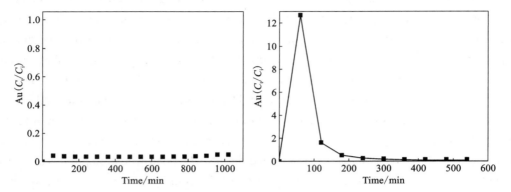

(initial gold concentration = 19 mg/L, temperature = 25 ℃, solution volume = 200 mL, flow rate = 1.0 mL/min, weight of adsorbents = 150 mg, bed depth = 1 cm, aqua regia concentration = 1.3 mol/L, ORP = 816 mV, hydrochloric acid concentration = 0.1 mol/L, thiourea concentration = 0.1 mol/L)

Fig. 12 Breakthrough curves of synthetic solution in fixed bed column by thiol-GP

3.13 Reducing agent on the precipitation yield of gold

Fig. 13 showed the effect of dosage of sodium borohydride on the recovery efficicency for gold, and it could be seen that even a very small aliquot of sodium borohydride could lead to the almost complete recovery of gold from the eluted solution, and more reducing agent was not necessary to increase the preciptiation efficiency, as shown in Fig. 13, and dosage of 1 g/L was quite enough for the effective precipitation gold.

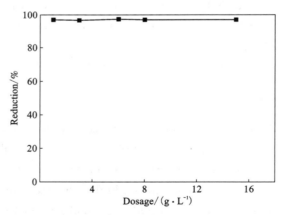

(initial gold concentration = 24.28 mg/L, temperature = 25 ℃, solution volume = 10 mL, contact time = 60 min, hydrochloric acid concentration = 0.1 mol/L, thiourea concentration = 0.1 mol/L)

Fig. 13 Effect of reducing agent dosage on equilibrium time at different S/L ratio of reducing agent

3.14 Practical separation and recovery of gold from actual leach liquor

Fig. 14 showed the breakthrough profiles of the column adsorption by real leach liquor of gold from the waste PCB in 1.3 M aqua regia. It could be found that the gold in the real leachate was also effectively adsorbed onto the column, and almost 11 factors condensed gold solution was obatined efficiently by using the mixing solution of 0.1 mol/L HCl and 0.1 mol/L CH_4N_2S for elution.

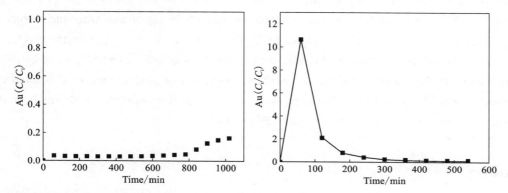

(initial gold concentration = 18.3 mg/L, solution volume = 200 mL, contact time = 200 min, temperature = 25 ℃, flow rate = 1.0 mL/min, weight of adsorbents = 150 mg, bed depth = 3.5 cm, aqua regia concentration = 1.3 mol/L, ORP = 816 mV, hydrochloric acid concentration = 0.1 mol/L, thiourea concentration = 0.1 mol/L)

Fig. 14 Breakthrough curve of real leach liquor of waste PCB in fixed bed column by thiol-GP gel

Fig. 15 demonstrated the morphology and elemental composition of the precipitated particles obtaiend by adding into the eluted solution with sodium borohydride. The SEM and EDS photo showed that the eluted gold off the thiol-GP column was successfully recovered as the gold partilces by reduction. And the SEM photo showed that the gold particles were aggregated together into larger one that would be beneficial to the solid/liquid seapration. The impurities such as chloride and sodium were also found in the precipitated particles.

(temperature = 25 ℃, solution volume = 100 mL, contact time = 60 min, pH = 3, weight of reducing agent = 150 mg, hydrochloric acid concentration = 0.1 mol/L, thiourea concentration = 0.1 mol/L)

Fig. 15 SEM and EDS photos of the precipitation

4 Conclusions

The Thiol-GP gel was successfully prepared with the selective adsorption capability of gold directly in 1.3 M aqua regia, which probably offered a green, environmentally friendly, low-cost, and efficient method for recovering gold from the leach liqour (aqua regia) of waste circuit boards. The adsorption efficiency of gold was close to 100% on the thiol-GP gel that was quite close to the D201 resin, and the adsorption capacity was 42.59 mg/g that was much larger than that of D201 resin (3.33 mg/g). The waste PCB powder was leached by aqua regia, and then the leach liquor was diluted 10-fold for direct adsorptive recovery of gold selectively by thiol-GP gel. It was found that the thiol-GP gel had the selective adsorption of gold and the new material could be reused at least 3 times of adsorption/elution, showing good prospective in the real industrial application of recovering gold from the waste PCB board.

Funding

The present work was financially supported by the special fundamental funds by National Key Research and Development Plan Key Special Projects for the key technology and engineering demonstration for multi-component deep separation and extraction of sulfurbased nickel and cobalt slag (Grant No. 2019YFC1907403).

References

[1] BISWAS F B, RAHMAN I M M, NAKAKUBO K, et al. Highly selective and straight forward recovery of gold and platinum from acidic waste effluents using cellulose-based bio-adsorbent[J]. Journal of Hazardous Materials, 2021, 410: 124569-124580. https://doi.org/10.1016/j.jhazmat.2020.124569

[2] DAS GRAÇAS SANTOS N T, DA SILVA M G C, VIEIRA M G A. Development of novel sericin and alginate-based biosorbents for precious metal removal from wastewater[J]. Environmental Science and Pollution Research, 2019, 26: 28455-28469. https://doi.org/10.1007/s11356-018-3378-z

[3] EINASR R S, ABDELBASIR S M, KAMEL A H, et al. Environmentally friendly synthesis of copper nanoparticles from waste printed circuit boards[J]. Separation and Purification Technology, 2020, 230: 115860-115870. https://doi.org/10.1016/j.seppur.2019.115860

[4] FAN R Y, MIN H Y, HONG X X, et al. Plant tannin immobilized Fe_3O_4@SiO_2 microspheres: A novel and green magnetic bio-sorbent with superior adsorption capacities for gold and palladium[J]. Journal of Hazardous Materials, 2019, 364: 780-790. https://doi.org/10.1016/j.jhazmat.2018.05.061

[5] GOLEV A, CORDER G D, RHAMDHANI M A. Estimating flows and metal recovery values of waste printed circuit boards in Australian e-waste[J]. Minerals Engineering, 2019, 137: 171-176. https://doi.

org/10. 1016/j. mineng. 2019. 04. 017

[6] HASHEM M A, ELNAGAR M M, KENAWY I M, et al. Synthesis and application of hydrazo-no-imidazoline modified cellulose for selective separation of precious metals from geological samples[J]. Carbohydrate Polymers, 2020, 237: 116177-116184. https: //doi. org/10. 1016/j. jcis. 2020. 01. 094

[7] HSU E, BARMAK K, WEST A C, et al. Advancements in the treatment and processing of electronic waste with sustainability: A review of metal extraction and recovery technologies[J]. Green Chemistry, 2019, 21: 919-936. https: //doi. org/10. 1039/C8GC03688H

[8] HUANG B, HU M, TOSTE F D. Homogeneous gold redox chemistry: Organometallics, catalysis, and beyond[J]. Trends in Chemistry, 2020, 2: 707-720. https: //doi. org/10. 1016/j. trechm. 2020. 04. 012

[9] ISILDAR A, VAN HULLEBUSCH E D, LENZ M, et al. Biotechnological strategies for the recovery of valuable and critical raw materials from waste electrical and electronic equipment (WEEE): A review[J]. Journal of Hazardous Materials, 2019, 362: 467-481. https: //doi. org/10. 1016/j. jhazmat. 2018. 08. 050

[10] ISMAIL H, HANAFIAH M M. An overview of LCA application in WEEE management: Current practices, progress and challenges[J]. Journal of Cleaner Production, 2019, 232: 79-93. https: //doi. org/10. 1016/j. jclepro. 2019. 05. 329

[11] JOHNSON C R, FEIN J B. A mechanistic study of Au (III) removal from solution by Bacillus subtilis [J]. Geomicrobiology Journal, 2019, 36: 506-514. https: //doi. org/10. 1080/01490451. 2019. 1573279

[12] KORF N, LØVIK A N, FIGI R, et al. Multi-element chemical analysis of printed circuit boards: Challenges and pitfalls[J]. Waste Management, 2019, 92: 124-136. https: //doi. org/10. 1016/j. wasman. 2019. 04. 061

[13] KUMAR A, HOLUZKO M E, JANKE T. Assessing the applicability of gravity separation for recycling of non-metal fraction from waste printed circuit boards[J]. Advanced Sustainable Systems, 2020: 2000231-2000238. https: //doi. org/10. 1002/adsu. 202000231

[14] LIU H, ZHANG H, PANG J, et al. Superhydrophobic property of epoxy resin coating modified with octadecylamine and SiO_2 nanoparticles[J]. Materials Letters, 2019, 247: 204-207. https: //doi. org/10. 1016/j. matlet. 2019. 03. 128

[15] MURAKAMI H, NISHIHAMA S, YOSHIZUKA K. Separation and recovery of gold from waste LED using ion exchange method[J]. Hydrometallurgy, 2015, 157: 194-198. https: //doi. org/10. 1016/j. hydromet. 2015. 08. 014

[16] PÁEZ-VÉLEZ C, RIVAS R E, DUSSÁN J. Enhanced gold biosorption of Lysinibacillus sphaericus CBAM5 by encapsulation of bacteria in an alginate matrix[J]. Metals, 2019, 9: 818-827. https: //doi. org/10. 3390/met9080818

[17] PEREZ J P H, FOLENS K, LEUS K, et al. Progress in hydrometallurgical technologies to recover critical raw materials and precious metals from low-concentrated streams [J]. Resources, Conservation and Recycling, 2019, 142: 177-188. https: //doi. org/10. 1016/j. resconrec. 2018. 11. 029

[18] PIETRELLI L, FERRO S, VOCCIANTE M. Eco-friendly and cost-effective strategies for metals recovery from printed circuit boards[J]. Renewable and Sustainable Energy Reviews, 2019, 112: 317-323. https: //doi. org/10. 1016/j. rser. 2019. 05. 055

[19] SEABRA D, CALDEIRA-PIRES A. Destruction mitigation of thermodynamic rarity by metal recycling [J]. Ecological Indicators, 2020, 119: 106824-106832. https: //doi. org/10. 1016/j. ecolind. 2020. 106824

[20] SHEN N, CHIRWA E M N. Short-term adsorption of gold using self-flocculating microalga from wastewater and its regeneration potential by bio-flocculation[J]. Journal of Applied Phycology, 2019, 31:

1783-1792. https：//doi. org/10. 1007/s10811-018-1670-4

[21] SODHA A B, QURESHI S A, KHATRI B R, et al. Enhancement in iron oxidation and multi-metal extraction from waste television printed circuit boards by iron oxidizing Lepto-spirillum feriphillum isolated from coal sample[J]. Waste and Biomass Valorization, 2019, 10：671-680. https：//doi. org/10. 1007/s12649-017-0082-z

[22] SODHA A B, SHAH M B, QURESHI S A, et al. Decouple and compare the role of abiotic factors and developed iron and sulphur oxidizers for enhanced extraction of metals from television printed circuit boards [J]. Separation Science and Technology, 2019, 54：591-601. https：//doi. org/10. 1080/01496395. 2018. 1512616

[23] SODHA A B, TIPRE D R, DAVE S R. Optimisation of biohydrometallurgical batch reactor process for copper extraction and recovery from non-pulverized waste printed circuit boards[J]. Hydrometallurgy, 2020：191105170-105178. https：//doi. org/10. 1016/j. hydromet. 2019. 105170

[24] SRIVASTAVA R R, ILYAS S, KIM H, et al. Biotechnological recycling of critical metals from waste printed circuit boards[J]. Journal of Chemical Technology & Biotechnology, 2020, 95：2796-2810. https：//doi. org/10. 1002/jctb. 6469

[25] THAKUR P, KUMAR S. Metallurgical processes unveil the unexplored "sleeping mines" e-waste：A review[J]. Environmental Science and Pollution Research, 2020, 27：32359-32370. https：//doi. org/10. 1007/s11356-020-09405-9

[26] TIPRE D R, KHATRI B R, THACKER S C, et al. The brighter side of e-waste-A rich secondary source of metal[J]. Environmental Science and Pollution Research, 2021, 28：10503-10518. https：//doi. org/10. 1007/s11356-020-12022-1

[27] TOUZE S, GUIGNOT S, HUBAU A, et al. Sampling waste printed circuit boards：Achieving the right combination between particle size and sample mass to measure metal content[J]. Waste Management, 2020, 118：380-390. https：//doi. org/10. 1016/j. wasman. 2020. 08. 054

[28] WANG Z, LI P, FANG Y, et al. One-step recovery of noble metal ions from oil/water emulsions by chitin nanofibrous membrane for further recycling utilization[J]. Carbohydrate Polymers, 2019, 223：115064-115071. https：//doi. org/10. 1016/j. carbpol. 2019. 115064

[29] WATANABE K, WELLING T A J, SADIGHIKIA S, et al. Compartmentalization of gold nanoparticle clusters in hollow silica spheres and their assembly induced by an external electric field[J]. Journal of Colloid and Interface Science, 2020, 566：202-210. https：//doi. org/10. 1016/j. jcis. 2020. 01. 094

[30] XUE D, LI T, LIU Y, et al. Selective adsorption and recovery of precious metal ions from water and metallurgical slag by polymer brush graphene-polyurethane composite [J]. Reactive and Functional Polymers, 2019, 136：138-152. https：//doi. org/10. 1016/j. reactfunctpolym. 2018. 12. 026

[31] ZHANG S H, GU Y, TANG A, et al. Forecast of future yield for printed circuit board resin waste generated from major household electrical and electronic equipment in China[J]. Journal of Cleaner Production, 2021, 283：124575-124587. https：//doi. org/10. 1016/j. jclepro. 2020. 124575

Thermodynamic and Experimental Analyses of the Carbothermic Reduction of Tungsten Slag

Tungsten slag contains valuable elements, such as Bi, W, Fe, Nb, and Mn, and harmful elements, such as Pb. Here, the carbothermic reduction of tungsten slag was analyzed thermodynamically using FactSage thermochemical software and verified experimentally. CaO, SiO_2, and Al_2O_3 could not react with C and were the major constituents of the slag. WO_3, Fe_2O_3, Nb_2O_5, and MnO_2 were reduced and collected in the alloy, and Bi_2O_3 and PbO were reduced and formed Bi and Pb, respectively, which volatilized and collected in the fumes. Increased temperature and decreased reductant addition contributed to an increase in the recovery of Nb, while the recovery of W was less affected. At reduction conditions of 14%-16% reductant addition and 1500-1600 ℃, the recovery percentage of W, Fe, and Nb were over 90%, the recovery percentage of Mn was over 50%, and Bi and Pb were collected in the fumes.

1 Introduction

China is a major tungsten producer. However, a significant amount of tungsten slag is produced during its production process, and the amount of stacking tungsten slag from past production has reached 1 million tons. Although tungsten slag contains harmful elements such as Pb and has been listed in the *National Catalogue of Hazardous Wastes*, it also contains valuable elements such as Bi, W, Fe, Nb, and Mn[1]. Therefore, there is an urgent need to treat tungsten slag to recover these valuable metals. Hydrometallurgical techniques including acid leaching and solvent extraction have been used to treat tungsten slag to recover Bi, Pb, W, Fe, Nb, Mn, and other valuable elements.[2-6] These techniques have the advantages of high recovery and high metal purity, but their disadvantages are significant, including a prolonged procedure and the

---

Published in *JOM*, 2021, 73(6): 1853-1860. Authors: Liao Chunfa, Xie Sui, Wang Xu, Zhao Baojun, Cai Boqing, Wang Lianghui.

inability to cope with the effluent. Alkaline treatment techniques have been used to recover W from tungsten slag by alkaline roasting, but their economic efficiency is low. The flotation metallurgy process can be used for the preliminary separation of tungsten slag, but the recovery percentage of valuable metals is not high. Ceramsite prepared with the main raw materials of diatomite and tungsten slag can be employed to effectively adsorb Cu^{2+} from wastewater and ammonium from leachate.[7,8] However, valuable elements such as W and Nb cannot be recycled. The pyrometallurgical process involves the preparation of a master alloy by the carbothermic reduction method to recover valuable elements in tungsten slag, and has the characteristics of a high additional value alloy, a shorter procedure, and reduction.[9]

Nevertheless, the mechanism of carbothermic reduction of tungsten slag remains unclear. Consequently, to better understand the carbothermic reduction behavior of tungsten slag and to reveal the reduction mechanism, the effects of temperature and reductant addition on the constituents of the reduction products and the valuable metals recovery were systematically investigated using Fact-Sage thermochemical software. The thermodynamic calculations have been compared to experimental results.

2 Materials and Methods

2.1 Materials

As shown in Table 1, Fe_2O_3, MnO_2, CaO, Al_2O_3, Na_2O, and SiO_2 were the primary constituents in tungsten slag, with trace amounts of WO_3, Nb_2O_5, Bi_2O_3, and PbO.

Table 1 Chemical composition of tungsten slag   %

Fe_2O_3	MnO_2	CaO	Al_2O_3	SiO_2	WO_3	Nb_2O_5	Bi_2O_3	PbO	Na_2O	LOL
29.07	35.45	8.65	1.83	9.88	0.66	0.86	0.68	0.34	5.41	7.17

2.2 Calculations

The Equilib Module of FactSage 8.0, based on the minimization of the total Gibbs free energy, was used to predict the equilibrium conditions in the carbothermic reduction of tungsten slag.[10,11] For simplifying the calculation, the tungsten slag compositions have been simplified to Fe_2O_3, MnO_2, CaO, Al_2O_3, SiO_2, WO_3, Nb_2O_5, Bi_2O_3, and PbO. The databases selected in FactSage 8.0 were "FactPs" "Ftoxide" "SGnobl-FCC" and "SpMCBN". The solution phases selected in calculations were "Ftoxid-SLAGA" "Ftoxid-SPINB" "Ftoxid-MeO_A" "Ftoxid-WOLLA" "Ftoxid-bC2SA" "Ftoxid-aC2SA" "FToxid-OlivA" "FToxid-Mull" "FToxid-Rhod" "SpMCBN-LIQU" "SpMCBN-FCC1" "SpMCBNBCC1" "SpMCBN-HCP1" "SpMCBN-

CBCC" "SpMCBN-CUB1" "SpMCBN-30" "SpMCBNLAV1" "SpMCBN-LAV2" "SpMCBN-M23C" "SpMCBN-M7C3" "SpMCBN-M6C" "SpMCBNM5C2" "SpMCBN-MC_2" "SpMCBN-MS1" "SpMCBN-M3S1" "SpMCBN-M5S1" "SpMCBNBCC2" "SpMCBN-222" "SpMCBN-M2SI" "SpMCBN-MNTF" "SpMCBN-MOF7" "SpMCBNMU22" "SpMCBN-CNF3" "SGnobl-FCC" "SGnobl-HCP1" "SGnobl-LIQ1" and "SGnobl-RHOM".

2.3 Experiments

Carbothermic reduction experiments were conducted using a high-temperature vertical tube furnace, which consisted of a corundum tube and was heated by molybdenum disilicide. The experimental system is shown in Fig. 1. Graphite powder was provided by Sinopharm Chemical Reagent China. Tungsten slag was thoroughly mixed with graphite and then pressed into columns using a tablet press. The tungsten slag mixtures in Al_2O_3 crucibles were heated for approximately 4 h to reach the desired temperature, then held for 2 h under a nitrogen protective atmosphere, followed by furnace cooling. The compositions of the furnace slag and alloy were analyzed by X-ray fluorescence spectroscopy and inductively coupled plasma mass spectrometry.

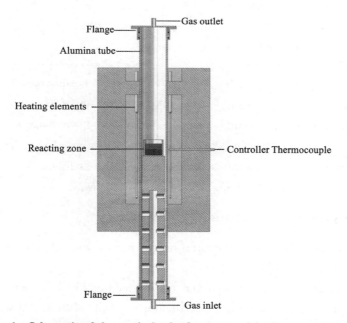

Fig. 1 Schematic of the vertical tube furnace used in the present study

3 Results and Discussion

3.1 The reduction sequence of oxides

The effect of reductant addition on the reduction percentage of oxides at 1550 ℃ is shown in Fig. 2.

As shown in Fig. 2, Bi_2O_3 partially decomposes and forms Bi and O_2 without the reductant. Pb and W begin to form when the reductant addition is 2% and 6%, respectively. Fe begins to form when the reductant addition is over 6%. When the reductant addition increases to 10%, Nb begins to form. At a reductant addition of 12%, FeO is completely reduced by the reductant. In addition, the reductions of Bi_2O_3, PbO, WO_3, and Nb_2O_5 are complete. As the reductant addition continues to increase, MnO begins to react with the reductant,

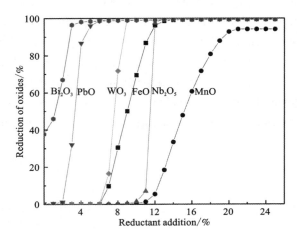

Fig. 2 Effect of reductant addition on the reduction percentage of oxides

and when the reductant addition reaches 20%, MnO is completely reduced. Therefore, during the reduction of tungsten slag, the reduction sequence of oxides is as follows:

$$Bi_2O_3 > PbO > WO_3 > FeO > Nb_2O_5 > MnO.$$

3.2 Reduction product compositions

Fig. 3(a) shows the effect of temperature on the reduction product compositions of Fe and Mn at a reductant addition of 18%.

Fe(bcc) appears at 700 ℃ and transforms into Fe(fcc) at 750 ℃. Then, Fe reacts with C to form a Fe-C alloy. FeO is entirely reduced at a temperature of 1100 ℃, at which point the alloy transforms into a molten alloy containing 4.4% carbon. As the temperature continues to increase, MnO begins to react with reductant. Because the reduction of MnO preferentially forms carbide, which dissolves into the molten alloy, the Mn and carbon contents in the molten alloy continuously increase. MnO reacts with carbon in the molten alloy once the reductant is exhausted, and thus the carbon content in the molten alloy decreases, while the Mn content continuously increases.

The reduction product compositions of W and Nb are affected by the reduction of FeO and MnO. The effect of temperature on these compositions with a reductant addition of 18% is shown in Fig. 3(b). W and Nb in the tungsten slag exist in the forms of $CaWO_4$ and Nb_2O_5,

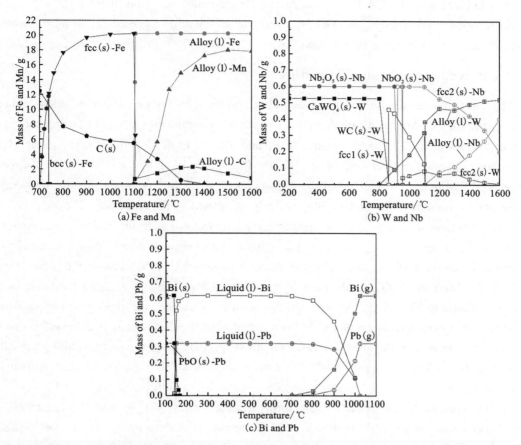

Fig. 3 Effect of temperature on the reduction product compositions at a reductant addition of 18%

respectively, at low temperatures. As the temperature increases to 860 ℃, $CaWO_4$ begins to be reduced to W, which further reacts with a reductant to form WC(s). The WC(s) partially reacts with the Fe-C alloy to form a Fe-W-C alloy. At temperatures above 920 ℃, Nb_2O_5 is reduced to NbO_2, which is further reduced and forms an Nb-C alloy as the temperature rises to 950 ℃. Subsequently, the WC(s) partially reacts with the Nb-C alloy to form an Nb-W-C alloy. Above 1100 ℃, the Fe-Mn-W-C alloy transforms into a molten alloy, and then WC(s) dissolves in the molten alloy. As the temperature further increases, the solubility of Nb in the molten alloy increases. Thus, the Nb-W-C alloy dissolves in the molten alloy and forms a Fe-Mn-W-Nb-C alloy.

As can be seen in Fig. 3(c), Bi_2O_3 is highly reactive. Bi_2O_3 and PbO are completely reduced to Bi-Pb alloy when the temperature increases to 139 ℃. The melting points of Bi and Pb are 271 ℃ and 326 ℃, while their boiling points are 1560 ℃ and 1740 ℃, respectively. Although their boiling points are high, the vapor pressures of Bi and Pb are also high at high temperatures, and the change in the saturated vapor pressure of Bi is close to that of Pb.[12] Therefore, Bi and Pb can be recovered in the fumes. At temperatures above 700 ℃, Bi and Pb start to volatilize, and at 1024 ℃, all the Bi and Pb are collected in the fumes.

As mentioned previously, W, Fe, Nb, and Mn are mainly recovered in the alloy, whereas Bi and Pb are collected in the fumes, and CaO, SiO_2, and Al_2O_3 primarily comprise the slag.

3.3 The recovery of valuable metals

The formation temperature of the molten alloy and the recovery percentage of valuable metals are related to the Fe, Mn, and carbon contents of the alloy, which are affected by temperature and the reductant addition. The melting points of Fe and Mn are 1538 ℃ and 1244 ℃, respectively. The molten formation temperature of the Fe-Mn-C alloy falls to below 1100 ℃ in the phase diagram of Fe-Mn-C.[13] The effect of the formation of low-melting-point alloys is the temperature and the carbon content in the molten alloy, which are related to the reductant addition.

The generation temperature of the molten alloy is correlated with the recovery percentage of Fe, as can be observed in Fig. 4(a). The molten transformation temperature of the alloy is required to be above 1500 ℃ at a reductant addition of less than 12% because the alloy consists mainly of Fe, and the FeO is only partially reduced. All the FeO is reduced as the reductant addition increases to 12%-13%. As the reductant addition increases, the generation temperature of the molten alloy decreases as Fe and C forms a low-melting-point alloy. It is noteworthy that the generation temperature of the alloy decreases rapidly to 1110 ℃ at a reductant addition greater than 13% owing to small quantities of MnO that react with C forming a Fe-Mn-C alloy with a low melting point.

MnO is reduced by the reductant and forms a molten alloy with Fe and C. Therefore, the recovery percentage of Mn reflects its degree of reduction. As shown in Fig. 4(b), the change in Mn recovery percentage presents as a hill shape, wherein increasing the temperature and reductant addition promotes the recovery of Mn.

The recovery percentage curves of Mn partially overlap at different reductant additions, as shown in Fig. 4(c). This suggests that the reduction of MnO is primarily controlled by temperature.

The reduction of MnO is divided into four steps, as shown in Fig. 4(c): (1) small quantities of MnO are reduced by the reductant and then react with Fe and C to form a low-melting-point alloy which melts at 1110 ℃; (2) MnO reacts with the reductant to form Mn_7C_3 and then dissolves in the molten alloy; (3) at above 1250 ℃, MnO reacts with the reductant to form Mn_2C_5 and then dissolves in the molten alloy; and (4) as the reductant is exhausted, the carbon in the molten alloy reacts with MnO.

The carbon content in the molten alloy is related to the carburization reaction and the reduction of MnO. As shown in Fig. 4(d), the change in carbon content first increases and then decreases with increasing temperature.

The change of the carbon content of the molten alloy is categorized into three stages. First, after the reduction of FeO by the reductant, Fe reacts with C and Mn to produce a low-melting-point and carbonsaturated alloy that melts at 1110 ℃. Second, the reaction of MnO and reductant

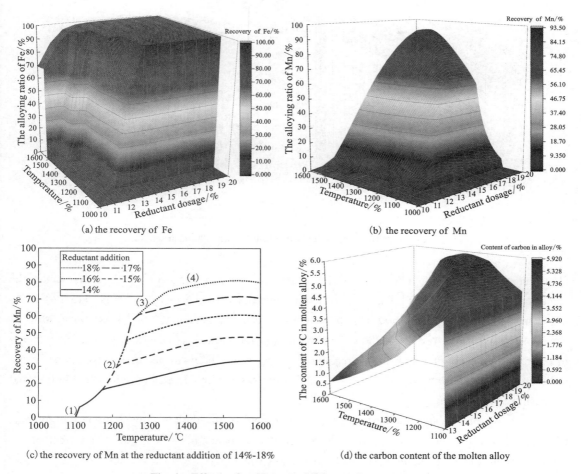

Fig. 4 Effects of reductant addition and temperature

produces carbide, which dissolves in the molten alloy,[14, 15] while the carbon content in the molten alloy also increases. Third, as the reductant is exhausted, the carbon in the molten alloy reacts with MnO, and the carbon content in the molten alloy decreases. The increase in reductant addition facilitates an increase in the recovery percentage of Mn but also results in an increase in the carbon content in the molten alloy.

Figure 5(a) shows that the recovery of W first decreases and then increases. From 1100 ℃ to 1300 ℃, the carbon content in the molten alloy increases, thereby decreasing the recovery of W. At 1300-1600 ℃, due to the consumption of carbon in the molten alloy, the carbon content in the molten alloy decreases, leading to an increase in the recovery of W. As the temperature increases beyond 1400 ℃, the recovery percentage of W begins to stabilize.

The Nb solubility in the molten alloy affects the recovery of Nb. The effect of reductant addition and temperature on the recovery of Nb in the alloy is shown in Fig. 5(b). As the temperature increases, the recovery percentage of Nb increases at the same reductant addition.

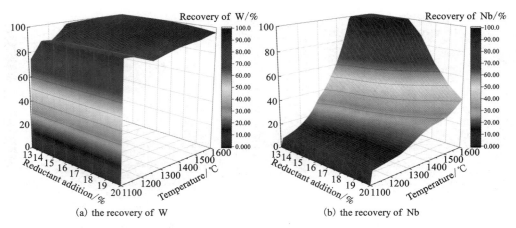

(a) the recovery of W   (b) the recovery of Nb

Fig. 5 Effects of reductant addition and temperature

However, with increasing reductant addition, the recovery percentage of Nb decreases under the same temperature conditions. The best reduction conditions for the recovery of Nb are 13%-15% reductant addition and 1500-1600 ℃, for which the recovery percentage of Nb is above 90%. It can be seen from Fig. 4(d) that the carbon content in the molten alloy increases with an increase in the reductant addition, and that increasing the temperature helps to decrease the carbon content in the molten alloy. However, the recovery of Nb increases with decreasing reductant addition and increasing temperature.

As shown previously, increasing the carbon content in the molten alloy inhibits the contents of W and Nb in the molten alloy, thus decreasing the recovery of W and Nb. The effect of the carbon content on the contents of W and Nb is shown in Fig. 6.

The inflection point of the change in the carbon content is the highest in the molten alloy. The reductant is completely consumed over the temperature range, and then carbon in the molten alloy reacts with MnO to produce Mn. Thus, the carbon content increases. With the reduction of MnO by the reductant, the gross mass of the molten alloy increases and the W content decreases. The carbon content decreases as the carbon in the molten alloy start to react with MnO, but the W content plateaus after a slight increase. However, there is a close affinity between the contents of Nb and C

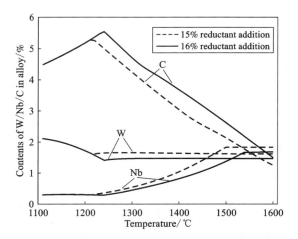

Fig. 6 Effect of the carbon content on the W and Nb contents

in molten alloy, in that an increase in the carbon content leads to a decrease in the Nb content in the molten alloy. The formation of a high-C alloy adversely affects the recovery of Nb.

In summary, the optimal reduction conditions to more efficiently recover W, Fe, Nb, and Mn are a reductant addition of 14%-16% and a temperature of 1500-1600 ℃. Over 90% of the W, Fe, and Nb and at least 50% of the Mn are recovered under these conditions.

4 Verifying experiments

To verify the computer calculations, the temperature and reductant addition were experimentally examined. The effects of reductant addition and temperature on the recovery of W, Nb, and Mn are shown in Fig. 7. The experiments of reductant addition were carried out at 1550 ℃.

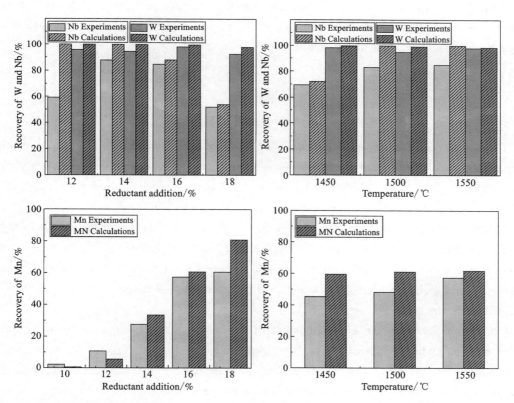

Fig. 7 Comparison of the recovery of W, Nb and Mn from calculations and experiments

The experimental results indicate that the recovery percentage of Nb becomes relatively stable with a 14%-16% reductant addition, but that the recovery of Nb decreases as the reductant addition increases to 18%. Increasing the temperature facilitates an increase in the recovery of Nb, whereas an increase in the reductant addition increases the recovery of Mn. The recovery percentage of W remains essentially stable as the reductant addition and temperature change. In addition, the alloy and slag compositions could not be determined for Bi and Pb. The Bi and Pb contents in the fumes are 14.5% and 25.5%, respectively. They are volatilized and collected in the fumes. These experimental results verify the thermodynamic analysis of the aforementioned carbothermic reduction of the tungsten slag.

The occurrence pattern of Mn in tungsten slag is simplified to MnO_2 in the thermodynamic calculations. However, the tungsten slag contains a lower valence oxide of Mn. Therefore, the discrepancy between the experimental and the predicted values is large for 10% and 12% reductant additions. Furthermore, MnO is more difficult to reduce because the silicate formation contains Mn. Moreover, the reduction of MnO is inhibited such that the activity of MnO is low at a low basicity of the slag. Further, the activity of MnO in the slag decreases with the reaction of MnO.[16, 17] As the recovery percentage of Mn reaches 60%, it tends to stabilize.[18] However, the carburization reaction continuously occurs in the alloy.[19] Therefore, the carbon content in alloy increases with the reductant addition, and thus the recovery percentage of Nb decreases. In addition, as the reductant addition increases to 20%, the particle size of the alloy drastically decreases, until it is difficult to separate the alloy from the slag.

5 Conclusion

The carbothermic reduction of tungsten slag has been investigated to inform the direction of its treatment, relating to both the advantageous and harmful compounds contained within it. Thermodynamic equilibrium conditions were calculated to investigate the reduction behavior and mechanism and to determine the appropriate tungsten slag reduction conditions.

(1) The reduction sequence of oxides is $Bi_2O_3 > PbO > WO_3 > FeO > Nb_2O_5 > MnO$ during the reduction of tungsten slag. WO_3, Fe_2O_3, Nb_2O_5, and MnO_2 are reduced to metals in this process and collected in the alloy. CaO, SiO_2, and Al_2O_3 cannot react with carbon and are the major constituents of the slag, whereas Bi_2O_3 and PbO are reduced to Bi and Pb, which volatilize and are collected in the fumes.

(2) The variation in the carbon content of the molten alloy is affected by the reduction process of MnO, which preferentially reacts with the reductant to form carbide. Consequently, the carbon content in the alloy first increases and then decreases. An increased temperature and reduced reductant addition are conducive to decreasing the carbon content of the alloy.

(3) The recovery of W and Nb is affected by the alloy's carbon content. This is beneficial for the recovery of Nb, which results in a decreased carbon content in the alloy. An increased temperature and decreased reductant addition are further conducive to increasing the recovery of Nb, but do not affect the recovery of W, which remains constant.

(4) Temperatures in the range of 1500-1600 ℃ and a reductant addition range of 14%-16% are the optimal carbothermic reduction conditions in tungsten slag. Under these conditions, the

recovery percentages of W, Fe, and Nb are over 90%, and that of Mn is more than 50%. Furthermore, Bi and Pb can be collected in the fumes.

Funding

This work was supported by the National Key Research and Development Program of China (No. 2019YFC1907404) and the Jiangxi Province Research and Development Program of China (No. 20192ACB70008). These sources were not involved in the study design, in the collection, analysis, and interpretation of data, in the writing of the report, nor in the decision to submit the article for publication.

References

[1] LIU H, LIU H L, NIE C, et al. Manag. 2020, 270: 110927.
[2] ZHONG X M, FU M S, QIN Y C. Adv. Mater. Res. 2014, 1033: 395.
[3] ZHONG X Z, FU M S, QIN Y C. J. Radioanal. Nucl. Chem. 2015, 304: 1099.
[4] NIE H P, WANG Y B, WANG Y L. Hydrometallurgy, 2018, 175: 117.
[5] NGUYEN V K, HA M G, SHIN S. J. Environ. Manag, 2018, 223: 852.
[6] DAI Y Y, ZHONG H, ZHONG H Y. Adv. Mater. Res, 2011, 236: 2554.
[7] JING Q X, WANG Y Y, CHAI L Y. Trans. Nonferrous Met. Soc, 2018, 28: 1053.
[8] JING Q X, WANG Y Y, CHAI L Y. Chin. J. Nonferrous Met, 2018, 28: 1033.
[9] WANG X, MA X D, LIAO C F. 11th International Symposium on High-Temperature Metallurgical Processing, 2020.
[10] WEN Y, TANG Q Y, CHEN J A. Int. J. Miner. Met. Mater, 2016, 23: 1126.
[11] NEZHAD S M M, ZABETT A. Calphad, 2016, 52: 143.
[12] KONG X F, YANG B, XIONG H. Trans. Nonferrous Met. Soc, 2014, 24: 1946.
[13] DJUROVIC D, HALLSTEDT B, APPEN J V. Calphad, 2011, 35: 479.
[14] HUANG W M. Metall. Mater. Trans. A, 1990, 21: 2115.
[15] RANKIN W J, WYNNYCKYJ J R. Metall. Mater. Trans, 1997, B 28: 307.
[16] MARISSA V R, ANTONIO R S, FEDERICO C A. Steel Res, 2002, 73: 378.
[17] SAFARIAN J, TRANELL G, KOLBEINSEN L. Metall. Mater. Trans. 2009, B 39: 702.
[18] EISSAA M, GHALIA S, AHMEDA A. Ironmak. Steelmak. 2012, 39: 419.
[19] SAFARIAN J, KOLBEINSEN L, TANGSTAD M. Metall. Mater. Trans. 2009, B 40: 929.

Removal and Recovery of Arsenic(Ⅲ) from Hydrochloric Acid Leach Liquor of Tungsten Slag by Solvent Extraction with 2-ethylhexanol

Arsenic(Ⅲ) removal and recovery from hydrochloric acid leach liquor of tungsten slag were systematically investigated by solvent extraction with 2-ethylhexanol. Because the iron in leach liquor could also be extracted by 2-ethylhexanol, the effects of various conditions on arsenic(Ⅲ) and iron extraction were investigated and the optimum conditions were determined. Single extraction efficiency of 84.9% for arsenic(Ⅲ) was achieved under the optimal conditions (10% (v/v) 2-ethylhexanol, 3.25 mol/L H^+, 5.91 mol/L Cl^-, O/A=2, 303.15 K, 5 min). After two-stage countercurrent extraction, the extraction efficiency of arsenic(Ⅲ) was 97.3% with only 0.75% of iron co-extracted. Using 1.0 mol/L of hydrochloric acid as stripping reagent, 97.4% of arsenic(Ⅲ) was stripped at O/A ratio of 2 by single stage. Furthermore, the possible extraction mechanism of arsenic with 2-ethylhexanol was investigated via combining experimental results and FT-IR analysis, and the structure of the extracted complex was determined to be $HAs(OH)_2Cl_2 \cdot ROH$. The results of this study propose a potential application process for arsenic removal and recovery from hydrochloric acid system.

1 Introduction

China supplies nearly 85% of the world's tungsten production[1]; however, it also suffers most from the pollution from tungsten metallurgy. In the production of tungsten, a huge amount of tungsten slag is produced through alkali leaching of tungsten concentrates[2]. Such slag contains valuable metals including tungsten (W, 1%-5%), tin (Sn, 0.4%-1.5%), iron (Fe, 15%-20%), manganese (Mn, 9%-15%), tantalum (Ta, 0.1%-0.5%), niobium (Nb, 0.3%-0.8%)

Published in *JOM*, 2022, 74(8): 3021-3029. Authors: Zeng Luqi, Yang Tianzu, Guo Lei, Liao Chenggang, Chen Ping, Liu Jiaxun, Huo Guangsheng.

and scandium (Sc, 0.01%-0.06%)[3], but it also contains toxic element arsenic (0.16%-2.69%), which has high leaching toxicity and environmental contamination risk. In 2016, tungsten slag was listed in the *National Hazardous Waste List of China* according to the latest environmental law. However, a million tons of tungsten slag has accumulated in dumps located near metallurgical plants in the past decades. Even worse, >70000 tons of tungsten slag is produced every year in China[4]. A growing treatment for tungsten slag has consequently been proposed by many researchers, including flotation, gravity separation, hydrometallurgical and pyrometallurgical methods. Among these methods, the hydrometallurgical method can recover almost all kinds of valuable metals (W, Sn, Fe, Mn, Ta, Nb, Sc) in the slag with a high recovery rate and moderate cost. For example, scandium (Sc) could be extracted by primary amine (N1923)[5], di-2,4,4-trimethylpentyl phosphinic acid (Cyanex 572)[6] and mixture of dialkyl phosphate (P204) and tributyl phosphate (TBP)[7]. Tungsten (W) could be extracted with a mixture of TBP and 2-ethylhexylphosphonic acid mono-(2ethylhexyl) ester (P507)[8], 2-octanol[9] and N1923[10]. However, all researchers were focused on valuable metal recovery, and the treatment of toxic element arsenic lacks corresponding recognition and comprehensive study. As a result, an arsenic and cheap metal (Fe, Ca, etc.) containing solution would be produced after recovery of the valuable metals. According to our process of leaching a tungsten slag with hydrochloric acid, a large amount of hydrochloric acid leach liquor containing 1-4 mol/L hydrogen ion, 0.5-5 g/L arsenic, 10-30 g/L iron and 10-30 g/L calcium was produced[11], while the permissible arsenic concentration for wastewater discharge is 0.5 mg/L in China. As a result, arsenic must be removed before further utilization of this arsenic-containing liquor.

Various processes, such as adsorption, chemical precipitation, electrocoagulation, membrane technology and solvent extraction, were developed to remove arsenic[12]. As for the above hydrochloric acid leach liquor with high concentration of arsenic, chemical precipitation seems to be a suitable method, since the arsenic can be co-precipitated with iron and calcium ions in the liquor as pH increases. However, the majority of arsenic in hydrochloric acid leach liquor exists in trivalent (III) oxidation state, and a pre-oxidation process should be conducted before co-precipitation, as pentavalent (V) arsenic is more easily immobilized with iron and calcium[13]. However, the iron (II) in the liquor will also be oxidized inevitably and a lot of arsenic-bearing residue generated, which can easily cause secondary pollution if improperly treated[14]. Though arsenic is harmful to the human body and environment, it is also considered critical to the economic and national security of the USA[15, 16]. Instead of precipitating arsenic as impure arsenites or arsenates, to recover arsenic as a product could not only avoid environmental pollution but also achieve the sustainable utilization of resources. Solvent extraction in particular offers the possibility for removal and recovery of arsenic with an arsenic rich aqueous product that is suitable for crystallization of As_2O_3[17, 18].

Many extractants have been proposed for arsenic extraction, such as organophosphorus reagents: TBP[18-20], hexabutylphosphoric triamide (HBPTA)[21], bis(2,4,4-trimethylpentyl) dithiophosphinic acid (Cyanex 301), mixture of straight chain alkylated phosphine oxides

(Cyanex 923), mixture of branched chain alkylated phosphine oxides (Cyanex 925)[22, 23]; aliphatic alcohols: 2-ethylhexanol and 2-ethylhexane-1, 3-diol[23, 24]; and mixtures: dibutyl butyl phosphonate (DBBP) + bis(2-ethylhexyl) phosphoric acid (D2EHPA)[25], 1, 2-octanedio + 2-ethylhexanol[18]. Among these, most of the research has focused on the extraction of arsenic (V) from sulfuric acid for recycling or further utilization of electrolytes in copper electrolysis process. Only thiosubstituted phosphorus-containing acids and aliphatic alcohols are reported to be able to extract arsenic (III) from hydrochloric acid solution. Cyanex 301 is among the most effective extractants for arsenic (III)[26], but arsenic (III) is difficult to strip from loaded Cyanex 301. Pinying[27] studied the extraction of arsenic from hydrochloric acid solution with 25% TBP + 25% di (2-ethylhexyl) dithiophosphoric acid (D2EHDTPA) in kerosene, and they suggested TBP extracted arsenic (III, V) through coordination bond, D2EHDTPA extracted arsenic (III) through substation and coordination bond but extracted arsenic (V) through simple molecular extraction. The extraction efficiency of arsenic was > 99%, but iron (III) in the solution must be reduced to iron (II) first to avoid iron (III) co-extraction. Orlandini[28] determined the distribution of 45 elements between 83% 2-ethylhexanol + 17% petroleum ether and hydrochloric acid solution. The results showed that 2-ethylhexanol has a high distribution (> 95%) of trivalent elements (As, Ga, Fe, Au) and poor extraction of bivalent elements (Fe(II), Ca), but no systematic study was presented for variations in the 2-ethylhexanol, H$^+$ and Cl$^-$ concentration. As mentioned above, the arsenic in hydrochloric acid leach liquor of tungsten slag is in trivalent (III) oxidation state, and the majority of iron is in bivalent (II) oxidation state. Thus, 2-ethylhexanol may be capable of extracting arsenic (III) selectively, and the other elements in liquor (e.g., Fe, Ca) have the possibility to be further recovered.

In this article, 2-ethylhexanol was employed for the removal and recovery of arsenic (III) from hydrochloric acid leach liquor of tungsten slag. Since the small amount of ferric iron in leach liquor could also be extracted, the effects of different parameters such as 2-ethylhexanol concentration, H$^+$ concentration, Cl$^-$ concentration, phase ratio, temperature and contact time on arsenic (III) and iron extraction were investigated. In addition, the extraction mechanism of arsenic (III) was revealed. The results indicated that this process could be a potential method for arsenic removal and recovery from hydrochloric acid.

2 Experimental section

2.1 Materials and reagents

The hydrochloric acid leach liquor was prepared by leaching a tungsten slag with 6.0 mol/L hydrochloric acid at 90 ℃ for 2.0 h according to our previous studies.[11] The major components of the liquor are shown in Table 1. According to Table 1, all of the arsenic and a small amount of iron in hydrochloric acid leach liquor are trivalent. The extractant 2-ethylhexanol ($C_8H_{18}O$, ⩾

99.0%) was supplied by Xilong Scientific Co., Ltd. The diluent sulfonated kerosene was purchased from Fuchen Chemical Reagents Co., Ltd. All the aqueous solutions were prepared in deionized water, and all the inorganic reagents were of analytical grade.

Table 1 Major components of the hydrochloric acid leach liquor   g/L

As_{total}	As(Ⅲ)	H^+	Cl^-	Fe_{total}	Fe^{2+}	Fe^{3+}	Ca^{2+}
3.06±0.01	3.06	3.25±0.05	209.60±0.63	22.72±0.03	20.58	2.14	25.14±0.05

2.2 Experimental and analysis

The extraction and stripping processes were conducted in the separatory funnel with PTFE plugs; the organic phase and aqueous phase were mixed and shaken in a temperature-controlled water bath shaker. Then, phase separation was realized because of gravity after standing for 10 min. In some experiments, sodium hydroxide or calcium chloride was added to the wastewater to adjust the concentration of H^+ or Cl^-, respectively. In addition, the arsenic(Ⅲ) concentration was increased to 15.1 g/L by dissolving arsenic trioxide into the hydrochloric acid leach liquor in saturation capacity method experiment. The saturation capacity method was carried out through mixing the same organic phase with fresh aqueous phase several times until the concentration of arsenic in aqueous phase was unchanged under the condition of 2% (v/v) 2-ethylhexanol, $c(H^+) = 3.25$ mol/L, $c(Cl^-) = 5.91$ mol/L, O/A = 2, $T = 303.15$ K, $t = 5$ min.

The total concentration of arsenic, arsenic(Ⅲ) and calcium in aqueous phase was determined via inductively coupled plasma-optical emission spectrometry (iCAP™ 7200 ICP-OES, Thermo Fisher Scientific), and a hydride generator was used when determining arsenic(Ⅲ) by hydrogenation method. The total concentration of iron and iron(Ⅲ) was determined via phenanthroline spectrophotometry. The concentration of Cl^- was measured via ion chromatograph (DIONEX ICS-90), and the concentration of H^+ was analyzed through titration method with 1.0 mol/L NaOH standard solutions. The concentrations of elements in the loaded organic phase were calculated according to mass balance. The infrared spectra of the extractant and loaded organic phase were measured using a Fourier transform infrared (FT-IR) spectrometer (Nicolet IS10 spectrometer, Thermo Scientific Corp) over the range of 4000-400 cm^{-1}. The distribution ratio (D), separation factor (β), extraction efficiency (E) and stripping efficiency (S) were calculated by Eqs. (1) and (4), respectively.

$$D = \frac{C_{org}}{C_{aq}} \quad (1)$$

$$\beta_{A/B} = \frac{D_A}{D_B} \quad (2)$$

$$E = \frac{D}{D + (V_{aq}/V_{org})} \times 100\% \quad (3)$$

$$S = \frac{C_{\text{aq, s}} \times V_{\text{aq, s}}}{C_{\text{org}} \times V_{\text{org}}} \times 100\% \tag{4}$$

where C_{aq} and C_{org} (g/L) stand for the concentration of elements in aqueous and loaded organic phase, respectively; V_{aq} and V_{org} (mL) represent the volume of the aqueous and loaded organic phase, respectively. $C_{\text{aq, s}}$ (g/L) represents the concentration of elements in strip liquor and $V_{\text{aq, s}}$ (mL) represents the volume of strip liquor.

3 Results and discussion

3.1 Extraction

3.1.1 Effect of 2-ethylhexanol concentration

The effect of 2-ethylhexanol concentration on arsenic (Ⅲ) and iron extraction was investigated against the volume percentage of 2-ethylhexanol in the organic phase. As shown in Fig. 1, the extraction efficiency of both elements increased with the increase of 2-ethylhexanol concentration. A maximum extraction efficiency of arsenic(Ⅲ) was observed at 30% (v/v) 2-ethylhexanol and remained at approximately 90.8% with further increase of the 2-ethylhexanol concentration. However, a part of iron was also extracted in the extraction process. The extraction efficiency increased slowly from 1.9% to 2.1% as the 2-ethylhexanol concentration increased from 5% to 10% (v/v) and then increased linearly from 2.1% to 14.1% as the 2-ethylhexanol concentration increased from 10% to 50% (v/v). This is because the small amount of iron(Ⅲ) in hydrochloric acid leach liquor could also be extracted by 2-ethylhexanol according to Orlandini's[28] study. Obviously, 2-ethylhexanol has a higher ability to extract arsenic(Ⅲ) than iron(Ⅲ) since the extraction efficiency of iron increased dramatically only after most of the arsenic(Ⅲ) had been extracted, and excess 2-ethylhexanol existed. As a result, the separation factor ($\beta_{\text{As/Fe}}$) remained at about 257 with 10% (v/v) 2-ethylhexanol and then decreased dramatically as the 2-ethylhexanol increased. Therefore, selective extraction of arsenic(Ⅲ) from the hydrochloric acid leach liquor by 2-ethylhexanol is feasible. The concentration of 2-ethylhexanol was determined to be 10% (v/v) with arsenic (Ⅲ) extraction efficiency of 84.9%, and only 2.1% of iron was co-extracted.

3.1.2 Effect of H$^+$ concentration

The effect of H$^+$ concentration on arsenic (Ⅲ) and iron extraction was studied by adding sodium hydroxide to adjust the concentration of H$^+$. As shown in Fig. 2, the H$^+$ concentration had a great effect on arsenic (Ⅲ) extraction. The extraction efficiency of arsenic (Ⅲ) increased dramatically from 43.5% to 84.1% as H$^+$ concentration increased from 0.5 mol/L to 3.25 mol/L, indicating that high acidity is conductive to arsenic(Ⅲ) extraction. This is because the increase of H$^+$ concentration could promote the formation of extractable species (HAs(OH)$_2$Cl$_2$) as we discussed in subsequent "Arsenic (Ⅲ) Complex Composition in the Hydrochloric Acid Leach

Liquor and Organic" section. The trend of the extraction of iron with the increase of H^+ concentration was the opposite to that of arsenic (Ⅲ), while the decline was slighter. Consequently, the separation factor ($\beta_{As/Fe}$) increased significantly from 17 to 251 as H^+ concentration increased from 0.5 mol/L to 3.25 mol/L. Usually, the increase of H^+ concentration would increase the iron extraction by extracting with neutral extractant, such as extracting iron (Ⅲ) with TBP in hydrochloric acid solution[29-31]. This opposite trend may also indicate that 2-ethlyhexanol has higher ability to extract arsenic(Ⅲ) or hydrochloric acid than iron(Ⅲ). When the H^+ concentration increased, the extracted iron (Ⅲ) was replaced by arsenic (Ⅲ) or hydrochloric acid. As listed in Table 1, the H^+ concentration in hydrochloric acid leach liquor is about 3.25 mol/L, which is beneficial to the extraction process. Therefore, the initial H^+ concentration of hydrochloric acid leach liquor was maintained for the subsequent extraction experiments.

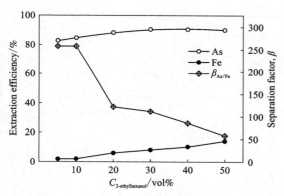

(Conditions: $c(H^+) = 3.25$ mol/L, $c(Cl^-) = 5.91$ mol/L, O/A = 2:1, T = 303.15 K, t = 5 min)

Fig. 1 Effect of 2-ethylhexanol concentration on arsenic (Ⅲ) and iron extraction

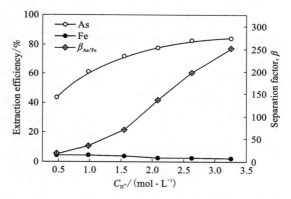

(Conditions: 10%(v/v) 2-ethylhexanol, $c(Cl^-) = $ 5.91 mol/L, O/A = 2:1, T = 303.15 K, t = 5 min)

Fig. 2 Effect of H^+ concentration on arsenic (Ⅲ) and iron extraction

3.1.3 Effect of Cl^- concentration

Similarly, the effect of Cl^- concentration on arsenic(Ⅲ) and iron extraction was investigated by adding calcium chloride to adjust the concentration of Cl^-. As shown in Fig. 3, the Cl^- concentration seemed to have a modest effect on arsenic(Ⅲ) extraction compared to H^+, and the extraction efficiency of iron increased with the increase of Cl^- concentration as well. The extraction efficiency of arsenic (Ⅲ) increased from 84.9% to 87.0% as Cl^- concentration increased from 5.91 to 6.80 mol/L, and the extraction efficiency of iron increased from 1.9% to 6.7%. Then, the extraction efficiency of both remained unchanged with further increase of Cl^- concentration. The slight increase of arsenic(Ⅲ) and iron may be explained by the fact that the initial concentration of Cl^-(5.91 mol/L) in hydrochloric acid leach liquor was sufficiently high. The effect of a lower concentration of Cl^- on arsenic(Ⅲ) extraction was not investigated, because

Cl⁻ is hard to remove from aqueous solution. However, the positive effect shown in Fig. 3 still indicated that high Cl⁻ concentration is good for arsenic(Ⅲ) and iron extraction because higher Cl⁻ concentration could also promote the formation of extractable chloro-complex species (HAs(OH)$_2$Cl$_2$, FeCl$_3$[29-31]). However, the co-extraction of iron decreased the separation factor of ($\beta_{As/Fe}$). Therefore, the initial Cl⁻ concentration of hydrochloric acid leach liquor was selected.

3.1.4 Effect of temperature

The effect of temperature on arsenic(Ⅲ) and iron extraction was studied at different temperatures of 298.15 K, 303.15 K, 308.15 K and 313.15 K. As shown in Fig. 4, the temperature had nearly no effect on arsenic(Ⅲ) extraction and little effect on iron extraction. The extraction efficiency of arsenic(Ⅲ) remained at 84.9% in the entire temperature range. While the extraction efficiency of iron increased gently from 2.0% to 2.3% as the temperature increased from 298.15 K to 313.15 K, the separation factor ($\beta_{As/Fe}$) decreased from 281 to 235 accordingly. It can be extrapolated that in this case the thermodynamic parameters did not affect the extraction efficiency profoundly. Therefore, the extraction process should be carried out at room temperature to avoid the extra cost of heat exchange.

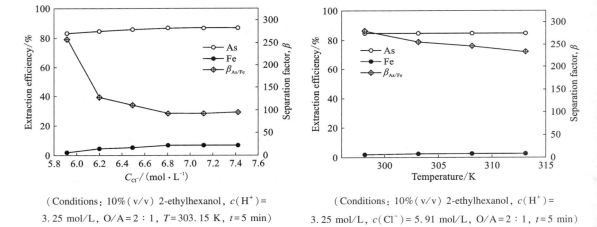

(Conditions: 10%(v/v) 2-ethylhexanol, $c(H^+)$ = 3.25 mol/L, O/A = 2 : 1, T = 303.15 K, t = 5 min)

Fig. 3 Effect of Cl⁻ concentration on arsenic(Ⅲ) and iron extraction

(Conditions: 10%(v/v) 2-ethylhexanol, $c(H^+)$ = 3.25 mol/L, $c(Cl^-)$ = 5.91 mol/L, O/A = 2 : 1, t = 5 min)

Fig. 4 Effect of temperature on arsenic(Ⅲ) and iron extraction

3.1.5 Effect of contact time

The effect of contact time on arsenic(Ⅲ) and iron extraction was investigated at different contact times varied from 0.5 to 20 min. As shown in Fig. 5, the extraction efficiency of arsenic(Ⅲ) increased from 82.4% to 84.9% as contact time increased from 0.5 to 2 min, and then the increase in subsequent time had no greater impact on the extraction of arsenic(Ⅲ). In the case of iron, the extraction efficiency decreased as the contact time went on and reached the minimum value after 5 min. As a result, the separation factor ($\beta_{As/Fe}$) increased from 89 to 259 as the contact time increased from 0.5 min to 5.0 min, and thereafter there was no change. This phenomenon also indicated that 2-ethlyhexanol has a higher ability to extract arsenic(Ⅲ) than iron; part of extracted iron could be replaced by arsenic(Ⅲ) as contact time goes on. Cleary 5 min is enough to attain the extraction equilibrium.

3.1.6 Batch continuous extraction test

Under optimal experimental conditions (10%(v/v) 2-ethylhexanol, 3.25 mol/L H^+, 5.91 mol/L Cl^-, O/A = 2, T = 303.15 K, t = 5 min), the arsenic(Ⅲ) extraction McCabe Thiele diagram was conducted and is presented in Fig. 6. It was found that two theoretical extraction stages were required for the extraction of arsenic(Ⅲ) at O/A = 2. The results of two-stage countercurrent extraction are shown in Table 2. The remaining arsenic(Ⅲ) in the raffinate was only about 0.08 g/L with the extraction efficiency of 97.3%. The co-extracted iron decreased from 2.1% to 0.75% compared to single-stage extraction, and calcium was not extracted. In addition, about 1.3% H^+, 1.4% Cl^- were co-extracted respectively because of the formation of extractable species, which will be explained in the following extraction mechanism section. In this case, the consumption of H^+ and Cl^- should be considered when the concentration of arsenic(Ⅲ) in solution is high, because the decrease of H^+ and Cl^- concentration will affect the extraction of arsenic(Ⅲ) as we discussed above. Therefore, arsenic(Ⅲ) was selectively removed from hydrochloric acid leach liquor and could be recovered as a pure product instead of precipitating arsenic as impure arsenites or arsenates. Though the concentration of arsenic in raffinate still does not meet the integrated pollutant discharge standard in China (<0.5 mg/L), the cost of further arsenic removal would be reduced greatly. On the other hand, the iron and calcium in arsenic-free leach liquor could be recovered after arsenic extraction, and the solution can be recycled without arsenic contamination.

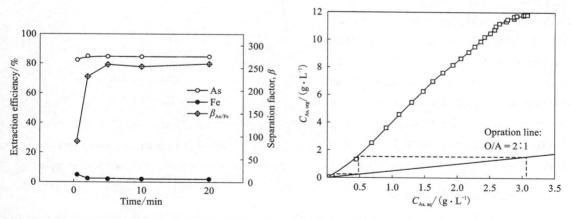

(Conditions: 10%(v/v) 2-ethylhexanol, $c(H^+)$ = 3.25 mol/L, $c(Cl^-)$ = 5.91 mol/L, O/A = 2, T = 303.15 K)

Fig. 5 Effect of contact time on arsenic(Ⅲ) and iron extraction

(Conditions: 10%(v/v) 2-ethylhexanol, $c(H^+)$ = 3.25 mol/L, $c(Cl^-)$ = 5.91 mol/L, O/A = 2, t = 5 min, T = 303.15 K)

Fig. 6 McCabe-Thiele diagram for arsenic(Ⅲ) extraction

Table 2 Results of two-stage countercurrent extraction

Element	As_{total}	Fe_{total}	Ca^{2+}	H^+	Cl^-
Feed solution/(g·L^{-1})	3.06±0.01	22.72±0.03	25.14±0.05	3.25±0.05	209.60±0.63
Raffinate/(g·L^{-1})	0.08±0.003	22.55±0.03	25.17±0.09	3.21±0.05	206.62±0.69
Loaded organic/(g·L^{-1})	1.49	0.08	—	0.02	1.49
Extraction efficiency/%	97.3	0.7	—	1.29	1.42

3.2 Stripping

Hydrochloric acid was used to strip the loaded organic phase from the two-stage countercurrent extraction process, and the effect of hydrochloric acid concentration on arsenic(Ⅲ) and iron stripping is presented in Fig. 7. The stripping efficiency of arsenic(Ⅲ) decreased from 97.4% to 65.7% as hydrochloric acid concentration increased from 0 to 4.0 mol/L, but the stripping efficiency of iron increased from 93.1% to 98.9%. This phenomenon is contrary to the extraction process as discussed in Sect. "Arsenic(Ⅲ) Complex Composition in the Hydrochloric Acid Leach Liquor and Organic", which also indicates that higher concentration of H$^+$ is beneficial for arsenic removal and iron separation. The stripping efficiency was maintained at about 97% as hydrochloric acid concentration varied from 0 to 1.0 mol/L. However, slight emulsification and long phase separation time occurred when the hydrochloric acid concentration was <0.5 mol/L. This may be due to the co-extraction of water in low acidity, since some big droplets appeared and disappeared gradually as time was extended without the formation of third phase. Therefore, 1.0 mol/L hydrochloric acid was selected as the optimal concentration for stripping arsenic(Ⅲ).

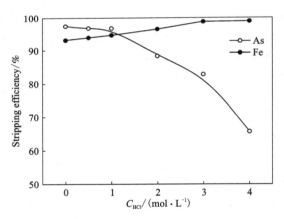

(Conditions: O/A=2:1, T=303.15 K, t=10 min)

Fig. 7 Effect of hydrochloric acid (HCl) concentration on arsenic (Ⅲ) and iron stripping

3.3 Extraction mechanism

3.3.1 Arsenic (Ⅲ) complex composition in the hydrochloric acid leach liquor and organic

In the system of hydrochloric acid solution, arsenic(Ⅲ) forms cationic and chloro-complexes because of its amphiprotic property, a gradual transition from H_3AsO_3 (= As(OH)$_3$), through As(OH)$_2^+$, As(OH)$_2$Cl, As(OH)Cl$_2$, to AsCl$_3$ was proposed by Arcand and Sella[32,33]. The general equilibrium may be written as Eq. (5), where $i = 0, 1, 2$. In addition, the dominant species in 3-8 mol/L hydrochloric acid were $As(OH)_2^+$ and $As(OH)_2Cl$ according to the calculation by Sella et al.[32]

$$As(OH)_{3-i}Cl_i + H^+ + Cl^- \rightleftharpoons As(OH)_{2-i}Cl_{i+1} + H_2O \quad (5)$$

Furthermore, the arsenic(III) was extracted to polar oxygenated solvents as $As(OH)_2Cl$, $As(OH)Cl_2$ at lower hydrochloric acid concentration whereas $AsCl_3$ was extracted at 10 mol/L hydrochloric acid according to Ma and Marcus's[34, 35] reviewers. As 2-ethylhexanol is also a polar oxygenated solvent, $As(OH)_2Cl$ or $As(OH)Cl_2$ would be the extracted species at our experimental conditions. To ascertain the mechanism of arsenic(III) extraction with 2-ethylhexanol, the effects of different 2-ethylhexanol, H^+, and Cl^- concentrations on arsenic(III) distribution ratio were studied by logarithm linear regression method based on above experimental data.

As shown in Fig. 8(a), the $\log[D]$ versus $\log[$2-ethylhexanol$]$ exhibited a poor linear fitting as the value of the correlation coefficient was low ($R^2 \approx 0.93$). This deviation may be caused by the effects of hydrogen bonding, molecular interactions, or the complex composition of the aqueous solution according to the explanation of Zhang's study[36]. In this case, the saturation capacity method was applied to investigate the association relationship between arsenic(III) and 2-ethylhexanol with 2% (v/v) 2-ethylhexanol. The saturation capacity of 2% (v/v) 2-ethylhexanol was about 10.09 g/L, and the molar ratio of 2-ethylhexanol : $As(III) = 0.135 : 0.128 \approx 1 : 1$, indicating that one mole of 2-ethylhexanol was required to extract one equivalent of arsenic(III). The $\log[D]$ versus $\log[H^+]$ and $\log[Cl^-]$ plots [Fig. 8(b) and (c)] were straight lines with a slope of 1 and 2, respectively, indicating the association of one H^+ and two Cl^- with arsenic(III) in the extracting species with 2-ethylhexanol. To sum up, the extracting species was $HAs(OH)_2Cl_2 \cdot ROH$, and the extraction reaction could be expressed as Eq. (6) or Eq. (7), where ROH represents the extractant 2-ethylhexanol.

$$As(OH)_2^+ + H^+ + 2Cl^- + ROH \rightleftharpoons HAs(OH)_2Cl_2 \Delta ROH \quad (6)$$

$$As(OH)_2Cl + H^+ + Cl^- + ROH \rightleftharpoons HAs(OH)_2Cl_2 \Delta ROH \quad (7)$$

3.4 FT-IR of the organics

The FT-IR spectra of the extractant and loaded organic phase from the two-stage batch continuous extraction in Sect. "Batch Continuous Extraction Test" were compared to further understand the extraction mechanism. As shown in Fig. 8(d), the absorption peaks of two organic phases at 1378/1460 and 2954 cm^{-1} correspond to the flexural vibration and stretching vibration absorption peaks of $-CH_3$, respectively, and the absorption peaks of two organic phases at 2854 and 2923 cm^{-1} correspond to the stretching vibration absorption peaks of $-CH_2$[37-39]. All these peaks were unchanged during the extraction. A significant shift and intensity increase of the absorption bands for O-H stretching vibration in the 3100-3700 cm^{-1} region and flexural vibration in 1550-1650 cm^{-1} were observed. The stretching vibration of C-O shifted from 1308 cm^{-1} to 1304 cm^{-1}. All these shifts in the spectra peaks indicate the involvement of O-H groups in

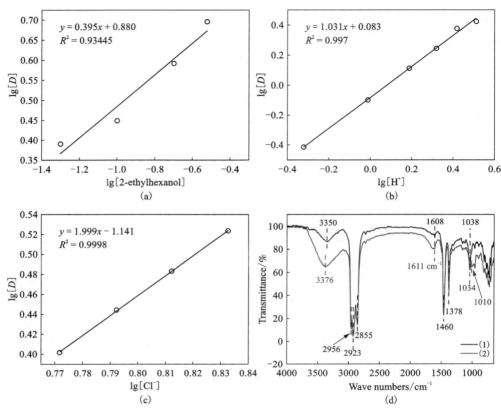

Fig. 8 Extraction mechanism: plots of (a) lg[D] against lg[2-ethylhexanol], (b) lg[D] against lg[H^+], (c) lg[D] against lg[Cl^-]; (d) FT-IR spectra of the extractant (1) and loaded organic phase (2)

complexation with arsenic(Ⅲ). Furthermore, the appearance of new bands at 1010 cm^{-1} might be attributed to As-O vibrations, which showed that oxygen-rich functional groups were mainly responsible for the extraction of arsenic species[40, 41]. Based on the above discussion, arsenic(Ⅲ) was probably extracted with 2-ethylhexanol by solvation of the hydrogen ion of complex arsenic(Ⅲ) acid species (HAs(OH)$_2$Cl$_2$), which is similar to extraction of iron from hydrochloric acid solution with TBP in the form of HFeCl$_4$[42].

4 Conclusion

A process for the removal and recovery of arsenic(Ⅲ) from hydrochloric acid leach liquor of tungsten slag by solvent extraction with 2-ethylhexanol was proposed. The results indicate that under the optimal single-stage extraction conditions-10% (v/v) 2-ethylhexanol, 3.25 mol/L H^+, 5.91 mol/L Cl^-, O/A of 2∶1, 303.15 K, 5 min, 84.9% arsenic(Ⅲ)—could be extracted with

2.1% co-extracted iron, and the separation factor ($\beta_{As/Fe}$) reached up to 257. After two-stage countercurrent extraction, 97.3% arsenic(Ⅲ) was extracted and the co-extracted iron decreased to 0.75%, and calcium was not extracted. Furthermore, 97.4% of arsenic(Ⅲ) could be stripped from the organic phase using 1.0 mol/L hydrochloric acid at O/A of 2∶1. Furthermore, the experimental results and FT-IR analysis confirmed that the mechanism of arsenic(Ⅲ) extraction by 2-ethylhexanol was possibly attributed to the solvation of the hydrogen ion of complex arsenic (Ⅲ) acid species ($HAs(OH)_2Cl_2$), and the composition of the arsenic(Ⅲ) containing extract may be $HAs(OH)_2Cl_2 \cdot ROH$.

Acknowledgments

The authors are thankful to the National Key Research and Development Program of China (2019YFC1907400) for financial support.

References

[1] MA X, QI C, YE L, et al. J. Clean. Prod. 2017, 149: 936. https://doi.org/10.1016/j.jclepro.2017.02.184.

[2] LIAO C, XIE S, WANG X, et al. JOM, 2021, 73: 1853. https://doi.org/10.1007/s11837-021-04671-5.

[3] LIU H, LIU H, NIE C, et al. J Environ Manage, 2020, 270: 110927. https://doi.org/10.1016/j.jenvman.2020.110927.

[4] JUNJIE L, DEWN H, KANGGEN Z, et al. Conserv. Utiliz. of Mineral Res. 2019, 39: 125.

[5] XU Y, DENG Z, LI W, et al. Jiangxi Nonferr. Met. 1997, 11: 32.

[6] NIE H, WANG Y, WANG Y, et al. Hydrometallurgy, 2018, 175: 117-123. https://doi.org/10.1016/j.hydromet.2017.10.026.

[7] ZHANG W, ZHANG T A, LV G, et al. JOM, 2018, 70: 2837-2845 https://doi.org/10.1007/s11837-018-3166-8.

[8] ZENG L, YANG T, YI X, et al. Hydrometallurgy, 2020. https://doi.org/10.1016/j.hydromet.2020.105500.

[9] LI Y, LV S, FU N, et al. JOM, 2020, 72: 373 https://doi.org/10.1007/s11837-019-03676-5.

[10] FU H, LI Y, CAO G, et al. JOM, 2018, 70: 2864. https://doi.org/10.1007/s11837-018-3167-7.

[11] HUO G, GUO L, YI X, et al. Treatment method of decomposition tin containing slag (soopat). 2019. http://www1.soopat.com/Patent/CN106180138A.

[12] ALKA S, SHAHIR S, IBRAHIM N, et al. J. Cleaner Prod, 2021. https://doi.org/10.1016/j.jclepro.2020.123805.

[13] ZENG T, DENG Z, ZHANG F, et al. Hydrometallurgy, 2021. https://doi.org/10.1016/j.hydromet.2021.105562.

[14] PANTUZZO F L, CIMINELLI V S. Water Res, 2010, 44: 5631. https://doi.org/10.1016/j.watres.2010.07.011.

[15] USGS, 2018. Interior releases 2018's final list of 35 minerals deemed critical to U. S. National security and the economy (Office of Communications and Publishing, 2018), https://www.usgs.gov/news/national-news-release/interior-releases-2018s-final-list-35-minerals-deemed-critical-us.

[16] MARTINS L S, GUIMARÃES L F, JUNIOR A B B, et al. Environ. J. Manage, 2021, 295: 113091.

[17] CAO P, LONG H, ZHANG M, et al. Chem Eng, 2021. https://doi.org/10.1016/j.jece.2021.105871.

[18] JANTUNEN N, VIROLAINEN S, LATOSTENMAA P, et al. Hydrometallurgy, 2019, 187: 101. https://doi.org/10.1016/j.hydromet.2019.05.008.

[19] DEMIRKIRAN A, RICE N M. ISEC, 2002: 892-895.

[20] NAVARRO P, ALGUACIL F J. Can. Metall. Q. 1996, 35: 133-141.

[21] TRAVKIN V F, GLUBOKOV Y M, MIRONOVA E V, et al. Sorption and Ion-Exchange Processes[J]. 2001, 74: 1614-1617.

[22] IBERHAN L, WISNIEWSKI M. J. Chem. Technol. Biotechnol, 2003, 78: 659-665. https://doi.org/10.1002/jctb.843.

[23] MB B, M W, J S. Journal of Radioanalytical and Nuclear Chemistry, 1998, 228: 57-61.

[24] BARADEL A, GUERRIERO R, MEREGALLI L, et al. JOM, 1986, 38: 32-37 https://doi.org/10.1007/BF03257918.

[25] BALLINAS M A D L, MIGUEL E R G D S, MUÑOZ M A, et al. Ind. Eng. Chem. Res., 2003, 42: 574-580.

[26] GUPTA B, BEGUM Z. Sep. Purif. Technol, 2008, 63: 77-85.

[27] PINYING L, ZHOULAN Y. J. CENT. -South Inst. Min. Metall, 1990, 21: 673-678.

[28] ORLANDINI K A, WAHLGREN M A, BARCLAY J. Anal. Chem, 1965, 37: 1148-1151.

[29] REDDY B R, SARMA P V R B. Hydrometallurgy, 1996, 43: 299.

[30] LEE M S, LEE G -S, SOHN K Y. Mater. Trans, 2004, 45: 1859.

[31] YI X, HUO G, TANG W. Hydrometallurgy, 2020. https://doi.org/10.1016/j.hydromet.2020.105265.

[32] SELLA C, MENDOZA R N, BAUER D. Hydrometallurgy, 1991, 27: 179-190.

[33] ARCAND G M. J. Am. Chem. Soc., 1957, 76: 1865-1870.

[34] MA R J. Yunnan Metallurgy (In Chinese), 1982, 6: 40-46.

[35] MARCUS Y. Coordin. Chem. Rev., 1967, 2: 195-238. https://doi.org/10.1016/S0010-8545(00)80205-2.

[36] ZHANG R, XIE Y, SONG J, et al. Hydrometallurgy, 2016, 160: 129. https://doi.org/10.1016/j.hydromet.2016.01.001.

[37] ZHOU X, ZHANG Z, KUANG S, et al. Hydrometallurgy, 2019, 185: 76. https://doi.org/10.1016/j.hydromet.2019.02.001.

[38] LE W, KUANG S, ZHANG Z, et al. Hydrometallurgy, 2018, 178: 54. https://doi.org/10.1016/j.hydromet.2018.04.005.

[39] KUANG S, ZHANG Z, LI Y, et al. J. Rare Earths, 2018, 36: 304 https://doi.org/10.1016/j.jre.2017.09.007.

[40] SHAKOOR M B, NIAZI N K, BIBI I, et al. Environ Int, 2019, 123: 567. https://doi.org/10.1016/j.envint.2018.12.049.

[41] ZHOU Z, LIU Y G, LIU S B, et al. Chem. Eng. J., 2017, 314: 223. https://doi.org/10.1016/j.cej.2016.12.113.

[42] KISLIK V S. Solvent extraction: classical and novel approaches. Elsevier, 2012.

Separation of Molybdenum and Tungsten from Iron in Hydrochloric-phosphoric Acid Solutions Using Solvent Extraction with TBP and P507

Abstract: A mixture of TBP and P507 was proposed for the separation of molybdenum (Mo) and tungsten (W) from iron (Fe) in a hydrochloric-phosphoric acid ($HCl-H_3PO_4$) solution that was obtained from the hydrochloric acid preleaching of low-grade scheelite. Various extractants with TBP were investigated, and P507 showed synergistic effects on Mo and W extraction and a strong antagonistic effect on Fe extraction. Then, the effects of various conditions on the separation of Mo and W from Fe with the mixture of TBP and P507 were investigated. Using 24% (v/v) TBP, 16% (v/v) P507 and 60% (v/v) kerosene as extractants, 90% Mo and 88% W were extracted with only 8% Fe co-extraction, and the separation factors of Mo/Fe and W/Fe were 153 and 118, respectively, through one-stage extraction under the following conditions: 1.7 mol/L H^+, A/O ratio of 3∶1, contact time of 10 min, and temperature of 15 ℃. After two-stage batch continuous extraction, 99% Mo, 94% W and 10% Fe were extracted. By washing with 2 mol/L hydrochloric acid, most of the Fe and other impurities were removed with negligible losses of Mo and W. Upon stripping with 5 mol/L ammonia at a 1∶1 organic-to-aqueous phase ratio, almost 100% of the Mo and W were recovered from the washed organic phase. Other impurities, such as P, Si, Ca, Al, and Mn, minimally affected the extraction of Mo and W.

Keywords: Molybdenum; Tungsten; Iron; Solvent extraction; Separation; TBP; P507

Published in *Hydrometallurgy*, 2020, 198: 105500. Authors: Zeng Luqi, Yang Tianzu, Yi Xintao, Chen Ping, Liu Jiaxun, Huo Guangsheng.

1 Introduction

Scheelite (CaWO$_4$) and wolframite ((Fe, Mn)WO$_4$) are the only commercial ores of tungsten (Lassner and Schubert, 1999; Zhao et al., 2011a). With the excessive depletion of easily accessible high-quality wolframite, scheelite has become the dominant mineral for tungsten extraction, which accounts for two thirds of the world's total reserves (Li et al., 2017; Yang et al., 2016; Zhao et al., 2011b). When scheelite is beneficiated via floatation, calcium-bearing minerals such as calcite, fluorite and apatite are commonly associated with scheelite in tungsten ores due to their similar surface properties and floatability (Kupka and Rudolph, 2018; Yang, 2018). The most typical technology for further separating scheelite from these calcium minerals is the conventional Petrov process. However, this process has high energy consumption because high temperature is required (Han et al., 2017; Yongxin and Changgen, 1983). In recent years, more selective collectors (Lyu et al., 2019; Yan et al., 2017; Zhao et al., 2019) for scheelite and more effective depressants (Chen et al., 2018; Dong et al., 2019; Kang et al., 2019) for other calcium-containing minerals at room temperature were investigated, but most remain in the theoretical study or laboratory stage. As a result, low-grade concentrates with content from 5% to 30% WO$_3$ remain in production to maintain the high recovery efficiency of tungsten and to reduce the production cost and energy consumption (Li et al., 2019). To remove apatite and calcite selectively, Huo et al. (2018) proposed a recycling preleaching process with hydrochloric acid (HCl) as shown in Fig. 1. This technology enriched the grade of scheelite effectively but generated a hydrochloric-phosphoric acid (HCl-H$_3$PO$_4$) preleaching solution, which contained small amounts of Mo and W(<2 g/L), over 50 g/L Ca, 15 g/L P, and 1 g/L Fe, among other components. The preleaching solution could be regenerated to HCl for circulating leaching via neutralization with calcium carbonate and reaction of calcium chloride solution with sulfuric acid (Al-Othman and Demopoulos, 2009). However, the molybdenum (Mo) and tungsten (W) in the solution must be recovered first because they would also be precipitated as Fe(Mo, W)O$_4$ or Ca(Mo, W)O$_4$ in the neutralization process.

Solvent extraction is widely used for Mo and W recovery. In a solution that contains phosphoric acid, W and Mo are mainly in the form of heteropoly acids ([P(W, Mo)$_{12}$O$_{40}$]$^{3-}$) (Bochet et al., 2009; Nagul et al., 2015). Several investigations have been conducted on the extraction of these heteropoly acids. Yang et al. (2016) investigated the extraction of tungsten from an acidic high-phosphorus solution with Aliquat 335 and realized 96% extraction efficiency when pH = 2; however, five stripping stages and high temperature (60 ℃) were required. Yatirajam used MIBK to separate Mo and W from impurity cations (e.g., Ca, Fe, Bi, Si, and As) as phosphotungstate (Yatirajam and Dhamija, 1977a), phosphotungsten blue (Yatirajam and Dhamija, 1977b) or phosphomolybdenum blue (Yatirajam and Ram, 1973) in an acid medium with a reducing agent. Li and Su (1990) used petroleum sulfoxide (PSO) and a diluent to extract

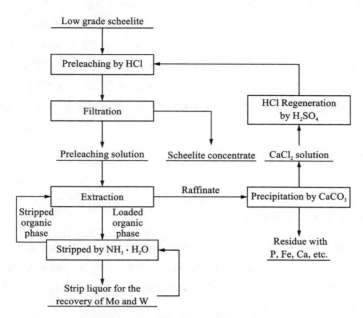

Fig. 1　Recycling preleaching process for low grade scheelite

a small amount of $[PW_{12}O_{40}]^{3-}$ from a sodium tungstate solution and suggested that the extraction species was $H_3PW_{12}O_{40} \cdot PSO$. Asakura et al. (2000) extracted $[P_2W_{17}O_{61}]^{10-}$ as $H_{10}P_2W_{17}O_{61}$ from an HNO_3 or HCl solution using TBP and established the experimental distribution equation of $[P_2W_{17}O_{61}]^{10-}$ in the nitrate system. Lakshmanan and Haldar (1970a) and Lakshmanan and Haldar (1969) demonstrated that 12-tungstophosphoric acid ($[PW_{12}O_{40}]^{3-}$) and 12-molybdophosphoric ($[PMo_{12}O_{40}]^{3-}$) could be extracted into neat, oxygenated organic solvents via ion pair solvation, and the extraction was profoundly influenced by the basicity and dielectric constant of the organic solvents. In addition, the extraction of these two heteropoly acids by TBP suggested that three molecules of TBP were associated with each acid in the organic phases (Lakshmanan and Haldar, 1970b). Liao and Zhao (2017) used TBP to extract 12-tungstophosphoric acid from an H_2SO_4-H_3PO_4 solution and found that iron (Fe) powder can reduce ferric (Fe^{3+}) to ferrous (Fe^{2+}) ions to inhibit the extraction of Fe^{3+} by TBP. However, the described operations either used a laboratory-synthesized solution without impurities or used a reducing agent to enhance the selectivity of the heteropoly acid ($[P(W, Mo)_{12}O_{40}]^{3-}$), which can't be applied directly in our HCl-H_3PO_4 preleaching solution, which contains various impurities. Fe^{3+} is easily extracted by the neutral solvents that are discussed above in a solution that contains a high concentration of chlorine ions (Cl^-), which may cause emulsification of the organic phase when stripping in an alkaline solution.

　　Recently, Zhang et al. (2019) reported that the mixture of an amine (N235) or neural solvent (TBP or TRPO) with P507 exhibited a significant antagonistic effect on Fe extraction and a significant synergistic effect on Mo extraction from nitric-sulfuric acid. Hence, the mixture of

TBP and organophosphorus or amine extractants may extract Mo and W from the $HCl-H_3PO_4$ preleaching solution selectively while the Fe remains in the solution. In this paper, various types of extractants, which include organophosphorus extractants (P204 and P507) and an amine extractant (N1923), in the TBP-base extraction system were used to separate Mo and W over Fe from the $HCl-H_3PO_4$ preleaching solution. The mixture of TBP and P507 yielded satisfactory results. Then, the influential parameters on extraction and stripping were investigated in detail.

2 Experimental

2.1 Materials and instruments

Tributyl phosphate (TBP), 2-ethylhexylphosphonic acid mono-(2-ethylhexyl) ester (P507), di(2-ethylhexyl) phosphoric acid (P204) and primary amine (N1923) were supplied by Shanghai Aladdin Reagent Co., Ltd. All the extractants were dissolved in kerosene that was purchased from Tianjin Hengxing Chemical Preparation Co., Ltd., to the required concentrations. The $HCl-H_3PO_4$ preleaching solution was prepared by leaching low-grade scheelite ore with 3.75 mol/L HCl at ambient temperature (15 ℃) for 1.5 h according to our previous studies. The chemical composition of the solution is presented in Table 1. Deionized water with a resistivity of 18.2 MΩ/cm was used in the experiments, and the other reagents were of analytical grade. The concentrations of the elements in the solution were analyzed via inductively coupled plasma-optical emission spectrometry ($ICAP^{TM}$ 7200 ICP-OES, Thermo Fisher Scientific).

Table 1 Chemical compositions of solutions and organic phases   g/L

	Mo	WO_3	Ca	P	Fe	Al	Si	Mn
$HCl-H_3PO_4$ solution	1.86	1.48	50.13	15.72	1.51	0.21	1.05	0.39
Raffinate[a]	0.02	0.09	50.30	15.14	1.35	0.19	0.96	0.39
Loaded extractant[b]	5.41	4.12	—	1.26	0.40	0.06	0.23	—
Washed extractant	5.38	4.12	—	0.15	0.07	0.06	0.29	—

a. average concentration of raffinate $L_1 \sim L_5$ in Section 3.5.
b. average concentration of loaded organic-phase $O_1 \sim O_2$ in Section 3.5; —: not detected.

2.2 Solvent extraction experiments

Separating funnels that were settled in a temperature-controlled water bath shaker were used to mix the organic phase and the acid solution. Then, phase separation was realized due to gravity after standing for 10 min. The concentrations of elements in raffinate were determined via ICP-OES, and the concentrations of elements in the loaded organic phase were deduced via mass-balance calculation. The distribution ratio (D), extraction efficiency (E) and separation fraction (β), and volume fraction (X) were calculated via Eqs. (1)-(4), respectively.

$$D = C_{org}/C_{aq} \tag{1}$$
$$E = D/(D + V_{aq}/V_{org}) \times 100\% \tag{2}$$
$$\beta_{A/B} = D_A/D_B \tag{3}$$
$$X_A = V_A/(V_A + V_B) \tag{4}$$

where C_{org} represents the concentrations of elements in the organic phases; C_{aq} represents the concentrations of elements in the raffinate; V_{org} and V_{aq} represent the volumes of the organic and aqueous phases, respectively; $\beta_{A/B}$ represents the separation factor of elements A and B; and V_A and V_B represent the volumes of TBP and P507, respectively.

3 Results and discussion

3.1 Effects of extractants with TBP

The effects of extractants with TBP on Mo, W and Fe extraction were investigated, as presented in Fig. 2. P507, P204 and N1923 all had synergistic effects with TBP on Mo extraction, which resulted in the remarkable increase of the extraction efficiency of the mixture compared with that of single TBP extraction, and according to the synergistic effect, the extractants followed the order P507 > P204 > N1923. However, the extractants with TBP behaved differently in W extraction. The extraction efficiency of W increased dramatically with TBP+N1923, while the mixture of TBP and P507 realized an extraction efficiency that exceeded that of single 20% (v/v) TBP but was slightly lower than that of 40% (v/v) TBP; hence, in terms of the synergistic effect, N1923>P507. In contrast, P204 exhibited an antagonistic effect when mixed with TBP, which led to a lower extraction efficiency of W than that of single 20% (v/v) TBP. For the extraction of Fe, the results demonstrate that P507, P204 and N1923 all promoted Fe extraction with TBP according to the following order: N1923>>P204>P507. However, the mixture of P507 and TBP only promoted the extraction of Fe slightly compared with single 20% (v/v) TBP, and the performance was far below that of single 40% (v/v) TBP. After comprehensive consideration, P507 was selected as a suitable extractant for the TBP-based extraction system to realize high extraction efficiencies for Mo and W but relatively low extraction efficiency for impurity Fe.

3.2 Effect of the composition of the extractant mixture

Solvent extraction experiments were conducted with various compositions of TBP and P507 to identify the optimal extraction composition for separating Mo and W from Fe under a total concentration of [TBP+P507]/[kerosene] = 40/60 (v/v). As shown in Fig. 3, the extraction behaviors of Mo, W and Fe differed substantially with the changing P507 volume fraction. The mixture of TBP and P507 showed a strong synergistic effect on Mo, and the extraction efficiency of Mo increased substantially compared to only TBP or P507. With increasing volume fraction of P507, the extraction efficiency reached a maximum of 97% at $X_{P507} = 0.5$ and, subsequently, gradually decreased. In addition, the extraction efficiency of Mo with single P507 (88%) was higher than that of single TBP (80%); hence, single P507 had higher extractability for Mo than single TPB. In addition, both cationic and neutral molecular forms of Mo were present. This is because $[PMo_{12}O_{40}]^{3-}$ in the HCl-H_3PO_4 solution could be decomposed into molybdenum cation ($[MoO_2]^{2+}$) at high acidity (Nagul et al., 2015). In this case, multiple mechanisms likely occurred during the extraction of Mo using a mixture of TBP and P507, and Mo can be extracted by TBP in the form of $PMo_{12}O_{40} \cdot 3[H_3O(H_2O)_3 \cdot 3TBP]$ (Lakshmanan and Haldar, 1970b) or by P507 in the form of $MoO_2 \cdot (HA_2)_2$ (Xia et al., 2015) (HA represents P507), as expressed in Eqs. (5)-(6). As a result, the synergistic effect of the TBP and P507 mixture on Mo extraction is more likely attributed to the multiple ionic forms of Mo, in addition to the interaction between TBP and P507.

$$9\overline{TBP} + 3H^+ + 12H_2O + PMo_{12}O_{40}^{3-} \rightleftharpoons \overline{PMo_{12}O_{40} \cdot 3[H_3O(H_2O)_3 \cdot 3TBP]} \quad (5)$$

$$2\overline{(HA)_2} + MoO_2^{2+} \rightleftharpoons \overline{MoO_2(HA_2)_2} + 2H^+ \quad (6)$$

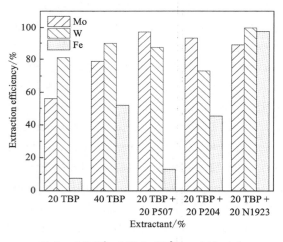

(1.7 mol/L H^+, A/O 1, 15 ℃, and 10 min)　　　(1.7 mol/L H^+, A/O 1, 15 ℃, and 10 min)

Fig. 2　Effects of extractants with TBP on Mo, W and Fe extraction　　　Fig. 3　Effects of the composition of the extractant mixture on Mo, W and Fe extraction

In the case of W extraction, the extraction efficiency of W slightly increased from 90% to 94% as X_{P507} increased from 0 to 0.4. Subsequently, the efficiency sharply decreased. Hence, the mixture of TBP and P507 has a weaker synergistic effect on W than on Mo, and the synergistic effect only occurs with sufficient TBP. In addition, single P507 has nearly no Mo extraction ability: only 9% W was extracted at $X_{P507} = 1.0$ because $[PW_{12}O_{40}]^{3-}$ was stable in the HCl-H_3PO_4 solution, in contrast to $[PMo_{12}O_{40}]^{3-}$, which was difficult to extract with an acidic organophosphorus extractant. Therefore, TBP acts as an extractant of W, as expressed in Eq. (7). The small synergistic effect of the TBP and P507 mixture at $X_{P507}<0.4$ may be attributed to the solvent effect through the substitution of the H_2O molecules in $PW_{12}O_{40} \cdot 3[H_3O(H_2O)_3 \cdot 3TBP]$ with P507, as expressed in Eq. (8).

$$9\overline{TBP} + 3H^+ + 12H_2O + PW_{12}O_{40}^{3-} \rightleftharpoons \overline{PW_{12}O_{40} \cdot 3[H_3O(H_2O)_3 \cdot 3TBP]} \quad (7)$$

$$\overline{PW_{12}O_{40} \cdot 3[H_3O(H_2O)_3 \cdot 3TBP]} + n\overline{HA} \rightleftharpoons$$
$$\overline{PW_{12}O_{40} \cdot 3[H_3O(H_2O)_{3-n} \cdot nHA \cdot 3TBP]} + nH_2O \quad (8)$$

For Fe extraction, the mixture of TBP and P507 showed a dramatic antagonistic effect. The extraction efficiency of Fe decreased substantially compared to those of single TBP and single P507, and the minimum extraction efficiency of Fe was 8% at $X_{P507} = 0.4$. Both TBP and P507 can extract Fe alone due to the complex reactions between Fe^{3+} and Cl^- that enable the coexistence of the neutral molecular form ($FeCl_3$), cationic forms (Fe^{3+}, $FeCl^{2+}$, $FeCl_2^+$) and even anionic form ($FeCl_4^-$) in the $HCl-H_3PO_4$ solution. The antagonistic effect may be explained by the hydrogen bonding between TBP and P507, which decreased the amounts of free TBP and P507 simultaneously and inhibited the extraction of $FeCl_3$ by TBP or of Fe cations by P507, as expressed in Eq. (9), according to (Zhang et al., 2019).

$$\overline{HA} + \overline{TBP} \rightleftharpoons \overline{HA \cdot TBP} \quad (9)$$

The separation factors of Mo/Fe and W/Fe both increased significantly with maximum values of 403 and 156, respectively, at $X_{P507} = 0.4$ due to the synergistic effects on Mo and W and the antagonistic effect on Fe. Hence, 6 : 4 was the effective volume ratio of TBP : P507 for separating Mo and W over Fe.

3.3 Effect of acidity

Fig. 4 shows that the acidity of the solution significantly affected the extraction process. The extraction of W decreased more rapidly than that of Mo as the acidity decreased. The extraction efficiency of W decreased from 93% to 62% as the acidity decreased from 1.7 mol/L H^+ to 1.0 mol/L H^+, while the extraction efficiency of Mo only decreased from 97% to 92%. This is because W was only present in the form of $[PW_{12}O_{40}]^{3-}$ and H^+ was required when extracted by TBP, while Mo was present in multiple forms and could be extracted by TBP and P507 simultaneously according to Section 3.2. In addition, the extraction efficiency of Fe slightly increased until the acidity was lower than 1.1 mol/L. As a result, the separation factors of Mo/Fe and W/Fe both decreased dramatically as the acidity decreased. Therefore, the initial preleaching solution was used directly for the solvent extraction.

3.4 Effect of A/O ratio

Fig. 5 presents the extraction effects of Mo, W and Fe with an A/O phase ratio range of 0.5 to 6. The extractions of Mo and W decreased monotonously with increasing A/O ratio, while the extraction of Fe initially decreased quickly and subsequently gently changed. This may be because there was excess extractant after the preferential extraction of Mo and W at an A/O ratio of 0.5. As a result, the separation factors of Mo/Fe and W/Fe both increased initially within an A/O ratio of 0.5 to 1 and subsequently decreased as the A/O ratio was further increased. Considering the loading and processing capacities of the extractant and to realize better separation of Mo and W over Fe, the A/O ratio was determined to be 3 : 1, at which 90% Mo, 88% W and 8% Fe were extracted and the separation factors of Mo/Fe and W/Fe were 153 and 118, respectively. A batch continuous extraction tests was adopted to realize higher extraction efficiency of Mo and W.

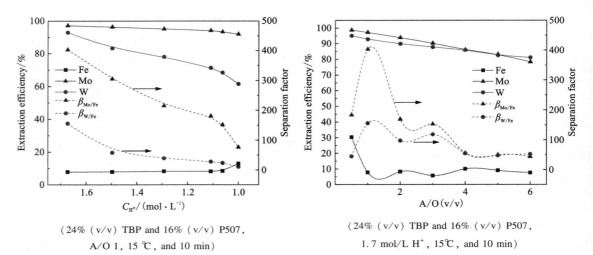

(24% (v/v) TBP and 16% (v/v) P507, A/O 1, 15 ℃, and 10 min)

Fig. 4 Effects of the acidity on Mo, W and Fe extraction

(24% (v/v) TBP and 16% (v/v) P507, 1.7 mol/L H^+, 15℃, and 10 min)

Fig. 5 Effects of the A/O ratio on Mo, W and Fe extraction

3.5 Batch continuous extraction test

Fig. 6 presents the extraction isotherms of Mo and W, which were obtained by changing the A/O ratio from 1 : 2 to 6 : 1. Based on the McCabe-Thiele rules, two theoretical extraction stages are required for the extraction of most of the Mo and W simultaneously at an A/O ratio of 3 via batch continuous extraction with 24% (v/v) TBP+16% (v/v) P507. A two-stage batch continuous simulation experiment was designed to evaluate the McCabe-Thiele prediction, as illustrated in Fig. 7. Each rectangle represents an extraction process; L and L with a subscript represent the initial liquid solution and raffinate, respectively, and O and O with a subscript represent the initial organic phase and the loaded organic phase, respectively. Using raffinate from the $L_1 \sim L_5$ operation, the recovery efficiencies of Mo, W and Fe were calculated, and the results are presented in Table 2. Approximately 99% and 94% of Mo and W, respectively, were extracted, while 10% Fe was coextracted. Raffinate of the corresponding average concentration of contained approximately 0.02 g/L Mo and 0.09 g/L WO_3, and the loaded organic phase contained 5.41 g/L Mo and 4.12 g/L WO_3, as presented in Table 1.

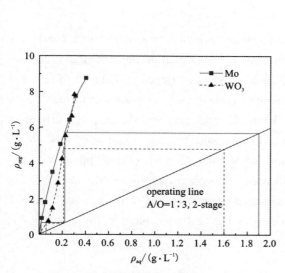

Fig. 6 McCable-Thiele plot for the extraction of Mo and W

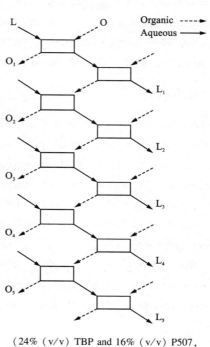

(24% (v/v) TBP and 16% (v/v) P507, 1.7 mol/L H^+, A/O 3, 15 ℃, and 10 min)

Fig. 7 Simulated batch continuous experiment

Table 2 Results of the two-stage simulated batch continuous extraction

Designed cycle	E(Mo)/%	E(W)/%	E(Fe)/%
L_1	99.04	93.80	14.02
L_2	98.92	94.27	12.39
L_3	98.93	94.13	7.43
L_4	98.86	93.59	6.40
L_5	98.91	94.24	11.94
Average	98.93	94.00	10.44

3.6 Washing of the loaded organic phase

With the extraction of Fe, additional impurities, such as P and Si, were coextracted, and the amounts of Ca, Al and Mn that were extracted were negligible, as shown in Table 1. The loaded organic phase was washed to efficiently separate Mo, W and impurities. Using 2 mol/L HCl as the washing liquor, 88% P and 84% Fe were washed under a 2 : 1 O/A ratio at 15 ℃ for 10 min, and the losses of Mo and W were negligible. However, Si was not washed. The chemical composition of the washed organic phase is listed in Table 1. Small amounts of P and Si remained in the washed organic phase due to the formation of heteropoly acids with Mo and W. In addition, the concentration of all impurities in the washed organic phase was very low, and their presence negligibly affected the Mo and W stripping process.

3.7 Stripping

Ammonia was used to strip the loaded organic phase from the two-stage batch continuous extraction in Section 3.5, and the effects of the ammonia concentration on the stripping of Mo and W are presented in Fig. 8. With deionized water, 59% of W was stripped, compared to only 6% of Mo; hence, W is easier to strip than Mo. As the concentration of ammonia was increased, the stripping efficiency of W increased to 100% with 3 mol/L ammonia, and the stripping efficiency of Mo increased to 100% with 5 mol/L ammonia. In addition, the consumption of OH^- in the stripping agent was very low since the concentrations of Mo and W in the organic phase were low so that the stripping liquor could be reused to increase the concentrations of Mo and W. The stripping efficiencies of Mo and W remained unchanged after recycling 4 times with 5 mol/L ammonia. Therefore, 5 mol/L ammonia was selected as the best concentration for stripping Mo and W.

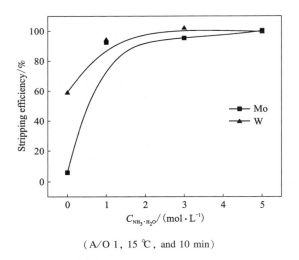

(A/O 1, 15 ℃, and 10 min)

Fig. 8 Effect of the ammonia concentration on the stripping of Mo and W

4 Conclusions

The separation of Mo and W from Fe in HCl-H_3PO_4 solutions using a mixture of TBP and P507 was feasible and profitable. Various extractants with TBP were investigated, and P507 showed strong synergistic effects on Mo and W extraction and a strong antagonistic effect on Fe extraction. The results demonstrate that 90% Mo and 88% W can be extracted with 8% Fe coextraction, and the separation factors of Mo/Fe and W/Fe were 153 and 118, respectively, after one stage of extraction using 24% (v/v) TBP, 16% (v/v) P507 and 60% (v/v) kerosene as the extractants under the following conditions: 1.7 mol/L H^+, A/O ratio of 3 : 1, contact time of 10 min, and temperature of 15 ℃. After two-stage batch continuous extraction, 99% Mo, 94% W and 10% Fe were extracted. Over 84% Fe could be washed using 2 mol/L HCl, and almost 100% of Mo and W could be stripped using 5 mol/L ammonia at a 1 : 1 A/O ratio. Additional impurities, such as P, Si, Ca, Al, and Mn, minimally affect the extraction of Mo and W. However, the mechanisms of the synergistic effects on Mo and W and the antagonistic effect on Fe using a mixture of TBP and P507 were discussed in terms of the states of these ions, which require further research.

Acknowledgment

The authors are thankful to the National Key Research and Development Program of China (2019YFC1907400) for financial support.

References

[1] AL-OTHMAN A, DEMOPOULOS G P. Gypsum crystallization and hydrochloric acid regeneration by reaction of calcium chloride solution with sulfuric acid[J]. Hydrometallurgy, 2009, 96: 95-102. https://doi.org/10.1016/j.hydromet.2008.08.010.

[2] ASAKURA T, DONNET L, PICART S. Extraction of hetero polyanions, P2W17O61 10-, P2W18O62 6-, SiW11O39 8- by TBP 246, 2000: 651-656.

[3] BOCHET C, DRAPER T, BOCQUET B, et al. 182 tungsten Mössbauer spectroscopy of heteropolytungstates[J]. Dalton Trans, 2009: 5127-5131. https://doi.org/10.1039/b904101j.

[4] CHEN W, FENG Q, ZHANG G, et al. Investigations on flotation separation of scheelite from calcite by using a novel depressant: Sodium phytate[J]. Miner. Eng, 2018, 126: 116-122. https://doi.org/10.1016/j.mineng.2018.06.008.

[5] DONG L, JIAO F, QIN W, et al. Selective flotation of scheelite from calcite using xanthan gum as depressant[J]. Miner. Eng, 2019, 138: 14-23. https://doi.org/10.1016/j.mineng.2019.04.030.

[6] HAN H, HU Y, SUN W, et al. Fatty acid flotation versus BHA flotation of tungsten minerals and their performance in flotation practice[J]. Int. J. Miner. Process, 2017, 159: 22-29. https://doi.org/10.1016/j.minpro.2016.12.006.

[7] HUO G, LIU Z, ZENG L, et al. Method for decomposing scheelite through cyclic hydrochloric acid of mother liquor, 2018: 108342597A.

[8] KANG J, KHOSO S A, HU Y, et al. Utilisation of 1-Hydroxyethylidene-1, 1-diphosphonicacid as a selective depressant for the separation of scheelite from calcite and fluorite [J]. Colloids Surf. A

Physicochem. Eng. Asp. 2019, 582: 123888. https://doi.org/10.1016/j.colsurfa.2019.123888.

[9] KUPKA N, RUDOLPH M. Froth flotation of scheelite: A review[J]. Int. J. Min. Sci. Technol. 2018, 28: 373-384. https://doi.org/10.1016/j.ijmst.2017.12.001.

[10] LAKSHMANAN V I, HALDAR B C. Studies on extractive behaviour of 12-heteropoly acids. I. Extraction of 12-tungstophosphoric acid[J]. J. Indian Chem. Soc. 1969, 46: 512.

[11] LAKSHMANAN V I, HALDAR B C. Studies on extractive behaviour of 12-heteropoly acids. 2. extraction of 12-molybdophosphoric, 12-tungstosilicic and 12-molybdosilicic acids[J]. J. Indian Chem. Soc. 1970a, 47: 231- +.

[12] LAKSHMANAN V I, HALDAR B C. Extractive behavior of 12-heteropoly acids. III. Extraction of 12-tungstophosphoric, 12-molybdophosphoric, 12-tungstosilicic, and 12-molybdosilicic acids by TBP in organic diluents[J]. J. Indian Chem. Soc. 1970b, 47: 72-78.

[13] LASSNER E, SCHUBERT W D. Tungsten: Properties, chemistry, technology of the element, alloys, and chemical compounds[M]. Berlin: Springer, 1999. https://doi.org/10.1007/978-1-4615-4907-9.

[14] LI D R, SU Y F. Separation of tungsten and phosphorus by solvent extraction with petroleum sulfoxide[J]. J. Chem. Ind. Eng. China, 1990, 1: 43-49.

[15] LI Y, YANG J, ZHAO Z. Recovery of tungsten and molybdenum from low-grade scheelite[J]. JOM, 2017, 69: 1958-1962. https://doi.org/10.1007/s11837-017-2440-5.

[16] LI Z, ZHANG G, ZENG L, et al. Continuous solvent extraction operations for the separation of W and Mo in high concentrations from ammonium solutions with acidified N1923[J]. Hydrometallurgy, 2019, 184: 39-44. https://doi.org/10.1016/j.hydromet.2018.12.015.

[17] LIAO Y, ZHAO Z. Effects of phosphoric acid and ageing time on solvent extraction behavior of phosphotungstic acid[J]. Hydrometallurgy, 2017, 169: 515-519. https://doi.org/10.1016/j.hydromet.2017.03.003.

[18] LYU F, SUN W, KHOSO S A, et al. Adsorption mechanism of propyl gallate as a flotation collector on scheelite: A combined experimental and computational study[J]. Miner. Eng. 2019, 133: 19-26. https://doi.org/10.1016/j.mineng.2019.01.003.

[19] NAGUL E A, MCKELVIE I D, WORSFOLD P, et al. The molybdenum blue reaction for the determination of orthophosphate revisited: Opening the black box[J]. Anal. Chim. Acta, 2015, 890: 60-82. https://doi.org/10.1016/j.aca.2015.07.030.

[20] XIA Y, XIAO L, XIAO C, et al. Direct solvent extraction of molybdenum(VI) from sulfuric acid leach solutions using PC-88A[J]. Hydrometallurgy, 2015, 158: 114-118. https://doi.org/10.1016/j.hydromet.2015.10.016.

[21] YAN W, LIU C, AI G, et al. Flotation separation of scheelite from calcite using mixed collectors[J]. Int. J. Miner. Process. 2017, 169: 106-110. https://doi.org/10.1016/j.minpro.2017.10.009.

[22] YANG X. Beneficiation studies of tungsten ores: A review[J]. Miner. Eng, 2018, 125: 111-119. https://doi.org/10.1016/J.MINENG.2018.06.001.

[23] YANG Y, XIE B, WANG R, et al. Extraction and separation of tungsten from acidic high-phosphorus solution[J]. Hydrometallurgy, 2016, 164: 97-102. https://doi.org/10.1016/j.hydromet.2016.05.018.

[24] YATIRAJAM V, DHAMIJA S. Extractive separation of tungsten as phosphotungstate[J]. Talanta, 1977a, 24: 52-55. https://doi.org/10.1016/0039-9140(77)80187-2.

[25] YATIRAJAM V, DHAMIJA S. Extractive separation and spectrophotometric determination of tungsten as phosphotungsten blue[J]. Talanta, 1977b, 24: 497-501. https://doi.org/10.1016/0039-9140(77)

80033-7.

[26] YATIRAJAM V, RAM J. Separation of molybdenum from interfering elements by extraction as phosphomolybdenum blue[J]. Talanta, 1973, 20: 885-890. https://doi.org/10.1016/0039-9140(73)80204-8.

[27] YONGXIN L, CHANGGEN L. Selective flotation of scheelite from calcium minerals with sodium oleate as a collector and phosphates as modifiers. I. Selective flotation of scheelite[J]. Int. J. Miner. Process. 1983, 10: 205-218. https://doi.org/10.1016/0301-7516(83)90011-X.

[28] ZHANG H, HUA J, LIU Z, et al. Improving the separation of molybdenum from iron in acidic sulfate solutions due to the antagonistic effect[J]. Hydrometallurgy, 2019, 186: 187-191. https://doi.org/10.1016/J.HYDROMET.2019.04.001.

[29] ZHAO Z, LI J, WANG S, et al. Extracting tungsten from scheelite concentrate with caustic soda by autoclaving process[J]. Hydrometallurgy, 2011a, 108: 152-156. https://doi.org/10.1016/j.hydromet.2011.03.004.

[30] ZHAO Z, LIANG Y, LIU X, et al. Sodium hydroxide digestion of scheelite by reactive extrusion[J]. Int. J. Refract. Met. Hard Mater., 2011b, 29: 739-742. https://doi.org/10.1016/j.ijrmhm.2011.06.008.

[31] ZHAO C, SUN C, YIN W, et al. An investigation of the mechanism of using iron chelate as a collector during scheelite flotation[J]. Miner. Eng. 2019, 131: 146-153. https://doi.org/10.1016/j.mineng.2018.11.009.

Process and Mechanism Investigation on Comprehensive Utilization of Arsenic-Alkali Residue

Abstract: Arsenic-alkali residue is a solid waste produced by the antimony smelting industry, which can pose a threat to the environment and human health. The common wet treatment process of arsenic-alkali residue has a low recovery of valuable elements, incomplete separation of arsenic and alkali, and also produces arsenic-alkali mixed salt, which cannot realize the completely harmless treatment of arsenic-alkali residue. In order to solve these problems, the oxidative water leaching process was used to treat arsenic-alkali residue, which realized the separation of arsenic and antimony. The leaching efficiencies of arsenic and antimony were 91.79% and 0.62%, respectively. The leaching residue could be returned to the antimony smelting system to recover antimony. Then the arsenic and alkali were directly separated from the arsenic-alkali mixed salt by carbothermal reduction, and 98.3% of arsenic was removed, and the non-toxic metallic arsenic with 99.9% purity was prepared. The alkali could be recovered from the slag after reduction, which solved the problem of harmless and recycling treatment of arsenic-alkali mixed salts. The mechanism of arsenic reduction pathway was studied through thermodynamic, phase, and arsenic valence state analyses.

Keywords: Arsenic-alkali residue; Resource recovery; Separation; Carbothermal reduction; Metallic arsenic

1 Introduction

Arsenic-alkali residue is a solid waste produced in the refining process of crude antimony. The main components of arsenic-alkali residue are sodium carbonate, sodium arsenate, and sodium antimonate[1]. Arsenic-alkali residue cannot simply be treated by solidification and

---

Published in *Journal of Central South University*, 2023, 30(3): 721-734. Authors: Gong Ao, Wu Xuangao, Li Jinhui, Wang Ruixiang, Xu Jiacong, Wen Shenghui, Yi Qin, Tian Lei, Xu Zhifeng.

landfilling, because it contains a large amount of sodium arsenate and high-solubility sodium carbonate. Serious arsenic contamination may occur if the arsenic dissolves in drinking water sources[2-4]. As a major antimony smelting country, China produces approximately 84.0% of the world's antimony[5], and produces approximately 5000 tons of arsenic-alkali residue each year. At present, the total reserves of arsenic-alkali residue in China have reached 200000 tons[6-7]. Therefore, China is facing serious environmental pollution because of arsenic-alkali residue.

Additionally, arsenic-alkali residue contains large amounts of alkali, arsenic, and antimony, which are important recoverable resources. Arsenic is widely used in glass clarifiers, batteries, medicine and semiconductors[8-9]. Antimony is one of China's four major strategic resources, and widely used in batteries, glass, flame retardants, thermally sensitive fibers and other products[10-11]. Therefore, the treatment of arsenic-alkali residue must consider the recycling potential of these valuable components to achieve eco-friendly resource recycling and promote the sustainable development of antimony smelting industry.

The main treatment methods of arsenic-alkali residue can be divided into pyrometallurgy, stabilization, and hydrometallurgy processes[12-14]. The pyrometallurgical process primarily uses the volatile characteristics of arsenic oxide to remove arsenic from arsenic-alkali residue in the form of As_2O_3 through oxidation roasting, or to recover arsenic in the form of metallic arsenic through continuous reduction refining[14-15]. However, the separation of arsenic is not complete, arsenic-containing dust will be generated, and cause secondary pollution. Therefore, the application of the pyrometallurgical treatment of arsenic-alkali residue is rare[16-17]. Stabilization treatment reduces the solubility of arsenic by adding solidifying agent to physically encapsulate arsenic, or by changing the form of arsenic in the arsenic-containing waste sludge[18-21]. However, stabilization treatment increases the amount of slag and raises the cost of landfill. It is impossible to recycle valuable components through this method, and it leads to resource waste. Furthermore, the residue after stabilization treatment still has the risk of secondary pollution due to arsenic dissolution.

At present, the main treatment process of arsenic-alkali residue is hydrometallurgy process. Since sodium arsenate and sodium carbonate are highly soluble in water, and sodium antimonate is not easily dissolved in water, the separation of arsenic and antimony can be achieved by water immersion. The leaching residue can be returned to the antimony smelting system for recovery. The treatment of high-arsenic leaching solution primarily involves chemical precipitation[22-23] or fractional crystallization[14, 24-25]. Under chemical precipitations, Fe-As and Ca-As precipitations are formed by adding precipitating agents such as iron salts or calcium salts, so as to realize the harmless treatment of high-arsenic leaching solution, but it is difficult to realize the resource utilization of arsenic and alkali, and there is still a risk of arsenic dissolution from the arsenic precipitation slag[22-23, 26].

Fractional crystallization uses the solubility difference between sodium arsenate and sodium carbonate at high temperature to achieve the separation of arsenic and alkali. However, in the actual production process, it is difficult to control the evaporation temperature. Hence, the separation effect of arsenic and alkali is poor and the crystallization product is generally a mixture of arsenic and alkali[14]. Some scholars converted Na_2CO_3 into $NaHCO_3$ with low solubility by adding CO_2 into the solution to improve the separation effect of arsenic and alkali[14, 24-25]. However, this process is only suitable for treating low arsenic solutions. The crystallized product still needs to be washed several times to ensure purity, and a large amount of wastewater is generated. In addition, there is overproduction of sodium arsenate in the market. Therefore, it is the conversion of arsenic into non-toxic and potentially commercial metallic arsenic that is required.

In this study, we designed a process scheme for comprehensive utilization and safe disposal of arsenic-alkali residue through the combination of pyrometallurgy and hydrometallurgy. We achieved the separation of arsenic and antimony through oxidation-water leaching, and utilized carbothermal reduction for the treatment of the crystalline product of the leaching solution, volatilizing and removing arsenic in the form of arsenic vapor. Through sublimation, we recovered non-toxic and potentially commercial metal arsenic, and the alkali can be recovered from the reduced residue. This makes it a safe and resource-efficient method to treat arsenic-alkali residue.

2 Materials and methods

2.1 Materials

The arsenic-alkali residue used in this investigation was obtained from Henan province, China. After drying at 80 ℃ for 8 h, the arsenic-alkali residue was crushed and separated with 75 μm sieve, and subsequently stored in a sealed plastic reagent bottle.

The sample was analyzed by X-ray fluorescence (XRF) for its general elements, and inductively coupled plasma optical emission spectrometer (ICP-OES) was used to determine the accurate assays of arsenic and antimony. The results were shown in Table 1, which indicate that arsenic-alkali residue mainly consisted of As, Sb, Na, and a few other elements. Compared with the general arsenic-alkali residue, the contents of arsenic and antimony are higher (17.61% and 16.05%, respectively), therefore, the hazard to the environment and the recyclability of valuable elements are higher.

Table 1 Main chemical compositions of arsenic-alkali residue   %

O	As	Sb	Na	Pb	Bi	S	Se	Mo	Cl	Si	Others
28.13	17.61	16.05	32.21	3.90	1.33	0.29	0.07	0.07	0.03	0.02	0.29

The main phase composition of arsenic-alkali residue detected by X-ray powder diffractometry (XRD) analysis is shown in Fig. 1. It primarily contains NaOH, Na_2CO_3, Na_3AsO_4,

$Na_2As_4O_{11}$, and $NaSbO_3$. Notably, the mixed alkali was used in antimony smelting, and arsenic-alkali residue was stored in the open air, and Na_2CO_3 and NaOH were observed in the arsenic-alkali residue.

For a better understanding of the distribution and existence of arsenic and antimony in the residue, chemical phase analysis was conducted, and the results were shown in Tables S1 and S2. Sodium antimonate (52.64%), sodium subantimonate (30.63%), and metallic antimony (16.18%) are the main antimony phases, while sodium arsenate (99.41%) is the main arsenic phase in the arsenic-alkali residue.

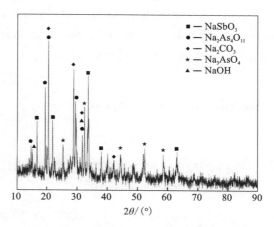

Fig. 1　XRD pattern of arsenic-alkali residue

2.2　Experimental procedures

As shown in Fig. 2, the experimental procedures recovering arsenic, antimony, and alkali from arsenic-alkali residue primarily included the following steps: 1) Leaching arsenic-alkali residue with oxidation-water to enrich antimony and leach out alkali and arsenic; the leaching residue can be directly returned to the antimony smelting system; 2) Evaporation and

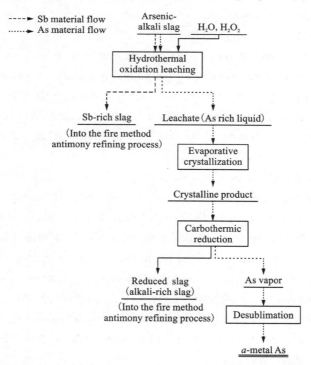

Fig. 2　Schematic flowsheet for the comprehensive utilization of arsenic-alkali residue

crystallization of the leaching solution to obtain the arsenic-alkali mixed salt; 3) The carbothermal reduction of crystallization product to recover arsenic and alkali. The operational conditions and the detailed methods were summarized in supplementary file.

3 Results and discussion

3.1 Separation of arsenic and antimony from arsenic-alkali residue by oxidation-water leaching

3.1.1 Determination of the optimum leaching conditions of arsenic-alkali residue

Fig. 1 shows that arsenic-alkali residue is mainly composed of $NaSbO_3$ with low solubility and Na_2CO_3 and Na_3AsO_4 with high solubility. And As and Sb can be separated by oxidation-water leaching. The effects of leaching time, leaching temperature, liquid-solid ratio, and oxidizer addition (H_2O_2) on the leaching efficiency of As and Sb were investigated, and the results are shown in Fig. 3.

Fig. 3(a) shows the effect of temperature on As and Sb leaching. When temperature increased from 20 ℃ to 90 ℃, the leaching efficiency of As increased from 88.35% to 91.79%, and the leaching efficiency of Sb decreased from 3.68% to 0.62%, indicating that the increase in temperature is favorable for the separation of As and Sb, and promotes the hydrolysis of antimony ions in the solution, thus promoting the separation of arsenic and antimony. Thus, 90 ℃ was selected as the optimal leaching temperature.

Fig. 3(b) shows the effect of time on As and Sb leaching. With the increase of the leaching time, the solubility of arsenic increased, and the antimony partially dissolved in the solution also had sufficient time to hydrolyze and precipitate, thereby improving the separation effect of arsenic and antimony. When the leaching time reached 2 h, the continued increases of time would not significantly improved the separation of arsenic and antimony. Thus, 2 h was selected as the optimal leaching time.

Fig. 3(c) shows the effect of liquid-solid ratio on the leaching of As and Sb. When the liquid-solid ratio increased from 1∶1 to 2∶1, the leaching efficiency of As increased from 89.38% to 91.79%. The increase of liquid-solid ratio has less effect on the leaching rate of antimony, illustrating that the increase of the liquid-solid ratio is favorable for the separation of As and Sb. On the one hand, the increase of the liquid-solid ratio reduces the viscosity of the solution and promotes the mass transfer process. On the other hand, it increases the solubility of sodium arsenate, thereby improves the separation of As and Sb. Although the further increase of the liquid-solid ratio is beneficial to the separation of arsenic and antimony, it also increases the water treatment capacity in subsequent evaporation and crystallization. Thus, the liquid-solid ratio of 2∶1 was identified as the optimal ratio.

Fig. 3(d) shows the effect of dosage of H_2O_2 on As and Sb leaching. The addition of H_2O_2

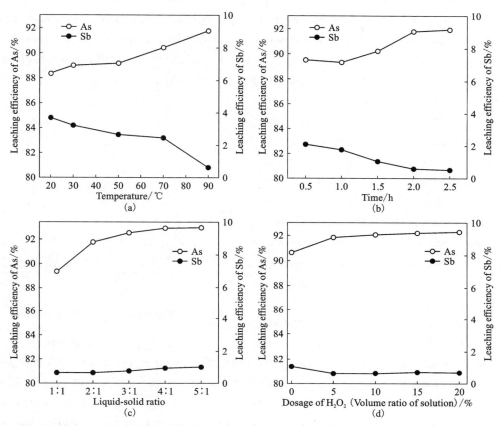

Figure 3 Effect of (a) leaching temperature (leaching time of 2 h, liquid-solid ratio of 2:1, H_2O_2 addition of 5%), (b) leaching time (leaching temperature of 90 ℃, liquid-solid ratio of 2:1, H_2O_2 addition of 5%), (c) liquid-solid ratio (leaching temperature of 90 ℃, leaching time of 2 h, H_2O_2 addition of 5%), and (d) dosage of H_2O_2 (leaching temperature of 90 ℃, leaching time of 2 h, liquid-solid ratio of 2:1) on leaching efficiency of As and Sb

significantly improved the separation of arsenic and antimony. As(V) is known to be more soluble than As(Ⅲ), while Sb(V) is less soluble than Sb(Ⅲ). Moreover, a part of As(Ⅲ) and Sb(Ⅲ) existed in arsenic-alkali residue, the addition of H_2O_2 increased the dissolution difference between arsenic and antimony As(Ⅲ/V) and Sb(Ⅲ/V) refer to trivalent or pentavalent As and Sb elements, respectively. However, it can be seen from Tables S1 and S2 that As in the experimental material primarily existed in the form of sodium arsenate, and the relative proportion of sodium antimonate was also small. 5% of H_2O_2 was sufficient for the oxidation of As(Ⅲ) and Sb(Ⅲ), and the continued increase in H_2O_2 addition had little effect on the separation of As and Sb. Therefore, 5% of H_2O_2 addition was identified as the optimal dosage in this experiment.

3.1.2 Characteristics of leaching residue and crystallization product

Through the exploration of the preliminary process conditions, we can achieve leaching rates of 91.79%, 0.62% for arsenic and antimony, respectively under optimal process conditions. The leaching residue and crystallization product were analyzed by ICP, scanning electron microscopyenergy-dispersive X-ray spectroscopy (SEM-EDS), XRD and X-ray photoelectron spectroscopy (XPS) to further explore the changes in the chemical composition, morphology, element distribution, and mineralogical phases after the leaching process.

ICP analysis revealed that the contents of arsenic and antimony in leached slag were 1.05% and 62.17%, respectively, which indicated that through the water leaching process, most of the arsenic was leached into the solution, while antimony was enriched in the leaching residue. Therefore, the separation of arsenic and antimony was achieved. According to the XRD pattern of leaching residue shown in Fig.4, the main phase in the leaching residue was $NaSbO_3$.

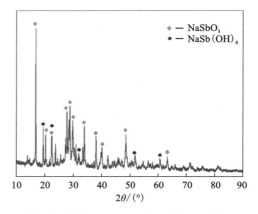

Fig. 4　XRD pattern of leaching residue

The SEM figures of arsenic-alkali residue and leaching residue are shown in Fig.5(a) and (b), respectively. By comparison, the particle of arsenic-alkali residue is fine and exists in the form of irregular spherical particles with smooth surface and agglomeration, which is an important reason for the insufficient arsenic leaching efficiency. The leaching residue becomes smaller in size and looser in structure, and its shape changes from the original spherical particles to irregular shapes such as needles and flocs, indicating that the original structure of arsenic-alkali residue was destroyed by water leaching[27]. Additionally, geometrically shaped crystals were found in the leach residue (marked in the red box in Fig.5(b)), which are $NaSb(OH)_6$ crystals formed by the hydrolysis of antimony salts, indicating that some antimony was first dissolved into the solution and then hydrolyzed into the leaching residue[28]. The element distribution scanning maps of arsenic-alkali residue and leaching residue are shown in Fig.5(c) and (d). The comparison shows that the contents of arsenic and antimony were both high in arsenic-alkali residue. The distribution areas of arsenic and antimony were consistent with elements O and Na, respectively, indicating that arsenic and antimony mainly exist in the form of sodium arsenate and sodium antimonate. The distribution of As is very sparse and antimony is more widespread in the leaching residue, indicating that most of the arsenic is leached into the solution, while antimony is enriched in the leaching residue, leading to the separation of arsenic and antimony. The leaching residue can be directly used for antimony smelting.

The leaching solution was crystallized by evaporation to obtain crystallization product rich in arsenic and alkali. The arsenic content is 18.87%, which is roughly the same as the total amount of arsenic in the leaching solution, and this indicates that the evaporation crystallization process is relatively safe, and arsenic is not volatile. The phase compositions of the samples were analyzed

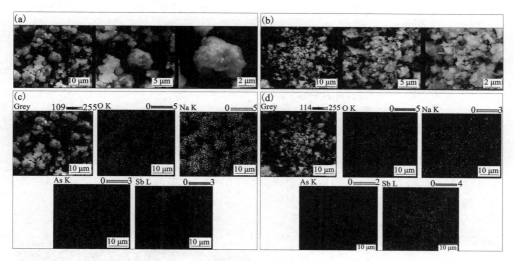

Fig. 5 SEM-EDS images of arsenic-alkali residue (a, c) and leaching residue (b, d)

by XRD, and the results are presented in Fig. 6(a). The crystallization product is mainly composed of NaOH, Na_2CO_3, Na_3AsO_4, and $Na_2As_4O_{11}$. XPS is a nondestructive surface analysis method for solid materials, and it provides information regarding the chemical composition of the surface of the sample[29-30]. The XPS survey spectrum is presented in Fig. 6(b). The presence of the As 3d core level peak as well as As Augre lines and As 3p peaks confirms the presence of arsenic. Significant peaks corresponding to O 1s, C 1s, and Na 1s are also observed. Fig. 6(c) shows the high-resolution scan of As 3d, and its binding energy is 45.3 eV. As reported, the binding energy of As 3d indicates the oxidation states of arsenic, with a value of 45.2-45.6 eV of As(V)—O bonds[31-32]. Thus, arsenic on the surface of crystallization products existed in the form of As(V). SEM-EDS image of crystallization products is presented in Fig. 6(d). As shown in the EDS analysis, the distribution of oxygen and sodium is consistent, and the distribution of arsenic is uniform, which proves the existence of sodium arsenate and alkali.

3.2 Preparation of metal arsenic by carbothermal reduction of crystalline products

3.2.1 Reduction mechanism analysis of crystalline products carbothermal reduction

For arsenic-alkali mixed salts, we attempted to remove arsenic through carbothermal reduction, and prepared non-toxic and commercially valuable metal arsenic. Therefore, the carbothermal reduction process was analyzed through thermodynamic analysis before the experiment. The specific thermodynamic analysis can be viewed in the supplementary file.

According to the analysis of the thermodynamics results, it can be concluded that the crystallization product has two reduction paths: $Na_2As_4O_{11}$ and Na_3AsO_4 are directly reduced to As_4, and $Na_2As_4O_{11}$ is first reduced to As_2O_3 and then reduced to As_4, As_4 is condensed to obtain

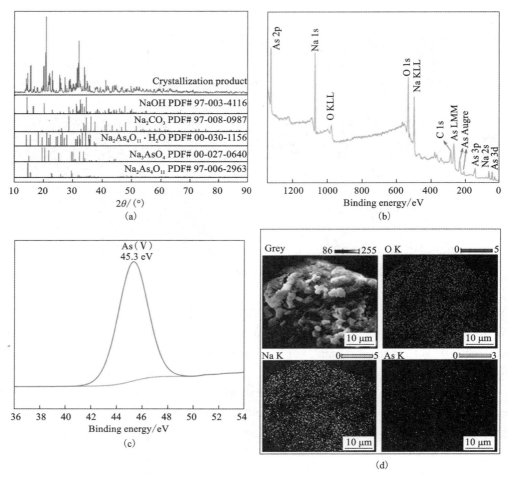

Fig. 6 (a) XRD pattern; (b) XPS survey spectra; (c) High-resolution scan for As 3d; (d) SEM-EDS image of the crystallization product

metal arsenic.

The sample with the addition of 12% carbon was subjected to thermogravimetry differential thermogravimetric (TG-DTG) analysis, and the results are presented in Fig. 7(a). From the TG-DTG curve, the mass of the sample decreases rapidly with the increase in temperature, which can be attributed to the rapid evaporation of free water or crystalline water in the sample, and the mass loss is 23.87%. Two mass loss peaks appeared at 560 ℃ and 720 ℃, respectively. Combined with the previous thermodynamic analysis, the reduction of $Na_2As_4O_{11}$ to As(Ⅲ) and As occurred at 560 ℃, and the reduction of Na_3AsO_4 to As occurred when the temperature reached 720 ℃. To further confirm the reduction process of products, the crystalline products with 12% carbon were roasted at 700 ℃ and 900 ℃, respectively, and the arsenic state changes in the roasted product were studied using XRD and XPS analysis techniques. The results are shown in Fig. 7(b) and (c). As shown in Fig. 7(b), the roasting products at different roasting temperatures all contain

Na_2CO_3, NaOH, and $NaAsO_2$, indicating that there was the reaction to generate $NaAsO_2$. The presence of Na_3AsO_4 was caused by the short roasting time and incomplete reaction. In addition, the diffraction peaks of As_2O_3 still existed at a roasting temperature of 700 ℃, and the diffraction peaks of As_2O_3 disappeared as the temperature rised, indicating that As_2O_3 had undergone reduction reaction. XRD analysis verifies our previous analysis on the reduction process of arsenic-alkali mixed salts. XPS analysis results [Fig. 7(c)] also support this conclusion. The As 3d spectra of roasted product obtained at 700 ℃ fitted to three peaks at 44.9, 44.3 and 43.9 eV[33-35], which correspond to Na_3AsO_4, $NaAsO_2$, and As_2O_3, respectively; the As 3d spectra of roasted product obtained at 900 ℃ fitted to three peaks at 44.9, 44.3 and 42.1 eV[33-35], which correspond to Na_3AsO_4, $NaAsO_2$, and As, respectively. Error of the binding energy of peaks due to calibration and noise is estimated at ±0.1 eV[36]. It is speculated that the existence of elemental arsenic was due to the short roasting time and part of elemental arsenic remained in the roasting slag.

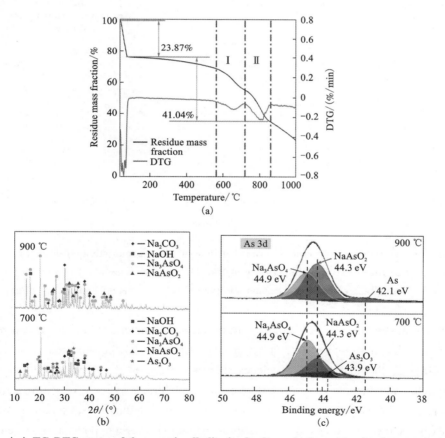

Fig. 7 (a) TG-DTG curve of the arsenic-alkali mixed salts carbothermal reduction process; (b) XRD pattern (roasting time of 10 min, carbon addition of 12%); (c) High-resolution scan for As 3d of the reduction product at different temperatures

3.2.2 Determination of the optimum roasting conditions of arsenic-alkali mixed salts

Through thermodynamic analysis, we have confirmed the feasibility of recovering metal arsenic by carbothermal reduction arsenic-alkali mixed salts. Moreover, we examined the effects of reaction temperature, reaction time, and amount of carbon addition on arsenic removal efficiency, and the results are shown in Fig. 8.

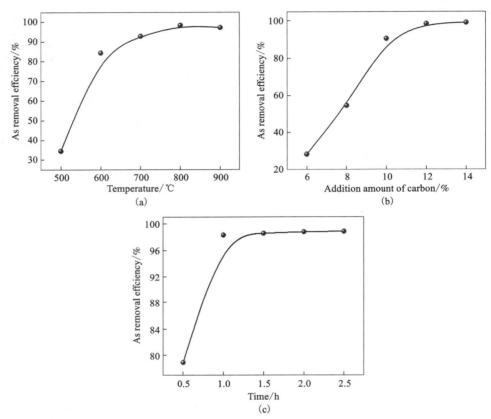

Fig. 8 Effect of (a) roasting temperature (roasting time of 1 h, carbon addition of 12%), (b) addition amount of carbon (roasting temperature of 800 ℃, roasting time of 1 h), and (c) roasting time (roasting temperature of 800 ℃, carbon addition of 12%) on removal efficiency of arsenic

Fig. 8(a) shows the effect of temperature on arsenic removal efficiency. It can be observed that temperature has a significant effect on arsenic removal. As the temperature increased from 500 ℃ to 800 ℃, the As removal efficiency increased from 34.45% to 98.3%, indicating that high temperature was conducive to arsenic removal, and further increase of temperature had little effect on arsenic removal, thus, 800 ℃ was selected as the optimal roasting temperature.

Fig. 8(b) shows the effect of the addition amount of carbon powder on arsenic removal efficiency. As a reducing agent, carbon powder provided electrons to As(Ⅴ), thus, the amount of carbon powder was crucial to the removal efficiency of arsenic. With the increase in carbon powder from 6% to 14% (mass fraction of crystallization products), the arsenic removal efficiency increased from 28.53% to 98.3%. However, increasing the amount of carbon powder did not promote the arsenic removal, indicating that the addition of carbon powder was sufficient to reduce As(V) to As at this time. Therefore, the optimal dosage of carbon powder was selected to be 12% in the experiment.

Fig. 8(c) shows the effect of roasting time on arsenic removal efficiency. It was found that the rate of arsenic removal was very fast, and removal efficiency is 98.3% after 1 h of reaction, further increasing the roasting time had only improved the arsenic removal efficiency marginally. Therefore, the roasting time of 1 h was selected as optimal time for the experiment.

3.2.3 Characteristics of reduction residue and product of arsenic

Based on the exploration of the preliminary process conditions, we can achieve 98.3% removal efficiency of arsenic under optimal conditions (roasting temperature of 800 ℃, roasting time of 1 h, carbon addition of 12%), which indicated effective separation of arsenic and alkali. The roasting residue and arsenic were analyzed by ICP, SEM-EDS, XRD, and XPS to further explore the changes in the chemical composition, morphology, element distribution, and mineralogical phases after the leaching process.

According to the ICP analysis, the content of arsenic in roasting residue is 0.57%, according to the XRD pattern of roasting residue shown in Fig. 9(a), the main phases in the roasting residue were NaOH and Na_2CO_3. No obvious arsenic-containing minerals are observed, demonstrating the

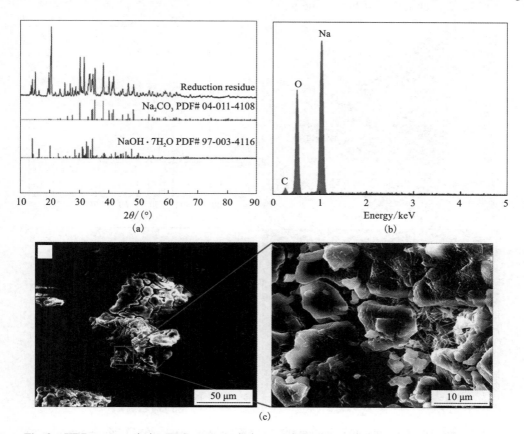

Fig. 9 XRD pattern (a), EDS analysis (b), and SEM image (c) of the reduction residue

highly efficient removal of arsenic from the arsenicalkali mixed salts. From the EDS analysis in Fig. 9(b), the main elements present in the reduced slag are Na and O, which further illustrates that arsenic has been removed from the arsenic-alkali mixed salts. The SEM image in Fig. 9(c) shows that the reduced slag is irregular granular with many cracks on the surface, which could be due to high-temperature sintering. The reduced slag is essentially composed of alkali and can be returned to the antimony refining process.

From the photograph of volatile product [Fig. 10(a)], it can be observed that the collected volatile products are flaky, and the surface has metallic luster. ICP analysis revealed that the arsenic content reached 99.9%, with very high purity. The XRD pattern of the volatile product in Fig. 10(b) also shows that the reduction product is pure metallic arsenic without other phases, and the metallic arsenic possesses a good crystalline rather than amorphous form. Thus, the product has commercial value. The SEM image of metallic arsenic in Fig. 10(c) shows that the surface of metallic arsenic is relatively smooth with bumps of different sizes. Fig. 10(d) presents the high resolution of the peak As 3d, which further shows that the reduction product is elemental arsenic, as it perfectly matches the reported binding energies of 41.8 eV of element arsenic[37-38]. Contrastingly, a small peak is found at 45.2 eV attributed to As_2O_3[39-40]. The metal arsenic is expected to have been oxidized before the test.

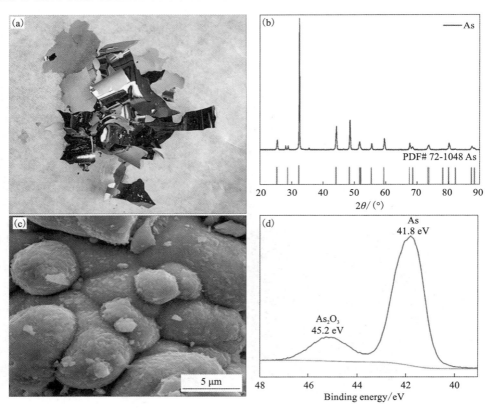

Fig. 10　(a) Entitative photograph; (b) XRD pattern; (c) SEM image and (d) high-resolution scan for As 3d of the volatile product

4 Conclusions

In this work, efficient recovery of arsenic, antimony, and alkali from arsenic-alkali residue was realized by oxidation-water leaching-carbothermal reduction, and the following conclusions were obtained:

First, the separation of arsenic and antimony from arsenic-alkali residue by oxidation-water leaching was experimentally studied. Under the conditions of the 2 : 1 liquid-solid, 90 ℃ leaching temperature, addition of 5% H_2O_2, and 2 h leaching time, the leaching rates of arsenic and antimony were 91.79% and 0.62%, respectively. The phase and morphology of the leaching residue were analyzed, and it was found that oxidation-water leaching destroyed the phase structure of the arsenic-alkali residue, such that the arsenic could be leached. Furthermore, it was found that some antimony was first dissolved into the solution during leaching and then hydrolyzed into the leaching slag.

Second, the high arsenic leaching solution was treated by evaporation crystallization method, and the crystalline product obtained constituted arsenic-alkali mixed salts. The feasibility of removing arsenic from the arsenic-alkali mixed salt by reduction and volatilization was analyzed by thermodynamic calculation. The arsenic reduction path in the crystalline product was as follows: $Na_2As_4O_{11}$ and Na_3AsO_4 were directly reduced to As_4, and $Na_2As_4O_{11}$ was first reduced to As_2O_3 and then reduced to As_4. The thermodynamic analysis results were verified through thermogravimetric analysis, reduction product phase analysis, and arsenic valence analysis.

Third, through carbothermal reduction, at the temperature of 800 ℃, 12% carbon addition, and 1 h reaction time, the removal efficiency of arsenic in the arsenic-alkali mixed salts was 98.3%. The obtained metal arsenic exists in the form of flakes with metallic luster, and the purity was more than 99.9%. The reduction residues were mainly $NaOH \cdot 7H_2O$ and Na_2CO_3, and the arsenic content was only 0.57%.

Appendix

Table S1 Antimony phase analysis results %

Antimony phase	Sodium antimonate	Sodium subantimonate	Metallic antimony	Antimony alloy	Total antimony
Content	52.64	30.63	16.18	0.55	100

Table S2 Arsenic phase analysis results   %

Arsenic phase	Sodium arsenate	Metallic arsenic	Total
Content	99.41	0.59	100

References

[1] ZHAO T C. Antimony[M]. Beijing: Metallurgical Industry Press, 1987.

[2] JIANG G H, MIN X B, KE Y, et al. Solidification/stabilization of highly toxic arsenic-alkali residue by MSWI fly ash-based cementitious material containing Friedel's salt: Efficiency and mechanism [J]. Journal of Hazardous Materials, 2022, 425: 127992. DOI: 10.1016/j.jhazmat.2021.127992.

[3] WANG D Y, REPO E, HE F S, et al. Dual functional sites strategies toward enhanced heavy metal remediation: Interlayer expanded Mg-Al layered double hydroxide by intercalation with L[J]. Journal of Hazardous Materials, 2022, 439: 129693. DOI: 10.1016/j.jhazmat.2022.129693.

[4] CHAI F, ZHANG R, MIN X B, et al. Highly efficient removal of arsenic (Ⅲ/Ⅴ) from groundwater using nZVI functionalized cellulose nanocrystals fabricated via a bioinspired strategy[J]. Science of the Total Environment, 2022, 842: 156937. DOI: 10.1016/j.scitotenv.2022.156937.

[5] HE M C, WANG X Q, WU F C, et al. Antimony pollution in China [J]. Science of the Total Environment, 2012, 421-422: 41-50. DOI: 10.1016/j.scitotenv.2011.06.009.

[6] GUO X J, WANG K P, HE M C, et al. Antimony smelting process generating solid wastes and dust: Characterization and leaching behaviors[J]. Journal of Environmental Sciences, 2014, 26(7): 1549-1556. DOI: 10.1016/j.jes.2014.05.022.

[7] LI J S, LIANG H Q. Treatment strategies study on the comprehensive utilization of arsenic-alkali residue in Xikuangshan area[J]. Hunan Nonferrous Metals, 2010, 26(5): 53-55, 76. (in Chinese)

[8] LONG H, HUANG X Z, ZHENG Y J, et al. Purification of crude As_2O_3 recovered from antimony smelting arsenic-alkali residue[J]. Process Safety and Environmental Protection, 2020, 139: 201-209. DOI: 10.1016/j.psep.2020.04.015.

[9] CHEN C H, LAI M, FANG F Z. Study on the crack formation mechanism in nano-cutting of gallium arsenide[J]. Applied Surface Science, 2021, 540: 148322. DOI: 10.1016/j.apsusc.2020.148322.

[10] WANG C L, WANG D, YANG R Q, et al. Preparation and electrical properties of wollastonite coated with antimony-doped tin oxide nanoparticles[J]. Powder Technology, 2019, 342: 397-403. DOI: 10.1016/j.powtec.2018.09.092.

[11] TANG G W, LIU W W, QIAN Q, et al. Antimony selenide core fibers[J]. Journal of Alloys and Compounds, 2017, 694: 497-501. DOI: 10.1016/j.jallcom.2016.10.043.

[12] LI L, ZHANG R J, LIAO B, et al. Separation of As from As and Sb contained smoke dust by selective oxidation[J]. The Chinese Journal of Process Engineering, 2014, 14(1): 71-77. (in Chinese)

[13] ZHANG N, FANG Z W, LONG H, et al. Stabilization of arsenic from arsenic alkali residue by forming crystalline scorodite[J]. The Chinese Journal of Nonferrous Metals, 2020, 30(1): 203-213. (in Chinese)

[14] TIAN J, SUN W, ZHANG X F, et al. Comprehensive utilization and safe disposal of hazardous arsenic-alkali slag by the combination of beneficiation and metallurgy[J]. Journal of Cleaner Production, 2021,

295: 126381. DOI: 10. 1016/j. jclepro. 2021. 126381.

[15] ZHANG W J, CHE J Y, XIA L, et al. Efficient removal and recovery of arsenic from copper smelting flue dust by a roasting method: Process optimization, phase transformation and mechanism investigation[J]. Journal of Hazardous Materials, 2021, 412: 125232. DOI: 10. 1016/j. jhazmat. 2021. 125232.

[16] XUE J R, LONG D P, ZHONG H, et al. Comprehensive recovery of arsenic and antimony from arsenic-rich copper smelter dust[J]. Journal of Hazardous Materials, 2021, 413: 125365. DOI: 10. 1016/j. jhazmat. 2021. 125365.

[17] ZHANG Y H, FENG X Y, QIAN L, et al. Separation of arsenic and extraction of zinc and copper from high-arsenic copper smelting dusts by alkali leaching followed by sulfuric acid leaching[J]. Journal of Environmental Chemical Engineering, 2021, 9(5): 105997. DOI: 10. 1016/j. jece. 2021. 105997.

[18] WANG X, DING J Q, WANG L L, et al. Stabilization treatment of arsenic-alkali residue (AAR): Effect of the coexisting soluble carbonate on arsenic stabilization[J]. Environment International, 2020, 135: 105406. DOI: 10. 1016/j. envint. 2019. 105406.

[19] COUSSY S, PAKTUNC D, ROSE J, et al. Arsenic speciation in cemented paste backfills and synthetic calcium-silicatehydrates[J]. Minerals Engineering, 2012, 39: 51-61. DOI: 10. 1016/j. mineng. 2012. 05. 016.

[20] LIANG Y J, MIN X B, CHAI L Y, et al. Stabilization of arsenic sludge with mechanochemically modified zero valent iron[J]. Chemosphere, 2017, 168: 1142-1151. DOI: 10. 1016/j. chemosphere. 2016. 10. 087.

[21] JIANG G H, MIN X B, KE Y, et al. Solidification/stabilization of highly toxic arsenic-alkali residue by MSWI fly ash-based cementitious material containing Friedel's salt: Efficiency and mechanism[J]. Journal of Hazardous Materials, 2022, 425: 127992. DOI: 10. 1016/j. jhazmat. 2021. 127992.

[22] TIAN J, WANG Y F, ZHANG X F, et al. A novel scheme for safe disposal and resource utilization of arsenicalkali slag[J]. Process Safety and Environmental Protection, 2021, 156: 429-437. DOI: 10. 1016/j. psep. 2021. 10. 029.

[23] SU R, MA X, LIN J R, et al. An alternative method for the treatment of metallurgical arsenic-alkali residue and recovery of high-purity sodium bicarbonate[J]. Hydrometallurgy, 2021, 202: 105590. DOI: 10. 1016/j. hydromet. 2021. 105590.

[24] LONG H, ZHENG Y J, PENG Y L, et al. Separation and recovery of arsenic and alkali products during the treatment of antimony smelting residues[J]. Minerals Engineering, 2020, 153: 106379. DOI: 10. 1016/j. mineng. 2020. 106379.

[25] LONG H, ZHENG Y J, PENG Y L, et al. Recovery of alkali, selenium and arsenic from antimony smelting arsenic-alkali residue[J]. Journal of Cleaner Production, 2020, 251: 119673. DOI: 10. 1016/j. jclepro. 2019. 119673.

[26] LEI J, PENG B, LIANG Y J, et al. Effects of anions on calcium arsenate crystalline structure and arsenic stability[J]. Hydrometallurgy, 2018, 177: 123-131. DOI: 10. 1016/j. hydromet. 2018. 03. 007.

[27] ZHAO F P, CHEN S X, XIANG H R, et al. Selectively capacitive recovery of rare earth elements from aqueous solution onto Lewis base sites of pyrrolic-N doped activated carbon electrodes[J]. Carbon, 2022, 197: 282-291. DOI: 10. 1016/j. carbon. 2022. 06. 033.

[28] YANG Y D, ZHANG Z T, LI Y H, et al. The catalytic aerial oxidation of As(III) in alkaline solution by Mn-loaded diatomite[J]. Journal of Environmental Management, 2022, 317: 115380. DOI: 10. 1016/j. jenvman. 2022. 115380.

[29] TIAN L, YU X Q, XU J C, et al. Preparation and study of tungsten carbide catalyst synergistically codoped

with Fe and nitrogen for oxygen reduction reaction[J]. Journal of Materials Research and Technology, 2021, 15: 7100-7110. DOI: 10.1016/j.jmrt.2021.11.134.

[30] WANG S L, MULLIGAN C N. Speciation and surface structure of inorganic arsenic in solid phases: A review[J]. Environment International, 2008, 34(6): 867-879. DOI: 10.1016/j.envint.2007.11.005.

[31] OUVRARD S, DE DONATO P, SIMONNOT M O, et al. Natural manganese oxide: Combined analytical approach for solid characterization and arsenic retention[J]. Geochimica et Cosmochimica Acta, 2005, 69(11): 2715-2724. DOI: 10.1016/j.gca.2004.12.023.

[32] DU Q, ZHANG S J, PAN B C, et al. Bifunctional resin-ZVI composites for effective removal of arsenite through simultaneous adsorption and oxidation[J]. Water Research, 2013, 47(16): 6064-6074. DOI: 10.1016/j.watres.2013.07.020.

[33] BANG S, JOHNSON M D, KORFIATIS G P, et al. Chemical reactions between arsenic and zero-valent iron in water[J]. Water Research, 2005, 39(5): 763-770. DOI: 10.1016/j.watres.2004.12.022.

[34] ZHANG S J, LI X Y, CHEN J P. An XPS study for mechanisms of arsenate adsorption onto a magnetitedoped activated carbon fiber[J]. Journal of Colloid and Interface Science, 2010, 343(1): 232-238. DOI: 10.1016/j.jcis.2009.11.001.

[35] SHAN C, DONG H, HUANG P, et al. Dualfunctional millisphere of anion-exchanger-supported nanoceria for synergistic As(Ⅲ) removal with stoichiometric H_2O_2: Catalytic oxidation and sorption[J]. Chemical Engineering Journal, 2019, 360: 982-989. DOI: 10.1016/j.cej.2018.07.051.

[36] MARTINSON C A, REDDY K J. Adsorption of arsenic(Ⅲ) and arsenic(Ⅴ) by cupric oxide nanoparticles[J]. Journal of Colloid and Interface Science, 2009, 336(2): 406-411. DOI: 10.1016/j.jcis.2009.04.075.

[37] BAHL M K, WOODALL R O, WATSON R L, et al. Relaxation during photoemission and LMM Auger decay in arsenic and some of its compounds[J]. The Journal of Chemical Physics, 1976, 64(3): 1210-1218. DOI: 10.1063/1.432320.

[38] FANTAUZZI M, ATZEI D, ELSENER B, et al. XPS and XAES analysis of copper, arsenic and sulfur chemical state in enargites[J]. Surface and Interface Analysis, 2006, 38(5): 922-930. DOI: 10.1002/sia.2348.

[39] YANG K, QIN W Q, LIU W. Extraction of metal arsenic from waste sodium arsenate by roasting with charcoal powder[J]. Metals, 2018, 8(7): 542. DOI: 10.3390/met8070542.

[40] YANG K, QIN W Q, LIU W. Extraction of elemental arsenic and regeneration of calcium oxide from waste calcium arsenate produced from wastewater treatment[J]. Minerals Engineering, 2019, 134: 309-316. DOI: 10.1016/j.mineng.2019.02.022.

The Catalytic Aerial Oxidation of As(Ⅲ) in Alkaline Solution by Mn-loaded Diatomite

Abstract: The oxidization of As(Ⅲ) to As(Ⅴ) is necessary for both the detoxification of arsenic and the removal of arsenic by solidification. In order to achieve high efficiency and low cost As(Ⅲ) oxidation, a novel process of catalytic aerial oxidation of As(Ⅲ) is proposed, using air as oxidant and Mn-loaded diatomite as a catalyst. Through systematic characterization of the reaction products, the catalytic oxidation reaction law of Mn-loaded diatomite for As(Ⅲ) was found out, and its reaction mechanism was revealed. Results show that Mn-loaded diatomite achieved a good catalytic effect for aerial oxidation of As(Ⅲ) and maintained high performance over multiple cycles of reuse, which was directly related to the structure of diatomite and the behavior of manganese. Under the conditions of a catalyst concentration of 20 g/L, an air flow rate of 0.3 m^3/h, a reaction temperature of 50 ℃ and an initial pH of 12.6, 96.04% As(Ⅲ) oxidation was achieved after 3 h. Furthermore, the efficiency of As(Ⅲ) oxidation did not change significantly after ten cycles of reuse. XPS analysis of the reaction products confirmed that the surface of the catalyst was rich in Mn(Ⅲ), Mn(Ⅳ) and adsorbed oxygen(O-H), which was the fundamental reason for the excellent performance of Mn-loaded diatomite in the catalytic oxidation of As(Ⅲ).

Keywords: Arsenic; Catalytic oxidation; Mn-loaded catalyst; Diatomite; Arsenic oxidation

1 Introduction

The current arsenic(As) pollution in water has become a problem of widespread concern due to its high toxicity. In many regions worldwide, various signs and symptoms of As poisoning have emerged, endangering both human health and sustainable development (Podgorski and Berg,

2020). At least 140 million people globally are exposed to environmental As concentrations above the World Health Organization limit (10 μg/L) (Bagchi, 2007; Siddiqui et al., 2019). According to redox potential of As, As mainly exists as As4 and As(V) in water (Navarrete-Magaña et al., 2021; Wei et al., 2020). In industrial applications, treatment techniques such as precipitation, adsorption, ion exchange and membrane treatment are usually used to remove As(V) in wastewater (D. Kim et al., 2020; Omwene et al., 2019; Park et al., 2016). However, compared with As(V), As(Ⅲ) is extremely tough to be removed from the solution, and the removal efficiency of As(Ⅲ) often greatly reduced comparatively (Chang et al., 2010; Huling et al., 2017; Masliy et al., 2020). At the same time, based on the stronger toxicity and higher mobility of As(Ⅲ), As(Ⅲ) is usually pre-oxidized to As(V) as much as possible to facilitate subsequent As removal and fixation (Hu et al., 2021).

Chemical oxidants such as chlorine, hydrogen peroxide and potassium permanganate are usually used for the oxidation of As(Ⅲ) (Li et al., 2010; Nazari et al., 2017; Sarkar and Paul, 2016; Siddiqui and Chaudhry, 2018; Zhang et al., 2007). Although these reagents can oxidize As(Ⅲ) efficiently, they also produce toxic by-products or residues, which will lead to secondary pollution problems (Siddiqui and Chaudhry, 2017). In addition, owing to the relatively high price of these oxidants, the operating costs are greatly increased. Although using readily available air to oxidize As(Ⅲ) is undoubtedly an attractive method, aerial oxidation of As(Ⅲ) has been plagued by the extremely slow oxidation rate and incomplete oxidation (Bissen and Frimmel, 2003; Wang et al., 2021). Therefore, it is not feasible to directly oxidize As(Ⅲ) using air or oxygen. Previous research has shown that the introduction of Fe(Ⅲ) oxyhydroxides, sulfite, activated carbon and other substances into the reaction system could improve the efficiency of As(Ⅲ) oxidation (Wang and Giammar, 2015; Wu et al., 2020; Zaw and Emett, 2002). The introduction of catalysts into the As(Ⅲ) aerial oxidation reaction system has become a commonly applied method for the improvement of oxidation efficiency. Kim et al. found that the use of Pt-TiO_2 as catalyst under air saturation conditions can effectively catalyze the aerial oxidation of As(Ⅲ), only 4% Pt loading on TiO_2 can realize efficient oxidation of As(Ⅲ), and the oxidation rate can reach 13.2 μmol/(min·gcat) (J. G. Kim et al., 2020). Previous researches have found that the introduction of potassium permanganate into alkaline As-containing solutions resulted in the phenomenon of over-stoichiometric oxidation. When the As/Mn molar ratio was increased to 21∶1, As(Ⅲ) could still be rapidly and completely oxidized (Li et al., 2014). It is speculated that the catalytic oxidation of manganese may lead to the hyper-chemical dose oxidation of As(Ⅲ) by potassium permanganate. At present, the mechanism of manganese catalyzed aerial oxidation of As(Ⅲ) remains unclear. Therefore, based on the purpose of clarifying the reaction mechanism of manganese on the catalytic oxidation of As(Ⅲ), the catalytic oxidation performance of manganese loaded catalyst on As(Ⅲ) was evaluated. Combined with the characterization of reaction products, the catalytic oxidation law was found out and the reaction mechanism was further revealed. This proposed method provides a novel approach for low-cost and high-efficiency oxidation of As(Ⅲ) and a new idea for the treatment of As-containing wastewater.

2 Materials and methods

2.1 Materials

The purity of the reagents used in the experiment were all analytical grade, including $MnSO_4 \cdot H_2O$, molecular sieves 13X and 4A, zeolite, activated carbon, diatomite and $NaAsO_2$ used for catalytic experiments, HCl and NaOH for pH adjustment, HNO_3 for sample digestion. The compressed catalyst used in the experiments was supplied using an air compressor (WSC21070F, Anzheng, China). The water used in the experiment was ultrapure water, which was provided by the ultrapure water equipment (Medium-1600, Hetai, China).

2.2 Preparation of the Mn-loaded catalyst

The selected carrier (molecular sieve 13X, molecular sieve 4A, zeolite, activated carbon, or diatomite) was activated using the manganese sulfate solution impregnation method. Firstly, 25 g carrier was added to 200 mL of 1 M manganese sulfate solution, and stirred continuously for 2 h at 50 ℃ and 300 r/min. After impregnation, liquidsolid separation was carried out and the carrier was washed. Then the solid phase product was dried at 110 ℃ for 12 h. The above process was repeated twice to produce the primary Mn-loaded carrier product.

2.3 The catalytic aerial oxidation of As(Ⅲ)

The catalytic aerial oxidation of As(Ⅲ) was carried out in a 1 L fourport glass reactor (Fig. S1). Other equipment used included a constant temperature water bath (DF-101S, Yuhua, China), a digital constant speed mixer (OHS-20, Titan, China), an air compressor (WSC21070F, Anzheng, China), a potential-pH meter (PB-10, Sartorius, China) and a low-speed freeze centrifuge (KDC-2046, USTC, China). The operational conditions and the details methods were summarized in Supporting Material.

2.4 Analysis and characterization methods

The chemical properties of Mn-loaded diatomite were studied by various characterization techniques such as X-ray diffraction (XRD), Scanning electron microscope (SEM) and X-ray photoelectron spectroscopy (XPS). Details were shown in the Supporting Material.

3 Results and discussion

3.1 Effect of different carriers on the aerial oxidation of As(Ⅲ)

The effects of different carriers on the aerial oxidation of As(Ⅲ) were shown in Fig. S2[①]. It can be evident from Fig. S2[①] that although different types of carriers show slightly different catalytic capacities for the aerial oxidation of As(Ⅲ), the difference was not significant compared with the blank control experiment and the solution pH value with different carriers did not change significantly. The extent of As(Ⅲ) oxidation was less than 0.5% and the pH change was less than 0.2 after 3 h of reaction. The above results indicate that the carriers (molecular sieve 13X, molecular sieve 4A, zeolite, activated carbon and diatomite) studied in this experiment had no catalytic effect on the aerial oxidation of As(Ⅲ). Furthermore, it was confirmed that the aerial oxidation of As(Ⅲ) was slow without catalyst.

3.2 Effect of different Mn-loaded catalysts on the aerial oxidation of As(Ⅲ)

As shown in Fig. 1(a), the effects of five types of Mn-loaded catalysts on As(Ⅲ) oxidation in solution were significantly different. Among them, Mn-loaded diatomite exhibited the best catalytic effect on the aerial oxidation of As(Ⅲ), and the oxidation extent of As(Ⅲ) can reach 92.94% after 3 h of reaction. The molecular sieve 13X carrier performed the worst, achieving only 50% As(Ⅲ) oxidation after 3 h of reaction.

As it was shown in Fig. 1(b), within 1.5 h of the initial reaction, the solution pH value decreased rapidly, with the rate of decrease then slowing down. The pH of solutions containing different Mn-loaded catalysts varied from 0.6 to 1.31. Within 2 h of the initial reaction, the pH changed greatly and then gradually stabilized, which was consistent with the trend of As(Ⅲ) oxidation extent. The change in solution pH during the oxidation of As(Ⅲ) is described in Eq. (1), in which the aerial oxidation of As(Ⅲ) occurs via a process of alkali consumption:

$$2AsO_2^- + O_2 + 4OH^- \longrightarrow 2AsO_4^{3-} + 2H_2O \tag{1}$$

It can be seen from Fig. 1(c) that there were obvious differences in the adsorption properties of different Mn-loaded catalysts for As, among which diatomite had the strongest adsorption capacity for As, while activated carbon had the weakest As adsorption capacity. Combined with the data shown in Fig. 1(a), these results showed that the decrease of As concentration in the system was closely related to the As(Ⅲ) oxidation effect. From the perspective of the oxidation extent of As(Ⅲ), the adsorption of As(Ⅲ) by the catalyst had a promotion effect on the efficiency of catalytic aerial oxidation of As(Ⅲ), which was also the reason why the catalytic oxidation of As(Ⅲ) by Mn-loaded diatomite is more significant.

[①] Omited.

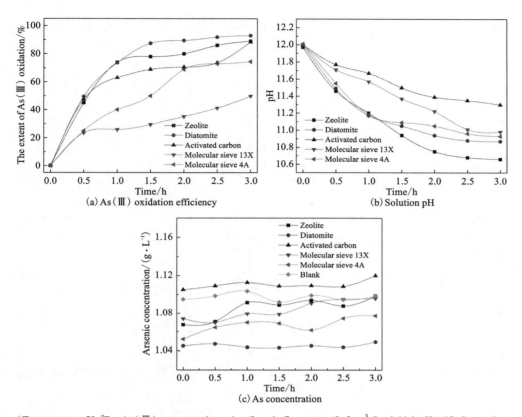

(Temperature = 50 ℃, As(Ⅲ) concentration = 1 g/L, air flow rate = 0.2 m³/h, initial pH = 12.0, catalyst concentration = 20 g/L)

Fig. 1 The effect of different Mn-loaded catalysts on the aerial oxidation of As(Ⅲ)

3.3 Catalytic aerial oxidation of As(Ⅲ) by Mn-loaded diatomite

The influence of various factors on the catalytic aerial oxidation of As(Ⅲ) by Mn-loaded diatomite was investigated under reaction temperatures of 20-80 ℃, initial alkali concentrations of 0.004-0.7 mol/L, air flow rates of 0-0.4 m³/h, initial As(Ⅲ) concentrations of 1-5 g/L, and catalyst concentration of 2-33 g/L (Fig. 2).

The effect of catalytic aerial oxidation of As(Ⅲ) by Mn-loaded diatomite was greatly affected by reaction temperature [Fig. 2(a)]. When the temperature was 20 ℃, the extent of As(Ⅲ) oxidation reached 82.36% after 3 h of reaction. With the increase of temperature, the rate of As(Ⅲ) oxidation further accelerated. When the temperature was raised to 80 ℃, the extent of As(Ⅲ) oxidation increased to 93.87% after the reaction for 3 h. Based on the consideration of As(Ⅲ) oxidation efficiency and energy consumption, an optimum reaction temperature of 50 ℃ was selected.

With the raise in initial pH value [Fig. 2(b)], the As(Ⅲ) oxidation rate initially raised and

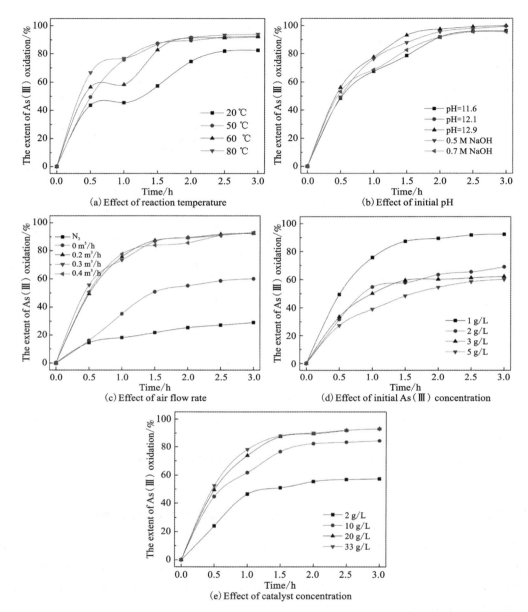

Fig. 2 The catalytic aerial oxidation of As(Ⅲ) by Mn-loaded diatomite

then decreased. When the initial solution pH was less than 12.1, with the raise of the initial pH of the solution, the rate of catalytic aerial oxidation of As(Ⅲ) by Mn-loaded diatomite gradually increased. However, when the initial pH of the solution was greater than 12.1, the oxidation rate of As(Ⅲ) showed a downward trend as the initial pH of the solution continued to increase. The oxidation reaction of As(Ⅲ) is a process that consumes OH^-, and the raise of pH value in the system will help promote the oxidation of As(Ⅲ). However, when the alkalinity is excessively high, the ability of OH⁻ and As(Ⅲ) to coordinate is strengthened and the morphology of As in solution is changed, making the oxidation of As(Ⅲ) more difficult (Liu and Qu, 2021; Mercer and Tobiason, 2008). In addition, the raise in solution pH reduces the stability of the catalyst, resulting in a decrease in catalytic performance. Therefore, considering the oxidation efficiency of As(Ⅲ), it is better to control the initial pH of the solution at 12.0-13.0.

When the reaction solution was in a N_2 atmosphere, after 3 h of reaction, about 29% of As(Ⅲ) was oxidized [Fig. 2(c)]. This was owing to the residual dissolved oxygen in the solution undergoes an oxidation reaction with As(Ⅲ). As the reaction progresses, the lessened of dissolved oxygen in the system slowed down the oxidation rate of As(Ⅲ). When the reaction solution was in an open system, the dissolved oxygen in the reaction system was replenished and the oxidation rate of As(Ⅲ) was improved. After 3 h of reaction, the extent of As(Ⅲ) oxidation reached 60.12%. As the air flow rate raised, the dissolved oxygen consumed by the As(Ⅲ) oxidation process was effectively replenished and the As(Ⅲ) oxidation rate raised rapidly. When the air flow rate was 0.3 m^3/h and the reaction time was 1.5 h, the oxidation extent of As(Ⅲ) reached 92.92%. However, when the air flow rate exceeded 0.3 m^3/h, continuing to increase the air velocity had little effect on the As(Ⅲ) oxidation rate. This may be the fact that when the air flow rate was 0.3 m^3/h, the dissolved oxygen content of the reaction system reached a saturated state, with further increases in the air flow rate unable to raise the concentration of dissolved oxygen and thus, no further improvement in the As(Ⅲ) oxidation efficiency could be achieved. Based on the consideration of As(Ⅲ) oxidation efficiency, an air flow rate of 0.3 m^3/h was selected as appropriate.

As the initial concentration of As(Ⅲ) increased, the extent of As(Ⅲ) oxidation gradually decreased [Fig. 2(d)]. With the concentration of As(Ⅲ) raised from 1 to 5 g/L, the oxidation extent of As(Ⅲ) decreased from 92% to 60.2% after 3 h of reaction. When the air flow rate and the amount of catalyst remained constant, the content of dissolved oxygen involved in As(Ⅲ) oxidation was fixed, with the amount of As(Ⅲ) catalytically oxidized was also unchanged and therefore, the increase of As(Ⅲ) concentration in the system will reduce its total oxidation extent. With the progression of the As(Ⅲ) oxidation reaction, the higher initial As(Ⅲ) concentration results in a higher level of alkali consumption, resulting in a rapid decline in pH value in the system, which will be unfavorable for the oxidation of As(Ⅲ). Therefore, a higher As(Ⅲ) concentration in the system results in a lower oxidation effect. For solutions with higher initial As(Ⅲ) concentrations, a certain amount of alkali can be added to improve the oxidation effect of As(Ⅲ).

The oxidation extent of As(Ⅲ) raised significantly with the increase of catalyst dosage [Fig. 2(e)]. When the catalyst concentration raised from 2 to 33 g/L, the oxidation extent of As(Ⅲ) raised from 57.29% to 92.91% after 3 h of reaction. In addition, it was found that when the reaction time exceeded a certain duration, the extent of As(Ⅲ) oxidation tended to stabilize under different catalyst concentration conditions. This phenomenon occurs owing to the As adsorption equilibrium of the catalyst. As the oxidation extent of As(Ⅲ) raised, the concentration of

As(III) in the system decreased, and the adsorption amount of As(III) on the catalyst surface was reduced, thus the oxidation extent of As(III) was lowered. With the oxidation of As(III), the concentration of As(V) in solution raised. As(V) is more easily adsorbed than As(III) (Zhu et al., 2009), which increases the amount of As(V) adsorbed on the catalyst surface, resulting in a decrease in the active sites on the catalyst surface and hindering the oxidation of As(III) (Tournassat et al., 2002). Therefore, so as to realize efficient oxidation of As(III), 20 g/L catalyst was selected as the optimal concentration.

3.4 Recycling performance of Mn-loaded diatomite

According to the above experimental results, it can be concluded that the optimal process conditions for As(III) oxidation using Mn-loaded diatomite as the catalyst are: an initial As(III) concentration of 1 g/L, catalyst concentration of 20 g/L, an air flow rate of 0.3 m^3/h, a reaction temperature of 50 ℃, an initial solution pH of 12.6 and a reaction time of 3 h. Under the above conditions, the recycling performance of Mn-loaded diatomite was investigated [Fig. 3(a)].

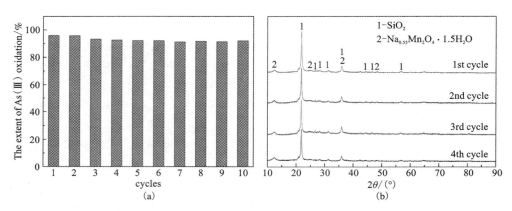

Fig. 3 (a) Effect of Mn-loaded diatomite on the oxidation efficiency of As(III), (b) XRD patterns of the reaction products during recycling experiments

The performance of Mn-loaded diatomite in the oxidation of As(III) did not change significantly after ten cycles of reuse, and the extent of As(III) oxidation still reached >90% [Fig. 3(a)], indicating that Mn-loaded diatomite has good recycling performance, which is beneficial to the reduction of As(III) oxidation costs.

For the sake of further investigating the stability of Mn-loaded diatomite in the cycle experiment, the reaction products generated in the previous four recycling experiments were characterized by XRD [Fig. 3(b)]. As it is shown in Fig. 3(b), there was no obvious change in the phase of manganese in the Mn-loaded diatomite during the recycling experiment, with manganese existing in the form of $Na_{0.55}Mn_2O_4 \cdot 1.5H_2O$. It showed that Mn-loaded diatomite has good cycle stability. In addition, no other manganese phases were found. Therefore, it can be inferred that $Na_{0.55}Mn_2O_4 \cdot 1.5H_2O$ is the catalytically active material.

3.5 Kinetics of catalytic aerial oxidation of As(Ⅲ) by Mn-loaded diatomite

Based on the experimental data at different reaction temperatures, the relationship between ln[As(Ⅲ)] and time was plotted (Fig. 4). It can be found that the reaction kinetics of catalytic aerial oxidation of As(Ⅲ) by Mn-loaded diatomite conforms to the characteristic first-order kinetic curve. Therefore, the As(Ⅲ) oxidation reaction can be expressed as shown in Eq. (2), with the rate constant k increasing with an increase in reaction temperature:

$$\ln[As(Ⅲ)] = -kt + b \tag{2}$$

where, k (min^{-1}) is the oxidation rate constant of As(Ⅲ); t is reaction time; and b (min^{-1}) is an integral constant.

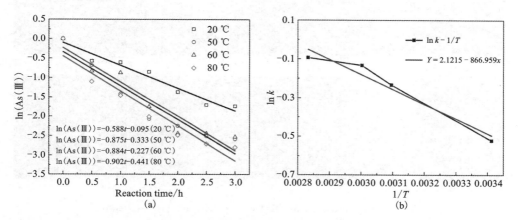

Fig. 4 Kinetics of catalytic aerial oxidation of As(Ⅲ) by Mn-loaded diatomite

According to the Arrhenius equation (Kang et al., 2020; Liu et al., 2020), the relationship between $\ln k$ and $1/T$ was obtained as $\ln k = -866.959x + 2.1215$, and the reaction activation energy was obtained as 7.208 kJ/mol. For chemical reactions, the reaction activation energy is between 5 and 20 kJ/mol, and it is generally controlled by the diffusion process (Abdel-Aal and Rashad, 2004; Zhang et al., 2021), which is roughly correspond with the experimental results observed in this study and previous research results (Li et al., 2014).

3.6 Reaction mechanism of catalytic aerial oxidation of As(Ⅲ) by Mn-loaded diatomite

The samples of diatomite, Mn-loaded diatomite and the reaction products were analyzed by XRD and SEM, so as to establish the reaction mechanism of catalytic aerial As(Ⅲ) oxidation by Mn-loaded diatomite, with the results shown in Fig. 5.

The main components of diatomite are SiO_2 and $SiO_2 \cdot xH_2O$, which do not contain Mn, As and S elements [Fig. 5(a)]. The particles exhibit a rich pore structure, providing a large number of active sites which are beneficial to manganese loading. Fig. 5(b) shows that the particle surface contains Mn and S elements, and the distribution of manganese is relatively uniform, indicating

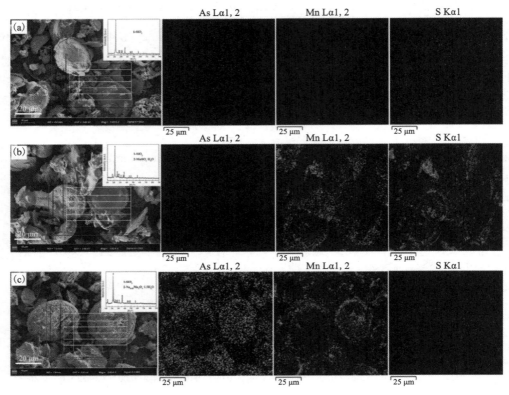

Fig. 5　XRD patterns and SEM-EDS results of: (a) diatomite;
(b) Mn-loaded diatomite; (c) reaction products

that manganese sulfate can be successfully loaded onto diatomite using the impregnation method. The morphology and composition of the support before and after the catalytic oxidation reaction did not change significantly (Fig. 5(b) and (c) show before and after the catalytic oxidation reaction, respectively), although the manganese-containing phase was transformed from manganese sulfate to $Na_{0.55}Mn_2O_4 \cdot 1.5H_2O$ and the manganese sulfate phase completely disappeared. At the same time, Fig. 5(c) shows that Mn and As are highly correlated, indicating that As is present on the surface of the reaction product in an adsorbed form. It is of note, that manganese underwent a speciation change from Mn(II) to Mn(IV), which has an important influence on the catalytic oxidation reaction. These results show that the diatomite structure is more suitable as a carrier of manganese, with this structure able to remain stable during impregnation and catalytic oxidation.

As a result of the study, Fig. 6 shows the XPS of the reaction products generated in the recycling experiment. As shown in the full range XPS spectra[Fig. 6(A)], Mn and S elements appear in addition to the carrier elements, indicating that $MnSO_4$ was successfully loaded onto diatomite. However, after catalytic oxidation reaction, the characteristic S element peak disappeared, with Na and As peaks appearing on the particle surface, indicating that the phase of manganese was transformed. Combined with the results of XRD analysis, these findings show that manganese was transformed from manganese sulfate to $Na_{0.55}Mn_2O_4 \cdot 1.5H_2O$.

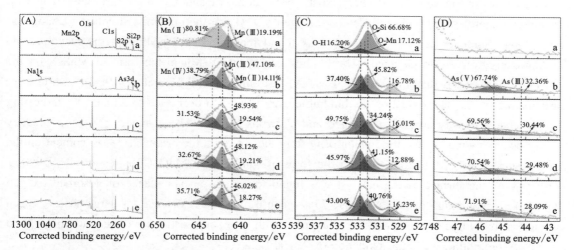

Fig. 6 XPS spectra of Mn-loaded diatomite and the reaction products generated during recycling experiments: (A) Full range XPS spectra; (B) Mn 2p3/2; (C) O 1s; (D) As 3d. Reaction cycles are indicated by: (a) Mn-loaded diatomite; (b) 1st cycle; (c) 2nd cycle; (d) 3rd cycle; (e) 4th cycle

It can be seen from Fig. 6(B)(a) that the peaks located at 641.5 and 642.7 eV correspond to Mn(Ⅲ) and Mn(Ⅱ), respectively (Ivanov et al., 1983; Strohmeier and Hercules, 1984). The proportions of Mn(Ⅲ) and Mn(Ⅱ) were 19.19% and 80.81%, respectively, indicating that while Mn(Ⅱ) is mainly in the form of $MnSO_4$, some Mn(Ⅱ) is converted to Mn(Ⅲ) on the surface of Mn-loaded diatomite. Fig. 6(B)(b), (c), (d) and (e) show that the peaks at 640.1, 642.2 and 643.4 eV correspond to Mn(Ⅱ), Mn(Ⅲ) and Mn(Ⅳ) (Penke et al., 2020; Yang et al., 2020), with the proportions of Mn(Ⅱ), Mn(Ⅲ) and Mn(Ⅳ) being 14.11%-19.54%, 46.02%-48.93% and 31.53%-38.79%, respectively. During recycling, the ratio of Mn(Ⅲ)/Mn(Ⅳ) on the catalyst remained >1.2, indicating that Mn on the surface of the catalyst mainly exists in the form of Mn(Ⅲ), with a smaller proportion of Mn(Ⅳ). The existence of Mn(Ⅲ) is closely related to the formation of oxygen vacancies (Dong et al., 2021), as shown in Eq. (3) as follows:

$$-Mn(Ⅳ) - O^{2-} - Mn(Ⅳ) \longrightarrow -Mn(Ⅲ) - \square - Mn(Ⅲ) - + 1/2O^{2-} \quad (3)$$

where the □ delegates the oxygen vacancy, higher oxygen vacancies are beneficial to the oxidation of organic pollutants. Therefore, the production of intermediate Mn(Ⅲ) was considered to be the rate-limiting step in organic pollutant oxidation (Ding et al., 2021; Hou et al., 2014). A higher Mn(Ⅲ) content on the catalyst surface indicates higher oxygen vacancies and therefore, stronger catalytic ability (Dong et al., 2021). During the recycling experiment, the change in valence state of manganese showed that mutual transformation occurs between

Mn(Ⅱ), Mn(Ⅲ) and Mn(Ⅳ) during the oxidation of As(Ⅲ), which enabled Mn-loaded diatomite to be effectively recycled and reused.

As shown in Fig. 6(C), the characteristic peaks at 532.1 and 532.8 eV were consistent with the binding energies of O-Si and adsorbed oxygen (O-H) in the diatomite structure (Chang and Tsai, 2004; Sprenger et al., 1990). It was found that after participating in the reaction, the binding energy of O-Mn changed. The binding energy of O-Mn in the form of $MnSO_4$ on the surface of Mn-loaded diatomite was 532.4 eV (Strohmeier and Hercules, 1984), while after participating in As(Ⅲ) oxidation the binding energy of O-Mn decreased to 530.0 eV, indicating that the charge density around O changed and electrons shifted. The change in O charge density mainly occurs owing to the introduction of O from the air, with the original O-Mn content on the particle surface being transformed from the form of $MnSO_4$ into MnO_2 (Wagner et al., 1980). Furthermore, the surface of the catalyst was found to contain an abundance of adsorbed oxygen (O-H), accounting for 37.40%-49.75%, confirming the existence of large-scale oxygen vacancies, providing the Mn-loaded diatomite with a strong catalytic oxidation ability and effectively promoting the aerial oxidation of As(Ⅲ).

As shown in Fig. 6(D), As(Ⅴ) and As(Ⅲ) were adsorbed on the surface of the catalyst, as shown by the characteristic peaks at 45.4 and 44.2 eV (Prucek et al., 2013), respectively. During the recycling experiments, the proportion of As(Ⅴ) on the surface of the catalyst increased gradually, while the ratio of As(Ⅴ)/As(Ⅲ) on the surface of the catalyst (71.97%/28.09% = 2.56) remained significantly lower than that of As(Ⅴ)/As(Ⅲ) in the solution, which indicates that the adsorption capacity of Mn-loaded diatomite for As(Ⅲ) is much higher than that of As(Ⅴ). Furthermore, no obvious accumulation of As(Ⅴ) occurred on the surface of the catalyst, which is beneficial as accumulation leads to a reduction in active sites and inhibits the oxidation of As(Ⅲ). Therefore, this ensures the good recycling performance of Mn-loaded diatomite following multiple cycles of reuse. The specific reaction process of catalytic air oxidation of As(Ⅲ) by Mn-loaded diatomite can be seen as Supplementary Materials.

4 Conclusions

Aerial oxidation of As(Ⅲ) is relatively slow under natural conditions, while the rate and extent of As(Ⅲ) oxidation are greatly improved after the addition of Mn-loaded diatomite. It is found that the active material for the oxidation of As(Ⅲ) is $Na_{0.55}Mn_2O_4 \cdot 1.5H_2O$, rather than the initially loaded manganese sulfate.

Under the optimum conditions of a catalyst concentration of 20 g/L, an initial As(Ⅲ) concentration of 1 g/L, an air flow rate of 0.3 m^3/h, reaction temperature of 50 ℃ and initial pH of 12.6, the extent of As(Ⅲ) oxidation reached 96.04% after 3 h, and the complete oxidation of As will be achieved by prolonging the time. Mn-loaded diatomite still exhibited good performance after ten cycles of reuse, with the extent of As(Ⅲ) oxidation continuing to reach 92.67%. XPS analysis results showed that the catalyst surface contained a large amount of Mn(Ⅲ) with high oxygen vacancies, enhancing the catalytic effect on aerial oxidation of As(Ⅲ). In addition, the adsorption capacity of the catalyst for As(Ⅲ) is much higher than that of As(Ⅴ), ensuring the good performance of recycled Mn-loaded diatomite. These results promote our understanding of the oxidation reaction law and mechanism for aerial oxidation of As(Ⅲ) catalyzed by Mn-loaded diatomite, helping to effectively overcome the problem of poor reaction kinetics for the aerial oxidation of As(Ⅲ). Catalytic oxidation of As(Ⅲ) with air by Mn-loaded diatomite has the advantages of high efficiency, low cost, easy preparation of catalyst and reusable, which makes it have a good prospect of industrial application.

Acknowledgments

This study is supported by the National Key R&D Program of China (SQ2019YFC190179), National Natural Science Foundation of China (51864019, 52004111), the Distinguished Professor Program of Jinggang Scholars in institutions of higher learning, Jiangxi Province, the Program of Qingjiang Excellent Young Talents, Jiangxi University of Science and Technology, and the Jiangxi Province Natural Science Foundation of China (20181BAB206019), Education Department of Jiangxi Province (GJJ170508).

References

[1] ABDEL-AAL E A, RASHAD M M. Kinetic study on the leaching of spent nickel oxide catalyst with sulfuric acid[J]. Hydrometallurgy, 2004, 74: 189-194. https://doi.org/10.1016/j.hydromet.2004.03.005.
[2] BAGCHI S. Arsenic threat reaching global dimensions[J]. Can. Med. Assoc. J. 2007, 177: 1344-1345. https://doi.org/10.1503/cmaj.071456.
[3] BISSEN M, FRIMMEL F H. Arsenic: A review. Part Ⅱ: Oxidation of arsenic and its removal in water treatment[J]. Acta Hydrochim. Hydrobiol, 2003, 31: 97-107. https://doi.org/10.1002/aheh.200300485.
[4] CHANG F, QU J, LIU R, et al. Practical performance and its efficiency of arsenic removal from groundwater using Fe-Mn binary oxide[J]. J. Environ. Sci. 2010, 22: 1-6. https://doi.org/10.1016/S1001-0742(09)60067-X.
[5] CHANG J K, TSAI W T. Effects of temperature and concentration on the structure and specific capacitance of manganese oxide deposited in manganese acetate solution[J]. J. Appl. Electrochem, 2004, 34: 953-961. https://doi.org/10.1023/B: JACH.0000040547.12597.bb.
[6] DING W, ZHENG H, SUN Y, et al. Activation of $MnFe_2O_4$ by sulfite for fast and efficient removal of arsenic(Ⅲ) at circumneutral pH: involvement of Mn(Ⅲ)[J]. J. Hazard Mater, 2021, 403: 123623. https://doi.org/10.1016/j.jhazmat.2020.123623.
[7] DONG C, WANG H, REN Y, et al. Layer MnO_2 with oxygen vacancy for improved toluene oxidation

activity[J]. Surface. Interfac, 2021, 22: 100897. https://doi.org/10.1016/j.surfin.2020.100897.

[8] HOU J, LI Y, MAO M, et al. Tremendous effect of the morphology of birnessite-type manganese oxide nanostructures on catalytic activity[J]. ACS Appl. Mater. Interfaces, 2014, 6: 14981-14987. https://doi.org/10.1021/am5027743.

[9] HU X, LIU Y, LIU F, et al. Simultaneous decontamination of arsenite and antimonite using an electrochemical CNT filter functionalized with nanoscale goethite[J]. Chemosphere, 2021, 274: 129790. https://doi.org/10.1016/j.chemosphere.2021.129790.

[10] HULING J R, HULING S G, LUDWIG R. Enhanced adsorption of arsenic through the oxidative treatment of reduced aquifer solids[J]. Water Res, 2017, 123: 183-191. https://doi.org/10.1016/j.watres.2017.06.064.

[11] IVANOV, EMIN B N, NEVSKAYA N A, et al. ChemInform abstract: Synthesis and properties of calcium and strontium HYDROXYMANGANATES(Ⅲ)[J]. Chemischer Informationsdienst, 1983, 14. https://doi.org/10.1002/chin.198325036.

[12] KANG K H, KIM J, JEON H, et al. Energy efficient sludge solubilization by microwave irradiation under carbon nanotube (CNT)-coated condition[J]. J. Environ. Manag, 2020, 259: 110089. https://doi.org/10.1016/j.jenvman.2020.110089.

[13] KIM D, MOON G, KOO M S, et al. Spontaneous oxidation of arsenite on platinized TiO_2 through activating molecular oxygen under ambient aqueous condition[J]. Appl. Catal. B Environ. 2020, 260: 118146. https://doi.org/10.1016/j.apcatb.2019.118146.

[14] KIM J G, KIM H B, YOON G S, et al. Simultaneous oxidation and adsorption of arsenic by one-step fabrication of alum sludge and graphitic carbon nitride (g-C3N4)[J]. J. Hazard Mater. 2020, 383: 121138. https://doi.org/10.1016/j.jhazmat.2019.121138.

[15] LI X, LIU C S, LI F, et al. The oxidative transformation of sodium arsenite at the interface of α-MnO_2 and water[J]. J. Hazard Mater. 2010, 173: 675-681. https://doi.org/10.1016/j.jhazmat.2009.08.139.

[16] LI Y, LIU Z H, LIU F, et al. Promotion effect of $KMnO_4$ on the oxidation of As(Ⅲ) by air in alkaline solution[J]. J. Hazard Mater. 2014, 280: 315-321. https://doi.org/10.1016/j.jhazmat.2014.08.008.

[17] LIU J, BAEYENS J, DENG Y, et al. The chemical CO_2 capture by carbonation-decarbonation cycles[J]. J. Environ. Manag, 2020, 260: 110054. https://doi.org/10.1016/j.jenvman.2019.110054.

[18] LIU R, QU J. Review on heterogeneous oxidation and adsorption for arsenic removal from drinking water[J]. J. Environ. Sci. 2021: 178-188. https://doi.org/10.1016/j.jes.2021.04.008.

[19] MASLIY A N, KUZNETSOV A M, KORSHIN G V. The intrinsic mechanism of catalyticoxidation of arsenite by hydroxyl-radicals in the $H_3AsO_3-CO_3^{2-}/HCO_3^--H_2O$ system: A quantum-chemical examination [J]. Chemosphere, 2020, 238: 124466. https://doi.org/10.1016/j.chemosphere.2019.124466.

[20] MERCER K L, TOBIASON J E. Removal of arsenic from high ionic strength solutions: Effects of ionic strength, pH, and preformed versus in situ formed HFO[J]. Environ. Sci. Technol, 2008, 42: 3797-3802. https://doi.org/10.1021/es702946s.

[21] NAVARRETE-MAGAÑA M, ESTRELLA-GONZÁLEZ A, MAY-IX L, et al. Improved photocatalytic oxidation of arsenic(Ⅲ) with WO_3/TiO_2 nanomaterials synthesized by the sol-gel method[J]. J. Environ. Manag, 2021, 282: 111602. https://doi.org/10.1016/j.jenvman.2020.111602.

[22] NAZARI A M, RADZINSKI R, GHAHREMAN A. Review of arsenic metallurgy: Treatment of arsenical minerals and the immobilization of arsenic[J]. Hydrometallurgy, 2017, 174: 258-281. https://doi.org/10.1016/j.hydromet.2016.10.011.

[23] OMWENE P I, ÇELEN M, ÖNCEL M S, et al. Arsenic removal from naturally arsenic contaminated ground water by packed-bed electrocoagulator using Al and Fe scrap anodes[J]. Process Saf. Environ. Protect, 2019, 121: 20-31. https://doi.org/10.1016/j.psep.2018.10.003.

[24] PARK J H, HAN Y S, AHN J S. Comparison of arsenic co-precipitation and adsorption by iron minerals and the mechanism of arsenic natural attenuation in a mine stream[J]. Water Res, 2016, 106: 295-303. https://doi.org/10.1016/j.watres.2016.10.006.

[25] PENKE Y K, YADAV A K, SINHA P, et al. Arsenic remediation onto redox and photo-catalytic/electrocatalytic Mn-Al-Fe impregnated rGO: Sustainable aspects of sludge as supercapacitor[J]. Chem. Eng. J. 2020, 390: 124000. https://doi.org/10.1016/j.cej.2019.124000.

[26] PODGORSKI J, BERG M. Global threat of arsenic in groundwater[J]. Science, 2020, 368: 845-850. https://doi.org/10.1126/science.aba1510.

[27] PRUCEK R, TUČEK J, KOLAŘÍK J, et al. Ferrate(VI)-induced arsenite and arsenate removal by in situ structural incorporation into magnetic iron(III) oxide nanoparticles[J]. Environ. Sci. Technol, 2013, 47: 3283-3292. https://doi.org/10.1021/es3042719.

[28] SARKAR A, PAUL B. The global menace of arsenic and its conventional remediation: A critical review [J]. Chemosphere, 2016, 158: 37-49. https://doi.org/10.1016/j.chemosphere.2016.05.043.

[29] SIDDIQUI S I, NAUSHAD M U, CHAUDHRY S A. A review on graphene oxide and its composites preparation and their use for the removal of As^{3+} nd As^{5+} from water under the effect of various parameters: Application of isotherm, kinetic and thermodynamics[J]. Process Saf. Environ. Protect, 2018, 119: 138-163. https://doi.org/10.1016/j.psep.2018.07.020.

[30] SIDDIQUI S I, CHAUDHRY S A. Arsenic removal from water using nanocomposites: A review[J]. CEE 4, 2017. https://doi.org/10.2174/2212717804666161214143715.

[31] SIDDIQUI S I, CHAUDHRY S A. Promising prospects of nanomaterials for arsenic water remediation: A comprehensive review[J]. Process Saf. Environ. Protect, 2019, 126: 60-97. https://doi.org/10.1016/j.psep.2019.03.037.

[32] SPRENGER D, BACH H, MEISEL W, et al. XPS study of leached glass surfaces[J]. J. Non-Cryst. Solids, 1990, 126: 111-129. https://doi.org/10.1016/0022-3093(90)91029-Q.

[33] STROHMEIER B R, HERCULES D M. Surface spectroscopic characterization of manganese/aluminum oxide catalysts[J]. J. Phys. Chem, 1984, 88: 4922-4929. https://doi.org/10.1021/j150665a026.

[34] TOURNASSAT C, CHARLET L, BOSBACH D, et al. Arsenic (III) oxidation by birnessite and precipitation of manganese(II) arsenate[J]. Environ. Sci. Technol, 2002, 36: 493-500. https://doi.org/10.1021/es0109500.

[35] WAGNER C D, ZATKO D A, RAYMOND R H. Use of the oxygen KLL Auger lines in identification of surface chemical states by electron spectroscopy for chemical analysis[J]. Anal. Chem, 1980, 52: 1445-1451. https://doi.org/10.1021/ac50059a017.

[36] WANG L, GIAMMAR D E. Effects of pH, dissolved oxygen, and aqueous ferrous iron on the adsorption of arsenic to lepidocrocite[J]. J. Colloid Interface Sci. 2015, 448: 331-338. https://doi.org/10.1016/j.jcis.2015.02.047.

[37] WANG Z, FU Y, WANG L. Abiotic oxidation of arsenite in natural and engineered systems: Mechanisms and related controversies over the last two decades (1999–2020)[J]. J. Hazard Mater, 2021, 414: 125488. https://doi.org/10.1016/j.jhazmat.2021.125488.

[38] WEI X, ZHOU Y, TSANG D C W, et al. Hyperaccumulation and transport mechanism of thallium and

arsenic in brake ferns (Pteris vittata L.): A case study from mining area[J]. J. Hazard Mater, 2020, 388: 121756. https://doi.org/10.1016/j.jhazmat.2019.121756.

[39] WU C, MAHANDRA H, RADZINSKI R, et al. Green catalytic process for in situ oxidation of Arsenic (III) in concentrated streams using activated carbon and oxygen gas[J]. Chemosphere, 2020, 261: 127688. https://doi.org/10.1016/j.chemosphere.2020.127688.

[40] YANG W H, SU Z, XU Z, et al. Comparative study of α-, β-, γ- and δ-MnO_2 on toluene oxidation: Oxygen vacancies and reaction intermediates[J]. Appl. Catal. B Environ, 2020, 260: 118150. https://doi.org/10.1016/j.apcatb.2019.118150.

[41] ZAW M, EMETT M T. Arsenic removal from water using advanced oxidation processes[J]. Toxicol. Lett, 2002, 133: 113-118. https://doi.org/10.1016/S0378-4274(02)00081-4.

[42] ZHANG G S, QU J H, LIU H J, et al. Removal mechanism of As(III) by a novel Fe-Mn binary oxide adsorbent: Oxidation and sorption[J]. Environ. Sci. Technol, 2007, 41: 4613-4619. https://doi.org/10.1021/es063010u.

[43] ZHANG H, CHEN G, CAI X, et al. The leaching behavior of copper and iron recovery from reduction roasting pyrite cinder[J]. J. Hazard Mater, 2021, 420: 126561. https://doi.org/10.1016/j.jhazmat.2021.126561.

[44] ZHU M, PAUL K W, KUBICKI J D, et al. Quantum chemical study of arsenic(III, V) adsorption on Mn-oxides: Implications for arsenic(III) oxidation[J]. Environ. Sci. Technol, 2009, 43: 6655-6661. https://doi.org/10.1021/es900537e.

Catalytic Oxidation Effect of MnSO$_4$ on As(Ⅲ) by Air in Alkaline Solution

Abstract: The catalytic oxidation effect of MnSO$_4$ on As(Ⅲ) by air in an alkaline solution was investigated. According to the X-ray diffraction (XRD), scanning electron microscope-energy dispersive spectrometer (SEM-EDS) and X-ray photoelectron spectroscopy (XPS) analysis results of the product, it was shown that the introduction of MnSO$_4$ in the form of solution would generate Na$_{0.55}$Mn$_2$O$_4\cdot$1.5H$_2$O with strong catalytic oxidation ability in the aerobic alkaline solution, whereas the catalytic effect of the other product MnOOH is not satisfactory. Under the optimal reaction conditions of temperature 90 ℃, As/Mn molar ratio 12.74∶1, air flow rate 1.0 L/min, and stirring speed 300 r/min, As(Ⅲ) can be completely oxidized after 2 h reaction. The excellent catalytic oxidation ability of MnSO$_4$ on As(Ⅲ) was mainly attributed to the indirect oxidation of As(Ⅲ) by the product Na$_{0.55}$Mn$_2$O$_4\cdot$1.5H$_2$O. This study shows a convenient and efficient process for the oxidation of As(Ⅲ) in alkali solutions, which has potential application value for the pre-oxidation of arsenic-containing solution or the detoxification of As(Ⅲ).

Keywords: Arsenic; Air oxidation; Catalytic oxidation; Manganese sulfate

1 Introduction

The arsenic-alkali residue is a typical by-product of the antimony refining process, which mainly contains 15%-20% arsenic, 18%-22% antimony, and a certain amount of caustic alkali. Arsenic usually exists in the form of sodium arsenite and sodium arsenate. Antimony, in addition to sodium antimonate and sodium antimonite, is also present in the form of free metal in a small amount (Wang et al., 2020a). The composition of arsenic-alkali residue is relatively simple, and it is rich in valuable metal antimony. However, separation of arsenic and antimony is difficult due to their similar properties. Therefore, the disposal of arsenic-alkali residue has always been a difficult problem in the antimony smelting industry.

To dispose arsenic-alkali residue, various distinctive schemes, mainly based on wet process, have been reported. Jin et al. (1999) recovered antimony and arsenic from the arsenic-alkali residue by hot water leaching → calcium oxide precipitation → sulfuric acid dissolution → SO_2 reduction → cooling crystallization process. The recovery extent of antimony reached 96%, and the purity of arsenic trioxide reached 95% (Jin et al., 1999). Long et al. (2020) reported the process of antimony enrichment with water leaching, twice alkali recovery with CO_2, and recovery of arsenic by acidification-SO_2 reduction. The recovery extents of arsenic, antimony and alkali in arsenic-alkali residue reached 79.4%, 98.2% and 87.8%, respectively (Long et al., 2020). Although these methods can achieve better separation of arsenic and antimony, they have the disadvantages of longer time requirement and higher cost, restricting their wide use. Considering the significant difference in the solubility of sodium arsenite and sodium antimonate in alkaline solution, the water leaching-oxidation method was also proposed to treat arsenic-alkali residue. In this method, the initial separation of arsenic and antimony was realized by the water leaching process, and then the leaching solution was treated by an oxidation process to complete the separation of arsenic and antimony (Wan et al., 2015; Deng et al., 2014). Due to the water leaching-oxidation process is simple and easy to implement, it has become the current mainstream technology for the treatment of arsenic-alkali residue.

In addition to the separation of arsenic and antimony, reducing arsenic toxicity and facilitating arsenic solidification are other purposes of the oxidation treatment of arsenic-alkali residue leaching solution. Therefore, oxidation treatment is one of the necessary steps for treating solutions containing As(Ⅲ) (Era et al., 2017; Wang et al., 2016). The main oxidants that have been reported for solutions of As(Ⅲ) mainly include hydrogen peroxide, chlorine, ozone, potassium permanganate, etc. (Hug and Leupin, 2003; Khuntia et al., 2014; Lee et al., 2011). Among them, hydrogen peroxide is the most widely used oxidant. Hydrogen peroxide in excess can be used as an effective arsenic oxidant in a wide pH range, especially in alkaline solutions, which can reduce the concentration of As(Ⅲ) in solutions by 10-15 times in a short time (Ritcey, 2005). The introduction of metal oxides as catalysts in the oxidation reaction system can further enhance the oxidation effect of hydrogen peroxide (Kim et al., 2015). Shan et al. (2019) have studied the effect of nano-CeO_2 on the oxidation of As(Ⅲ) by H_2O_2 and found that the utilization rate of H_2O_2 was greatly improved compared with the blank solution, reaching 90.5%-122% stoichiometric reaction (Shan et al., 2019). Thus the oxidation efficiency of the above strong oxidants is high towards As(Ⅲ) and their oxidation effect on As(Ⅲ) is ideal, but it is indisputable that their high consumption increases the overall cost. Besides, most oxidants still poses problem of introducing impurities. Therefore, developing low-cost As(Ⅲ) oxidation methods is one of the most sought-after research directions for the treatment of arsenic-containing materials.

Air is undoubtedly the ideal oxidant, which is cheap and easy to obtain without introducing impurities. Oxidation of As(Ⅲ) by air is thermodynamically feasible, but its oxidation efficiency is low due to kinetic reasons (Bissen and Frimmel, 2003). Bisceglia et al. (2005) found that the oxidation rate of As(Ⅲ) by oxygen is very slow, with a half-life of about one year (Bisceglia et al., 2005). Kim and Nriagu (2000) took natural water containing 46-62 μg/L arsenic (As(Ⅲ) >72%) as the research sample and showed that after 5 days of aerial oxidation, only 54% As(Ⅲ) could be oxidized (Kim and Nriagu, 2000). Therefore, it has become the consensus of most researchers to introduce catalysts into the aerial oxidation system of As(Ⅲ) to improve the reaction efficiency (Huling et al., 2017; Kim et al., 2020; Liu and Qu, 2021). Under oxygen atmosphere, the maximum oxidation capacity of Fe-Mn nodules reached 3.71 mg/g when initial concentration of As(Ⅲ) was 10 mg/L and the mass ratio of Fe-Mn nodules to As(Ⅲ) was 100∶1 (Rady et al., 2020). In previous studies, we found that adding potassium permanganate to alkaline arsenic-containing solutions can significantly promote As(Ⅲ) oxidation by air (Li et al., 2014). This indicates that manganese-based oxides may have a catalytic effect during aerial oxidation of As(Ⅲ). In order to confirm this conjecture, $MnSO_4$ was introduced into the reaction system for aerial oxidation of arsenic-alkali residue leaching solution in this study to synthesize a new fresh manganese-based oxide in one step. Through the systematic characterization of the products, the reaction law and catalytic oxidation mechanism were identified. Thus this method not only realizes the oxidation of arsenic, but also is beneficial to the subsequent solidification of arsenic by scorodite. The results are expected to provide new methods and ideas for the oxidation of arsenic-alkali residue leaching solution and even the pre-oxidation of arsenic-containing waste solution.

2 Materials and methods

2.1 Materials

The leaching solution of arsenic-alkali residue used in the experiment was collected from antimony smelters in China, the initial pH was 13.4, and the mass concentrations of total arsenic (As_T), As(Ⅲ), As(Ⅴ), total antimony (Sb_T), Se, Fe, Zn, and Al were 31.46, 0.65, 30.81, 0.19, 0.02, 0.03, 0.01 and 0.04 g/L, respectively. All of reagents used in this work were of analytical grade.

2.2 Experimental equipment and procedures

The schematic diagram of the experimental device is shown in Appendix A Fig. S1[①]. The oxidation experiment of arsenic-alkali residue leaching solution was carried out in a 2 L oxidation

---

[①] Omited.

reactor equipped with a microporous aeration device. The equipment used in the experiment consisted of a constant temperature water bath (DF-101S, Yuhua, China), digital display constant speed stirrer (OHS-20, Titan, China), air compressor (WSC21070F, Anzheng, China) and low-speed refrigerated centrifuge (KDC-2046, Zonkia, China).

Firstly, a 1.2 L arsenic-alkali residue leaching solution was taken, and its pH was adjusted with NaOH or HCl solution according to the experimental conditions. This solution was heated to the target temperature and stirred. At the same time, air was pumped into the solution and the flow rate was adjusted. Then a certain amount of $MnSO_4$ was added to the solution to start the oxidation experiment. During the experiment, 10 mL of the slurry was taken out every 15 min (water was replenished the mark 5 min before sampling), and the liquid-solid separation was carried out immediately by centrifuge (4000 r/min). The supernatant was collected to analyze the concentration of As(III) in the supernatant. At the end of the experiment, the solid product was filtered, washed, dried, and then sent for analysis.

To determine the oxidation extent of As(III) in solution, it was first separated from the solution by extracting with benzene, the concentration of As(III) was then determined by the volumetric titration method in hydrochloric acid medium. Potassium bromate was used as the titrator, and methyl orange was used as the indicator (Kew et al., 1952; Sharma et al., 2016). Each sample was tested three times, and the average value of the three test results was used as the final concentration of As(III) in the sample to calculate the oxidation extent of As(III). The oxidation extent of As(III) was computed by Eq. (1):

$$\alpha = \frac{3C_1 \times (V_1 - V_0)}{C_{As} \times V} \times 100\% \qquad (1)$$

where α is the oxidation extent of As(III), C_{As} (mol/L) is the average value of the initial concentration of As(III) in the leaching solution determined three times, and C_1 (mol/L) is the concentration of potassium bromate standard solution. V_0 (mL) represents the volume of potassium bromate standard solution consumed in the blank experiment, V_1 (mL) is the volume of potassium bromate standard solution consumed by the sample, and V (mL) represents the volume of the leaching solution of the sample.

3 Results and discussion

3.1 Oxidation of As(III) by air in the presence of manganese sulfate

3.1.1 Effect of manganese sulfate addition methods

The effect of manganese sulfate addition methods on the efficiency of the As(III) oxidation process was investigated, and the results are shown in Fig. 1. It is evident from Fig. 1(a) that the addition methods of manganese sulfate had a significant influence on its catalytic efficiency towards oxidation of As(III) present in the leaching solution. Without the addition of manganese sulfate into the As(Ⅲ) leaching solution, the efficiency of the As(Ⅲ) oxidation was poor when only air was introduced, affording only 0.44% oxidation after 4 h of reaction and proved that As(Ⅲ) could not be oxidized well only through the air. When manganese sulfate was added in the form of a solution, the aerial oxidation efficiency significantly improved, and the oxidation extent of As(Ⅲ) reached 99% after 2 h. However, the addition of manganese sulfate in the form of solid did not improve the oxidation rate of As(Ⅲ) much, yielding only 37.23% oxidation extent of As(Ⅲ) after 4 h of reaction.

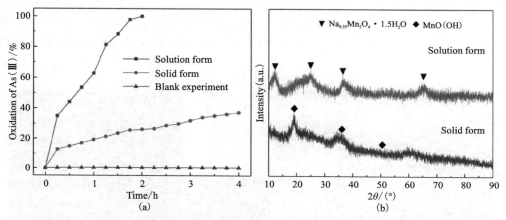

Fig. 1 (a) Effect of manganese sulfate addition methods on the oxidation efficiency of As(Ⅲ) and (b) X-ray diffraction (XRD) of reaction products obtained under different adding methods. Reaction conditions: temperature 90 ℃, As/Mn molar ratio 12.74 : 1, air flow rate 1.0 L/min, stirring speed 300 r/min, and initial pH = 13.4

Fig. 1(b) shows the X-ray diffraction (XRD) patterns of the solid phase products formed after adding manganese sulfate in different forms. The addition of manganese sulfate solution afforded mainly $Na_{0.55}Mn_2O_4 \cdot 1.5H_2O$ (Joint Committee on Powder Diffraction Standards (JCPDS) 43-1456) as a solid product. However, manganese mainly exists in the form of MnOOH (JCPDS 18-0804) when solid manganese sulfate was applied for the oxidation (Bargar et al., 2005; Tu, 1994). Hence, the difference in product phase leads to the entirely different efficiency of manganese sulfate for aerial oxidation of As(Ⅲ). In addition, the redox potential of Mn(Ⅲ) is lower than that of Mn(Ⅳ), resulting in a slower electron transfer rate from Mn(Ⅲ) to As(Ⅲ) compared to the electron transfer from Mn(Ⅳ) and As(Ⅲ) (Hou et al., 2016; Zhu et al., 2009), and hence the oxidation ability of $Na_{0.55}Mn_2O_4 \cdot 1.5H_2O$ is stronger than the crystalline Mn(Ⅲ) oxide (Chiu and Hering, 2000; Lan et al., 2018).

Fig. 2 shows the Scanning electron microscope-energy dispersive spectrometer (SEM-EDS) analysis results of the solid phase product formed after adding manganese sulfate. The morphology and particle size of the solid phase products obtained from various manganese sulfate addition methods were completely different. When manganese sulfate was added in the form of a solution, the resulting solid phase products were irregularly shaped dense particles with relatively smooth surfaces, and a wide particle size distribution. The addition of manganese sulfate in the form of solid produced the solid phase dendritic floc with a rough surface and a large number of dendritic crystals as products. The average particle size was 3-5 μm, and the particle size was relatively uniform. The components of solid products obtained by different feeding methods were similar, mainly manganese-containing compounds, however, with certain microscopic composition differences, especially the content of arsenic. There was almost no arsenic on the surface of the obtained solid product from the catalytic reaction when manganese sulfate in the form of the solution was utilized, whereas a small amount of arsenic (0.22%-0.30%) was present on the surface of different particles when manganese sulfate solid was added (Table 1). This may be due to the different properties of solid products obtained under different manganese sulfate addition methods.

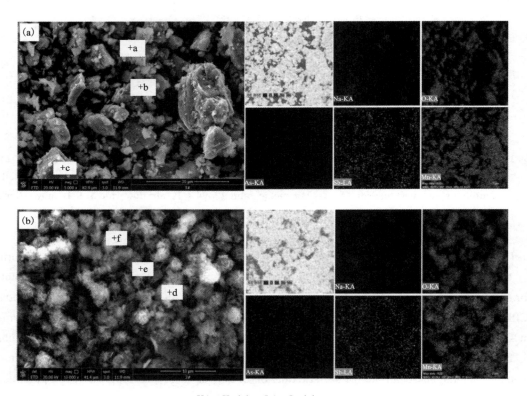

KA—K alpha; LA—L alpha.

Fig. 2 Scanning electron microscope-energy dispersive spectrometer (SEM-EDS) analysis results of solid products obtained by different addition methods: (a) solution form and (b) solid form

Table 1 Elemental content in mass within each region

Region	Mass fraction of elements/%				
	Na	O	As	Sb	Mn
a	7.81	50.41	—	0.94	40.86

Continue the Table 1

Region	Mass fraction of elements/%				
	Na	O	As	Sb	Mn
b	3.90	46.65	—	0.90	48.56
c	3.89	52.73	—	0.93	42.45
d	6.82	52.99	0.25	0.43	39.52
e	2.14	50.97	0.30	1.38	45.20
f	4.85	51.35	0.22	0.26	43.32

a, b, c, d, e, and f are EDS fixed-point analysis shown in Fig. 2.

According to the solution and the solid phase composition, arsenic mainly exists in the adsorbed form in the solid phase products. When using solid manganese sulfate, the obtained product particles were loose flocs with a large number of dendrites on the surface [Fig. 2(b)], having abundant surface active potential, enabling good adsorption of arsenic (Guo et al., 2015; Hou et al., 2017). The product particles obtained by the manganese sulfate in the form of a solution were coarse and dense [Fig. 2(a)], and the adsorption capacity of arsenic was poor. Therefore, the difference in the morphology and particle size of different solid phase products resulted in varying adsorption performance for arsenic.

3.1.2 Effect of As/Mn molar ratio

The effect of As/Mn molar ratio on the catalytic oxidation of As(Ⅲ) was also studied, and the results are displayed in Fig. 3.

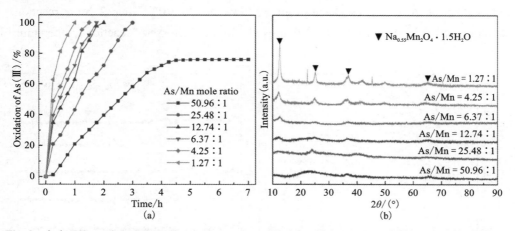

Fig. 3 (a) Effect of As/Mn molar ratio on the oxidation efficiency of As(Ⅲ) and (b) XRD of reaction products under different As/Mn molar ratios. Reaction conditions: temperature 90 ℃, air flow rate 1.0 L/min, stirring speed 300 r/min, and initial pH 13.4

It is evident from Fig. 3(a) that as the As/Mn molar ratio increased, the oxidation efficiency of As(Ⅲ) gradually decreased. As(Ⅲ) was completely oxidized after 2 h of reaction when the

As/Mn molar ratio was 12.74 : 1. For the As/Mn molar ratio higher than 25.48 : 1, the oxidation effect of As(Ⅲ) is poor and As(Ⅲ) could not be completely oxidized. The oxidation rate of As(Ⅲ) slowed down upon increasing the As/Mn molar ratio to 50.96 : 1, and after 5 h reaction, the extent of As(Ⅲ) oxidation was only 77% which did not improve further over time. For a lower As/Mn molar ratio, the amount of catalyst per unit volume would be more, and the more active sites will be available for reaction, resulting in the faster oxidation rate of As(Ⅲ).

Fig. 3(b) exhibits the XRD patterns of the solid phase products obtained under different As/Mn molar ratios. The crystallinity of the obtained solid product $Na_{0.55}Mn_2O_4 \cdot 1.5H_2O$ decreased with the increase of As/Mn molar ratio, evident from Fig. 3(b). When the As/Mn molar ratio exceeded 25.48 : 1, the characteristic diffraction peak of $Na_{0.55}Mn_2O_4 \cdot 1.5H_2O$ could not be observed. Usually, a higher crystallization degree of the catalyst provides less active sites, and therefore the catalytic efficiency decreases. Even though $Na_{0.55}Mn_2O_4 \cdot 1.5H_2O$ had poor crystallinity, its amorphous nature increased its arsenic adsorption capacity, thus reducing the catalytically active sites (Rathi et al., 2020; Wang et al., 2017). The adsorption of arsenic on manganese-based oxides had been confirmed by Manning et al. (Manning et al., 2002; Wei et al., 2019). Therefore, a large amount of arsenic adsorption on the catalytically active Mn species when using the high As/Mn molar ratio reduces the catalyst reaction interface, resulting in the suppression of the oxidation of As(Ⅲ).

3.1.3 Effect of reaction temperature

The effect of reaction temperature on As(Ⅲ) oxidation was further investigated. The corresponding results are shown in Fig. 4. As the reaction temperature rises, the oxidation of As(Ⅲ) also increases [Fig. 4(a)]. When the temperature increased from 30 ℃ to 75 ℃, the oxidation extent of As(Ⅲ) escalated from 60.44% to 90.70% after 2 h of reaction. The crystallization of $Na_{0.55}Mn_2O_4 \cdot 1.5H_2O$ improved with the increase of reaction temperature [Fig. 4(b)].

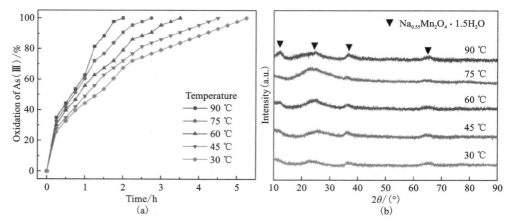

Fig. 4 (a) Effect of reaction temperature on the oxidation efficiency of As(Ⅲ) and (b) XRD of reaction products under different reaction temperatures. Reaction conditions: As/Mn molar ratio 12.74 : 1, air flow rate 1.0 L/min, stirring speed 300 r/min, and initial pH 13.4

According to the previous experimental results, the crystallization of $Na_{0.55}Mn_2O_4 \cdot 1.5H_2O$ directly impact the aerial catalytic oxidation of As(Ⅲ). The presence of manganese in the form of amorphous or less crystalline $Na_{0.55}Mn_2O_4 \cdot 1.5H_2O$ leads to enhanced arsenic adsorption and decreased reaction interface of the catalyst, thereby reducing its oxidation efficiency towards arsenic (Lafferty et al., 2010b). Moreover, temperature not only determines the oxidation extent of As(Ⅲ), but also affects the recyclability of the catalyst. Considering the energy consumption and oxidation efficiency, the optimum reaction temperature should not be less than 60 ℃.

3.1.4 Effect of the initial pH of the solution

The influence of the initial pH of the reaction solution on As(Ⅲ) oxidation was accessed, and the results are shown in Fig. 5. It can be seen from Fig. 5(a) that the oxidation rate of As(Ⅲ) in the leaching solution increased rapidly as the initial pH increased. Although beyond pH 13, the continued increase in pH had little effect on As(Ⅲ) oxidation. The increment in pH from 12 to 13 enhanced the oxidation extent of As(Ⅲ) from 63.79% to 95.35% after 2 h reaction. Further increase in pH to 13.4 led to 100% oxidation of As(Ⅲ) was achieved.

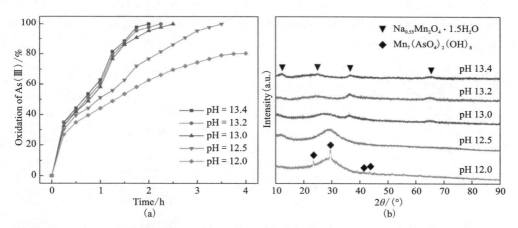

Fig. 5 (a) Effect of initial pH of the solution on the oxidation efficiency of As(Ⅲ) and (b) XRD of reaction products under different initial pH conditions. Reaction conditions: temperature 90 ℃, As/Mn molar ratio 12.74 : 1, air flow rate 1.0 L/min, and stirring speed 300 r/min

Furthermore, the solid products obtained under different pH conditions were significantly different [Fig. 5(b)]. $Na_{0.55}Mn_2O_4 \cdot 1.5H_2O$ with certain crystallinity is obtained when the initial pH is more than 13. In contrast, for an initial pH of less than 13, no characteristic diffraction peak of $Na_{0.55}Mn_2O_4 \cdot 1.5H_2O$ was observed, and the obtained diffraction patterns were mainly of amorphous substances. At pH 12, the characteristic peak of $Mn_7(AsO_4)_2(OH)_8$ (JCPDS17-0748) appeared in the XRD pattern, again confirming that the crystallinity of $Na_{0.55}Mn_2O_4 \cdot 1.5H_2O$ had a significant effect on its catalytic efficiency towards the oxidation of As(Ⅲ). Since the adsorption of arsenic on amorphous or less crystalline $Na_{0.55}Mn_2O_4 \cdot 1.5H_2O$ was enhanced, even when the pH was too low, manganese combined with arsenic to form a precipitate of manganese arsenate, reducing the production of $Na_{0.55}Mn_2O_4 \cdot 1.5H_2O$ significantly, which diminishes its effect on the catalytic aerial oxidation of As(Ⅲ) (Tournassat et al., 2002).

3.1.5 Effect of air flow rate

The effect of air flow rate on the oxidation of As(Ⅲ) was examined and the results are shown in Fig. 6. It can be seen from Fig. 6(a) that the oxidation extent of As(Ⅲ) increased significantly upon accelerating the air flow rate. Even though the oxidation rate of As(Ⅲ) was relatively slow when the air was not forced into the reaction system, As(Ⅲ) could still be completely oxidized within 9.5 h. At an air flow rate of 1.0 L/min, the complete oxidation of arsenic was achieved in 2 h, because the oxygen present in the air rapidly diffused to the catalyst surface through the liquid surface, assisting in continuous oxidation of As(Ⅲ). However, due to the small gas-liquid contact interface, the As(Ⅲ) oxidation reaction is affected by the oxygen diffusion rate, and therefore the oxidation rate of As(Ⅲ) is slow when air is not forced or passed continuously, whereas the injection of air into the solution system significantly improves the oxygen diffusion and increasing the concentration of dissolved oxygen in the solution. At the same time, the sufficient mixing of the gas-liquid-solid multiphase medium is promoted, and thus the oxidation rate of As(Ⅲ) significantly rises (Rady et al., 2020).

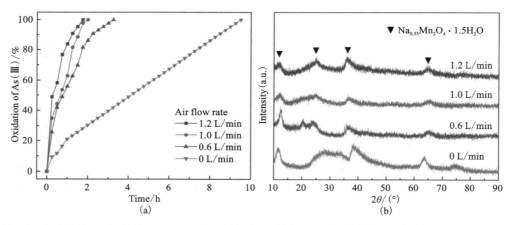

Fig. 6 (a) Effect of air flow rate on the oxidation efficiency of As(Ⅲ) and (b) XRD of reaction products under different air flow rates. Reaction conditions: temperature 90 ℃, As/Mn molar ratio 12.74 : 1, stirring speed 300 r/min, and initial pH 13.4

Fig. 6(b) demonstrates that $Na_{0.55}Mn_2O_4 \cdot 1.5H_2O$ was present in the products obtained under different air flow rates. Nevertheless, the diffraction peak broadened when the reaction system was not aerated, and this broadening was apparent in the range of $2\theta = 25°-35°$, indicating the presence of a certain amount of amorphous or low crystalline material in the product. When the dissolved oxygen was insufficient in the system, Mn(Ⅱ) cannot be completely converted to Mn(Ⅳ), forcing the consumption of lattice oxygen in $Na_{0.55}Mn_2O_4 \cdot 1.5H_2O$ to promote the conversion of Mn(II) to Mn(IV) (Luo et al., 2008), which will change the crystal form of the product and reduce the number of catalysts, resulting in the inhibition of As(III) oxidation. After the concentration of dissolved oxygen increases, lattice oxygen in $Na_{0.55}Mn_2O_4 \cdot 1.5H_2O$ oxidizes Mn(II) into Mn(IV) with the formation of oxygen vacancy and the consumed lattice oxygen is replenished by dissolved oxygen (Cheng et al., 2019; Aydin et al., 2000). Therefore, for the catalytic aerial As(III) oxidation, the air flow rate of 1.0 L/min was considered suitable.

3.2 Antimony precipitation behavior during the aerial catalytic oxidation of As(III)

In the process of aerial catalytic oxidation of As(III), it is also accompanied by the oxidation of Sb(III), and Sb(III) is oxidized prior to As(III). After Sb(III) is oxidized to Sb(V), which is rapidly transformed into $NaSb(OH)_6$ precipitate. The chemical reactions involved in this process are described in Eqs. (2)-(3) as follows (Liu et al., 2019; Sun et al., 2018; Zhang et al., 2021):

$$MnO_2 + Sb(OH)_3 + 4H_2O + OH^- \longrightarrow Sb(OH)_6^- + Mn(OH)_2 \qquad (2)$$

$$2Sb(OH)_3 + 2H_2O + 2OH^- + O_2 \xrightarrow{MnO_2} 2Sb(OH)_6^- \qquad (3)$$

During the experiment, it was found that the precipitation extent of Sb(V) was seriously affected by the standing time of the oxidation solution. The results are shown in Fig. 7. As seen from Fig. 7(a), with the increase of standing time, the precipitation rate of antimony is higher. When the standing time is 0 h, the precipitation extent of antimony is only about 8.66%. With the extension of standing time, the precipitation extent of antimony increases gradually. When the standing time is 12 h, the precipitation rate of antimony reaches 80.69%. It can be attributed to the low concentration of antimony ions, resulting in slow nucleation and growth (Leuz and Johnson, 2005). With the extension of standing time, $NaSb(OH)_6$ gradually grew and agglomerated to form aggregates. As shown in Fig. 7(b), although the obtained product at different standing time both was $NaSb(OH)_6$, the particle size obtained by standing for 12 h was larger than that obtained by standing for 6 h, and the particle agglomeration was more obvious.

3.3 Recycling of reaction product $Na_{0.55}Mn_2O_4 \cdot 1.5H_2O$

The catalytic oxidation behavior of As(III) by the reaction product $Na_{0.55}Mn_2O_4 \cdot 1.5H_2O$ during five successive reuse cycles was studied to investigate the recycling performance of $Na_{0.55}Mn_2O_4 \cdot 1.5H_2O$, the reaction product $Na_{0.55}Mn_2O_4 \cdot 1.5H_2O$ was recovered by stirring in deionized water at 200 r/min for 10 min, followed by centrifugation. The recycling performance of $Na_{0.55}Mn_2O_4 \cdot 1.5H_2O$ is shown in Fig. 8. The catalytic oxidation extent of As(III) to As(V) by the reused $Na_{0.55}Mn_2O_4 \cdot 1.5H_2O$ remained constant in the first three cycles and decreased slightly after the 4th cycle. After five regeneration cycles, the oxidation extent of As(III) was only decreased by 3.1% compared with the first cycle, illustrating that $Na_{0.55}Mn_2O_4 \cdot 1.5H_2O$ can be repeatedly used.

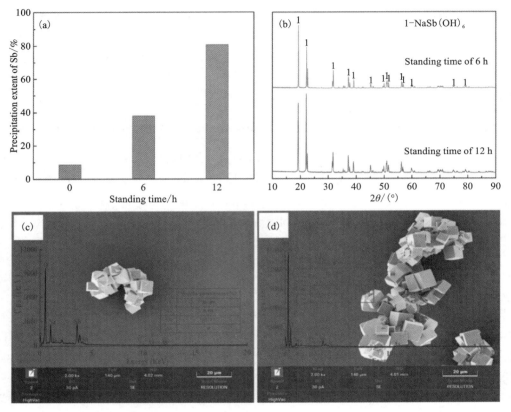

Fig. 7 (a) Effect of standing time on precipitation extent of antimony and (b) XRD and SEM-EDS of precipitation products: standing time of (c) 6 h and (d) 12 h. Cps: counts per second

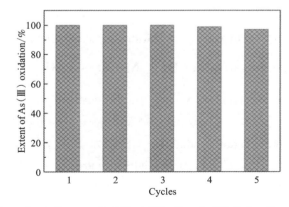

Fig. 8 Recycling of reaction product $Na_{0.55}Mn_2O_4 \cdot 1.5H_2O$ for five cycles. Reaction conditions: temperature 90 ℃, As/Mn molar ratio 12.74∶1, air flow rate 1.0 L/min, stirring speed 300 r/min, initial pH 13.4, and reaction time 2 h

3.4 Mechanism of air oxidation of As(Ⅲ) catalyzed by manganese sulfate

The activity of the catalyst depends on the atomic state exposed to the surface of the catalyst. In order to further identify the interaction between As(Ⅲ) and manganese, the reaction product obtained under the optimized experimental conditions of 90 ℃ reaction temperature, 12.74 : 1 As/Mn molar ratio, 1.0 L/min air flow rate, and 300 r/min stirring speed for 2 h was characterized by X-ray photoelectron spectroscopy (XPS), and Gaussian curve fitting was used to describe the oxidation state of elements, as shown in Fig.9.

It can be seen from Fig.9(b) that the Mn2p3/2 peak can be divided into three peaks at 640.8, 641.9 and 643.0 eV, respectively, corresponding to Mn(Ⅱ), Mn(Ⅲ), and Mn(Ⅳ) (Di Castro et al., 1990; Ivanov-Emin et al., 1983; Yang et al., 2020). Through peak area calculation, it is determined that the atomic ratios of Mn(Ⅱ), Mn(Ⅲ), and Mn(Ⅳ) are 8.96%, 39.88% and 51.16%, respectively, which indicates that the manganese on the surface of the reaction product mainly exists in the form of MnO_2 and MnOOH. At the same time, the existence of Mn(Ⅱ) confirmed the cyclic behavior of different valences of manganese in the oxidation process of As(Ⅲ).

O1s spectra are used to identify the types of surface oxygen species in oxides. The fitting of O1s spectra [Fig.9(c)] shows that oxygen mainly exists in three forms, and the main peak at 529.8 eV is the lattice oxygen (O_{latt}) formed by combining with Mn species. The shoulder peak at 530.9 eV is adsorbed oxygen (O_{ads}), such as O_2^{2-} or O^- in the form of hydroxyl OH^- and oxygen vacancies. The weakest peak at 532.5 eV is surface residual water (O_{sur}) (Chang and Tsai, 2004; Sharma et al., 2008). The atomic ratios of O_{latt}, O_{ads}, and O_{sur} are 61.91%, 33.69% and 4.4%, respectively, indicating that although the surface of the reaction product contained a large amount of lattice oxygen (O_{latt}), there are still many hydroxyl OH^- and superoxide radicals in the form of oxygen vacancies on the surface. The discovery of these superoxide radicals confirmed the aerial oxidation of As(Ⅲ) catalyzed by $Na_{0.55}Mn_2O_4 \cdot 1.5H_2O$. By fitting the spectra of As3d [Fig.9(d)], it can be found that the peaks at 45.1 and 44.3 eV correspond to As(Ⅴ) and As(Ⅲ) (Ouvrard et al., 2005; Wen et al., 2014), respectively, and the atomic ratios of As(Ⅲ) and As(Ⅴ) are 12.49% and 87.51%, respectively. The As(Ⅲ)/As(Ⅴ) ratio is much higher than the initial As(Ⅲ)/As(Ⅴ) ratio of the solution. Thus, during the oxidation process of As(Ⅲ), first it is adsorbed on the catalyst surface and then oxidized in situ. The generated As(Ⅴ) is desorbed from the catalyst surface. Fig.9(e) shows that the peaks at 539.2 and 539.8 eV correspond to Sb(Ⅲ) and Sb(Ⅴ) (Li et al., 2018; Wu et al., 2019), respectively, and their atomic ratios are 27.35% and 72.65%, respectively, consistent with the oxidation behavior of arsenic.

According to the above results, when $MnSO_4$ was added to the arsenic-alkali residue leaching solution, $Na_{0.55}Mn_2O_4 \cdot 1.5H_2O$ and MnOOH were formed under aerobic conditions, but only

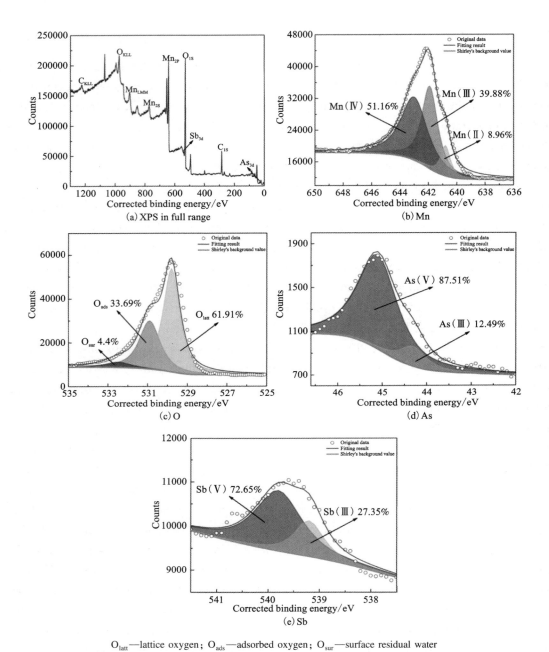

O_{latt}—lattice oxygen; O_{ads}—adsorbed oxygen; O_{sur}—surface residual water

Fig. 9　X-ray photoelectron spectroscopy (XPS) of reaction product

$Na_{0.55}Mn_2O_4 \cdot 1.5H_2O$ had a catalytic effect. During the reaction process (Appendix A Fig. S2①), after adding $MnSO_4$ in the form of solution in alkaline arsenic leaching solution, free Mn(Ⅱ) ions react with alkali to form $Mn(OH)_2$, and then with dissolved oxygen to form

---

① Omitted.

$Na_{0.55}Mn_2O_4 \cdot 1.5H_2O$ (Aydin et al., 2000; Nishi et al., 2020; Wu et al., 2012). During the heterogeneous oxidation reaction, As(Ⅲ) and dissolved oxygen in solution were adsorbed to the reactive active sites at the edge (edge site) and vacancy sites of the $Na_{0.55}Mn_2O_4 \cdot 1.5H_2O$ particle layer. With electron transfer, the direct oxidation rate of dissolved oxygen and As(Ⅲ) was accelerated, and As(Ⅲ) was rapidly oxidized to As(Ⅴ), while part of Mn(Ⅳ) was reduced to $Mn(OH)_2$ and retained in $Na_{0.55}Mn_2O_4 \cdot 1.5H_2O$ after participating in the oxidation reaction of As(Ⅲ) (Lafferty et al., 2010a, 2010b; Nishi et al., 2020; Villalobos et al., 2014; Wang et al., 2020b; Wei et al., 2019). Subsequently, dissolved oxygen reacts with $Mn(OH)_2$ again to form $Na_{0.55}Mn_2O_4 \cdot 1.5H_2O$. In addition, the abundant hydroxyl OH^- on the new ecological MnO_2 surface was conducive to the adsorption of dissolved oxygen and As(Ⅲ) (Huang et al., 2015; Wei et al., 2021), which further promotes the direct oxidation reaction of dissolved oxygen and As(Ⅲ). Therefore, in aerobic alkaline solution, the excellent catalytic oxidation ability of $MnSO_4$ for As(Ⅲ) is mainly attributed to the indirect oxidation of As(Ⅲ) by $Na_{0.55}Mn_2O_4 \cdot 1.5H_2O$ and the direct oxidation on its surface.

4 Conclusions

In conclusion, the addition mode of manganese sulfate showed that its introduction in solution or solid form into the aerial oxidation reaction system of arsenic-alkali residue leaching solution would form $Na_{0.55}Mn_2O_4 \cdot 1.5H_2O$ and MnOOH, respectively. However, only $Na_{0.55}Mn_2O_4 \cdot 1.5H_2O$ had strong catalytic oxidation ability for As(Ⅲ). The As(Ⅲ) could be completely oxidized under suitable conditions of 12.74∶1 As/Mn molar ratio, air flow rate of 1.0 L/min, stirring speed of 300 r/min, reaction temperature of 90 ℃ and time of 2 h. The excellent catalytic oxidation performance for As(Ⅲ) was mainly attributed to the indirect oxidation of As(Ⅲ) by the product $Na_{0.55}Mn_2O_4 \cdot 1.5H_2O$, and not through direct oxidation by dissolved oxygen. This study shows a convenient and efficient process for the oxidation of As(Ⅲ) in alkali solution, which has potential application value for the pre-oxidation of arsenic-containing solution or the detoxification of As(Ⅲ). Hereafter, it is suggested that the arsenic-containing wastewater should be treated by catalytic oxidation firstly, then the alkali should be separated and recovered by membrane technology, and finally the arsenic should be solidified in the form of scorodite by ferric salt precipitation.

Acknowledgments

This work was supported by the National Key R&D Program of China (No. SQ2019YFC190179), National Natural Science Foundation of China (Nos. 51864019 and 52004111), the Distinguished Professor Program of Jinggang Scholars in institutions of higher learning, Jiangxi province, the Program of Qingjiang Excellent Young Talents, Jiangxi University of Science and Technology, and the Jiangxi Province Natural Science Foundation of China (No. 20181BAB206019).

References

[1] AYDIN S, TUFEKCI N, ARAYICI S, et al. Catalytic effects of high Mn(IV) concentrations on Mn(II) oxidation[J]. Water. Sci. Technol, 2000, 42: 387-392.

[2] BARGAR J R, TEBO B M, BERGMANN U, et al. Biotic and abiotic products of Mn(II) oxidation by spores of the marine bacillus sp. strain SG-1[J]. Am. Mineral, 2005, 90: 143-154.

[3] BISCEGLIA K J, RADER K J, CARBONARO R F, et al. Iron(II)-catalyzed oxidation of arsenic(III) in a sediment column[J]. Environ. Sci. Technol, 2005, 39: 9217-9222.

[4] BISSEN M, FRIMMEL F H. Arsenic: A review. Part II: Oxidation of arsenic and its removal in water treatment[J]. Acta. Hydrochim. Hydrobiol, 2003, 31: 97-107.

[5] CHANG J K, TSAI W T. Effects of temperature and concentration on the structure and specific capacitance of manganese oxide deposited in manganese acetate solution[J]. J. Appl. Electrochem, 2004, 34: 953-961.

[6] CHENG Y, HUANG T, LIU C, et al. Effects of dissolved oxygen on the start-up of manganese oxides filter for catalytic oxidative removal of manganese from groundwater[J]. Chem. Eng. J., 2019, 371: 88-95.

[7] CHIU V Q, HERING J G. Arsenic adsorption and oxidation at manganite surfaces. 1. Method for simultaneous determination of adsorbed and dissolved arsenic species[J]. Environ. Sci. Technol, 2000, 34: 2029-2034.

[8] DENG W H, CHAI L Y, DAI Y J. Industrial experimental study on comprehensive recoverying valuable resources from antimony smelting arsenic alkali residue[J]. Hunan Nonferrous Metals, 2014, 30: 24-27.

[9] DI CASTRO V, POLZONETTI G, CONTINI G, et al. XPS study of MnO_2 minerals treated by bioleaching [J]. Surf. Interface Anal, 1990, 16: 571-574.

[10] ERA Y, HIRAJIMA T, SASAKI K, et al. Microbiological As(III) oxidation and immobilization as scorodite at moderate temperatures[J]. Solid State Phenom, 2017, 262: 664-667.

[11] GUO S, SUN W, YANG W, et al. Superior As(III) removal performance of hydrous MnOOH nanorods from water[J]. RSC Adv., 2015, 5: 53280-53288.

[12] HOU J, LUO J, HU Z, et al. Tremendous effect of oxygen vacancy defects on the oxidation of arsenite to arsenate on cryptomelane-type manganese oxide[J]. Chem. Eng. J., 2016, 306: 597-606.

[13] HOU J, LUO J, SONG S, et al. The remarkable effect of the coexisting arsenite and arsenate species ratios on arsenic removal by manganese oxide[J]. Chem. Eng. J., 2017, 315: 159-166.

[14] HUANG Y, HUANG C, YANG Q, et al. Preliminary mechanisms for arsenic removal by natural ferruginous manganese ore[J]. Mater. Res. Innovations, 2015, 19S5-1313-S5-1317.

[15] HUG S J, LEUPIN O. Iron-catalyzed oxidation of arsenic(III) by oxygen and by hydrogen peroxide: pH-dependent formation of oxidants in the Fenton reaction[J]. Environ. Sci. Technol., 2003, 37: 2734-2742.

[16] HULING J R, HULING S G, LUDWIG R. Enhanced adsorption of arsenic through the oxidative treatment of reduced aquifer solids[J]. Water Res., 2017, 123: 183-191.

[17] IVANOV-EMIN B N, NEVSKAYA N A, ZAITSEV B E, et al. ChemInform abstract: Synthesis and properties of calcium and strontium hydroxymanganates (Ⅲ)[J]. Chemischer Informationsdienst, 1983, 14.

[18] JIN Z, JIANG K, WEI X, et al. New process for treating antimony arsenic alkali residue[J]. Nonferrous Metals (Extractive Metallurgy), 1999: 11-14.

[19] KEW D J, AMOS M D, GREAVES M C. Notes. The bromate titration of tervalent arsenic[J]. Analyst, 1952, 77: 488.

[20] KHUNTIA S, MAJUMDER S K, GHOSH P. Oxidation of As(Ⅲ) to As(Ⅴ) using ozone microbubbles [J]. Chemosphere, 2014, 97: 120-124.

[21] KIM D, BOKARE A D, KOO M S, et al. Heterogeneous catalytic oxidation of As(Ⅲ) on nonferrous metal oxides in the presence of H_2O_2[J]. Environ. Sci. Technol, 2015, 49: 3506-3513.

[22] KIM D, MOON G, KOO M S, et al. Spontaneous oxidation of arsenite on platinized TiO_2 through activating molecular oxygen under ambient aqueous condition[J]. Appl. Catal., 2020, B 260: 118146.

[23] KIM M J, NRIAGU J. Oxidation of arsenite in groundwater using ozone and oxygen[J]. Sci. Total Environ, 2000, 247: 71-79.

[24] LAFFERTY B J, GINDER-VOGEL M, SPARKS D L. Arsenite oxidation by a poorly crystalline manganese-oxide 1. Stirred-flow experiments[J]. Environ. Sci. Technol, 2010a, 44: 8460-8466.

[25] LAFFERTY B J, GINDER-VOGEL M, ZHU M, et al. Arsenite oxidation by a poorly crystalline manganese-oxide. 2. Results from X-ray absorption spectroscopy and X-ray diffraction[J]. Environ. Sci. Technol, 2010b, 44: 8467-8472.

[26] LAN S, YING H, WANG X, et al. Efficient catalytic As(Ⅲ) oxidation on the surface of ferrihydrite in the presence of aqueous Mn(Ⅱ)[J]. Water Res, 2018, 128: 92-101.

[27] LEE G, SONG K, BAE J. Permanganate oxidation of arsenic (Ⅲ): Reaction stoichiometry and the characterization of solid product[J]. Geochim. Cosmochim. Acta, 2011, 75: 4713-4727.

[28] LEUZ A -K, JOHNSON C A. Oxidation of Sb(Ⅲ) to Sb(Ⅴ) by O_2 and H_2O_2 in aqueous solutions. Geochim[J]. Cosmochim. Acta, 2005, 69: 1165-1172.

[29] LI W, FU F, DING Z, et al. Zero valent iron as an electron transfer agent in a reaction system based on zero valent iron/magnetite nanocomposites for adsorption and oxidation of Sb(Ⅲ)[J]. J. Taiwan Inst. Chem. Eng, 2018, 85: 155-164.

[30] LI Y, LIU Z, LIU F, et al. Promotion effect of $KMnO_4$ on the oxidation of As(Ⅲ) by air in alkaline solution[J]. J. Hazard. Mater, 2014, 280: 315-321.

[31] LIU R, QU J. Review on heterogeneous oxidation and adsorption for arsenic removal from drinking water [J]. J. Environ. Sci., 2021, 110(12): 178-188.

[32] LIU W, LIU H, ZHANG D, et al. Pressure oxidation dissolution of antimony trioxide in KOH solution for preparing sodium pyroantimonate[J]. JOM, 2019, 71: 4631-4638.

[33] LONG H, ZHENG Y, PENG Y, et al. Recovery of alkali, selenium and arsenic from antimony smelting arsenic-alkali residue[J]. J. Clean. Prod, 2020, 251: 119673.

[34] LUO J, ZHANG Q, GARCIA-MARTINEZ J, et al. Adsorptive and acidic properties, reversible lattice oxygen evolution, and catalytic mechanism of cryptomelane-type manganese oxides as oxidation catalysts [J]. J. Am. Chem. Soc., 2008, 130: 3198-3207.

[35] MANNING B A, FENDORF S E, BOSTICK B, et al. Arsenic(Ⅲ) oxidation and arsenic(Ⅴ) adsorption reactions on synthetic birnessite[J]. Environ. Sci. Technol, 2002, 36: 976-981.

[36] NISHI R, KITJANUKIT S, NONAKA K, et al. Oxidation of arsenite by self-regenerative bioactive birnessite in a continuous flow column reactor[J]. Hydrometallurgy, 2020, 196: 105416.

[37] OUVRARD S, DE DONATO P, SIMONNOT M O, et al. Natural manganese oxide: Combined analytical approach for solid characterization and arsenic retention[J]. Geochim. Cosmochim. Acta, 2005, 69: 2715-2724.

[38] RADY O, LIU L, YANG X, et al. Adsorption and catalytic oxidation of arsenite on Fe-Mn nodules in the presence of oxygen[J]. Chemosphere, 2020, 259: 127503.

[39] RATHI B, JAMIESON J, SUN J, et al. Process-based modeling of arsenic(Ⅲ) oxidation by manganese oxides under circumneutral pH conditions[J]. Water Res, 2020, 185: 116195.

[40] RITCEY G M. Tailings management in gold plants[J]. Hydrometallurgy, 2005, 78: 3-20.

[41] SHAN C, LIU Y, HUANG Y, et al. Non-radical pathway dominated catalytic oxidation of As(Ⅲ) with stoichiometric H_2O_2 over nanoceria[J]. Environ. Int., 2019, 124: 393-399.

[42] SHARMA A, SHARMA G, NAUSHAD M, et al. Estimation of arsenic(Ⅲ) in organic arsines and its complexes using potassium bromate and potassium iodate as oxidants[J]. J. Chil. Chem. Soc., 2016, 61: 2940-2948.

[43] SHARMA R K, RASTOGI A C, DESU S B. Manganese oxide embedded polypyrrole nanocomposites for electrochemical supercapacitor[J]. Electrochim. Acta, 2008, 53: 7690-7695.

[44] SUN Q, LIU C, ALVES M E, et al. The oxidation and sorption mechanism of Sb on δ-MnO_2[J]. Chem. Eng. J., 2018, 342: 429-437.

[45] TOURNASSAT C, CHARLET L, BOSBACH D, et al. Arsenic(Ⅲ) oxidation by birnessite and precipitation of manganese(Ⅱ) arsenate[J]. Environ. Sci. Technol. 2002, 36: 493-500.

[46] TU S. Transformations of synthetic birnessite as affected by pH and manganese concentration[J]. Clays Clay Miner, 1994, 42: 321-330.

[47] VILLALOBOS M, ESCOBAR-QUIROZ I N, SALAZAR-CAMACHO C. The influence of particle size and structure on the sorption and oxidation behavior of birnessite: I. Adsorption of As(Ⅴ) and oxidation of As(Ⅲ)[J]. Geochim. Cosmochim. Acta, 2014, 125: 564-581.

[48] WAN W, CHEN W, HUANG S, et al. Pilot-scale study on separation of arsenic and antimony in arsenic-alkali residue[J]. China Nonferrous Metall, 2015, 44: 32-36.

[49] WANG H, WANG Y, SUN Y, et al. A microscopic and spectroscopic study of rapid antimonite sequestration by a poorly crystalline phyllomanganate: Differences from passivated arsenite oxidation[J]. RSC Adv., 2017, 7: 38377-38386.

[50] WANG X, DING J, WANG L, et al. Stabilization treatment of arsenic-alkali residue (AAR): Effect of the coexisting soluble carbonate on arsenic stabilization[J]. Environ. Int., 2020a, 135: 105406.

[51] WANG Y, DUAN J, LI W, et al. Aqueous arsenite removal by simultaneous ultraviolet photocatalytic oxidation-coagulation of titanium sulfate[J]. J. Hazard. Mater, 2016, 303: 162-170.

[52] WANG Y, LIU H, WANG S, et al. Simultaneous removal and oxidation of arsenic from water by δ-MnO_2 modified activated carbon[J]. J. Environ. Sci., 2020b, 94: 147-160.

[53] WEI W, GUO K, KANG X, et al. Complete removal of organoarsenic by the UV/permanganate process via HO· oxidation and in situ-formed manganese dioxide adsorption[J]. ACS EST Eng., 2021, 1: 794-803.

[54] WEI Z, WANG Z, YAN J, et al. Adsorption and oxidation of arsenic by two kinds of β-MnO_2[J]. J. Hazard. Mater, 2019, 373: 232-242.

[55] WEN Z, ZHANG Y, DAI C, et al. Synthesis of ordered mesoporous iron manganese bimetal oxides for arsenic removal from aqueous solutions[J]. Microporous Mesoporous Mater, 2014, 200: 235-244.

[56] WU K, LIU R, LIU H, et al. Enhanced arsenic removal by in situ formed Fe-Mn binary oxide in the aeration-direct filtration process[J]. J. Hazard. Mater, 2012, 239-240: 308-315.

[57] WU T, SUN Q, FANG G, et al. Unraveling the effects of gallic acid on Sb(Ⅲ) adsorption and oxidation on goethite[J]. Chem. Eng. J. 2019, 369: 414-421.

[58] YANG W, SU Z, XU Z, et al. Comparative study of α-, β-, γ- and δ-MnO_2 on toluene oxidation: Oxygen vacancies and reaction intermediates[J]. Appl. Catal. , 2020, B 260: 118150.

[59] ZHANG L, GUO X, TIAN Q, et al. Selective removal of arsenic from high arsenic dust in the NaOH-S system and leaching behavior of lead, antimony, zinc and tin[J]. Hydrometallurgy, 2021, 202: 105607.

[60] ZHU M, PAUL K W, KUBICKI J D, et al. Quantum chemical study of arsenic(Ⅲ, Ⅴ) adsorption on Mn-Oxides: Implications for arsenic(Ⅲ) oxidation[J]. Environ. Sci. Technol, 2009, 43: 6655-6661.

Enhanced Removal of Organic Matter from Oxygen-pressure Leaching Solution by Modified Anode Slime

Abstract: Organic matter has become a typical harmful impurity, posing a serious problem and catastrophic effects in the hydrometallurgical extraction of zinc. The highly efficient and inexpensive method for the removal of organics from the leaching solution is of great practical significance for the efficient operation of zinc hydrometallurgy. For this reason, the removal of organic matter from oxygen-pressure leaching solution using modified anode slime was investigated by TOC analysis, FTIR spectroscopy, XRD, GCMS, and other characterization methods, and its mechanism was also discussed. The results show that the removal of organic matter using anode slime modified with sulfuric acid was significantly improved, and the removal of TOC reached 57.83% for anode slime modified at a temperature of 80 ℃ with a ratio of sulfuric acid to anode slime of 14 mL/100 g, which was more than 2.7 times that of untreated anode slime. The good removal effect of organic matter using the modified anode slime could be attributed to the transformation of inert α-MnO_2 to active MnOOH by concentrated sulfuric acid-mediated modification. It was also found that the acid-solid ratio and temperature exhibited a significant effect on the modification of anode slime. The XPS analysis results of Mn 2p, Mn 3s, and O 1s, combined with the XRD patterns of the modified products indicated that the modification of anode slime led to the conversion of MnO_2 in anode slime to MnOOH in the presence of H_2SO_4. However, MnO_2 was further converted into high crystallinity MnOOH, even $MnSO_4$ when the temperature exceeded 80 ℃ or the acid-solid ratio exceeded 18 mL/100 g, resulting in the decreased degradation performance for organic matter. The characterization results of FTIR spectroscopy and GCMS show that modified anode slime does not exhibit a degradation effect on all types of organic substances in leaching solution; however, it is very effective for some alcohols, olefins, and esters.

Keywords: Zinc hydrometallurgy; Organic matter; Modification; Anode slime; Oxidative degradation

---

Published in *Journal of Cleaner Production*, 2023, 404: 136886. Authors: Li Yuhu, Liu Ran, Yang Yudong, Yang Sijie, Zhao Yi.

1 Introduction

Organic matter is not only a major water pollutant, but also a typical harmful impurity often encountered in hydrometallurgy. Unlike inorganic impurities, the behavior of organic compounds is more complex and their adverse impact is more far-reaching. Therefore, the study of their removal methods has attracted significant attention and become a research hotspot. The oxygen-pressure leaching process of zinc sulfide concentrate offers the advantages such as high resource utilization and environmental-friendliness, and has become the most representative zinc hydrometallurgy process (Gu et al., 2010). This process directly uses zinc concentrate as a raw material, which easily introduces the flotation agent remaining in the zinc concentrate into the production system. At the same time, organic substances such as calcium lignosulfonate are used as sulfur dispersants during the leaching process (Kolmachikhina et al., 2021). In contrast, the zinc concentrate needs to be roasted in the conventional zinc hydrometallurgy process and no additives required, which strictly restricts the introduction of organic substances. Consequently, the concentration of organic matter in the leaching solution is significantly higher than that in the conventional hydrometallurgical zinc leaching solution. However, organic matter is an extremely harmful impurity causing deleterious effects on the zinc hydrometallurgy systems. Excessive organic matter not only causes a decrease in current efficiency and plate-burning during electrolysis, but also leads to corrosion of the anode and an increase of lead in the cathode zinc (Lu et al., 2018; Majuste et al., 2017). On the other hand, the zinc hydrometallurgy process is a closed loop system, which leads to continuous accumulation of the concentration of organic substances in the production system, resulting in increasing harm. Therefore, removal of organics from smelting system via cost-effective and high-efficiency process has become increasingly urgent.

The reported methods for removing organic impurities present in solution mainly include the air flotation method (Pawliszak et al., 2018), membrane separation method (Pelalak et al., 2018; Zhang et al., 2022), activated carbon adsorption method (Plaza et al., 2010), oxidation method (Matilainen and Sillanpää, 2010), and other processes. In the air flotation method, gas is introduced into the oily wastewater so that impurities adhere to the fine bubbles, the bubbles then float to the water surface to form scum, and the organic matter can be removed by separating the scum. For example, Shi et al. (2017) conducted a study on the removal of natural organic matter (NOM) from drinking water and found that the removal rate of NOM reached 17.7% by air flotation. However, air flotation is suitable only for water-insoluble organic impurities that can be complexed, precipitated, or flocculated, and is not effective in removing water-soluble organics. A multistage integrated membrane system was proposed to remove organic matter from wastewater, and the concentration of total organic carbon (TOC) was reduced from 5.4 to 0.6 mg/L (Chu et al., 2022). Separation via the membrane method in actual production offers the advantages such as high separation efficiency, low energy consumption, and green environmental protection (Padaki et al., 2015). However, this method still encounters many problems, such as high processing cost and poor stability (Baker, 2010). Liu et al. (2022) proposed a method for the removal of sulfonated chloropyridine in solution by activated carbon adsorption method. The removal rate of sulfonated chloropyridine reached 95.88% via activated carbon pretreated by microwave irradiation, which was twice the adsorption capacity of untreated activated carbon. Lu et al. (2022) realized simultaneous removal of Cr and organic matter sustainably by a Cr-Fenton-like reaction, and the removal rates of TOC after 8 h running reached 41.34%.

Various methods for the removal of organic matter present in solution have been developed to date; however, these methods face challenges such as complex processes and high costs. With the large-scale use of organic additives such as sulfur dispersants and flocculants and the lack of removal capacity of organic matter from the present zinc hydrometallurgy system, organic matter continues to accumulate, leading to its increasingly significant negative impact on zinc electrowinning. At present, the removal of organic matter from zinc hydrometallurgy system mainly relies on activated carbon adsorption method (Park et al., 2019). Lin et al. (2018) and Wang et al. (2019) studied the removal of tannins from zinc leaching solution by activated carbon adsorption successively, and its removal efficiency for tannins reached 85%. As mentioned above, the cost of activated carbon is relatively high, and activated carbon adsorption is suitable only for the removal of organic compounds containing hydroxyl and carboxyl groups, which makes this method ineffective for the removal of organic compounds in oxygen-pressure leaching solution (OPLS). On the other hand, the anode slime produced in the zinc hydrometallurgical system is rich in manganese dioxide (MnO_2) and exhibits good oxidation performance. The use of anode slime as an oxidant to remove organics from the OPLS can be considered as a feasible method, as it can not only remove organics, but also effectively use anode slime to reduce treatment cost. However, the main species of Mn in anode slime is inert α-MnO_2, which shows poor reactivity and low degradation of organic matter (Ohzuku et al., 1984). In addition to the relatively stable nature of organics in the leaching solution, the complex composition and high salinity of the leaching solution also pose a significant challenge to the removal of organics. Therefore, at present, feasible method to remove organic substances from the oxygen-pressure leaching system is not available.

To realize the low-cost removal of organic matter from zinc hydrometallurgy system, in this study, a method for oxidative degradation of organic matter using modified anode slime was proposed. The effects of reaction conditions on the modification of anode slime and its removal effect on organic matter were investigated, and the reaction mechanism was also revealed. The by-product anode slime was used as a costeffective and high-efficient raw material and the sulfuric acid modification process was used to obtain the modified anode slime with good removal effect toward organic substances. This method not only solves leaching solution by the existing process, but also realizes low-cost removal of organic matter and effective utilization of the by-product anode slime. More importantly, the oxidizing ability of oxidants is not the only factor that determines the degradation of organic compounds, the physical properties of oxidants also need to be evaluated. The research results are expected to provide theoretical and technical support for the low-cost removal of organic matter from zinc hydrometallurgy system. This study provides a reference for the development of organic pollutants removal technology from water systems.

2 Experimental

2.1 Material

The solution used in the experiment was collected from a zinc hydrometallurgy smelter by oxygen-pressure leaching process. It contained 144.42 g/L Zn with an acidity of 4.92 g/L and a TOC concentration of 157.76 mg/L. Fig. 1 shows the Fourier transform infrared (FTIR) spectra of the OPLS. Fig. 1 exhibits that the FTIR bands at 1160 cm^{-1} correspond to the symmetrical stretching absorption of the ester group, and peaks located below 1000 cm^{-1} belong to stretching vibrations of $-CH_2$ and $-CH_3$ on the carbon chains of various alkanes and alkenes. Based on the above-mentioned results, it can be inferred that the organic substances in OPLS mainly consist of stable esters and long carbon-chain hydrocarbons.

The anode slime was obtained from the zinc electrolysis system, and its main components are listed in Table 1. These components mainly include manganese and lead, and their main crystalline phases are α-MnO_2 and $PbSO_4$, respectively. The other reagents H_2SO_4, NaOH, etc., used in the experiment were all analytical reagents.

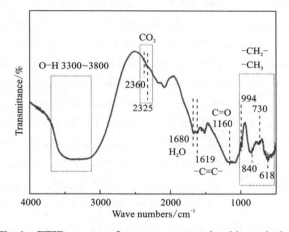

Fig. 1 FTIR spectra of oxygen-pressure leaching solution

Table 1 Main components of anode slime used in the experiment

Element	Mn	Pb	S	Sr	Ca	Zn	F
Content/%	36.664	8.982	4.936	1.552	0.718	0.539	0.522

2.2 Experimental procedures

Modification experiment: Anode slime (100 g) was weighed in a corundum crucible, and an accurately measured amount of concentrated sulfuric acid was added to the crucible according to the set conditions. The materials were then mixed homogeneously and the crucible was placed in an oven. It was maintained at a specific temperature for 2 h to effectively carry out the modification process. The modification product of anode slime was obtained after the reaction.

Evaluation experiment of the modified anode slime: Leaching solution (OPLS, 100 mL) was first measured in a 200 mL beaker, which was then placed in a constant-temperature water bath, stirred, and heated to a set temperature. After it reached the set temperature, the specific amount of modified anode slime was added to the beaker, which was then maintained at the set constant temperature for a certain time. After the completion of the reaction, the postreaction liquid was collected by filtration, and the composition and content of the organic matter were sampled for characterization.

2.3 Analysis and characterization methods

A TOC analyzer (Metash, TOC-2000) was used to characterize the content of the organic compounds in the solution. The X-ray diffraction (XRD, Rigaku-TTR Ⅲ, Cu/K$_\alpha$, $\lambda = 0.15406$ nm) was used to characterize the species of the solid samples. FTIR spectroscopy (Nicolet, 1800) and gas chromatography-mass spectroscopy (GCMS-QP2020) were used to analyze the type and content of organic substances in the solution. X-ray photoelectron spectroscopy (Escalab 250Xi, Thermo Scientific) was used to characterize the valence states of Mn and the distribution of O on the surface of modified anode slime.

In order to effectively conduct FTIR and GCMS characterization, dichloromethane extraction method is used to enrich the organic matter in the sample before sample characterization, and then the extracted concentrates is dropwise added to KBr compression tablets. After dichloromethane volatilizes, a thin film sample containing the organic matter is obtained, which can be used for FTIR characterization (Maria, 2012). The extracted concentrates can be directly used for GCMS characterization.

3 Results and discussion

3.1 Removal effect of activated carbon on organic matter

Activated carbon consists of an abundant void structure and a large specific surface area and is extensively used in the removal of organic matter in hydrometallurgy (Patawat et al., 2020). To determine the effect of activated carbon on the removal of organic matter from the OPLS, experiments on the effect of the dosage of activated carbon on the removal of organic matter were conducted under normal temperature for 2 h, and the corresponding results are shown in Fig. 2. The results reveal that when the amount of activated carbon varies from 5 to 25 g/L, the concentration of TOC in the final solution is maintained at ~175 mg/L, indicating that the activated carbon has no removal effect on the organic matter present in the OPLS. Activated carbon usually shows a good removal effect on aromatics, phenols, and humus (Ozkaya, 2006; Sato et al., 2007), while the organic substances present in OPLS mainly include esters and alkanes, thus the removal of TOC is poor.

3.2 Removal effect of unmodified anode slime on organic matter

The main component of anode slime is MnO_2 with strong oxidizing properties, and it is often used as an oxidant for the degradation of organic matter (Xu et al., 2019). To investigate the effect of unmodified anode slime on the removal of organic matter in the OPLS, the experiments using unmodified anode slime were carried out at 80 ℃ for 2 h for comparative analysis, and the results are shown in Fig. 3.

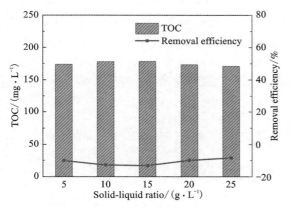

Fig. 2 Removal effect of activated carbon on organics in the OPLS

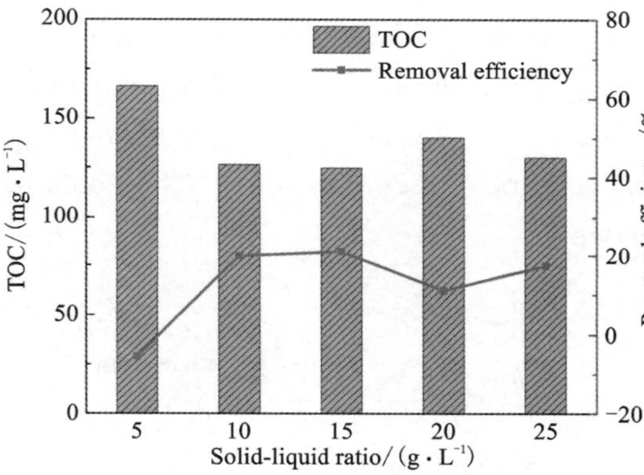

Fig. 3 Removal effect of unmodified anode slime on organic matter in the OPLS

Fig. 3 demonstrates that the unmodified anode slime exhibits a certain removal effect on TOC in the OPLS. When the amount of unmodified anode slime was increased from 5 to 10 g/L, the concentration of TOC decreased from 166.36 to 126.41 mg/L. When the amount of anode slime was further increased, the concentration of TOC remained almost unchanged, and the removal rate of TOC became ~20%. Therefore, although the unmodified anode slime showed a certain removal effect on TOC in the OPLS, the removal effect was rather poor, and the removal of TOC reached only ~20%. Since the standard electrode potential of MnO_2/Mn^{2+} is up to 1.23 eV, most organic substances can be degraded; however, practically it is not possible. This may be attributed to the fact that manganese in anode slime mainly exists as $\alpha\text{-}MnO_2$, which shows relatively complete crystallization and poor activity, resulting in low oxidative degradation of organic matter (Kanungo et al., 1981).

3.3 Removal effect of modified anode slime on organic matter

Improving the reactivity of anode slime is the key to enhancing its removal effect of TOC from the OPLS. To this end, the effect of concentrated sulfuric acid-mediated modification conditions and reaction parameters on the removal of TOC using modified anode slime in the OPLS was investigated.

3.3.1 Effect of the acid-solid ratio on the removal of TOC

The effect of the acid-solid ratio on the modification of anode slime was investigated under the conditions of modified anode slime dosage of 20 g/L, temperature of 80 ℃, and time of 1 h, and the corresponding results are shown in Fig. 4. Fig. 4(a) exhibits that the removal of TOC with the modified anode slime increased slightly with the increase in the acid-solid ratio from 2 mL/100 g to 6 mL/100 g, and the removal of TOC increased from 24.01% to 31.97%. When the acid-solid ratio was increased to 14 mL/100 g, the removal of TOC rapidly increased to 57.83%. It exhibited little effect on the degradation promotion of the modified anode slime for a further increase of the acid-solid ratio to 18. The removal of TOC remained at ~57%, and the concentration of TOC in the final solution was approximately 65 mg/L. However, the removal of the organic matter using the obtained modified anode slime decreased rapidly when the acid-solid ratio exceeded 18 mL/100 g. The removal of TOC decreased from 57.83% to 39.49% with the acid-solid ratio of 26 mL/100 g. This could be attributed to the transformation of the manganese phase.

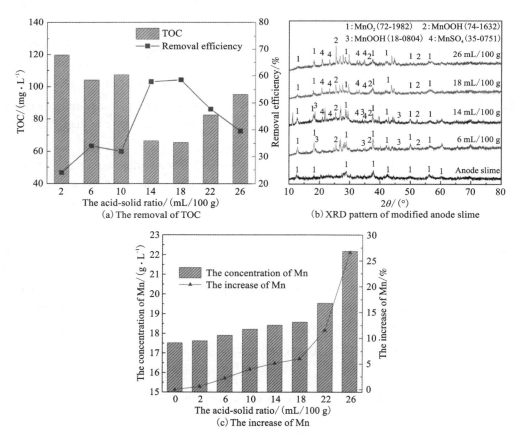

Fig. 4 Effect of the acid-solid ratio on the removal of TOC using modified anode slime

Fig. 4(b) illustrates that the diffraction peak of α-MnO_2 in the product gradually weakens, and that of MnOOH appears simultaneously with the increase in the acid-solid ratio, indicating the conversion of inert α-MnO_2 in the anode slime to highly active MnOOH (Yang et al., 2005; Zhang et al., 2005), as revealed by Reaction (1). This significantly improves the oxidative degradation effect of the modified anode slime on organic matter. However, when the acid-solid ratio exceeded 14 mL/100 g, the characteristic diffraction peak of $MnSO_4$ appeared in the XRD pattern of the product, indicating the conversion of MnO_2 and MnOOH into $MnSO_4$, when the acid-solid ratio exceeded 14 mL/100 g, as indicated by Reaction (2). Although MnOOH is still present in the product, its reaction activity becomes poor, which is attributed to the gradual increase in the diffraction intensity of its characteristic peak with the increase of the acid-solid ratio. In this event, the degradation effect of the product on organic matter is no longer improved. On the other hand, it was observed that the introduction of manganese was inevitable in the removal of TOC with the modified anode slime, and the change trend of manganese was different from that of TOC. The concentration of manganese in the solution after reaction increased with the increase of acid-solid ratio, in particular, when the acid-solid ratio exceeded 18 mL/100 g, which again confirms the influence of the acid-solid ratio on manganese phase during the modification of anode slime [Fig. 4(c)]. Insoluble MnO_2 and MnOOH were converted into soluble $MnSO_4$ at a high acid-solid ratio, resulting in an increase in manganese dissolution. The characteristic diffraction peak intensity of MnOOH becomes stronger with the increase of the acid-solid ratio; nonetheless, its removal effect toward organic matter is worse, which indicates that the reaction activity of the product significantly impacts the degradation of organic matter except for the manganese phase. Therefore, considering the removal effect of TOC and acid consumption, the acid-solid ratio for the modification of anode slime is determined to be 14 mL/100 g.

$$3MnO_2 + H_2SO_4 \longrightarrow 2MnOOH + MnSO_4 + O_2 \uparrow \qquad (1)$$

$$2MnO_2 + 2H_2SO_4 \longrightarrow 2MnSO_4 + O_2 \uparrow + 2H_2O \qquad (2)$$

3.3.2 Effect of modification temperature on modification of anode slime

The effect of modification temperature on the removal of organic matter using the modified anode slime was investigated under the conditions of modified anode slime dosage of 20 g/L, temperature of 80 ℃, and time of 1 h, and the results are shown in Fig. 5.

The results illustrate that the removal of TOC with the modified anode slime first increased and then decreased with the increase in the temperature. In contrast, the amount of manganese introduced in the system continued to increase, in particular, when the temperature exceeded 140 ℃. When the temperature was increased from 20 to 80 ℃, the removal of TOC gradually increased from 41.95% to 57.83% [Fig. 5(a)], and the concentration of Mn in final liquor changed to below 0.9 g/L [Fig. 5(c)]. However, the removal of TOC gradually decreased with a further increase in temperature, and the removal of TOC decreased rapidly at temperatures above 200 ℃, while the introduction of Mn increased rapidly. The removal of TOC was only ~23.45% at 360 ℃, which is almost the same as that of the unmodified anode slime; however, the increase of Mn reached more than 47%. The effect of modification temperature on TOC removal and Mn

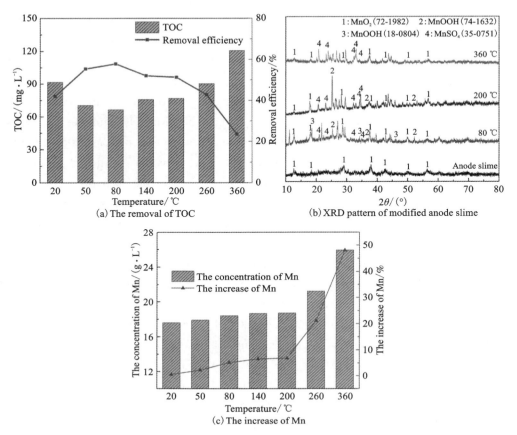

Fig. 5　Effect of temperature on the modification of anode slime

introduction from anode slime could be attributed to the transformation of the manganese phase.

$$4MnOOH + 4H_2SO_4 \longrightarrow 4MnSO_4 + O_2(g)\uparrow + 6H_2O \qquad (3)$$

Fig. 5(b) shows that the manganese phase in the anode slime gets converted from stable α-MnO_2 to highly active MnOOH with increasing modification temperature, which promotes the degradation performance of the modified anode slime. However, MnOOH and H_2SO_4 further reacted and dissolved to form $MnSO_4$, as indicated by Reaction (3), when the modification temperature exceeded 80 ℃, reducing the degradation effect of the modified anode slime on organic matter. When the temperature was increased to 360 ℃, MnO_2 reacted directly with H_2SO_4 to form $MnSO_4$, resulting in poor removal of TOC, but also the excessive introduction of Mn. Therefore, considering the removal of TOC and the introduction of Mn, the optimized modification temperature of anode slime was determined to be 80 ℃.

3.3.3　Effect of modified anode slime dosage on the removal of TOC

The effect of modified anode slime dosage on the removal of organic matter was investigated at a temperature of 80 ℃ for 1 h, and the results are shown in Fig. 6. The removal of TOC continues to increase with the increase in the dosage of modified anode slime. When the dosage of modified anode slime was increased from 5 to 20 g/L, the removal of TOC increased from 35.42% to 51.90%, and the concentration of TOC in the final solution decreased from 101.87 to 75.88 mg/L. In contrast, the TOC content in the solution did not change much with the further increase in the dosage of modified anode slime. This may be attributed to the presence of a variety of organics in the OPLS, and the modified anode slime could only oxidize and degrade part of the organic matter, and was not suitable for high-stability organic matter, such as straightchain alkanes. Therefore, considering the removal of TOC, the optimized dosage of modified anode slime was determined to be 20 g/L.

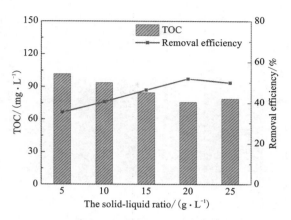

Fig. 6 Effect of the dosage of modified anode slime on the removal of TOC

3.3.4 Effect of reaction temperature on the removal of TOC

The effect of reaction temperature on the removal of organic matter was investigated under conditions of a reaction time of 1 h and a dosage of anode slime of 20 g/L. Fig. 7 shows the corresponding results, revealing that the removal of TOC gradually increased with an increase in temperature to 80 ℃; however, the removal rate of TOC remained almost unchanged at higher temperatures. When the temperature was increased from 30 to 80 ℃, the removal of TOC increased from 42.79% to 51.91%, and the concentration of TOC in the reaction solution decreased from 90.25 to 75.88 mg/L. In general, the higher the temperature, the better the degradation effect; however, the removal of TOC in the OPLS using modified anode slime was not the same. It also confirms that the types of organic substances in the OPLS are relatively complex, and some organic substances are quite stable, which are difficult to oxidize and degrade. Although an increase in temperature generally benefits the removal of TOC, considering the rather small improvement in the removal of TOC at above 80 ℃, the reaction temperature was determined to be 80 ℃.

3.3.5 Effect of reaction time on the removal of TOC

The effect of reaction time on the removal of organic matter was investigated under conditions including temperature of 80 ℃ and anode slime dosage of 20 g/L. Fig. 8 shows the corresponding results, revealing that when the reaction time was increased from 15 to 60 min, the removal of TOC increased from 40.11% to 57.09%, and the concentration of TOC in the reaction solution decreased from 94.03 to 67.37 mg/L. However, a further increase in reaction time exhibited a little effect on the removal of TOC. Considering the removal of TOC and energy consumption, the reaction time was determined to be 60 min.

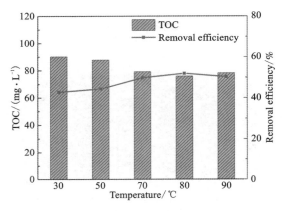

Fig. 7　Effect of reaction temperature on the removal of TOC

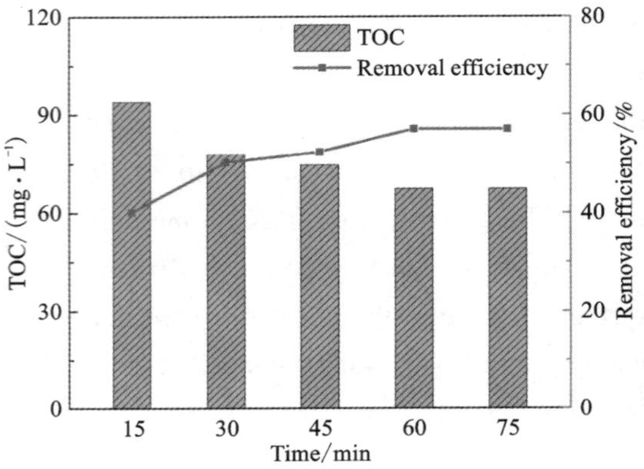

Fig. 8　Effect of reaction time on the removal of TOC

3.4　Removal mechanism of modified anode slime for organics

3.4.1　Characterization of modified anode slime

To reveal the removal mechanism of the modified anode slime toward organics in the OPLS, X-ray photoelectron spectroscopy (XPS) was employed to characterize the modified anode slime obtained at different modification temperatures. The corresponding results are shown in Figs. 9 and 10. The XPS spectrum of the Mn 2p orbital for the modified anode slime is presented in Fig. 9. The characteristic peaks with binding energies of 641.12, 642.6, and 644.49 eV in the Mn $2p_{3/2}$ spectrum represent Mn^{2+}, Mn^{3+}, and Mn^{4+}, respectively (Wang et al., 2015).

Fig. 9(a) shows that the contents of Mn^{4+}, Mn^{3+}, and Mn^{2+} for the unmodified anode slime are 37.49%, 61.10%, and 1.41%, respectively. Noteworthy, the valence state of the modified anode slime changes significantly. When the modification temperature was increased to 80 ℃, the Mn^{4+} content of the obtained product decreased, the content of Mn^{3+} increased significantly, and the Mn^{2+} content increased slightly. In contrast, when the modification temperature was increased to 200 ℃, the Mn^{4+} and Mn^{3+} content of the modified product decreased, while the content of Mn^{2+} increased significantly. Moreover, the decreasing trend of manganese valence of the modified product became more obvious with a further increase in temperature. When the modification temperature reached 360 ℃, the Mn^{4+} and Mn^{3+} contents of the product decreased to 32.09% and 55.34%, respectively, and the Mn^{2+} content increased to 12.57%. Moreover, the fitting of the spectra gradually deteriorated with increasing modification temperature, in particular, the characteristic peaks of Mn $2p_{3/2}$ with a lower binding energy appear obvious shift. The satellite peaks of Mn^{2+} appear between the characteristic peaks of Mn $2p_{3/2}$, which reduces the fitness of the spectral line. This phenomenon also confirms the formation of Mn^{2+} in the modified anode slime at temperatures above 80 ℃ (Banerjee and Nesbitt, 2001). It can also be confirmed from the XPS analysis results of Mn 3s shown in Fig. 9(f) that the gap in binding energy between the

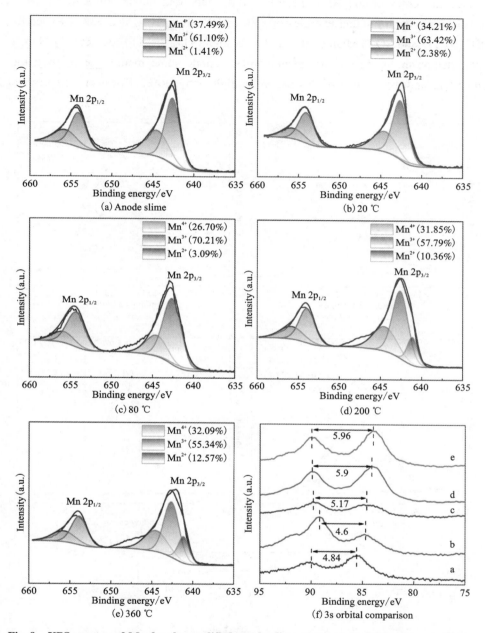

Fig. 9　XPS spectra of Mn for the modified anode slime obtained at different temperatures

Mn 3s doublets increases with increasing modification temperature. The larger the difference in the energy gap between the doublet binding on the Mn 3s orbital, the lower the average valence state of Mn. It indicates that the average valence state of Mn of the modified anode slime gradually decreases with increasing modification temperature (Googlev et al., 2016).

In order to further clarify the transformation law of Mn during the modification process of anode slime, the O 1s orbital XPS spectrum of the modified anode slime was fitted, and the characteristic peaks with binding energies of 530.1, 531.3, 532.1, and 532.4 eV correspond to O^{2-}, OH^-, H_2O, and SO_4^{2-} (Mian et al., 2019; Wang et al., 2015), respectively. Fig. 10 shows the results, demonstrating that O of the unmodified anode slime mainly comes from the surface adsorbed water, and the content of this part is as high as 59.36%. However, the source of O on

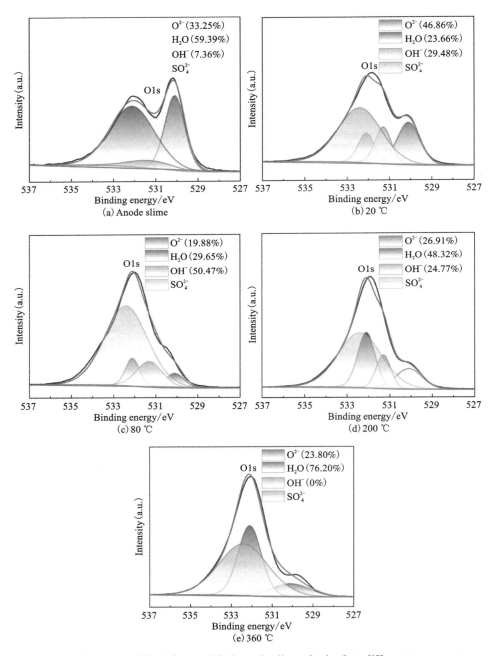

Fig. 10 XPS spectra of O 1s for modified anode slime obtained at different temperatures

the surface of the modified anode slime changes significantly. The results indicate that O in the form of adsorbed water gradually decreases, while O in the form of OH^- gradually increases and reaches 50.47% at 80 °C. At the same time, the content of O present in the form of O^{2-} reaches a minimum value of only 19.88%. However, the content of OH^- decreases rapidly to 24.77% at 200 °C, which even disappears completely at 360 °C.

Therefore, based on the orbital XPS analysis results of Mn 2p, Mn 3s, and O 1s, combined with the XRD patterns of the modified products, the modification process of the anode slime can be inferred. That is, MnO_2 in the anode slime gets converted to MnOOH with the increase in the amount of sulfuric acid and temperature, resulting in the improvement of the degradation performance of the anode slime. The further conversion of MnOOH into $MnSO_4$ begins when the temperature exceeds 80 °C, and MnO_2 is directly converted into $MnSO_4$ when the temperature is above 360 °C. Combined with the removal effect of different modified anode slimes on the organic matter in the leaching solution, it was found that the removal effect of the organic matter is related to the formation of Mn(Ⅲ), and the more favorable it is for the generation of MnOOH, the more beneficial it is for the removal of organic matter.

3.4.2 Characterization of organic matter

The above-mentioned results confirm that only a part of the organic matter could be removed from the OPLS with the modified anode slime, which indicates that the modified anode slime is not effective for the removal of all types of organics. To determine the types of organic matter that could be easily degraded with the modified anode slime, the samples collected before and after the reaction were characterized by FTIR spectroscopy and GCMS, and the corresponding results are presented in Fig. 11 and Table 2.

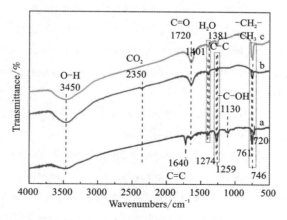

a—Raw solution; b—Modified anode slime at 50 °C; c—Modified anode slime at 360 °C.

Fig. 11 FTIR spectra of modified anode slime

Table 2　GCMS analysis results of organic matter in solution before and after reaction

No.	Component name	Molecular weight	Original content/%	After content/%
1	Hexanedioic acid, bis(2-ethylhexyl) ester	370.31	15.69	12.39
2	Diisooctyl phthalate	390.28	5.71	4.82
3	1,3-Benzenedicarboxylic acid, bis(2-ethylhexyl) ester	390.28	2.08	1.62
4	Hexacosane	366.42	4.31	1.17
5	Docosane	310.36	4.18	1.66
6	Heptacosane	380.44	3.55	2.07
7	Heneicosane	296.34	3.29	3.6
8	Hexadecane	226.27	2.59	1.17
9	Octadecane	254.3	2.15	2.3
10	Heptadecane	240.28	1.79	1.86
11	Octacosane	394.45	1.49	0.82
12	Pentacosane	352.41	1.42	1.26
13	Tetracosane	338.39	1.2	1.94

Fig. 11 illustrates that the changes in the organic species before and after the reaction are quite significant. Except for the hydroxyl vibration peak at 3450 cm^{-1}, the asymmetric stretching double peak of CO_2 at 2350 cm^{-1}, the absorption peaks of water of crystallization for KBr at 1401 and 1381 cm^{-1}, and the C—C stretching vibration peaks of aromatic rings at 1274 and 1259 cm^{-1} remain unchanged; and the —C=C— double bond vibration peak at 1640 cm^{-1} and the stretching vibration peak of C—OH in alcohols or aromatic compounds adjacent to 1130 cm^{-1} disappear. It indicates that the oxidative degradation of alcohols and olefins occurred after the modified anode slime was added to the OPLS.

To further identify the degradation reaction processes of organic compounds using modified anode slime, GCMS was used to characterize the solution before and after the reaction. Table 2 presents the results, revealing that the main organic compounds in the OPLS were large molecules including long carbon chain alkanes and ester organic compounds. After the treatment with modified anode slime, the content of the three ester organics was reduced by 21.03%, 15.58% and 22.12%, respectively. Moreover, the content of alkane organic compounds with molecular weights above 300 decreased, while that of alkane organic compounds with molecular weights below 300 remained basically unchanged or even slightly increased. This result indicates that some esters and alkane organic compounds with larger molecular weights in the solution underwent degradation to produce organic compounds with smaller molecular weights or even degraded to CO_2, which reduced the TOC content in the solution.

4 Conclusions

In this study, the organic matter was removed from oxygen-pressure leaching solution of ZnS concentrate using modified anode slime, the effects of reaction conditions on the modification of anode slime and the removal of TOC were investigated, and its reaction mechanism was revealed. Based on the results, the following conclusions can be drawn.

(1) Organic matter in the OPLS could not be removed with activated carbon. In contrast, unmodified anode slime exhibited a certain removal effect on organic matter; nonetheless, the removal of TOC was less than ~20%.

(2) The modification of anode slime with concentrated sulfuric acid could significantly improve the removal of organic matter, and the acid-solid ratio and temperature were the key factors involved in the modification of anode slime. The removal of TOC reached 57.83%, and the concentration of TOC decreased from 157.76 to 66.52 mg/L under an acid-solid ratio of 14 mL/100 g and a modification temperature of 80 ℃, while the introduction of Mn was less than 0.9 g/L.

(3) The good removal effect of organic matter with the modified anode slime could be attributed to the transformation of inert α-MnO_2 to active MnOOH with concentrated sulfuric acidmediated modification. The content of Mn^{3+} on the surface of modified anode slime particles increased from 61.10% to 70.21%, while the content of OH^- increased from 7.36% to 50.47%. Modified anode slime exhibited good degradation effect on the alcohols, olefins, and esters present in the OPLS.

(4) The selection of oxidants is the key to the degradation of organic compounds; however, in addition to the oxidation ability, the influence of physical properties such as reaction activity cannot be ignored. Undeniably, a lot more systematic explorations are further demanded to investigate modification, activation, doping, and other treatment methods that can improve the degradation performance of oxidants.

Acknowledgments

This study is supported by the National Key R&D Program of China (SQ2019YFC190179), National Natural Science Foundation of China (51864019, 52004111), the Jiangxi Provincial Natural Science Foundation (20224BAB214040), the Program of Qingjiang Excellent Young Talents, Jiangxi University of Science and Technology, and the "Double Thousand Plan" talent project of Jiangxi Province (jxsq2018106051).

References

[1] BAKER R W. Research needs in the membrane separation industry: Looking back, looking forward[J]. J. Membr. Sci., 2010, 362: 134-136. https://doi.org/10.1016/j.memsci.2010.06.028.

[2] BANERJEE D, NESBITT H W. XPS study of dissolution of birnessite by humate with constraints on reaction mechanism. Geochem[J]. Cosmochim. Acta, 2001, 65: 1703-1714. https://doi.org/10.1016/S0016-7037(01)00562-2.

[3] CHU H, MA J, LIU X, et al. Spatial evolution of membrane fouling along a multi-stage integrated membrane system: A pilot study for steel industry brine recycling[J]. Desalination, 2022, 527: 115566. https://doi.org/10.1016/j.desal.2022.115566.

[4] GOOGLEV K A, KOZAKOV A T, KOCHUR A G, et al. Determining the valence state of manganese ions in complex oxides $La_{1-x}Ca_xMnO_3$ ($x = 0.5, 0.7, 0.85,$ and 0.9) based on Mn2p and Mn3s X-ray photoelectron spectra[J]. Bull. Russ. Acad. Sci. Phys., 2016, 80: 638-640. https://doi.org/10.3103/S1062873816060125.

[5] GU Y, ZHANG T, LIU Y, et al. Pressure acid leaching of zinc sulfide concentrate[J]. Trans. Nonferrous Metals Soc. China, 2010, 20: s136-s140. https://doi.org/10.1016/S1003-6326(10)60028-3.

[6] KANUNGO S B, PARIDA K M, SANT B R. Studies on MnO_2-Ⅱ. Relationship between physicochemical properties and electrochemical activity of some synthetic MnO_2 of different crystallographic forms[J]. Electrochim. Acta, 1981, 26: 1147-1156. https://doi.org/10.1016/0013-4686(81)85092-X.

[7] KOLMACHIKHINA E B, BLUDOVA D I, LUGOVITSKAYA T N, et al. Behavior of indium during the pressure leaching of a zinc concentrate in the presence of lignosulfonate[J]. Russ. Metall. 2021: 1387-1393. https://doi.org/10.1134/S0036029521110070.

[8] LIN G, CHENG S, WANG S, et al. Process optimization of spent catalyst regeneration under microwave and ultrasonic spray-assisted[J]. Catal. Today, 2018, 318: 191-198. https://doi.org/10.1016/j.cattod.2017.09.042.

[9] LIU J, CUI J, GAO J, et al. Study on continuous adsorption/microwave-activated carbon for removing sulfachloropyridazine[J]. Chem. Phys. Lett., 2022, 792: 139417 https://doi.org/10.1016/j.cplett.2022.139417.

[10] LU J, DREISINGER D, GLÜCK T. Cobalt electrowinning-a systematic investigation for high quality electrolytic cobalt production[J]. Hydrometallurgy, 2018, 178: 19-29. https://doi.org/10.1016/j.hydromet.2018.04.002.

[11] LU K, GAO M, SUN B, et al. Simultaneous removal of Cr and organic matters via coupling Cr-Fenton-like reaction with Cr flocculation: The key role of Cr flocs on coupling effect[J]. Chemosphere, 2022, 287: 131991. https://doi.org/10.1016/j.chemosphere.2021.131991.

[12] MAJUSTE D, BUBANI F C, BOLMARO R E, et al. Effect of organic impurities on the morphology and crystallographic texture of zinc electrodeposits[J]. Hydrometallurgy, 2017, 169: 330-338. https://doi.org/10.1016/j.hydromet.2017.02.013.

[13] MARIA C. Application of FTIR spectroscopy in environmental studies. In: Akhyar Farrukh, M. (Ed.)[J]. Advanced Aspects of Spectroscopy. InTech, 2012. https://doi.org/10.5772/48331.

[14] MATILAINEN A, SILLANPÄÄ M. Removal of natural organic matter from drinking water by advanced

oxidation processes[J]. Chemosphere, 2010, 80: 351-365. https://doi.org/10.1016/j.chemosphere.2010.04.067.

[15] MIAN M M, LIU G, FU B, et al. Facile synthesis of sludge-derived MnO_x-Nbiochar as an efficient catalyst for peroxymonosulfate activation[J]. Appl. Catal. B Environ, 2019, 255: 117765 https://doi.org/10.1016/j.apcatb.2019.117765.

[16] OHZUKU T, HIGASHIMURA H, HIRAI T. XRD studies on the conversion from several manganese oxides to β-manganese dioxide during acid digestion in $MnSO_4$-H_2SO_4 system[J]. Electrochim. Acta, 1984, 29: 779-785. https://doi.org/10.1016/0013-4686(84)80014-6.

[17] OZKAYA B. Adsorption and desorption of phenol on activated carbon and a comparison of isotherm models [J]. J. Hazard Mater., 2006, 129: 158-163. https://doi.org/10.1016/j.jhazmat.2005.08.025.

[18] PADAKI M, SURYA MURALI R, ABDULLAH M S, et al. Membrane technology enhancement in oil-water separation: A review[J]. Desalination, 2015, 357: 197-207. https://doi.org/10.1016/j.desal.2014.11.023.

[19] PARK J E, KIM E J, PARK M -J, et al. Adsorption capacity of organic compounds using activated carbons in zinc electrowinning[J]. Energies, 2019, 12: 2169. https://doi.org/10.3390/en12112169.

[20] PATAWAT C, SILAKATE K, CHUAN-UDOM S, et al. Preparation of activated carbon from Dipterocarpus alatus fruit and its application for methylene blue adsorption[J]. RSC Adv., 2020, 10: 21082-21091. https://doi.org/10.1039/D0RA03427D.

[21] PAWLISZAK P, ZAWALA J, ULAGANATHAN V, et al. Interfacial characterisation for flotation: 2. Air-water interface[J]. Curr. Opin. Colloid Interface Sci. 2018, 37: 115-127. https://doi.org/10.1016/j.cocis.2018.07.002.

[22] PELALAK R, HEIDARI Z, SOLTANI H, et al. Mathematical model for numerical simulation of organic compound recovery using membrane separation[J]. Chem. Eng. Technol, 2018, 41: 345-352. https://doi.org/10.1002/ceat.201700445.

[23] PLAZA M G, GARCÍA S, RUBIERA F, et al. Post-combustion CO_2 capture with a commercial activated carbon: Comparison of different regeneration strategies[J]. Chem. Eng. J., 2010, 163: 41-47. https://doi.org/10.1016/j.cej.2010.07.030.

[24] SATO S, YOSHIHARA K, MORIYAMA K, et al. Influence of activated carbon surface acidity on adsorption of heavy metal ions and aromatics from aqueous solution[J]. Appl. Surf. Sci., 2007, 253: 8554-8559. https://doi.org/10.1016/j.apsusc.2007.04.025.

[25] SHI Y, YANG J, MA J, et al. Feasibility of bubble surface modification for natural organic matter removal from river water using dissolved air flotation[J]. Front. Environ. Sci. Eng., 2017, 11: 10. https://doi.org/10.1007/s11783-017-0954-2.

[26] WANG T, LIN G, GU L, et al. Role of organics on the purification process of zinc sulfate solution and inhibition mechanism[J]. Mater. Res. Express, 2019, 6: 106588. https://doi.org/10.1088/2053-1591/ab3dc5.

[27] WANG Y, SUN H, ANG H M, et al. 3D-hierarchically structured MnO_2 for catalytic oxidation of phenol solutions by activation of peroxymonosulfate: Structure dependence and mechanism[J]. Appl. Catal. B Environ., 2015, 164: 159-167. https://doi.org/10.1016/j.apcatb.2014.09.004.

[28] XU B, LU H, CAI W, et al. Synergistically enhanced oxygen reduction reaction composites of specific surface area and manganese valence controlled α-MnO_2 nanotube decorated by silver nanoparticles in Al-air batteries[J]. Electrochim. Acta, 2019, 305: 360-369. https://doi.org/10.1016/j.electacta.2019.

03.062.

[29] YANG R, WANG Z, DAI L, et al. WITHDRAWN: synthesis and characterization of single-crystalline nanorods of α-MnO_2 and γ-MnOOH[J]. Mater. Chem. Phys, 2005. https://doi.org/10.1016/j.matchemphys.2005.02.019.S0254058405001392.

[30] ZHANG T, WANG Q, LUAN W, et al. Zwitterionic monolayer grafted ceramic membrane with an antifouling performance for the efficient oil-water separation[J]. Chin. J. Chem. Eng., 2022, 42: 227-235. https://doi.org/10.1016/j.cjche.2021.03.049.

[31] ZHANG Y C, QIAO T, HU X Y, et al. Simple hydrothermal preparation of −MnOOH nanowires and their low-temperature thermal conversion to β-MnO_2 nanowires[J]. J. Cryst. Growth, 2005, 280: 652-657. https://doi.org/10.1016/j.jcrysgro.2005.04.017.

Thermodynamic Analysis and Process Optimization of Zinc and Lead Recovery from Copper Smelting Slag with Chlorination Roasting

Abstract: An efficient chlorination roasting process for recovering zinc (Zn) and lead (Pb) from copper smelting slag was proposed. Thermodynamic models were established, illustrating that Zn and Pb in copper smelting slag can be efficiently recycled during the chlorination roasting process. By decreasing the partial pressure of the gaseous products, chlorination was promoted. The Box-Behnken design was applied to assessing the interactive effects of the process variables and optimizing the chlorination roasting process. $CaCl_2$ dosage and roasting temperature and time were used as variables, and metal recovery efficiencies were used as responses. When the roasting temperature was 1172 ℃ with a $CaCl_2$ addition amount of 30% and a roasting time of 100 min, the predicted optimal recovery efficiencies of Zn and Pb were 87.85% and 99.26%, respectively, and the results were validated by experiments under the same conditions. The residual Zn- and Pb-containing phases in the roasting slags were $ZnFe_2O_4$, Zn_2SiO_4, and PbS.

Keywords: Chlorination roasting; Copper smelting slags; Thermodynamic models; Optimization; Zn and Pb recovery; Box-Behnken design

1 Introduction

As critical nonferrous metals, zinc (Zn) and lead (Pb) are widely used in machinery manufacturing, battery production, and galvanization[1]. In China, most smelters currently produce Zn and Pb with Pb-Zn symbiotic ore as the raw material by direct reduction and fuming processes[2]. After years of mining, the gradual depletion of high-grade Pb-Zn ore has forced smelters to recycle metallurgical secondary resources[3]. Moreover, the energy consumption of

---

Published in *Transactions of Nonferrous Metals Society of China*, 2021, 31(12): 3905−3917. Authors: Zhang Beikai, Guo Xueyi, Wang Qinmeng, Tian Qinghua.

extracting metals from secondary resources is lower than that of extracting metals from primary resources[4].

Currently, the research in this field is focused on the comprehensive utilization of metallurgical slag[5, 6]. HU et al.[7] developed an integrated process for the recovery of Zn from Zn hydrometallurgy residue by roasting and water leaching in sequence. The results indicated that, under optimal conditions, the Zn recovery efficiency reached 90.9% and the leaching liquor could be directly returned to the Zn smelting. Xin et al.[8] studied the recovery of Zn, Pb, and Sb from antimony smelting slag using a fuming furnace. The results showed that Zn, Pb, and Sb in antimony smelting slag could be recovered by the fuming furnace process, with metal recoveries of 88%, 95%, and 60%, respectively. Peng et al.[9] presented a beneficiation-metallurgy combination process to recover Zn and Fe and enrich In and Ag from high-iron-bearing Zn calcine. The results demonstrated that 90% of Zn was extracted, and 83% of Fe was recovered during the process. Approximately 86% of In entered the iron concentration, whereas Ag mainly entered the tailings, with a recovery efficiency of 76%.

Copper smelting slag contains 1%-4% Zn and 0.2%-0.6% Pb[10]. However, it is difficult to recover Zn and Pb from copper smelting slag by traditional metallurgical processes because the slag has a stable structure and complicated multi-component system[11, 12]. Chlorination roasting is a mature metallurgical method that can effectively separate and recover valuable metals from the smelting slag[13]. A pyrometallurgical process allows certain raw material components to react with the chlorinating agent to generate volatile chlorides at temperatures lower than their melting points[14]. Chlorinating agents generally include solid chlorinating agents ($CaCl_2$, NaCl, $MgCl_2$, NH_4Cl, etc.) and gaseous chlorinating agents (Cl_2, HCl, etc.)[15]. Although the gaseous chlorinating agent has a better chlorination effect, it is more likely to corrode the production pipeline[16]. Therefore, a solid chlorinating agent is preferable for use in chlorination roasting. Considerable research efforts have been devoted to the recovery of valuable metals from metallurgical slag by chlorination roasting. Wang et al.[17] presented a novel process based on one-step chlorination roasting to simultaneously extract gold and Zn from refractory carbonaceous gold ore using NaCl as a chlorination agent. The results showed that the recovery efficiencies of gold and Zn were 92% and 92.56%, respectively, under optimal conditions. Qin et al.[18] investigated the effect of chlorination roasting with pyrite on gold and silver recovery efficiencies from gold tailings. The results demonstrated that the addition of pyrite during the chlorination process promoted the recovery of gold and silver from gold tailings, and 98.56% of gold and 87.92% of silver were recovered under optimal conditions. Compared with other solid chlorinating agents, $CaCl_2$ is an ideal chlorinating agent because of its excellent removal performance on heavy metals at low cost[19]. During chlorination roasting, Ca from $CaCl_2$ is transformed into CaO, and the roasting slag can be further processed to prepare cement[20]. Consequently, $CaCl_2$ was used as the chlorinating agent in the present study.

Herein, an optimized chlorination roasting method was presented to recover Zn and Pb simultaneously from copper smelting slag, and the roasting mechanism was analyzed in detail. The

Box-Behnken design is an efficient approach for fitting second-order polynomials to response surfaces and uses relatively small numbers of observations to estimate the parameters[21]. In the present study, the chlorination roasting process was modeled using the Box-Behnken design to obtain optimum conditions for recovering Zn and Pb and investigate the interaction among the critical process parameters such as $CaCl_2$ dosage and roasting temperature and time.

2 Experimental

2.1 Materials

The raw material used in the present study was copper smelting slag obtained from a copper smelter in Shandong Province, China, after dilution using a flotation method. The elemental content of the copper smelting slag was analyzed using an inductively coupled plasma emission spectrometer (ICP-OES), and the results are given in Table 1. The Zn and Pb contents were 2.80% and 0.45%, respectively. The phase composition of the copper smelting slag was determined using X-ray diffraction (XRD). The XRD pattern in Fig. 1(a) indicates that fayalite (Fe_2SiO_4), magnetite (Fe_3O_4), and zinc ferrite ($ZnFe_2O_4$) were the significant minerals present in the copper smelting slag. Because of the complex composition and structure of copper slags, the minor minerals are encapsulated in the significant minerals; therefore, they cannot be accurately characterized by XRD[22]. Chemical phase analysis is based on the difference in the solubility and dissolution rate of various minerals in chemical solvents, by using a selective dissolution method to accurately identify the mineral composition of raw materials[23]. Zn and Pb occurrences in the different phases of the copper smelting slag are shown in Tables 2 and 3, respectively. The Zn in the copper smelting slag was mainly Zn_2SiO_4, $ZnFe_2O_4$, ZnO, and ZnS, whereas only 0.36% of Zn was present as $ZnSO_4$. The Pb in the copper smelting slag was mainly PbS, $PbSiO_3$, $PbSO_4$, and PbO, whereas only 2.08% of Pb was present as metallic Pb. From the energy dispersive spectroscopy (EDS) mapping of the copper smelting slag shown in Fig. 1(b), the Zn in the slag was encapsulated in fayalite and iron oxides, and the Pb in the slag was mainly combined with S, which was consistent with the afore-mentioned analysis results.

Table 1 Chemical composition of copper smelting slag %

Zn	Pb	Cu	Fe	O
2.80	0.45	0.55	41.23	36.10
Si	Ca	Na	K	S
11.75	1.76	1.36	0.58	0.41

Table 2 Occurrence of Zn in different phases of copper smelting slag

Phase	Content/(kg·t^{-1})	Distribution/%
Zn_2SiO_4	13.1	46.74
$ZnFe_2O_4$	9.75	34.79
ZnO	3.15	11.23
ZnS	1.93	6.88
$ZnSO_4$	0.1	0.36
Total	28.03	100

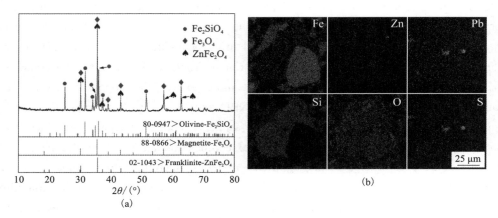

Fig. 1 XRD pattern (a) and EDS mappings (b) of copper smelting slag

Table 3 Occurrence of Pb in different phases of copper smelting slag

Phase	Content/(kg·t^{-1})	Distribution/%
PbS	1.78	39.58
$PbSiO_3$	1.41	31.25
$PbSO_4$	0.66	14.59
PbO	0.56	12.5
Pb	0.094	2.08
Total	4.5	100

2.2 Experimental design

The Box-Behnken design is a response surface methodology suitable for experiments with limited factors[24]. In the present study, the Box-Behnken design was employed to evaluate the influence of the CaCl₂ dosage and the roasting temperature and time on the metal recovery efficiency. Thus, the optimum operating conditions for recycling were determined. The $CaCl_2$ dosage and roasting temperature and time are critical parameters during the chlorination roasting process. Specifically, without sufficient chlorinating agents during the roasting process, the Zn and Pb encapsulated in significant minerals cannot be converted into chlorides. The thermodynamic calculations show that the chlorination reaction of Zn and Pb was restricted at low temperatures. It is necessary to provide sufficient roasting time to ensure an adequate reaction. The chlorination reaction process might be adversely affected by excessive experimental parameters. For instance, the materials can be melted under an excessively high roasting temperature, resulting in the solidification of heavy metals, which can be converted into oxides during the long chlorination roasting time, and the cost can be increased using excessive chlorinating agent[25, 26].

Three levels for each parameter were chosen: $CaCl_2$ dosage (10%, 20%, and 30%), roasting temperature (700, 950, and 1200 ℃), and roasting time (30 min, 60 min, and 90 min). The Zn and Pb recovery efficiencies were chosen as the response variables. The variables and their levels are presented in Table 4. The polynomial equation modeling of the response variable as a function of the operating parameters studied was expressed as follows[27]:

$$Y = \beta_0 + \sum_{i=1}^{n} \beta_i x_i + \sum_{i=1}^{n} \beta_{ii} x_i^2 + \sum_{i<j}^{n} \beta_{ij} x_i x_j + k \quad (1)$$

where Y is the predicted response (Zn and Pb recovery efficiencies), n is the number of the patterns, i and j are the index numbers for the patterns, β_0 is the constant coefficient, β_i is the linear coefficient, β_{ii} is the quadratic coefficient, β_{ij} is the interaction coefficient, x_i and x_j are the coded independent variables, and k is the random error.

Table 4 Independent variables and their levels used for response surface methodology

Parameter	Code	Level		
		−1	0	1
$w(CaCl_2)/\%$	x_1	10	20	30
Temperature/℃	x_2	700	950	1200
Time/min	x_3	30	60	90

2.3 Experimental set-up and procedures

All Box-Behnken design experiments were conducted in a high-temperature tube furnace. A 20 g sample of dried copper smelting slag was ground and mixed with different $CaCl_2$ dosages, loaded in a corundum porcelain boat, and placed in a tube furnace. The temperature of the furnace was heated to a preset temperature at a heating rate of 10 ℃/min, and air was then introduced at a flow rate of 100 mL/min. The raw materials were entirely roasted by holding them for a specified duration. Eventually, the roasting slags were recovered and ground into powder. The air was cut off when the furnace was cooled to room temperature (25-30 ℃). During the experiment, the volatilized flue gas was gathered through a gas scrubber containing 200 mL of 10% NaOH at the tube furnace outlet. The Zn and Pb recovery efficiencies are presented in Eq. (2):

$$R = \frac{w_g \cdot m_g - w_s \cdot m_s}{w_g \cdot m_g} \tag{2}$$

where R is the recovery efficiency (%), w_g is the mass fraction of an element in the copper smelting slag, m_g is the mass of the copper smelting slag in the porcelain boat, w_s is the mass fraction of an element in the roasting slag, and m_s is the mass of the roasting slag.

2.4 Characterization and analyses

The experimental design and analyses were performed using Design-Expert 12. A high-temperature smelting quartz tube furnace (Shanghai Jujing, China, SGL-1700C) was used as the experimental equipment, and a corundum porcelain boat (120 mm in length, 60 mm in width, and 20 mm in height) was used as the material container in the present study. Chemical analysis of the experimental materials and roasting slags was performed using an ICP-OES (Optima 7300 V, Perkin Elmer, USA). The occurrence states of Zn and Pb in the copper smelting slag were determined by chemical phase analysis (China Changsha Institute of Mining and Metallurgy). The morphological characteristics of the roasting slags were determined using a scanning electron microscope (SEM) equipped with EDS. An XRD analyzer (D8 Discover 2500) using a PANalytical X'Pert X-ray diffractometer (Cu K_α radiation) was used. FactSage™ 7.1 software was used to establish the detailed thermodynamic modeling.

3 Results and discussion

3.1 Volatilization principle and thermodynamic analysis

To determine the feasibility of recovering Pb and Zn by chlorination roasting, the saturated vapor pressures of their chlorides, sulfides, and oxides must be calculated. The variation in the saturated vapor pressure as the temperature changes can be determined using the Antoine equation[28] [Eqs. (3)-(7)]. The results are shown in Fig. 2 (the relevant calculation data for ZnO are unknown; however, it can be confirmed from the literature that ZnO is not volatile[29]). The saturated vapor pressure of chlorides was higher than that of sulfides and oxides. Therefore, Zn and Pb in the copper smelting slag could be separated and recovered during the chlorination roasting process.

$$\lg P_{PbCl_2} = 10000T^{-1} - 6.65\lg T + 33.52 \tag{3}$$

$$\lg P_{ZnCl_2} = 8415T^{-1} - 5.035\lg T + 28.545 \tag{4}$$

$$\lg P_{PbS} = -11597T^{-1} + 12.57 \tag{5}$$

$$\lg P_{PbO} = -13300T^{-1} - 0.81\lg T - 0.00043T + 16.93 \quad (6)$$

$$\lg P_{ZnS} = -13846T^{-1} + 12.7 \quad (7)$$

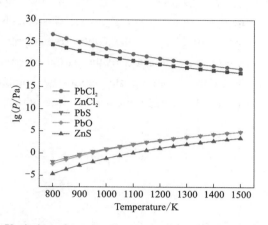

Fig. 2 Variation of saturated vapor pressure versus temperature

A thermodynamic reaction model of chlorination roasting was developed to understand the reaction process[30]. Combined with the preceding process mineralogy analysis of copper smelting slag, when using $CaCl_2$ as the chlorine donor, possible reactions in chlorination roasting process are shown in Eqs. (8)-(15). Details of thermodynamic modeling were undertaken using FactSage™ 7.1 software[31].

$$Zn_2SiO_4 + 2CaCl_2 + SiO_2 \rightleftharpoons 2ZnCl_2(g) + 2CaSiO_3 \quad (8)$$

$$ZnFe_2O_4 + CaCl_2 + SiO_2 \rightleftharpoons ZnCl_2(g) + CaSiO_3 + Fe_2O_3 \quad (9)$$

$$ZnS + CaCl_2 \rightleftharpoons ZnCl_2(g) + CaS \quad (10)$$

$$ZnO + CaCl_2 + SiO_2 \rightleftharpoons ZnCl_2(g) + CaSiO_3 \quad (11)$$

$$PbSiO_3 + CaCl_2 \rightleftharpoons PbCl_2(g) + CaSiO_3 \quad (12)$$

$$PbS + CaCl_2 \rightleftharpoons PbCl_2(g) + CaS \quad (13)$$

$$PbO + CaCl_2 + SiO_2 \rightleftharpoons PbCl_2(g) + CaSiO_3 \quad (14)$$

$$PbSO_4 + CaCl_2 \rightleftharpoons PbCl_2(g) + CaSO_4 \quad (15)$$

The conventional thermodynamic model is based on the constant partial pressure of gaseous products[32]. However, according to Eqs. (16) and (17), the Gibbs free energy change of the reaction in Eqs. (8)-(15) can be changed by controlling the partial pressure of gaseous products[33]:

$$\Delta G^\ominus = -RT\ln K^\ominus = -RT\ln \frac{P_{ZnCl_2}}{P^0} \quad (16)$$

$$\Delta G^\ominus = -RT\ln K^\ominus = -RT\ln \frac{P_{PbCl_2}}{P^0} \quad (17)$$

During the chlorination roasting process in the experiment, the air was blown into the experimental furnace to make the gaseous products ($ZnCl_2$ and $PbCl_2$) leave the reaction zone with the flowing gas, thereby decreasing the partial pressure of the gaseous products in the system and facilitating the positive direction of the reaction in Eqs. (8)-(15). If the partial pressures of the gaseous products were 1, 1/2, 1/10, 1/100, 1/1000, and 1/10000 of the total pressure, the Gibbs free energy change of the chlorination roasting process was calculated and is shown in Figs. 3(a-h). At a roasting temperature of 850 ℃, the Gibbs free energy changes (ΔG) of these reactions were all positive. Therefore, as the effect of the partial pressure of gaseous products was considered, Zn and Pb in the copper smelting slag could be completely recovered from the chlorination roasting process[34]. In all the experiments, air was pumped at a constant rate (100 mL/min) into the furnace.

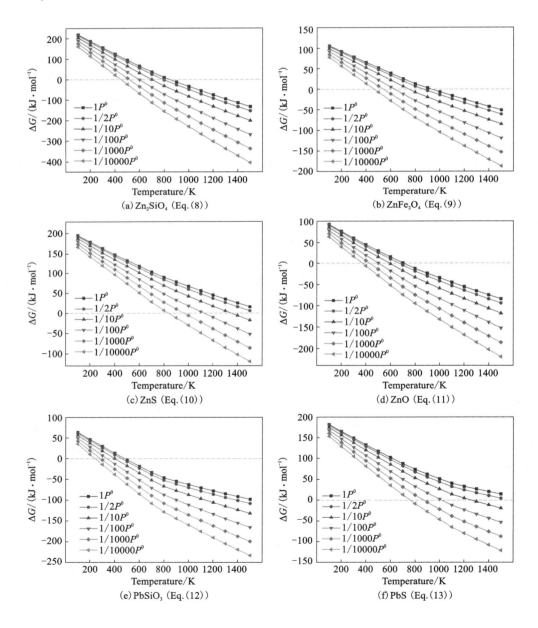

(a) Zn_2SiO_4 (Eq.(8)) (b) $ZnFe_2O_4$ (Eq.(9))
(c) ZnS (Eq.(10)) (d) ZnO (Eq.(11))
(e) $PbSiO_3$ (Eq.(12)) (f) PbS (Eq.(13))

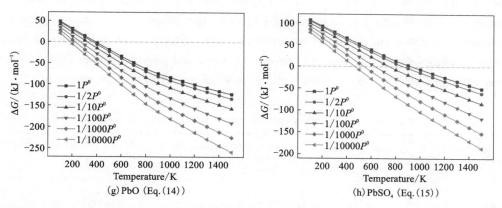

Fig. 3 Thermodynamic modeling of chlorination roasting at different partial pressures of gaseous products

3.2 Statistical analysis

3.2.1 Model selection

Mathematical polynomial models can be divided into first-order, two-factor interactions, quadratic, cubic, and so on[35]. To evaluate the reliability of experimental results and the credibility of mathematical model, the experimental results must be tested for significance[36]. The model statistics are listed in Table 5. The adjusted R^2 values of the quadratic model corresponding to Zn and Pb were 0.9818 and 0.9972, respectively, indicating that 98.18% and 99.72% of the response value changes of the recovery ratios for Zn and Pb were attributed to the three factors and the model fit well with the actual situation. The value of $R^2_{\text{adj(Zn)}} = 0.9583$ was close to $R^2_{\text{(Zn)}} = 0.9818$ in the quadratic model of the recovery ratio for Zn, whereas the value of $R^2_{\text{adj(Pb)}} = 0.9936$ was close to $R^2_{\text{(Pb)}} = 0.9972$ in the quadratic model of the recovery ratio for Pb, indicating the high significance of the models[37]. Therefore, recovery ratio models are recommended for use in the quadratic model.

Table 5 Analysis of different response surface models

Metal	Source	Standard deviation	R-squared	Adjusted R-squared
Zn	Linear	11.82	0.8255	0.7852
	2FI	13.34	0.8288	0.7261
	Quadratic	5.20	0.9818	0.9583
Pb	Linear	14.47	0.6206	0.5331
	2FI	16.40	0.6252	0.4003
	Quadratic	1.69	0.9972	0.9936

Table 6 gives the experimental data design matrix and the responses proposed by the Box-

Behnken design with three factors at three levels. The plots in Fig. 4 depict the correlation between the predicted values obtained from Eq. (1), and the actual values of the experimental observations. The actual and predicted values were evenly distributed on both sides of the linear regression, indicating a good agreement between the actual and predicted values for the Zn and Pb recovery efficiencies. Thus, the quadratic model could accurately predict the test results[38].

Table 6 Box-Behnken experimental design matrix and results

No.	$w(CaCl_2)$ /%	Temperature /℃	Time /min	Zn recovery efficiency/%		Pb recovery efficiency/%	
				Observed	Predicted	Observed	Predicted
1	10	700	60	1.77	−4.26	42.50	42.29
2	30	700	60	29.04	27.24	53.46	51.85
3	10	1200	60	45.37	47.17	88.05	89.66
4	30	1200	60	80.18	86.21	96.85	97.06
5	10	950	30	38.35	38.76	91.25	89.49
6	30	950	30	73.65	69.82	99.41	99.05
7	10	950	90	31.46	35.29	93.80	94.16
8	30	950	90	75.18	74.77	99.81	101.57
9	20	700	30	15.41	21.04	44.43	46.40
10	20	1200	30	80.12	77.92	98.04	98.19
11	20	700	90	21.25	23.45	55.66	55.51
12	20	1200	90	82.61	76.98	98.24	96.27
13	20	950	60	67.76	65.38	98.33	97.90
14	20	950	60	64.59	65.38	97.59	97.90
15	20	950	60	64.36	65.38	97.59	97.90
16	20	950	60	64.97	65.38	97.79	97.90
17	20	950	60	65.20	65.38	98.19	97.90

3.2.2 Analysis of variance (ANOVA)

The ANOVA results for the regression model are given in Table 7. When the P value was lower than 0.05, the model and chosen variable were remarkable[39]. As seen in Table 7, the P values of the recovery efficiency models were lower than 0.0001, indicating that the credibility of the model was high. In the quadratic model of Zn, x_1, x_2, x_1^2, and x_2^2 were significant model items because their P values were lower than 0.05, whereas other linear, quadratic, and interactive terms were insignificant. Similarly, x_1, x_2, x_3, x_2x_3, x_1^2, and x_2^2 were significant model items in the quadratic model of Pb. The higher the F values are, the more significant the influence of the variables is[36]; therefore, the notable degrees for influencing the Zn recovery efficiency were the

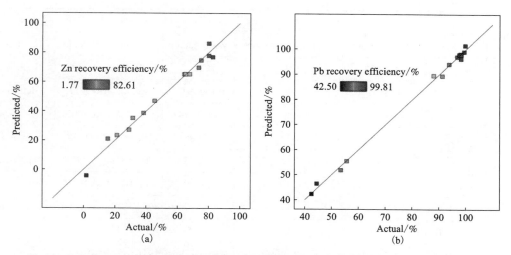

Fig. 4 Relationship between predicted values and actual values defined for experimental region of Zn recovery efficiency (a) and Pb recovery efficiency (b)

$CaCl_2$ dosage and roasting temperature. All independent variables were notable for influencing the Pb recovery efficiency. In addition, the coefficients of variation for Zn and Pb were 9.81% and 1.98%, respectively. The adequate precisions for Zn and Pb were 22.6717 and 45.794, respectively, confirming that the models can be used to predict Zn and Pb recovery efficiencies.

Table 7 Analysis of variance for response surface quadratic model

Metal	Source	Analysis					
		Sum of squares	Degree of freedom	Mean square	F value	P value	Conclusion
Zn	Model	10209.68	9	1134.41	41.90	<0.0001	Significant
	x_1	2488.65	1	2488.65	91.92	<0.0001	Significant
	x_2	6094.63	1	6094.63	225.10	<0.0001	Significant
	x_1^2	484.38	1	484.38	17.93	0.0039	Significant
	x_2^2	1018.02	1	1018.02	37.60	0.0005	Significant
	Residual	189.52	7	27.07			
	C.V.: 9.81%			Adequate precision: 22.6717			

Continue the Table 7

Metal	Source	Analysis					
		Sum of squares	Degree of freedom	Mean square	F value	P value	Conclusion
Pb	Model	7156.55	9	795.17	279.17	<0.0001	Significant
	x_1	143.91	1	143.91	50.52	0.0002	Significant
	x_2	4284.14	1	4284.14	1504.08	<0.0001	Significant
	x_3	25.85	1	25.85	9.07	0.0196	Significant
	$x_2 x_3$	30.42	1	30.42	10.68	0.0137	Significant
	x_1^2	34.30	1	34.30	12.04	0.0104	Significant
	x_2^2	2595.70	1	2595.70	911.30	<0.0001	Significant
	Residual	19.94	7	2.85			
	C.V.: 1.98%			Adequate precision: 45.7940			

3.2.3 Effect of roasting parameters and interactions

Based on the significant regression terms and their corresponding estimated coefficients, the regression equations for the metal recovery efficiency for each coded factor are given in Eqs. (18) and (19), describing the functional relationship between the response values and impact factors:

$$Y_{Zn} = 65.38 + 17.64x_1 + 27.6x_2 + 0.3712x_3 + 1.88x_1x_2 + 2.11x_1x_3 - 0.8375x_2x_3 - 10.74x_1^2 - 15.55x_2^2 + 0.0208x_3^2 \quad (18)$$

$$Y_{Pb} = 97.9 + 4.24x_1 + 23.14x_2 + 1.8x_3 - 0.54x_1x_2 - 0.5375x_1x_3 - 2.76x_2x_3 - 2.85x_1^2 - 24.83x_2^2 + 1.02x_3^2 \quad (19)$$

Based on the quadratic response functions, 3D surface plots and contour plots were used to illustrate the interaction of factors on response[40]. Figs. 5(a-c) and 5(d-f) show the response surfaces for Zn and Pb recovery, respectively. As shown in Figs. 5(a, d), the $CaCl_2$ dosage and roasting temperature positively affected Zn and Pb recovery efficiencies with a roasting time of 60 min. Increasing the $CaCl_2$ dosage resulted in a gradual increase in metal recovery, especially for Zn. Similarly, the roasting temperature is an important parameter for improving metal recovery efficiency. However, when the temperature exceeded 1041 ℃, increasing the roasting temperature reduced the Pb recovery efficiency. The roasting slags obtained at 700, 950, 1000, and 1050 ℃ were further analyzed using SEM, as shown in Fig. 6. At 700-1000 ℃, the mineral surface gradually dissociated with increasing temperature, owing to the destruction of the frame of the copper smelting slag at high temperatures. At 700-1000 ℃, the metal recovery efficiency increased with increasing temperature, attributed to the further exposed metals enhancing the mass transfer process of the chlorination reaction[18]. When the roasting temperature was increased to 1050 ℃, the slag became dense, caused by sintering at high temperatures; therefore, the porosity of the mineral was reduced, hindering the chlorination reaction[41]. Thermodynamically, the chlorination reaction could be enhanced by increasing the temperature. From the response surface methodology results, at 700-1200 ℃, the Zn recovery efficiency increased with increasing temperature, even if the mineral was sintered. Owing to the high Zn content in minerals, the adverse effect of sintering was lower than the favorable influence of thermodynamics. Conversely, at 1041-1200 ℃, the Pb recovery efficiency decreased with increasing temperature because the sintering of the mineral hindered the chlorination reaction.

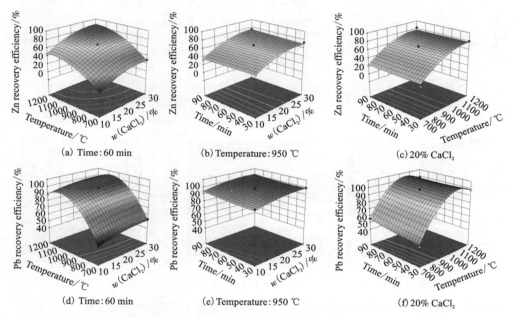

Fig. 5 Surface 3D plots for Zn recovery efficiency (a-c) and Pb recovery efficiency (d-f)

Fig. 6 SEM images of roasting slag roasted at different temperatures

As shown in Figs. 5(b, e), an increase in the roasting time increased Zn and Pb recovery efficiencies; therefore, extending the roasting time promoted the extent of the chlorination reaction and increased the metal recovery efficiency. Figs. 5(c, f) confirmed that the temperature effect on the recovery rates of Zn and Pb was slightly different. In the range of 700-1200 ℃, the Zn recovery efficiency increased as the temperature increased. For the Pb recovery efficiency, when the temperature was higher than 1079 ℃, the Pb recovery gradually decreased as the temperature increased, and the reason was the same as above.

Meanwhile, Figs. 7(a-c) and 7(d-f) show the contour plots for Zn and Pb recovery, respectively. The contour shapes of the roasting temperature and $CaCl_2$ dosage were close to ellipses in Figs. 6(a, d), demonstrating that their interaction was significant. The contour shapes of the other factors were similar to straight lines, implying that their interaction was small[42].

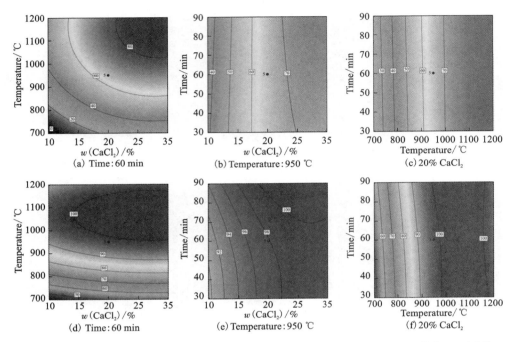

Fig. 7 Contour 2D plots for Zn recovery efficiency (a-c) and Pb recovery efficiency (d-f)

3.2.4 Optimization

The optimum roasting conditions and the predicted recovery efficiencies were determined using Design-Expert12, as given in Table 8. Three experiments were conducted under optimal conditions to verify the optimized results. The average Zn and Pb recovery efficiencies obtained under optimized conditions were 87.85% and 99.26%, respectively. The results showed that the predicted values were in good agreement with the observed data, indicating that the model was statistically significant.

Table 8 Optimum roasting conditions and results

$w(CaCl_2)$ /%	Temperature /℃	Time /min	Zn recovery efficiency/%		Pb recovery efficiency/%	
			Observed	Predicted	Observed	Predicted
30	1172	100	87.85	88.57	99.26	99.99

Due to the small amount of Zn and Pb components in the roasting slag, Zn- and Pb-containing phases could not be accurately detected by XRD. Therefore, SEM-EDS was utilized to analyze the micromorphology of the roasting slags when roasted at 1172 ℃ for 100 min, and

CaCl$_2$ dosage was 30%. The results are shown in Fig. 8(a) and Table 9. The mineral existed in three typical areas (gray, white, and black), and white areas were enclosed in dark parts. The EDS-point results showed that Fe, Si, and O mainly occurred in the dark parts, whereas Zn mainly occurred in the white areas. Therefore, the residual Zn-containing phase was a ternary (Fe-Zn-O) or quaternary compound (Fe-Zn-Si-O), and residual Pb-containing phase could not be accurately determined, because of the Pb traces in the slag.

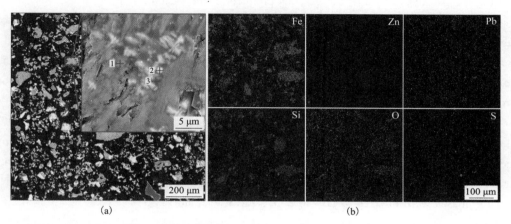

Fig. 8 SEM-EDS images of roasting slag under optimum roasting conditions (a) and EDS mapping of roasting slag under optimum roasting conditions (b)

Table 9 Analytical results of roasting slag under optimum roasting conditions   %

Point in Fig. 8(a)	Fe	Si	O	Zn	Pb	Ca	S
1	39.61	21.09	26.55	5.31	1.91	1.06	0.34
2	17.87	16.31	15.59	26.08	2.04	2.98	10.64
3	18.62	26.25	30.62	2.35	2.11	12.96	1.15

EDS mapping of the roasting slags under the optimum roasting conditions is shown in Fig. 8(b). The elemental distributions of Fe, Si, and O were dispersive and homogeneous. The residual Zn was almost encapsulated in these elements, and the elemental distributions of Pb and S were analogous. In conclusion, the residual Zn- and Pb-containing phases might be ZnFe$_2$O$_4$, Zn$_2$SiO$_4$, and PbS. The raw material compounds could not be thoroughly chlorinated, which was consistent with thermo-dynamic calculation results.

4 Conclusions

(1) The thermodynamic reaction model of chlorination roasting indicated that Zn- and Pb-containing phases in copper smelting slag could be efficiently separated and recovered by decreasing the partial pressure of the gaseous products in the system.

(2) The Box-Behnken design was applied to assessing the interactive effects of the process variables and optimizing the chlorination roasting process. The ANOVA results demonstrated that the regression equations were appropriate for optimizing the independent variables within the range. The 3D surface plot and contour plot results indicated a notable influence of $CaCl_2$ dosage and roasting temperature on Zn and Pb recovery.

(3) The optimal chlorination roasting conditions for recovering Zn and Pb were 30% of $CaCl_2$ at 1172 ℃ for 100 min. Under these conditions, the predicted Zn and Pb recovery efficiencies reached 87.85% and 99.26%, respectively, which matched well with the actual observed values. After the optimal chlorination roasting process, the residual Zn- and Pb-containing phases were inferred to be $ZnFe_2O_4$, Zn_2SiO_4, and PbS.

Acknowledgments

The authors are grateful for the financial supports from the National Natural Science Foundation of China (Nos. 51620105013, 51904351), Innovation-Driven Project of Central South University, China (No. 2020CX028), Natural Science Fund for Distinguished Young Scholar of Hunan Province, China (No. 2019JJ20031), and the National Key R&D Program of China (No. 2019YFC1907400).

References

[1] PENG R Q. Metallurgy of lead and zinc [M]. Beijing: Science Press, 2003. (in Chinese)
[2] LI W F, ZHAN J, FAN Y F, et al. Research and industrial application of a process for direct reduction of molten high-lead smelting slag[J]. JOM, 2017, 69(4): 784-789.
[3] PAN D A, LI L L, TIAN X, et al. A review on lead slag generation, characteristics, and utilization[J]. Resources Conservation & Recycling, 2019, 146: 140-155.
[4] GUO X Y, TIAN Q H, LIU Y, et al. Progress in research and application of non-ferrous metal resources recycling[J]. The Chinese Journal of Nonferrous Metals, 2019, 29(9): 65-107. (in Chinese)
[5] ZHANG X U, YANG L S, LI Y H, et al. Estimation of lead and zinc emissions from mineral exploitation based on characteristics of lead/zinc deposits in China[J]. Transactions of Nonferrous Metals Society of

China, 2011, 21(11): 2513-2519.
[6] YANG S J, ZHANG L W, YU D H. Intensive development and comprehensive utilization of metallurgical slag[J]. Applied Mechanics & Materials, 2012, 174-177: 1424-1428.
[7] HU M, PENG B, CHAI L Y, et al. High-zinc recovery from residues by sulfate roasting and water leaching [J]. JOM, 2015, 67(9): 2005-2012.
[8] XIN P F, WEI J Y, XU L, et al. The recovery of Pb and Zn in antimony smelting slag [M]. PbZn 2020: 9th International Symposium on Lead and Zinc Processing. San Diago, USA: Springer-TMS, 2020.
[9] PENG B, PENG N, LIU H, et al. Comprehensive recovery of Fe, Zn, Ag and In from high iron-bearing zinc calcine[J]. Journal of Central South University, 2017, 24: 1082-1089.
[10] ZHANG J L, YANG X, ZHANG J K, et al. Influence of slag contents on sedimentation separation of slag and matte at high temperature[J]. The Chinese Journal of Nonferrous Metals, 2019, 29(8): 1712-1720. (in Chinese)
[11] WANG H Y, SONG S X. Separation of silicon and iron in copper slag by carbothermic reduction-alkaline leaching process[J]. Journal of Central South University, 2020, 27(8): 2249-2258.
[12] SUN W, SU J F, ZHANG G, et al. Separation of sulfide lead-zinc-silver ore under low alkalinity condition [J]. Journal of Central South University, 2020, 27(8): 2249-2258.
[13] LI G S, ZOU X L, CHENG H W, et al. A novel ammonium chloride roasting approach for the high-efficiency Co-sulfation of nickel, cobalt, and copper in polymetallic sulfide minerals[J]. Metallurgical and Materials Transactions B, 2020, 21: 2769-2784.
[14] DING J, HAN P W, LU C C, et al. Utilization of gold-bearing and iron-rich pyrite cinder via a chlorination-volatilization process[J]. International Journal of Minerals, Metallurgy, and Materials, 2017, 24(11): 1241-1250.
[15] MUKHERJEE T K, GUPTA C K. Base metal resource processing by chlorination[J]. Mineral Processing and Extractive Metallurgy Review, 1983, 1(1-2): 111-153.
[16] JENA P K, BROCCHI E A. Base metal extraction through chlorine metallurgy[J]. Mineral Processing and Extractive Metallurgy Review, 2008, 16(4): 211-237.
[17] WANG H J, FENG Y L, LI H R, et al. Simultaneous extraction of gold and zinc from refractory carbonaceous gold ore by chlorination roasting process[J]. Transactions of Nonferrous Metals Society of China, 2020, 30(4): 1111-1123.
[18] QIN H, GUO X Y, TIAN Q H, et al. Pyrite enhanced chlorination roasting and its efficacy in gold and silver recovery from gold tailing[J]. Separation and Purification Technology, 2020, 250: 117168.
[19] LI H Y, PENG J H, LONG H L, et al. Cleaner process: Efficacy of chlorine in the recycling of gold from gold containing tailings[J]. Journal of Cleaner Production, 2021, 287: 125066.
[20] ZHU M L, XIAO N, TANG L C, et al. Preparation of new cementitious material by reduction and activation of copper slag and its application in mine filling[J]. The Chinese Journal of Nonferrous Metals, 2020, 30(11): 2736-2745. (in Chinese)
[21] WHITTINGHILL D C. A note on the robustness of Box-Behnken designs to the unavailability of data[J]. Metrika, 1998, 48(1): 49-52.
[22] HAO X D, LIU X D, YANG Q, et al. Comparative study on bioleaching of two different types of low-grade copper tailings by mixed moderate thermophiles[J]. Transactions of Nonferrous Metals Society of China, 2018, 28(9): 1847-1853.
[23] DU K, WANG C Y, WANG L. Distribution of main elements and phase characteristics of copper converter

slag[J]. Journal of Central South University (Science and Technology), 2018, 49(11): 2649-2655. (in Chinese)

[24] FATMA U, FARUK K. Modelling of relation between synthesis parameters and average crystallite size of Yb₂O₃ nanoparticles using Box-Behnken design[J]. Ceramics International, 2020, 46(17): 26800-26808.

[25] ZHOU H, XU J N, ZHOU M X, et al. Study on immobilization of heavy metals in a waste SCR catalyst during high temperature melting treatment[J]. Journal of Chinese Society of Power Engineering, 2020, 40(6): 492-501.

[26] JAAFAR I, GRIFFITHS A J, HOPKINS A C, et al. An evaluation of chlorination for the removal of zinc from steelmaking dusts[J]. Minerals Engineering, 2011, 24(9): 1028-1030.

[27] MYERS R H, MONTGOMERY D C, ADWESON-COOK C M. Response surface methodology: Process and product optimization using designed experiments [M]. New York: John Wiley and Sons, Inc, 2008.

[28] DONG Z W, XIONG H, DENG Y, et al. Separation and enrichment of PbS and Sb₂S₃ from jamesonite by vacuum distillation[J]. Vacuum, 2015, 121: 48-55.

[29] DAI Y N, YANG B, MA W H, et al. Advances on vacuum metallurgy of nonferrous metals [J]. Engineering Sciences, 2004, 2(3): 12-19, 29.

[30] YANG X M, LI J Y, WEI M F, et al. Thermodynamic evaluation of reaction abilities of structural units in Fe-O binary melts based on the atom-molecule coexistence theory [J]. Metallurgical and Materials Transactions B, 2016, 47(1): 174-206.

[31] BALE C W, BÉLISLE E, CHARTRAND P, et al. FactSage thermochemical software and databases, 2010—2016[J]. Calphad, 2016, 54: 35-53.

[32] HUANG X H. Principles on ferrous metallurgy [M]. Beijing: Metallurgical Industry Press, 2013. (in Chinese)

[33] LI H G. Metallurgical principle [M]. Beijing: Science Press, 2005. (in Chinese)

[34] DONG Z W, XIA Y, GUO X Y, et al. Direct reduction of upgraded titania slag by magnesium for making low-oxygen containing titanium alloy hydride powder[J]. Powder Technology, 2020, 368: 160-169.

[35] MA Z Y, YANG H Y. Microwave assisted leaching of selenium from copper anode slime optimized by response surface methodology[J]. Journal of Central South University: Science and Technology, 2015, 46(7): 2391-2397. (in Chinese)

[36] HUO Q, LIU X, CHEN L J, et al. Treatment of backwater in bauxite flotation plant and optimization by using Box-Behnken design [J]. Transactions of Nonferrous Metals Society of China, 2019, 29(4): 821-830.

[37] NAIR A T, AHAMMED M M. Coagulant recovery from water treatment plant sludge and reuse in post-treatment of UASB reactor effluent treating municipal wastewater[J]. Environmental Science and Pollution Research, 2014, 21(17): 10407-10418.

[38] YANG S Y, LI Y, JIA D Y, et al. The synergy of Box-Behnken designs on the optimization of polysaccharide extraction from mulberry leaves[J]. Industrial Crops and Products, 2017, 99: 70-78.

[39] GHASEMI S, FARHADIZADEH A R, GHOMI H. Effect of frequency and pulse-on time of high power impulse magnetron sputtering on deposition rate and morphology of titanium nitride using response surface methodology[J]. Transactions of Nonferrous Metals Society of China, 2019, 29(12): 2577-2590.

[40] XU Z P, GUO X Y, LI D, et al. Optimization of tellurium and antimony extraction from residue generated in alkaline sulfide leaching of tellurium-bearing alkaline skimming slag using central composite design[J]. Mining Metallurgy and Exploration, 2020, 37: 493-505.

[41] ZHOU Y N, YAN D H, LI L, et al. Volatile characteristics of lead and zinc during co-processing hazardous waste in sintering machines[J]. Acta Scientiae Circumstantiae, 2015, 35(11): 3769-3774. (in Chinese)

[42] CIFUENTES B, FIGUEREDO M, COBO M. Response surface methodology and aspen plus integration for the simulation of the catalytic steam reforming of ethanol[J]. Catalysts, 2017, 7(1): 15-34.

A Method of High-quality Silica Preparation from Copper Smelting Slag

High-quality silica was prepared from copper smelting slag through a method of in situ modification. The effects of the addition of an amount of polyethylene glycol-6000 as a modifier, the modification temperature and the modified endpoint pH on the particle size and specific surface area of the silica were systematically studied. It has been shown that the particle size, specific surface area, and the interstices between the particles were greatly affected by the modification temperature and the pH of the modification endpoint. Optimal conditions are: modifier 10% as solute mass, modification temperature 40 ℃, and pH of modification endpoint 8.5. Under these conditions, the silicon sinking rate was as high as 97.82%, the prepared silica particles had good dispersibility, the average particle size was 20 nm, the particle morphology was spherical, and the specific surface area was as high as 244.67 m^2/g, which was superior to A-grade standard of HG/T 3061-1999 and ISO 5794-1: 2005(E), and could be directly used in the rubber industry.

1 Introduction

Copper smelting slag comes mainly from pyrometallurgical processes, such as bottom blowing furnace smelting, flash furnace smelting, the Noranda method, the ISASMELT process, electric furnaces, etc.[1-4] The amount of slag produced by smelting processes is very large, usually about twice the amount of the produced metals.[5-7] As a common method, the copper slag is crushed and screened, and then valuable metals are extracted by two flotation processes. The amount of extracted metals accounts for a small part of the total amount of slag, and the slag tailings yield is also huge.[8-10] Key components in the tailings are iron and silicon. The grade of iron is generally about 40%, much higher than the average grade of iron ore, and the grade of silicon is as high as about 20%, which is an ideal silicon source.[11] At present, the disposal of the tailings is mainly being dumped in an area, with only a small part being used in the cement concrete industry. Therefore, the silicon and iron elements in copper smelting slag have not been efficiently utilized and do not generate large economic values.[12,13]

After the tailings are mixed with a certain amount of sodium hydroxide and smelted at low temperature, the silicon in the tailings is enriched in the form of silicate which can be dissolved in water and then separated from the iron in the tailings. Thus, the leached liquid is an ideal source for preparing silica.

When silica is used as a reinforcing agent, the tensile and impact properties of the polymer can be obviously enhanced. At present, the preparation methods of silica include the gas phase method, the precipitation method, the microemulsion method, the sol-gel method, etc. The gas phase and precipitation methods are the two main methods in industrial production.[14,15] Because the gas phase method requires a high quality of raw materials and complicated equipment, it has the disadvantages of operation, high cost, difficult manipulation, low yield, etc.[16] In situ-modified precipitation preparation is a method in which the modifier and raw materials are prepared together and modified simultaneously in the process of producing silica (Fig. 1). Compared with the gas phase method, the precipitation method simplifies the process, shortens the reaction time and at the same time reduces the cost.[17-19]

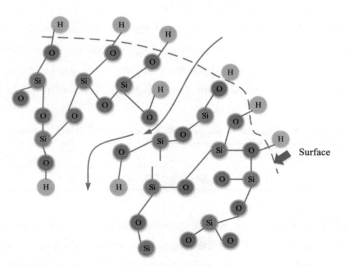

Fig. 1 Schematic of the preparation of silica by the precipitation method

The modifier used has the property of a high molecular weight and can form hydrogen bonds with vertex oxygen. It is adsorbed on the surface of colloid particles and wound to form a high molecular layer to wrap the reactant. It also controls the dehydration speed of the vertex O of the tetrahedron to prepare silica with high dispersibility and high specific surface area. Polyethylene glycol (PEG) is a nonionic neutral surface modifier,[20] and will change from zigzag to a meander spatial network structure when dissolved in water, thus forming steric effects to control the reaction process of silica. Since PEG has different relative molecular masses, it is also important to select the PEG with an appropriate relative molecular mass. According to the research of some scholars, PEG modification of silica has two effects: a network aggregation effect, and a steric effect.[21] When the relative molecular mass of PEG is less than 6000, its steric effect is stronger than its network aggregation effect, resulting in a smaller tendency of aggregation among the particles. When its relative molecular mass is higher than 6000, the trend of ion aggregation gradually increases due to the network aggregation effect being greater than the steric effect.[22]

The effect of the addition of an amount of PEG-6000 modifier, the modification temperature and the modification endpoint pH on the properties of silica have been explored in detail in our studies, and the more suitable process conditions obtained, which provided the basis for industrial application of a new silica preparation method.

2　Materials and methods

2.1　Raw material

The main components (%) of copper smelting slag are Cu 0.36, Fe 47.94, SiO_2 37.35, Zn 3.47, Al 2.25, CaO 1.64, and MgO 1.05, and the main phases in copper smelting slag are Fe_2SiO_4 and Fe_3O_4. The raw material used in our experiments is the leaching solution obtained from copper smelting slag provided by a domestic company after low-temperature alkaline smelting and water immersion separation. Its elemental composition is shown in Table 1.

Table 1　Element of copper tailings smelting lixivium (CME)　　g/L

Element	Fe	Si	Zn	Al	Ca	Mg	Cu
Contents	0	8.41	0.54	0.38	0	0	0

As can be seen from Table 1, Si is mainly present in the leachate. The content of the silicon element is 8.41 g/L, and the content of the amphoteric metals, Zn and Al, are lower, 0.54 g/L and 0.38 g/L, respectively. The Zn and Al ions in the solution can be removed through subsequent processes. The leachate was titrated with 1 mol/L hydrochloric acid with phenolphthalein as indicator. The alkalinity of the solution was 1.14 mol/L.

2.2　Characterization

A cold field-emission-corrected transmission electron microscope (TECNAI G2 60-300; USA) was used to analyze the morphology, particle size, dispersibility, and electron diffraction of the particles. A thermogravimetry (TG) and differential thermal analyzer (DTA) (SDTQ600; USA) was used to study the desorption, condensation and crystallization water of the silica samples. The particle size was measured by a nano-particle size and Zeta potential analyzer (Nano-zs; UK). The specific surface area and pore size analyzer (Monosorb Autosorb; Kontha; USA) was used to analyze the specific surface area of the sample. The aggregation degree of the silica prepared by precipitation will affect its performance, and the aggregation degree can be measured by the size of the space between the particles. The volume of DBP (dimethyl phthalate) absorbed by the silica was used to measure the size of the void, which is related to the specific surface area and particle size of the samples.[23]

2.3 Experimental principle

According to the principle that carbonic acid generated after carbon dioxide is dissolved in water as a weak acid, the pH reduction rate of the system can be effectively controlled by the reaction of the carbonic acid with sodium silicate, so that the agglomeration of silica particles is inhibited. The reaction formulae are shown in Eqs. (1) and (2).

$$Na_2O \cdot mSiO_2 + nH_2O + CO_2 \Longleftrightarrow Na_2CO_3 + mSiO_2 \cdot nH_2O \quad (1)$$

$$SiO_2 \cdot nH_2O \Longleftrightarrow SiO_2 \cdot n'H_2O + (n - n')H_2O \quad (2)$$

Various silicate ions in sodium silicate solution have their corresponding equilibrium, and changes in system conditions (temperature, solution pH value, solute concentration) will affect their equilibrium. Since sodium silicate can be regarded as a multi-tetrahedral structure with a vertex of —ONa, when acid is added to the system, —ONa will be rapidly converted into —OH to generate a particle resulting in a loss of surface activity and the reduction of specific surface area of the particles. By using a modifier with a high molecular weight, a high molecular weight layer is formed on the surface of the colloidal particles. This layer wraps the particles and contains certain electric charges, which push the particles away from each other due to the electrostatic force. At the same time, the addition of a modifier can react with active hydroxyl groups on the surface of the silica particles, thus reducing their surface hydrophilicity. By using the modifier with a high molecular weight and due to the property of hydrogen bonding with the vertex oxygen, the modifier is adsorbed on the surface of the colloidal particles and forms a polymer layer which wraps the reactant and controls the dehydration rate of tetrahedron vertex O, so that silica with high dispersion and high specific surface area is prepared.

3 Results and discussion

3.1 Impact of modifier amount on the properties of silica

PEG-6000 is an organic polymer with hydrophilic groups and lipophilic groups in its structure. The hydrophilic groups have polarity and the lipophilic groups have the opposite. When the PEG-6000 contacts particles with polarity, the lipophilic groups are adsorbed on the surface of the particles, while the hydrophilic groups are arranged on the outside to reduce their surface tension. The low surface tension allows organic molecules to penetrate between aggregated particles to wrap them and so separate them from each other. However, too much PEG-6000 will also increase the steric hindrance between the particles in the solution, weaken their diffusion ability, slow the reaction speed, or cause serious particle agglomeration in local areas. Therefore, it is particularly important to select an appropriate amount of modifier in the silica preparation process.

As can be seen from Fig. 2(a), (b), (c), (d), and (e), the particle size distribution of silica is relatively concentrated at 0% and 10%, dispersed at about 900 nm at 0% and at about 600 nm at 10%. The addition amounts are 5% and 15%, and the particle size distribution is more dispersed. When the addition amount is 20%, the particle size distribution is 150 nm and 400-1200 nm. As can be seen from Fig. 2(f), when the amount of PEG-6000 is increased, the average particle size of the silica tends to first increase and then to decrease. The results show that the addition of the modifier can effectively reduce the agglomeration of the silica particles when the amount exceeds 5%.

As can be seen from Fig. 2(g), with the increase of the amount of PEG-6000 added, the reaction time shows a trend of first decreasing and then increasing, reaching a minimum when the addition amount is 10%. When the addition amount is 0-5%, the effect of the amount of modifier on the reaction rate is not obvious, while, when the amount of modifier is within 5%-10%, the reaction time obviously decreases. It is presumed that the coating effect of dimethyl phthalate modifier enables the particles in the solution to quickly aggregate and react, thus improving the

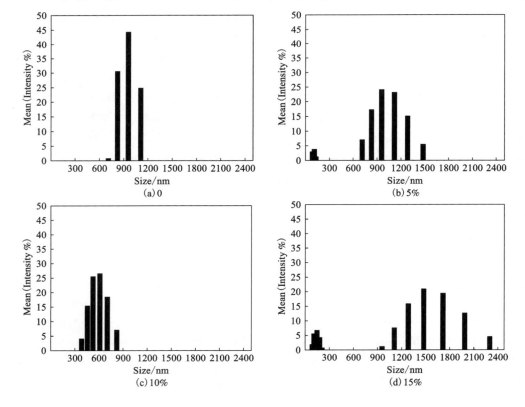

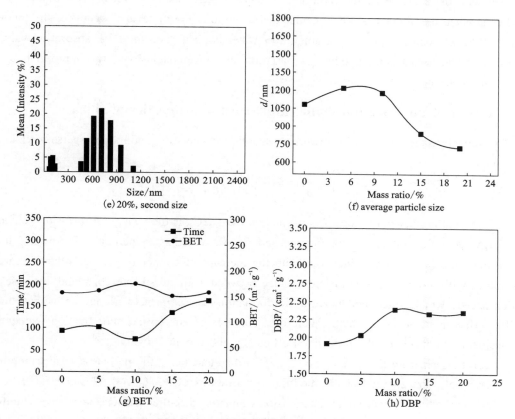

Fig. 2　The impact of the amount of modifier on silica

reaction rate. However, the reaction time obviously increases after the amount of modifier added exceeds 10%, which indicates that the increase in the amount of modifier leads to an increase in the viscosity of the solution, thus hindering the diffusion of dimethyl phthalate particles in the reaction and reducing the reaction rate. The Brunauer-Emmett-Teller (BET) amalysis of silica shows a trend of first increasing and then decreasing, reaching an extreme value with 10% addition.

The effect of PEG-6000 addition on silica preparation can be divided into two stages. Stage 1 is when the concentration of modifier is 0-10% and stage 2 is when the concentration is over 10%. With the addition of the modifier, the viscosity of the solution increases, resulting in an increase in steric effects and an increase in the effect of preventing the products from continuing to agglomerate and grow, thus increasing the specific surface area of the silica. However, when the modifier is more than 10%, the viscosity of the solution exceeds the critical value of silica modification. Due to the high steric hindrance and slow ionic reaction, the Si-O bond of silica is formed, resulting in the formation of dense silica particles on the surface of the sample and the decrease of the specific surface area of the silica.

As can be seen from Fig. 2(h), the DBP absorption value of silica first increases and then decreases with the increase of the modifier. It can be seen that the critical point of adding the modifier is when the amount is 10%. Under this condition, the molecular weight of the modifier bonded by Si-O bond is equal to the amount of water bonded, and the DBP absorption value of white carbon black reaches the maximum. Based on the above conclusions, it is appropriate to add 10% of the modifier.

3.2 Impact of modification temperature on the properties of silica

Theoretically, the relationship between preparation temperature and monodisperse particle size is as follows:[24]

$$J = J_0 \exp\left(\frac{-\Delta G_D}{KT}\right) \times \exp\left(\frac{-\Delta G^*}{KT}\right) \qquad (3)$$

J_0 is the initial nucleation rate, J is nucleation rate, ΔG_D is the change of diffusion activation free energy, K is the Boltzmann coefficient, and T is the reaction temperature. From this, it can be seen that the nucleation rate increases with the increase of temperature, i.e., the increase of temperature will promote the collision contact between sodium silicate ions and increase the reaction speed, which is also conducive to the formation of a larger network and an increase in the specific surface area of the product. Therefore, finding the appropriate modification temperature has considerable impacts on the production efficiency and product quality.

As can be seen from Fig. 3(a), (b), (c), (d), and (e), the particle size distribution of silica is relatively concentrated at the modifier temperatures of 30 ℃, 40 ℃ and 50 ℃, and is about 1200 nm, which indicates that the collision contact of the silica particles is relatively uniform under these three temperatures. When the modification temperature is at 20 ℃ and 80 ℃, the particle size distribution is relatively dispersed, which means that the collision nucleation of the silica particles is not uniform under these two temperature conditions. As can be seen from Fig. 3 (f), the particle size of the silica shows a trend of a rapid increase at first and then a slow decrease with the increase of the modification temperature, and the average particle size reaches its maximum at 50 ℃.

The reaction temperature in different ranges has different impacts on the particle size of the silica. In the low reaction temperature range, the increase of temperature intensifies the collision contact between the sodium silicate ions, which is conducive to the formation of larger network particles, thus increasing the particle size. In the high reaction temperature range, as the temperature increases, the solubility of CO_2 gradually decreases, resulting in a decrease in the concentration of carbonic acid that reacts with sodium silicate ions in the solution, thus weakening the network accumulation and reducing the particle size. It can be seen from Fig. 3 that, in the range of 20-50 ℃, the former predominates, so that the particle size of the silica increases rapidly. When the temperature exceeds 50 ℃, the latter predominates, resulting in the particle size of the silica slowly decreasing.

From Fig. 3(g), it can be seen that the BET value of the silica gradually decreases with the increase of temperature, with the extreme value being 244.67 m²/g at 20 ℃, which indicates that

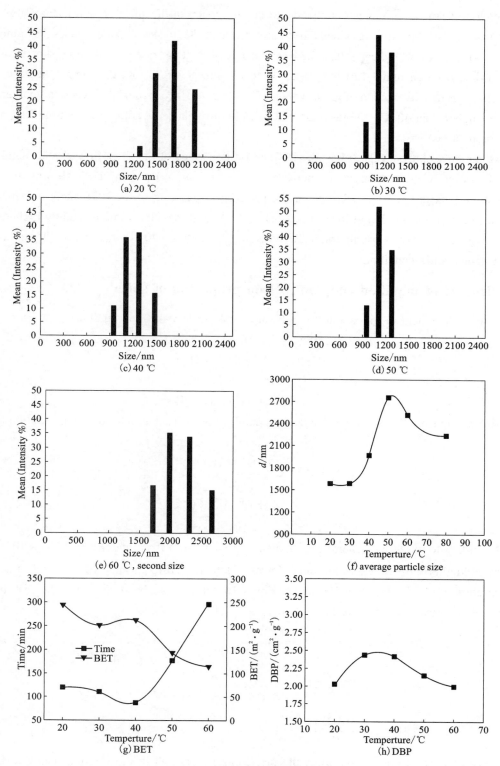

Fig. 3 The impact of modified temperature on silica

the increase of temperature is not conducive to the dispersion of the silica particles. Fig. 3(g) also shows that the reaction time gradually decreases from 20 ℃ to 40 ℃ and reaches its minimum value at 40 ℃, which indicates that the reaction rate was at its fastest at 40 ℃, shortening the reaction time from an initial 120 min at 20 ℃ to 87 min at 40 ℃. At the same time, the BET value also reaches its second highest value of 212.29 m^2/g. After the reaction temperature becomes higher than 40 ℃, the decrease of CO_2 solubility plays a leading role and the BET value of the product decreases.

As can be seen from Fig. 3(h), the DBP absorption value of the silica first increases and then decreases as the reaction temperature increases. At 30 ℃ and 40 ℃, the DBP absorption values are close to 2.44 cm^2/g and 2.42 cm^2/g, respectively. Referring to the BET analysis, since the reaction rate at 40 ℃ is 23 min lower than that at 30 ℃ and the BET value is slightly higher, and considering the energy consumption and industrial feasibility, 40 ℃ has been chosen as the optimal temperature condition.

3.3 Impact of modified endpoint pH on properties of silica

As the silicic acid in the solution is a binary weak acid, ionization equilibrium exists:

$$HSiO_3 \longrightarrow HSiO_3^- + H^+ \quad (4)$$

$$HSiO_3^- \longrightarrow SiO_3^{2-} + H^+ \quad (5)$$

$$CO_2 + H_2O \longrightarrow H_2CO_3 \quad (6)$$

$$H_2SiO_3 + H_2O \longrightarrow SiO_2 \cdot H_2O \quad (7)$$

The relationship between the ionization equilibrium constant and ion concentration is as follows:

$$K_{\alpha1}^{\ominus} = \frac{[H_3O^+][HA^-]}{[H_2A]} \quad (8)$$

$$K_{\alpha2}^{\ominus} = \frac{[H_3O^+][A^{2-}]}{[HA^-]} \quad (9)$$

According to the Blue Chemistry Manual, $K_{\alpha1}^{\ominus} = 1.70 \times 10^{-10}$ and $K_{\alpha2}^{\ominus} = 1.58 \times 10^{-12}$. In ionization equilibrium, Si exists in three forms: H_2SiO_3, $HSiO_3^-$, and SiO_3^{2-}[25]. If the total concentration is C, the following formulae Eqs. (10)-(12) can be derived for the distribution coefficients of various existing forms:[26]

$$\delta H_2SiO_3 = \frac{[H_2SiO_3]}{C} = \frac{[H_3O^+]^2}{[H_3O^+]^2 + K_{\alpha1}^{\ominus}[H_3O^+] + K_{\alpha1}^{\ominus} \times K_{\alpha2}^{\ominus}} \quad (10)$$

$$\delta HSiO_3^- = \frac{[HSiO_3^-]}{C} = \frac{K_{\alpha1}^{\ominus}[H_3O^+]}{[H_3O^+]^2 + K_{\alpha1}^{\ominus}[H_3O^+] + K_{\alpha1}^{\ominus} \times K_{\alpha2}^{\ominus}} \quad (11)$$

$$\delta SiO_3^{2-} = \frac{[SiO_3^{2-}]}{C} = \frac{K_{\alpha1}^{\ominus} \times K_{\alpha2}^{\ominus}}{[H_3O^+]^2 + K_{\alpha1}^{\ominus}[H_3O^+] + K_{\alpha1}^{\ominus} \times K_{\alpha2}^{\ominus}} \quad (12)$$

As can be seen from Fig. 4, when the pH is close to 14, Si in the solution mainly exists in the form of SiO_3^{2-}. With the decrease of pH, the concentration of SiO_3^{2-} gradually decreases, and

H_2SiO_3 appears when pH reaches about 12.5, and its concentration increases with the decrease of pH. When pH is close to 8, Si mostly exists in the form of H_2SiO_3.

At a stirring speed of 360 rad/min, CO_2 gas is introduced at a flow rate of 0.6 L/min to react. Aeration is stopped when the pH of the solution reaches a certain value, and stirring is continued for 10 min.

As can be seen from Fig. 5 (a), (b), (c), (d), and (e), the particle size distribution of silica is concentrated at pH values of 8.5 and 9, and the particle size distribution is about 900 nm. When the pH value takes other values, the particle size distribution is above 1000 nm, indicating that it is relatively uniform in the pH range of 8.5-9. As can be seen from Fig. 5(f), with the increase of the pH value at the end of the reaction, the average particle size of the silica first decreases and then increases, and the minimum value of 859 nm is obtained when the pH value is 8.5. According to the analysis in Fig. 4, the lower the pH value, the greater the tendency of silicic acid formation in the solution, i.e., the more stable the product. Therefore, when the pH value is greater than 8.5, the particle size of the silica gradually increases with the increase of the pH value. However, when the pH value is between 8 and 8.5, the smaller pH value leads to the agglomeration of silicic acid and the increase of the particle size of the silica.

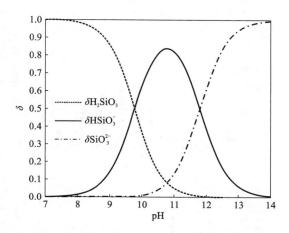

Fig. 4　Ion distribution curve (7-14) of silicon

As can be seen from Fig. 5(g), the reaction time of the silica gradually decreases with the increase of the pH at the end of the reaction, while the BET analysis tends to first increase and then decrease, reaching an extreme value of 236.42 m^2/g at the pH of 8.5. When the pH value is 8.5, the average secondary particle size of the silica reaches a minimum value, and the specific surface area of the larger particles with a network structure composed of small particles is larger, so that the specific surface area is the largest at this pH value.

From Fig. 5(h), it can be seen that the DBP value of the silica particles first increases and then decreases with the increase of the endpoint pH value. In the range of pH 8.0-9.0, the DBP value presents an inverse parabola shape with large numerical changes, which indicates that the DBP value is more sensitive to pH value changes in the range of pH 8.0-9.0 and reaches a maximum of 2.67 cm^2/g when the pH is 8.5. To sum up, pH 8.5 is taken as the optimal condition.

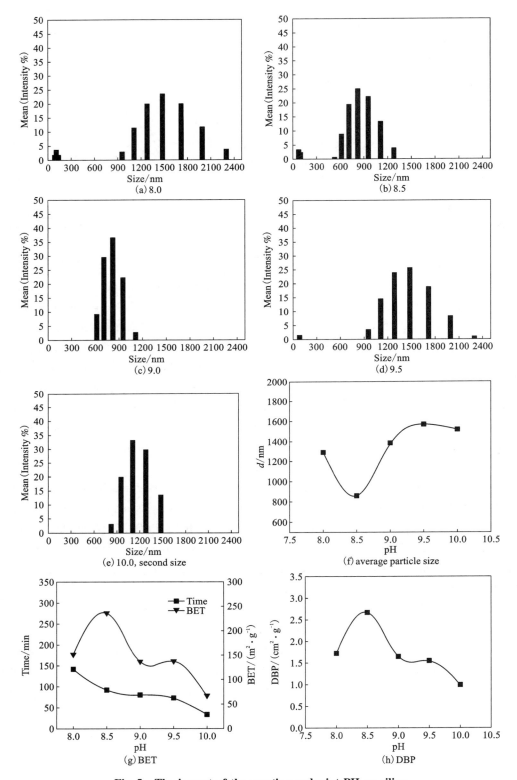

Fig. 5 The impact of the reaction endpoint PH on silica

3.4 Characterization and analysis of silica

The silica was prepared according to the optimized conditions, and the silicon deposition rate was as high as 97.82%. From the X-ray diffraction pattern of silica in Fig. 6(a), it can be seen that the silica gradually forms a steamed bun peak in the range $2\theta=15°\text{-}36°$, and the intensity of the peak is smaller and the width is larger, which indicates that the substance may be amorphous or caused by fine crystallites. The wet silica (white flocculent precipitate) was analyzed by electron selective diffraction (SAED). It can be seen from Fig. 6(b) that the diffraction pattern has no lattice and is an annular halo, and thus it can be determined that the silica is amorphous.

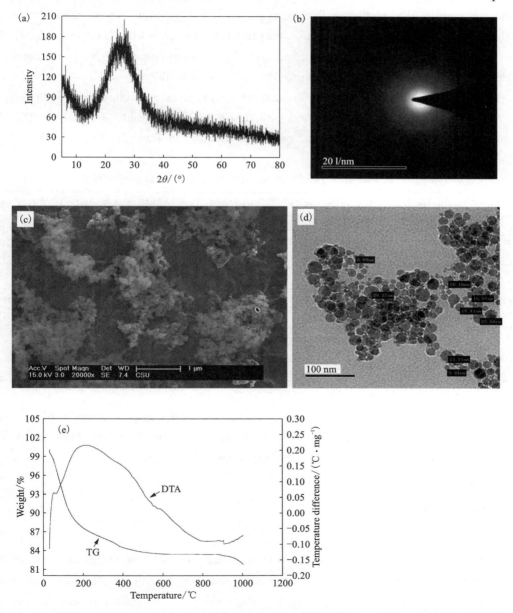

Fig. 6 (a) XRD, (b) SAED, (c) FE-SEM morphology, (d) TEM morphology, and (e) TG-DTA

From Fig. 6(c), it can be observed that the prepared silica particles are in contact with each other and cross-linked into a network structure, which is called a secondary structure. The chain branch structure of the silica is an important factor that can be used as a reinforcing agent in the rubber industry. In order to accurately understand the micro-morphology and primary particle size of the silica particles, TEM analysis was carried out on the samples, and the results are shown in Fig. 6(d). The silica particles TG curve that there is a large weight loss between 70 ℃ and 190 ℃, with a mass loss of 8.83%. It is presumed that the loss of adsorbed water on the surface of the silica is represented by a large endothermic peak on the DTA curve. Then, the mass prepared through comprehensive experiments are spherical with a diameter of about 20 nm. Some particles are connected to each other through aggregation and collision, and the microstructure is a spatial network. From Fig. 6(e), it can be seen from the change of the TG curve gradually slows down, indicating that the weight loss after 200 ℃ is mainly produced by condensation and dehydration of silicon hydroxyl on the surface, and tends to be flat at 500 ℃.

The specific surface area, SiO_2 purity, and dibutyl phthalate absorption value of the prepared silica product were analyzed and compared with HG/T 3061-1999 and ISO 5794-1: 2005(E). The analysis of the physical and chemical properties is shown in Table 2.

Table 2 Analysis of physical and chemical properties of high-quality silica

Test items	Data of HG/T 3061-1999	Data of ISO 5794-1: 2005(E)	Test results
BET/($m^2 \cdot g^{-1}$)	Type A≥191	—	244.67
Purity of SiO_2/%	≥90	≥90	91.61
DBP/($cm^3 \cdot g^{-1}$)	2.00-3.50	—	2.48
Total copper content/(mg·kg^{-1})	≤30	≤30	0
Total manganese content/(mg·kg^{-1})	≤50	≤50	0
Total iron content /(mg·kg^{-1})	≤1000	≤500	0

As can be seen from Table 2, the SiO_2 content in the sample is as high as 91.61%, and all the indices reach the HG/T 3061-1999 A-grade standard. Therefore, the high-quality silica prepared in this research can be used as a rubber compounding agent.

4 Conclusion

A single factor rotation method was used to systematically study the effects and laws of PEG-6000 modifier dosage, modification temperature and modification endpoint pH on the properties of silica powder prepared from copper smelting slag tailings. The following conclusions were obtained.

The particle size and specific surface area of the produced silica particles are greatly affected by the modification temperature and the modification endpoint pH, while the amount of the modifier that is added has a relatively weak influence on it. Under the synergistic effect of particle size and specific surface area, the DBP of the silica shows the following rules: the amount of modifier added, the modification temperature and the modification endpoint pH value all show the highest values at the critical point, and the DBP absorption value is greatly influenced by the modification temperature and the modification endpoint pH value.

When EPG-6000 is used as the modifier, in the conditions of modifier dosage of 10% of solute mass, modification temperature of 40 ℃, and a modification endpoint pH value of 8.5, the silicon deposition rate is as high as 97.82%, and the obtained highquality silica particles have good dispersibility, average particle diameter of 20 nm, spherical morphology, and specific surface area of 244.67 m^2/g, which is superior to the A-grade standard of HG/T 3061-1999 and ISO 5794-1: 2005(E), and can be directly used in the rubber industry.

Acknowledgments

The authors appreciate the financial support from the National Natural Science Foundation of China (No. 51620105013 and No. 51904351), Innovation-Driven Project of Central South University Hunan (No. 2020CX028) and Natural Science Fund for Distinguished Young Scholar of Hunan Province, China (No. 2019JJ20031).

References

[1] DU K, LI H, ZHANG M. JOM-US, 2017, 69: 2379.
[2] LI Y, CHEN Y, TANG C, et al. Mater, 2017, 322: 402.
[3] GUO Z, ZHU D, PAN J, et al. JOM-US, 2016, 68: 2341.
[4] COURSOL P, CARDONA VALENCIA N, MACKEY P, et al. JOM-US, 2012, 64: 1305.
[5] MAST R E, KENT G H. JOM-US, 1955, 7: 877.
[6] GUO Z, PAN J, ZHU D, et al. Prod, 2018, 199: 891.
[7] GORAI B, JANA R K. Resour. Conserv. Recycl. 2003, 39: 299.
[8] WANG Q, WANG S, TIAN M, et al. Int. J. Min. Met. Mater, 2019, 26: 301.
[9] WANG Q, GUO X, TIAN Q, et al. Metals, 2017, 7: 502.
[10] WANG Q, GUO X, TIAN Q, et al. Metals, 2017, 7: 302.
[11] CAO J, WANG B. China Min. Mag., 1994, 3: 17.
[12] ZHOU T, ZHANG C, WEI N, et al. Ceram. Soc., 2014, 33: 691.
[13] BIAN R, DU W. China Nonferr. Metall, 2012, 2: 8.
[14] TUROV V V, GUN'KO V M, TSAPKO M D, et al. Appl. Surf. Sci., 2004, 229: 197.

[15] ZHOU L, YIN L, ZHOU K, et al. Mater. Rev., 2003, 11: 56.
[16] CHEN Y, ZHAO S, WANG S. Inorg[J]. Chem. Ind. 2013, 45: 41.
[17] LOPEZ J F, PEREZ L D, LOPEZ B L. J. Appl. Polym. Sci., 2011, 122: 2130.
[18] AN D, WANG Z, ZHAO X, et al. Colloids Surf., 2010, A 369: 218.
[19] SONG Y, SONG L, LU A, et al. Ceram. Soc., 2013, 5: 674.
[20] WANG T, CHEN Q, ZHOU P, et al. Fine Chem., 2000, 17: 438.
[21] KONG Q, QIAN H, LI S. Inorg. Chem. Ind., 2009, 41: 33.
[22] HE K, CHEN H. Chem. React. Eng. Technol, 2006, 22: 181.
[23] QUAN P, FANG Q. Contemp. Chem. Ind., 2014, 7: 1168.
[24] GAO H, YANG Y. Chem. Ind. Times, 2010, 24: 16.
[25] PRASERTSRI S, RATTANASOM N. Polym. Test, 2012, 31: 593.
[26] MA X K, LEE N, OH H, et al. Colloids Surf., 2010, A 358: 172.

Recovery of Zinc and Lead from Copper Smelting Slags by Chlorination Roasting

Copper smelting slags are important secondary resources containing varieties of valuable metals. In this paper, $CaCl_2$ was used as the chlorination agent to recycle Zn and Pb from copper smelting slags by chlorination roasting. The effects of temperature, holding time, the dosage of $CaCl_2$, and the roasting atmosphere on the metal recovery ratio were investigated. The results showed that the recovery ratio of Zn and Pb reached 74.74% and 94.72%, respectively, under typical conditions. When the roasting temperature was above 1100 ℃, the material began to sinter or even melt with the increase of temperature, which was not conducive to the recovery of metals. An excessively long roasting time will convert chlorides into oxides and reduce the metal recovery ratio. The thermodynamic calculation results indicated that, when the temperature was higher than 850 ℃, the calcium introduced by the $CaCl_2$ would be integrated with Si and Fe in the slags.

1 Introduction

Copper smelting slags are pyrometallurgical residues produced from the copper smelting process.[1] According to statistics, 2.2 t copper smelting slags are produced for every 1 t electrolytic copper production.[2] In 2019, the production in China of electrolytic copper was 9.8 million tons, and the copper smelting slags were 21.5 million tons, which contained more than 650000 t Zn and 100000 t Pb.[3] At present, the utilization of copper smelting slag resources is grossly inadequate, and a mass of copper smelting slags are directly scrapped and consequently lead to the underutilization of heavy metal elements and to environmental contamination.[4, 5]

Because copper smelting slags are important secondary resources of non-ferrous metals, their comprehensive utilization has become a research hotspot in recent years. Zhang et al.[6] used

---

Published in *JOM*, 2021, 73(6): 1861−1870. Authors: Guo Xueyi, Zhang Beikai, Wang Qinmeng, Li Zhongchen, Tian Qinghua.

copper smelting slags as a micronized sand to prepare highperformance concrete, which not only reclaimed and reused copper smelting slags but also improved the performance of the concrete. Han et al. [7] developed a combination of flotation high-performance liquid chromatography solvent extraction to effectively utilize secondary copper resources. Under optimal extraction conditions, the recovery rate of copper in the stripping solution was 91.3%, and the recovery rate of iron was 98.6%. Videla et al. [8] investigated the effect of ultrasonic treatment on the recovery ratio of copper through the tailings flotation process. The application of ultrasound in the flotation process could increase the copper recovery ratio by 3.5%. At present, the research hotspots in this field mainly focus on the utilization of Cu and Fe in copper smelting slags and the preparation of building materials. [6-16] There are few studies concerning the recovery of other valuable metals in copper smelting slags, such as Zn and Pb. Copper smelting slags have a stable structure and a very complicated multi-component system, which make it difficult to achieve the separation and recovery of the valuable metals by conventional methods. [9, 10] In the traditional smelting waste treatment process, most of the copper smelting slags are made into building materials or directly landfilled. [11] The valuable metals in the slags are not effectively recovered and result in a serious waste of resources.

The chlorination process can recover valuable metals from the solid waste resources. This process mainly makes use of the characteristics of metal chlorides, such as low melting point, high volatility and easy solubility in water of metal chlorides, so that the metal elements are volatilized from the accompanying system in the form of chlorides. [12, 13] According to the difficulty level of the formation of metal chlorides and the differences in properties, the metal elements can be selectively chlorinated and volatilized by controlling the reaction temperature and product vapor pressure to achieve the purpose of separating and recovering the metal elements. [14-16] According to different chlorinated media, the chlorination process can be classified as anhydrous or aqueous. [17] The recovery technology of metallic oxide sources is anhydrous chlorination and the sulfide sources go through both anhydrous and aqueous chlorination. Ordinarily, the chlorinating agents for both kinds of chlorination are gaseous chlorine, hydrogen chlorides, and alkali metal chlorides such as $CaCl_2$, $MgCl_2$, $FeCl_2$, and NaCl. Owing to the $CaCl_2$ having a good ability of volatilizing metals and a rather low price, the recycling of heavy metals such as Au, Ag, Pb, Cd, and Cu from smelting slags in chlorination roasting with the addition of $CaCl_2$ has been widely studied.

This paper introduces a method of using $CaCl_2$ as a chlorination agent to chlorinate and recover Zn and Pb in copper smelting slags. This research conducted a process of mineralogical analysis of raw materials and roasting slags. By a single-factor rotation method, we systematically studied the effects of roasting temperature (700-1300 ℃), roasting time (20-120 min), the dosage of $CaCl_2$(5%-30%), and roasting atmosphere (different atmospheres and flow rates) on the recovery ratio of Zn and Pb in copper smelting slags.

2 Experimental

2.1 Experimental procedures

All roasting experiments in this study were carried out in a high-temperature tube furnace. First, 20 g of dried copper smelting slags were ground and mixed with a certain mass ratio of $CaCl_2$ for 15 min, loaded into a corundum porcelain boat, and then fed into a tube furnace. The mixture was heated to the preset temperature at a heating ratio of 10 °C/min, and then a certain flow of gas was introduced. By holding it for a time, the raw materials were fully roasted. Finally, the furnace was cooled to room temperature and the gas was cut off and the roasting slags were taken out and ground into powder. During the experiment, the volatilized flue gas was collected through a gas scrubber containing 200 mL of 10% NaOH at the outlet of the tube furnace. The recovery/volatilization ratio of Zn and Pb metals is determined by Eq. (1).

$$R = \frac{w_g \times m_g - w_s \times m_s}{w_g \times m_g} \quad (1)$$

where R is the recovery/volatility efficiency (%), w_g is the mass fraction of an element in the copper smelting slags, m_g is the weight of copper smelting slags in the porcelain boat, w_s is the mass fraction of an element in the roasting slags, and m_s is the weight of roasting slags.

2.2 Experimental facilities and analytical methods

A high-temperature tube furnace (SGL-1700C; Shanghai Jujing) was used as the experimental equipment, and a corundum porcelain boat (120 mm in length, 60 mm in width, and 20 mm in height) was used as the experimental container for this research. Inductively coupled plasma emission spectrometer (ICP-OES; Optima 7300 V; Perkin Elmer, USA) was used to determine the content of the various elements in the copper smelting slags and roasting slags. All the solid samples were dried in a vacuum oven at 80 °C for 12 h, then characterized by a scanning electron microscope (SEM) equipped with energy disperse spectroscopy (EDS), and an X-ray diffraction (XRD) analyzer (D8 Discover 2500), using a PANalytical X'Pert X-ray diffractometer (Cu K_α radiation). During the analysis of the chemical composition of the samples, each sample was tested twice, and the average value taken as the final result. To better understand the chlorination roasting mechanism, detailed thermodynamic modeling was carried out with the help of FactSage™ 7.1 software.

2.3 Material

The copper smelting slags as the raw materials used in the experiment were copper flotation tailings obtained from a copper smelter in Shandong Province, China. The mineralogical analysis of the copper smelting slags sample was carried out after grinding and drying. The element content of the copper smelting slags was analyzed by ICP-OES, and the results are shown in Table 1, which indicate that the Zn and Pb contents were 2.80% and 0.45%, respectively. The phase composition of the copper smelting slags were determined by XRD and the results are shown in Fig. 1. As seen, the main components in the materials were Fe_3O_4 and Fe_2SiO_4. Due to the low content of other components in the slags, they cannot be accurately displayed in the XRD pattern. The micromorphology and element content were characterized by SEM-EDS and the results are shown in Fig. 2, from which it can be seen that the minerals exist in the form of multiparticles or irregular particles, and exist in three typical forms: black (point 1, the main components of this area are iron oxide, Zn silicate, and silicon oxide), white (point 2, the main components of this area are Pb sulfate, Pb silicate, and Pb sulfide) and gray (point 3, the main components of this area are Zn ferrite, Zn sulfate, and Zn sulfide).

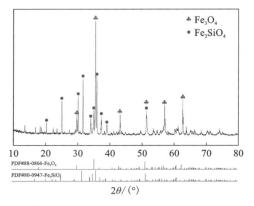

Fig. 1 The XRD pattern of the copper smelting slags

Table 1 Chemical composition of the copper smelting slags

Element	Zn	Pb	Cu	Fe	O	Si	Ca	Na	K	Cl	S
Contents/%	2.80	0.45	0.55	41.23	36.10	11.75	4.91	0.87	0.58	0.059	0.79

Fig. 2 SEM-EDS images of the copper smelting slags

3 Results and discussion

3.1 Reaction principle and thermodynamic analysis

The thermodynamic reaction model of chlorination roasting was conducted to assist in understanding the reaction process.[18] Combining the foregoing process mineralogy analysis of the copper smelting slags, the possible reactions in the chlorination roasting process are shown in Eqs. (2)-(10), where Eqs. (2)-(6) are the chlorination process of different phases of Zn, and Eqs. (7)-(10) are the chlorination process of different phases of Pb. Details of the thermodynamic modeling were carried out with the help of the FactSage™ 7.1 software. The standard Gibbs free energy change ($\Delta G^\ominus$, kJ/mol) of each reaction varies with temperature, as shown in Fig. 3.

$$Zn_2SiO_4 + 2CaCl_2 + SiO_2 \rightleftharpoons 2ZnCl_2(g) + 2CaSiO_3 \quad (2)$$
$$ZnFe_2O_4 + CaCl_2 + SiO_2 \rightleftharpoons ZnCl_2(g) + CaSiO_3 + Fe_2O_3 \quad (3)$$
$$ZnS + CaCl_2 \rightleftharpoons ZnCl_2(g) + CaS \quad (4)$$
$$ZnO + CaCl_2 + SiO_2 \rightleftharpoons ZnCl_2(g) + CaSiO_3 \quad (5)$$
$$ZnSO_4 + CaCl_2 \rightleftharpoons ZnCl_2(g) + CaSO_4 \quad (6)$$
$$PbSiO_3 + CaCl_2 \rightleftharpoons PbCl_2(g) + CaSiO_3 \quad (7)$$
$$PbS + CaCl_2 \rightleftharpoons PbCl_2(g) + CaS \quad (8)$$
$$PbO + CaCl_2 + SiO_2 \rightleftharpoons PbCl_2(g) + CaSiO_3 \quad (9)$$
$$PbSO_4 + CaCl_2 \rightleftharpoons PbCl_2(g) + CaSO_4 \quad (10)$$

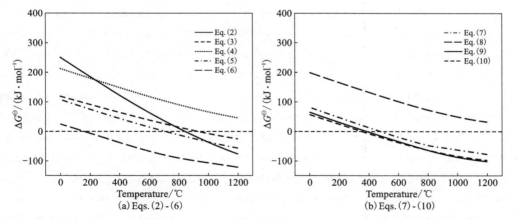

Fig. 3 The Gibbs free energies ($\Delta G^\ominus$) as a function of different temperatures for these chemical reactions

As seen from Fig. 3, when the temperature above 950 ℃, only the Gibbs free energy of Eqs. (4) and (8) are positive, which means that the reaction cannot happen.[19] Therefore, using $CaCl_2$ as the chlorination agent cannot completely achieve the chlorination roasting process of Zn and Pb in copper smelting slags. However, it is remarkable that the thermodynamic model is based on the constant partial pressure of the gas phases ($ZnCl_2$ and $PbCl_2$).[20] According to Eqs. (11) and (12), the Gibbs free energy of the reactions of Eqs. (2)-(10) can be reduced to a less negative value by controlling the partial pressure of the gas phase product.[21] Therefore, during the chlorination roasting process, air or an inert gas (such as argon) was blown into the experimental furnace to make the gas phase reaction products ($ZnCl_2$ and $PbCl_2$) leave the reaction zone with the flowing gas, thereby reducing the partial pressure of the reaction products in the system and facilitating the positive direction of the reactions of Eqs. (2)-(10).[22]

$$\Delta G^{\ominus} = -RT\ln K^{\ominus} = -RT\ln \frac{P_{ZnCl_2}}{P^{\ominus}} \tag{11}$$

$$\Delta G^{\ominus} = -RT\ln K^{\ominus} = -RT\ln \frac{P_{PbCl_2}}{P^{\ominus}} \tag{12}$$

3.2 Effect of reaction conditions on zinc and lead recovery

The $\Delta H > 0$ kJ/mol in each reaction of Eqs. (2)-(10) means that the chlorination is an endothermic process, and that the reaction efficiency is related to temperature.[19] In order to verify the effectiveness of the chlorination roasting at the roasting temperature to recover Zn and Pb, the raw material (20 g copper smelting slags + 6 g $CaCl_2$) was roasted at 700-1300 ℃ for 1 h with air flow rate of 100 mL/min, and the results are shown in Fig. 5(a). As seen, the metal recovery ratio of Zn and Pb increased significantly as the temperature increased from 700 to 1100 ℃. This was conceivably because the increase of temperature will destroy the structure of the copper smelting slags and thereby enhance the chemical reactions.[23] Among them, the metal recovery ratio of Zn and Pb was greatly affected by temperature in the range of 700-850 ℃. The metal recovery ratios of Zn and Pb were 74.74% and 94.72%, respectively, at 850 ℃. However, the metal recovery ratios of Zn and Pb are less affected by temperature above 850 ℃. Most of the slags begin to melt under high-temperature conditions, resulting in the densification of the tailings.[24] The roasting slags at 850 ℃ [shown in Fig. 5(a)] are loose and easy to sample, indicating that the material is not melted at this temperature. The morphology of the roasting slags at 1300 ℃ [shown in Fig. 5(a)] is significantly different from the roasting slags under a low temperature. The roasting slags are attached to the porcelain boat and are a solid block. The reason is that the roasting temperature is too high which leads to the melting of the raw materials.

The SEM images of the roasting slags after chlorinating at 850 ℃, 1100 ℃, 1200 ℃, and 1300 ℃ are shown in Fig. 5(b). This shows that, in the range of 1100-1300 ℃, as the temperature increases, the calcined product gradually sinters, resulting in a decrease in material porosity, and $CaCl_2$ particles cannot fully contact Zn and Pb in the copper smelting slags, which reduces the kinetic conditions of the chlorination reaction. The phenomenon of material sintering also has a certain solidification effect on heavy metals.[25] When the temperature is above 1300 ℃, the materials begin to melt and transform into a liquid phase, which seriously impacts the metal recovery ratio. In addition, the study by Zhou et al.[26] has shown that extremely high roasting temperatures will increase the consumption of chlorination agents and reduce the purity of gas phase products. In the industrial production process, the increase of the roasting temperature will increase the production and operation costs. Giving consideration to the metal recovery ratio, material cost, and energy consumption, the temperature of chlorination roasting should not exceed 850 ℃.

Fig. 5(c) shows the effect of chlorination roasting time on the metal recovery ratio of Zn and Pb when the raw material (20 g copper smelting slags + 6 g $CaCl_2$) was roasted at 850 ℃ for 20-120 min with air flow rate of 100 mL/min. It can be seen that, in the range of 15-60 min, increasing the roasting time has a significant effect on the improvement of the metal recovery ratio. The metal recovery ratio of Zn and Pb has increased from 57.75% to 74.74% and from 84.09% to 94.72% respectively, when increasing the roasting time from 15 min to 60 min. The roasting time should be extended appropriately, because a short roasting time may cause the chlorination reaction to be incomplete. When the roasting time is longer than 60 min, the effect of improving the metal recovery ratio of Pb is not significant, and the metal recovery ratio of Zn will decrease. The evolution of the phases was investigated by XRD when the copper smelting slags were roasted at different times under the same other conditions. The results are shown in Fig. 4, with the main phases in the roasting slags including Fe_2O_3, $ZnFe_2O_4$, and a few ZnO after the roasting process. Due to the progress of chlorination, the diffraction peak intensity of metallic oxides decreased along with roasting time before 60 min. However, when the roasting time exceeded 60 min, the peak intensity increased along with the roasting time. Combining the previous analysis results, it is confirmed that excessively long chlorination roasting times will convert chlorides into oxides and hence reduce metal recovery ratios, and the results are consistent with the study of Jaafar et al.[27]

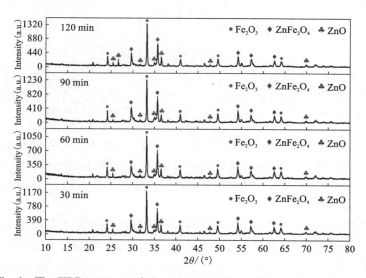

Fig. 4 The XRD patterns of the roasting slags at different roasting times

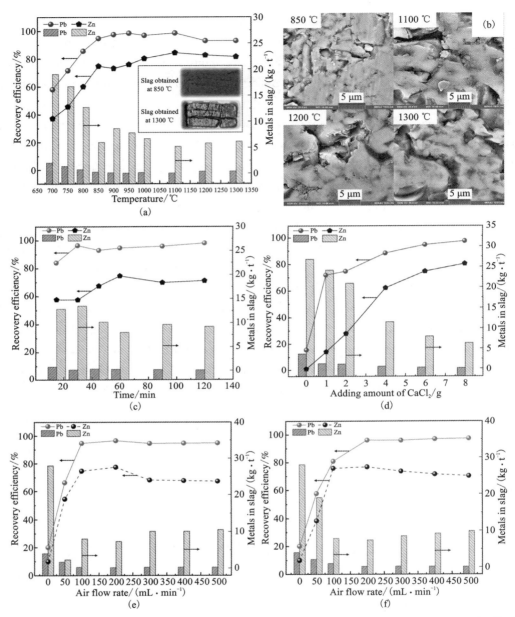

Fig. 5 Effects of (a) roasting temperature, (c) roasting time, (d) CaCl$_2$ addition amount, (e-f) air/argon flow rate on the recovery of Zn and Pb. (b) SEM images of samples chlorinated at different temperatures: 600 ℃, 900 ℃, 1000 ℃, and 1050 ℃

CaCl$_2$ is a chlorination agent with a wide range of sources and a good selective chlorination effect. It has low energy consumption and low pollution in roasting treatments.[28] Fig. 5(d) shows the effect of the dosage of CaCl$_2$ on the metal recovery ratios of Zn and Pb when 20 g of copper smelting slags roasted at 850 ℃ for 60 min with air flow rate of 100 mL/min. As seen from Fig. 5(d), the metal recovery ratios of Zn and Pb are greatly affected by the amount of $CaCl_2$ added. When no $CaCl_2$ was added, the recovery ratios of Zn and Pb could still reach 3.68% and 0.94%, respectively. The reason is that their compounds have a higher saturated vapor pressure at high temperature. As the amount of $CaCl_2$ added increases, the metal recovery ratios of Zn and Pb rise. When the amount of $CaCl_2$ addition is 6 g, the metal recovery ratios of Zn and Pb are 74.74% and 94.72%, respectively. Increasing the amount of $CaCl_2$ can increase the contact ratio between the chlorine donor and the Zn and Pb phases, promote the chlorination process, and speed up the reaction. When the amount of $CaCl_2$ added exceeds 30% of the material, the increase of the metal recovery ratios of Zn and Pb slows down. The reason is that the content of Zn and Pb in the copper smelting slags is less than 5%. At this time, the chlorine donor and the contents of Zn and Pb in the copper smelting slags are basically saturated. Therefore, we can determine that increasing the amount of $CaCl_2$ can significantly increase the metal recovery ratios of Zn and Pb. However, the amount of $CaCl_2$ should not exceed 30% of the mass of the mineral material.

In order to study the effect of different roasting atmospheres on the metal recovery ratios, air and argon were selected for roasting experiments at different flow rates. Figs. 5(e) and 5(f) show the effect of the roasting atmosphere on the metal recovery ratio of Zn and Pb when the raw material (20 g copper smelting slags + 6 g $CaCl_2$) roasted at 850 ℃ for 60 min with different roasting atmospheres. As seen from Figs. 5(e) and 5(f), the metal recovery ratios of Zn and Pb increase with the increase of airflow rate. This is because the flowing gas brings the gas phase products out of the reaction zone, reduces the partial pressure of the $ZnCl_2$ and $PbCl_2$ vapor, and promotes the positive progress of Eqs. (2)-(10). In addition, the airflow rate can increase the contacting probability between $CaCl_2$, and the Zn and Pb in the copper smelting slags, and enhance the mass transfer process. However, when the airflow rate is higher than 60 mL/min, the metal recovery ratio of Zn shows a downward trend. The reason is that gas molecules occupy the reaction site and conflict with the chlorination reaction, leading the chlorination rate of the Zn phase to decrease as the gas pressure increases.[29]

During smelting production, gas corrosion in the pipeline is prone to occur.[17] Therefore, the behavior of hydrochloric acid in the chlorination roasting process should be considered via the following reactions:

$$2CaCl_2 + O_2 = 2CaO + 2Cl_2 \tag{13}$$

$$H_2O + Cl_2 = HCl + HClO \tag{14}$$

$$2MeS + 8HClO = 2MeCl_2 + 2H_2SO_4 + 4HCl \tag{15}$$

$$MeO + 2HCl = MeCl_2 + H_2O \tag{16}$$

The affinity of calcium ions for oxygen ions is greater than for chloride ions, therefore, part of the calcium chloride will react with oxygen during the roasting process to generate calcium oxide and chlorine gas. The introduced air in the experiments contains a small amount of moisture, so hydrochloric acid and hypochlorous acid will be produced in the system. They will react with metallic oxides and metal sulfides, respectively, to produce metal chlorides. The acid gas generated is consumed during the roasting process, so the influence of acid gas on the experiments can be ignored.

3.3 Mechanism of chlorination roasting of copper smelting slags

The experiments in argon can also explain the chlorination roasting mechanism of Zn and Pb in copper smelting slags. The chlorination can be divided into direct or indirect chlorination.[30] Taking $CaCl_2$ as a chlorination agent to treat metal oxides as an example, the direct chlorination mechanism can be represented by Eq. (17), and the solid chlorination agent directly interacts with the metal oxide to produce the corresponding chloride. The indirect chlorination mechanism can be expressed by Eqs. (18) and (19), and the chlorination agent decomposes under the action of oxygen to produce chlorine, and the chlorine reacts with metal oxide to produce corresponding chlorides. Before heating, the furnace should be passed over the argon at 500 mL/min for 5 min for prepurging. As shown in Fig. 5(e), the chlorination and volatilization of Zn and Pb can still be achieved in argon atmosphere, and the effect is similar to that in an air atmosphere. It can be explained that the Zn and Pb in the copper smelting slags do not need to participate with oxygen during the chlorination roasting process. Therefore, the type of chlorination is direct chlorination. The study by Sun et al.[31] showed that, at temperatures higher than the melting point of $CaCl_2$, liquid $CaCl_2$ in the $ZnO-Fe_2O_3-CaCl_2$ system directly reacts with Zn containing substances to produce gaseous $ZnCl_2$. The results of this study are the same as the conclusion.

$$CaCl_2 + MeO \Longrightarrow CaO + MeCl_2(s/l/g) \tag{17}$$

$$2CaCl_2 + O_2(g) \Longrightarrow 2CaO + 2Cl_2(g) \tag{18}$$

$$2MeO + 2Cl_2(g) \Longrightarrow 2MeCl_2(s/l/g) + O_2(g) \tag{19}$$

3.4 Characterization and analysis of roasting slags

Copper smelting slags have a complex composition including many other elements. Therefore, it is significant to consider the content of impurity elements in the chloride fumes. The trend of other elements can be judged by studying the changes of slag composition during the chlorination roasting process. Using 20 g of copper tailings, air flow rate of 100 mL/min, $CaCl_2$ addition amount of 6 g, and roasting time of 60 min, we have analyzed the elemental composition of raw materials and roasting slags, respectively. The results are shown in Fig. 6(a), and most of the critical ingredients in copper smelting slags decreased after roasting due to their volatility and the entrainment of chlorinated volatiles. As can be seen from Fig. 6(c), in the roasting slags, the contents of Fe, Si, Al, S, Cu, Mu, Ti, and As had slightly decreased and the contents of Ca had significantly increased due to the introduction of calcium chloride. This indicates that the chloride fumes contained a few impurities. In coming research on recycling, zinc and lead will be separated and recovered from the chloride fumes.

Under the same conditions, from the SEM image and EDS surface scanning analysis in Fig. 6 (b), it can be illustrated that the Fe in the roasting slags exists in the form of iron oxides. In addition, the residual Zn and Pb in the roasting slags presents a scattered distribution and are

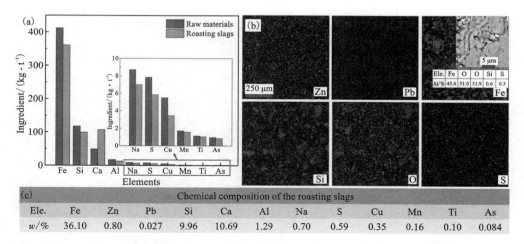

Fig. 6 (a) The content of other elements in the roasting slags, (b) SEM images and EDS map scanning analysis, (c) chemical composition of the roasting slags

encapsulated in fayalite and iron oxides. Combining the XRD analysis results and the 5 μm microscale electron microscopy analysis, it can be concluded that the Fe in the roasting slags mainly exists in the form of Fe_2O_3, and that the roasting slags can be further studied for recovering iron.[32] The iron oxidation reaction can be represented as follows:

$$Fe_3O_4 + CaCl_2 + SiO_2 = Fe_2O_3 + FeCl_2(g) + CaSiO_3 \quad (20)$$

3.5 Equilibrium consideration

To further understand the thermodynamic feasibility of the chlorination roasting process, a thermodynamic model was developed using the equilibrium composition module of FactSage™ 7.1. The initial calculation components include 20 g copper smelting slags and 4 g $CaCl_2$. The relationship between the equilibrium amounts of the species and the temperature of the reaction between the copper smelting slags and $CaCl_2$ under sufficient oxygen is shown in Fig. 7(a), (b). It can be seen that gaseous $ZnCl_2$ and $PbCl_2$ have been produced when the temperature is greater than 400 ℃, and peaks of their amount appear at about 600 ℃. When the temperature reaches 850 ℃, 1000 ℃, and 1200 ℃, respectively, the calcium introduced by the $CaCl_2$ will begin to gradually transform into $Ca_3Si_2O_7(s)$, $Ca_2SiO_4(s)$, and $CaFe_2O_4(s)$ by reacting with the Si and Fe in the materials. The thermodynamic calculation results can be used to further determine the composition types of the roasting slags, and to analyze the mechanism of calcium migration during chlorination roasting. In addition, the results can show the predicted phases and their proportions at different temperatures.

During the chlorination roasting process, oxygen partial pressure will affect the stability of different phases in the copper smelting slags and thus impact the volatilization of metal chlorides. Figs. 7(c) and 7(d) show the predominant region diagram of the Zn-Pb-O system and the

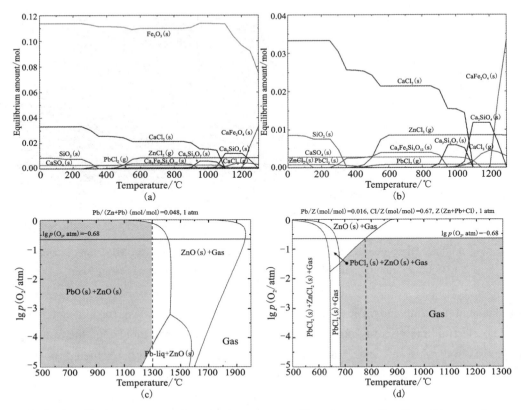

Fig. 7 Equilibrium amounts of species in copper smelting slags as a function of temperature for reaction during chlorination roasting (a, b), predominant region diagram of Zn-Pb-O system (c), predominant region diagram of Zn-Pb-O-Cl system (d)

Zn-Pb-O-Cl system, which can illustrate the correlation between the composition of the zinc-lead phase equilibrium system and the oxygen partial pressure and temperature. Because the volume fraction of oxygen in the air is about 21%, lg $p(O_2$, atm) in this study is -0.68. Fig. 7(c) shows the phase equilibrium of the Pb-Zn-O system when $n_{Pb} : n_{Zn+Pb} = 0.048$ (e.g., copper smelting slags), which illustrates that the system consists of PbO(s) and ZnO(s) in the range of 500-1300 ℃ in air atmosphere. In this state, there is no liquid phase generated during the chlorination roasting process, which is conducive to the volatilization of metal chlorides. Fig. 7 (d) shows the phase equilibrium of the Pb-Zn-O-Cl system when $n_{Pb} : n_{Zn+Pb+Cl} = 0.016$ and $n_{Cl} : n_{Zn+Pb+Cl} = 0.67$ (e.g., copper smelting slags, $n_{Cl} = 2n_{Pb+Zn}$), which illustrates that the system is a gas phase composed of $ZnCl_2$ and $PbCl_2$ when the temperature is in the range of 780-1300 ℃ in air atmosphere. The above phase equilibrium calculation results indicate that $ZnCl_2$ and $PbCl_2$ can be efficiently formed under the conditions of this study, and that the thermodynamic calculation results are consistent with the experimental results.

4 Conclusion

This paper provides a new process for recovering Zn and Pb from copper smelting slags by a method of chlorination roasting. The factors of the influence and mechanism of the metal recovery ratio of Zn and Pb were studied, concluding as follows:

The result of thermodynamic calculations showed that the method of chlorination roasting can recover Zn and Pb from copper smelting slags. The experimental results showed that, when the roasting temperature was below 1100 ℃, the metal recovery ratios of Zn and Pb increased as the roasting temperature increased. When the temperature was higher than 1300 ℃, the minerals began to melt and the mineral structure became dense, resulting in a decrease in the metal recovery ratios of Zn and Pb. If the roasting time is too short, this will result in an incomplete chlorination reaction, and the increasing of holding time will increase the metal recovery ratios of Zn and Pb, but an excessively long roasting time will convert chlorides into oxides and reduce the metal recovery ratios. When the dosage of $CaCl_2$ is less than 30%, the effect of improving the metal recovery ratios of Zn and Pb is obvious with the increase of the dosage of $CaCl_2$. The experimental results of the roasting atmosphere show that the chlorination reaction type of Zn and Pb in the copper smelting slags is a direct chlorination process. The metal recovery ratios of Zn and Pb reached to 74.74% and 94.72%, respectively, when the roasting temperature was 850 ℃ for 60 min with an airflow rate of 100 mL/min and the dosage of $CaCl_2$ is 30%. In addition, it will not release acid gas during the roasting process. The residual Zn and Pb were encapsulated in fayalite and iron oxides, and the Fe mainly existed in the form of Fe_2O_3 in the roasting slags. It can be seen from the results of the thermodynamic calculations that the calcium introduced by the $CaCl_2$ will transform into $Ca_3Si_2O_7(s)$, $Ca_2SiO_4(s)$, and $CaFe_2O_4(s)$ by integrating with the Si and Fe in the materials during chlorination roasting. The phase equilibrium calculation results further verify the possibility of recovering Zn and Pb metals under the experimental conditions.

Acknowledgments

The authors appreciate the financial support from the National Natural Science Foundation of China (No. 51620105013 and No. 51904351), Innovation-Driven Project of Central South University Hunan (No. 2020CX028) and Natural Science Fund for Distinguished Young Scholar of Hunan Province, China (No. 2019JJ20031). Project (2019YFC1907400) supported by the National Key R&D Program of China.

References

[1] CHENG X, CUI Z, CONTRERAS L, et al. JOM, 2019, 71: 1897.
[2] WANG Q, LI Z, LI D, et al. JOM, 2020, 72: 2676.
[3] National Bureau of Statistics. Statistical Communique of the People's Republic of China on the 2019 National Economic and Social Development. (China Statistical Yearbook, 2020), http://www.stats.gov.cn/tjsj/zxfb/202002/t20200228_1728913.html. Accessed 28 February 2020.
[4] LIU S, LI Q, SONG J. Powder Technol, 2018, 330: 105.
[5] GUO Z, ZHU D, PAN J, et a. JOM, 2018, 70: 533.
[6] ZHANG Y, SHEN W, WU M, et al. Constr. Build. Mater, 2020, 244: 118312.
[7] HAN B, ALTANSUKH B, HAGA K, et al. Mater, 2018, 352: 192.
[8] VIDELA A R, MORALES R, SAINT-JEAN T, et al. Miner. Eng., 2016, 99: 89.
[9] SUN Y, ZHANG J, WANG Y, et al. Rare Metal Technology, ed. ALAM S, KIM H, NEELAMEGGHAM N R, et al. (The Minerals, Metals & Materials Society, 2016), 2016: 87-94.
[10] WANG H, SONG S. J. Cent. South Univ. (Engl. Ed.), 2020, 27: 2249.
[11] JUNG M, CHOI Y, JEONG J. Environ. Geochem. Health, 2011, 33: 113.
[12] WANG H, FENG Y, LI H, et al. Powder Technol, 2019, 355: 191.
[13] WANG H, FENG Y, LI H, et al. Trans. Nonferrous Met. Soc. China, 2020, 30(4): 1111.
[14] MUKHERJEE T K, MENON P R, SHUKLA P P, et al. JOM, 1985, 37: 29.
[15] BAI S, BI Y, DING Z, et al. J. Alloys Compd, 2020, 840: 155722.
[16] QIN H, GUO X, TIAN Q, et al, Sep. Purif. Technol, 2020, 250: 117168.
[17] MUKHERJEE T K, GUPTA C K. Miner. Process. Extr. Metall. Rev., 1983, 1: 111.
[18] YU W, TANG Q, CHEN J, et al. Int. J. Miner., Metall. Mater., 2016, 23: 1126.
[19] HUANG X. Principles on ferrous metallurgy, 4th ed. Beijing: Metallurgical Industry Press, 2013: 4-18.
[20] LI H. Metallurgical Principle, 2nd ed. Beijing: Science Press, 2005: 206-208.
[21] ZHA G, YANG B, YANG C, et al. JOM, 2019, 71: 2413.
[22] DONG Z, XIA Y, GUO X, et al. Powder Technol, 2020, 368: 160.
[23] WANG Q, GRAYDON J W, KIRK D W. J. Chongqing Univ., 2003, 26: 73.
[24] LEI C, CHEN T, YAN B, et al. Rare Met. (Beijing, China) (2015) doi: https://doi.org/10.1007/s12598-015-0499-0.
[25] HAN S, WANG X, ZHOU R. Gang Tie Fan Tai, 1993, 14: 39.
[26] ZHOU Y, YAN D, LI L, et al. Acta Sci. Circumst, 2015, 35: 3769.
[27] JAAFAR I, GRIFFITHS A J, HOPKINS A C, et al. Miner. Eng., 2011, 24: 1028.
[28] JIANG G, WU P, WANG Z, et al. Nonferrous Met. Sci. Eng., 2016, 6: 43.
[29] WANG C, HU X, MATSUURA H, et al. ISIJ Int., 2007, 47: 370.
[30] XING Z, CHENG G, YANG H, et al. Miner. Eng., 2020, 154: 106404.
[31] SUN J, HAN P, LIU Q, et al. Trans. Indian Inst. Met., 2019, 72: 1053.
[32] LI K, PING S, WANG H, et al. Int. J. Miner. Metall. Mater., 2013, 20: 1035.

Recovery of Copper, Lead and Zinc from Copper Flash Converting Slag by the Sulfurization-reduction Process

This research proposed an efficient and environmentally friendly technology for recovering valuable metals from copper flash converting slag. In a reducing atmosphere, pyrite is used as the sulfurizing agent to efficiently and selectively sulfurize and recover Cu, Pb and Zn of copper flash converting slag. Cu of copper flash converting slag is mainly presented in the form of copper oxide and metallic copper. Pb mainly exists in the form of lead oxide, and Zn mainly exists in the form of zinc ferrite and zinc silicate. The feasibility of the technology is confirmed by thermodynamic analysis and experimental study. The optimum slag cleaning conditions were determined as follows: reductive agent coke dosage of 6%, 2% pyrite addition of converter slag weight and smelting at 1350 ℃ for 3 h. Under the optimum conditions, the recoveries of Cu, Pb and Zn were 98.3%, 99.56% and 83.08%, respectively. The grade of matte was 52.44%. The cleaned slag contained 1.10% Cu, 0.09% Pb and 1.25% Zn. The proportions of zinc entering the vapor, slag and matte phases were 62.07%, 16.92% and 21.01%, respectively. The proportions of lead entering the vapor, slag and matte phases were 2.16%, 0.41% and 97.43%, respectively.

1 Introduction

Nonferrous metals (Cu, Pb, Zn) remaining in copper converting slag have become a global problem in copper pyrometall urgy.[1] By the end of 2020, more than 30 copper smelters were put into operation in China with an annual capacity of refined copper of 10225 kt, ranking first in the world.[2, 3] After more than 100 years of development, copper converter technology includes the Peirce-Smith converter,[4, 5] Kennecott-Outotec flash converter,[6] bottom blowing converter,[7] Ausmelt converter,[8] Mitsubishi converter,[9] and so on.[10] The chemical compositions of the

---

Published in *JOM*, 2023, 75(4): 1107-1118. Authors: Li Zhongchen, Tian Qinghua, Wang Qinmeng, Guo Xueyi.

converting slag contain approximately 4%-27% Cu, 2%-11% Pb and 2%-6% Zn, which is an important resource.[11] At present, most of the copper converting slag is treated and recovered by returning to the smelting process. Nevertheless, most copper of flash converting slag exists in the form of Cu_2O, which increases the processing difficulty of the copper smelting process.[12, 13] Impurities such as Pb/Zn enter copper matte during the smelting process, reducing the quality of copper matte and increasing the impurity removal pressure in subsequent processes.[14] Therefore, it is of economic and environmental significance for the copper smelting industry to effectively recover valuable components from copper converting slag and develop new technologies for comprehensively using copper slag resources.[1, 15]

The flash converter adopts calcium ferrite slag, and the calcium-iron ratio is controlled at 0.3%-0.4%.[8, 16] The furnace operates in a strong oxidizing atmosphere, the copper in the slag mainly exists in the form of Cu_2O, and the content of Cu in the slag is generally >20%. The high content of Cu_2O in the furnace improves its ability to dissolve Fe_3O_4 so that it will not be precipitated, so the content of Fe_3O_4 is generally >15% in the flash converter process.[17] The content of SiO_2 in flash converting slag is generally controlled at 1.5%-3.0%, which affects the lead removal effect of the converting furnace.

In recent years, many scholars have performed extensive research on the recovery of valuable metals from copper converting slag, mainly hydrometallurgy and pyrometallurgy.[11, 18, 19] Du et al.[20] carried out reduction treatment on topblown converting slag. By adding SiO_2 and coke, the content of Cu in the slag was decreased from 19.85% to approximately 3%, and the content of Fe_3O_4 decreased to 7.0%. Dosmukhamedov et al.[1] sulfurized copper converter slag and explored the influence of copper concentrate, copper-zinc concentrate and matte of the blast furnace on the content of copper and magnetite in converter slags. The content of Cu in the slag was decreased from 2.5%-7% to 1.3%-2.3%, and the content of iron (Ⅲ) oxide in the slag decreased from 16%-18% to 10%-13% after short term treatment of the converter slag with oxygen flash furnace matte containing 40%-45% Cu. Topcu et al.[21] reduced copper losses to the converting slag in the converter stage by adding calcined colemanite. The content of Cu in the slag was decreased from 4.45% to 12% by adding 2% calcined colemanite at 1250 ℃ for 3 h.

Gargul et al.[22] used citric acid solutions to selectively leach Cu and Pb from copper flash converting slag. The Pb concentration was decreased from 3.05% to 0.41%-0.6% and the analogous values for Cu were 12.44% (before leaching) and 11.5%-11.8% (after leaching). Meshram et al.[23] recovered Cu, Ni and Co from waste copper converter slag by organic acids. In 9-10 h at 308 K and 15% pulp density with 2 N citric acid using <45 μm particles, 99.1% Cu, 89.2% Ni and 94% Co were recovered. Aracena et al.[24] carried out in 1.2 m high columns with a cross-section diameter of 7.5×10^{-2} m, using 2.0 kg of converter slag. Using an ammonia system, the recovery values reached 87.7% for Cu, with almost no impurities.

At present, the main research focus is on copper converter slag, but with the development of the copper industry, the flash converter is the future development trend, so flash converting slag is studied in this paper. In this study, the valuable metals in flash converting slag were recovered by sulfurization-reduction method. The novel method was proposed to recover copper and eliminate other hazardous metals simultaneously from copper converting slag. The physical and chemical property composition change rules of copper converting slag during cleaning, fuming and thermal modulation were studied. The multiphase distribution rules of Cu, Pb and Zn in copper converting slag were clarified, thus realizing the cleaning of copper converting slag. By adding different contents of sulfurizing agent and a reducing agent to copper converting slag and controlling the reaction temperature, the distribution of various valuable elements to copper matte, slag and vapor was explored. The optimum addition combination and process conditions were explored to realize Cu recovery and Pb/Zn fumes.

2 Experimental

2.1 Materials

Copper flash converting slag was taken from an enterprise in Henan Province. The raw materials were crushed and dried for composition and chemical phase analysis. The chemical compositions of the raw materials are shown in Table 1. The copper flash converting slag contains 26.62% Cu, 8.84% Pb and 3.04% Zn. Pyrite contains 46.62% Fe and 46.61% S. Coke ash chemical composition, proximate analysis and ash fusibility analysis were determined by GB/T 212-2008 and GB/T 219-2008, and the results are shown in Table 1.

Table 1 Chemical analysis of experimental raw materials   %

Materials	Composition									
Flash converting slag	Fe_{total}	Cu	Pb	Zn	As	CaO	SiO_2	Ni	S	Co
	31.68	26.62	8.84	3.04	0.27	8.54	0.9	0.29	0.18	0.12
Pyrite	Fe_{total}	S	Zn	CaO	MgO	Al_2O_3	SiO_2	Bi		
	46.62	46.61	0.12	1.41	0.12	0.12	3.82	0.07		
Coke	Industrial analysis/%				Chemical composition of the ash					
	M_{ad}	A_{ad}	V_{ad}	FC_{ad}	TFe	Al_2O_3	CaO	MgO	S	P
	0.87	14.10	7.22	77.81	0.30	9.18	0.378	0.105	0.55	0.35

The distribution of copper/iron/lead/zinc in associated minerals in converting slag was determined based on the solubility difference of various phases in a solvent, and the results are given in Table 2, which shows that 82.19% of copper is present in the form of copper oxide, and 6.01% of copper exists in metallic copper. Because flash converting is a strong oxidation process, most of the copper in the slag exists in the form of oxide. Iron mainly exists in the form of magnetite, ferric oxide and ferrous oxide, accounting for 51.39%, 16.07% and 23.45% of the total iron, respectively. The high content of Fe_3O_4 leads to excessive copper in slag. Lead mainly exists in the form of lead oxide, accounting for 92.08% of the total lead, and zinc mainly exists in the form of zinc ferrite and zinc silicate, accounting for 89.47% and 5.59% of the total zinc, respectively.

Table 2 Distribution of copper/iron/lead/zinc in associated minerals   %

Element	Mineral				
Copper	Copper oxide	Metallic copper	Copper sulfide	Combined copper oxide	Cu_{total}
Content	21.88	1.60	0.68	2.46	26.62
Fraction	82.19	6.01	2.56	9.24	100
Iron	Magnetite	Metallic iron	Fayalite	Ferric oxide	Ferrous oxide	Fe_{total}
Content	16.28	1.28	1.60	5.09	7.43	31.68
Fraction	51.39	4.04	5.05	16.07	23.45	100
Lead	Lead sulfide	Lead oxide	Lead sulfate	Plumbojaro	Pb_{total}
Content	0.33	8.14	0.02	0.35	8.84
Fraction	3.73	92.08	0.23	3.96	100
Zinc	Zinc ferrite	Zinc silicate	Zinc oxide	Zinc sulfate	Zn_{total}
Content	2.72	0.17	0.02	0.12	3.04
Fraction	89.47	5.59	0.66	3.95	100

The phase compositions and microstructure of the copper flash converting slag were determined by X-ray diffraction (XRD, D/max-2550, Cu K_α radiation, from Nippon Science Co., Ltd.) analyses and SEM-EDS (SIRION200 from FEI Co., USA). The corresponding results are presented in Fig. 1. The result indicated that the converter slag was mainly comprised of calcium iron oxide ($Ca_{0.15}Fe_{2.85}O_4$), calcium iron oxide ($Ca_4Fe_9O_{17}$), cuprite (Cu_2O), copper (Cu), delafossite ($CuFeO_2$) and litharge (PbO), which was the same as the chemical composition analysis. Fig. 1(b) illustrates the microstructure and element compositions of the converter slag. Three mineral phases were identified with the help of EDS analysis: calcium iron oxides, cuprite (Cu_2O) and copper (Cu).

2.2 Experimental methods

In the experimental process, 200 g copper flash converting slag and a certain proportion of composite additives (pyrite and coke) were fully mixed in an agate mortar and then put into an alumina crucible. The alumina crucible was put into a vertical tube furnace (as shown in Fig. 2), heated at a rate of 10 ℃/min under inert atmosphere Ar and kept at a constant temperature (1250-

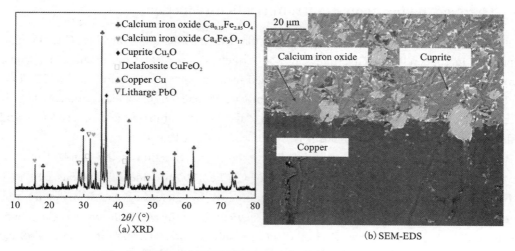

Fig. 1 XRD and SEM-EDS patterns of the converting slag

1400 ℃) for 3 h to ensure full smelting. After the reaction, the experimental sample was cooled to room temperature under the protection of an inert atmosphere. Then, the experimental sample was taken out of the furnace body and weighed. After the crucible was broken, the copper matte and slag phases appeared in the experimental sample, and the element content was detected by ICP analysis (PS-6, Barid Co., USA). Because of the high content of copper in the copper matte phase, the content of copper in the copper matte phase was determined by chemical titration.

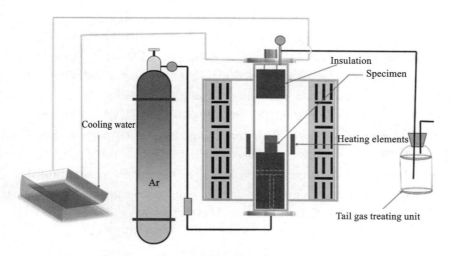

Fig. 2 Schematic diagram of the experimental apparatus

2.3 Thermodynamic analysis of sulfurization-reduction

In the previous analysis, we determined the existing state of valuable Cu, Pb and Zn components in copper flash converting slag. Nevertheless, to express the reaction mechanism as accurately as possible and simplify the complicated metallurgical reaction process, we used Cu_2O, PbO and $ZnFe_2O_4$ to express the existing forms of Cu, Pb and Zn dissolved in molten slag. In the sulfurization-reduction process, Cu_2O, PbO and $ZnFe_2O_4$ are sulfurized and reduced to sulfide or metallic states. The reaction mechanism is expressed by Eq. (1)-(12).

Under the same atmospheric pressure, Fig. 3 shows the variation in the standard free energy with temperature for the sulfurization-reduction of Cu/Fe/Pb/Zn oxides. Equations (1) to (12) can be promoted in the sulfurization-reduction process by increasing the temperature. In the sulfurization-reduction process, Cu_2O in molten slag can be sulfurized-reduced into Cu_2S and Cu. $ZnFe_2O_4$ in the slag is reduced to ZnS and ZnO by sulfurization. On the one hand, ZnS can be volatilized into the gas phase. On the other hand, ZnO reacts with the reducing agent to generate elemental zinc, which is volatilized into the gas phase. PbO in molten slag is reduced to metallic lead, which enters the copper matte phase or volatilizes into the gas phase.

$$Cu_2O(l) + FeS(l) = Cu_2S(l) + FeO(l) \quad (1)$$
$$Cu_2O(l) + C(s) = 2Cu(l) + CO(g) \quad (2)$$
$$2Cu(l) + FeS(l) = Cu_2S(l) + Fe(l) \quad (3)$$
$$PbO(l) + C(s) = Pb(g) + CO(g) \quad (4)$$
$$PbO(l) + C(s) = Pb(l) + CO(g) \quad (5)$$
$$PbO(l) + FeS(l) = PbS(g) + FeO(l) \quad (6)$$
$$PbO(l) + FeS(l) = PbS(l) + FeO(l) \quad (7)$$
$$ZnFe_2O_4(l) + FeS(l) + C(s) = ZnS(g) + CO(g) + 3FeO(l) \quad (8)$$
$$ZnFe_2O_4(l) + FeS(l) + C(s) = ZnS(l) + CO(g) + 3FeO(l) \quad (9)$$
$$ZnFe_2O_4(l) + C(s) = ZnO(l) + CO(g) + 2FeO(l) \quad (10)$$
$$ZnO(l) + C(s) = Zn(g) + CO(g) \quad (11)$$
$$Fe_3O_4(l) + C(s) = 3FeO(l) + CO(g) \quad (12)$$

In the cleaning process of copper flash converting slag, Fe_3O_4 plays a critical role.[25] The low content of Fe_3O_4 can decrease the viscosity and improve the fluidity of molten slag. Therefore, considering that there is 51.39% Fe_3O_4 in the copper flash converting slag, it is necessary to reduce the content of Fe_3O_4 by adding a reducing agent and sulfurizing agent.

$$Fe_3O_4(s) + C = 3FeO(l) + CO(g) \quad (13)$$
$$3Fe_3O_4(s) + FeS(l) = 10FeO(l) + SO_2(g) \quad (14)$$
$$G^\ominus = -RT\ln K^\ominus \quad (15)$$
$$K^\ominus = \frac{\alpha_{FeO}^{10} \times P_{SO_2}}{\alpha_{Fe_3O_4}^3 \times \alpha_{FeS}} \quad (16)$$
$$\alpha_{Fe_3O_4}^3 \times \alpha_{FeS} = \alpha_{FeO}^{10} \times \frac{P_{SO_2}}{e^{-\frac{G^\ominus}{RT}}} \quad (17)$$

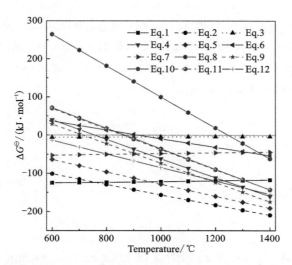

Fig. 3 Variation of standard Gibbs free energy $\Delta G^{\ominus}$ of several reactions with temperature obtained by thermodynamic calculations using Factsage 7.1 software

When the temperature, P_{SO_2} and α_{FeS} are fixed, the relationship between α_{FeO} and $\alpha_{Fe_3O_4}$ can be expressed by combining Eq. (13)-(17). Therefore, the relationships between α_{FeO} and $\alpha_{Fe_3O_4}$ in the slag with different values of α_{FeS} were calculated by Eq. (17). In the case of $P_{SO_2} = 0.25$ atm, the relationships between α_{FeO} and $\alpha_{Fe_3O_4}$ in the molten slag with different values of α_{FeS} were calculated by Eq. (17), and the results are presented in Fig. 4(a). With increasing αFeS, the activity of Fe_3O_4 is obviously decreased, which indicates that Fe_3O_4 can be easily reduced to FeO by adding the sulfurizing agent. Therefore, adding a reducing agent and sulfurizing agent can reduce the content of Fe_3O_4 in copper converting slag, thus reducing the loss of copper in slag.

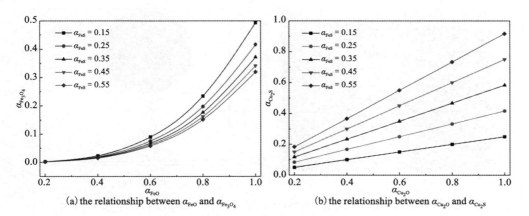

Fig. 4 Activity relationship

Cu of copper flash converting slag is mainly presented in the form of copper oxide. It is necessary to consider the chemical reaction between Cu_2O and the reaction system additives. The relationship between α_{Cu_2O} and α_{Cu_2S} with the activity of the sulfurizing agent was investigated, as shown in Fig. 4(b). When the sulfurizing agent is constant, α_{Cu_2S} is increased with an increase in α_{Cu_2O}. With an increasing in α_{FeS}, α_{Cu_2S} is obviously increased, which shows that Cu_2O can easily become Cu_2S by adding the sulfurizing agent.

$$\alpha_{Cu_2O} \times \alpha_{FeS} = \frac{\alpha_{FeO} \times \alpha_{Cu_2S}}{e^{\frac{-G^\ominus}{RT}}} \tag{18}$$

3 Results and discussion

3.1 Experimental study on the station of copper converting slag

According to the above experimental process, the whole heating and cooling stage was carried out in an argon atmosphere (150 mL/min). After cooling to room temperature, the experimental samples were removed. The picture shows the physical picture of the experimental sample after melting.

Fig. 5 shows that small copper particles can aggregate into large copper bulk after standing for 3 h at 1350 ℃. However, since most of the copper in the copper converting slag exists in the form of oxide, only the simple copper in the copper converting slag can be collected and recovered through the standing experiment. If we want to recover Cu_2O from copper converting slag, we can recover valuable metals from copper converting slag by the appropriate reducing or sulfurizing agent.

Fig. 5 Section map of copper converting slag after standing at 1350 ℃ for 3 h

3.2 Effect of pyrite addition on the separation and recovery of copper, lead and zinc

In the process of sulfurization and reduction, the copper or copper matte produced by the reaction is dispersed in the slag as tiny droplets. When the gravity of tiny droplets of copper or copper matte suspended or immersed in the slag is greater than the buoyancy, the droplets sink in the slag, and the average speed of the drop can be expressed by the Stokes formula[25]:

$$v = \frac{2gr^2}{9\eta_s}(\rho_M - \rho_S) \qquad (19)$$

In Eq. (19), v = the average speed of droplet descent, m/s; r = the radius of the falling droplet, m; η_s = viscosity of slag, Pa/s; ρ_M, ρ_S = the density of copper (or copper matte) and slag, respectively, kg/m^3; g = acceleration of gravity, m/s^2. In the molten state, the density of metallic copper is approximately 8920 kg/m^3, and the density of copper matte is 4000-5000 kg/m^3. In the same slag system, the slag viscosity and slag density are certain, so under the same droplet radius, the sedimentation rate of copper is approximately twice that of copper matte.

The recovery of Cu and the grade of Cu in copper cleaned slag and matte by adding a sulfurizing agent (pyrite) are shown in Fig. 6(a). With the addition of the sulfurizing agent, the grade of matte is increased from 46.68% to 52.44% and then decreased from 52.44% to 47.69% at the same amount of reducing agent, smelting temperature and constant temperature time. With the addition of the sulfurizing agent, the content of Cu in molten slag is first decreased and then increased. When the content of pyrite is 0%-2%, the content of Cu in slag is decreased from 3.16% to 1.10%. When the content of pyrite is 2%-4%, the content of Cu in slag is increased from 1.10% to 2.60%.

With the addition of the sulfurizing agent, Cu_2O in slag can be sulfurized-reduced into Cu_2S and Cu, and then Cu_2S can combine with FeS to form the copper matte phase. However, with the excessive amount of FeS, the proportion of Cu_2S is increased, which leads to a decrease in the proportion of metallic copper. As Eq. (19) shows, the settling effect of metallic copper is better than that of copper matte, so adding an excessive sulfurizing agent makes more copper mixed in cleaned slag, resulting in an increase of copper in slag. The experiment also found that the slag and matte can be easily separated by adding a proper amount of sulfurizing agent.

The recovery of Pb/Zn and the content of Pb/Zn in cleaned slag are shown in Fig. 6(b). With increasing sulfurizing agent content, the content of Pb in slag is decreased significantly from 1.36% to 0.09% and then increased to 1.06%. Lead mainly exists in the form of lead oxide in converting slag. With the addition of the reducing agent and the sulfurizing agent, lead in converting slag forms liquid lead in the copper matte or gaseous lead in the vapor phase. However, with an excessive sulfurizing agent, S will be supersaturated in the smelting system, which will deteriorate slag properties and make it difficult for lead sulfide produced by lead to volatilize into the gas phase.

The content of Zn in slag is decreased from 1.96% to 1.25% and then increased to 1.70%

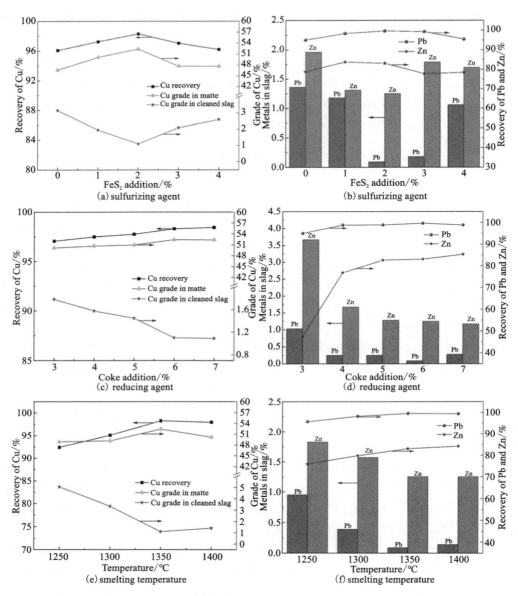

Fig. 6　Effect of compound additives on the separation and recovery of copper, lead and zinc from copper slag

with the addition of the sulfurizing agent, in which Zn mainly exists in the form of zinc silicate and zinc ferrite. With the addition of sulfurizing agent and reducing agent, the structure of zinc silicate and zinc ferrite is destroyed, so the zinc element is sulfurized and reduced to elemental zinc form and then volatilized. However, if the sulfurizing agent is excessive, on the one hand, the content of iron oxide in slag is decreased, leading to an increase of the viscosity of the molten slag; on the other hand, it leads to the increase of the content of ZnS, so that the sulfurization-reduction effect of zinc becomes worse. In summary, the optimum amount of sulfurizing agent was 2% of the total mass of copper slag.

3.3 Effect of reducing agent addition on the separation and recovery of copper, lead and zinc

The recovery of copper by adding the reducing agent (coke) and the grade of copper in cleaned slag and copper matte is shown in Fig. 6(c). Under the same amount of sulfurizing agent, smelting temperature and constant time, the grade of matte is gradually increased and then tends to be stable with the addition of coke. When the reducing agent is 3%-6%, the grade of matte is increased from 50.21% to 52.44%. When the reducing agent is added at 6%-7%, the grade of matte is unchanged. The content of Cu in the slag is gradually decreased and then stabilized. When the reducing agent is 3%-6%, the content of copper in slag is decreased from 1.79% to 1.10%, and the recovery of Cu is reached 98.30%.

The recovery of Pb/Zn and the content of Pb/Zn in cleaned slag are shown in Fig. 6(d). The recovery of Pb is gradually increased with increasing coke and then tends to be stable. With the addition of 3%-4% reducing agent, the recovery of Pb is increased from 94.93% to 98.81%. With the addition of 4%-7% reducing agent, the recovery of Pb is not increased obviously, only from 98.81% to 99.59%. The recovery of Zn is increased from 47.51% to 85.31% in the range of 3%-7% reducing agent because zinc mainly exists in the form of zinc ferrite. With the addition of the reducing agent, more iron oxides are reduced to FeO, the structure of zinc ferrite is destroyed, and the zinc oxide contacts the reducing agent more fully, resulting in a gradual increase in the recovery of Zn. However, adding more reducing agents lead to a waste of heat and resources. On the other hand, it can lead to the formation of "Fe", which deteriorates the properties of molten slag and is not conducive to the stratification of slag and matte.[26] Therefore, 6% reducing agent is selected as the optimal condition, under which the recovery of Zn is 83.08%, and the content of zinc in cleaned slag is 1.25%.

When the addition of the reducing agent is 6% to 7%, the total mass still decreases after the reaction, which shows that at this stage, the reducing agent reduces the high-valent iron oxide to the low-valent iron oxide. Combined with Fig. 7(a), the formation of zinc ferrite is related to the Fe_2O_3 content, so the structure of zinc ferrite is destroyed more thoroughly with the addition of a reducing agent, so the above analysis is verified.

Zn volatilizes into the vapor phase mainly by reducing to gaseous zinc, so the effect of reducing agent addition on the three-phase distribution of zinc was investigated separately. As shown in Fig. 7(b), in the range of 3%-5% of reducing agent, the proportion of zinc entering the gas phase is obviously increased, from 39.89% to 61.82%. In the range of 5%-7% of reducing agent, the proportion of zinc entering the gas phase is increased from 61.82% to 62.10%.

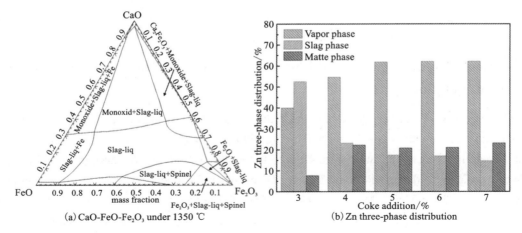

Fig. 7 Effect of reductant addition on the three-phase distribution of zinc

3.4 Effect of smelting temperature on the separation and recovery of copper, lead and zinc

The relationship between melt density and viscosity changing with temperature is shown in Eq. (20) and Eq. (21):

$$\eta = A_\eta \exp\left(\frac{E_\eta}{RT}\right) \quad (20)$$

$$\rho_T = \rho_m - \alpha(T - T_m) \quad (21)$$

In Eq. (20) and Eq. (21), ρ_T = density of melt at a certain temperature T, kg/m³; ρ_m = density of melt at melting temperature T_m, kg/m³; α = constant related to melt properties; A_η = constant; E_η = activation energy of the viscous flow. Eq. (20) and Eq. (21) show that the density and viscosity of melt are negatively correlated with temperature; that is, the density and viscosity decrease with increasing temperature.

The effect of temperature change on copper recovery and copper grade in copper cleaned slag and copper matte is shown in Fig. 6(e). At 1250 ℃, there are obvious small copper particles in the upper slag phase, and the delamination effect is not good, which shows that the temperature is too low, the slag viscosity is too high, and the small copper matte particles have difficulty aggregating into large copper matte particles, resulting in a poor copper matte sedimentation effect. However, with increasing temperature, the slag and matte separation effect is good because molten copper slag has good fluidity and low viscosity at high temperatures, which is beneficial to copper.

Under the same addition amount of reducing agent, sulfurizing agent and constant temperature time, the grade of matte is increased from 48.90% to 52.44% and then decreased to 50.16% with increasing temperature. The recovery of Cu and the content of Cu in cleaned slag is gradually decreased and then stabilized. In the range of 1250-1350 ℃, the content of Cu in the cleaned slag is decreased from 5.06% to 1.10%. In the range of 1350-1400 ℃, the content of Cu in the cleaned slag remained unchanged.

The recovery of Pb/Zn and the content of Pb/Zn in the cleaned slag are shown in Fig. 6(f). The recovery of Pb/Zn is gradually increased with increasing temperature and then tends to be stable. In the range of 1250-1350 ℃, the recovery of Pb is increased from 95.70% to 99.59%, and the recovery of Zn is increased from 75.95% to 84.21%. In the range of 1350-1400 ℃, the recovery of Pb/Zn is unchanged. With increasing temperature, the content of Pb/Zn in the cleaned slag is first decreased and then stabilized, and the content of Pb/Zn is decreased from 0.95% to 0.09% and from 1.83% to 1.25%, respectively. Therefore, 1350 ℃ was chosen as the best temperature for the reaction system.

3.5 Optimizing conditions for the separation and recovery of copper, lead and zinc

Under the optimum conditions, the distribution regularities of Pb/Zn are different from those of Cu. The former can volatilize into the gas phase, so it is necessary to consider the three-phase distribution regularities of Pb/Zn. Through a single-factor experiment, the optimum reaction conditions were determined as follows: reductive agent coke dosage of 6%, 2% pyrite addition of converter slag weight and smelting at 1350 ℃ for 3 h. The recovery and three-phase distribution of Pb/Zn are shown in Fig. 8(a). The recoveries of Pb/Zn were 99.59% and 83.08%, respectively. The proportions of lead entering the vapor, slag and matte phases were 2.16%,

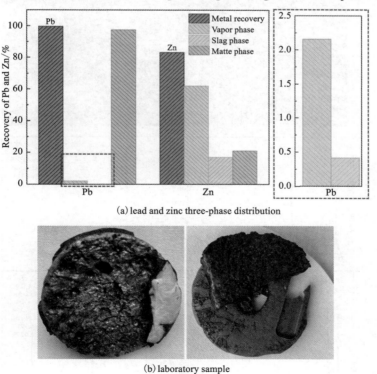

(a) lead and zinc three-phase distribution

(b) laboratory sample

Fig. 8 Optimization conditions

0.41% and 97.43%, respectively. The proportions of zinc entering the vapor, slag and matte phase were 62.07%, 16.92% and 21.01%, respectively. Through experiments, it was found that lead significantly entered the matte phase.

Under optimized conditions, Fig. 9(a) presents the SEM-EDS images of cleaned slag. In area A, the main elements are Fe and O, with contents of 72.45% and 21.06%, respectively, and a small amount of Cu and Zn, which are converted into atomic ratios and combined with XRD [Fig. 9(b)]. The phase in this area is FeO, indicating that with the addition of a reducing agent, high-valent iron oxide is reduced to low-valent iron oxide. In area B, the main elements are Fe, O, Ca and Al, with contents of 29.25%, 27.57%, 33.06% and 9.35%, respectively, with small amounts of Cu and Zn. Combined with XRD, the phase in this area is $Ca_2(Al, Fe)O_5$. In area C, Zn, Fe, S and O are the main elements, with 47.4%, 29.51%, 14.05% and 4.26%, respectively. Combined with XRD, ZnS exists in this area. In area D, the main elements are Cu, Fe and S, the contents of which are 63.46%, 23.89% and 4.82%, respectively, indicating that there are copper matte inclusions in the cleaned slag, which proves that with the increasing of sulfurizing agent, small particles of copper matte do not aggregate into large particles of copper matte and enter the copper matte phase.

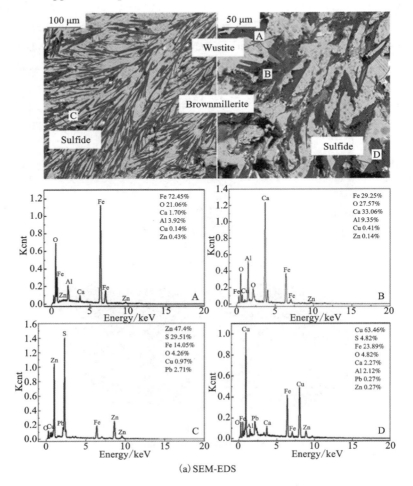

(a) SEM-EDS

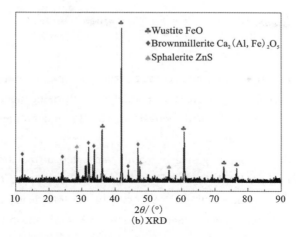

Fig. 9 SEM-EDS and XRD pattern of the cleaned slag

Under the optimized conditions, four different matte morphologies were identified in Fig. 10. In area I, the content of copper is 93.33%, which is mainly the Cu phase. In area II, the contents of copper and iron are 8.04% and 89.23%, respectively, which are mainly the Cu–Fe phase. In area III, Zn, Cu, Fe and S are the main elements, with contents of 36.70%, 2.57%, 9.69% and 46.58%, respectively, which are mainly the Zn–Cu–Fe–S phase, indicating that Zn partially enters copper matte. In area IV, the contents of Cu and Pb are 73.43% and 22.08%, respectively, which are mainly the Cu–Pb phase, indicating that lead can enter into the matte phase.

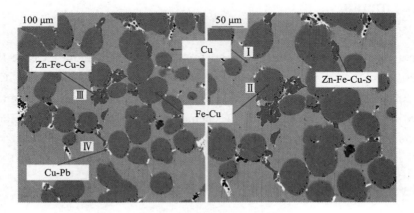

Fig. 10 SEM-EDS of the copper matte

4 Conclusion

In this article, the occurrence forms of copper, lead and zinc in copper flash converting slag are analyzed in detail, and the effects of the amount of sulfurizing agent, reducing agent and smelting temperature on the separation and recovery of Cu, Pb and Zn are systematically studied. The following conclusions are obtained:

(1) Cu of copper flash converting slag is mainly presented in the form of copper oxide and metallic copper, accounting for 82.19% and 6.01% of the total copper, respectively. Pb mainly exists in the form of lead oxide accounting for 92.08% of the total lead, and Zn mainly exists in the form of zinc ferrite and zinc silicate, accounting for 89.47% and 5.59% of the total zinc, respectively.

(2) The optimum reaction conditions in the sulfurization-reduction reaction system were as follows: the mass of sulfurizing agent and the reducing agent was 6% and 2% of the total mass of copper slag, respectively. The melting temperature was 1350 ℃, and the constant temperature was 3 h. Under the optimum conditions, the recoveries of Cu, Pb and Zn were 98.3%, 99.56% and 83.08%, respectively. The grade of matte was 52.44%. The cleaned slag contained 1.10% Cu, 0.09% Pb and 1.25% Zn. The proportions of zinc entering the vapor, slag and copper matte phases were 62.07%, 16.92%, and 21.01%, respectively. The proportions of lead entering the vapor, slag and matte phases were 2.16%, 0.41% and 97.43%, respectively.

Acknowledgments

The authors appreciate the financial support from the National Natural Science Foundation of China (51620105013, and U20A20273), the Innovation Driven Program of Central South University (2020CX028), the National Key Research and Development Program of China (2019YFC1907400) and the Science and Technology Innovation Program of Hunan Province (2021RC3005).

References

[1] DOSMUKHAMEDOV N, EGIZEKOV M, ZHOLDASBAY E, et al. JOM, 2018, 70: 2400.
[2] LIU Z, XIA L. Min. Proc. Ext. Met., 2019, 128: 117.
[3] WANG Q, LI Z, LI D, et al. JOM, 2020, 72: 2676.
[4] TASKINEN P, AKDOGAN G, KOJO I, et al. Min. Proc. Ext. Met., 2019, 128: 58.
[5] DING L, ZHANG L, BERNITSAS M M, et al. Renew. Energ., 2016, 85: 1246.

[6] YU F, XIA L, ZHU Y, et al. Metall. Mater. Trans. B., 2021, 52: 3468.
[7] GUO X, ZHANG B, WANG Q, et al. JOM, 2021, 73: 1861.
[8] WOOD J, HOANG J, HUGHES S. JOM, 2017, 69: 1013.
[9] PIATAK N M, PARSONS M B, SEAL R R. Appl. Geochem, 2015, 57: 236.
[10] ZHANG H, WANG Y, HE Y, et al. Miner. Eng., 2021, 160: 106661.
[11] SUN Y, CHEN M, CUI Z, et al. Metall. Mater. Trans., 2020, B 51: 426.
[12] MURAVYOV M, PANYUSHKINA A, BULAEV A, et al. Miner. Eng., 2021, 170: 107040.
[13] LI M, ZHOU J, TONG C, et al. Metall. Mater. Trans, 2018, B 49: 1794.
[14] YANG X, HAN Z, GUO Y, et al. China Nonferrous Metallurgy, 2018, 6: 8.
[15] SARFO P, DAS A, WYSS G, et al. Waste Manage, 2017, 70: 272.
[16] JANSSON J, TASKINEN P, KASKIALA M. Can. Metall. Quart, 2014, 53: 1.
[17] KOLCZYK E, MICZKOWSKI Z, CZERNECKI J. Metalurgija, 2017, 56: 139.
[18] ZHAI X, LI N, ZHANG X, et al. Trans. Nonferrous Met. Soc. China, 2011, 21: 2117.
[19] BOYRAZLI M, ALTUNDOGAN H S, TÜMEN F. Can. Metall. Quart, 2006, 45: 145.
[20] DU C, CHEN G, LUO Y. China Nonferrous Metallurgy, 2016, 2: 13.
[21] TOPÇU A M, RÜŞEN A, DERIN B. Mater J. Res. Technol, 2019, 8: 6244.
[22] GARGUL K, BORYCZKO B, BUKOWSKA A, et al. Arch. Civ. Mech. Eng., 2019, 19: 648.
[23] MESHRAM P, PRAKASH U, BHAGAT ABHILASH L, et al. Minerals, 2020, 10: 290.
[24] ARACENA A, FERNÁNDEZ F, JEREZB O, et al. Hydrometallurgy, 2019, 188: 31-37.
[25] GUO Z, PAN J, ZHU D, et al. J. Clean. Prod., 2018, 199: 891.
[26] LI Y, CHEN Y, TANG C, et al. J. Hazard. Mater, 2017, 322: 402.

Mechanism and Kinetics for the Carbothermal Reduction of Arsenic-alkali Mixed Salt Produced from Treatment of Arsenic-alkali Residue

Abstract: Arsenic-alkali mixed salt (AAS) is a hazardous by-product of the treatment of arsenic-alkali residue (AAR). Although direct carbothermal reduction can enable the harmless and resourceful disposal of AAS, the specific reaction pathway and mechanism of the reduction process remain unclear. To gain further insight into the reaction process, the non-isothermal kinetics and reaction mechanism of AAS carbothermal reduction were investigated. The results indicate that AAS reduction involves three pathways: (1) direct reduction of $Na_2As_4O_{11}$ to As(g), (2) reduction of $Na_2As_4O_{11}$ to As_2O_3 and subsequent reduction to As(g), and (3) direct reduction of Na_3AsO_4 to As(g). Calculations using the Flynn-Wall-Ozawa (FWO) and integral master-plot methods indicate that the kinetic model conforms to the two-dimensional diffusion function, with an average activation energy of 164.30 kJ/mol and a pre-exponential factor of 1.764×10^6 L/s. These results are significant for further in-depth research into the carbothermal reduction of AAS.

Keywords: Arsenic-bearing waste; Carbothermal reduction; Non-isothermal kinetics; Kinetic triplets; Thermal analysis

1 Introduction

Arsenic is a well-known toxic element that is generally found in association with various minerals. With the development of the smelting industry, the increasing amounts of arsenic-containing waste pose risks to the environment and human health[1,2]. As a waste material produced during the process of refining crude antimony, the harmless treatment of AAR has been widely studied[3-6].

AAR is an important secondary resource containing arsenic, antimony, and alkali[7,8]. Due to

---

Published in *Thermochimica Acta*, 2023(728): 179591. Authors: Gong Ao, Wu Xuangao, Li Jinhui, Wang Ruixiang, Chen Lijie, Tian Lei, Xu Zhifeng.

the high solubility and toxicity of sodium arsenate, the leakage of AAR poses a severe threat to human life and health[9, 10]. China has the world's richest antimony resources, and although antimony production has declined in recent years, China ranked first, with 67% of the global production from 2015 to 2019[11, 12]. Consequently, it is difficult to decrease stockpiles of AAR. According to various reports, the AAR reserve in China amounts to approximately 200000 tons, increasing at a rate of 5000 tons annually[4, 13]. Therefore, the effective disposal of AAR is of considerable significance for the sustainable development of the antimony smelting industry.

Currently, the most common method for harmless and resourceful treatment of AAR employs hydrometallurgy, which utilizes the difference in the solubilities of sodium arsenate and sodium antimonate in water to separate arsenic and antimony. After leaching, antimony remains in the residue in the form of sodium antimonate, while sodium arsenate and alkali enter the solution to generate an arsenic-alkali solution[14-16]. The arsenic-alkali solution has been subjected to evaporative crystallization to obtain AAS, which can be used as a glass-clarifying agent. This resulted in the overproduction of AAS, which accumulated in large stockpiles owing to the shrinking market.

The solidification process is one of the most common methods used to prevent arsenic leakage from arsenic-containing waste residues and is considered by the United States Environmental Protection Agency to be the most effective technology for treating toxic wastes[17, 18]. Numerous studies have been conducted on the stabilization treatment of arsenic-alkali waste residues[19, 20]. However, the solidification process increases the volume of the treated solids, requiring a large amount of land for landfilling, reduces available human living space, and squanders resources such as arsenic in the waste residue. In contrast, the reduction of arsenic-containing waste residues is an effective waste-treatment strategy. Reduction treatment greatly reduces the volume of waste residue, and toxic arsenic compounds can be reduced to non-toxic and commercially valuable zero-valent arsenic. AAS is an arsenic-containing waste residue suitable for treatment using this method. After reduction roasting, the total amount of residue is greatly reduced: for one ton of AAS after reduction roasting, the total amount of roasting residue is only approximately 400 kg, and the main components of the roasting residue are sodium carbonate and sodium hydroxide, which can be recycled further. Arsenic is recovered in the form of element arsenic, which can be used in semiconductor materials after further purification.

The reduction treatment of AAS has been studied in recent years[18, 21], where the effective disposal of AAS has been achieved through the carbothermal reduction process. Arsenic was recovered in the form of metallic arsenic with commercial value, with a recovery ratio of more than 99%. Although better recovery of arsenic has been achieved, there have been no studies on the non-isothermal kinetics of the carbothermal reduction of AAS, and the carbothermal reduction pathway and mechanism of AAS are still unclear. Consequently, there is a lack of theoretical guidance for AAS reduction. The amount of carbon powder added in current treatments is generally too high, and the reaction temperature is also too high to ensure a good reduction effect (e.g., to prevent the production of A_2O_3, leading to a decrease in the purity of elemental arsenic).

Therefore, in-depth study of the non-isothermal kinetics of the carbothermal reduction of AAS is a key step in the harmless and resourceful treatment of AAS.

Evaluation of the non-isothermal kinetics, which provides important guidance for practical production, has been widely used in various fields, such as for determining the thermal stability of materials[22], pyrolysis behavior[23,24], chlorination roasting[25], and reduction roasting[26,27]. By studying the non-isothermal kinetics, the actual reaction pathways and key control steps can be determined more clearly, process conditions can be optimized, unnecessary waste of resources can be avoided, and theoretical guidance for practical production can be provided. To further optimize the AAS reduction conditions, it is necessary to conduct in-depth study of the non-isothermal kinetics of the AAS carbothermal reduction, explore the reduction intermediates and reduction pathways, and determine the main kinetic parameters and key control steps in the reduction process. This can provide guidance for understanding the AAS carbothermal reduction process and optimizing the reduction process parameters.

2 Experimental methods

2.1 Experimental material

AAS was obtained via direct evaporation of an arsenic-alkali solution, which was obtained by mixing the filtrate and wash-water from the leaching of AAR. The concentrations of arsenic, antimony, and alkali in the arsenic-alkali solution were 40.41, 0.23, and 106.48 g/L, respectively. After thorough mixing, the chemical composition of the AAS was analyzed using inductively coupled plasma-optical emission spectroscopy (ICP-OES), and the results are summarized in Table 1.

Table 1 Major chemical elements in AAS   %

Element	O	As	Na	S	Cl	Other
Content	41.94	18.87	38.10	0.21	0.03	0.85

As shown in Table 1, the As and Na contents in AAS reached 18.87% and 38.10%, respectively. The phase and microstructure of AAS were further analyzed using X-ray diffraction (XRD) and scanning electron microscopy-energy dispersive spectroscopy (SEM-EDS), as shown in Fig 1. XRD analysis [Fig. 1(a)] showed that AAS was mainly composed of NaOH, Na_2CO_3, $Na_2As_4O_{11}$, and Na_3AsO_4. The NaOH present in the raw material melts during high-temperature roasting, which inevitably impacts the reduction of the arsenic-containing phase, which is more difficult than the reduction of sodium arsenate reported by Yang et al.[18]. The SEM image in Fig. 1(b) indicates that AAS was present as irregularly shaped particles with relatively smooth surfaces. EDS analysis [Fig. 1(c)] indicated that the main elements on the surface of AAS in different regions were As, O, and Na, and the elemental proportions were similar, indicating a relatively uniform distribution of the elements.

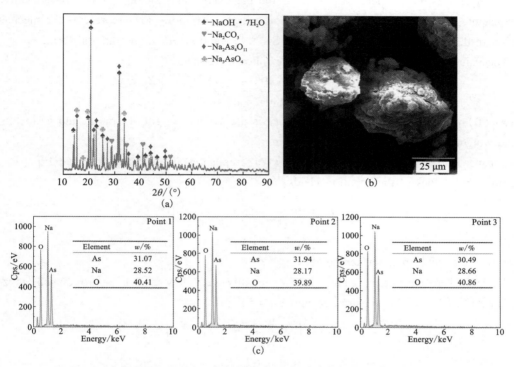

Fig. 1 (a) XRD pattern, (b) SEM image, (c) EDS spectrum of AAS

2.2 Experimental methods

To investigate the kinetics of the carbothermal reduction of AAS, thermogravimetric (TG) and differential thermogravimetric (DTG) experiments were conducted using a synchronous thermal analyzer (NETZSCH STA449 F5, Germany). The AAS and toner were thoroughly mixed in a C/As molar ratio of 1.5, after which an appropriate mixture was added to the alumina crucible. The TG experiments were conducted under argon atmosphere. The flow rate of argon gas was 120 mL/min. The experiments were carried out in the temperature range of 25-1000 ℃ at heating rates of 2, 5, 10, 15, and 20 K/min, respectively.

3 Kinetic analysis methods

Generally, the kinetic formula for heterogeneous reaction processes under non-isothermal conditions is as follows[28-30]:

$$\frac{\mathrm{d}\alpha}{\mathrm{d}t} = \beta \frac{\mathrm{d}\alpha}{\mathrm{d}T} = \frac{A}{\beta} \exp\left(\frac{-E}{RT}\right) f(\alpha) \tag{1}$$

where α is the conversion rate at different temperatures, %; t is the time, min; β is the heating rate, K/min; T is the temperature, K; A is the pre-exponential factor, 1/S; R is the universal gas constant, J/(mol·K); E is the activation energy, kJ/mol; $f(\alpha)$ represents the mechanism function in differential form, given in Table S1[25, 31]; and $G(\alpha)$ is the integral form.

The value of α can be obtained from the following equation:

$$\alpha = \frac{w(0) - w(t)}{w(0) - w(\infty)} \tag{2}$$

where $w(0)$ is the initial sample weight, $w(t)$ is the sample weight at time t, and $w(\infty)$ is the sample weight after completion of the reaction.

Because $G(\alpha)$ is an integral form of $f(\alpha)$, replacing $f(\alpha)$ with $G(\alpha)$ in Eq. (1) and integrating both sides of the equation yields:

$$G(\alpha) = \int_0^\alpha \frac{1}{f(\alpha)} \mathrm{d}\alpha = \frac{A}{\beta} \int_{T_0}^T e^{\frac{-E}{RT}} \mathrm{d}T \tag{3}$$

$$T = T_0 + \beta t \tag{4}$$

where T_0 is the initial reaction temperature.

The initial reaction temperature T_0 is generally very low and is approximated as zero; therefore, Eq. (3) can be transformed into Eq. (5).

$$G(\alpha) = \int_0^\alpha \frac{1}{f(\alpha)} \mathrm{d}\alpha \approx \frac{A}{\beta} \int_0^T e^{\frac{-E}{RT}} \mathrm{d}T \tag{5}$$

$$\text{Let } u = E/RT \tag{6}$$

By substituting Eq. (6) into Eq. (5), Eq. (7) can be obtained as:

$$G(\alpha) = \int_0^\alpha \frac{1}{f(\alpha)} \mathrm{d}\alpha = \frac{AE}{\beta R} \int_u^{+\infty} \frac{e^{-u}}{u^2} \mathrm{d}u = \frac{AE}{\beta R} \cdot P(u) \tag{7}$$

where $P(u)$ denotes the temperature integral. Eq. (7) shows that the key to solving $G(\alpha)$ is solving $P(u)$; however, $P(u)$ is a non-convergence integral, and an accurate analytical formula cannot be obtained. Nevertheless, $P(u)$ has many approximate solutions, such as the Frank-Kameneskii, Coats-Redfern, Doyle, Šesták-Šatava-Wendlandt, and Tang-Liu-Zhang-Wang-Wang approximations, for which the temperature integration is selected according to the range of u values.

3.1 Calculation of activation energy

In analyzing the thermal kinetics using the multi-scan rate method, multiple thermal analysis curves are acquired at different heating rates. Because data from multiple thermal analysis curves are typically used at the same conversion rate, this is also known as the equal-conversion rate method. The multi-scan rate non-isothermal method yields reliable activation energies without involving kinetic mode functions and has significant potential for extensive applications.

The FWO method is a common multiple scan rate method[31, 32] that has been widely used in thermal analysis kinetic studies, such as roasting and reduction. Therefore, the FWO method was used to determine the activation energy of the carbothermal reduction process. The equation for the FWO method is follows:

$$\lg \beta = \lg\left(\frac{AE}{RG(\alpha)}\right) - 2.315 - 0.4567\frac{E}{RT} \tag{8}$$

The $\ln \beta$ and $1/T$ values obtained at different heating rates were linearly fitted. The activation energy at different conversion rates can be calculated from the slope of the fitted line (α ranged from 0.1 to 0.9 with a step of 0.05).

3.2 Integral master-plot method for determining the optimal kinetic model function

The integral master-plot method employs an α value of 0.5, which is substituted into Eq. (7) to obtain $G(\alpha_{0.5}) = (AE)/(\alpha R) \times P(u_{0.5})$, which is then divided by Eq. (7) to obtain Eq. (9). Because the influence of the activation energy and pre-exponential factor were eliminated, the obtained mechanism function was the most objective[31, 33].

$$G(u)/G(u_{0.5}) = P(u)/P(u_{0.5}) \tag{9}$$

The temperature integral $P(u\alpha)$ adopts the Šesták-Šatava-Wendlandt approximation; the specific equation is as follows:

$$P_{ssw}(u) = \frac{e^{-u}}{u}\left(\frac{674.567 + 57.412u - 6.055u^2 - u^3}{1699.066 + 841.655u + 49.313u^2 - 8.02u^3 - u^4}\right) \tag{10}$$

By substituting the average apparent activation energy obtained in the previous stage into the right-hand side of Eq. (9), the experimental curve for α and $P(u)/P(u_{0.5})$ can be obtained. By substituting the various mechanism functions listed in Table S1 into the left-hand side of Eq. (9), the theoretical curve for α and $G(u)/G(u_{0.5})$ can be obtained. By comparing the experimental and theoretical curves, we identified the theoretical curve that was closest to the trend of the experimental curve. The kinetic model corresponding to this theoretical curve was the optimal kinetic mechanism function in the AAS carbothermal reduction process.

After the mechanism function was determined, the pre-exponential factor could be solved. By taking logarithms on both sides of Eq. (7), Eq. (11) was obtained. By linearly fitting $\ln(E/\beta R)$ $+\ln[P(u)]$ and $\ln[G(\alpha)]$, the pre-exponential factor can be calculated.

$$\ln(E/\beta R) + \ln[P(u)] = \beta\ln A + \ln[G(\alpha)] \tag{11}$$

4 Results and discussion

4.1 Analysis of the AAS carbothermal reduction process

The TG-DTG curves of the AAS carbothermal reduction process at a heating rate of 10 K/min are shown in Fig. 2. The TG-DTG curves indicate that the carbothermal reduction process proceeded in two stages. The first weight loss stage began at room temperature and ended at approximately 400 K. This stage mainly involved the removal of free water and crystal water from AAS, with a weight loss percentage of 27.25%. The second evident weight loss stage was between 400 and 1250 K, with a weight loss percentage of approximately 45.40%. The weight loss rate was the fastest from 800 to 1250 K, and the DTG curve exhibited three peaks at 971,

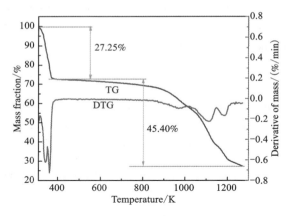

Fig. 2 TG-DTG curves of AAS carbothermal reduction

1113, and 1184 K. The reduction of AAS was mainly concentrated in this temperature range.

Through TG-DTG analysis, the AAS reduction reaction was found to be relatively complete when the temperature exceeded 1250 K; therefore, a carbothermal reduction test was conducted at this temperature. The reduction residue and volatile product were analyzed through XRD and SEM-EDS, and the results are shown in Fig. 3.

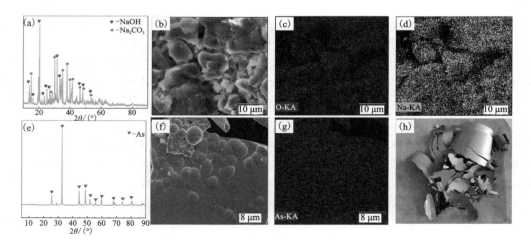

Fig. 3 (a) XRD pattern, (b) SEM image, (c) EDS element mapping of O, (d) EDS element mapping of Na in the reduction residue; (e) XRD pattern, (f) SEM image, (g) EDS element mapping of As, (h) entitative photograph of the volatile product

The XRD pattern [Fig. 3(a)] no longer showed the presence of the arsenic-containing phase. Further, by ICP detection, the arsenic content in the reduced residue was only 0.62%, indicating that the arsenic reduction was more complete. The XRD peaks of the reduced residue matched well with those of NaOH and Na_2CO_3, indicating that the main components of the carbothermal reduction residue in AAS were NaOH and Na_2CO_3. The SEM image of the reduced residue is shown in Fig. 3(b), where the surface is dense and smooth, and there are cracks between the particles, which is characteristic of high-temperature roasting. The elemental surface scan map [Fig. 3(c) and Fig. 3(d)] indicates that the surface elements of the reduced residue were mainly O and Na with consistent distribution areas, confirming the XRD analysis results. The XRD patterns of the volatile products are shown in Fig. 3(e). The diffraction peaks of the volatile product were sharp and strong, and all diffraction peaks corresponded to those of the standard arsenic, indicating that the volatile product was α-As with high crystallinity and purity. The arsenic content of the volatile products was analyzed using ICP. The results showed that the arsenic content exceeded 99.9%. As indicated in the SEM image [Fig. 3(f)] and elemental distribution scanning map [Fig. 3(g)], the volatile products were flaky, with a smooth surface and large overall size, and arsenic was distributed throughout the analyzed area. Amorphous elemental arsenic has been reported[9, 18], where the microscopic morphology is mainly dominated by small, spherical particles, averaging around 1 μm, which is significantly different from the morphology of elemental arsenic with good crystallinity in flake-form. Fig. 3(h) presents a photograph of the collected volatile products. The collected α-As was present in the form of flakes with a shiny metallic surface luster.

4.2 Kinetics and apparent activation energy analyses

Fig. 4(a) and Fig. 4(b) show the TG and conversion rate curves of the AAS carbothermal reduction process at different heating rates, respectively. As the temperature increases, the reduction reaction of AAS begins, the sample quality gradually decreases, and the conversion rate gradually increases and finally stabilizes. The variations in the TG and conversion rate curves at different heating rates were consistent, indicating that the change in the heating rate had no effect on the AAS reduction process; that is, it had no effect on solving the kinetic mechanism function.

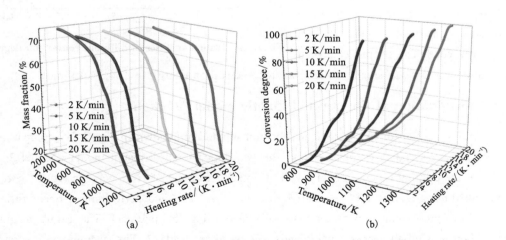

Fig. 4 (a) TG curves and (b) conversion rate curves of AAS carbothermal reduction

Using the FWO method, the apparent activation energy at each conversion rate was calculated by linear fitting; the fitting results are shown in Fig. 5.

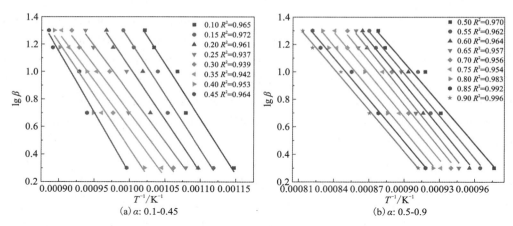

Fig. 5 Iso-conversional plots based on the FWO method

The relationship between the activation energy and conversion rate is shown in Fig. 6. As the conversion rate increased, the activation energy first slightly decreased, then gradually increased, and finally stabilized, which may be mainly due to the different reactions in the AAS reduction process. However, the activation energy showed an overall increasing trend, indicating that more energy was required to break the original chemical bonds as the reaction proceeded. The overall change in

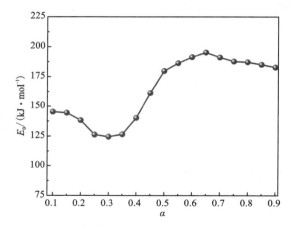

Fig. 6 Activation energies at different conversion degrees

the activation energy was small, with an average activation energy of 164.30 kJ/mol.

4.3 Determination of the optimal kinetic model function

The fitting results for the integral master plot method are shown in Fig. 7. By comparing the experimental curves and theoretical curves, the two-dimensional diffusion model (Model 2) was found to be the optimal kinetic model, with an integral formula of $G(\alpha) = \alpha + (1-\alpha)\ln(1-\alpha)$.

By substituting $G(\alpha) = \alpha + (1-\alpha)\ln(1-\alpha)$ into Eq. (11) and performing a linear fitting on Eq. (11), the pre-exponential factor was calculated using the intercept. The fitting results are shown in Fig. 8, and the kinetic parameters are listed in Table 2. The average pre-exponential factor was calculated as 1.764×10^6 1/s. A low pre-exponential factor ($<10^9$ 1/s) indicates that the reaction is mainly a surface reaction[34], which is consistent with the mechanism indicated by the mechanism function.

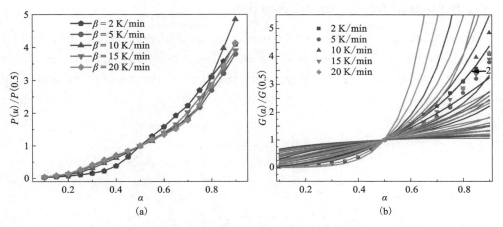

Fig. 7 (a) Experimental master-plots of AAS carbothermal reduction and (b) comparative analysis of theoretical and experimental master-plots for AAS carbothermal reduction

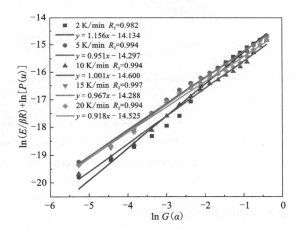

Fig. 8 Linear fitting results from Eq. (11)

Table 2 Calculated kinetic parameters

$\beta/(\text{K} \cdot \text{min}^{-1})$	$E_a/(\text{kJ} \cdot \text{mol}^{-1})$	A/s^{-1}	R^2
2		1.375×10^6	0.982
5		1.618×10^6	0.994
10	164.30	2.191×10^6	0.994
15		1.604×10^6	0.997
20		2.033×10^6	0.994
Average		1.764×10^6	0.992

4.4 Verification of kinetic model function

To verify its reliability, mechanism function 2 was substituted into Eq. (7) to determine the theoretical relationship between α and T. If the theoretical relationship curves are consistent with the experimental data points and the change trends are similar, the obtained mechanism function has a high degree of credibility. Fig. 9 shows a comparison of the theoretical curves and experimental data points.

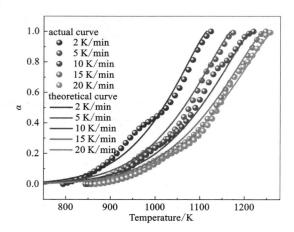

Fig. 9 Verification of AAS carbothermal reduction mechanism function

As shown in Fig. 9, the experimental data points exhibit the same trend as that of the theoretical curve. At a heating rate of 2 K/min, the experimental data points were slightly offset from the theoretical curve; however, the remaining curves were in good agreement, indicating that the determined mechanism function was accurate and reliable.

4.5 Mechanism of the reduction process

When carbon is the reducing agent, CO is also produced and participates in the reaction. The possible reactions of the arsenic-containing phase during the carbon reduction roasting of AAS are as follows:

$$Na_2As_4O_{11} + 2C = 2As_2O_3 + Na_2CO_3 + CO_2(g) \qquad (12)$$

$$Na_2As_4O_{11} + 4CO(g) = 2As_2O_3 + Na_2CO_3 + 3CO_2(g) \qquad (13)$$

$$Na_2As_4O_{11} + 5C = Na_2CO_3 + As_4(g) + 4CO_2(g) \qquad (14)$$

$$Na_2As_4O_{11} + 10CO(g) = Na_2CO_3 + As_4(g) + 9CO_2(g) \qquad (15)$$

$$2Na_3AsO_4 + C = Na_2CO_3 + As_2O_3 + 2Na_2O \qquad (16)$$

$$2Na_3AsO_4 + 2CO(g) = 2Na_2CO_3 + As_2O_3 + Na_2O \qquad (17)$$

$$4Na_3AsO_4 + 5C = 5Na_2CO_3 + As_4(g) + Na_2O \qquad (18)$$

$$4Na_3AsO_4 + 10CO(g) = 6Na_2CO_3 + As_4(g) + 4CO_2(g) \qquad (19)$$

$$2As_2O_3 + 3C = As_4(g) + 3CO_2(g) \qquad (20)$$

$$2As_2O_3 + 3CO(g) \rightleftharpoons As_4(g) + 3CO_2(g) \tag{21}$$

The relationship between the standard Gibbs free energy and temperature in Eqs. (12)-(21) is shown in Fig. 10. Generally, the reduction of high-valence compounds follows the stepwise reduction law, but the Gibbs free energy for the reduction of Na_3AsO_4 to As_2O_3 is always greater than zero in the range of 0-1000 ℃. The temperature required for the reduction of $Na_2As_4O_{11}$ to As_2O_3 must exceed 900 ℃. In contrast, the Gibbs free energy for the direct reduction of $Na_2As_4O_{11}$ and Na_3AsO_4 by CO to arsenic is less than zero. Therefore, from the thermodynamic perspective, $Na_2As_4O_{11}$ and Na_3AsO_4 are more readily directly reduced to As.

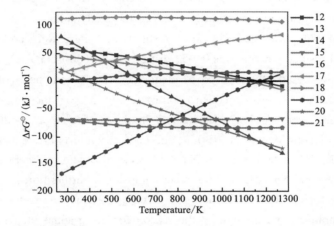

Fig. 10 Standard Gibbs free energy versus temperature plot for Eqs. (12)-(21)

To further explore the carbothermal reduction pathway of AAS and verify the kinetic analysis data, the reduction roasting products at different α (0.3 and 0.65) were obtained by applying the activation energy variation law with α in Fig. 6; the XRD patterns of the roasting products are shown in Fig. 11.

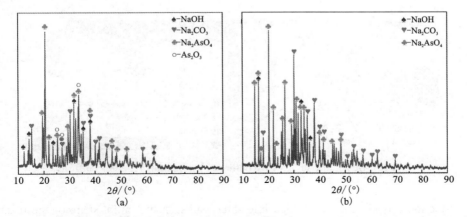

Fig. 11 XRD patterns of reduced residue at temperatures of (a) $\alpha = 0.3$ and (b) $\alpha = 0.65$

A comparison of Fig. 1(a) and Fig. 11(a) shows that there were no $Na_2As_4O_{11}$ peaks in the profile of the reduced residue when α was 0.3; diffraction peaks of Na_3AsO_4 and As_2O_3 were detected. The detection of As_2O_3 also indicates the reduction of $Na_2As_4O_{11}$ to trivalent arsenic, whereas the diffraction peak intensity of As_2O_3 was relatively low, indicating that its content was relatively small; therefore, this stage was dominated by the direct reduction of $Na_2As_4O_{11}$ to elemental arsenic. No diffraction peak of elemental arsenic was detected because of the complete volatilization of elemental arsenic into arsenic vapor, which was released from the reaction system. Meanwhile, as shown in Fig. 10, the Gibbs free energy for the direct reduction of $Na_2As_4O_{11}$ to elemental arsenic did not vary significantly with temperature, indicating that temperature has little effect on this reaction in terms of thermodynamics.

A comparison of Fig. 1(a) and Fig. 11(b) shows that there were no As_2O_3 peaks in the profile of the reduced residue when α was 0.65; diffraction peaks of Na_3AsO_4 were detected. This indicates that As_2O_3 was gradually reduced to elemental arsenic and then volatilized out of the reaction system as the reaction proceeded. In contrast to Fig. 3(a), the diffraction peaks of the arsenic-containing phase were no longer detectable in the profile of the reduced slag after the complete reaction, indicating that when α>0.3, the main reaction is the reduction of Na_3AsO_4 to form elemental arsenic. Diffraction peaks of Na_3AsO_4 were also detected in Fig. 10(b) because of the short reaction time, and Na_3AsO_4 did not react completely. Furthermore, as shown in Fig. 10, the Gibbs free energy for the direct reduction of Na_3AsO_4 to elemental arsenic gradually increased with increasing temperature, which may be the reason for the gradual increase in the activation energy when α>0.3. However, because the direct reduction of $Na_2As_4O_{11}$ and Na_3AsO_4 to monolithic arsenic is not thermodynamically difficult, the difference in the Gibbs free energy was not significant; therefore, the activation energy of the system also changed slightly, fluctuating between 124.54 kJ/mol and 195.2 kJ/mol.

Based on the thermodynamic analysis, non-isothermal kinetic analysis, and experimental studies, the AAS carbothermal reduction reaction pathway is summarized in Fig. 12.

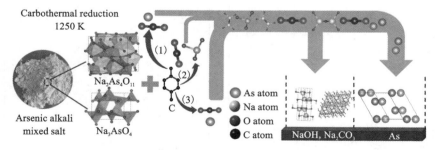

Fig. 12 Schematic of AAS carbothermal reduction reaction pathway

The reduction process can be divided into three paths. In the initial stage of the reaction, the chemical bond in the $Na_2As_4O_{11}$ molecule was broken at high temperature and the molecule was reduced to As_2O_3 or element arsenic, while As_2O_3 was further reduced to element arsenic. The initial activation energy was 145.31 kJ/mol. As the reaction progressed, the change in the activation energy gradually stabilized at approximately 124.54 kJ/mol. The corresponding reaction paths at this stage were Path 1 and Path 2, where Path 1 took precedence over Path 2; that is, the reduction of $Na_2As_4O_{11}$ occurred by preferential direct reduction to produce elemental arsenic (corresponding to $0.1<\alpha<0.35$). As the reaction continued, the reduction of Na_3AsO_4 to $As(g)$ was the main reaction in this phase, with an average activation energy of 184.77 kJ/mol, which corresponds to reaction Path 3 (corresponding to $0.35<\alpha$). At the end of the reaction, the reduction residues were NaOH and Na_2CO_3, indicating that complete reduction of the arsenic-containing phase in AAS could be achieved. Elemental arsenic eventually forms gaseous arsenic, which evaporates from the reaction zone into the condensation zone and crystallizes to form metallic arsenic. From the above analysis, the reduction of sodium arsenate is dominated by direct reduction to elemental arsenic, indicating that it is not necessary to dispense too much toner to achieve purity of metallic arsenic during actual production, which can prevent any effects of the residual toner on recovery of the alkali. The reduction process is mainly controlled by two-dimensional diffusion, and the temperature can be increased or the particle size can be reduced to increase mass transfer and improve the reaction rate.

5 Conclusions

Through thermodynamic analysis, non-isothermal kinetic analysis, and experimental studies, in-depth research on the mechanism of AAS carbothermal reduction was conducted. The research results provide theoretical guidance for the improvement of AAS carbothermal reduction technology. The conclusions are summarized as follows:

TG-DTG and XRD analyses of the residue reduced at different temperatures and non-isothermal kinetic studies showed that the AAS carbothermal reduction process was dominated by the direct reduction of $Na_2As_4O_{11}$ and Na_3AsO_4 to elemental As, whereas a small amount of $Na_2As_4O_{11}$ was reduced to As_2O_3 and then to As. The activation energy and mechanism function of the AAS carbothermal reduction were determined using the FWO and integral master-plot methods, respectively. Through calculations, the mechanism function of the reaction process was determined to be $G(\alpha)=\alpha+(1-\alpha)\ln(1-\alpha)$. The average activation energy was 163.40 kJ/mol, and the pre-exponential factor was 1.764×10^6 1/s. The carbothermal reduction process proceeds via two-dimensional diffusion. The correctness of the mechanism function was verified by comparing the theoretical and experimental curves for α and T. The non-isothermal kinetic study indicates that it is not necessary to excessively increase the amount of reducing agent during AAS carbothermal reduction, and mass transfer during the reaction should be strengthened to improve the reaction rate, which provides some theoretical guidance for future process optimization.

Acknowledgment

This work was supported and funded partly by the National Key R&D Program of China (No. 2019YFC1907405), the Key Project of Jiangxi Natural Science Foundation (20232ACB204016), the National Nature Science Foundation of China (Nos. 52064021, 52074136, 51974140, and 52064018), the Jiangxi Provincial Cultivation Program for Academic and Technical Leaders of Major Subjects (No. 20204BCJL23031), the Jiangxi Province Science Fund for Distinguished Young Scholars (No. 20202ACB213002), the Merit-based postdoctoral research in Jiangxi Province (No. 2019KY09), the Program of Qingjiang Excellent Young Talents, Jiangxi University of Science and Technology (No. JXUSTQJBJ2020004), the Key Projects of Jiangxi Key R&D Plan (No. 20212ACB204015), the Double Thousand Plan in Jiangxi Province (jxsq2019201040).

References

[1] ALKA S, SHAHIR S, IBRAHIM N, et al. Arsenic removal technologies and future trends: A mini review [J]. J. Clean. Prod, 2021: 278. 123805. https://doi.org/10.1016/j.jclepro.2020.123805.

[2] YANG K, QIN W, LIU W. Extraction of elemental arsenic and regeneration of calcium oxide from waste calcium arsenate produced from wastewater treatment[J]. Miner. Eng., 2019, 134: 309-316. https://doi.org/10.1016/j.mineng.2019.02.022.

[3] SU R, MA X, LIN J, et al. An alternative method for the treatment of metallurgical arsenic-alkali residue and recovery of high-purity sodium bicarbonate[J]. Hydrometallurgy, 2021, 202: 105590. https://doi.org/10.1016/j.hydromet.2021.105590.

[4] GUO X, WANG K, HE M, et al. Antimony smelting process generating solid wastes and dust: Characterization and leaching behaviors[J]. J. Environ. Sci., 2014, 26: 1549-1556. https://doi.org/10.1016/j.jes.2014.05.022.

[5] LONG H, ZHENG Y, PENG Y, et al. Recovery of alkali, selenium and arsenic from antimony smelting arsenic-alkali residue[J]. J. Clean. Prod., 2020, 251: 119673. https://doi.org/10.1016/j.jclepro.2019.119673.

[6] LONG H, ZHENG Y, PENG Y, et al. Separation and recovery of arsenic and alkali products during the treatment of antimony smelting residues[J]. Miner. Eng., 2020, 153: 106379. https://doi.org/10.1016/j.mineng.2020.106379.

[7] LONG H, HUANG X, ZHENG Y, et al. Purification of crude As_2O_3 recovered from antimony smelting arsenic-alkali residue[J]. Process Saf. Environ. Prot., 2020, 139: 201-209. https://doi.org/10.1016/j.psep.2020.04.015.

[8] ZHAO Z, WANG Z, XU W, et al. Arsenic removal from copper slag matrix by high temperature sulfide-reduction-volatilization[J]. J. Hazard. Mater., 2021, 415: 125642. https://doi.org/10.1016/j.jhazmat.2021.125642.

[9] NIE H, CAO C, XU Z, et al. Novel method to remove arsenic and prepare metal arsenic from copper electrolyte using titanium (Ⅳ) oxysulfate coprecipitation and carbothermal reduction[J]. Sep. Purif. Technol, 2020, 231: 115919. https://doi.org/10.1016/j.seppur.2019.115919.

[10] ZHONG D, LI L. Separation of arsenic from arsenic—antimony-bearing dust through selective oxidation—sulfidation roasting with CuS[J]. Trans. Nonferrous Met. Soc. China, 2020, 30: 223-235. https://doi.org/10.1016/S1003-6326(19)65194-0.

[11] DEMBELE S, AKCIL A, PANDA S. Technological trends, emerging applications and metallurgical strategies in antimony recovery from stibnite[J]. Miner. Eng., 2022, 175: 107304. https://doi.org/10.1016/j.mineng.2021.107304.

[12] HE M, WANG X, WU F, et al. Antimony pollution in China[J]. Sci. Total Environ, 2012, 421-422: 41-50. https://doi.org/10.1016/j.scitotenv.2011.06.009.

[13] LI J S, LIANG H Q. Treatment strategies study on the comprehensive utilization of arsenic-alkali residue in Xikuangshan Area[J]. Hunan Nonferrous Metals, 2010, 26: 53-55. (in chinese)

[14] TAN C, LI L, ZHONG D, et al, Separation of arsenic and antimony from dust with high content of arsenic by a selective sulfidation roasting process using sulfur[J]. Trans. Nonferrous Met. Soc. China., 2018, 28: 1027-1035. https://doi.org/10.1016/S1003-6326(18)64740-5.

[15] ZENG G S, LI H, CHEN S H, et al. Leaching kinetics and seperation of antimony and arsenic from arsenic alkali residue[J]. Adv. Mater. Res., 2011, 402: 57-60. https://doi.org/10.4028/www.scientific.net/AMR.402.57.

[16] QIU Y H, LU B Q, CHEN B Z, et al. Commercial scale test of anti-pollution control technique for slag of arsenic and soda[J]. J. Cent. South Univ. Technol, 2005, 36: 234-237. (In Chinese)

[17] LEIST M, CASEY R J, CARIDI D. The management of arsenic wastes: Problems and prospects[J]. J. Hazard. Mater, 2000, 76: 125-138. https://doi.org/10.1016/S0304-3894(00)00188-6.

[18] YANG K, QIN W, LIU W. Extraction of metal arsenic from waste sodium arsenate by roasting with charcoal powder[J]. Metals., 2018, 8: 542. https://doi.org/10.3390/met8070542.

[19] JIANG G, MIN X, KE Y, et al. Solidification/stabilization of highly toxic arsenic-alkali residue by MSWI fly ash-based cementitious material containing Friedel's salt: Efficiency and mechanism[J]. J. Hazard. Mater, 2022, 425: 127992. https://doi.org/10.1016/j.jhazmat.2021.127992.

[20] WANG X, DING J, WANG L, et al. Stabilization treatment of arsenic-alkali residue (AAR): Effect of the coexisting soluble carbonate on arsenic stabilization[J]. Environ. Int., 2020, 135: 105406. https://doi.org/10.1016/j.envint.2019.105406.

[21] GONG A, WU X, LI J, et al. Process and mechanism investigation on comprehensive utilization of arsenic-alkali residue[J]. J. Cent. South Univ., 2023, 30: 721-734. https://doi.org/10.1007/s11771-023-5253-4.

[22] GHADIKOLAEI S S, OMRANI A, EHSANI M. Non-isothermal degradation kinetics of ethylene-vinyl acetate copolymer nanocomposite reinforced with modified bacterial cellulose nanofibers using advanced isoconversional and master plot analyses[J]. Thermochim. Acta., 2017, 655: 87-93. https://doi.org/10.1016/j.tca.2017.06.014.

[23] LIU X, LV S, TANG J, et al. Thermal decomposition investigation on heat-resistant poly(dimethylsilylene ethynylenemethylphenylene-methylenemethylphenyleneethynylene) resins[J]. Thermochim. Acta., 2023, 724: 179489. https://doi.org/10.1016/j.tca.2023.179489.

[24] RUAN S, XING P, WANG C, et al. Thermal decomposition kinetics of arsenopyrite in arsenic-bearing

refractory gold sulfide concentrates in nitrogen atmosphere[J]. Thermochim. Acta., 2020, 690: 178666. https://doi.org/10.1016/j.tca.2020.178666.

[25] TIAN L, GONG A, WU X, et al. Non-isothermal kinetic studies of rubidium extraction from muscovite using a chlorination roasting-water leaching process[J]. Powder Technol, 2020, 373: 362-368. https://doi.org/10.1016/j.powtec.2020.06.015.

[26] LIU Y, LV X, YOU Z, et al. Kinetics study on non-isothermal carbothermic reduction of nickel laterite ore in presence of Na_2SO_4[J]. Powder Technol, 2020, 362: 486-492. https://doi.org/10.1016/j.powtec.2019.11.103.

[27] LV X, LV W, YOU Z, et al. Non-isothermal kinetics study on carbothermic reduction of nickel laterite ore [J]. Powder Technol, 2018, 340: 495-501. https://doi.org/10.1016/j.powtec.2018.09.061.

[28] MATTA G, COURTOIS N, CHAMPENOIS J B, et al. Kinetic analysis of pyrolysis and thermal oxidation of bitumen[J]. Thermochim. Acta., 2023, 724: 179514. https://doi.org/10.1016/j.tca.2023.179514.

[29] SHOKRI A, FOTOVAT F. Kinetic analysis and modeling of paper-laminated phenolic printed circuit board (PLP-PCB) pyrolysis using distributed activation energy models (DAEMs)[J]. Thermochim. Acta., 2023, 724: 179513. https://doi.org/10.1016/j.tca.2023.179513.

[30] SONG Q, ZHANG L, XU Z. Kinetic analysis on carbothermic reduction of GeO_2 for germanium recovery from waste scraps[J]. J. Clean. Prod., 2019, 207: 522-530. https://doi.org/10.1016/j.jclepro.2018.09.212.

[31] HU R Z, SHI Q Z. Thermal analysis kinetics[J]. Beijing: Science Press, 2008.

[32] SHARMA P, SINGH K, PANDEY O P. Investigation on oxidation stability of V2AlC MAX phase[J]. Thermochim. Acta., 2021, 704: 179010. https://doi.org/10.1016/j.tca.2021.179010.

[33] ALVES J L F, DA SILVA J C G, MUMBACH G D, et al. Assessing the potential of the invasive grass Cenchrus echinatus for bioenergy production: A study of its physicochemical properties, pyrolysis kinetics and thermodynamics[J]. Thermochim. Acta., 2023, 724: 179500. https://doi.org/10.1016/j.tca.2023.179500.

[34] MAIA A A D, DE MORAIS L C. Kinetic parameters of red pepper waste as biomass to solid biofuel[J]. Bioresour. Technol, 2016, 204: 157-163. https://doi.org/10.1016/j.biortech.2015.12.055.

Valuable Metals Substance Flow Analysis in High Pressure Acid Leaching Process of Laterites

Abstract: Substance flow analysis (SFA), an analytical tool, was applied in a high pressure acid leaching (HPAL) process of laterites. The results show that although the HPAL process has become the mainstream process for the treatment of laterites, a large amount of solid waste discharge has caused great harm to the environment and restricted its large-scale development. The annual treatment capacity of laterites by this HPAL process is 321×10^4 t/a, and 300×10^4 t of high pressure leaching residue, 10×10^4 t of sulfate residue, 1.6×10^4 t of iron and aluminum residue, and 0.08×10^4 t of acid leaching residue are discharged every year. Nickel, cobalt, and manganese are used as the raw material for the preparation of the precursor, and the masses that finally flow into the precursor preparation process are 2.70×10^4 t/a, 0.24×10^4 t/a, and 0.29×10^4 t/a, respectively, and the proportions are 77.14%, 75.00%, and 13.12%, respectively. Scandium finally flows into the scandium extraction process is 40.00 t/a, and the proportion is 37.70%. A total of 98.11% of iron and 99.86% of aluminum can be selectively removed by high pressure acid leaching. Some recommendations for improving emission control and resource recycling for a high pressure acid leaching process of laterites are put forward in the conclusions of this study.

Keywords: Substance flow analysis (SFA); Laterite nickel ores; HPAL process; Emission control; Resource recycling

1 Introduction

Nickel is commonly present in two principal ore types—sulfide and laterite ores[1-3]. Recently, as the demand for nickel continues to grow and sulfide deposits deplete[4-9], an increasing portion of nickel production has come from laterite deposits. The mineralization process

---

Published in *Journal of Central South University*, 2023, 30(6): 1776-1786. Authors: Tian Qinghua, Dong Bo, Guo Xueyi, Wang Qingao, Xu Zhipeng, Li Dong.

of laterites is complicated, which leads to large differences in the contents of elements and the mineral phase composition of each ore layer. Therefore, the extraction process of each ore layer is different. Until now, nickel has been recovered from high iron-containing laterites (limonite, nontronite/smectite) by hydrometallurgical processing, while pyrometallurgy is generally used to extract nickel from low iron-containing and high magnesium-containing saprolite and garnierite[10, 11]. The specific chemical composition and extraction method of different layers of laterites are shown in Table 1[12-15]. High pressure acid leaching (HPAL) and atmospheric acid leaching (AL) are the two prevailing technologies for hydrometallurgical processing of laterites[16-21]. Although AL processes have some advantages of lower capital cost and simpler process equipment than HPAL, they also suffer from several weaknesses[22-24]. The most important advantage offered by HPAL over AL is that under high temperature reaction conditions, most of the leached iron and aluminum are reprecipitated in the form of hematite and alunite, respectively, regenerating almost all of the acid that was consumed in leaching these metals. This results in significantly lowered acid consumption. Moreover, faster reaction kinetics at higher temperatures (230 ℃ to 270 ℃ versus ambient to 90 ℃) offer reduced residence time (i.e., from approximately 12 h in the case of AL to 0.5-1.5 h) and hence increased productivity[25, 26]. With the continuous improvement of autoclave manufacturing technology and the improvement of material properties, the advantages of the HPAL process have become increasingly prominent. Although the ultimate goal of the extraction of laterites is to produce the greatest economic benefits, the environmental pollution caused by the extraction of heavy metals still needs to be considered. Therefore, determining the migration paths of heavy metals and promoting the future development of the HPAL process to an environmentally friendly process is very important.

Table 1 Chemical composition and extraction method of different layers of laterites

Different layers	Chemical mass fraction/%					Recovery method
	Ni	Co	Fe	Cr	MgO	
Overburden	<0.8	<0.1	>50	<0.8	<0.5	Discard
Limonitic ore zone	0.8-1.5	0.1-0.2	40-50	0.8-3.0	0.5-5	Hydrometallurgy
Transition ore zone	1.5-1.8	0.02-0.1	25-40	0.8-1.5	5-15	Pyrometallurgy or hydrometallurgy
Saprolitic ore zone	1.8-3.0	0.02-0.1	10-25	0.8-1.6	15-35	Pyrometallurgy
Bed rocks	0.25	0.01-0.02	<5	0.1-0.8	35-45	No mining

Substance flow analysis (SFA) is a common method for evaluating the flows and stocks of one substance or a group of substances in a time-defined and space-defined system[27-30]. It is based upon the principle of mass balance and views economic processes through the lens of systems science[31-33]. Given a well-defined system boundary, SFA seeks to quantify all inflows, internal flows, stocks, and outflows of a particular chemical element, in this case phosphorus[34-36]. It is mainly used to study specific material flows, such as iron, copper, zinc, manganese, etc., which are of great significance to the national economy, as well as toxic and harmful material flows, such as arsenic, lead, mercury, fluorine, and other industrial sectors, such as iron and steel, chemical industry and forestry[28, 37-41]. Thus far, life cycle assessment (LCA) studies of laterite processing have determined the energy consumption, greenhouse gas emissions, and solid waste burden of various pyrometallurgical and hydrometallurgical routes. Norgate and Jahanshahi[42] described how the laterite ore grade affects the total energy consumption and GHG emissions. Janelle Zhiyun Khoo and Nawshad Haque et al.[43] provided a fair comparison between three different laterite processing technologies, quantified selected environmental impact categories and compared and analysed each impact category for three different nickel feedstocks for stainless steel production. However, as an ore extraction system, metal flows will significantly affect its application and development, but thus far, there is no relevant literature reported on the material flow direction of elements in the entire HPAL process. This paper conducts a detailed material flow analysis and research on the recovery of valuable elements in the process of treating laterites by the HPAL process, aiming to clarify the flow direction and recycling status and provide theoretical guidance for improving emission control and resource recycling for a HPAL process of laterite nickel ores.

2 Experimental method and data

2.1 Methodology

SFA should be implemented in six basic steps[44, 45]: 1) definition of research objective and selection of monitoring indicators; 2) system definition including scope, boundaries, and time frame; 3) identification of relevant flows, processes, and stocks; 4) design of material or substance flow chart; 5) mass balancing; and 6) illustration and interpretation of results, and conclusions. A stepwise framework for SFA is shown in Fig. 1[46]. Compared to applying SFA on a global, regional or local scale, applying SFA to an HPAL process of an enterprise is more microcosmic and more specific. The entire HPAL process is composed of a series of processes. Correspondingly, the SFA model of an HPAL process is composed of two parts: a substance flow chart of a single process and a substance flow chart of the entire process[46].

2.1.1 Substance flow chart model of a single process

If we trace all the materials input to and output from a single process within the system boundary defining the workshop of process j, then the substance flow model of process j is composed of six kinds of substance flows[46], as follows: 1) input upstream products substance flow, p_{j-1}; 2) input material flow, α_j; 3) circular substance flow, β_j; 4) emission substance flow, γ_j; 5) stock substance flow, θ_j; and 6) output substance, p_j. Since the HPAL process is a continuous production process, no product is stored temporarily, and the stock substance flow θ_j is

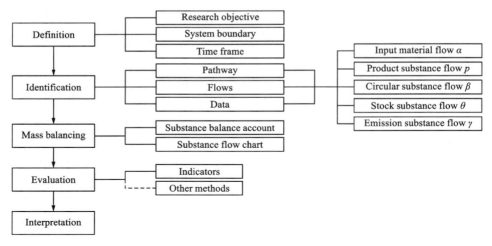

Fig. 1 Framework for applying SFA

not under calculation in this study. Because the HPAL process is an open circuit production process, only wastewater is recycled, and the circular substance flow β_j is not calculated in this study. Therefore, the four kinds of substance flows are shown in Fig. 2. According to the principle of conservation of mass, for a single process j, there is:

$$p_{j-1} + \alpha_j = p_j + \gamma_j \tag{1}$$

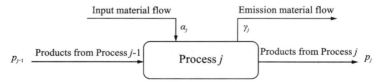

Fig. 2 Substance flow chart of a single process

2.1.2 Substance flow chart model of the entire system

If we combine the substance flow charts of all the single process units, we then have the substance flow chart of the entire system, as shown in Fig. 3. There are three types of substance flows in an overall system, as listed below.

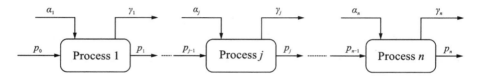

Fig. 3 Substance flow chart of the entire system

Input material flow, α, is:

$$\alpha = \sum_{j=1}^{n} \alpha_j \qquad (2)$$

Emission substance flow, γ, is:

$$\gamma = \sum_{j=1}^{n} \gamma_j \qquad (3)$$

Product substance flow, p, is:

$$p = \sum_{j=1}^{n} p_j \qquad (4)$$

2.1.3 Mass balance

The data required for establishing a mass balance account consist of the flow quantity of each substance flow, M_i, t/a, and the concentration or percentage of objective substance in each substance flow, C_i, %. Based on the total amount and objective element content of each substance flow, the concentration of the objective element in each substance flow, m_i, can be obtained by:

$$m_i = M_i \times C_i \quad j = 1, 2, \cdots, k \qquad (5)$$

According to the conservation of mass principle, the amount of objective element in all the input materials is equal to the amount in all the output materials. Thus for process j, we have

$$\sum_{i}^{k} m_{i,\text{input}} = \sum_{i}^{k} m_{i,\text{output}} \qquad (6)$$

Applying the principle of conservation of mass for the entire system (including all the processes), we have:

$$\sum_{j=1}^{n} \sum_{i=1}^{k} m_{i,\text{input}} = \sum_{j=1}^{n} \sum_{i=1}^{k} m_{i,\text{output}} \quad j = 1, 2, \cdots, n \qquad (7)$$

When establishing a mass balance account, we first calculate the mass balance for each process j according to Eq. (5) and Eq. (6); then, the mass balance of the entire system (including all processes) can be obtained according to Eq. (7).

Balancing mass flows is always a difficult task for any SFA study. Since the input and output substance flows are tested and measured for each of the processes, the degree of mass balance failure is evaluated by the percent of loss for each process according to Eq. (8). A review of many former studies reveals that balance differences of 10% between input and output are common and are usually not significant for the conclusions[47].

$$\text{percent of metal loss} = \frac{\text{metal}_{\text{input}} - \text{metal}_{\text{output}}}{\text{metal}_{\text{input}}} \times 100\% \qquad (8)$$

2.1.4 Evaluation indicators of SFA

For the purpose of explaining the relationship between the different substance flows and their production efficiency, two parameters were the cumulative direct yield (ω_j) and emission rate (E_j) of process j, the proportion of product (p_j) or waste (γ_j) from process j and the input raw material flow (α_1), %, as follows. According to Eq. (9) and Eq. (10), we can clarify the flow direction of objective elements in the system and then propose effective improvement methods.

$$\omega_j = \frac{p_j}{\alpha_1} \qquad (9)$$

$$E_j = \frac{\gamma_j}{\alpha_1} \qquad (10)$$

2.2 Data collection and processing

2.2.1 Investigated HPAL process

Based on the production data of nonferrous metals smelters in Hebei, China, a one-year valuable metal SFA, with January 2019 to December 2019 as the reference time, was investigated. Due to the restriction that laterites cannot be directly imported, this entire HPAL process is divided into two parts: nickel and cobalt hydroxide preparation, and purification and impurity removal. The first part of the product is nickel and cobalt hydroxide (produced abroad), and the second part of the product (produced in China) is scandium-rich solution and nickel-cobalt-manganese solution. The entire HPAL process consists of eight subprocesses, namely, high pressure acid leaching (HPAL), neutralization of iron with limestone (NIL), neutralization precipitation (NP), reductive leaching (RL), neutralization of iron and aluminum (NIA), acid leaching (AL), scandium extraction process (SE), and precursor preparation process (PP). The HPAL process as an ore extraction system was centered on valuable metal extraction; therefore, the system boundary of the study is defined as up to the scandium extraction process and precursor preparation process. The simplified normal flow process and process parameters of the HPAL process are shown in Fig. 4 and Table 2, respectively.

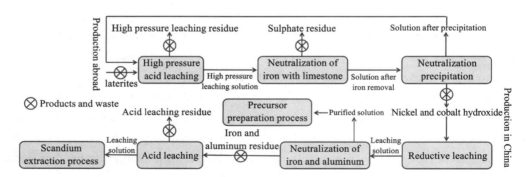

Fig. 4 Standardized flow chart of HPAL process and sampling sites

Table 2 Process parameters of the HPAL process

Process unit	Process parameters
HPAL	245-255℃, 4.3-4.8 MPa, 40-60 min, 200-280 kg/t dry ore
NIL	30% limestone slurry, pH = 3.6-5.0
NP	10% NaOH solution, pH = 7.6-8.5

Continue the Table 2

Process unit	Process parameters
RL	70 ℃, Na_2SO_3/Mn = 0.4-0.8 mol/mol, pH = 1.5-2.0, 2 h
NIA	pH = 5, 2 h, 60-80 ℃
AL	100 g/L H_2SO_4, 60-70 ℃, 1-1.5 h

2.2.2 Sampling and analysis

Ni, Co, Mn, Sc, Fe, and Al were selected as the research objective substances in this study. The system boundary is the entire HPAL process, as introduced above. Such data were usually obtained from statistics and production reports. The mass flow data from January 2019 to December 2019 were exported for calculation. However, this study focused on the internal flows and valuable metals in the entire HPAL process, which were usually beyond statistics and quality inspection plans. Therefore, all valuable metal concentration data were obtained by sampling tests. A valuable metal sampling test plan was created based on the sampling plan and mass flow statistics of the company. Because the first part of the HPAL process is produced abroad, considering the difficulty of sampling from the solution and data accuracy, solid samples were sampled in the study. The sampling sites are indicated in Fig. 4. One year of onsite analysis work was implemented to collect data on valuable metal flows. The frequency for each sampling site was 30 days, and five replicates were analysed to ensure the reliability of the test. First, the solid samples were digested through a four-acid digestion procedure, including nitric, hydrochloric, and hydrofluoric acid. Then, Ni, Co, Mn, Sc, Fe, and Al concentrations were determined with an atomic emission spectrometer (ICP-AES, IRIS Intrepid II XSP, Thermo Electron Corporation, USA) after pretreatment. The final result was the average value.

The annual production of the intermediate products and solid wastes produced in each subprocesses of laterite treatment and the annual average content of elements therein are shown in Table 3. Owing to the limitations of the analytical method, only the gross contents of valuable metals were obtained, and the influence of chemical forms was not considered. Data were calculated by mass balance. This paper mainly studies the material flow direction of valuable metals (Ni, Co, Mn, Sc, Fe, Al) in the entire HPAL process of laterites; therefore, some small streams, such as overflow of water and loss in the transfer process, were beyond the scope of the study. Considering their small quantity and minimal effect on the results, these small streams were neglected.

Table 3 The annual production of the intermediate products and solid wastes produced in each subprocess of laterite treatment and the annual average content of elements

Intermediate products and solid wastes	Content/%					
	Ni	Co	Mn	Sc	Fe	Al
α_1 Laterite nickel ore: 321×10^4 t/a	1.09	0.10	0.69	33 g/t	41.63	2.20
γ_1 High pressure leaching residue: 300×10^4 t/a	0.06	0.01	0.07	2 g/t	43.70	2.35
γ_2 Sulfate residue: 10×10^4 t/a	0.73	0.06	0.29	0.06	25.00	—
P_3 Nickel and cobalt hydroxide: 8×10^4 t/a	39.46	3.37	5.19	0.05	0.38	0.12
γ_5 Iron and aluminum residue: 1.6×10^4 t/a	28.51	2.08	8.30	0.25	1.88	0.63
γ_6 Acid leaching residue: 0.08×10^4 t/a	8.00	—	10.00	—	—	—

3 Results and discussion

3.1 Ni, Co, Mn, Sc, Fe, and Al substance flow chart of the HPAL process

According to this part of the methodology (2.1.3 Mass balance), the mass balance account of process j and the entire system is built. The Ni, Co, Mn, Sc, Fe, and Al substance flow chart, shown in Fig. 5, was sketched based upon metal substance flow data and the process flow

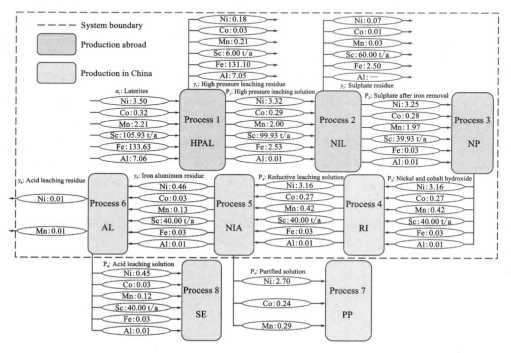

Fig. 5 Ni, Co, Mn, Sc, Fe, and Al substance flow chart of the HPAL process (10^4 t/a)

chart for HPAL. The HPAL process is very simple. There are 11 strands of substance flows altogether, and each of them has been identified with both name and flow code in this article. It should be noted that the Ni, Co, Mn, Sc, Fe, and Al loss of each process was also included in this mass balance account. The percentages of Ni, Co, and Mn losses from the entire HPAL process are 2.57%, 3.12%, and 70.14%, respectively, and the main loss occurs in the neutralization precipitation process, while there is no loss of other metals. According to the criteria mentioned previously, the percentages of Ni and Co loss are acceptable. Since the process and material flow are relatively simple, the material balance error of each process is small. The reason for the large loss of manganese is that the wastewater produced in this process is recycled as process water. In actual production, it is difficult to monitor and sample, which is regarded as an inevitable loss.

3.2 Evaluation indicators of the HPAL process

The evaluation results, based on the indicators defined in section "2.1.4 Evaluation indicators of SFA" above, are shown in Fig. 6. According to Table 3, the annual treatment capacity of laterite nickel ores by this HPAL process is 321×10^4 t/a. Due to the extremely low content of valuable elements such as nickel and cobalt in laterite nickel ores, a large amount of solid waste is discharged every year. The specific data are as follows: high pressure leaching residue ($\gamma_1 = 300 \times 10^4$ t/a), sulphate residue ($\gamma_2 = 10 \times 10^4$ t/a), iron and aluminum residue ($\gamma_5 = 1.6 \times 10^4$ t/a), acid leaching residue ($\gamma_6 = 0.08 \times 10^4$ t/a). The quality and proportion of Ni, Co, Mn, Sc, Fe, and Al in the above solid wastes can be further clarified from Fig. 5 and Fig. 6.

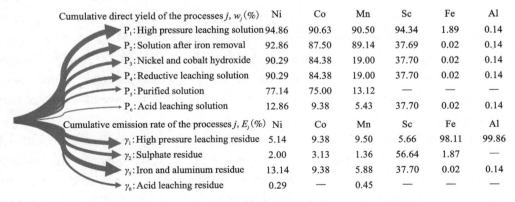

Cumulative direct yield of the processes j, w_j (%)	Ni	Co	Mn	Sc	Fe	Al
P_1: High pressure leaching solution	94.86	90.63	90.50	94.34	1.89	0.14
P_2: Solution after iron removal	92.86	87.50	89.14	37.69	0.02	0.14
P_3: Nickel and cobalt hydroxide	90.29	84.38	19.00	37.70	0.02	0.14
P_4: Reductive leaching solution	90.29	84.38	19.00	37.70	0.02	0.14
P_5: Purified solution	77.14	75.00	13.12	—	—	—
P_6: Acid leaching solution	12.86	9.38	5.43	37.70	0.02	0.14
Cumulative emission rate of the processes j, E_j (%)	Ni	Co	Mn	Sc	Fe	Al
γ_1: High pressure leaching residue	5.14	9.38	9.50	5.66	98.11	99.86
γ_2: Sulphate residue	2.00	3.13	1.36	56.64	1.87	—
γ_5: Iron and aluminum residue	13.14	9.38	5.88	37.70	0.02	0.14
γ_6: Acid leaching residue	0.29	—	0.45	—	—	—

Fig. 6 Cumulative direct yield and emission rate of process j

Nickel, cobalt, and manganese are used as the raw material for the preparation of the precursor, and the masses that finally flow into the precursor preparation process are 2.70×10^4 t/a, 0.24×10^4 t/a, and 0.29×10^4 t/a, respectively, and the proportions are 77.14% (w_5-Ni), 75.00% (w_5-Co), and 13.12% (w_5-Mn), respectively. Scandium with ultrahigh added value finally flows into the scandium extraction process is 40.00 t/a, and the proportion is 37.70%

(w_6-Co). 98.11% of iron (E_1-Fe) and 99.86% of aluminum (E_1-Al) can be selective removed by high pressure acid leaching.

The HPAL process has become the mainstream process for the treatment of low-grade laterite nickel ores, but a large number of solid waste discharges have caused great harm to the environment. At the same time, the harsh environmental policy year by year will also restrict the large-scale application of the process.

In the high pressure acid leaching (HPAL) process, the main purpose is to selectively leach valuable metals such as nickel and cobalt from laterite nickel ores and to remove most of the impurities such as iron and aluminum. Fig. 6 shows that the cumulative direct yields of Ni, Co, Mn, Sc, Fe, and Al in this process are 94.86%, 90.63%, 90.50%, 94.34%, 1.89%, and 0.14%, respectively, and the cumulative emission rates are 5.14%, 9.38%, 9.50%, 5.66%, 98.11%, and 99.86%, respectively. A total of 5.14% of the nickel flows into the high pressure leaching residue (E_1–Ni = 5.14%) because part of the nickel exists in the lattice of iron-bearing minerals such as goethite, which is difficult to deeply leach[48, 49], and another part of the nickel is due to incomplete washing in the stage of liquid-solid separation. A total of 9.38% of the cobalt flows into the high pressure leaching residue (E_1–Co = 9.38%) because cobalt in laterite nickel ores mainly occurs in manganese minerals. Compared with iron minerals, this mineral has strong acid resistance and is not easy to leach[49, 50]. Although this process has achieved relatively good selective leaching effects, the total amount of high-value elements such as nickel, cobalt, and scandium in the high pressure leaching residue each year is 0.18×10^4 t, 0.03×10^4 t, and 6 t, respectively, which causes a great waste of resources. High pressure leaching residue is currently mainly processed in deep-sea landfills. However, as foreign environmental protection policies become more stringent, this method will be eliminated. We have always treated the high pressure leaching residue as solid waste, but 98.11% of the iron (E_1–Fe = 98.11%) in the raw material flows into the residue every year (the total iron in the residue is 131×10^4 t/a, and the iron grade can reach 43.70%), which is a high-grade iron-containing raw material. China's iron resources are scarce and have long been dependent on imports, so it is urgent to solve the technical barriers of weak reduction of hematite to pelletizing.

In the neutralization of iron with limestone (NIL) process, low-cost limestone is used as a neutralizer to remove iron. Although the iron removal effect and filtration performance have achieved relatively good results, a large amount of sulfate residue is produced. As shown in Fig. 5 and Fig. 6, the cumulative direct yields of Ni, Co, Mn, Sc, Fe, and Al in this process are 92.86%, 87.50%, 89.14%, 37.69%, 0.02%, and 0.14%, respectively, and the cumulative emission rates are 2.00%, 3.13%, 1.36%, 56.64%, 1.87%, and 0.00%, respectively. It can be seen from Fig. 5 and Table 3 that to remove 2.50×10^4 t/a of iron, this process produces 10×10^4 t/a of sulfate residue, which not only increases the amount of solid waste treatment but also causes a large amount of waste of valuable metals. 2.00% of the nickel and 3.13% of the cobalt flow into the sulfate residue due to the adsorption of iron hydroxide[51]. It is noteworthy that 56.64% of scandium precipitates into the sulfate residue due to the similar chemical properties of iron and scandium[22], resulting in a great waste of resources, and it is difficult to selectively extract scandium from sulfate residue. Therefore, it is recommended to improve this process to reduce the loss of valuable metals nickel and cobalt, especially scandium, and to reduce the amount of solid waste generated from the source.

In the neutralization precipitation (NP) process, the pH of the solution after iron removal is adjusted to enrich the nickel and cobalt in the intermediate product, nickel, and cobalt hydroxide, to be transported back to the country. In the reductive leaching (RL) process, a reducing agent is added, and the acidity of the solution is adjusted to dissolve all the nickel and cobalt hydroxide. There is no waste residue in the above two processes. In the neutralization of iron and aluminum (NIA) process, to concentrate all scandium into iron and aluminum residues, it is necessary to adjust the pH value to a higher value, which leads to 13.14% of nickel (E_5-Fe = 13.14%) and 9.38% of cobalt (E_5-Co = 9.38%) flowing into the iron and aluminum residues. Therefore, iron and aluminum residues are not solid waste but the main raw material for extracting scandium. In the acid leaching (AL) process, 12.86% of nickel (w_6-Ni = 12.86%) and 9.38% of cobalt (w_6-Co = 9.38%) flow into the scandium extraction process, which leads to the complexity of the impurity removal of the scandium extraction process and the purity of subsequent scandium products. Moreover, the enrichment and extraction of nickel and cobalt in wastewater after scandium extraction is also an important problem to be overcome.

4 Conclusions

Based on the production data of nonferrous metal smelters in Hebei, China, an integrated SFA model of an HPAL process was developed to investigate Ni, Co, Mn, Sc, Fe, and Al flows.

(1) The annual treatment capacity of laterites by this HPAL process is 321×10^4 t/a, and 300×10^4 t of high pressure leaching residue, 10×10^4 t of sulfate residue, 1.6×10^4 t of iron and aluminum residue, and 0.08×10^4 t of acid leaching residue are discharged every year.

(2) Nickel, cobalt, and manganese are used as the raw material for the preparation of the precursor, and the masses that finally flow into the precursor preparation process are 2.70×10^4 t/a, 0.24×10^4 t/a, and 0.29×10^4 t/a, respectively and the proportions are 77.14%, 75.00%, and 13.12%, respectively.

(3) Scandium with ultrahigh added value finally flows into the scandium extraction process is 40.00 t/a, and the proportion is 37.70%.

(4) A total of 98.11% of iron and 99.86% of aluminum can be selectively removed by high pressure acid leaching.

References

[1] SENANAYAKE G, CHILDS J, AKERSTROM B D, et al. Reductive acid leaching of laterite and metal oxides: A review with new data for Fe(Ni, Co)OOH and a limonitic ore[J]. Hydrometallurgy, 2011, 110(1-4): 13-32. DOI: 10.1016/j.hydromet.2011.07.011.

[2] MCDONALD R G, WHITTINGTON B I. Atmospheric acid leaching of nickel laterites review Part I. Sulphuric acid technologies[J]. Hydrometallurgy, 2008, 91(1-4): 35-55. DOI: 10.1016/j.hydromet.2007.11.009.

[3] MESHRAM P, ABHILASH, PANDEY B D. Advanced review on extraction of nickel from primary and secondary sources[J]. Mineral Processing and Extractive Metallurgy Review, 2019, 40(3): 157-193. DOI: 10.1080/08827508.2018.1514300.

[4] KURSUNOGLU S, KAYA M. Atmospheric pressure acid leaching of Caldag lateritic nickel ore[J]. International Journal of Mineral Processing, 2016, 150: 1-8. DOI: 10.1016/j.minpro.2016.03.001.

[5] MACCARTHY J, NOSRATI A, SKINNER W, et al. Atmospheric acid leaching mechanisms and kinetics and rheological studies of a low grade saprolitic nickel laterite ore[J]. Hydrometallurgy, 2016, 160: 26-37. DOI: 10.1016/j.hydromet.2015.11.004.

[6] NASAB M H, NOAPARAST M, ABDOLLAHI H. Dissolution optimization and kinetics of nickel and cobalt from iron-rich laterite ore, using sulfuric acid at atmospheric pressure[J]. International Journal of Chemical Kinetics, 2020, 52(4): 283-298. DOI: 10.1002/kin.21349.

[7] MAKUZA B, TIAN Q H, GUO X Y, et al. Pyrometallurgical options for recycling spent lithium-ion batteries: A comprehensive review[J]. Journal of Power Sources, 2021, 491: 229622. DOI: 10.1016/j.jpowsour.2021.229622.

[8] ZHENG G L, ZHU D Q, PAN J, et al. Pilot scale test of producing nickel concentrate from low-grade saprolitic laterite by direct reduction-magnetic separation[J]. Journal of Central South University, 2014, 21(5): 1771-1777. DOI: 10.1007/s11771-014-2123-0.

[9] GUO X Y, ZHANG C X, TIAN Q H, et al. Liquid metals dealloying as a general approach for the selective extraction of metals and the fabrication of nanoporous metals: A review[J]. Materials Today Communications, 2021, 26: 102007. DOI: 10.1016/j.mtcomm.2020.102007.

[10] DE ALVARENGA OLIVEIRA V, DE JESUS TAVEIRA LANA R, DA SILVA COELHO H C, et al. Kinetic studies of the reduction of limonitic nickel ore by hydrogen[J]. Metallurgical and Materials Transactions B, 2020, 51: 1418-1431. DOI: 10.1007/s11663-020-01841-9.

[11] LI J H, LI D S, XU Z F, et al. Selective leaching of valuable metals from laterite nickel ore with ammonium chloride-hydrochloric acid solution[J]. Journal of Cleaner Production, 2018, 179: 24-30. DOI: 10.1016/j.jclepro.2018.01.085.

[12] DALVI A D, BACON W G, OSBORNE R C. The past and the future of nickel laterites[C]//PDAC 2004 International Convention, Trade Show & Investors Exchange. Toronto: The Prospectors and Developers Association of Canada, 2004: 1-27.

[13] AGACAYAK T, ZEDEF V, ARAS A. Kinetic study on leaching of nickel from Turkish lateritic ore in nitric acid solution[J]. Journal of Central South University, 2016, 23(1): 39-43. DOI: 10.1007/s11771-016-3046-8.

[14] CHEN S L, GUO X H, SHI W T, et al. Extraction of valuable metals from low-grade nickeliferous laterite ore by reduction roasting-ammonia leaching method[J]. Journal of Central South University of Technology, 2010, 17(4): 765-769. DOI: 10.1007/s11771-010-0554-9.

[15] SHAO S, MA B Z, WANG X, et al. Nitric acid pressure leaching of limonitic laterite ores: Regeneration of HNO_3 and simultaneous synthesis of fibrous $CaSO_4 \cdot 2H_2O$ by-products[J]. Journal of Central South University, 2020, 27(11): 3249-3258. DOI: 10.1007/s11771-020-4463-2.

[16] WHITTINGTON B I, MUIR D. Pressure acid leaching of nickel laterites: A review[J]. Mineral Processing & Extractive Metallurgy Review, 2000, 21(6): 527-599. DOI: 10.1080/08827500008914177.

[17] RUBISOV D H, KROWINKEL J M, PAPANGELAKIS V G. Sulphuric acid pressure leaching of laterites-universal kinetics of nickel dissolution for limonites and limonitic/saprolitic blends[J]. Hydrometallurgy, 2000, 58(1): 1-11. DOI: 10.1016/s0304-386x(00)00094-3.

[18] WHITTINGTON B I, JOHNSON J A, QUAN L P, et al. Pressure acid leaching of arid-region nickel laterite ore–Part Ⅱ. Effect of ore type[J]. Hydrometallurgy, 2003, 70(1-3): 47-62. DOI: 10.1016/s0304-386x(03)00044-6.

[19] ÖNAL M A R, TOPKAYA Y A. Pressure acid leaching of Çaldağ lateritic nickel ore: An alternative to heap leaching[J]. Hydrometallurgy, 2014, 142: 98-107. DOI: 10.1016/j.hydromet.2013.11.011.

[20] LIU K, CHEN Q Y, HU H P, et al. Pressure acid leaching of a Chinese laterite ore containing mainly maghemite and magnetite[J]. Hydrometallurgy, 2010, 104(1): 32-38. DOI: 10.1016/j.hydromet.2010.04.008.

[21] GUO X Y, SHI W T, LI D, et al. Leaching behavior of metals from limonitic laterite ore by high pressure acid leaching[J]. Transactions of Nonferrous Metals Society of China, 2011, 21(1): 191-195. DOI: 10.1016/s1003-6326(11)60698-5.

[22] KAYA Ş, TOPKAYA Y A. Extraction behavior of scandium from a refractory nickel laterite ore during the pressure acid leaching process[M]//Rare Earths Industry. Elsevier, 2016: 171-182. DOI: 10.1016/B978-0-12-802328-0.00011-5.

[23] KING M G. Nickel laterite technology–Finally a new dawn?[J]. JOM, 2005, 57(7): 35-39. DOI: 10.1007/s11837-005-0250-7.

[24] YAN K, LIU L, ZHAO H, et al. Study on extraction separation of thioarsenite acid in alkaline solution by CO_3^{2-}-type tri-n-octylmethyl-ammonium chloride[J]. Frontiers in Chemistry, 2021, 8: 592837. DOI: 10.3389/fchem.2020.592837.

[25] JOHNSON J A, MCDONALD R G, MUIR D M, et al. Pressure acid leaching of arid-region nickel laterite ore Part Ⅳ: Effect of acid loading and additives with nontronite ores[J]. Hydrometallurgy, 2005, 78(3-4): 264-270. DOI: 10.1016/j.hydromet.2005.04.002.

[26] WHITTINGTON B I, JOHNSON J A. Pressure acid leaching of arid-region nickel laterite ore Part Ⅲ: Effect of process water on nickel losses in the residue[J]. Hydrometallurgy, 2005, 78(3-4): 256-263. DOI: 10.1016/j.hydromet.2005.04.003.

[27] YU C J, LI H Q, JIA X P, et al. Heavy metal flows in multi-resource utilization of high-alumina coal fly ash: A substance flow analysis[J]. Clean Technologies and Environmental Policy, 2015, 17(3): 757-766. DOI: 10.1007/s10098-014-0832-6.

[28] STANISAVLJEVIC N, BRUNNER P H. Combination of material flow analysis and substance flow analysis: A powerful approach for decision support in waste management[J]. Waste Management & Research, 2014, 32(8): 733-744. DOI: 10.1177/0734242x14543552.

[29] EL-BAZ A A, EWIDA M K T, SHOUMAN M A, et al. Material flow analysis and integration of watersheds and drainage systems: I. Simulation and application to ammonium management in Bahr El-Baqar drainage system[J]. Clean Technologies & Environmental Policy, 2004, 7(1): 51-61. DOI: 10.1007/s10098-004-0258-7.

[30] MA D C, HU S Y, ZHU B, et al. Carbon substance flow analysis and CO_2 emission scenario analysis for China[J]. Clean Technologies & Environmental Policy, 2012, 14(5): 815-825. DOI: 10.1007/s10098-012-0452-y.

[31] MEGLIN R, KYTZIA S, HABERT G. Regional circular economy of building materials: Environmental and economic assessment combining material flow analysis, input-output analyses, and life cycle assessment [J]. Journal of Industrial Ecology, 2022, 26(2): 562-576. DOI: 10.1111/jiec.13205.

[32] KWONPONGSAGOON S, WAITE D T, MOORE S J, et al. A substance flow analysis in the southern hemisphere: Cadmium in the Australian economy[J]. Clean Technologies & Environmental Policy, 2007, 9(3): 175-187. DOI: 10.1007/s10098-006-0078-z.

[33] GARCIA-HERRERO I, MARGALLO M, ONANDÍA R, et al. Connecting wastes to resources for clean technologies in the chlor-alkali industry: A life cycle approach[J]. Clean Technologies and Environmental Policy, 2018, 20(2): 229-242. DOI: 10.1007/s10098-017-1397-y.

[34] GUO X Y, ZHANG J X, TIAN Q H. Modeling the potential impact of future lithium recycling on lithium demand in China: A dynamic SFA approach[J]. Renewable and Sustainable Energy Reviews, 2021, 137: 110461. DOI: 10.1016/j.rser.2020.110461.

[35] KWONPONGSAGOON S, BADER H P, SCHEIDEGGER R. Modelling cadmium flows in Australia on the basis of a substance flow analysis[J]. Clean Technologies & Environmental Policy, 2007, 9(4): 313-323. DOI: 10.1007/s10098-007-0095-6.

[36] SADENOVA M A, UTEGENOVA M E, KLEMEŠ J J. Synthesis of new materials based on metallurgical slags as a contribution to the circular economy[J]. Clean Technologies and Environmental Policy, 2019, 21: 2047-2059. DOI: 10.1007/s10098-019-01761-6.

[37] GHIRARDINI A, ZOBOLI O, ZESSNER M, et al. Most relevant sources and emission pathways of pollution for selected pharmaceuticals in a catchment area based on substance flow analysis[J]. The Science of the Total Environment, 2021, 751: 142328-142328. DOI: 10.1016/j.scitotenv.2020.142328.

[38] SHAFIQUE M, RAFIQ M, AZAM A, et al. Material flow analysis for end-of-life lithium-ion batteries from battery electric vehicles in the USA and China[J]. Resources, Conservation and Recycling, 2022, 178: 106061. DOI: 10.1016/j.resconrec.2021.106061.

[39] WANG D, TANG Y T, SUN Y, et al. Assessing the transition of municipal solid waste management by combining material flow analysis and life cycle assessment[J]. Resources, Conservation and Recycling, 2022, 177: 105966. DOI: 10.1016/j.resconrec.2021.105966.

[40] TIMMERMANS V, VAN HOLDERBEKE M. Practical experiences on applying substance flow analysis in Flanders: Bookkeeping and static modelling of chromium[J]. Journal of Cleaner Production, 2004, 12(8-10): 935-945. DOI: 10.1016/j.jclepro.2004.02.035.

[41] ZHONG W Q, DAI T, WANG G S, et al. Structure of international iron flow: Based on substance flow analysis and complex network[J]. Resources Conservation and Recycling, 2018, 136: 345-354. DOI: 10.1016/j.resconrec.2018.05.006.

[42] NORGATE T, JAHANSHAHI S. Assessing the energy and greenhouse gas footprints of nickel laterite processing[J]. Minerals Engineering, 2011, 24(7): 698-707. DOI: 10.1016/j.mineng.2010.10.002.

[43] KHOO J Z, HAQUE N, WOODBRIDGE G, et al. A life cycle assessment of a new laterite processing technology[J]. Journal of Cleaner Production, 2017, 142: 1765-1777. DOI: 10.1016/j.jclepro.2016.11.111.

[44] LIU M Z, WEN J X, FENG Y, et al. A benefit evaluation for recycling medical plastic waste in China based on material flow analysis and life cycle assessment[J]. Journal of Cleaner Production, 2022, 368: 133033. DOI: 10.1016/j.jclepro.2022.133033.

[45] GONG H Q, MENG F L, WANG G H, et al. Toward the sustainable use of mineral phosphorus fertilizers for crop production in China: From primary resource demand to final agricultural use[J]. Science of the Total Environment, 2022, 804: 150183. DOI: 10.1016/j.scitotenv.2021.150183

[46] BAI L, QIAO Q, LI Y P, et al. Substance flow analysis of production process: A case study of a lead smelting process[J]. Journal of Cleaner Production, 2015, 104: 502-512. DOI: 10.1016/j.jclepro.2015.05.020.

[47] WANG X X, WANG A J, ZHONG W Q, et al. Analysis of international nickel flow based on the industrial chain[J]. Resources Policy, 2022, 77: 102729. DOI: 10.1016/j.resourpol.2022.102729.

[48] LIU K, CHEN Q Y, HU H P. Comparative leaching of minerals by sulphuric acid in a Chinese ferruginous nickel laterite ore[J]. Hydrometallurgy, 2009, 98(3-4): 281-286. DOI: 10.1016/j.hydromet.2009.05.015.

[49] LUO J, LI G H, RAO M J, et al. Atmospheric leaching characteristics of nickel and iron in limonitic laterite with sulfuric acid in the presence of sodium sulfite[J]. Minerals Engineering, 2015, 78: 38-44. DOI: 10.1016/j.mineng.2015.03.030.

[50] SU C, GENG Y, ZENG X L, et al. Uncovering the features of nickel flows in China[J]. Resources, Conservation and Recycling, 2023, 188: 106702. DOI: 10.1016/j.resconrec.2022.106702.

[51] FARIS N, WHITE J, MAGAZOWSKI F, et al. An investigation into potential pathways for nickel and cobalt loss during impurity removal from synthetic nickel laterite pressure acid leach solutions via partial neutralisation[J]. Hydrometallurgy, 2021, 202(14): 105595. DOI: 10.1016/j.hydromet.2021.105595.

图书在版编目(CIP)数据

典型冶金固废有价金属清洁提取与资源化利用／郭学益等著. —长沙：中南大学出版社，2023.12
 ISBN 978-7-5487-5632-3

Ⅰ.①典… Ⅱ.①郭… Ⅲ.①冶金工业—固体废物处理—研究 Ⅳ.①X756.5

中国国家版本馆 CIP 数据核字(2023)第 227420 号

典型冶金固废有价金属清洁提取与资源化利用
DIANXING YEJIN GUFEI YOUJIA JINSHU QINGJIE TIQU YU ZIYUANHUA LIYONG

郭学益　王亲猛　田庆华　谢铿　管建红　徐志峰　著

□出 版 人	林绵优	
□责任编辑	史海燕	
□责任印制	李月腾	
□出版发行	中南大学出版社	
	社址：长沙市麓山南路	邮编：410083
	发行科电话：0731-88876770	传真：0731-88710482
□印　　装	湖南省众鑫印务有限公司	
□开　　本	787 mm×1092 mm 1/16　□印张 48.75　□字数 1450 千字	
□版　　次	2023 年 12 月第 1 版　□印次 2023 年 12 月第 1 次印刷	
□书　　号	ISBN 978-7-5487-5632-3	
□两册定价	328.00 元	

图书出现印装问题，请与经销商调换